CLINICAL EVALUATION OF MUSCLE FUNCTION

The Authors:

M. Lacôte, M.C.M.K.
Past Director, A.D.E.R.F. School of Physical Therapy,
Hôpital Necker, Enfants Malades, Paris.
Faculty Member, Ecole Française d'Orthopédie et Massage, Paris

A. M. Chevalier, M.C.M.K.
Director of Physical Therapy,
Hôpital St.-Michel, Paris.
Faculty Member, A.D.E.R.F. School of Physical Therapy,
Faculty Member, Ecole Française d'Orthopédie et de Massage,
Faculty Member, Ecole d'Assas, Paris

A. Miranda, M.C.M.K.
Director of Physical Therapy,
Neuro-Orthopedics Unit (Dr. Duval-Beaupère),
Department of Professor Held,
Hôpital R. Poincaré, Garches

J. P. Bleton, M.C.M.K.
Director of Physical Therapy,
Neurology Department of Professor Rondot,
Hôpital Ste.-Anne, Paris.
Faculty Member, A.D.E.R.F. School of Physical Therapy,
Hôpital Necker, Enfants Malades, and
Post Graduate School of Physical Therapy (I.N.K.), Paris

Ph. Stevenin, M.C.M.K.
Director, National Institute of Physical Therapy and
Post Graduate School of Physical Therapy (I.N.K.), Paris.
Faculty Member, A.D.E.R.F. School of Physical Therapy,
Hôpital Necker, Enfants Malades, Paris

Original Drawings and Photographs: A. Miranda

Translator: D. Thomas, M.C.M.K., R.P.T.
Post graduate trainee, Proprioceptive Neuromuscular Facilitation, Vallejo, California.
Former Staff Physical Therapist, Rancho Los Amigos Hospital.
Post graduate trainee in Functional Electrical Stimulation, University of Ljubljana, Yugoslavia

CLINICAL EVALUATION OF MUSCLE FUNCTION

M. Lacôte
A. M. Chevalier
A. Miranda
J. P. Bleton
P. Stevenin

Translated by
D. and J. Thomas

Second Edition

CHURCHILL LIVINGSTONE
EDINBURGH LONDON MELBOURNE AND NEW YORK 1987

CHURCHILL LIVINGSTONE
Medical Division of Longman Group UK Limited

Distributed in the United States of America by Churchill Livingstone Inc., 1560 Broadway, New York, N.Y. 10036, and by associated companies, branches and representatives throughout the world.

First and Second Editions published in French under the title *Evaluation Clinique de la Fonction Musculaire*

Second Edition published in English

ISBN 0-443-03720-5

British Library Cataloguing in Publication Data
Clinical evaluation of muscle function.
1. Muscles——Examination
I. Lacôte, M. II. Evaluation clinique de la fonction musculaire. *English*
616.7'40754 RC925.7

Typeset, printed and bound in Great Britain by
William Clowes Limited, Beccles and London

PREFACE

The field of indications for physical therapy has developed widely. Greater patho-physiological knowledge has led to the formulation of coherent therapeutic strategies and considerable improvement in neurological patients.

Physical therapy itself is within the scope of this more efficient treatment and needs testing systems in pathology as well as in therapeutic procedure.

To give a more general view, an attempt has been made here to associate the analytical with the global and the peripheral nervous system with the central nervous system.

This book is divided into three sections:

—The first concentrates on the testing and evaluation of facial muscles, an area little detailed up to now in scientific literature.

—The second section covers the evaluation of lower motor neuron lesions of trunk and extremity muscles. It is based on tests conducted by American researchers during polio epidemics, on experience acquired with numerous patients using these same tests, and on data from recent publications on biomechanics.

—The last section deals in a new way with disturbances encountered in central neurological affections. A study of the various clinical reports has led to the development of a rating system appropriate to each.

This technical work is designed for therapists' daily use. More than a guide for evaluation of function, it offers the student a view of physiology and anatomy as seen through normal and pathological movement.

Due to the range of subjects treated, this book can neither be complete, nor pretend to describe all of the different tests. Those represented were chosen on the basis of methods used in the departments of the respective authors.

ACKNOWLEDGMENTS

I wish to thank:
—Monsieur le Docteur J. P. Fombeur, Chef du Service Otorhino-Laryngologie et Chirurgie Maxillo-Faciale et Réparatrice (Hôpital Saint-Michel, Paris) who instigated this project and has given me advice and encouragement for many years,
—Madame le Docteur Dobler, Attachée O.R.L.,
—Madame le Docteur Levistre, Assistante Audiométriste,
—Monsieur le Docteur Seguin, Assistant O.R.L.,
—Monsieur le Docteur Geneviève, Assistant O.R.L.,
—Monsieur le Docteur Lecomte, Attaché O.R.L.
for their help and kindness.

A. M. Chevalier

I would like to express:
—my sincere thanks to Professeur J. P. Held, Chef de Service, Titulaire de la Chaire de Rééducation, Hôpital Raymond Poincaré, Garches and to Professeur A. Grossiord whose advice and whose articles I drew upon for part of this work,
—my gratitude to Docteur Duval-Beaupère whose competence and initiative have been inspirational and who particularly motivated me in improving my physical therapy techniques.

A. Miranda

I would like to thank the staff, doctors, and personnel of the Neurology Department, Hôpital Sainte-Anne, Paris, whose support and encouragement were essential in realizing this work. Most particularly I wish to mention:
—Monsieur le Professeur P. Rondot, whose advice and teaching have been invaluable; most of the photographs were taken in his department and several are from his private collection,
—Monsieur le Professeur Agrege J. de Recondo and Monsieur le Docteur M. Ziegler who have closely followed this work, counselling and permitting me to draw upon their knowledge,
—Mademoiselle N. Laboure, Head Nurse of the Neurology Department who provided the use of her department and optimal working conditions.

J. P. Bleton

The authors give their special thanks to physical therapists C. Day and D. Mirabaud of the Neuro-Orthopedics Unit (Docteur Duval-Beaupere), Department of Professeur Held, Hôpital Raymond Poincare, Garches, for their valuable and efficient participation.

Our appreciative thanks also go to the many patients who were kind enough to take part in this work.

The publishers wish to express their thanks to Mrs Moira Banks, Lecturer, Department of Physiotherapy, The Queen's College, Glasgow, for her help and advice during the preparation of this translation for publication.

Churchill Livingstone

TRANSLATOR'S NOTE

Reference books are put through extreme tests when the subject is repeatedly studied and indispensable to the basic structure of higher learning. Many are read and then placed aside, still new, while others, pages worn, covers reinforced, continue answering students' questions and practitioners' dilemmas. Anatomy is such a subject. In 1924, Henri Rouviere, Professor of Anatomy at the Faculté de Medecine de Paris, Member of the Academie de Medecine, published a lifetime's work. His book has become a classic in the library of the French-speaking student of anatomy. Like Gray's for the English-speaking student, the Rouviere is one of the first books acquired as the student begins his studies and, years later, it furnishes the ultimate answer when consulted as questions arise pertaining to basic anatomy.

In this work the authors have requested that Rouviere's anatomical references be translated directly into English. Thus insertions on the spinal column, for example, and other anatomical descriptions may not be precisely those to which the English-speaking reader is accustomed. It is hoped the differences may serve to enhance general understanding.

Dominique Thomas

CONTENTS

SECTION ONE: Evaluating Facial Motor Function in Patients with Peripheral and Central Lesions 13
A. M. Chevalier and Collaborators

Lateral view of facial muscles innervated by the facial nerve 14
Introduction 15

Single muscle testing 17
Testing schedules 18
Testing procedure and equipment 21
Motor points and morphological landmarks 22
Evaluation form: muscle activity 23
Evaluation form: muscle tone 24

Muscles of facial expression or facial skin muscles 25
Eyebrow and eyelid muscles 26
Muscles of the nose 46
Muscles of the lips 55

Muscles controlling the organs of facial cavities 87
Muscles of the eyes 89
Muscles of the tongue 92
Muscles of the ear 96

Mandibular motor muscles 97

Quick tests 108

Principal facial expressions 110

Bibliography 116

SECTION TWO: Evaluating Motor Function in Trunk and Limbs of Patients with a Peripheral Nerve Lesion 117
A. Miranda and Collaborators

Part 1: Introduction, Testing Protocol and Material 119

Introduction 120

Grading system for manual testing individual muscles 125

Evaluation form 127

Quick tests 137

Part 2: Individual Muscle Testing 149

Position chart for testing different grades of individual muscles 151

Head and neck muscles 159

Upper extremity muscles 173

Trunk muscles 273

Lower extremity muscles 317

Bibliography 400

SECTION THREE: Evaluating Motor Function in Disorders of the Central Nervous System 401
J. P. Bleton and Collaborators

Qualitative evaluation of muscle and its function in peripheral and central nervous system lesions in adults 402

Evaluation of hand function 409

Deficiency syndrome in pyramidal lesions 431

Pyramidal hypertonia or spasticity of cerebral origin 436

Spasticity of spinal origin 442

Synkineses or associated reactions 444

Evaluation of gait in spinal cord disorders 451

Evaluation of motor function in spasmodic paraparesia 453

Evaluation of posture and gait in hemiplegia 460

Evaluation of hemiplegic upper extremity motor functions 471

Motor function evaluation in parkinsonian syndromes. The effect on manual dexterity 478

Evaluation of gait in parkinsonian syndromes and functional consequences 483

Ataxia. Evaluation of motor coordination and equilibrium disturbances 486

Abnormal movements 497

Apraxia 505

Bibliography 512

SECTION FOUR: Motor Points 515

SECTION FIVE: Innervation: Schematic Nerve Distribution 527
M. Lacôte and A. Miranda

Plexus and nerves 529

Summary of radicular nerve distribution 545

Sensory distribution 549

Bibliography 561

Nerves index 562

MUSCLES INDEX 563

SECTION ONE

Evaluating Facial Motor Function in Patients with Peripheral and Central Lesions

A. M. Chevalier and Collaborators

LATERAL VIEW OF FACIAL MUSCLES INNERVATED BY THE FACIAL NERVE (VII)

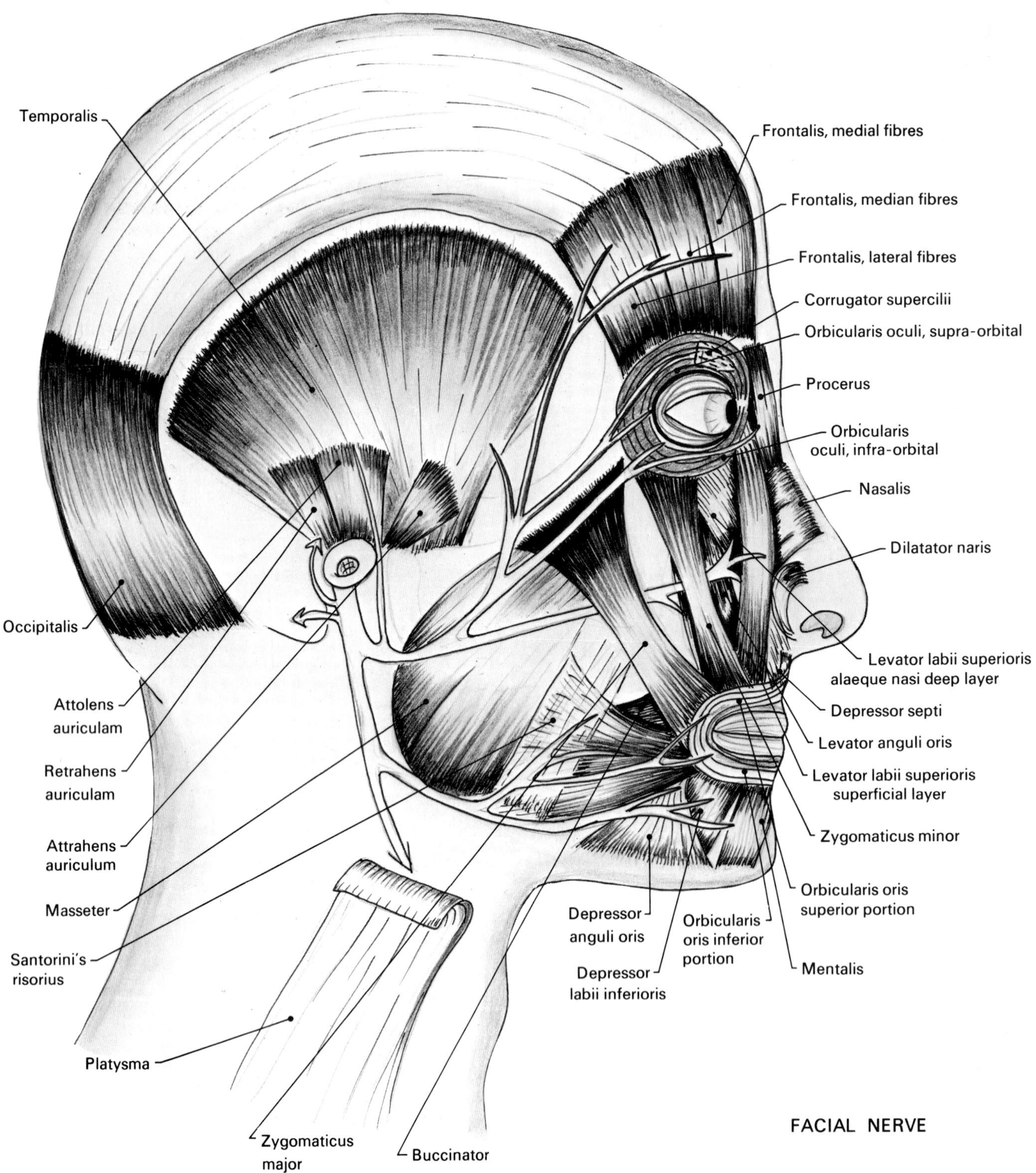

Note: Locations of facial nerve branches vary from one subject to another. The above corresponds to the anatomy of 40% of the cases studied.

INTRODUCTION

Throughout life one sees oneself reflected in the expressions of others. At around two or three years of age a person discovers his face in the mirror at the same time as he discovers his body. Then he 'forgets' his face, since he does not see it, until puberty when he starts searching for his personality.

A person has a certain idea of his own moral and physical identity. He confirms himself through relations with those around him. His facial expression reflects mental function although the muscles responsible go unrecognized; he does not see himself when speaking to others.

With a lesion to these integrated patterns of spontaneous expression, the manifestation of a person's personality in relation to others is disrupted. The results are unaesthetic change and the inability to use facial expression to translate mental function. In addition, every effort at facial expression accentuates facial asymmetry.

The study of facial movements has been divided into three chapters:

—Muscles of facial expression or facial skin muscles

—Muscles controlling the organs of facial cavities

—Mandibular motor muscles.

The term 'synkineses' has been used throughout to describe an unintentional movement accompanying a voluntary movement.

SINGLE MUSCLE TESTING

TESTING SCHEDULES

A facial muscle lesion can be of central or peripheral origin. Its evaluation and the testing schedules depend upon the origin of the lesion.

IN CENTRAL LESIONS

The inferior facial nerve is most specifically involved in motor tract affections at a capsular and cortical level of the corticonuclear pathway at the base of the ascending frontal convolution. These paralyses recede by themselves leaving few or no after-effects.
Test: As soon as possible and every two weeks thereafter.

IN PERIPHERAL LESIONS

The facial nerve can be affected at different levels along its pathway. The facial nerve pathway is composed of three sections.

1st SECTION: INTRA-CRANIAL

Belongs to the auditory-facial bundle.
The facial nerve is often damaged during excision of a neuroma of the auditory nerve (VIII). It can be bruised, partially sectioned and sutured, totally sectioned and sutured, totally sectioned and anastomosed, most often to the hypoglossal nerve, or sometimes to the contralateral facial nerve. The testing schedule will depend upon the nature of the facial nerve lesion.

If the facial nerve is sectioned

a. *Reconnected, sutured to its distal end*

Test: As soon as possible after surgery.
—Every six months in the first year.
—Every three months in the second year.
—Once per month in the third year.

b. *Anastomosed to the hypoglossal nerve*

Test: As soon as possible after surgery.
—Every three months in the first year.
—Once per month in the second and third years.
Note: Mobility of the tongue should figure in the evaluation of the facial muscles.

c. *Anastomosed to the contralateral nerve*

Test: As soon as possible after surgery.
—Every three months in the first year.
—Once per month in the second and third years.

If the facial nerve is bruised or stretched

Test: As soon as possible after surgery.
—Once per month until recovery is complete.

2nd SECTION: INTRA-PETROUS

The facial nerve crosses the petrous portion of the temporal bone, passing through a rigid osseous canal:

—site of traumatic lesions (fragment of fractured petrous bone).
—site of viral affections (zona).
—site of idiopathic affections.
—site of Bell's palsy.

a. *Traumatic lesions*

Test: As soon as possible after the patient regains consciousness (in lesions where the petrous bone is fractured), then after surgical decompression of the facial nerve.

—Once per month in the first year, and in the second and third years if the paralysis has not receded.

b. *Viral affection*

Test: As soon as possible after the onset, and then depending upon its intensity.

—Preferably every three months in the first year.
—Once a month in the second, third and fourth years.

c. *Idiopathic affection*

Test: As soon as possible, then weekly during the first months of the affection.

—If tests show steady improvement, they may be given every two weeks until complete recovery, which may take up to two or three months.
—If tests show no improvement, they may be given every three months during the first year, and monthly in subsequent years.
—If tests show a regression, the surgeon should be notified so that he may decide if an operation is necessary to decompress the facial nerve.

(i) Facial nerve not decompressed surgically.
Test: Weekly until total recovery, over a period which may last up to four years.

(ii) Facial nerve decompressed surgically.
Test: Same as for traumatic lesion.

d. *Bell's palsy*

May be located in the second or third section; the stapedial reflex confirms its precise level.
Test: Same as for idiopathic facial paralyses, or those originating in the third section.

3rd SECTION: EXTRA-PETROUS OR INTRA-PAROTIDIAN

The facial nerve leaves the stylo-mastoid foramen passing between the two lobes of the parotid gland. This is the site of lesions caused by stretching the facial nerve during thorough exeresis of the inner lobe of the parotid. The lesion, partial or temporary, is essentially of the cervico-facial branch, but does not exclude minimal lesion of the temporo-facial.
Test: The day after surgery, then every three days until total recovery (usually about one month). However, if there is a haematoma which is taking a long time to be reabsorbed, testing may be extended to once a week over a three- to six-month period.

CONCLUSION

Testing must be done as rapidly as possible after facial nerve lesion. The testing schedule depends upon the level of the lesion and its intensity. Recovery can be hoped for up to four years. Afterwards, surgeons may undertake corrective surgery on remaining sequelae, except for synkineses, which will partially regress but will not completely disappear.

Note: All facial paralyses for which testing shows nearly or no improvement over a three-month period will recede but with synkineses.

TESTING PROCEDURE AND EQUIPMENT

Muscles of the face should be tested preferably in the morning, since the sound hemiface muscles will not yet have stretched to a maximum the muscles of the paralysed half.
Testing should always be done by the same person, in a calm, well-lit room. No more than one other person should be present, and he should be there only with the patient's permission.
The patient should be seated in a comfortable chair, facing the examiner, who is also seated.
Facial testing must first be explained by the examiner. The testing depends upon the patient's level of comprehension and his degree of concentration. The explanation is essential for the efficiency of the testing since facial expression is innate.
Facial expression is the reflection of sensitive, passionate or intellectual expression. The wealth of facial expressions is closely related to the intensity and sensitivity of psychic function. In order to reconstruct global expression, it is necessary to incorporate all the essential and analytical elements of which it is composed.

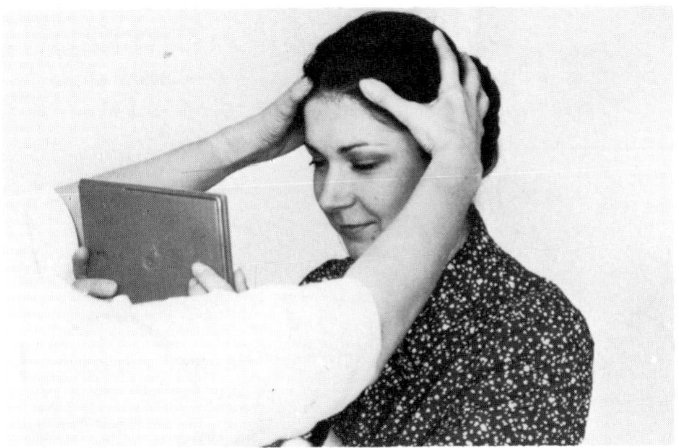

In the testing room there should be:

—a portable lamp directing light across the patient's face in such a way that the slightest movement of the skin is detected.
—a small mirror for the patient to hold between himself and the examiner.
—an evaluation form for muscle activity.
—an evaluation form for muscle tone.
—tongue blades.
—disposable rubber gloves.
—cardboard tubes of various diameters.
—compresses.
—calipers.
—a metric ruler.

MOTOR POINTS AND MORPHOLOGICAL LANDMARKS

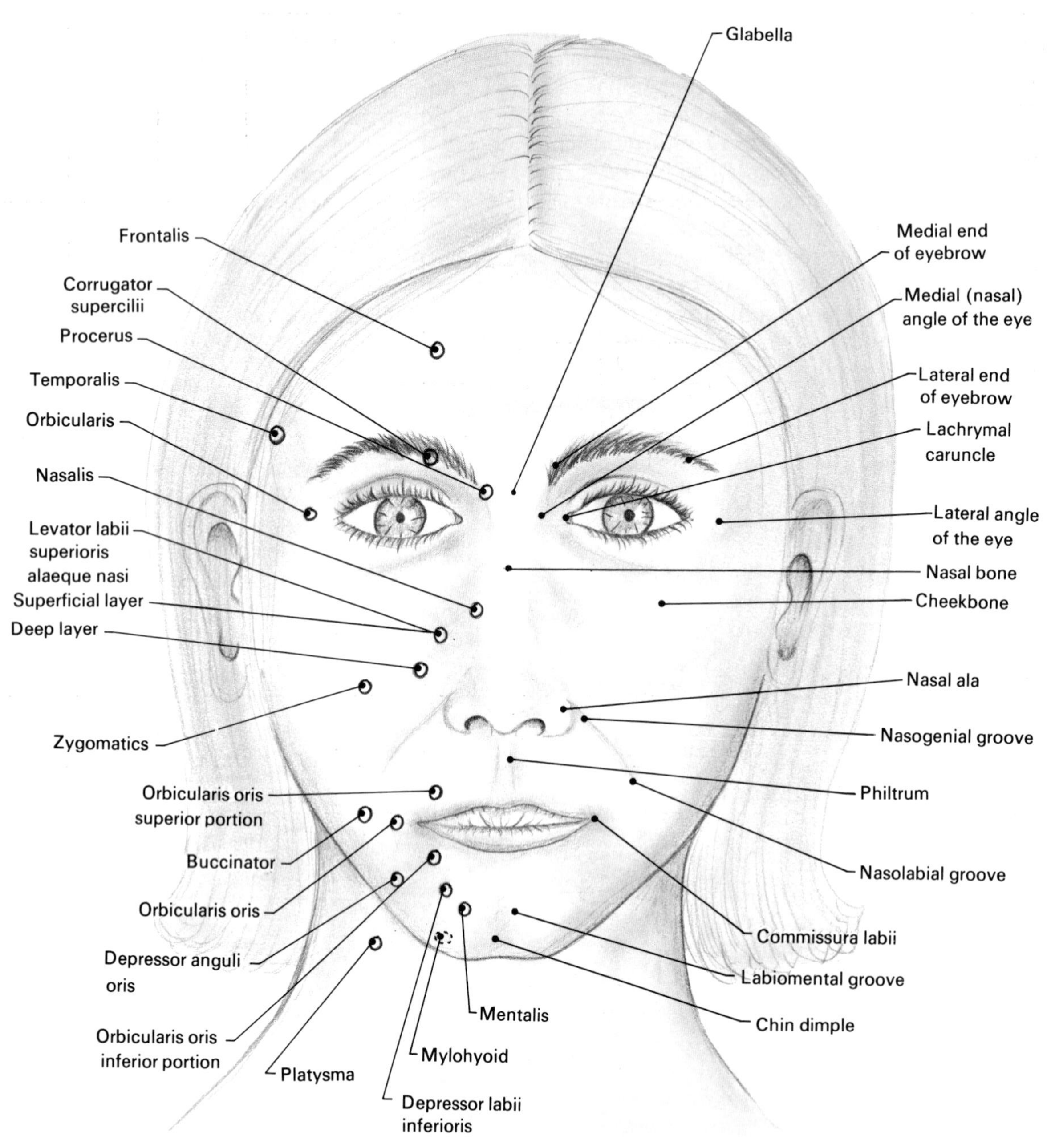
Glabella
Frontalis
Corrugator supercilii
Procerus
Temporalis
Orbicularis
Nasalis
Levator labii superioris alaeque nasi
Superficial layer
Deep layer
Zygomatics
Orbicularis oris superior portion
Buccinator
Orbicularis oris
Depressor anguli oris
Orbicularis oris inferior portion
Platysma
Depressor labii inferioris
Mylohyoid
Mentalis
Medial end of eyebrow
Medial (nasal) angle of the eye
Lateral end of eyebrow
Lachrymal caruncle
Lateral angle of the eye
Nasal bone
Cheekbone
Nasal ala
Nasogenial groove
Philtrum
Nasolabial groove
Commissura labii
Labiomental groove
Chin dimple

EVALUATION FORM: MUSCLE ACTIVITY

Surname . . .	Christian name . . .	date	date	date	date	date	date	date
Forehead eyelids & eyebrows	occipito frontalis corrugator supercilii orbicularis oculi: supra-orbital portion infra-orbital portion							
Objective signs—In mm—yes–no	Bell's phenomenon palpebral inocclusion ectropion Dupuis-Dutemps-Cestan sign sign of Souques' eyelashes							
Nose	procerus nasalis dilatator naris							
Lips	orbicularis oris superior part inferior part levator anguli oris (caninus) levator labii superioris depressor septi zygomaticus minor (cry) zygomaticus major (smile) risorius buccinator depressor anguli oris							
Chin	depressor labii inferioris mentalis							
Neck	platysma							

EVALUATION FORM: MUSCLE TONE

Flaccid facial paralysis	total − 2	partial − 1	normal 0
forehead wrinkles disappear			
lateral end of eyebrow droops			
nose deviates towards sound side			
nasogenial groove disappears			
nasolabial groove disappears			
deviation & drooping of commissura labii			
upper lip tucked in			
lower lip tucked in			
cheek sags			

Paralysis with hypertonia sequels	+ 1	+ 2	+ 3
lateral end of eyebrow abnormally elevated			
nasogenial groove exaggerated			
commissura labii deviated up and outwards			

Principal synkineses	+ 1	+ 2	+ 3
mouth/eye movement of orbicularis oris associated with the closing of the eye			
eye/mouth movement of orbicularis oculi (volitional or spontaneous), with deviation of commissura labii up and outwards			
Emotional synkinesis	yes		no

MUSCLES OF FACIAL EXPRESSION OR FACIAL SKIN MUSCLES

—All are grouped around the facial orifices.
—All have a fixed bony origin and a mobile cutaneous insertion.
—All are constrictors or dilators.
—All reflect mental function.
—All are essential to facial expression.
—Gravity has no effect on these muscles.
—All are innervated by the facial nerve (VII) except the levator palpebrae superioris which is innervated by the oculomotor nerve (III).
—Compared to other nerves of the organism, the facial nerve, essentially motor, has the capacity to regenerate itself for about four years.
—The face has a vertical axis and is not symmetrical, humans adhering to the law of asymmetry.

EYEBROW AND EYELID MUSCLES

OCCIPITO-FRONTALIS (EPICRANIUS)

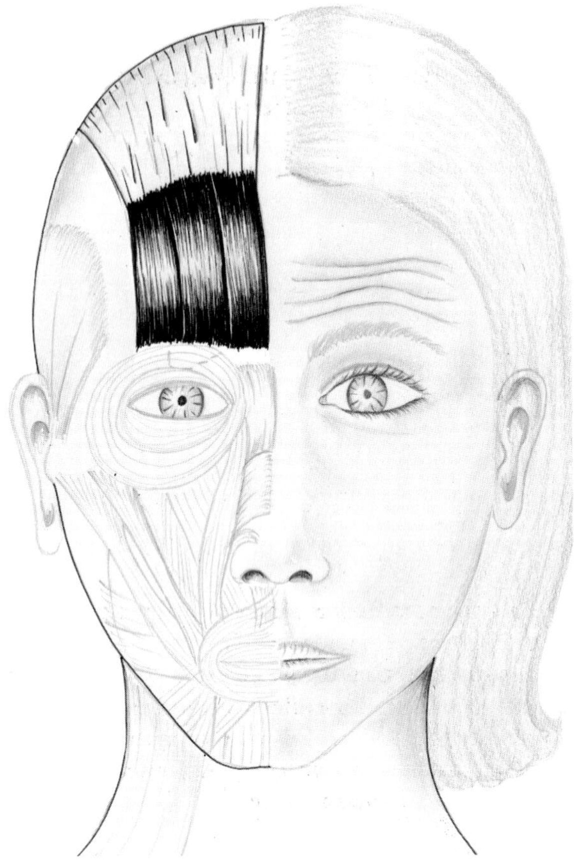

Relaxed state. Expression.

This digastric muscle in contact with the skull is composed of:

—the occipitalis at the back.

—the frontalis in front.

The two muscles are attached to each other by the epicranial aponeurosis.

OCCIPITALIS

This muscle is distinctly separated from the other occipitalis and attached to it by an extension of epicranial aponeurosis.

Origin Lateral two-thirds of the superior nuchal line and the mastoid process by aponeurotic fibres.

Insertion Posterior edge of the epicranial aponeurosis (galea aponeurotica).

Innervation The horizontal branch of the posterior auricular twig which itself branches off the facial nerve a couple of millimetres beneath the stylomastoid foramen.

FRONTALIS

This is a flat muscle forming one single muscular mass together with the other frontalis.

Composed of:

—Lateral fibres which interlace with the supra-orbital portion of the orbicularis oculi.

—Median fibres which interlace with the supra-orbital portion of the orbicularis oculi and the corrugator supercilii.

—Medial fibres which are an extension of the procerus fibres.

Origin Anterior edge of epicranial aponeurosis. It is attached to the other frontalis on the median line.

Insertion Deep surface of skin in the eyebrow–intereyebrow region (glabella).

Innervation Frontal division, superior branch of the temporofacial nerve.

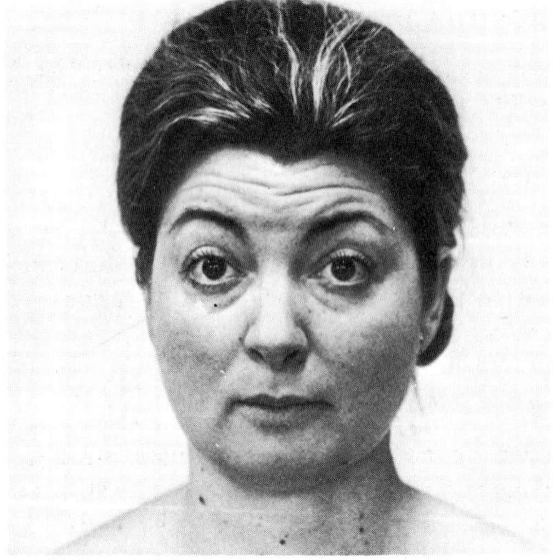

Expression.

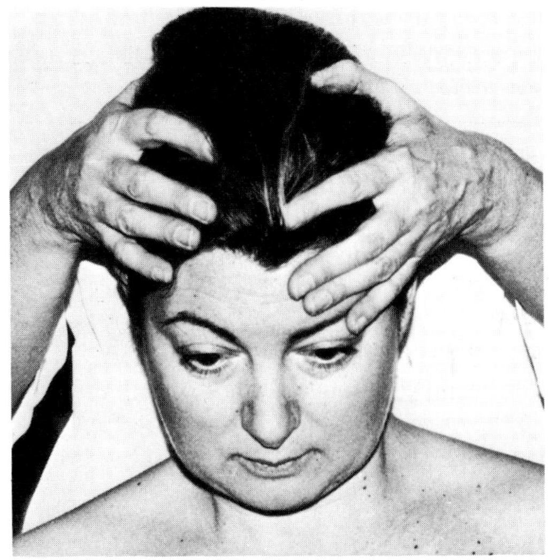
Starting position.

Function Raises the eyebrows forming horizontal wrinkles perpendicular to muscular fibres of the frontalis.

Agonists Medial fibres of the contralateral frontalis and occipitalis.

Antagonists Ipsilateral corrugator supercilii, orbicularis oculi and procerus.

Expression Surprise or attention.

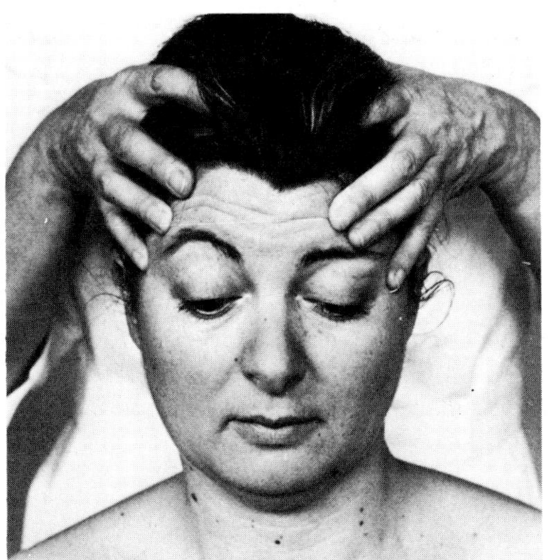
Final position.

MUSCLE TESTING

Starting position

Bring the head forward avoiding compensatory extension of head and neck. With the palms of the hands firmly on both parietal regions slide the skin forward so that as the affected frontalis starts to contract, the pull by the sound occipitalis obliquely down and back transmitted by the epicranial aponeurosis, can be avoided.

Test

Ask the patient to begin raising his eyebrows slowly and gradually. If he has difficulty understanding the test, looking up may help.

Possibility of error

The skin may be pulled diagonally upwards by the medial belly of the sound frontalis.

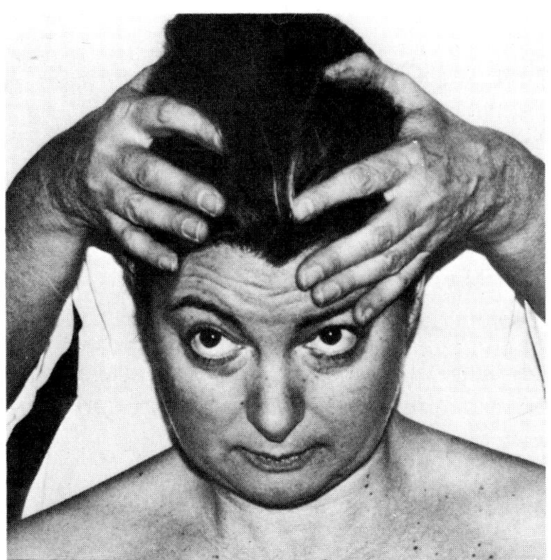
Final position.

Deficiency

The skin of the forehead is immobile. Wrinkles have completely disappeared. The affected hemi-forehead seems to have a larger vertical diameter than the non-affected.

Grading

0 The contraction is invisible, even under strong, glancing light. When the movement is requested, no contraction can be felt by the examiner's finger placed 2 cm above the superior lateral part of the eyebrow on the lateral belly of the tested frontalis.

Grade **0** (L F P).

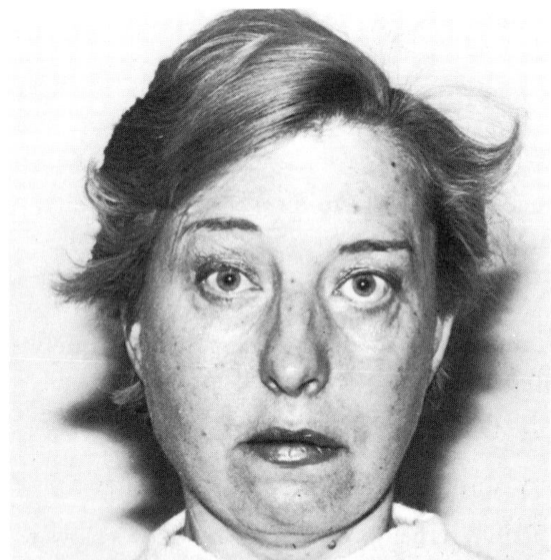

Grade **1** (L F P).

Grade **2** (L F P).

1 During the contraction a slight movement is observed in the surface of the skin, particularly around the lateral fibres of the frontalis.

2 The skin moves more. The first horizontal wrinkles appear at the lateral end of the eyebrow. The movement should be repeated five times. The affected muscle tires quickly compared to the sound side. The movement is slow and incomplete.

3 The skin of the forehead is distinctly more mobile. The number and depth of the wrinkles increase. The movement should be repeated ten times. It is accomplished to the full extent but is not synchronized with the sound side.

4 The skin of the forehead moves easily, amply, symmetrically and in synchronization with the sound side. The movement is fully integrated into voluntary facial expression.

TESTING MUSCLE TONE

– 2 *Atonia*

Wrinkles are completely absent. The forehead appears larger in its vertical diameter. The upper angle of the eyebrow sags.

– 1 *Hypotonia*

The upper angle of the eyebrow no longer sags. Wrinkles are beginning to form but their furrows are not yet as deep as on the sound side.

0 *Normalization of tone*

Wrinkles are deeper, symmetrical and similar to those on the sound side.

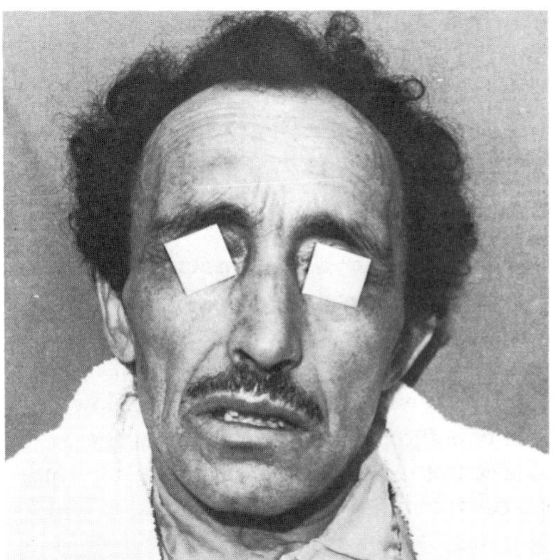

Grade **3** (L F P).

+ **1** *Moderate hypertonia*

When relaxed, the face is slightly asymmetrical. Wrinkles form more accentuated furrows which appear higher than on the sound side.

+ **2** *Major hypertonia*

The frontalis remains spasmodic. The subject is unable to relax deep forehead wrinkles. Forehead asymmetry is accentuated.

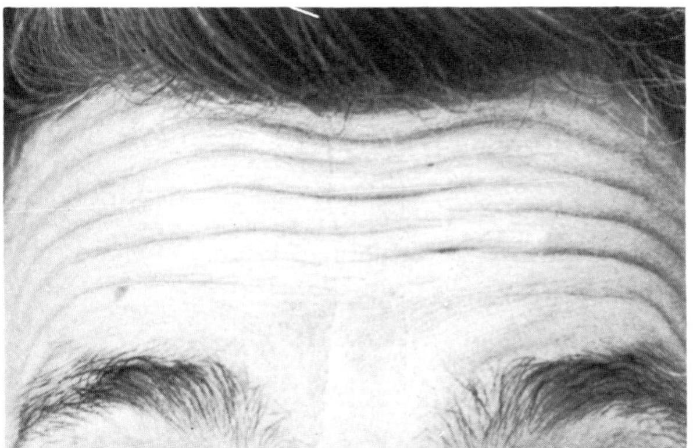

Grade **4.**

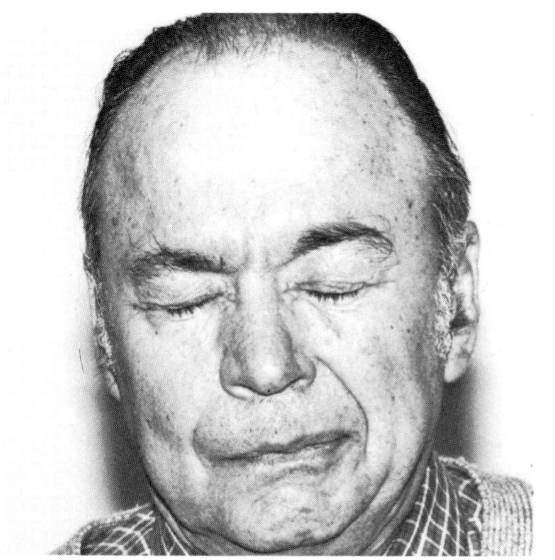

Synkinesis + 1, upon forced closing of the eye (L F P).

SYNKINESES

0 *Synkinesis is absent*

There is no disorderly frontalis movement, seen particularly:
—When the eyes close.
—when the mouth moves.

+ **1** *Synkinesis can be inhibited voluntarily*

The synkinesis appears:
—as the eye closes spontaneously, or as it is forced closed,
—as the mouth constricts or dilates.
The frontalis contracts. The subject manages to control voluntarily a minimal unconscious elevation of the lateral end of the eyebrow while looking in the mirror.

+ **2** *Synkinesis can be inhibited by digital pressure*

As the eye is closed or as the mouth moves, digital pressure is applied to demonstrate the possibility of correcting the disorder. The pathological movement is inhibited by digital pressure which is placed on the frontalis while the fibres are in a maximally eccentric position.

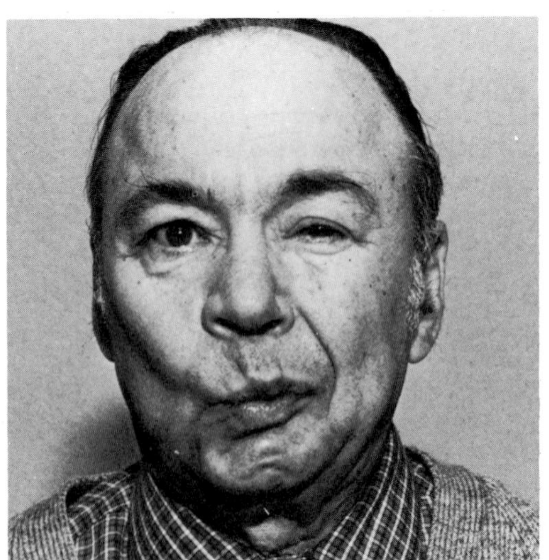

Forehead synkinesis + 1 with contraction of orbicularis oris (L F P).

—Either as the eyes close, while holding the frontalis: Ask the subject, when his face is relaxed, to begin to close his eyes slowly. The subject should maintain the position after digital pressure has gradually been released. The patient, who cannot control himself in front of the mirror with his eyes closed, keeps his finger in place touching the frontalis as a reminder of its immobility. The synkinesis should disappear.

—Or as the mouth moves, while holding the frontalis:
Ask the patient, with his face relaxed, to slowly, gradually begin to purse his lips. Gently release digitial pressure. A balance should establish itself between the frontalis in its relaxed position and the beginning of the constriction of the lips.
The patient controls himself before the mirror.
The synkinesis should disappear.

+ **3** *Synkinesis cannot be inhibited*

The frontalis contracts as the eyes close as well as when the mouth constricts no matter what efforts are made at inhibition.

Hemispasm: The involuntary contraction of the frontalis is part of the hemifacial spasm, pulling the sound side towards the damaged side.

CORRUGATOR SUPERCILII

A thin, flat muscle running from the medial part of the arcus superciliaris to its most medial part.

Relaxed state. Expression.

Origin Fleshy fibres from inner extremity of superciliary ridge.

Insertion After passing below the frontalis and the orbital part of the supra-orbital portion of the orbicularis oculi, it inserts into the deep surface of the skin of the eyebrow at two-thirds or at middle of orbit.

Innervation Superior branch of the temporo-facial.

Function Lifts medial end of eyebrow forward, accentuating promontory of the medial third of the arcus superciliaris. Draws lateral two-thirds of the eyebrow down and in.

Agonists Ipsilateral procerus and orbicularis oculi, particularly the supra-orbital portion.

Antagonists Medial and median portions of the ipsilateral frontalis.

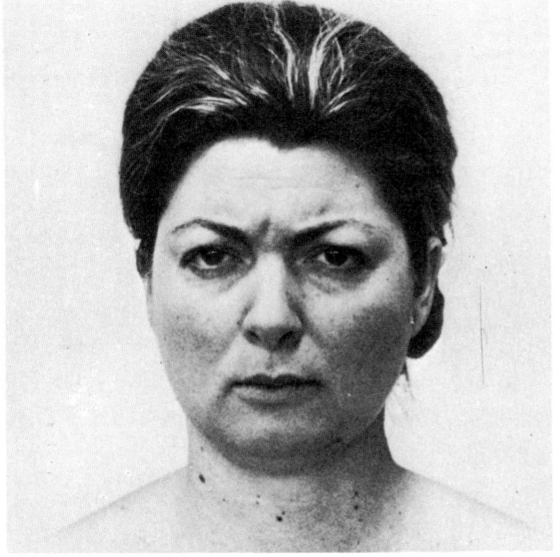

Expression.

Expression Severity, disapproval. A thinker's forehead (Rouille). Protects the eyes under bright, blinding light. The muscle of pathos, grief and suffering (Duchenne de Boulogne).

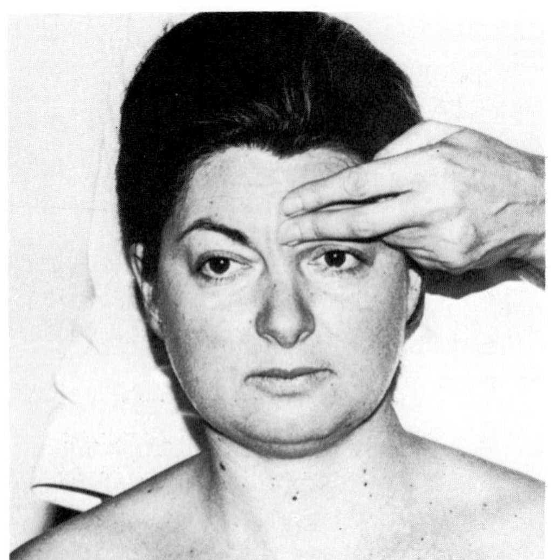

Starting position.

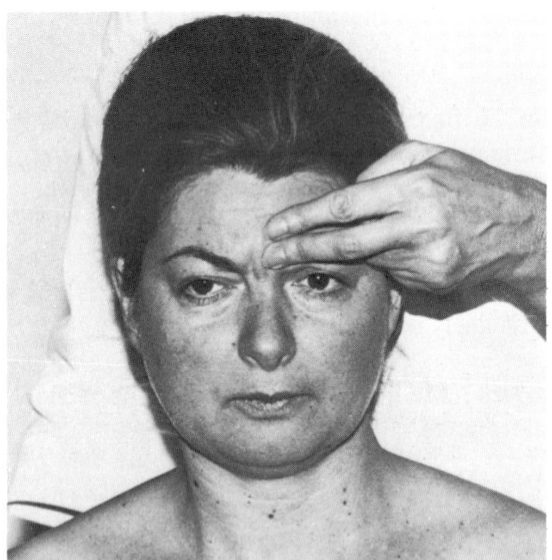

Final position.

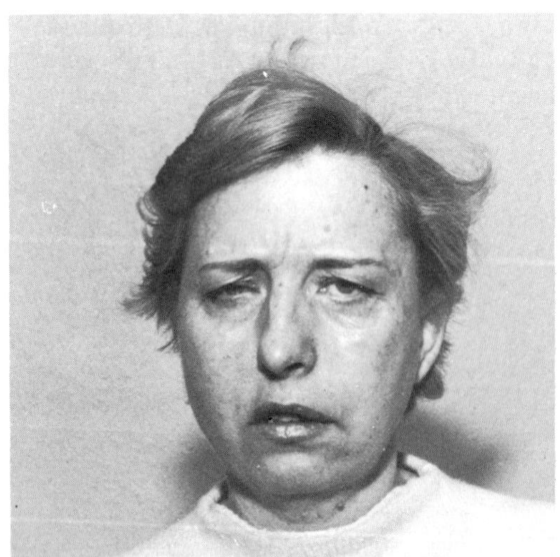

Possibility of error (L F P).

MUSCLE TESTING

Starting position

On a relaxed face, use digital pressure on the glabella to hold the median axis of the face, thus immobilizing the sound side.

Test

Ask the subject to frown, slowly and progressively, bringing his eyebrows towards the bridge of the nose.

Possibility of error

If the median axis is not well held, too strong or too rapid a contraction of the sound corrugator supercilii pulls across the median line. A passive wrinkle appears on the side of the affected muscle.

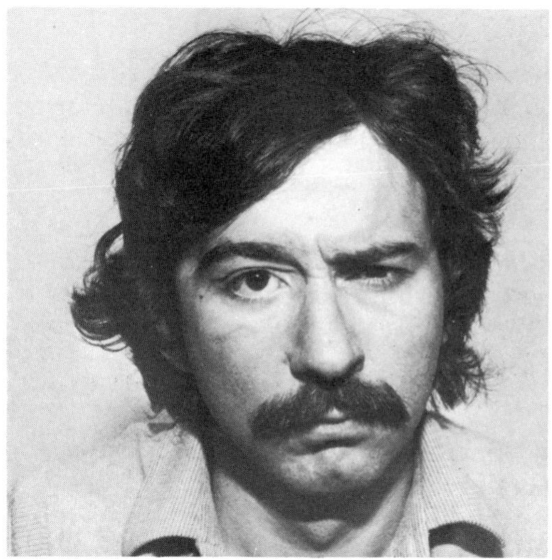

Grade **0** (R F P).

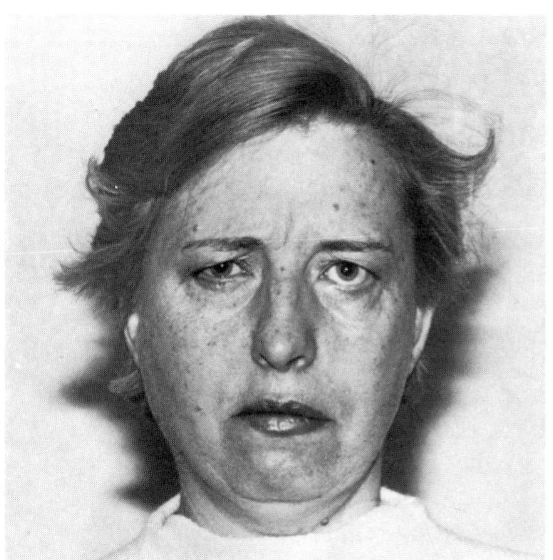

Grade **1** (L F P).

Grade **2** (L F P).

Deficiency

The skin of the fossa located above the eyebrow remains immobile. No vertical wrinkle forms near the median axis of the face.

Grading

0 The contraction is invisible, even under bright, glancing light. No contraction can be felt under the medial end of the eyebrow as voluntary movement is attempted.

1 During the contraction, a slight movement is perceived in the surface of the skin of the fossa over the eyebrow.

2 The skin is more mobile and the fossa deepens. During the contraction a first vertical wrinkle appears over the nasal angle of the eye. The movement should be repeated five times. Compared to the sound side, the muscle tires quickly. The movement is slow and incomplete.

3 The skin of the eyebrow moves more neatly and rapidly. The number of wrinkles increases to two or three usually, and they deepen. The movement should be repeated ten times fully, but it is not yet synchronized with the sound side.

4 The skin of the eyebrow moves easily, amply, symmetrically and in synchronization with the sound side. The movement is globally integrated into voluntary facial expression.

TESTING MUSCLE TONE

– 2 *Atonia*

Wrinkles are non-existent. The medial end of the eyebrow does not project forwards. On certain subjects, the medial end of the eyebrow is lowered, under that of the sound side.

Grade **3** (L F P).

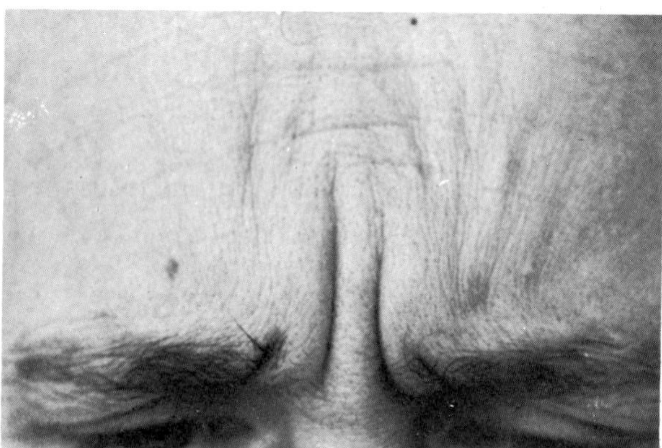

Grade **4.**

− **1** *Hypotonia*

The medial end of the eyebrow is the same height as the end on the sound side. Wrinkles are beginning to form but their furrows are not as deep.

0 *Normalization of tone*

When the face is relaxed, wrinkles on the affected side are symmetrical and similar to those on the sound side.

+ **1** *Moderate hypertonia*

When the face is relaxed, the wrinkle closest to the median axis, is accentuated.

+ **2** *Major hypertonia*

The corrugator supercilii remains spasmodic. Wrinkles are deep and cannot be relaxed; they may pull the skin of the glabella region, giving a marked asymmetry to the face.

SYNKINESES

0 *Synkinesis is absent*

There is no disorderly movement of the corrugator supercilii.

+ **1** *Synkinesis can be inhibited voluntarily*

The subject, when looking in a mirror, manages to control the unconscious contraction of the corrugator supercilii.

+ **2** *Synkinesis can be inhibited with digital pressure*

Synkinesis appears as the orbicularis oculi contracts. On a relaxed face, maintain the corrugator supercilii in a fully stretched position. Orient digital pressure against the direction of pathological contraction. Ask the patient to slowly and gently close his eyes. Gradually release digital pressure. The subject who cannot control the movement in front of the mirror should keep his finger lightly over the skin to remind himself of its immobility. The synkinesis should disappear.

+ **3** *Synkinesis cannot be inhibited*

The corrugator supercilii contracts as eyelids close spontaneously or are forced closed, in spite of all efforts at inhibition.

Hemispasm The involuntary contraction of the corrugator supercilii participates in the hemifacial spasm, pulling the sound side towards the damaged one.

PROCERUS

Thin, elongated muscle located on either side of the median axis of the face on the superior, lateral part of the nasal bone.

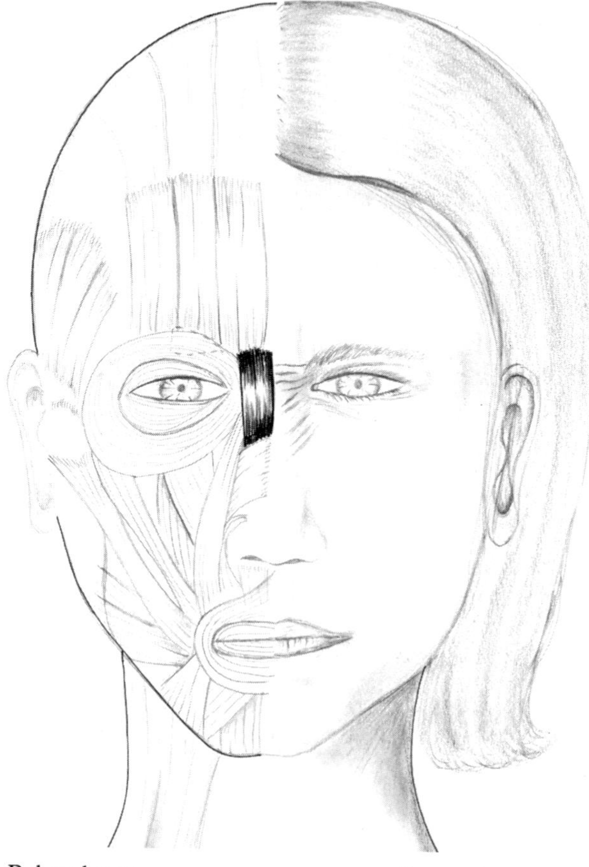

Relaxed state. Expression.

Origin Aponeurosis covering the lower portion of the nasal bone.

Insertion Deep surface of the skin between the two eyebrows after decussating with fibres of the frontalis.

Innervation Temporo-facial branch, infra-orbital palpebral twig.

Function Draws down skin of glabella. Raises skin over nasal bone. Forms horizontal wrinkles.

Agonists Contralateral procerus, nasalis, levator anguli oris and levator labii superioris.

Antagonists Medial fibres of the ipsilateral occipito frontalis, and the depressor septi on both sides.

Expression Threatening, aggression (Duchenne de Boulogne).

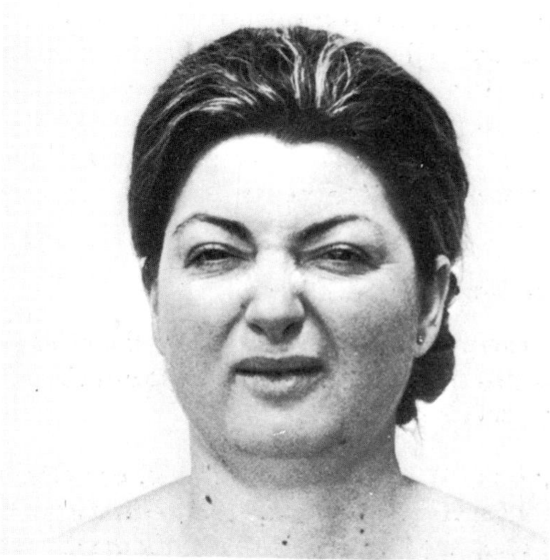

Expression.

MUSCLE TESTING

Starting position

On a relaxed face, use digital pressure to dislocate the median axis of the nose, pushing it over to the affected side and holding it there. This will prevent skin traction by the sound side.

Test

Ask the subject to raise the skin of his nose, wrinkling it slowly and gradually. If the subject has trouble with the movement, the corrugator and nasalis may help since they are facilitating muscles.

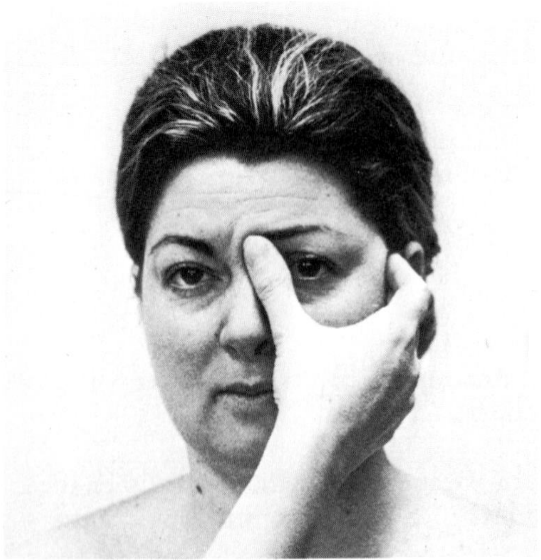

Starting position.

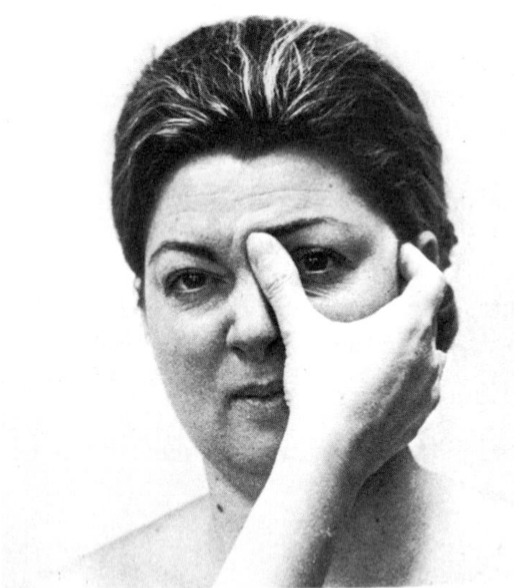

Final position.

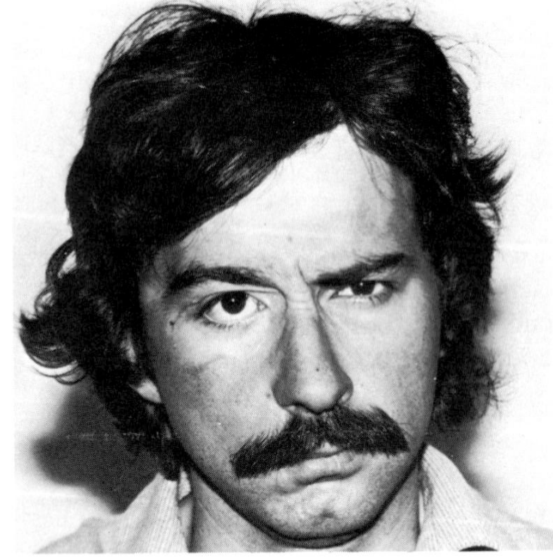

Grade **1** (R F P).

Possibility of error

The skin over the nasal bone may be pulled by the sound procerus.

Deficiency

The skin is immobile on the tested side. Wrinkles have completely disappeared.

Grading

0 The contraction is invisible, even under bright, glancing light. The contraction cannot be felt on the lateral face of the nose over the nasal angle of the eye when the movement is requested.

1 During the contraction, a slight movement in the surface of the skin is perceived particularly on the lateral face of the nose just over the nasal angle of the eye.

2 The skin is more mobile. The first transverse wrinkles cross the nose during the contraction. The movement should be repeated five times. The muscle tires quickly. The movement slows when compared to the sound side, and is incomplete.

3 The skin is clearly more mobile. The number of wrinkles has increased as has their depth. The movement should be repeated ten times to its full extent, but it is not yet synchronized with the sound side.

4 The skin over the nasal bone wrinkles neatly. The movement is ample, symmetrical and in synchronization with sound side. It is globally integrated into voluntary facial expression.

TESTING MUSCLE TONE

– 2 *Atonia*

Wrinkles have completely disappeared over the procerus on the tested side.

– 1 *Hypotonia*

The skin over the nasal bone is less pulled by the sound side. Wrinkles are beginning to form.

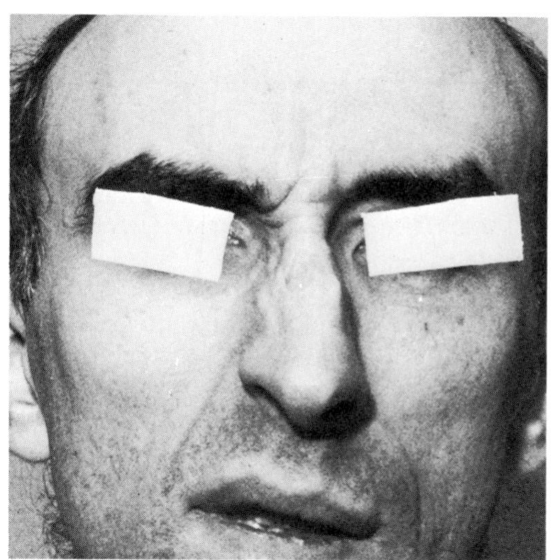

Grade **3** (L F P).

0 *Normalization of tone*

Wrinkles are symmetrical, similar to those on the sound side.

+1 *Moderate hypertonia*

Rare or exceptional.

+2 *Major hypertonia*

Rare or exceptional.

SYNKINESES

0 *Synkinesis is absent*

No disorderly movement of the procerus is noticed.

+1 *Synkinesis can be inhibited voluntarily*

Synkinesis appears generally with movement of dilatator naris and of nasalis pars transversa.

+2 Rare and exceptional.

+3 Rare and exceptional.

Hemispasm The involuntary contraction of the procerus participates in the hemifacial spasm, pulling the sound side towards the damaged side.

ORBICULARIS OCULI

A wide, thin muscle composed of concentric fibres, including a palpebral portion and an orbital portion.

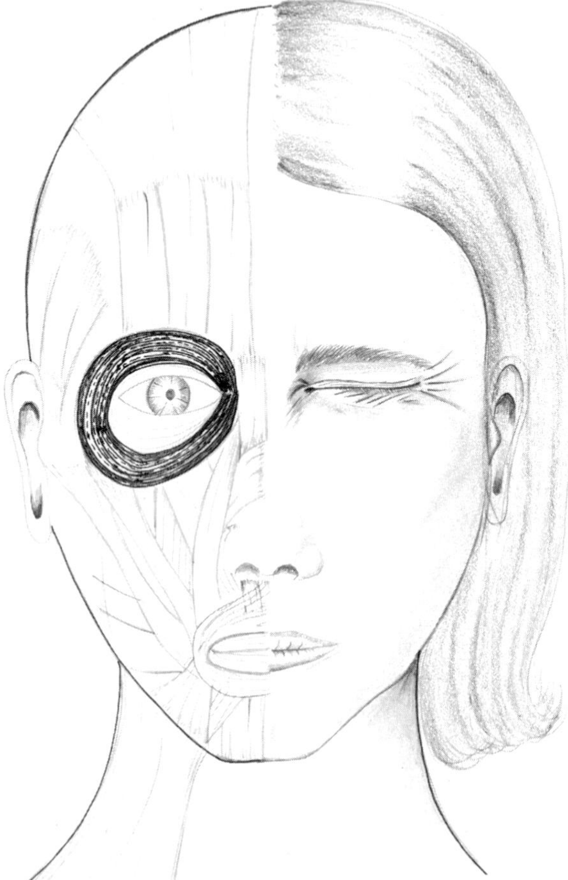

Relaxed state. Expression.

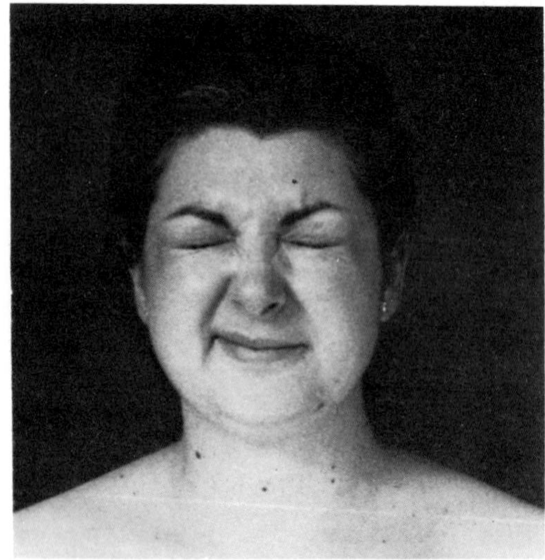

Expression.

PALPEBRAL PORTION

Origin Lateral portion of the internal palpebral ligament and posterior ridge of the lachrymal bone.

Insertion Lateral palpebral raphe.

ORBITAL PORTION

Origin Nasal portion of the frontal bone, frontal process of maxilla, and anterior surface and edge of medial palpebral ligament.

Insertion Muscular fibres form an ellipse which widens towards the lateral angle of the eye.

Innervation Supra-orbital portion: superior twig of superior branch of temporo-facial.
Infra-orbital portion: inferior twig of superior branch of temporo-facial.

Function Closes the eyelids. Closing is possible for the supra-orbital portion when the levator palpebrae is at rest (blinking the eyes). This is an alternating series of movements whose role is cleaning, humidifying and protecting the eye. It is an automatic movement which may also be voluntary. Horner's muscle favours the drainage of tears. It compresses the lachrymal canals and the lachrymal sac.

Agonists

a. In spontaneous closing: none.
b. In forced closing: the corrugator, procerus, nasalis and the zygomatics.

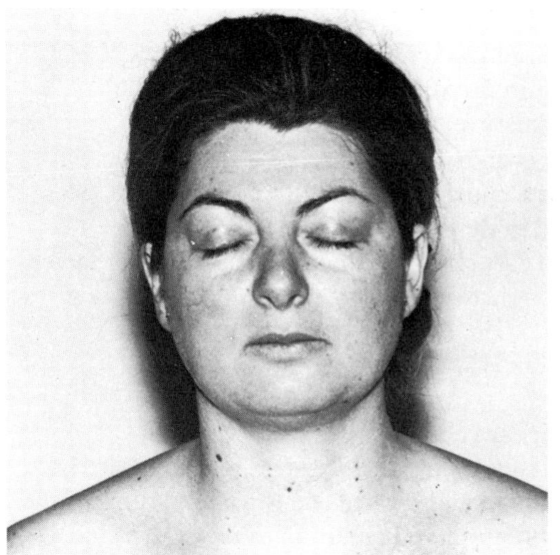

Normally closed.

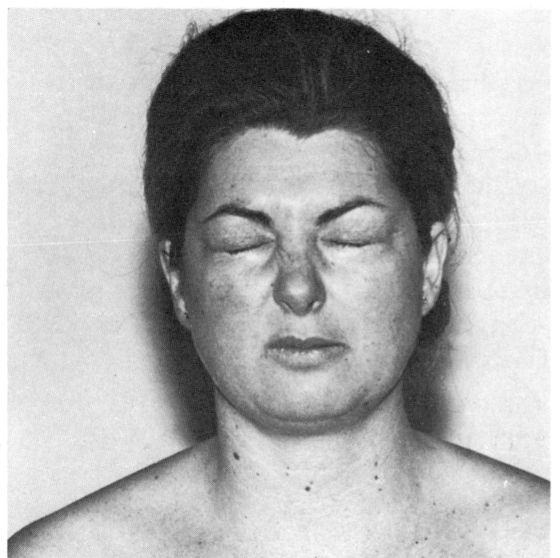

Tightly closed.

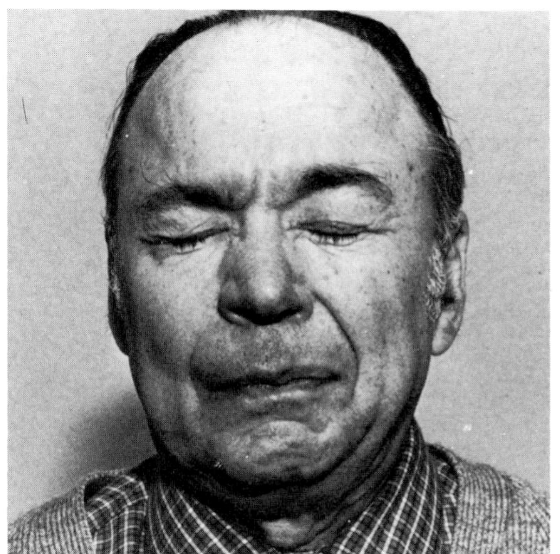

Test showing difference between orbicularis oculi supra-orbital and infra-orbital portions (L F P).

Antagonists

a. Upper eyelid: levator palpebrae superioris and the frontalis oppose forced closing.
b. Lower eyelid: orbicularis oris in its concentric position opposes forced closing.

Expression Spontaneous movement of protection from dazzling light or aggression, denotes worry or preoccupation (Rouille).

MUSCLE TESTING

Starting position

Keep the head in a neutral position, with the face relaxed. Eyes are open, looking straight ahead.

Test

a. *Palpebral portion*:

Ask subject to close his eyes slowly. Lashes should move toward nasal angle of eye.

b. *Orbital portion*:

Ask subject to close his eyes tightly. The lids wrinkle forming deep folds at the lateral angle of the eye. Lashes should nearly disappear in palpebral fissure. Compare with sound side. Note difference in force between supra- and infra-orbital portions of the orbicularis. The two parts are tested separately.

Possibility of error

None.

Grading

0 • Upper eyelid:
The contraction is invisible, even under bright, glancing light. Palpate over lateral angle of the eye.
• Lower eyelid:
The contraction is invisible, even under bright, glancing light. Palpate under lateral angle of eye. Major palpebral inocclusion (non-closing gap): measure in millimetres.

At **0**, it is more than 5 mm.

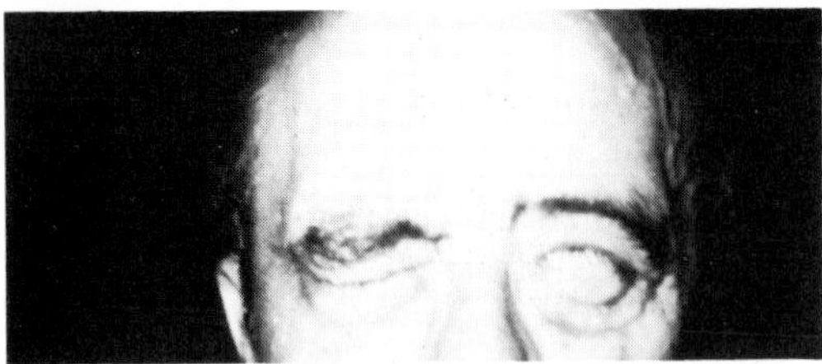
Inocclusion superior to 1 cm (L F P).

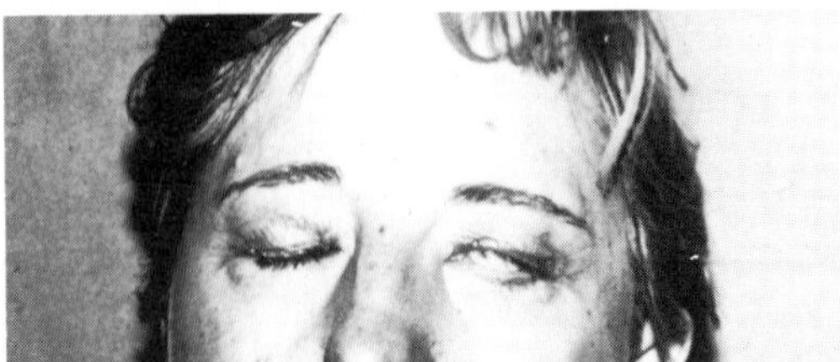
Inocclusion superior to 5 mm (L F P).

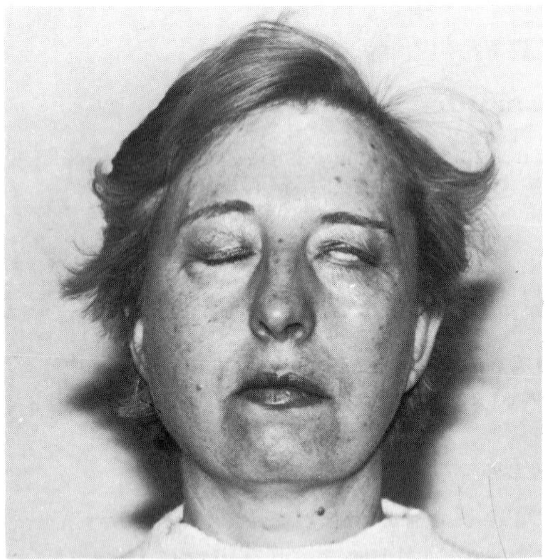
Inocclusion 5 mm (L F P).

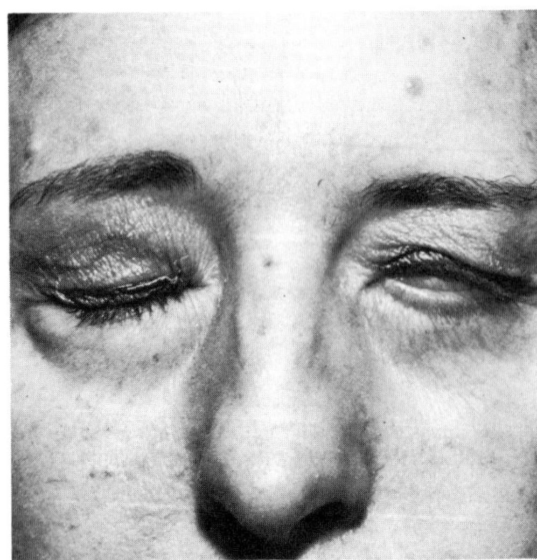
Inocclusion less than 5 mm (L F P).

1 • Upper eyelid:
During the contraction a slight movement is perceived in the surface of the skin above the palpebral fissure.
• Lower eyelid:
During the contraction a slight movement is perceived in the surface of the skin below the palpebral fissure. Fasciculation appears under lower eyelashes.

Inocclusion is 5 mm.

2 • Upper eyelid:
The skin is more mobile. It forms minute wrinkles slanting down and inwards on the lateral third of the eyelid. Lashes move in relation to each other and orient themselves towards the nasal angle of the eye. The movement should be repeated five times. The muscle tires quickly. The movement is slow and incomplete.
Levator palpebrae dominates as soon as orbicularis oculi tires, lifting the eyelid slightly (Dupuis-Dutemps-Cestan sign).
• Lower eyelid:
The skin is more mobile, forming wrinkles at the lateral corner of the eye. At the nasal angle wrinkles are numerous and minute. The nasal angle of the eye closes covering the lachrymal caruncle. Lashes move more and are oriented towards the nasal angle of the eye.
The movement should be repeated five times. The muscle tires quickly. The movement is slow and incomplete.
When orbicularis oculi is at **2**, eyelids close without force and lashes touch. The subject sees no light through his closed lids.

Inocclusion is 0 mm.

3 • Upper eyelid:
The eyelid definitely moves more. Wrinkles are deeper. The movement should be repeated ten times fully but it is not synchronized with the sound side.
• Lower eyelid:
The eyelid is quite mobile. The infra-orbital portion projects outwards. The nasal angle of the tested eye is symmetrical to that of the sound side.
The movement should be repeated ten times but is not yet synchronized with the sound side.
When orbicularis oculi is at **3**:
—at spontaneous closing there is no deficiency.
—at forced closing there is a difference in penetration of lashes on the tested side compared to the sound side (sign of Souques' lashes).

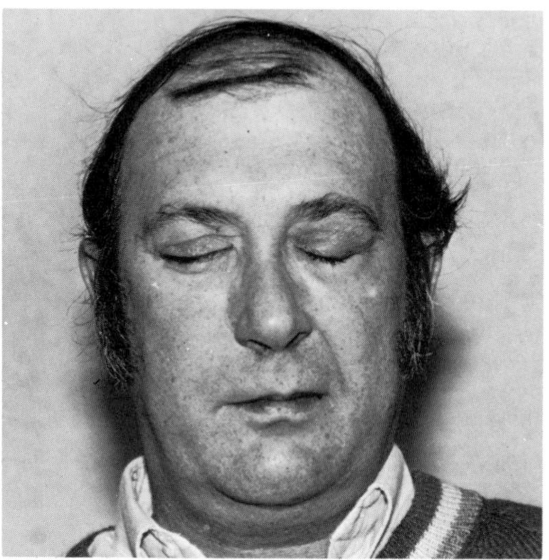

Grade **0** (R F P), palpebral occlusion.

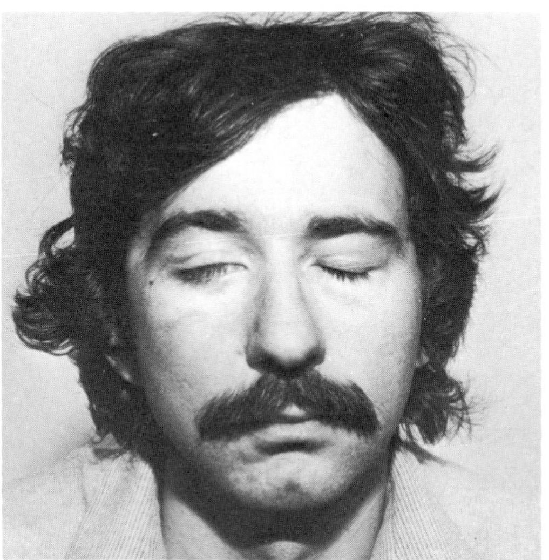

1st stage: Eye closes (R F P).

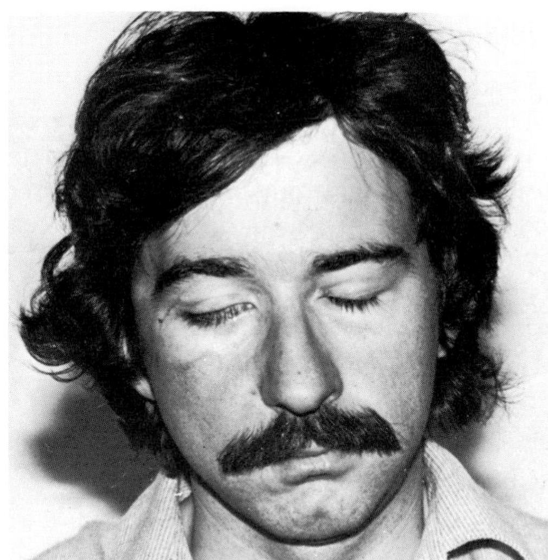

2nd stage: Upper lid lifts after eye closes (R F P). Dupuis-Dutemps-Cestan sign.

4 • Upper eyelid:
The tested eyelid folds in the same manner as the sound one.
Wrinkles in the lateral angle of the eye form a crow's foot similar to the one formed by the infra-orbital portion of the orbicularis.
• Lower eyelid:
The lower eyelid folds firmly, pulling the cheek up, causing it to protrude. The movement is symmetrical, identical in extent and synchronized with the sound side.
It is globally integrated into voluntary facial expression. When orbicularis oculi is at **4**, the lashes disappear into the palpebral fissure on the tested side to the same depth as on the sound side.

TESTING MUSCLE TONE

– 2 *Atonia*

• Upper eyelid:
Minute wrinkles have completely disappeared. The lid sags over its eyelash rim, disturbing vision.
• Lower eyelid:
Wrinkles have disappeared. The lower lid sags heavily, the eyelash rim hangs out, forming an ectropion.

– 1 *Hypotonia*

• Upper eyelid:
The upper lid no longer sags over the eyelash rim. As the eye closes, the upper lid pulls more strongly against the levator palpebrae.
• Lower eyelid:
There is no ectropion of the lower lid and wrinkles are beginning to form.

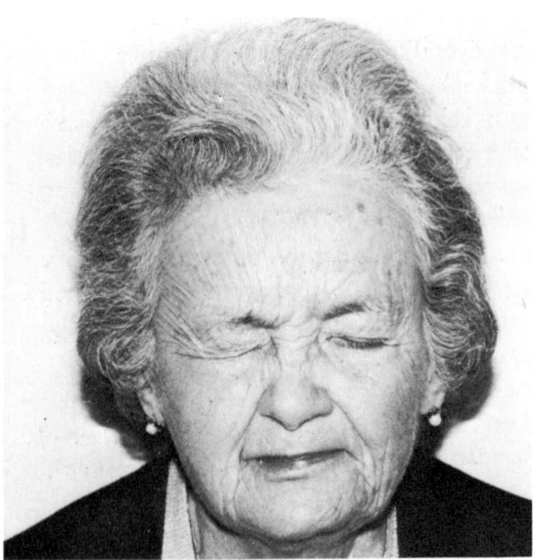

Sign of Souques's lashes (L F P).

Atonia of infra-orbital portions with ectropion (R F P).

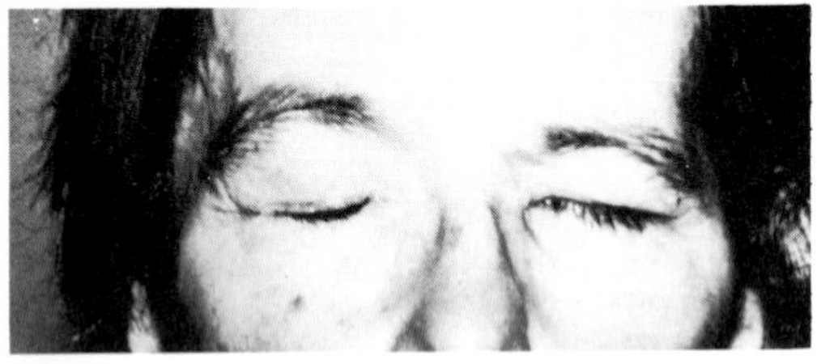

Hypotonia of upper and lower lids (L F P).

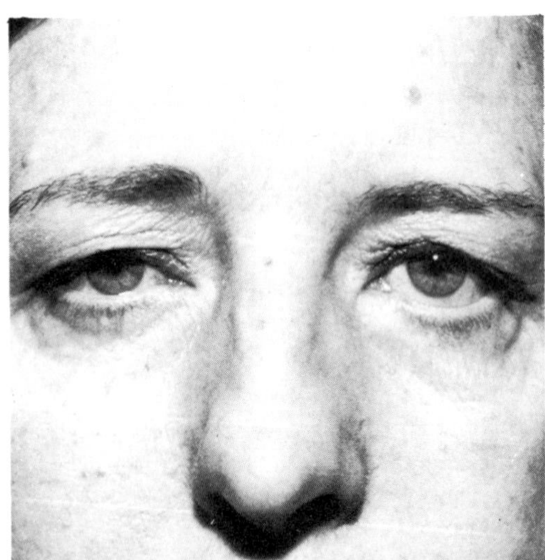

Hypotonia of lower lid (L F P).

0 *Normalization of tone*

• Upper eyelid:
Wrinkles on the tested side are symmetrical and similar to those on the sound side.
• Lower eyelid:
Same.

+1 *Moderate hypertonia*

Concerns both the upper and lower lids. The eye becomes smaller in both its diameters, compared to the sound side.

+2 *Major hypertonia*

Concerns both the upper and lower lids. The eye is much smaller, opening only a few millimetres vertically and about one centimetre in length, considerably impairing the subject's field of vision.

SYNKINESES

0 *Synkinesis is absent*

No disorderly movement of the orbicularis oculi is perceived.

+1 *Synkinesis can be inhibited voluntarily*

Since zygomaticus major links the infra-orbital portion of orbicularis oculi to the superior portion of orbicularis oris, the orbicularis oris can be fixed in its most shortened position by tightening the lips in an unamused smile.
This position places zygomaticus major in a maximally elongated position.
Synkinesis of the orbicularis oculi is completely thwarted by this starting position.

+2 *Synkinesis can be inhibited with digital pressure*

With the subject in the same position as described above, press down and in just over the nasogenial groove. Keep the fibres of zygomaticus major in a fully stretched position. Ask subject to hold his eye open and gradually release digital pressure.

Irrepressible synkinesis: + 3 (L F P).

A balance between the orbicularis oculi and the orbicularis oris should be achieved. The patient should be able to control the movement before the mirror. The eye should not close and the upper lip should not rise.

+ **3** *Synkinesis cannot be inhibited*

Called 'mouth-eye synkinesis'.
Orbicularis oris contracts entirely each time the eye is closed with force, creating a functional hindrance which is quite disagreeable for the patient.

Hemispasm Involuntary, global contraction of the orbicularis oculi passively pulls the zygomaticus major which in turn pulls the upper lip. This movement participates in the hemifacial spasm.

LEVATOR PALPEBRAE SUPERIORIS

Relaxed state. Expression.

A flat, trapezoidal muscle (see page 88).

Origin Common tendinous ring and inferior aspect of the body of the sphenoid in front of and above the optic canal.

Insertion In three layers.
—Superficial: blends with deep surface of palpebral septum and enters skin of upper eyelid.
—Middle: anterior surface of superior tarsus.
—Deep: superior fornix of the conjunctiva.

Innervation Common oculomotor nerve (III).

Function Raises upper eyelid; opens the eye when the orbicularis oculi is at rest. The movement is alternating and successive.

Agonist Smooth Muller muscle fibres.

Antagonists In spontaneous closing, the supra-orbital portion of the orbicularis oculi; in forced closing, the supra- and infra-orbital portions of orbicularis oculi, the corrugator supercili, procerus and the zygomatics.

Expression Fright.

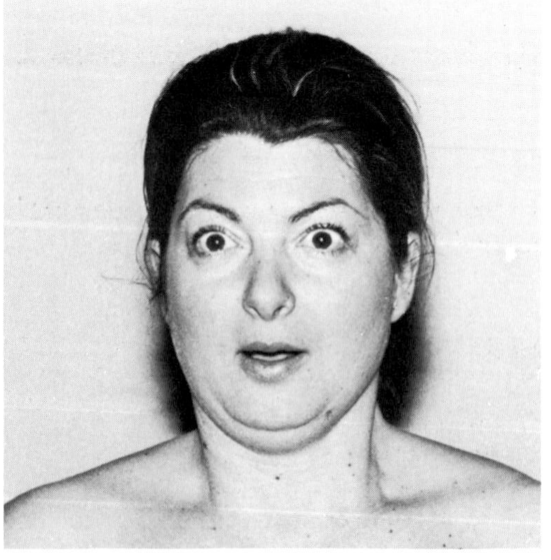

Expression.

MUSCLE TESTING

Starting position

Eye is spontaneously closed without force.

Test

Ask subject to raise his eyelids gradually until his eyes are looking entirely upwards.

Possibility of error

None.

Deficiency

When there is a lesion of the oculomotor nerve, it is impossible to completely raise the upper eyelid (ptosis).

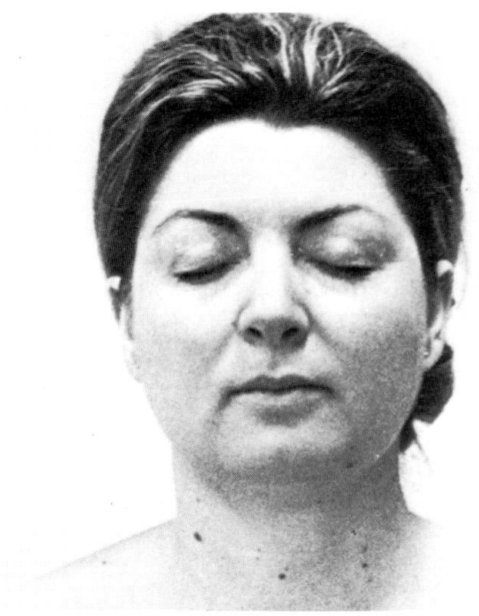

Starting position.

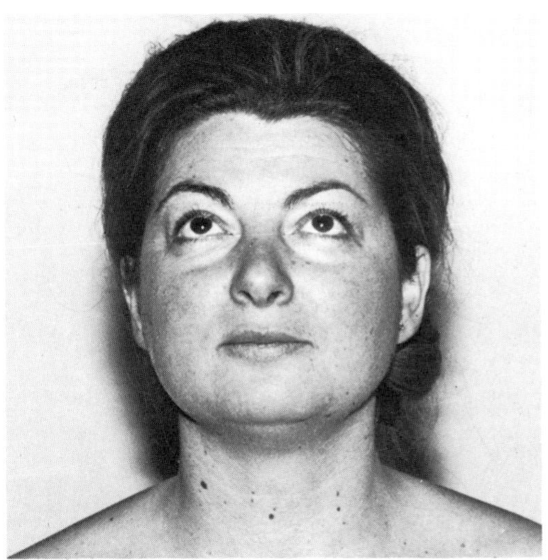

Final position.

Grading

0 The contraction is not visible, no trace of it is perceived in the upper eyelid.

1 At the starting position, a faint contraction is observed.

2 The eyelid is more mobile.
The movement should be repeated five times. When compared to the sound side, it is slow and incomplete as the muscle tires.

3 The upper eyelid moves more rapidly.
The movement should be repeated ten times fully but is not yet synchronized with the sound side.

4 The upper eyelid rises normally. The movement is extensive, symmetrical and in synchronization with the sound side. It is fully integrated into global facial expression. Batting the eyelids is the simultaneous, instant response to resting and to movements of the orbicularis oculi.
The movement may be performed automatically or voluntarily.

TESTING MUSCLE TONE

−2 *Atonia*

Major ptosis of the upper eyelid which is heavy and wrinkle free.

−1 *Hypotonia*

The upper lid is less heavy. The subject manages to lift it to about mid-iris.

0 *Normalization of tone*

The tested lid rises normally and resembles the lid on the sound side.

Hypertonia

Never encountered. Facial hypertonia is present only in muscles innervated by the facial nerve.

SYNKINESES

Never encountered.
Same for **hemispasms**.

MUSCLES OF THE NOSE

There are three:
—Two dilators:
- nasalis pars transversa (transverse part).
- dilatator naris (nasalis, alar part).

—One constrictor: depressor septi.

NASALIS PARS TRANSVERSA

A thin, flat, triangular muscle running from the centre of the longitudinal axis of the nose to the canine fossa.

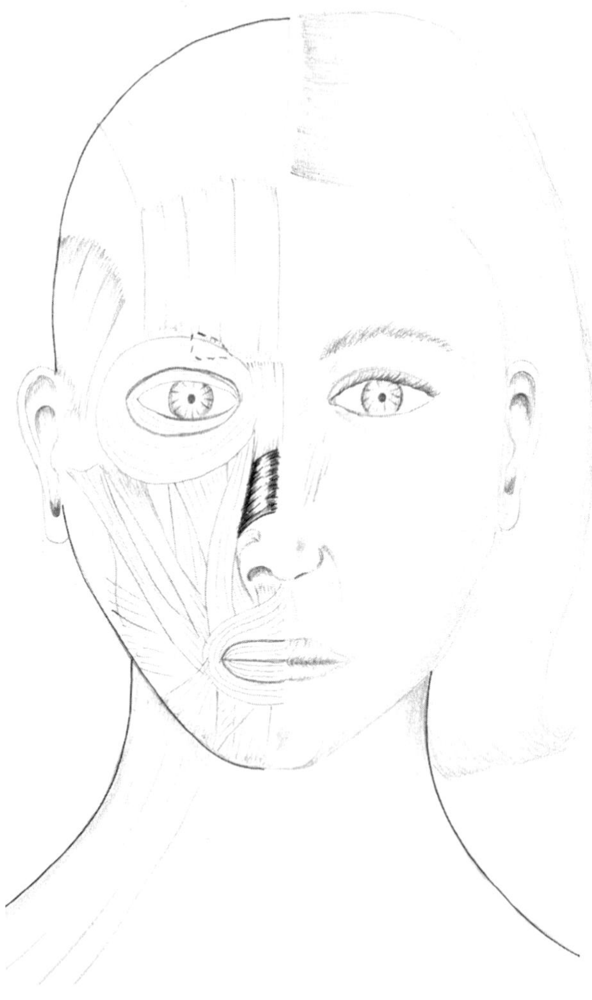

Relaxed state. Expression.

Origin The aponeurotic line along the ridge of the nose.

Insertion Inferior fibres are oriented towards and insert into the deep surface of skin.
Superior fibres blend into the lateral fibres of the depressor septi.

Innervation Transverse division of the infra-orbital branch of the temporo-facial.

Function Dilator fibres dilate the nostrils with dilator naris. Alar fibres lift the nasal ala up and forward. Transverse fibres may compress the nostrils working with the depressor septi.

Agonists *For dilating fibres*: ipsi- and contralateral levator anguli oris, dilatator naris, levator labii superioris.
For constricting fibres: lateral fibres of the ipsilateral depressor septi.

Antagonists Ipsi- and contralateral depressor septi.

Expression Participates in the sniff-test (works synergistically with the diaphragm). Marks the end of apnoea prolonged to the limit of asphyxia.

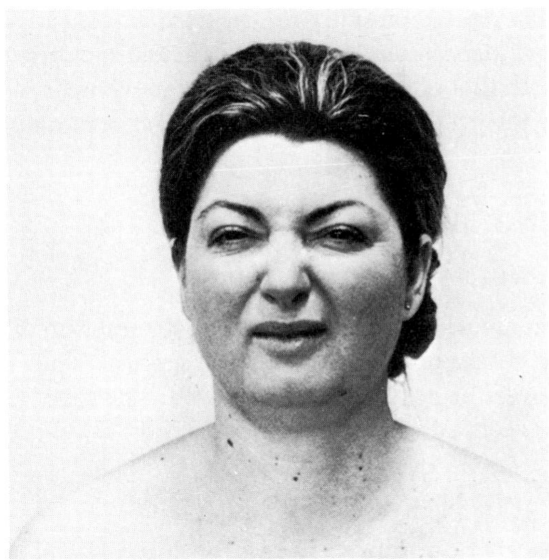

Expression.

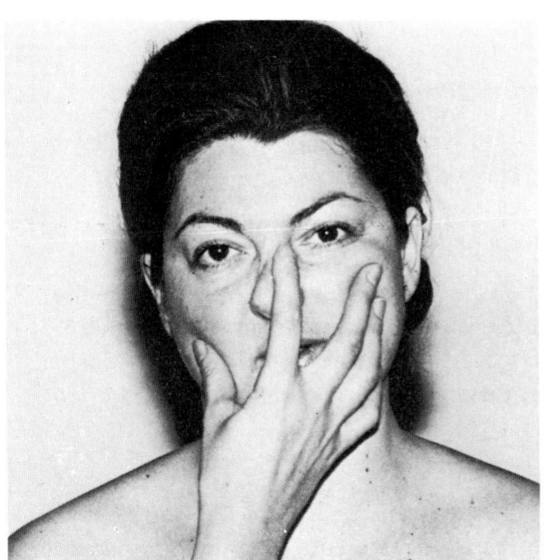

Starting position.

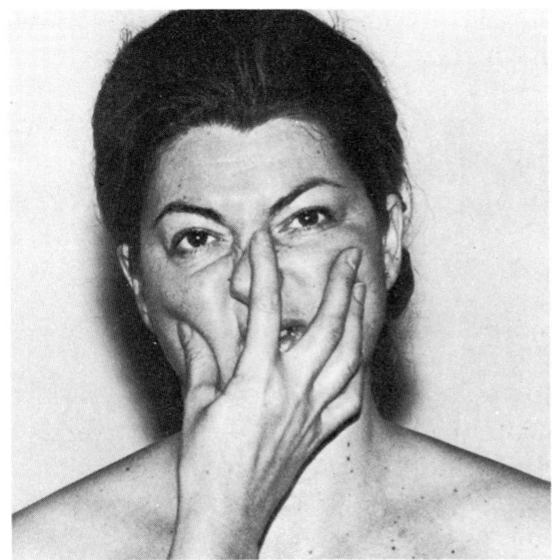

Final position.

MUSCLE TESTING

Starting position

Use digital pressure to dislocate the sound nasalis, pushing it to the median axis of the nose and holding it there.

Test

Alar fibres:

Raise nasal alae slightly by sniffing several times consecutively.
Note: The nasalis pars transversa is tested as a dilator.

Transverse fibres:

Lower nasal alae.

Possibility of error

Contralateral nasalis may draw skin of the nose all the way over to the sound side.

Deficiency

If alar fibres are paralysed, the skin just above the nostril is flattened against the nasal septum where it hampers breathing. The tip of the nose veers off towards the sound side, flattening the nostril and shortening its transverse diameter.

Grading

0 The contraction is invisible, even under bright, glancing light. It cannot be felt in the middle of the dorso-lateral portion of the nose when the movement is requested.

1 During the contraction a slight movement in the surface of the skin is perceived, particularly in the middle of the dorso-lateral portion of the nose where minute wrinkles form.

2 The skin is more mobile. Wrinkles stand out clearly on the dorso-lateral portion of the nose during the contraction. The movement should be repeated five times. The muscle tires quickly when compared to the sound side. The movement is slow and incomplete.

3 The skin is even more mobile. The number of wrinkles and their depth increases.
The movement should be repeated ten times to its full extent, though it is not yet synchronized with the sound side.

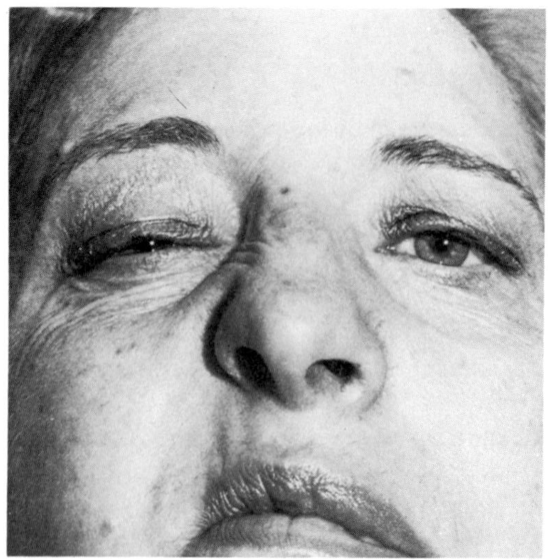
Grade **0** (L F P).

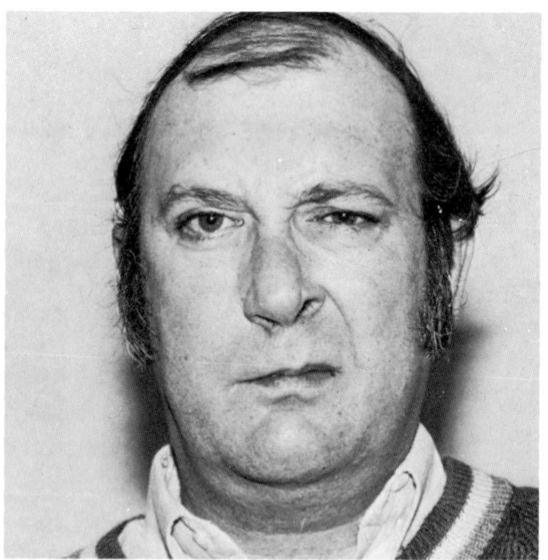
Atonia: − 2 (R F P).

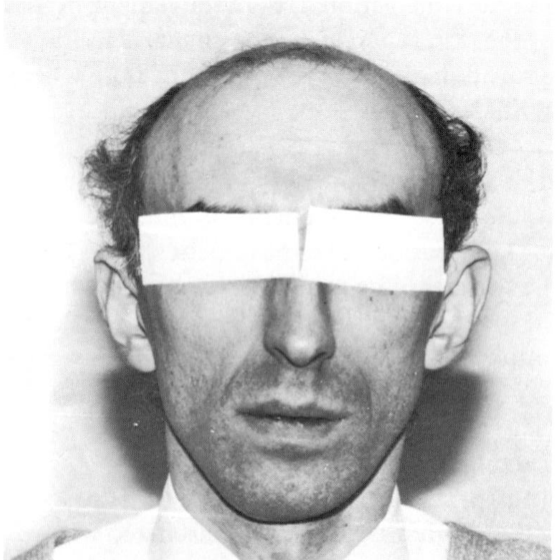
Normalization of tone: 0 (L F P).

4 The dorso-lateral skin over the nose moves easily. The movement is ample, symmetrical and synchronized with that of the sound side. The movement is globally integrated into voluntary facial expression.

TESTING MUSCLE TONE

− 2 *Atonia*

Wrinkles are non-existent. The dorso-lateral skin of the nose is flattened against the nasal septum. The groove over the nostril has disappeared. The tip of the nose veers off towards the sound side.

− 1 *Hypotonia*

Wrinkles are beginning to form. The dorso-lateral skin of the nose is less drawn towards the sound side. The nose is straightened and reintegrates itself as the axis of facial symmetry.

0 *Normalization of tone*

Wrinkles are symmetrical, deeper, and similar to those on the sound side.

+ 1 *Moderate hypertonia*

When the face is relaxed, on the affected side the nostril is elevated and wrinkles are moderately accentuated. Moderate hypertonia slightly deepens the nasogenial groove over the canine fossa.

+ 2 *Major hypertonia*

The nasalis pars transversa is spasmodic. Wrinkles are prominent. All elements described in moderate hypertonia are accentuated. The upper lip is elevated on the affected side.

Major hypertonia: + 2 (L F P).

SYNKINESES

0 *Synkinesis is absent*

There is no disorderly movement of the nasalis pars transversa.

+ 1 *Synkinesis can be inhibited voluntarily*

While looking in a mirror, the subject manages voluntary control of the muscle.

+ 2 *Synkinesis can be inhibited with digital pressure*

Very difficult to do because the levator anguli oris (caninus) usually participates. However, this synkinesis is rare and exceptional.

+ 3 *Synkinesis cannot be inhibited with digital pressure*

Rare and exceptional.

Hemispasm The involuntary contraction of the nasalis pars transversa participates in the hemifacial spasm drawing the sound side towards the affected side.

DILATATOR NARIS (nasalis alar part, or pars alaris)

A small, flat, triangular muscle starting at the nasolabial groove, widening over the nostril till it ends at the lateral margin of the nostril.

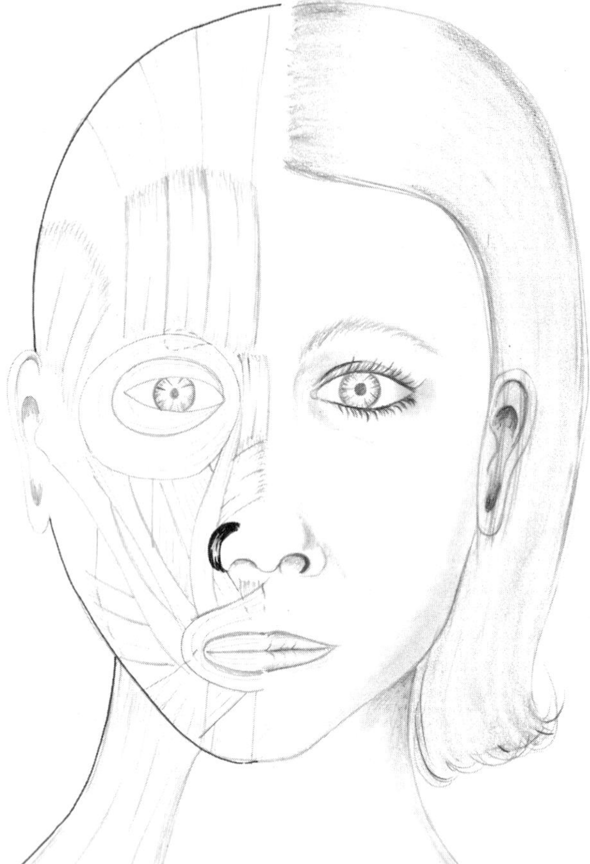

Origin Lateral cartilage of nasal ala.

Insertion Deep surface of skin in nasolabial groove at lower extremity of nasal ala.

Innervation Dilator division, infra-orbital branch of the temporo-facial.

Function Enlarges transverse diameter of nasal aperture, drawing nasal ala up and outwards.

Agonists Ipsilateral alar fibres of the nasalis pars transversa, levator labii superioris and levator anguli oris.

Antagonists Ipsilateral, inferior transverse fibres of the nasalis pars transversa and depressor septi.

Expression Joyful astonishment, craving or desire.

Relaxed state. Expression.

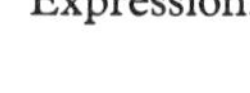

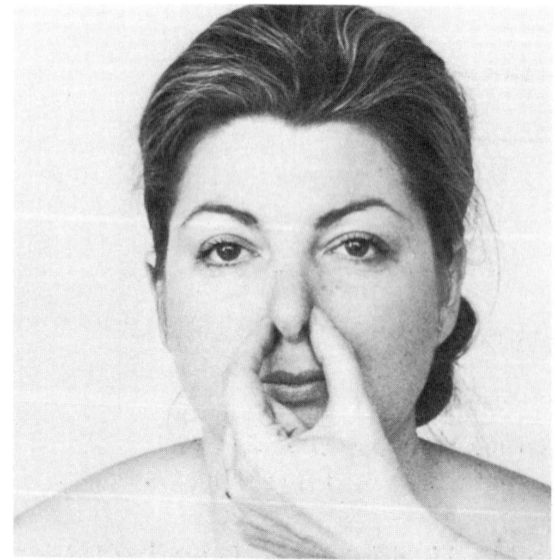

Starting position.

MUSCLE TESTING

Starting position

Hold nasal alae just over the nostrils, between thumb and forefinger of same hand without exerting pressure.

Test

Ask patient to breathe in deeply, dilating his nostrils as much as possible.

Possibility of error

The ipsilateral levator anguli oris might intervene, passively raising the nostril.

Deficiency

There is no trace of pressure against the finger on the tested side.

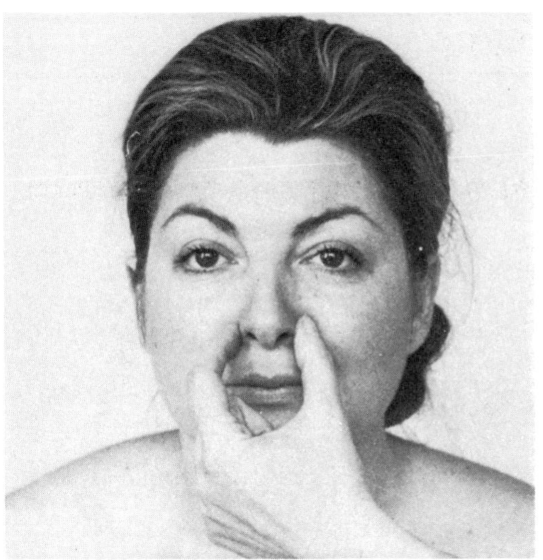

Final position.

Grading

0 The contraction is invisible, even under bright, glancing light, and it cannot be felt.

1 During the contraction the groove over the nostril deepens.

2 A slight lifting of the nasal ala is felt beneath the finger.

3 The nasal ala widens, the nostril clearly dilating, but the movement is not synchronized with that of the sound side.

4 The nasal ala moves easily. The nostril dilates amply, symmetrically and in synchronization with the sound side. The movement is globally integrated into voluntary facial expression.

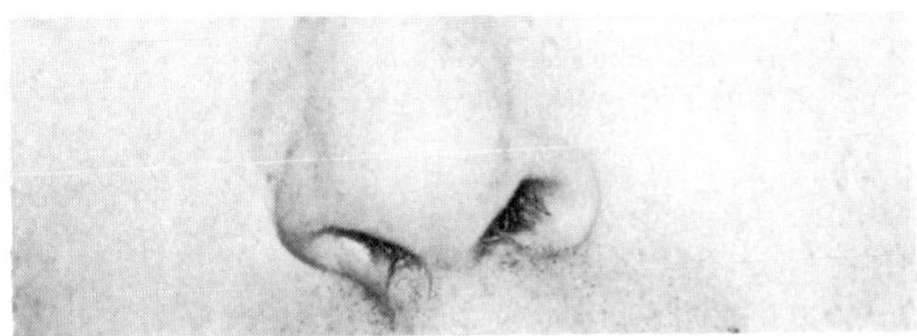

Atonia: − 2 (R F P).

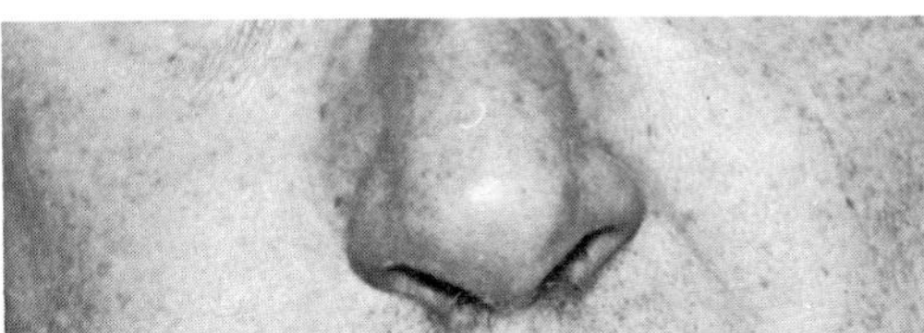

Hypotonia: − 1 (R F P).

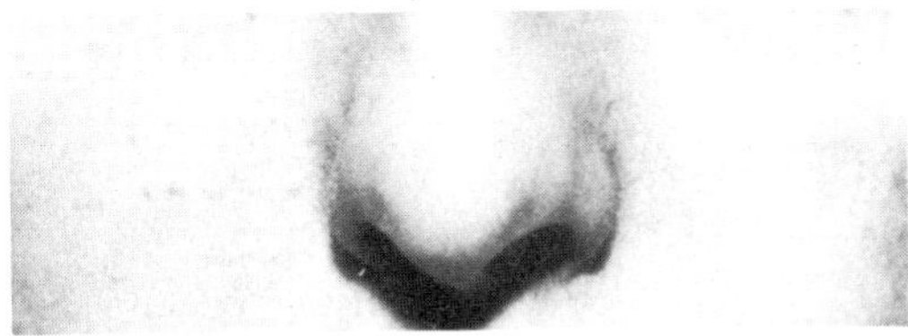

Normalization of tone: 0 (R F P).

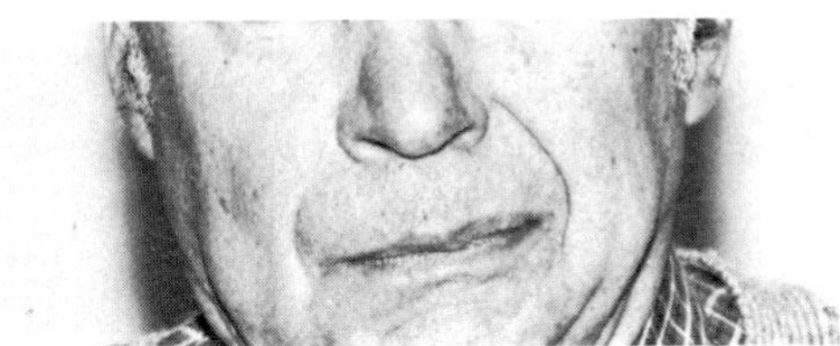

Left hemispasm.

TESTING MUSCLE TONE

− 2 *Atonia*

The nostril has lost its shape.

− 1 *Hypotonia*

The nostril is less flattened, less sagging. The groove over the nostril is beginning to form.

0 *Normalization of tone*

The nostrils are symmetrical in shape.

+ 1 *Moderate hypertonia*

Rare.

+ 2 *Major hypertonia*

Rare.

SYNKINESES

0 *Synkinesis is absent*

There is no disorderly nostril movement.
Grades + 1, + 2, and + 3 do not exist.

Hemispasm The nasal ala rises passively, drawn by the spasm of levator anguli oris.

DEPRESSOR SEPTI (myrtus)

A flat, four-sided muscle running from the posterior margin of the nostril towards the lip.

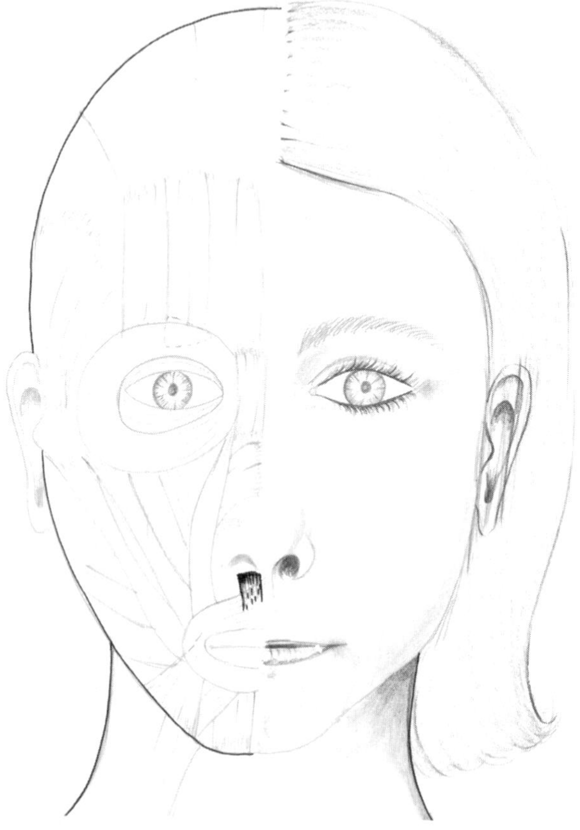

Relaxed state. Expression.

Origin Incisive fossa of the maxilla and alveolar protrusion of the canine tooth.

Insertion Nasal septum and posterior edge of nasal ala.

Innervation Depressor septi twig of inferior branch of temporo-facial.

Function Lowers nasal ala and reduces transversal diameter of nasal aperture. Lowers upper lip.

Agonists Contralateral depressor septi and ipsilateral lateral fibres of the nasalis pars transversa.

Antagonists Ipsilateral levator anguli oris, levator labii superioris, dilatator naris, and alar fibres of the nasalis pars transversa.

Expression Permits navigating a razor over the upper lip, 'the barber's muscle'.

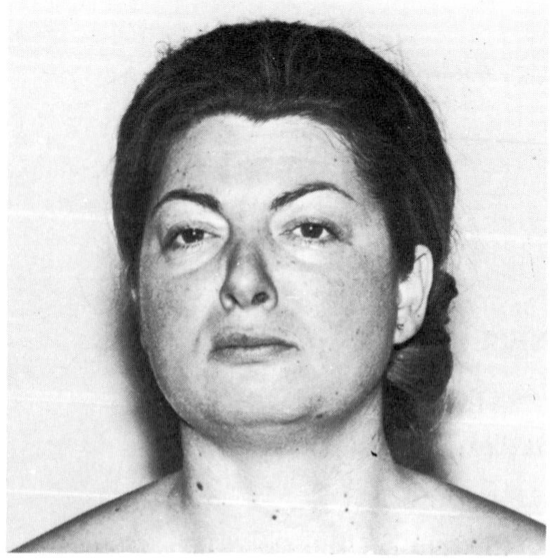

Function.

MUSCLE TESTING

Starting position

On the relaxed face, use digital pressure on the sound upper lip to dislocate the median axis of the face, pushing it towards the affected side and holding it there. The muscular fibres to be tested are thus in a shortened position.

Test

Ask subject to draw his upper lip down gradually and tuck it under his teeth.

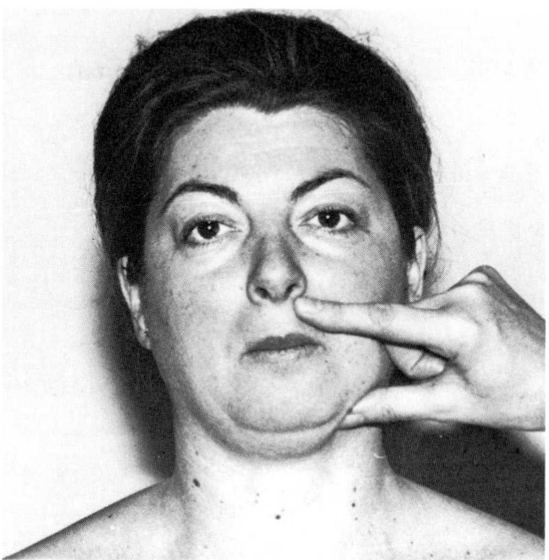

Starting position.

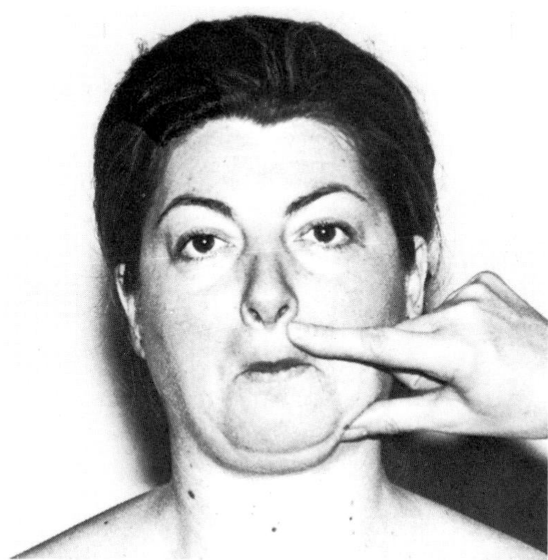

Final position.

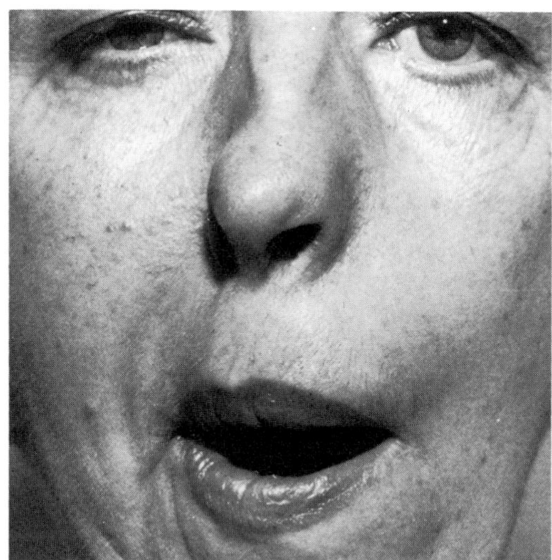

Grade **0** (L F P).

Possibility of error

If the nasolabial groove is not well held, too fast or too strong a contraction of the contralateral depressor septi will pull upon the tested side. Skin movement gives the illusion of a contraction.

Deficiency

The surface of the skin from beneath the lateral infra-nasal region to the upper lip does not move downwards.

Grading

0 The contraction is invisible, even under bright, glancing light. No contraction can be felt on the antero-lateral part of the upper lip beneath the nostril on the tested side.

1 During the contraction a slight downward movement is perceived in the surface of the skin.

2 The inter-nasolabial space increases, the skin is more mobile. The movement should be repeated five times. The muscle tires quickly and when completely exhausted, is pulled by the sound muscle. The movement is slow and incomplete.

3 The inter-nasolabial skin moves more and faster. The movement should be repeated ten times fully but it is not yet synchronized with the sound side.

4 The inter-nasolabial skin moves easily. The movement is ample, symmetrical and in synchronization with the sound side. It is globally integrated into voluntary facial expression.

TESTING MUSCLE TONE

– 2 *Atonia*

The edges defining the fossa in the centre of the inter-nasolabial space have disappeared.

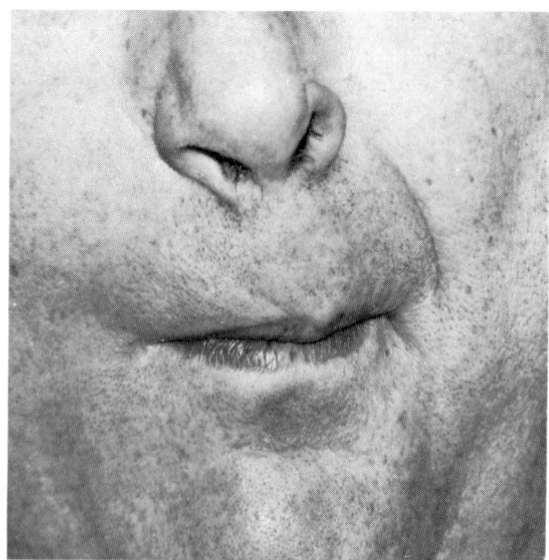

Grade **1** (R F P).

− **1** *Hypotonia*

The edges of the fossa are visible but less so than on the sound side.

0 *Normalization of tone*

Relief is symmetrical.

+ **1** *Moderate hypertonia*

Rare or non-existent.

+ **2** *Major hypertonia*

Rare or non-existent.

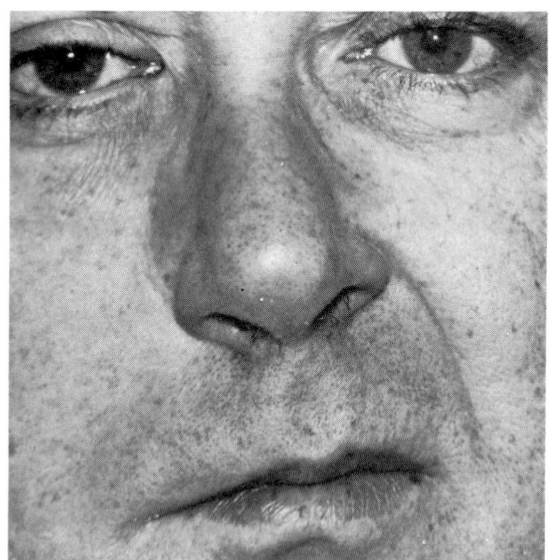

Hypotonia: − 1 (R F P).

SYNKINESES

Synkineses of the depressor septi are non-existent.

Hemispasm None.

MUSCLES OF THE LIPS

The dilators are:
—levator labii superioris alaeque nasi.
—levator anguli oris.
—zygomaticus minor.
—zygomaticus major.
—buccinator.
—risorius.
—depressor labii inferioris.
—mentalis.
—depressor anguli oris.
—platysma.

The constrictors are:
—orbicularis oris, superior and inferior portions.
—compressor labiorum.

LEVATOR LABII SUPERIORIS ALAEQUE NASI

There are two portions: the superficial and the deep.

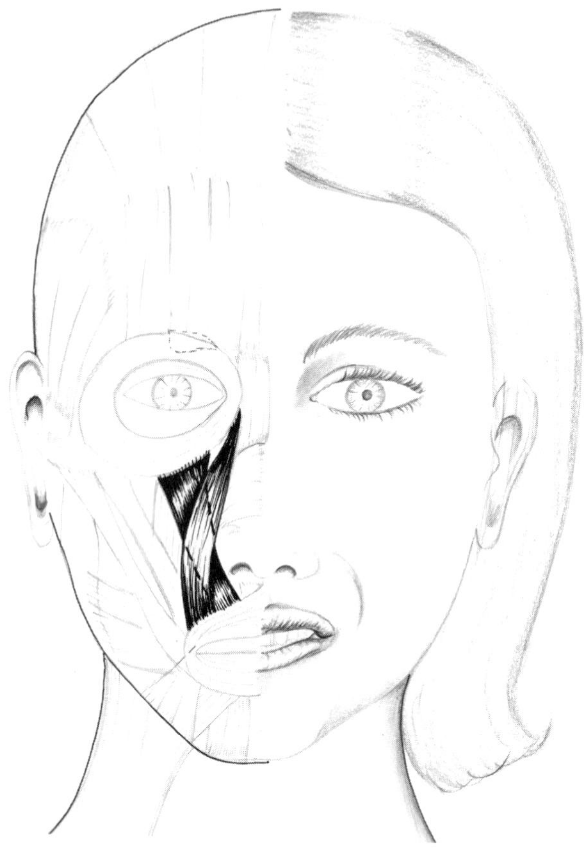

Relaxed state. Expression.

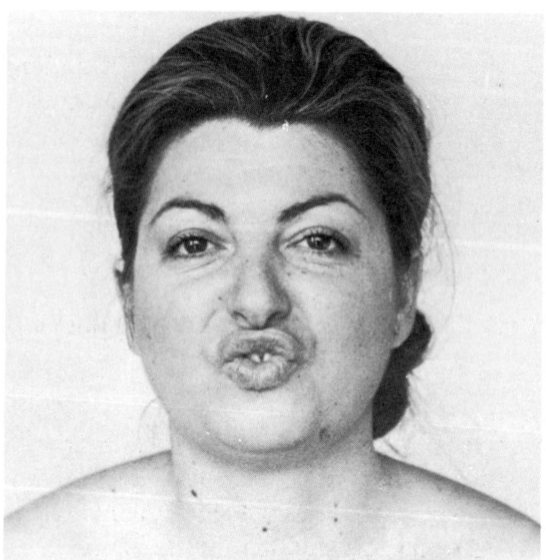

Function.

SUPERFICIAL PORTION

A long, thin muscle located in the nasogenial groove, running from the medial edge of the orbit down to the upper lip.

Origin Lateral surface of frontal process of the maxillary, it is covered by the infra-orbital portion of the orbicularis oculi.

Insertion Deep surface of skin at the posterior edge of nasal ala, and the deep surface of skin of the upper lip.

Innervation Levator division of infra-orbital branch of temporo-facial nerve.

Function Raises nasal ala and upper lip which lifts and projects outward uncovering upper incisors up to the gum.

Agonists Ipsilateral levator labii deep portion, and superior portion of orbicular oris.

Antagonists Ipsilateral depressor septi and the orbicularis oris in its position of maximal contraction.

Expression Displeasure, dissatisfaction.

DEEP PORTION

This muscle is covered by the superficial portion.

Origin Medial half of the margin of the orbit above the infra-orbital foramen.

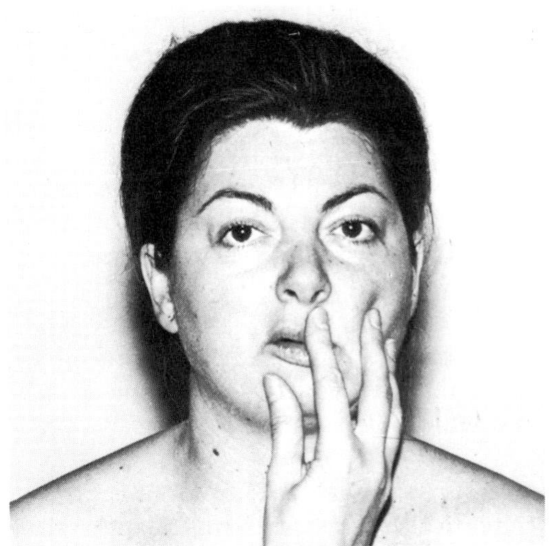

Starting position.

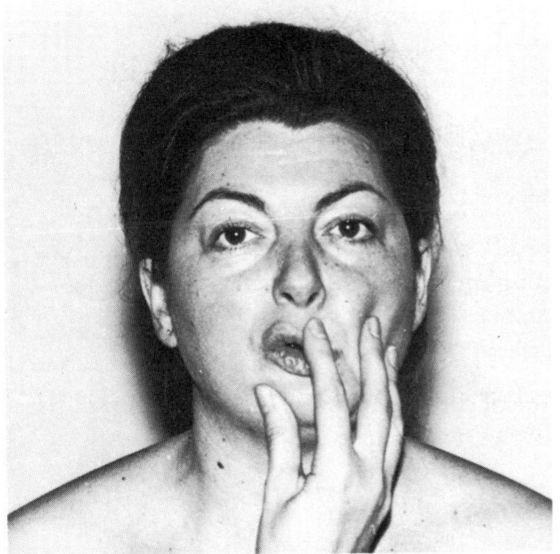

Final position.

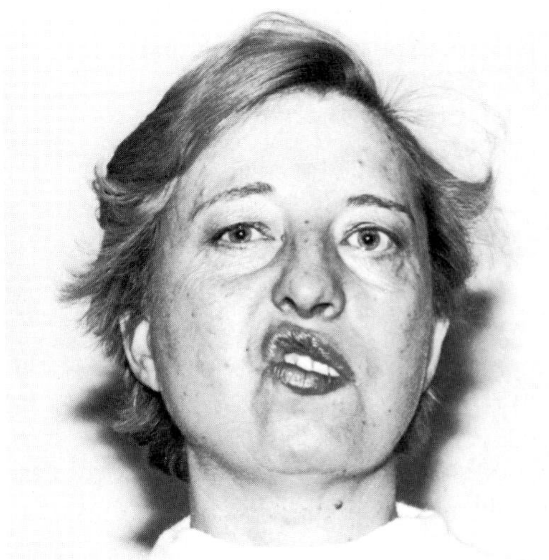

Grade **0** (L F P).

Insertion After crossing the fibres of the levator anguli oris, it inserts into the deep surface of skin at inferior edge of nasal ala and upper lip.

Innervation Same as for superficial portion.

Function Same as for superficial portion.

Agonists Ipsilateral superficial portion of the levator labii superioris and superior portion of the orbicularis oris.

Antagonists Ipsilateral depressor septi and the orbicularis oris in its position of maximal contraction.

Expression Same as for the superficial portion of the levator labii superioris.
Note: The two portions of levator labii superioris alaeque nasi are tested together.

MUSCLE TESTING

Starting position

On the relaxed face, use digital pressure to hold the contralateral levator in a maximally stretched position. Pressure is placed over the upper lip pushing it down and inward towards the median axis of the face.

Test

Ask subject to raise his lip curling it backwards so as to expose his incisor teeth.

Possibility of error

The upper lip may be raised laterally without curling backwards by the levator anguli oris which exposes the canine tooth.
The upper lip may be drawn up and out by contraction of zygomaticus minor.

Deficiency

The skin and the rim of the upper lip are immobile.

Grading

0 The contraction is invisible, even under bright, glancing light. On the tested side, no contraction can be felt over the upper lip when the movement is requested. The upper lip should push the finger forward and upwards.

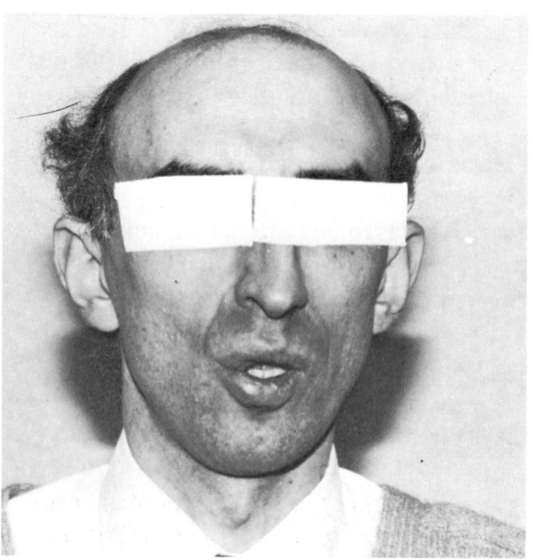
Grade **2** (L F P).

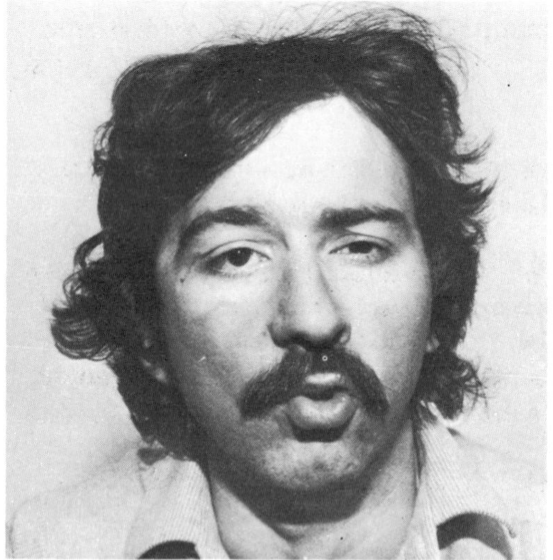
Grade **4** (R F P).

1 During the contraction a slight movement is perceived in the skin's surface, raising the hair (moustache) over the upper lip.

2 The upper lip rises more clearly, curling backwards. The movement should be repeated five times. The muscle tires quickly and is then pulled towards the sound side. The movement is slow and incomplete.

3 The upper lip moves more readily. The movement should be repeated ten times fully but it is not yet synchronized with the sound side.

4 The upper lip moves easily. The movement is ample, symmetrical and in synchronization with the sound side. It is globally integrated into voluntary facial expression.

TESTING MUSCLE TONE

− 2 *Atonia*

The upper lip on the tested side is lower, tucked in, and lies clinging against the incisors on that side.

− 1 *Hypotonia*

The outline of the upper lip is higher, the rim of the lip is more clearly defined.

0 *Normalization of tone*

Relief is symmetrical.

+ 1 *Moderate hypertonia*

Very rare and not to be confused with that of the levator anguli oris, which is frequent.

+ 2 *Major hypertonia*

Rarer still.

SYNKINESES

Synkineses of the levator labii superioris alaeque nasi are exceptional.

Hemispasm The levator labii superioris alaeque nasi sometimes participates in the hemifacial spasm.

LEVATOR ANGULI ORIS

Muscle passing from the maxillary canine fossa to the upper lip.

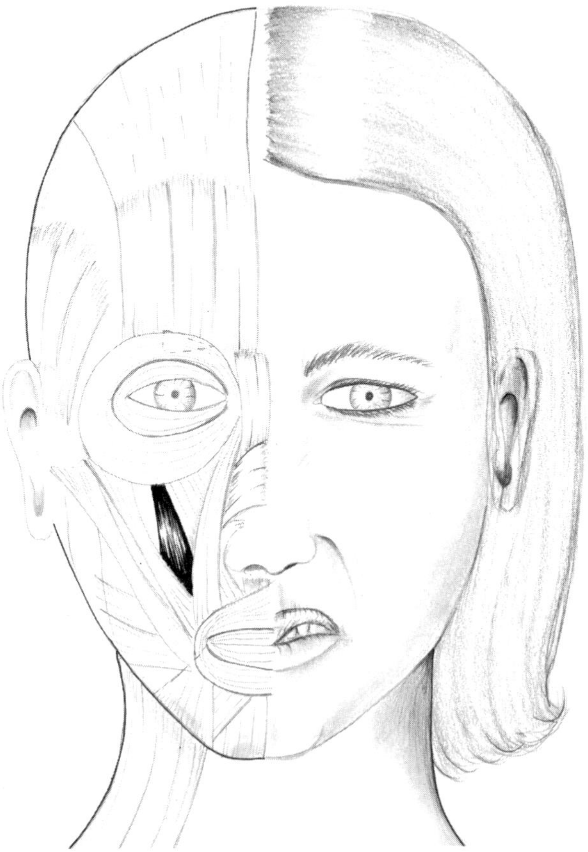

Relaxed state. Expression.

Origin Immediately below the infra-orbital foramen in the canine fossa.

Insertion Into the deep surface of the skin at the commissura labii, fibres intermingle with those of orbicularis oris, depressor anguli oris, and the zygomatics.

Innervation Canine twig, inferior infra-orbital branch of the temporo-facial, forming a naso-genial plexus.

Function Raises the commissura labii, deepens the nasogenial groove, reveals the canine tooth.

Agonists Nasalis pars transversa, zygomatics and levator labii superioris.

Antagonists Orbicularis oris contracted and the depressor septi.

Expression Pride, self-esteem, mockery, sniggering.

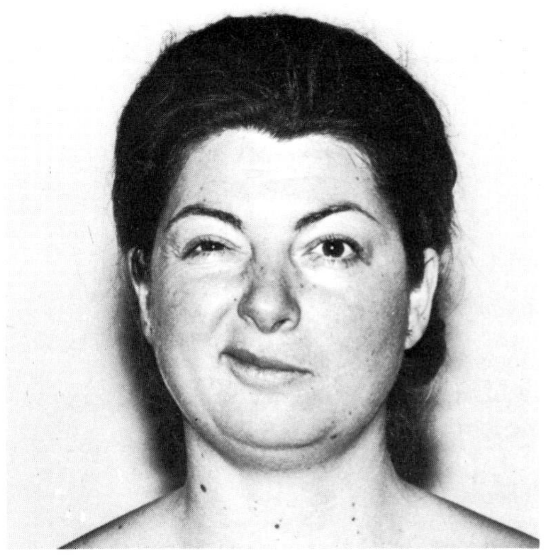

Expression.

MUSCLE TESTING

Starting position

On a relaxed face, hold the contralateral levator anguli oris with digital pressure beneath the nostril. Direct the pressure down and in towards the median axis of the face.

Test

Ask the subject to lift the side of his upper lip, exposing his canine tooth.

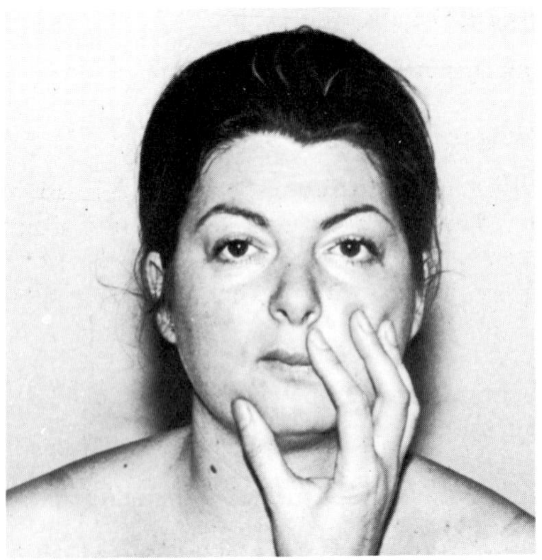

Starting position.

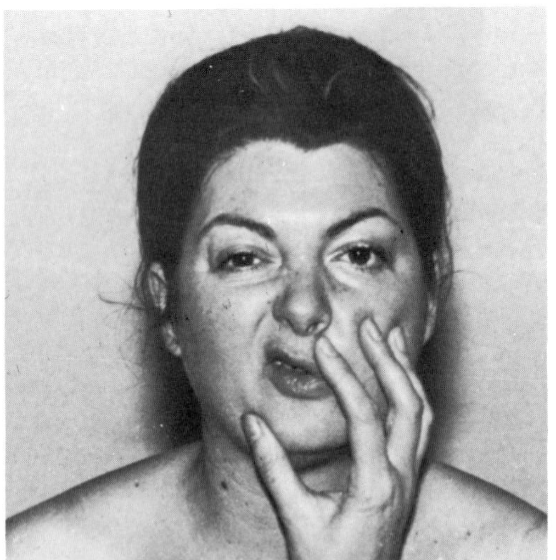

Final position.

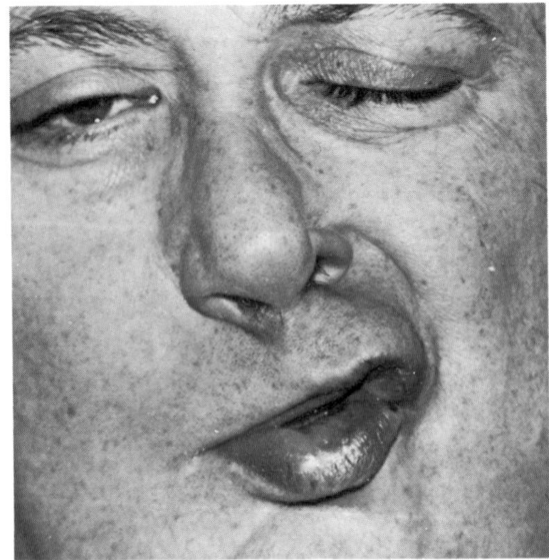

Grade **0** (R F P).

Possibility of error

The upper lip may rise, curling itself forward as the levators contract.
The upper lip may be drawn laterally upward without curling as the zygomatics contract.

Deficiency

The upper lip is immobile. It clings to the upper teeth. The angle formed by the nasogenial groove and the nasal ala is absent.

Grading

0 The contraction is invisible, even under bright, glancing light. No contraction is felt laterally above the upper lip.

1 During the contraction, a slight movement is seen in the skin's surface at the angle formed by the nasal ala and the nasogenial groove, barely raising the upper lip.

2 The upper lip rises more distinctly, uncovering the canine tooth. The movement should be repeated five times. The muscle tires quickly and when exhausted, is pulled over by the sound side. The movement is slow and incomplete.

3 The upper lip moves more rapidly. The canine tooth is more visible. The movement should be repeated ten times fully. It is not yet synchronized with the sound side.

4 The upper lip moves more easily and exposes the canine tooth. The movement is ample, symmetrical and in synchronization with the sound side. It is globally integrated into voluntary facial expression.

TESTING MUSCLE TONE

– 2 *Atonia*

Often associated with atonia of the levator labii superioris alaeque nasi. The lip clings specifically to the canine tooth.

– 1 *Hypotonia*

The angle formed by the nasogenial groove and the nasal ala begins to deepen.

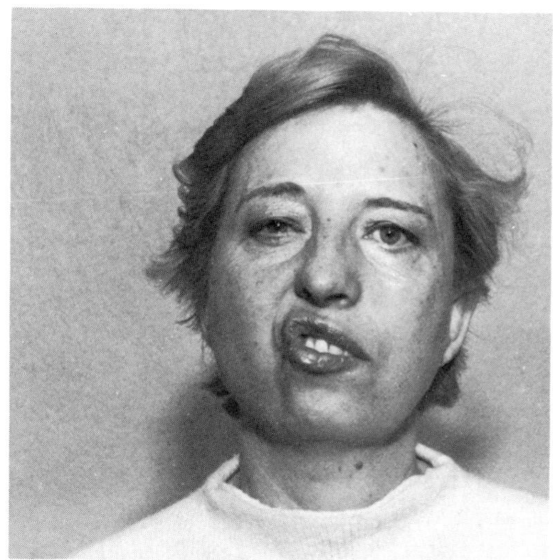

Atonia: − 2 (L F P).

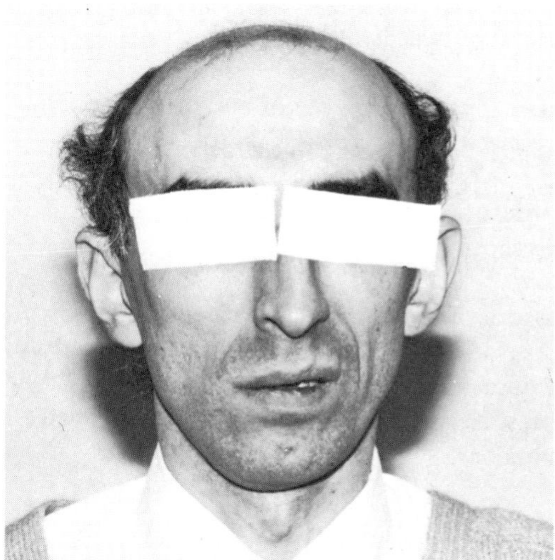

Hypertonia: + 2 (L F P).

0 *Normalization of tone*

Nasogenial grooves are symmetrical.

+ 1 *Moderate hypertonia*

On the relaxed face, there is a slight lateral elevation of the tested lip. The angle formed by the nostril and the nasogenial groove is slightly accentuated. The nasogenial groove is a little higher than on the sound side.

+ 2 *Major hypertonia*

The face is distinctly asymmetrical, the sound side being drawn toward the pathological side. The angle of the nose on the tested side is closed, deep, and elevated when compared with that of the sound side.

SYNKINESES

0 *Synkinesis is absent*

There is no disorderly movement of the levator anguli oris particularly when
—the lips are being constricted.
—the lips are being dilated.

+ 1 *Synkinesis can be inhibited voluntarily*

The subject manages to control a minimal rising of the lip when looking in a mirror and while constricting or dilating his mouth.

+ 2 *Synkinesis can be inhibited with digital pressure*

On a relaxed face, the muscle fibres of the levator anguli oris are held in a maximally stretched position. The upper, lateral portion of the upper lip is pushed in the direction opposite to that of the pathological movement.
Ask the subject to purse his lips gradually tighter and tighter. Gradually release digital pressure. A balance should be achieved between the orbicularis oris and the levator anguli oris, which should not rise.
The same is repeated for dilation of the lips.

+ 3 *Synkinesis cannot be inhibited*

The levator anguli oris contracts tightly as soon as the mouth moves.

Hemispasm The hemifacial spasm includes the global, involuntary contraction of the levator anguli oris.

ZYGOMATICUS MINOR

Parallels lateral edge of deep portion of levator labii superioris.

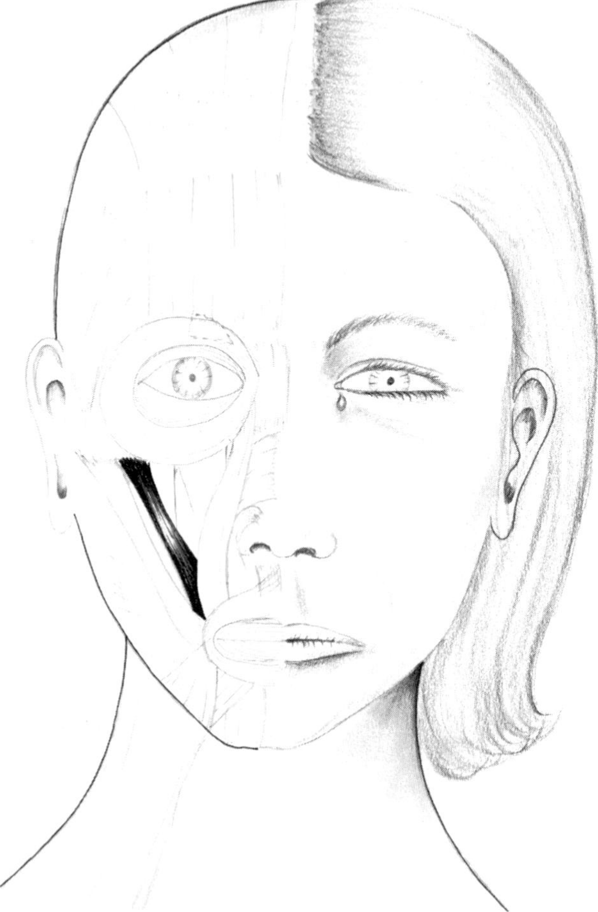

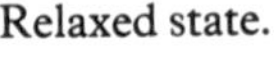

Relaxed state. Expression.

Origin Middle part of outer surface of malar bone.

Insertion The deep surface of skin of the upper lip laterally to levator labii superioris.

Innervation Infra-orbital twig of the temporo-facial branch.

Function Raises the upper lip projecting it upwards and outwards.

Agonists Zygomaticus major and inferior portion of the orbicularis oculii.

Antagonists Ipsilateral orbicularis oris and depressor septi.

Expression In grief, as with moderate sobbing (Duchenne de Boulogne), it works synergistically with the depressor anguli oris.
In laughter, it works synergistically with the zygomaticus major.

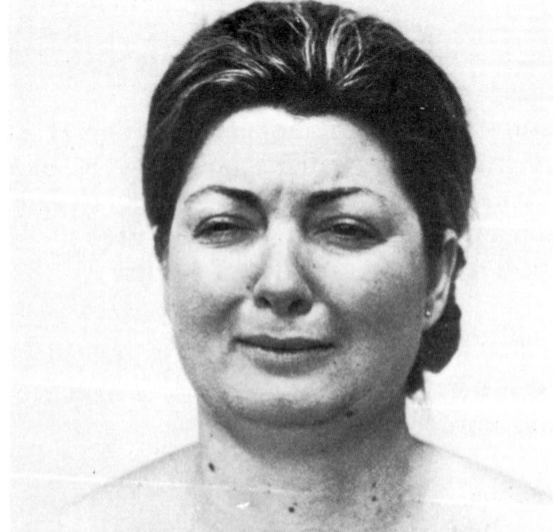

Expression.

MUSCLE TESTING

Starting position

On a relaxed face, apply digital pressure to the sound side above the nasogenial groove, pushing it down and inwards, dislocating the median axis of the face towards the affected side. The muscle to be tested is thus in its shortened position.

Test

Ask the subject to raise his cheek slightly as though he were crying a little. The contraction of the zygomaticus minor is indicated by the dimple above and outside the angle of the nose.

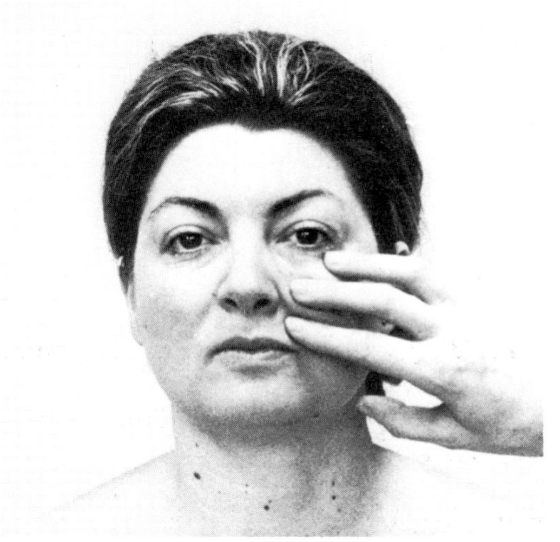

Starting position.

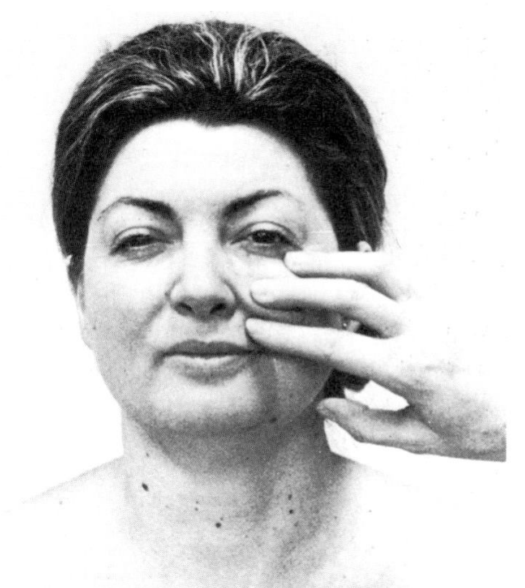

Final position.

Possibility of error

The zygomaticus major may lift the cheek and raise the upper lip, pushing it upwards and outwards. The dimple may be formed by passive traction of zygomaticus minor as the orbiculus oculi furnishes a strong contraction.

Deficiency

The cheek is immobile. No movement is perceived in the surface of the skin above the nasogenial groove.

Grading

Grading will be done jointly with zygomaticus major in laughter since it is difficult to separate the two muscles. However, the presence or absence of the dimple above that of zygomaticus major, will be noted.

TESTING MUSCLE TONE

Muscle tone of zygomaticus minor will be tested with that of zygomaticus major.

SYNKINESES

Synkineses will be seen with the zygomaticus major as will hemispasm.

ZYGOMATICUS MAJOR

Follows the outer margin of zygomaticus minor.

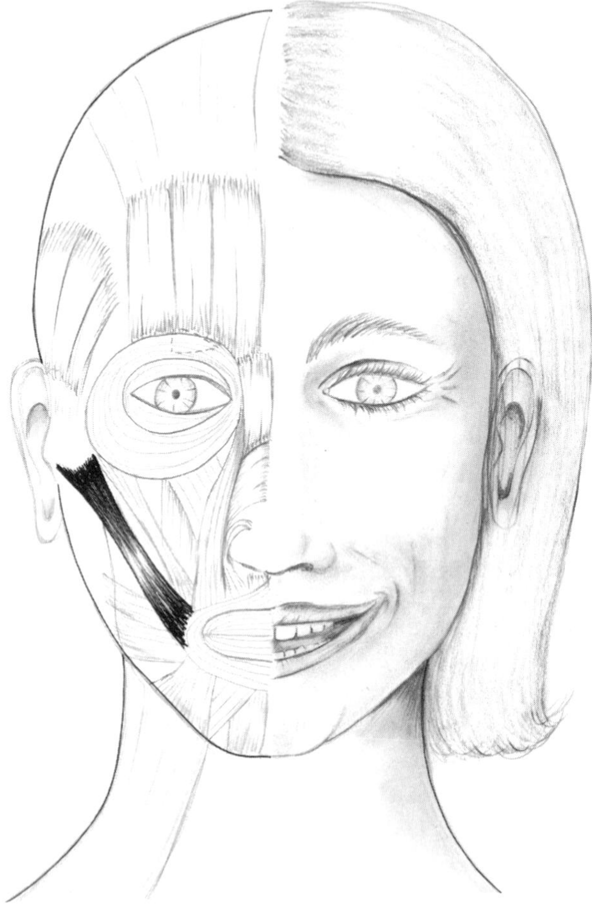

Relaxed state. Expression.

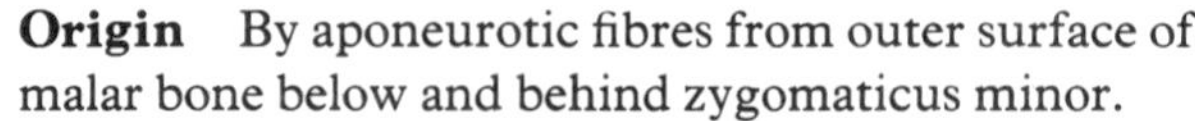

Origin By aponeurotic fibres from outer surface of malar bone below and behind zygomaticus minor.

Insertion After a slanting course down and inwards, it crosses the buccinator which is a deep muscle. It attaches to the deep surface of skin and mucous membrane at the commissura labii.

Innervation Zygomatic twigs of the infra-orbital branch of the temporo-facial.

Function Raises the commissura labii, drawing it upwards and outwards.

Agonists Zygomaticus minor and inferior portion of the orbicularis oculi.

Antagonists Orbicularis oris, depressor septi, depressor anguli oris, and platysma.

Expression Laughter, a broad smile, joy (Duchenne de Boulogne).

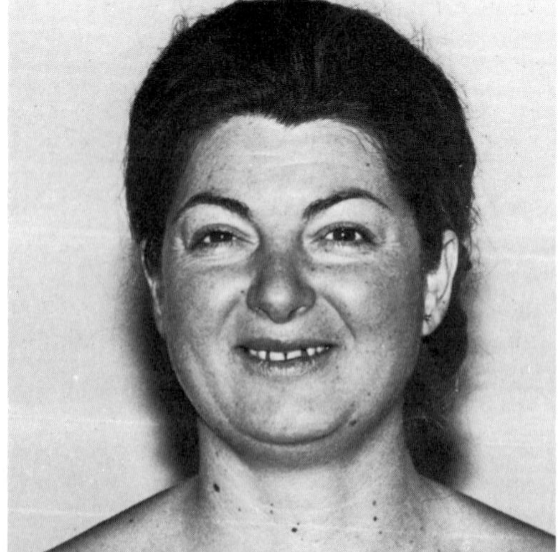

Expression.

MUSCLE TESTING

Starting position

On a relaxed face, apply digital pressure at the commissura labii on the sound side, pushing it towards the median axis of the face. The muscular fibres to be tested are thus in their shortened position.

Test

Ask subject to smile, revealing his teeth.

Possibility of error

The risorius may pull the commissura labii, but only outwards.

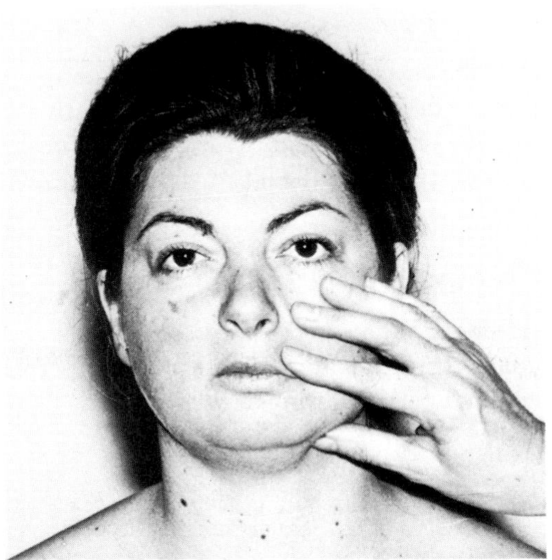

Starting position.

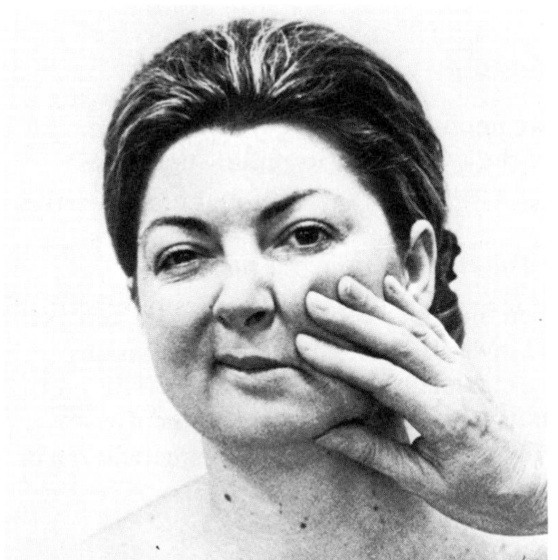

Final position.

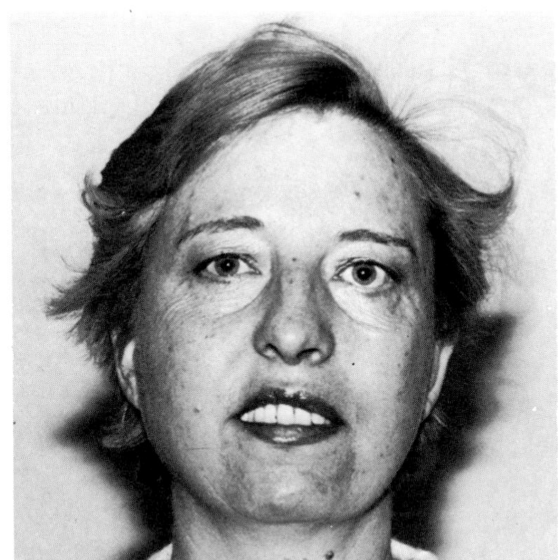

Grade **2** (L F P).

Deficiency

There is no movement in the commissura labii on the tested side. Skin is immobile at the cheekbone. The indentation which forms the nasogenial groove is non-existent.

Grading

0 The contraction is invisible, even under bright, glancing light. No contraction can be felt under the malar bone, beneath the cheekbone.

1 During the contraction a slight movement is perceived in the skin's surface at the commissura labii. The nasogenial groove begins to form.

2 The commissura labii rises more clearly upwards and outwards. The movement should be repeated five times. It slows when compared to the sound side, and is incomplete as the muscle tires.

3 The nasogenial groove is deeper. The commissura labii lifts higher. The skin over the cheekbone rises more rapidly. The movement should be repeated ten times fully but it is not yet synchronized with the sound side.

4 The cheek rises in an easy movement which is ample, symmetrical, and in synchronization with the sound side. It is globally integrated into voluntary facial expression.

TESTING MUSCLE TONE

− 2 *Atonia*

The cheek loses its form. It hangs limply and forms a bag at the bottom. The nasogenial groove sags and the commissura labii is pulled down.

− 1 *Hypotonia*

The form of the cheek has lifted slightly, the bagging at the bottom is diminished. The commissura labii is not pulled so far down. The nasogenial groove is beginning to appear.

0 *Normalization of tone*

The contour of the cheeks is symmetrical, the nasogenial groove is redefined. The commissura labii is at the same height on each side.

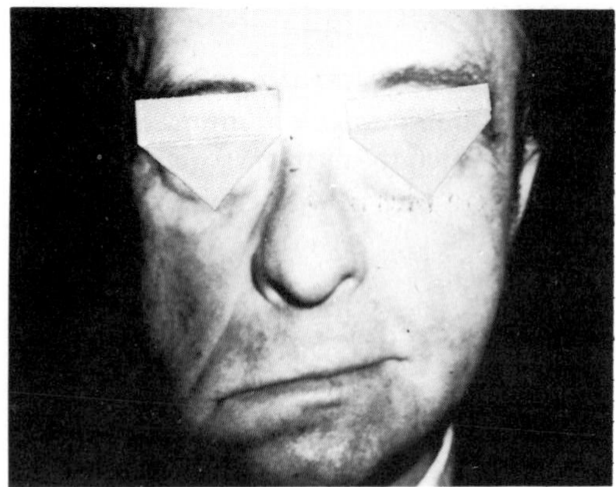

Atonia: − 2 (R F P).

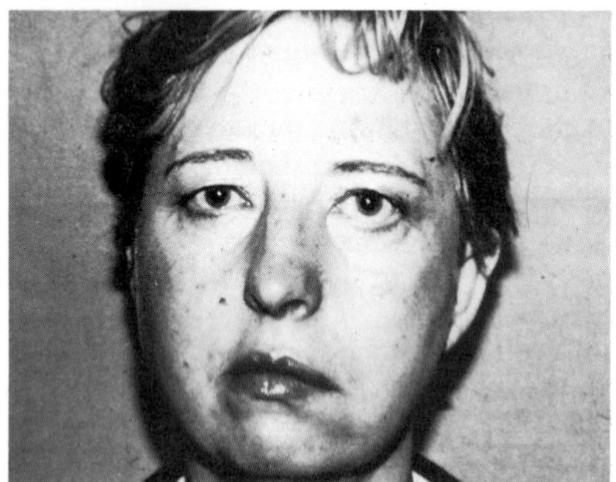

Hypotonia: − 1 (L F P).

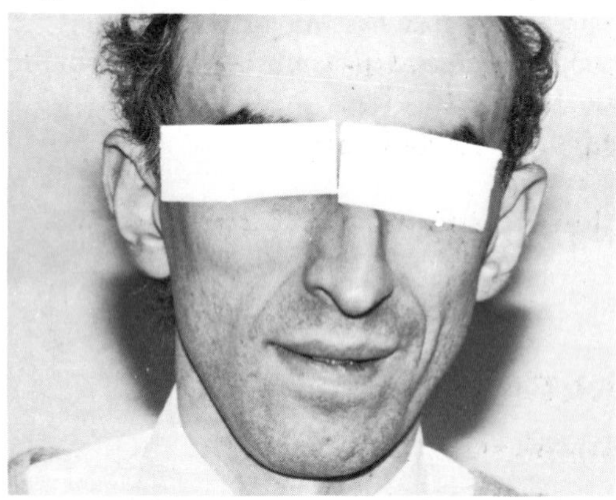

Moderate hypertonia: + 1 (L F P).

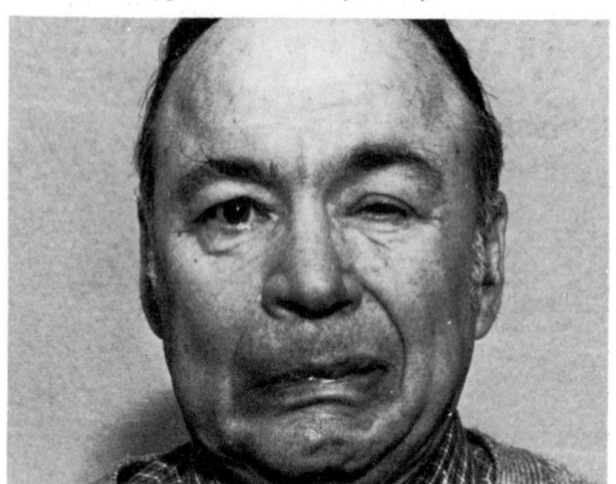

Irrepressible synkinesis: + 3 (L F P).

+ 1 *Moderate hypertonia*

The form of the cheek protrudes more. The naso-genial groove is deeper and angles up and slightly outwards. The commissura labii is higher than on the sound side, stretching the lips towards the tested side.

+ 2 *Major hypertonia*

The sound side is pulled by the pathological side. Hypertonia of the zygomaticus major includes lifting of the cheek and partial closing of the eye.

SYNKINESES

0 *Synkinesis is absent*

No disorderly movement of the zygomaticus major is perceived at the spontaneous closing of the eye.

+ 1 *Synkinesis is inhibited voluntarily*

This is accomplished by fixing the opening and the closing of the eye without its pulling upon the commissura labii.

+ 2 *Synkinesis is inhibited by digital pressure*

Pressure on the supra-commissura dimple, directed down and inwards, places fibres of zygomaticus major in a maximally stretched position. This position permits a better contraction of orbicularis oris, which will inhibit contraction of zygomaticus major.

+ 3 *Synkinesis cannot be inhibited*

The entire zygomaticus major contracts as soon as the eye is opened or closed spontaneously or voluntarily.

Hemispasm The involuntary, global contraction of both zygomatics is part of the hemifacial spasm.

BUCCINATOR

Deep muscle of the cheek, U-shaped, and open towards the front, it occupies the interval between maxilla and mandible.

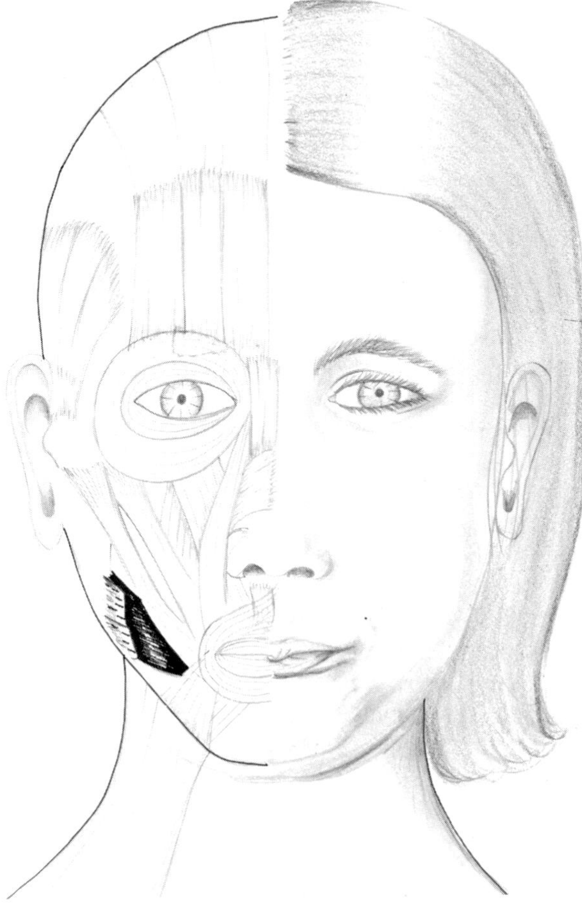

Relaxed state. Expression.

Origin
a. Alveolar process of maxilla and mandible opposite sockets of last three molars,
b. Anterior border of pterygo mandibular raphe.

Insertion Superior and inferior fibres interlace before inserting into commissura labii and deep surface of skin of lateral third of the vestibule of the mouth.

Innervation Buccinator twig of superior buccal branch of the cervico-facial.

Function Draws commissura labii backwards, lengthening the mouth opening.
Compresses the oral cavity, pushes strongly against the molars, aids in mastication, and permits high-pitched whistling.

Agonists Risorius and the masseters.

Antagonist Orbicularis oris.

Expression Satisfaction (Duchenne de Boulogne).

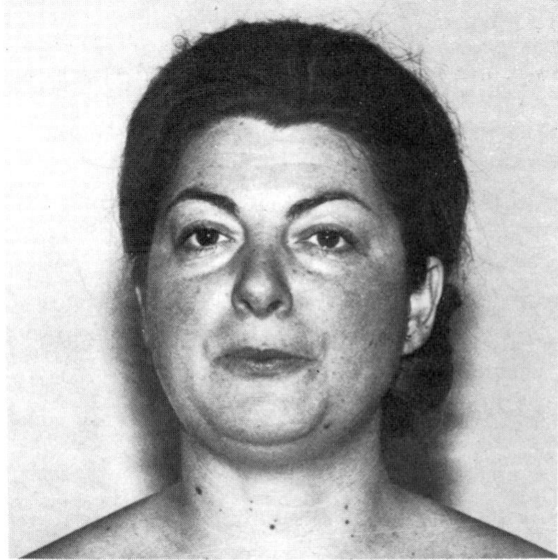

Expression.

MUSCLE TESTING

Starting position

On a relaxed face, push the lips on the sound side with two fingers, dislocating this side towards the median axis of the face. The sound side is thus immobilized, placing the muscle to be tested in its shortened position.

Test

Ask subject to compress his lips, tightening his cheeks against his molars.

Possibility of error

The commissura labii may be pulled outwards and backwards by the risorius. Masseters may give an illusion of the contraction.

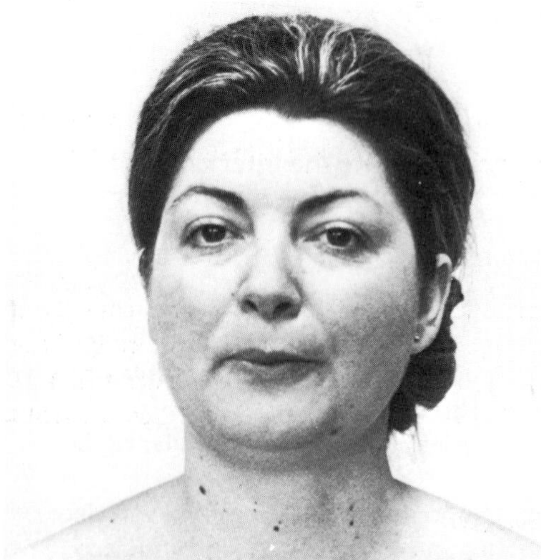

A piercing, high-pitched whistle.

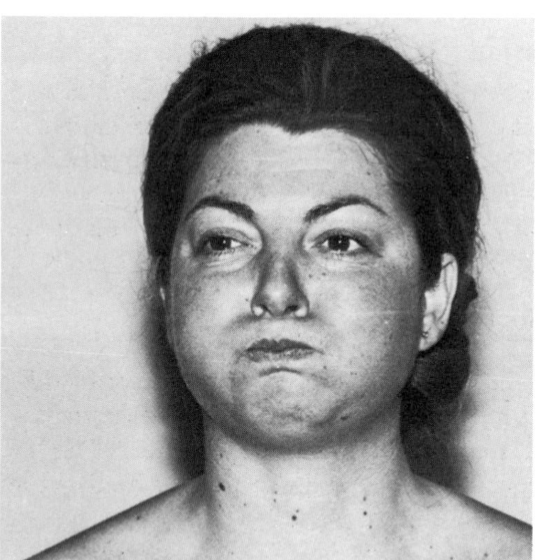

Inflating cheeks.

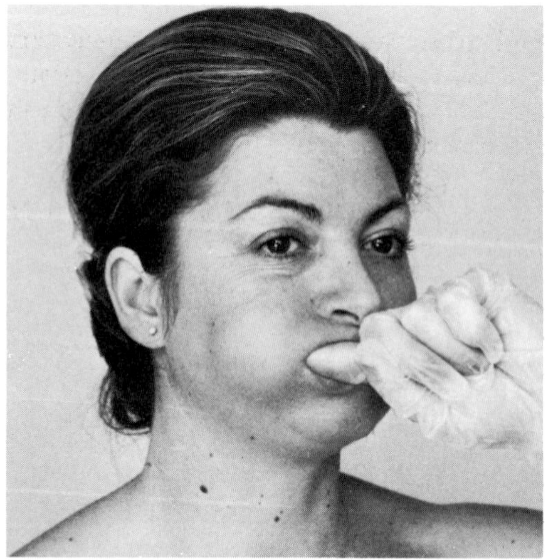

Starting position, Endo-buccal test: stretch.

Deficiency

This is felt from within the mouth. With maxilla and mandible held tightly together, slide the index along the interior face of the cheek. The cheek is flaccid, incapable of compressing itself against the palmar aspect of the index.
Ask subject to inflate his cheeks, then to compress them. The tested cheek remains distended (sign of the smoker).

Grading

0 The contraction is invisible, even under bright, glancing light. No contraction can be felt endo-buccally. Alimentary bolus cannot be centred in the buccal cavity. Food remains between cheek and mandible. The tongue must be used to dislodge it.

1 During the contraction, a slight depression is formed along the commissural line. When the lips are stretched and compressed, a slight contraction is perceived in the cheek from within the mouth. Alimentary bolus still cannot be gathered in the centre of the buccal cavity, requiring intervention of tongue for cleaning the mouth.

2 The cheek compresses more tightly against the molars. This contraction is clearly felt against the pulp of the index within the mouth. The movement should be repeated five times. It slows when compared to the sound side, and is incomplete as the muscle tires. Alimentary bolus is more effectively assembled in the centre of the buccal cavity although a slight difficulty still remains. The subject is able to produce a high-pitched whistle.

3 The lips and cheeks compress more forcefully. The dimple outside the commissura labii is deeper. Within the mouth, the cheek resists stretching by the index against which it presses strongly. The movement should be repeated ten times fully but it is not yet synchronized with the sound side. Alimentary bolus can be perfectly centred in the oral cavity without the aid of the tongue.

4 The cheeks compress identically. Dimples are symmetrical, a piercing whistle is possible. Movement is ample, symmetrical and in synchronization with the sound side. It is globally integrated into voluntary facial expression.

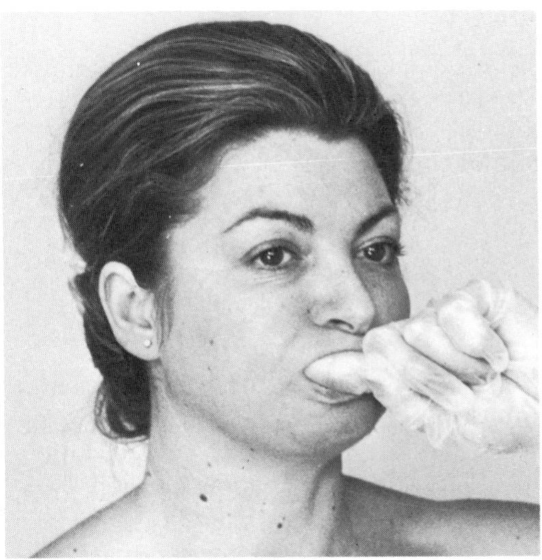
Final position, Endo-buccal test, compression.

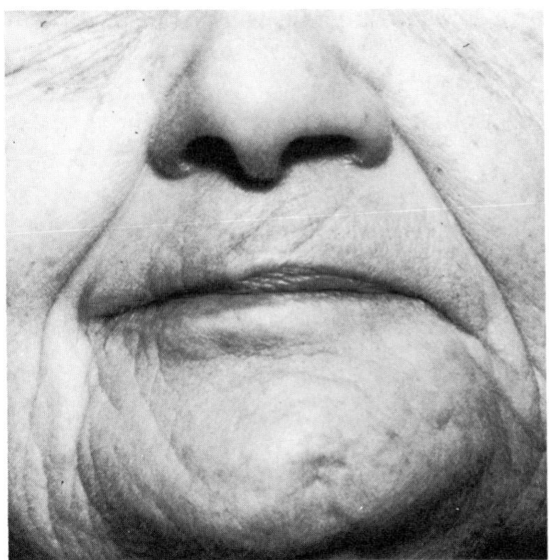
Hypotonia: − 1 (L F P).

TESTING MUSCLE TONE

− 2 *Atonia*

The cheek puffs out at each buccal expiration (sign of the smoker). The compression of stretched lips is impossible.

− 1 *Hypotonia*

The cheek is less distended following buccal expiration, the 'sign of the smoker' is still present but is less important. The subject is incapable of emitting a high-pitched whistle. Compression of lips is possible but weak.

0 *Normalization of tone*

Compression of the lips is possible. The subject is able to tighten his cheeks and whistle loudly.

+ 1 *Moderate hypertonia*

Commissura labii on the tested side is pulled sideways, lengthening the lip. The nasogenial groove is deeper. Hypertonia of the buccinator is particularly evident in endo-buccal palpation. A thin, internal cord is felt under the nasogenial groove. This cord is easily stretched.

+ 2 *Major hypertonia*

The commissura labii is distinctly pulled outwards. The cheek clings tightly to the teeth. The patient bites the inside of his cheek when chewing. The intra-buccal cord is very important and hinders the opening of the mouth.

SYNKINESES

Since this is a deep muscle, it is difficult to detect and grade synkineses. However, during the contraction of the orbicularis oris, endo-buccal palpation reveals an irrepressible synkinesis.

Hemispasm The global, involuntary contraction of the buccinator participates in the hemifacial spasm.

RISORIUS

Muscle varying in size and in form, located in the middle region of the cheek.

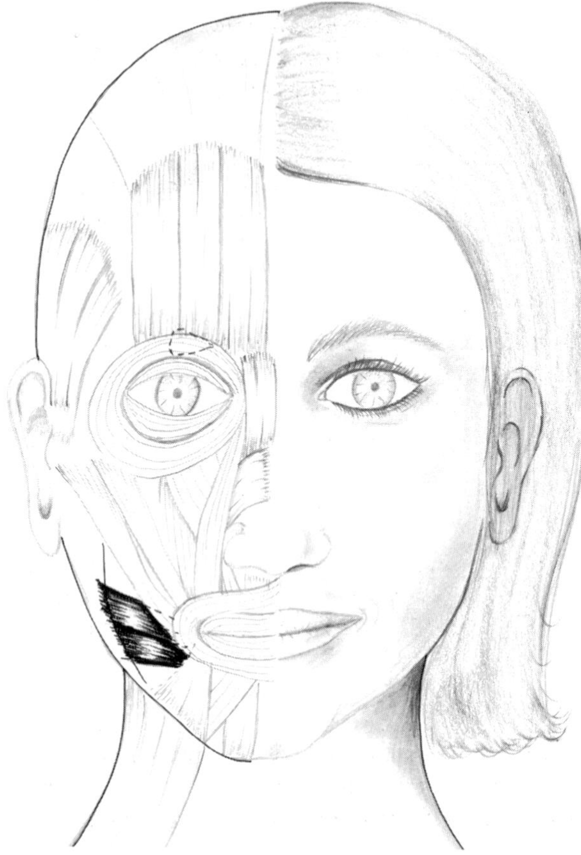

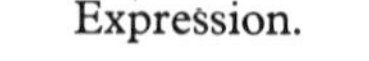

Relaxed state.

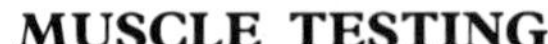

Expression.

Origin Posteriorly, from aponeurotic fibres of the masseter muscle.

Insertion The skin at the commissura labii.

Innervation Risorius twig of inferior, maxillary-buccal branch of the cervico-facial.

Function Draws the commissura labii outwards and backwards.

Agonists Ipsilateral superior fibres of the platysma.

Antagonist Orbicularis oris.

Expression Bargaining, denotes irony, gives the enigmatic smile of the Mona Lisa.

MUSCLE TESTING

Starting position

On a relaxed face, using two fingers, dislocate the sound side of the lips pushing them towards the median axis and holding them there. The sound side is thus immobilized and the muscle to be tested is held in its shortened position.

Test

Ask subject to stretch his lips outwards and backwards horizontally.

Possibility of error

The commissura labii may be drawn upwards and backwards as the zygomaticus major contracts.
The commissura labii may also be lightly stretched, the cheeks tightened against the molars as the buccinator contracts.
The commissura labii may be drawn down and back as the depressor anguli oris contracts.

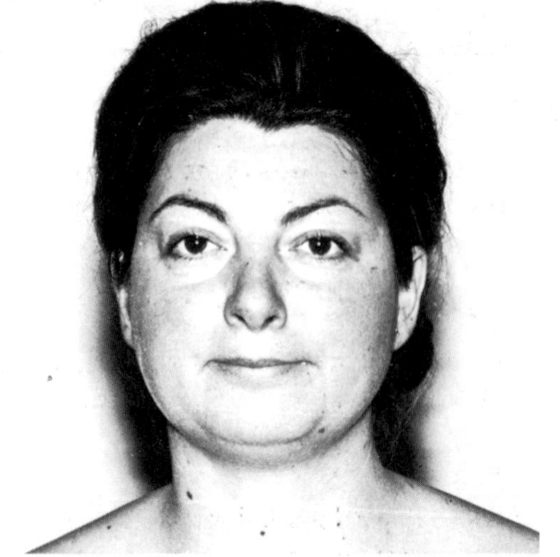

Expression.

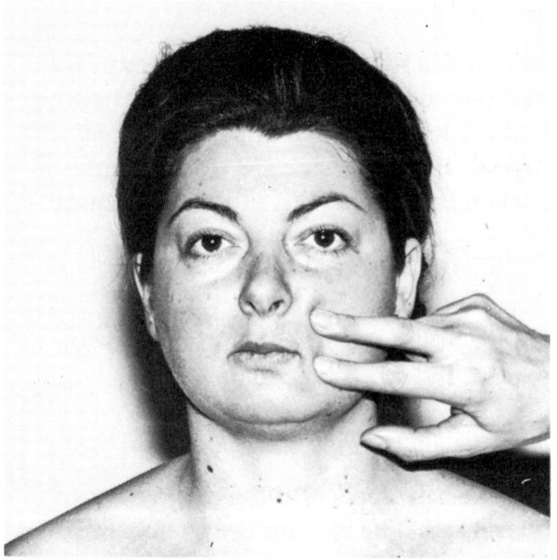

Starting position.

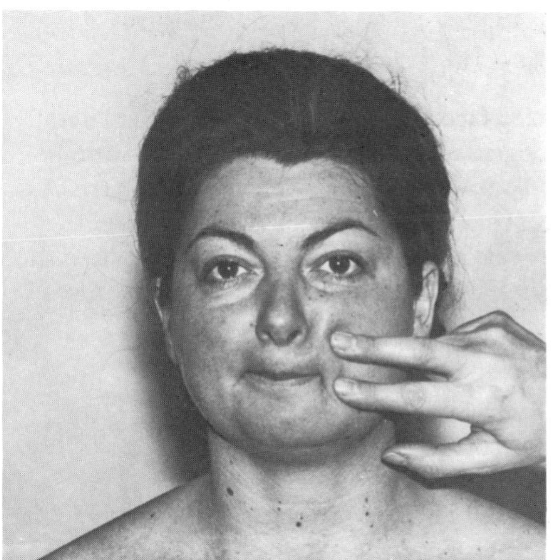

Final position.

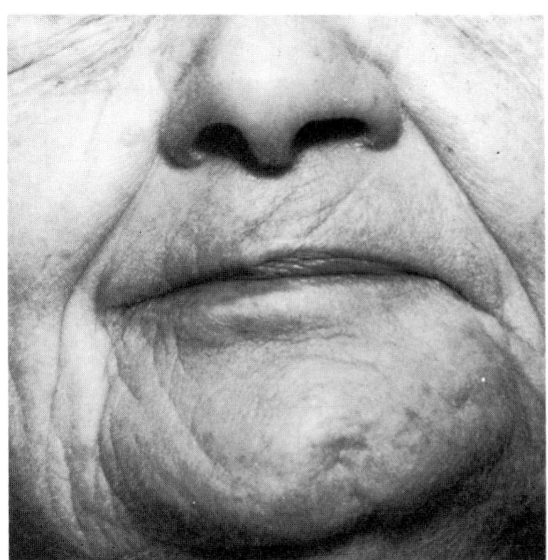

Atonia: − 2 (L F P).

Deficiency

There is no horizontal movement in the skin's surface beside the commissura labii.

Grading

0 The contraction is invisible, even under bright, glancing light. No contraction can be palpated posteriorly to the nasogenial groove in line with the commissura labii.

1 During the contraction, a slight horizontal mobility is perceived just laterally to the commissura labii.

2 The commissura labii stretches more distinctly, forming a little wrinkle at the inferior end of the nasogenial groove. The movement should be repeated five times. It slows when compared to the sound side and is incomplete as the muscle tires.

3 The commissura labii moves more rapidly. The indentation formed in the nasogenial groove is deeper. The movement should be repeated ten times fully but it is not yet synchronized with the sound side. It is globally integrated into voluntary facial expression.

4 The commissura labii moves out and back symmetrically on each side. The indentations are identical. Movement is ample, symmetrical and in synchronization with the sound side. It is integrated into global, voluntary facial expression.

TESTING MUSCLE TONE

− 2 *Atonia*

The commissura labii sags.

− 1 *Hypotonia*

The commissura labii has risen slightly and the inferior portion of the nasogenial groove begins to reappear.

0 *Normalization of tone*

The commissura labii is symmetrical with that of the sound side. The nasogenial grooves are identical.

+ 1 *Moderate hypertonia*

The commissura labii is slightly stretched out on the tested side. The nasogenial indentation is accentuated. Lips are thinner and are stretched outwards.

Hypertonia: + 2 (L F P).

Attempt at balancing orbicularis oris and risorius (L F P).

Irrepressible synkinesis: + 3 (L F P).

+ 2 *Major hypertonia*

The commissura labii is distinctly pulled outwards. The inferior part of the cheek bulges, accentuating the depth of the lower portion of the nasogenial groove. Lips on the tested side are very thin and stretched when compared with those of the sound side.

SYNKINESES

0 *Synkinesis is absent*

+ 1 *Synkinesis can be inhibited voluntarily*

Synkinesis generally appears at the contraction of the orbicularis oris. The risorius completely impedes its contraction, causing buccal asymmetry. The subject should begin to contract the orbicularis oris and find the balance point between the two.

+ 2 *Synkinesis can be inhibited with digital pressure*

On a relaxed face, apply digital pressure just laterally to the nasogenial groove on the tested side. Push horizontally towards the median axis of the face. Ask the subject to contract the orbicularis oris slightly. Gradually release pressure. The balance between the orbicularis oris and the risorius should be achieved.

+ 3 *Synkinesis cannot be inhibited*

At the slightest contraction of the orbicularis oris, the risorius contracts firmly.

Hemispasm The involuntary global contraction of the risorius participates in the hemifacial spasm.

DEPRESSOR LABII INFERIORIS (quadratus labii inferioris)

A small, quadrilateral muscle located at the lateral margin of the chin and the lower lip.

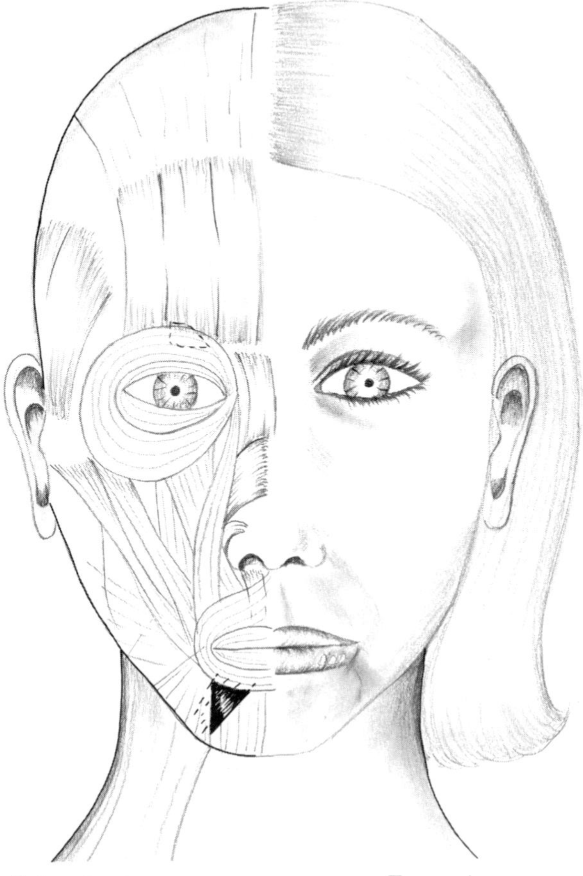

Relaxed state. Expression.

Origin Anterior third of external, oblique line of the mandible.

Insertion Teguments of the lower lip, fibres blending with those of the contralateral depressor on the median axis of the face.

Innervation Mandibular twig of cervicofacial.

Function Draws the lower lip down and out, slightly everting the outer rim of the lower lip.

Agonists Ipsilateral and contralateral mentalis, depressor anguli oris and platysma.

Antagonists Ipsilateral zygomaticus major and orbicularis oris.

Expression Irony, pouting.

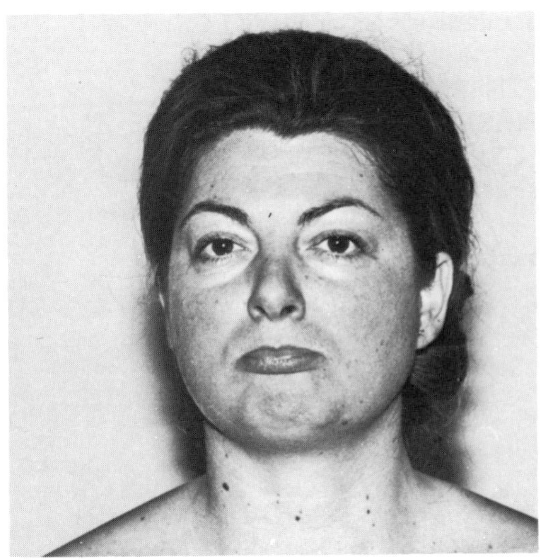

Expression.

MUSCLE TESTING

Starting position

On a relaxed face, dislocate the middle of the chin on the sound side by pushing it towards the tested side, horizontally inwards, and holding it there.

Test

Ask subject to draw lower lip down and out, slightly everting outer rim on the tested side.

Possibility of error

The commissura labii may be drawn obliquely down and backwards by the depressor anguli oris and the platysma. The mentalis may pull the depressor labii, passively everting the lower lip.

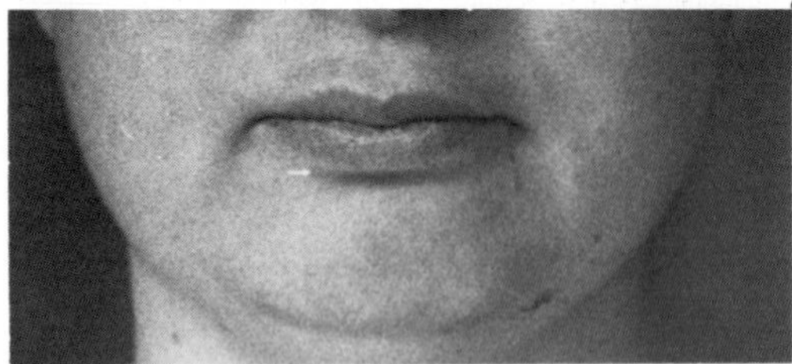
Starting position.

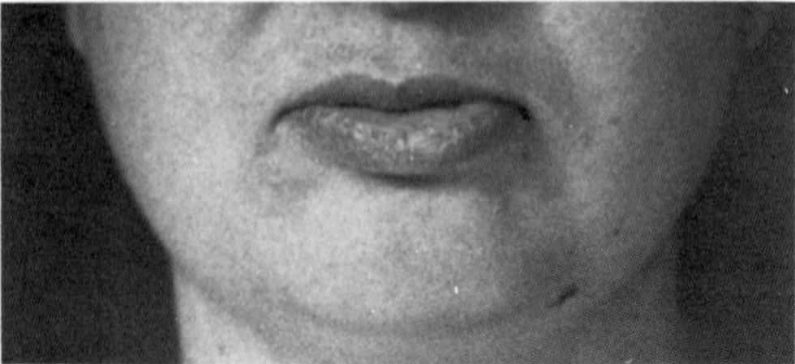
Final position.

Grade **0**: No contraction (L F P).

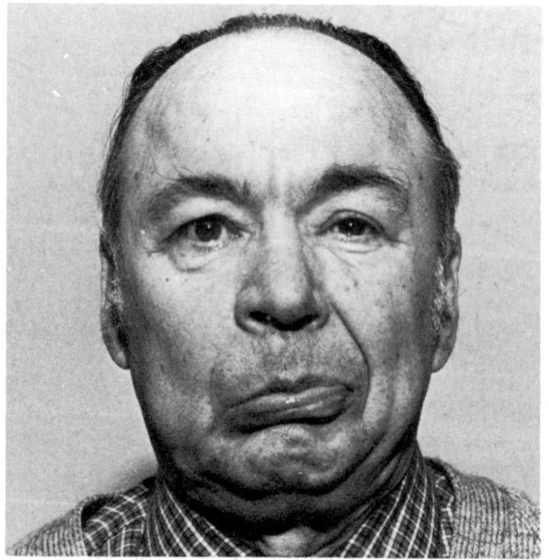
Grade **2**: Lower lip drawn down and everted (L F P).

Deficiency

The rim of the lower lip remains immobile and does not curl outwards.

Grading

0 The contraction is invisible, even under bright, glancing light. No contraction can be felt laterally to the mentalis.

1 During the contraction a slight movement in the surface of the skin is perceived in the lower lip inside the commissura labii.

2 The lower lip everts, lowering itself a little. The movement should be repeated five times. It is slow when compared to the sound side, and incomplete as the muscle tires.

3 The lower lip everts farther, creating a deeper, lateral indentation which outlines the mentalis. The movement should be repeated ten times fully, but it is not yet synchronized with the sound side.

4 The lower lip curls out easily. Lateral indentations symmetrically outline the mentalis. The movement is ample, accomplished easily, it is symmetrical and in synchronization with the sound side. It is globally integrated into voluntary facial expression.

TESTING MUSCLE TONE

Testing the tone of such a tiny muscle is difficult because its tone is masked by others (ipsilateral depressor anguli oris and mentalis).

SYNKINESES

Synkinesis is rare.

Hemispasm Rare.

MENTALIS (levator labii inferioris or levator menti)

Small, conical fasciculus located between the depressor labii inferioris and the median axis of the face.

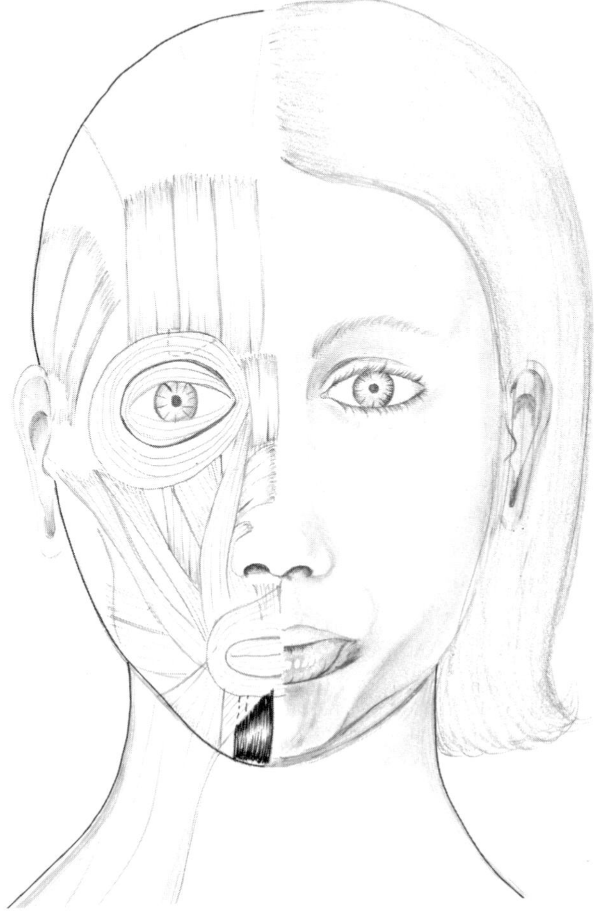

Relaxed state. Expression.

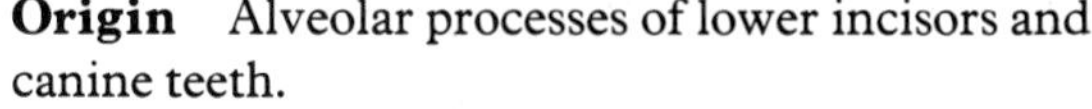

Origin Alveolar processes of lower incisors and canine teeth.

Insertion Spreads tuft-like into integument of chin.

Innervation Mandibular branch of the cervico-facial.

Function Raises the chin then the lower lip.

Agonists Ipsi- and contralateral depressor labii inferioris, contralateral mentalis.

Antagonists Ipsi- and contralateral depressor anguli oris and platysma.

Expression Doubt, indecision and pouting.

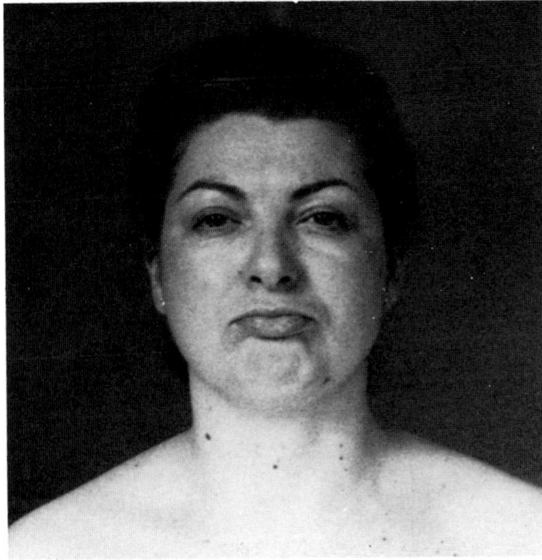

Expression.

MUSCLE TESTING

Starting position

On a relaxed face, use transverse digital pressure to dislocate the sound mentalis by pushing it across the median zone of the chin and holding it there.

Test

Ask subject to thrust out his chin and raise it towards his lower lip.

Possibility of error

The movement may be partially started by the depressor labii inferioris.

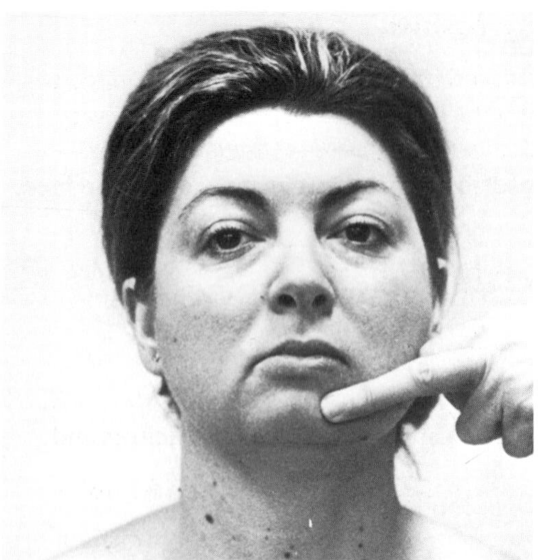

Starting position.

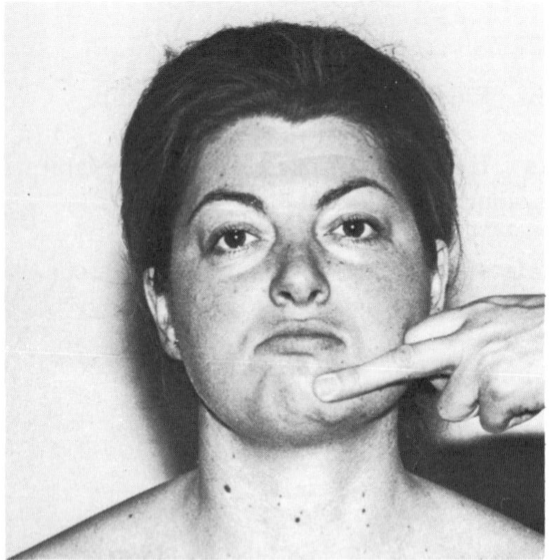

Final position.

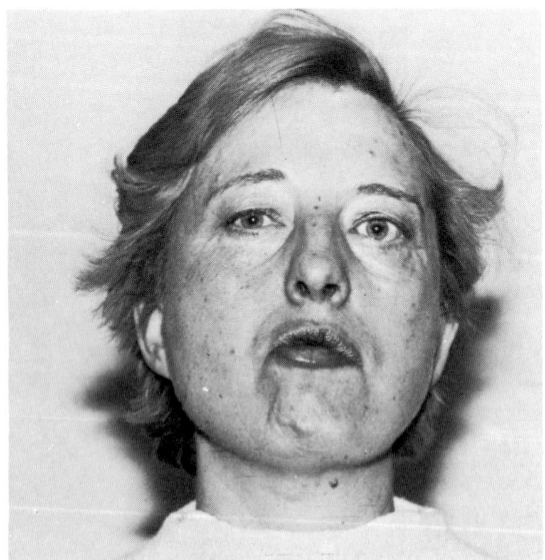

Grade **0**: No contraction (L F P).

Deficiency

Muscular relief of the chin is absent. The sound mentalis pulls the pathological towards the sound side.

Grading

0 The contraction is invisible, even under bright, glancing light. No contraction can be felt on the anterior part of the chin near the median axis of the face.

1 During the contraction, a slight depression forms close to the median axis of the chin.

2 The chin depresses laterally, the contraction is clearly felt under the finger tip. The lower lip rises. The movement should be repeated five times. It is slow when compared with the sound side, and is incomplete as the muscle tires.

3 Chin and lower lip move more distinctly, the labiomental groove is deeper. The movement should be repeated ten times fully, but it is not yet synchronized with the sound side.

4 The chin rises easily. The movement is ample, symmetrical and in synchronization with the sound side. The movement is globally integrated into voluntary facial expression.

TESTING MUSCLE TONE

– 2 *Atonia*

The para-median muscular relief on the tested side is absent. It deviates, drawn by the sound mentalis. The labiomental groove, common in men, has disappeared.

– 1 *Hypotonia*

The para-median muscular relief is formed and is less pulled by the sound mentalis. The labiomental groove appears.

0 *Normalization of tone*

The labiomental groove is at its normal depth. On certain individuals a vertical groove separates the two muscles.

+ 1 *Moderate hypertonia*

The mentalis rises slightly, a cluster of little depressions puckering its lower portion.

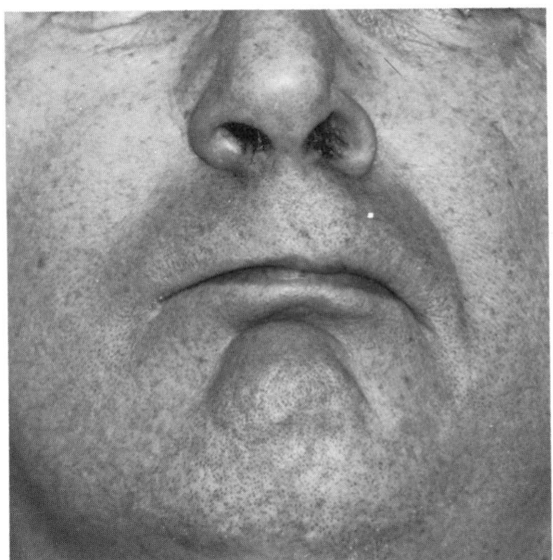

Grade **3**: Labiomental groove (R F P).

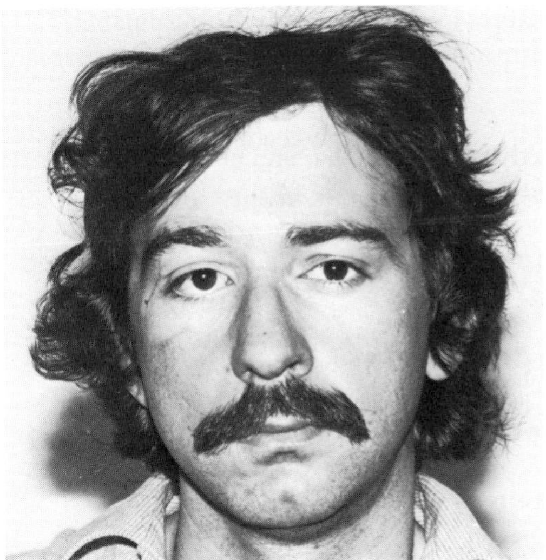

Normalization of tone: 0 (R F P).

Irrepressible synkinesis: + 3 (L F P).

+ 2 *Major hypertonia*

The mentalis on the tested side has risen, pulling the sound side with it. The labiomental groove is very deep. The muscular mass of the mentalis protrudes, its relief is asymmetrical.

SYNKINESIS

Synkinesis appears with lip movement.

0 Synkinesis is absent.

+ 1 Voluntary inhibition of synkinesis is impossible.

+ 2 Inhibition of synkinesis by digital pressure is also impossible.

+ 3 Synkinesis cannot be inhibited.
The mentalis contracts strongly at the slightest lip movement.

Hemispasm The mentalis participates in the hemifacial spasm.

DEPRESSOR ANGULI ORIS

A wide, flat, triangular muscle running from the mandible to the commissura labii.

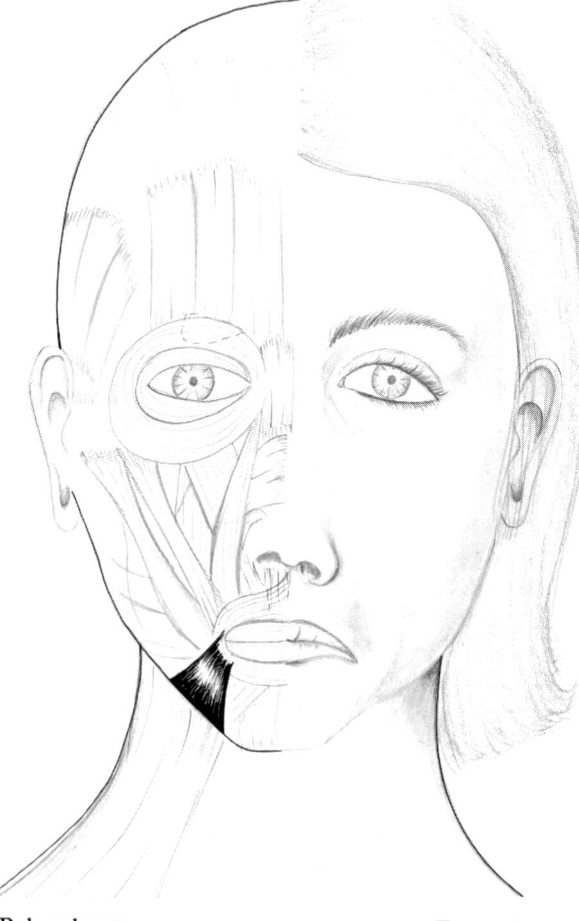

Relaxed state. Expression.

Origin Anterior part of the external, oblique line of the mandible below origin of depressor labii inferioris.

Insertion

a. At the commissura labii, superficial fibres mingle with the zygomatics.
b. Deep fibres intermingle with the buccinator.
c. Sometimes fibres reach the cartilage of the nasal ala.

Innervation Mandibular and buccal branches of the cervicofacial.

Function Pulls commissura labii downwards and outwards.

Agonists Platysma, depressor labii inferioris.

Antagonists Orbicularis oris and the zygomatics.

Expression Fright, grief, deep sobbing.

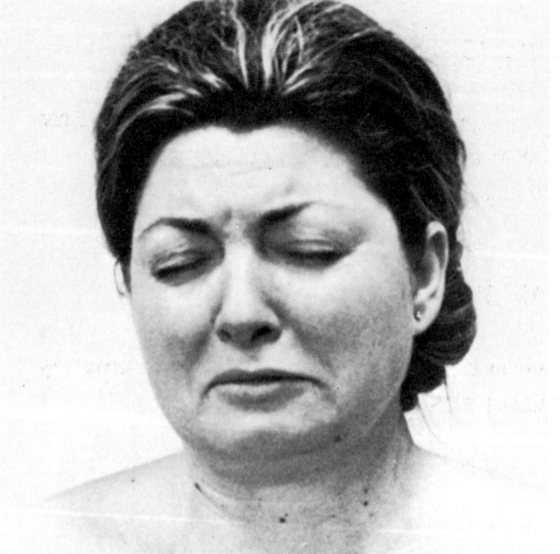

Expression.

MUSCLE TESTING

Starting position

On a relaxed face, the fingers as close to the median axis as possible, dislocate the chin zone by pushing it towards the affected side.

Test

Ask patient to pull the commissura labii downwards and outwards. The movement should be modulated so that the platysma does not contract.

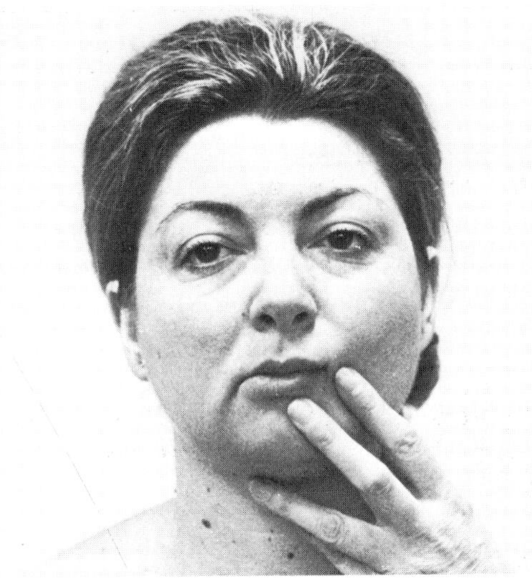

Starting position.

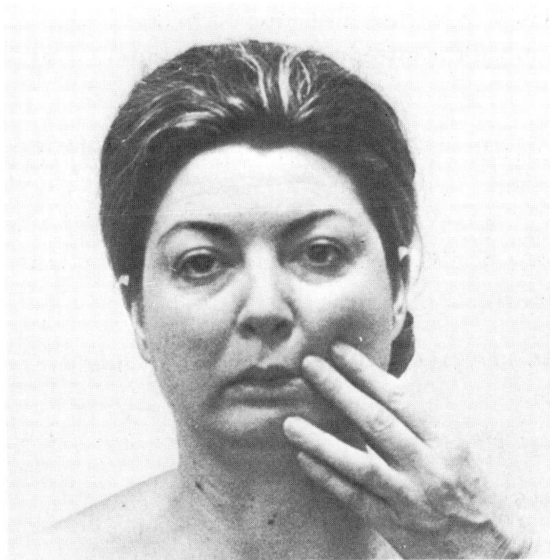

Final position.

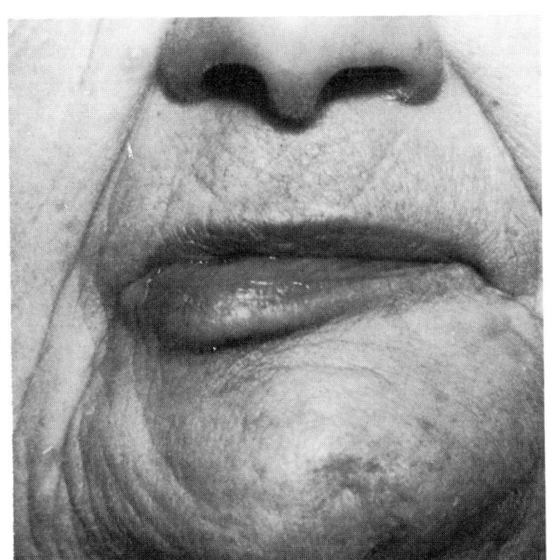

Grade **0**: Deficiency (L F P).

Possibility of error

Commissura labii may be pulled by the ipsilateral platysma.
Commissura labii may be pulled down by the depressor labii inferioris.

Deficiency

Commissura labii cannot be pulled down.

Grading

0 The contraction is invisible, even under bright, glancing light.
No contraction can be felt beneath and outside the commissura labii.

1 During the contraction a slight movement in the skin's surface is perceived just beneath the commissura labii. The rim of the lower lip drops, the fleshy part widening.

2 The commissura labii pulls down a little farther, little wrinkles may appear on the inferior part of the lower jaw. The movement should be repeated five times. It is slow when compared with the sound side, and incomplete as the muscle tires.

3 The commissura labii drops farther and more neatly. The movement should be repeated ten times fully, but is not yet synchronized with the sound side.

4 The commissura labii drops easily. The movement is ample, symmetrical and in synchronization with the sound side. It is globally integrated into voluntary facial expression.

TESTING MUSCLE TONE

The tone of the depressor anguli oris is masked by that of the middle fibres of the platysma. It is difficult to test.

SYNKINESES

Synkinesis and hemispasm are rare and intimately associated with those of the platysma.

PLATYSMA

A broad, thin muscle running from the upper thorax to the mandible and cheek.

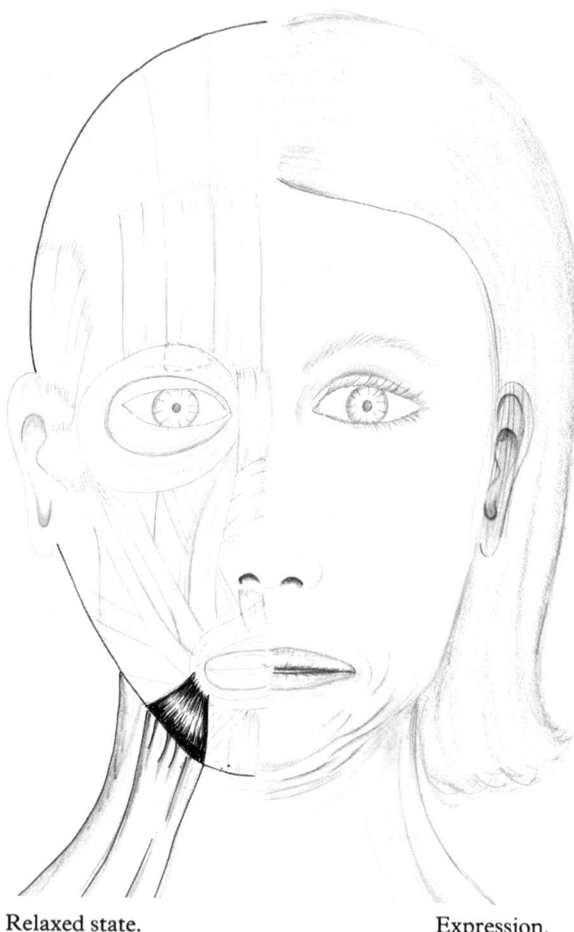

Relaxed state. Expression.

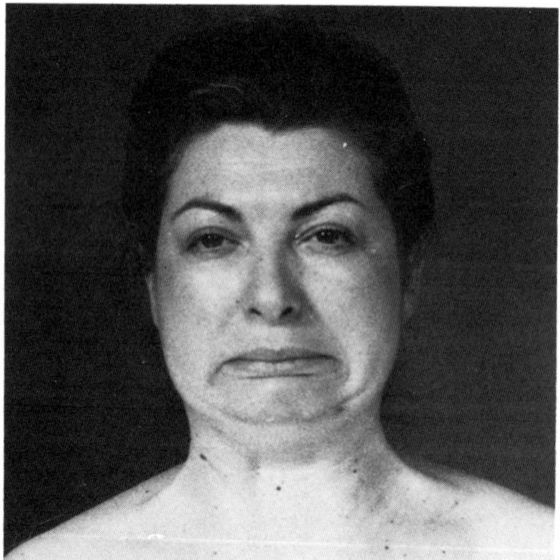

Expression.

Origin Thin, fibrous bands from superficial aponeurosis covering the acromion and upper part of the deltoid, clavicular and pectoral regions.

Insertion

a. Anterior fibres interlace with fibres of the contralateral platysma and insert into integuments of the symphysis menti.
b. Middle fibres insert into inferior edge of the mandible where they intermingle with those of depressor anguli oris and depressor labii inferioris.
c. Posterior (or lateral) fibres pass over the mandible where some mingle with fibres of the depressor anguli oris while others insert into skin and subcutaneous tissue of the lower part of the face and commissura labii.

Innervation Twigs of the cervical branch of the cervico facial.

Function Pulls down skin of the chin and commissura labii. Tenses skin of the neck causing ridges to form while raising it slightly.

Agonists Depressor anguli oris, depressor labii inferioris.

Antagonist Orbicularis oris.

Expression Torture, terror, fright.

MUSCLE TESTING

Starting position

On a relaxed face, hold median zone of the face near chin on non-tested side, with fingers.
Dislocate sound side by pushing it towards the affected, shortening the depressor anguli oris, a position which favours contraction of the platysma.

Test

Ask subject to do his best at pulling commissura labii downwards and outwards.

Possibility of error

None.

Deficiency

Skin of the neck remains immobile and cannot be made taut. There is no augmentation in transverse diameter of the neck.

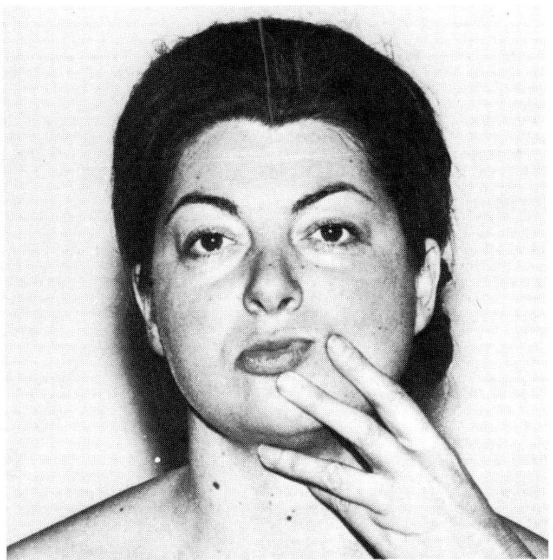
Starting position.

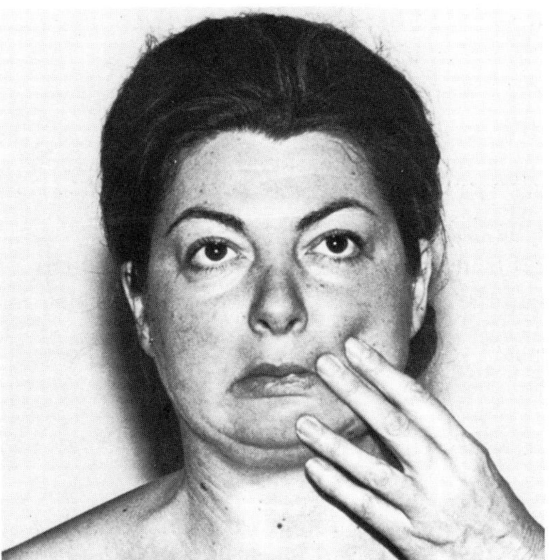
Final position.

Grade **0**: Deficiency (L F P).

Grading

0 The contraction is invisible, even under bright, glancing light. No contraction can be felt beneath the depressor anguli oris at the lower aspect of the mandible.

1 During the contraction there is a slight movement in the surface of skin over the infero-lateral portion of the neck.

2 Skin of the neck rises more clearly. Certain fibres, anterior, middle or inferior, may stand out beneath the skin. The movement should be repeated five times. It is slow when compared with the sound side, and is incomplete as the muscle tires.

3 Transverse diameter of the neck increases. Muscular fibres stand out much farther. The movement should be repeated ten times fully, but it is not yet synchronized with the sound side.

4 Muscular fibres individualize, forming real cords which lift the skin of the neck. The movement is ample, symmetrical and in synchronization with the sound side. It is globally integrated into voluntary facial expression.

TESTING MUSCLE TONE

Tone of the platysma is difficult to test because it is large and flat. Only its possibilities of contraction may be determined.

SYNKINESES

Synkinesis and hemispasm are rare and directly associated with those of the depressor anguli oris.

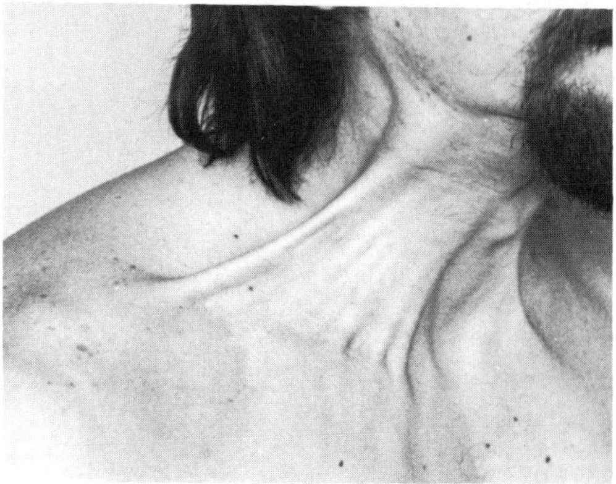
Grade **4**.

Main constrictor of the mouth, this muscle is elliptical and gives thickness to the lips. It is composed of two parts, one lateral or peripheral, the other medial or central.

Relaxed state. Expression.

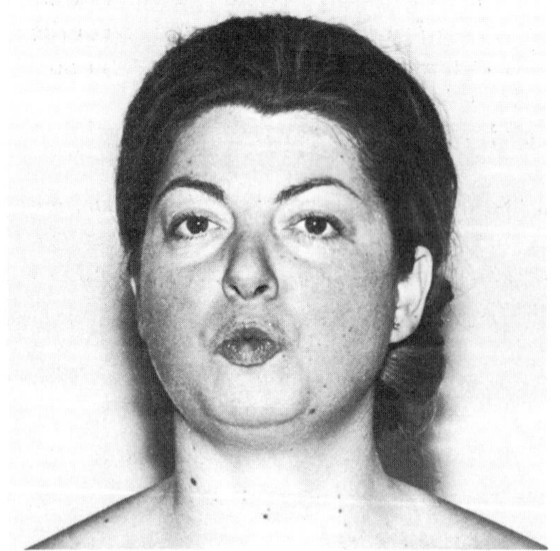

Expression.

LATERAL (OR PERIPHERAL) PART

Composed of extrinsic and intrinsic fibres.

Extrinsic fibres

These are an intermingling of the inserting fibres of depressor anguli oris, buccinator and levator anguli oris mainly, and of the rest of the inter-maxillary region muscles.

Intrinsic fibres

These belong to the incisor muscles, there are two for each lip:
a. Musculus incisivus superior, the fibres of which originate in the alveolar border of the maxilla opposite the lateral incisor tooth.
b. Musculus incisivus inferior, the fibres of which originate in the mandible laterally to the mentalis. They arch outward and insert into integuments of the commissura labii.

MEDIAL (OR CENTRAL) PART

Composed of fibres which follow the rim of the lips and interlace with fibres of the compressor labiorum.

Innervation

Upper lip: superior buccal branch of the cervico-facial.
Lower lip: inferior buccal branch of the cervico-facial.

Agonist Compressor labiorum.

Antagonists Levator anguli oris, zygomatics, risorius, depressor anguli oris, depressor labii inferioris, mentalis, and platysma. All these muscles, and particularly the risorius, are dilators of the buccal orifice.

Function Closes, contracts and compresses the lips.

Expression Reserve.

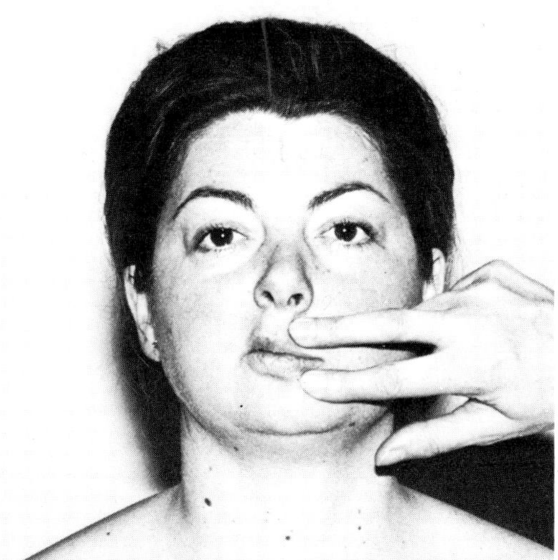

Starting position.

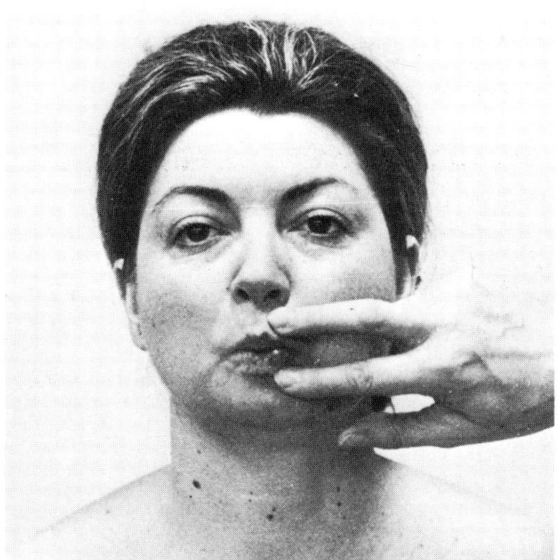

Final position.

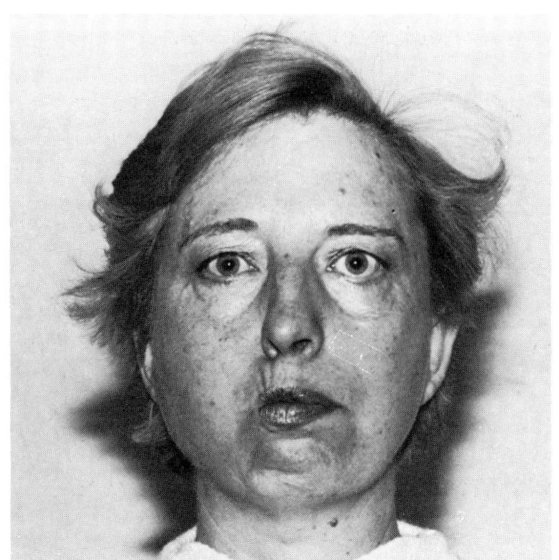

Grade **0**: Deficiency (L F P).

MUSCLE TESTING

Starting position

On a relaxed face use two fingers, one over the upper lip and the other under the lower lip, to dislocate the sound side of the face by pushing it towards the pathological one and holding it there.

Test

Ask subject to purse his lips in an unamused smile.

Possibility of error

None.

Deficiency

The commissura labii remains immobile on the tested side, it will not approach the median axis.

Grading

0 The contraction is invisible, even under bright, glancing light. No contraction can be felt on the fleshy part of either lip.

1 During contraction, the commissura labii moves slightly toward the median axis. The fleshy part of one or both lips thickens as the horizontal axis shortens.

2 The commissura labii approaches the median axis of the mouth a little more. Little wrinkles form around the rims of the upper and lower lips. The movement should be repeated five times. It is slow when compared with the sound side, and incomplete as the muscle tires.

3 The commissura labii draws much closer to the median axis of the face. The fleshy part of one of the lips becomes much thicker. The movement should be repeated ten times fully, but it is not yet synchronized with the sound side.

4 The upper and lower lips move easily. Movement is ample, symmetrical, and in synchronization with the sound side. It is globally integrated into voluntary facial expression.

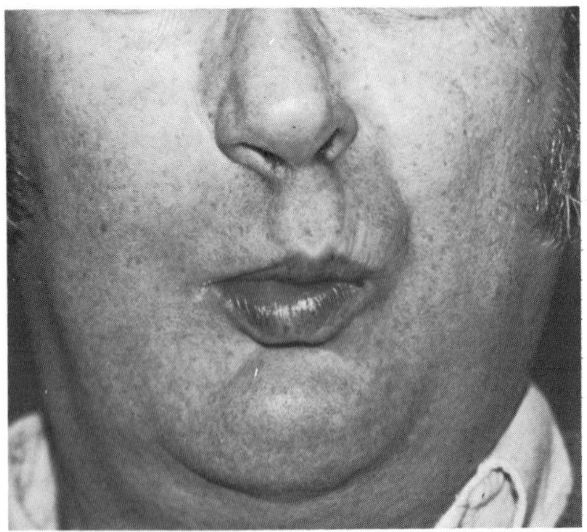

Orbicularis oris, superior portion: Grade **2** (R F P).
Orbicularis oris, inferior portion: Grade **1**.

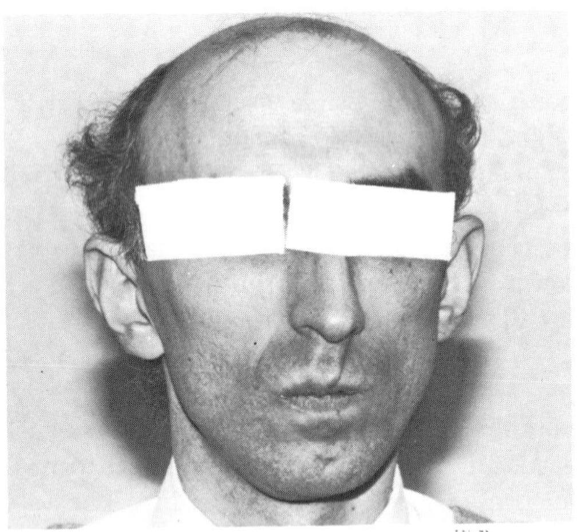

Orbicularis oris, superior portion: Grade **2** (L F P).
Orbicularis oris, inferior portion: Grade **3**.

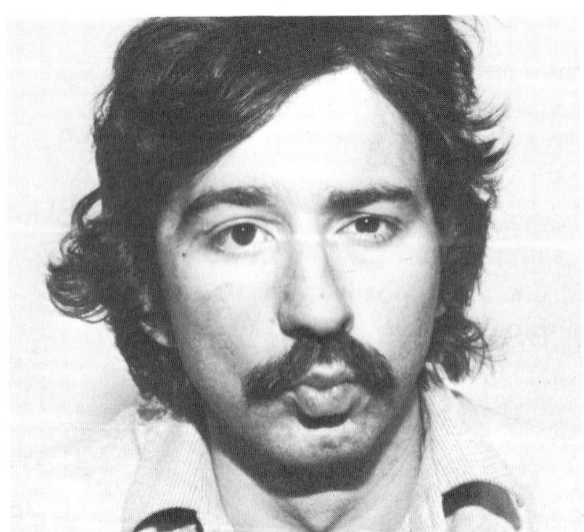

Orbicularis oris, superior portion: Grade **4** (R F P).
Orbicularis oris, inferior portion: Grade **4**.

TESTING MUSCLE TONE

− 2 *Atonia*

Muscular relief in the fleshy part of the lips is diminished. The pathological side appears swallowed. The commissura labii on the pathological side is farther from the median axis than on the sound side.

− 1 *Hypotonia*

Muscular relief becomes more prominent. The commissura labii is closer to the median axis and the fleshy part of the lip is thicker.

0 *Normalization of tone*

Buccal relief is symmetrical.

+ 1 *Moderate hypertonia*

Never encountered since antagonistic muscles are numerous.

+ 2 *Major hypertonia*

Never encountered.

SYNKINESES

Synkineses and hemispasm of the orbicularis oris are non-existent.

COMPRESSOR LABIORUM

Name given to anterior-posterior fibres situated around the opening of the mouth. These fibres mingle with those of the internal portion of the orbicularis oris. This suction muscle is particularly developed in newborn babies.

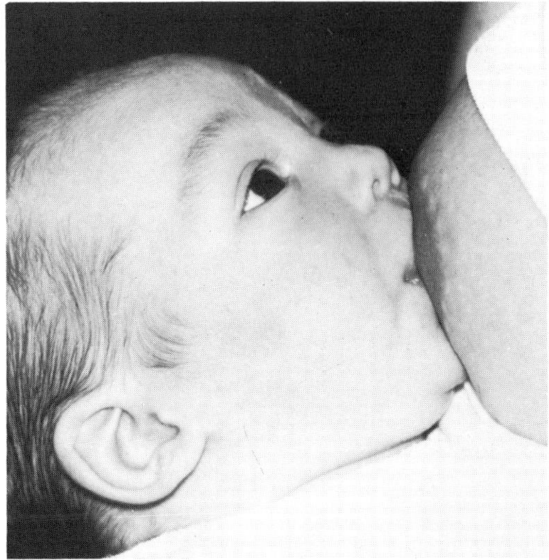

Function.

Innervation Same as the orbicularis oris.

Function Compresses the lips from front to back. Works in synchronization with the orbicularis oris, buccinator, masseters and tongue muscles.

Expression That of savouring.

Grading

The examiner will note whether or not nursing or suction is possible.

MUSCLES CONTROLLING THE ORGANS OF FACIAL CAVITIES

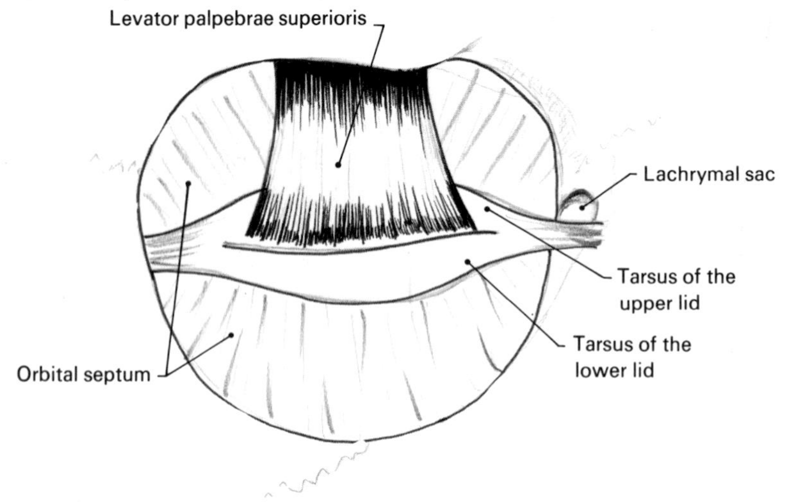

MUSCLE OF THE UPPER LID

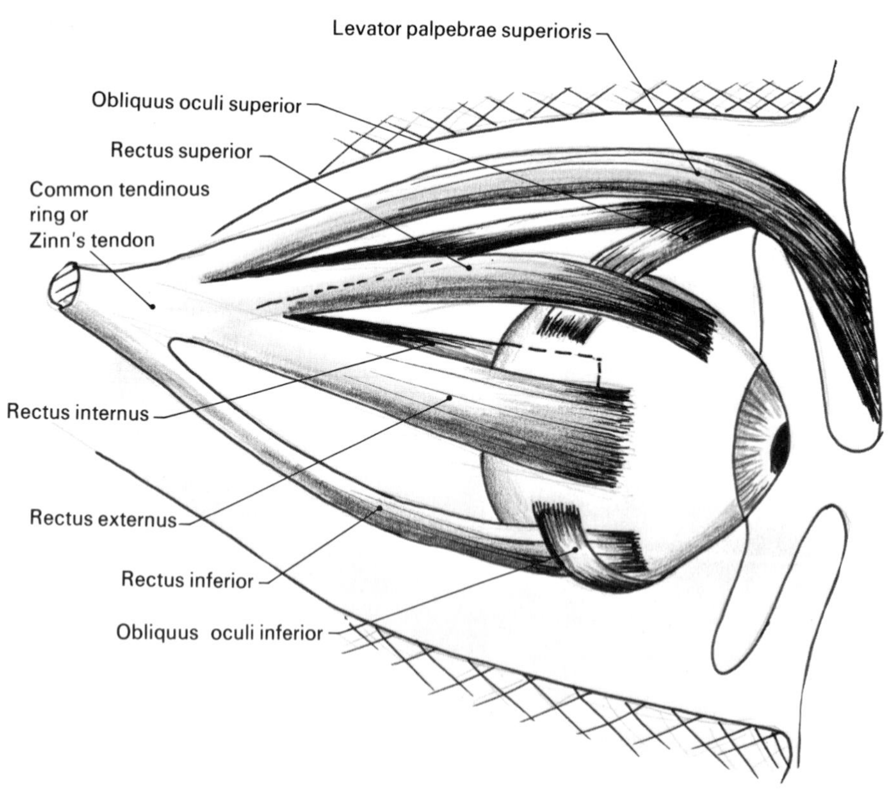

MUSCLES OF THE ORBIT

MUSCLES OF THE EYES

The synchronized action of these muscles lead the eye through the six main directions of sight. There are six muscles in all. The two which are not innervated by the oculomotor nerve (III) are:
—obliquus oculi superior, innervated by the trochlear nerve (IV).
—rectus externus, innervated by the abducent nerve (VI).
All these muscles mobilize the eyeballs and widen the visual field without head movement.
Two muscles are needed for each movement.
Each eye has four rectus muscles and two obliques.

RECTUS SUPERIOR

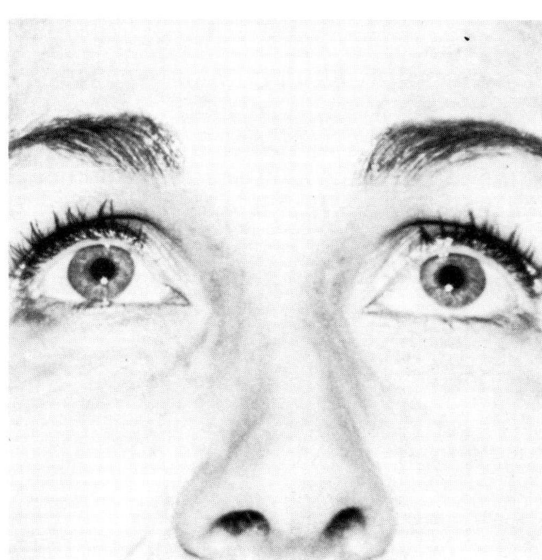

A flat, ribbon-like muscle.

Origin Upper margin of the optic foramen (common tendinous ring: Zinn's tendon).

Insertion Through a tendinous expansion into the sclera about 8 mm from the margin of the cornea in a vertical plane just medially to the vertical axis of the eyeball.

Innervation Branch of the superior division of the oculomotor (III).

Function Draws the cornea up and slightly inwards.

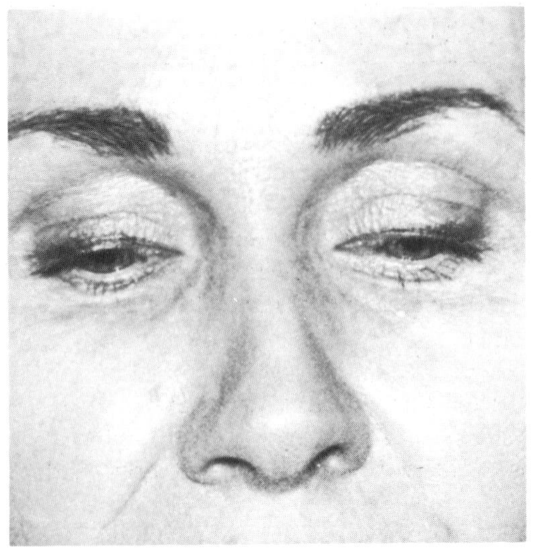

RECTUS INFERIOR

Origin Inferior portion of Zinn's tendon.

Insertion Into the sclera about 6 mm from the margin of the cornea in a vertical plane, just medially to the vertical axis of the eyeball.

Innervation Branch of inferior division of the oculomotor (III).

Function Draws the cornea down and slightly inwards.

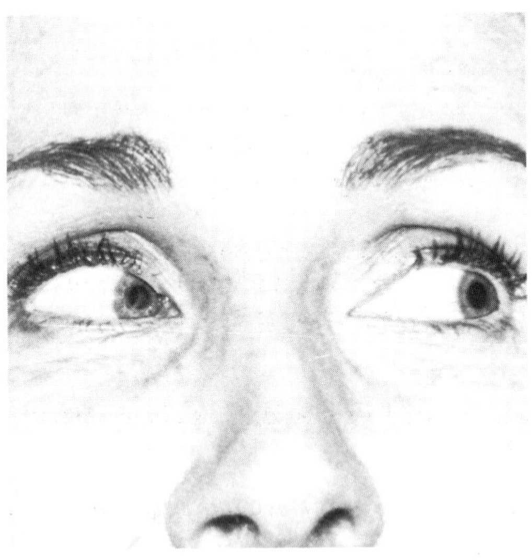

RECTUS MEDIALIS

Origin From median portion of Zinn's tendon.

Insertion Into the sclera about 5 mm from the margin of the cornea in the transverse plane of the eyeball.

Innervation Branch of inferior division of the oculomotor (III).

Function Draws the cornea inward.

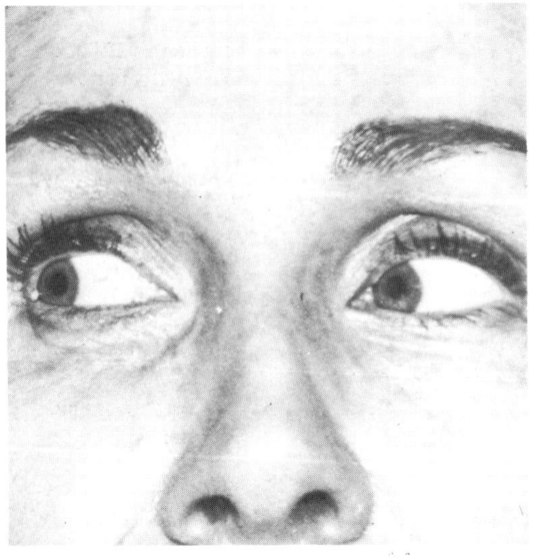

RECTUS LATERALIS

Origin From the lateral portion of Zinn's tendon.

Insertion Into the sclera about 7 mm from the margin of the cornea in the transverse plane of the eyeball.

Innervation The abducent nerve (VI).

Function Draws the cornea out.

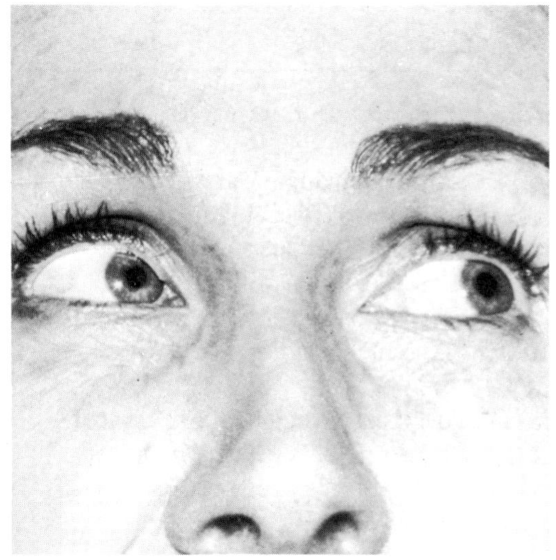

OBLIQUUS SUPERIOR

Origin From Zinn's tendon above the margin of the optic foramen, above and medially to the origin of the rectus superior.

Insertion After passing through a fibrocartilaginous pulley, the tendon is reflected back, outwards and down where it inserts into the sclera between the superior and lateral recti.

Innervation The trochlear nerve (IV).

Function Draws the eyeball forward, down and inward.

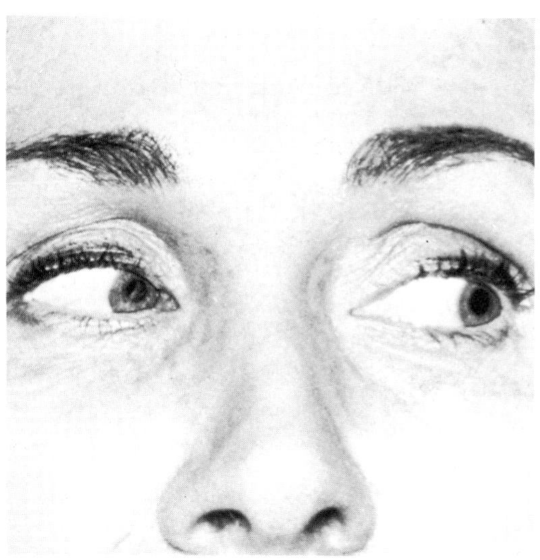

OBLIQUUS INFERIOR

Origin Posteriorly and laterally to the nasolachrymal canal on the floor of the orbit.

Insertion After passing backwards, outwards, and upwards around the eyeball, it inserts into the outer part of the sclera between the inferior and lateral recti.

Innervation Inferior division of the oculomotor (III).

Function Draws the eyeball up, forward and inward.

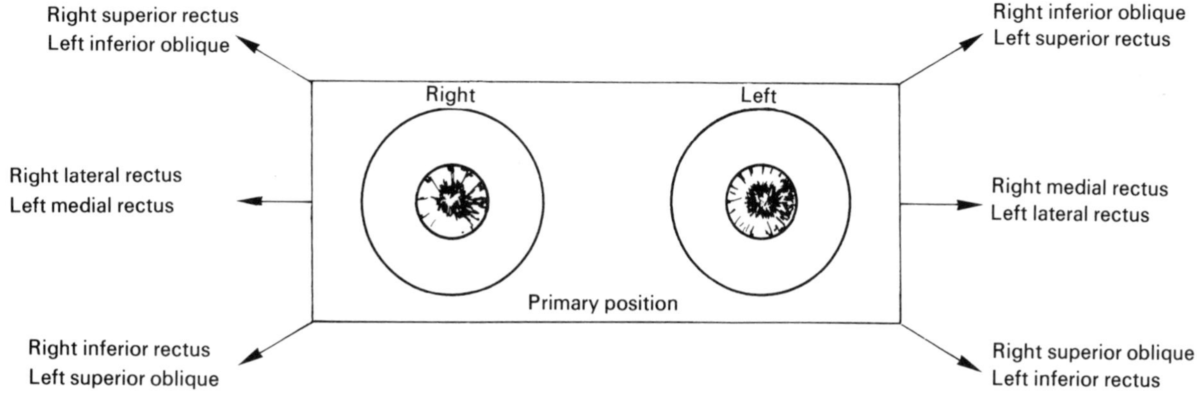

Diagram from Chusid, J.G., Correlative Neuroanatomy and Functional Neurology, 14th ed., Los Altos, Calif., Lange Medical Publications, 1970, page 94.

MUSCLES OF THE TONGUE

The tongue is a mobile, fleshy organ which fills the oral cavity, pressing against the palate when the mouth is closed. It is the essential organ of taste and of speech; it plays a major role in suction, chewing, in salivation and in preparing alimentary bolus, in swallowing, in phonation and in grasping.
It is composed of seventeen muscles: eight symmetrical pairs and one odd, transversal, superficial muscle. They are all covered by a mucous membrane (a continuation of the buccal mucous membrane which forms the lingual frenulum under the tongue).
On the superior portion of the tongue, this mucous membrane is covered with tactile and gustatory papillae (taste buds). The hypoglossal nerve (XII) is responsible for the tongue's great variety of movement, the glosso-pharyngeal (IX) for taste sensitivity and the lingual (VII) for tactile sensitivity.

GENIOGLOSSUS

Extrinisic muscle vertically situated on either side of the median line. With its contralateral muscle, it forms the median, fibrous septum.

Origin Upper genial tubercle of the mandible.

Insertion Superior fibres: tip of tongue.
Middle fibres: entire under-surface of tongue.
Lower fibres: superior margin of the body of the hyoid bone.

Innervation Muscular branches of the hypoglossal (XII).

Function Depresses and projects the tongue forward retracting the tip backward. It raises the hyoid bone.

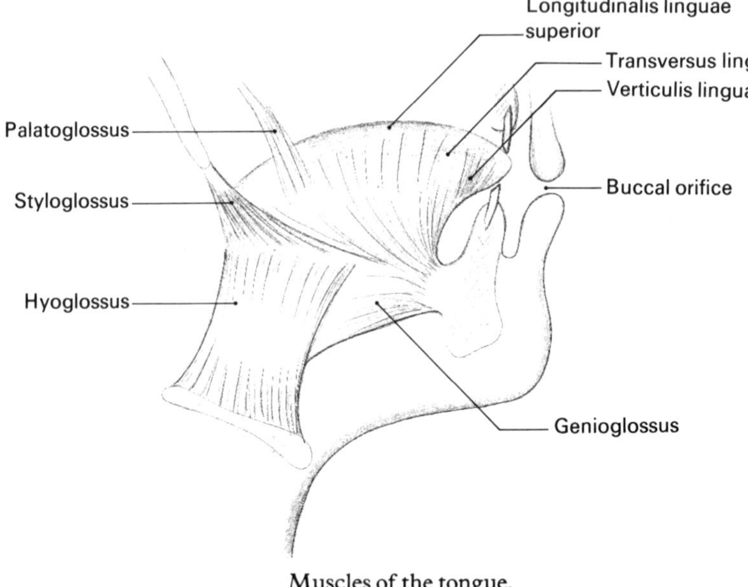

Muscles of the tongue.

HYOGLOSSUS

A thin, flat, quadrilateral, extrinsic muscle posterior and lateral to the genioglossus.

Origin Side and body of the greater cornu of the hyoid bone.

Insertion Side of the tongue.

Innervation Muscular branches of the hypoglossal (XII).

Function Depresses, retracts and draws down the sides of the tongue.

STYLOGLOSSUS

An extrinsic muscle, it is situated laterally to the hyoglossus.

Origin Anterior border of styloid process.

Insertion Side of tongue, its fibres mingling with those of the hyoglossus.

Innervation Muscular branches of the glosso-pharyngeal (IX).

Function Raises the root of the tongue and widens it.

LONGITUDINALIS LINGUAE INFERIOR

Lateral intrinsic muscle oriented on a sagittal plane.

Origin Under-surface of tongue between genioglossus and hyoglossus.

Insertion Tip of the tongue where it blends with the styloglossus.

Innervation Muscular branches of the hypoglossal (XII).

Function Retracts the tongue and curls the tip downwards.

PHARYNGOGLOSSUS

Name given to the constrictor pharyngis superior fibres which insert into the tongue, forming an extrinsic muscle laterally to the genioglossus.

Origin Superior constrictor of pharynx.

Insertion Side of tongue.

Innervation Branches of the glossopharyngeal (IX) and vagus (X).

Function Flattens and widens the tongue, draws it backwards and upward, acts in swallowing.

PALATOGLOSSUS

The most external, lateral muscle, this is an extrinsic muscle composed of small, fleshy fascicules running from the anterior surface of the soft palate to the side of the tongue.

Origin Anterior surface of the soft palate.

Insertion Dorsum and side of tongue, blending with styloglossus and transversus linguae.

Innervation Branches of hypoglossal (XII).

Function Retracts tongue upwards and backwards, narrows fauces.

AMYGDALOGLOSSUS

Inconstant, extrinsic muscle situated in a sagittal plane.

Origin Fibrous capsule covering the lateral surface of the tonsils.

Insertion Side of tongue.

Innervation Branch of hypoglossal (XII).

Function Acts in swallowing.

TRANSVERSUS LINGUAE

Intrinsic muscle, its fibres mingle with the extrinsic, transverse fibres of the palatoglossus and the superior constrictor of the pharynx. The transverse fibres intersect fibres of the genioglossus, longitudinalis linguae inferior, hyoglossus and the styloglossus.

Origin Median fibrous septum.

Insertion Dorsum and side of tongue.

Innervation Branch of hypoglossal (XII).

Function Elongates and narrows the tongue.

LONGITUDINALIS LINGUAE SUPERIOR

Intrinsic muscle consisting of a thin muscular layer, it covers the whole dorsum of the tongue.

Origin Submucous fibres beginning at the back of the tongue.

Insertion Tip of tongue.

Innervation Branch of the hypoglossal (XII).

Function Retracts the tongue, curls the tip and sides upwards.

TESTING MUSCLES OF THE TONGUE

Testing tongue muscles is done from a basis of global movements such as the following:
—extension: thrusting the tongue all the way out as far as it will go (protrusion).
—retraction: pulling the tongue straight back in as far as possible.
—widen, narrow, elongate, deviate or move to the right or left.
—to hollow, trough-like, or to hump.

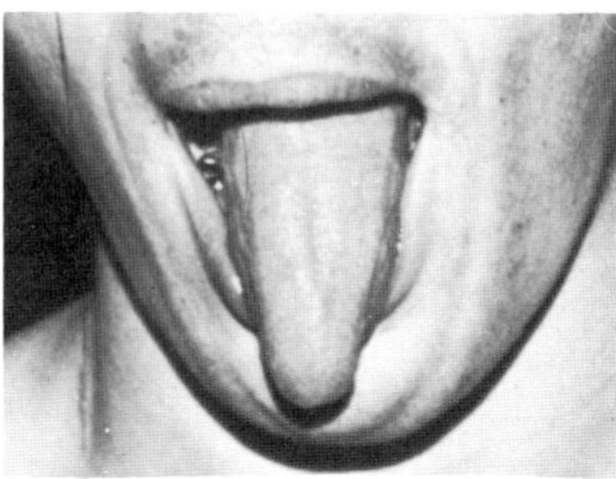

Extension.

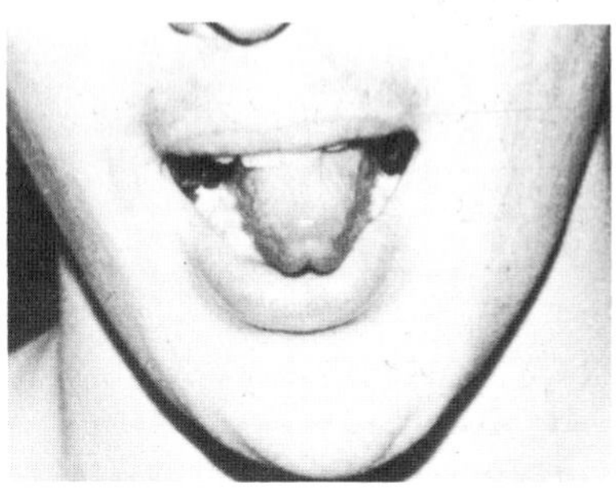

Retraction.

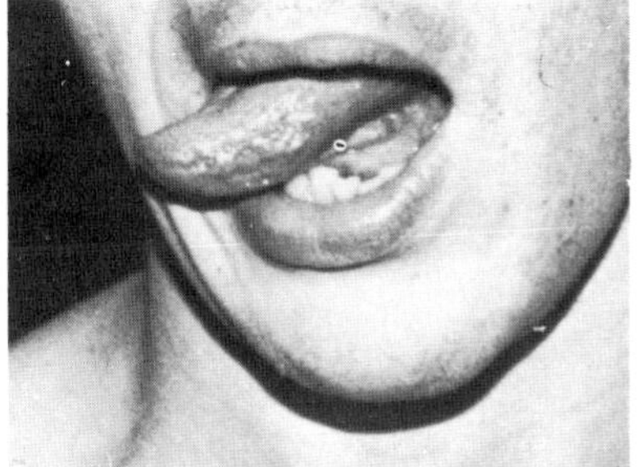

Right deviation.

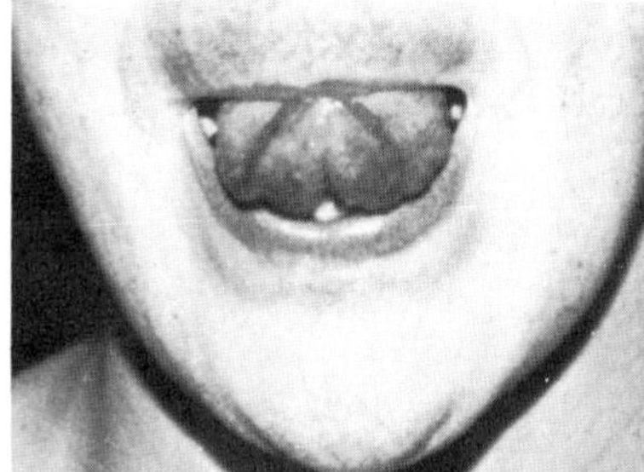

Tipping.

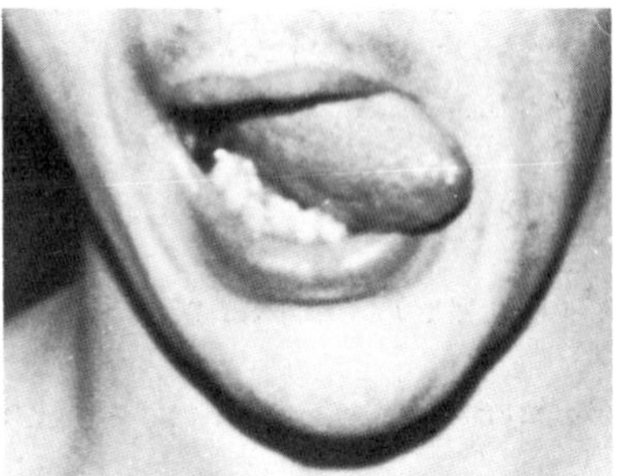

Left deviation.

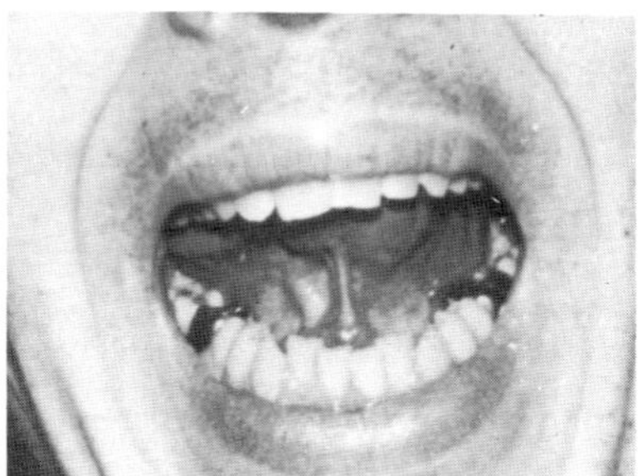

Retraction raising the root.

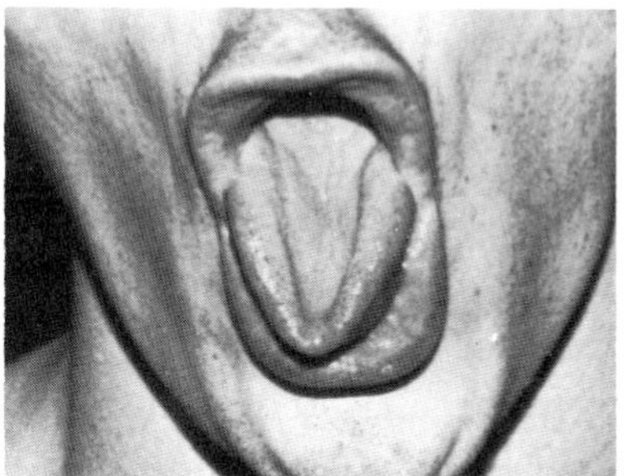

Trough.

MUSCLES OF THE EAR

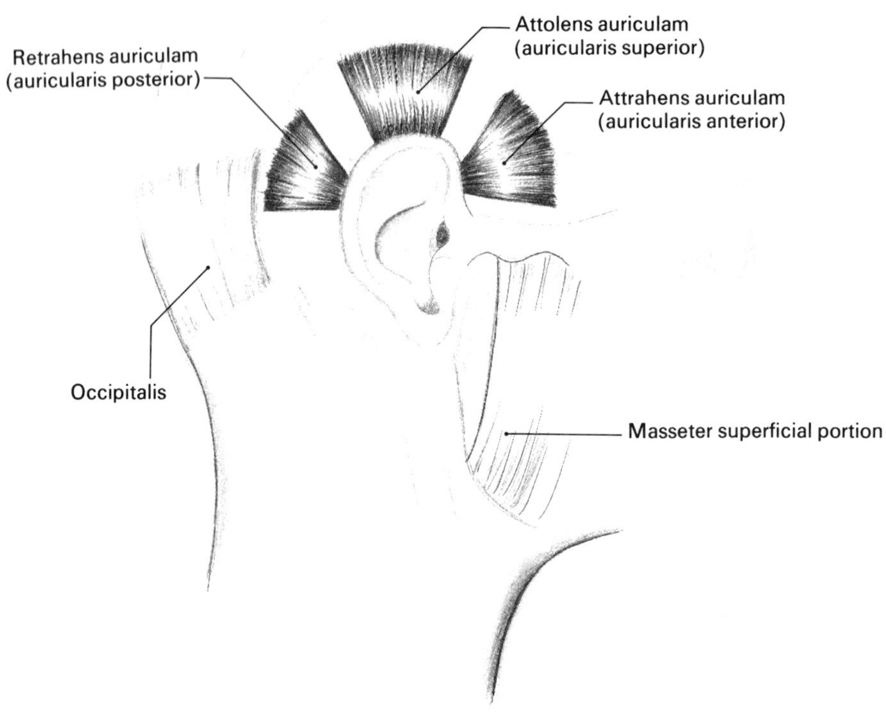

They are three in number and are extrinsic.
These muscles are rudimentary in man. Few people are capable of moving their ears (auricles), and then only backwards.
These muscles are:

ATTRAHENS AURICULAM (auricularis anterior)

Situated anteriorly to the auricle, this muscle is fan-shaped.
It originates from the lateral margin of the aponeurosis of the occipito-frontalis, and inserts into the front of the helix and the anterior margin of the concha.

ATTOLLENS AURICULAM (auricularis superior)

Situated superiorly to the auricle, this muscle is also fan-shaped. It originates in the aponeurosis of the occipito-frontalis and converges to insert into the upper portion of the cranial surface of the pinna.

RETRAHENS AURICULAM (auricularis posterior)

Originates in the mastoid portion of the temporal bone and inserts into the lower portion of the cranial surface of the concha.

MANDIBULAR MOTOR MUSCLES

THE PTERYGOID MUSCLES

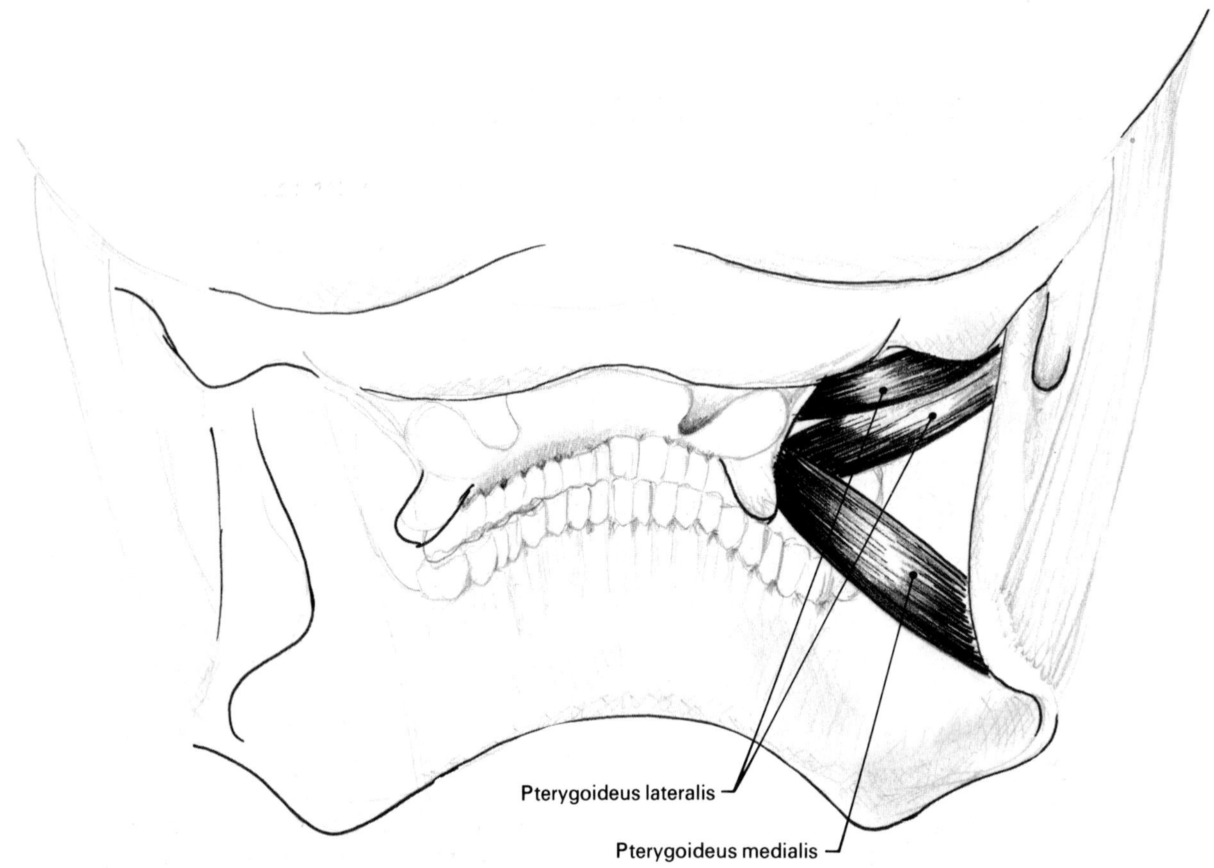

POSTERIOR VIEW

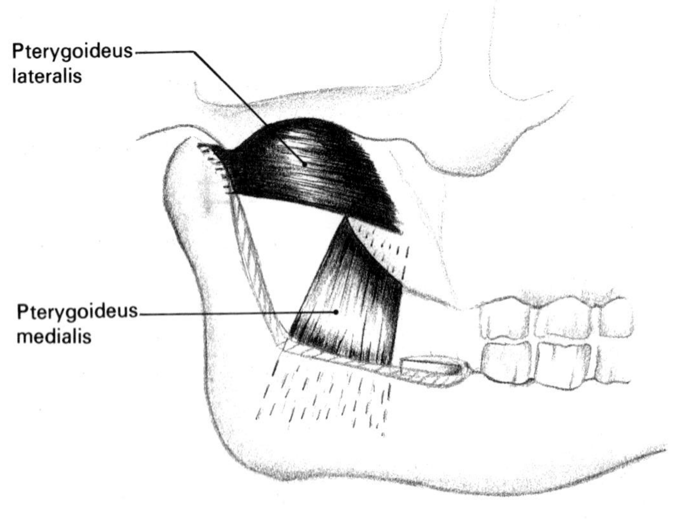

LATERAL VIEW

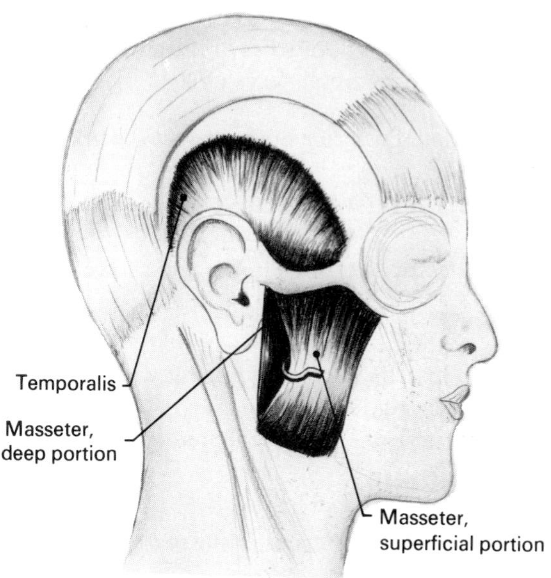

There are four
—the masseter
—the pterygoideus lateralis
—the temporalis
—the pterygoideus medialis

MASSETER

This is a short, thick muscle composed of three portions:
—two principal: the superficial, and the middle,
—one accessory: the deep (not always present).

SUPERFICIAL PORTION

Origin Anterior three-quarters of the lower border of the zygomatic arch of maxilla.

Insertion The angle and lower half of the outer surface of the mandibular ramus.

Innervation Masseteric nerve from the anterior branch of the mandibular division of the trigeminal (V).

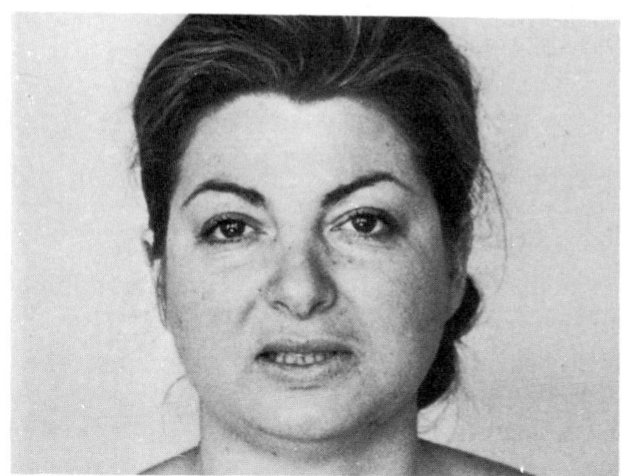

Bilateral function.

MIDDLE PORTION

Origin Posterior third of the lower border, and the whole inner surface of the zygomatic arch.

Insertion Upper half of the ramus and outer surface of the coronoid process of the mandible.

Innervation As the superficial portion.

Agonists Ipsi- and contralateral temporalis, and the pterygoideus medialis.

Antagonists Pterygoideus lateralis and the supra-hyoid muscles.

Function Elevates the mandible against the maxilla, forcefully clenches the two unilaterally or bilaterally. Acts in mastication and in phonation.

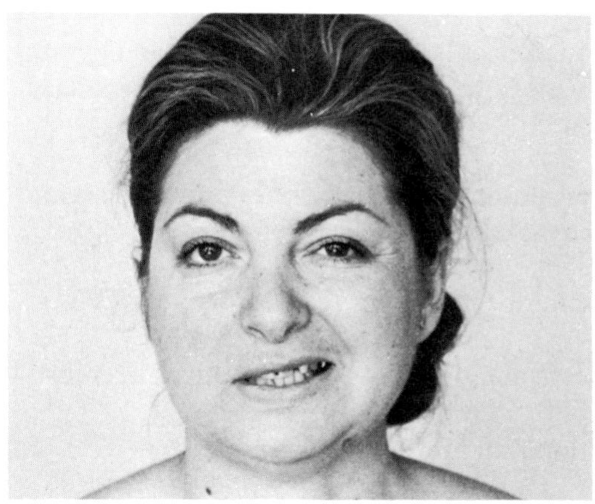

Unilateral function.

TEMPORALIS

Broad, flat, radiating muscle.

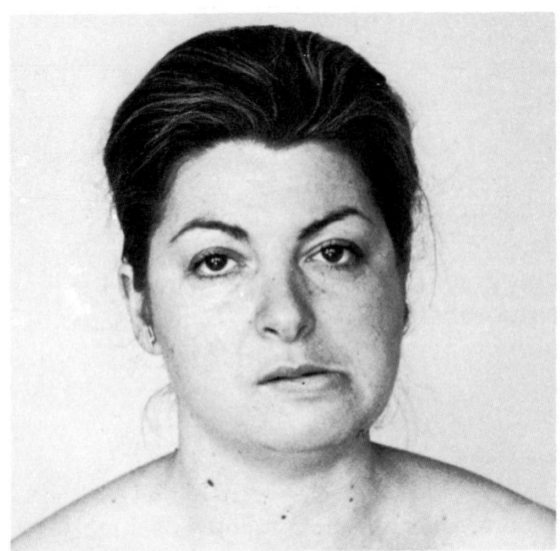
Pterygoideus lateralis: Unilateral contraction, Mandibular deviation.

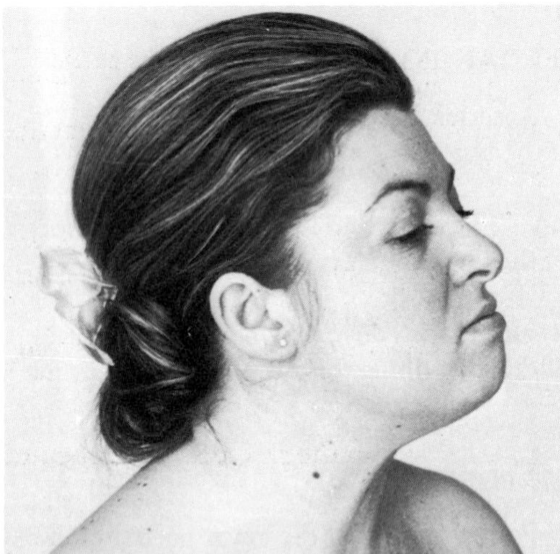
Pterygoideus lateralis: Bilateral contraction, Mandibular protrusion.

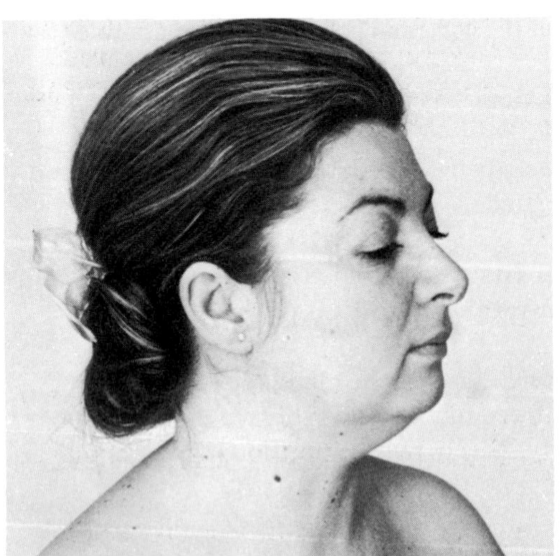
Pterygoideus medialis: Bilateral contraction, Mandibular retropulsion.

Origin From the whole of the temporal fossa to the great wing of the sphenoid and temporal fascia.

Insertion Fibres converge into a thick, flat tendon which inserts into the entire coronoid process of the jaw.

Innervation Deep temporal branches of the anterior division, mandibular trunk, trigeminal nerve (V).

Function Mastication. Posterior fibres retract the mandible.

PTERYGOIDEUS LATERALIS (externus)

A short, thick muscle composed of an upper and a lower head.

Origin

a. *Upper head*: outer, inferior surface of greater wing of the sphenoid, and from pterygoid ridge.
b. *Lower head*; outer surface of the lateral pterygoid plate.

Insertion

a. *Upper head*: into depression in front of the neck of mandibular condyle.
b. *Lower head*: anterior margin of interarticular fibrocartilage.

Innervation Lateral pterygoid branch, anterior division, mandibular trunk of the trigeminal nerve (V).

Agonist Masseter.

Antagonists Pterygoideus medialis, supra- and infra-hyoid muscles.

Function Moves mandible laterally if one muscle acts alone. Acts in mastication (rotary motion of grinding), and in phonation. If both muscles act simultaneously, they cause the mandible to protrude forward.

PTERYGOIDEUS MEDIALIS (internus)

Deepest muscle of mastication.

Origin Medial surface of lateral pterygoid plate and pyramidal process of palatine bone, and a small portion of maxillary tuberosity.

Insertion Lower, posterior portion, inner side of the ramus and the angle of the mandible.

Innervation Same as for the pterygoideus lateralis (V).

Agonists Masseter, temporalis and pterygoideus lateralis.

Antagonists Supra- and infra-hyoid muscles.

Function Retracts and elevates the mandible.

MASTICATORS: GLOBAL TESTING

Movements revolve around the two condyles and are of three types: rotation, sliding, and laterality or deviation. For the purposes of this section, the following tests will imply sound bones and joints.

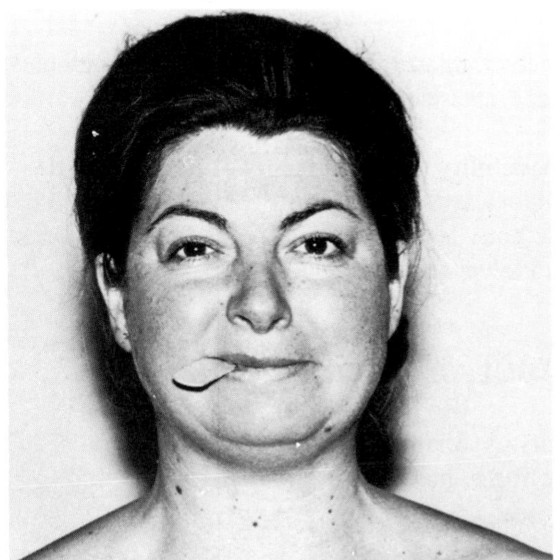

Tight, unilateral bite.

RAISING THE MANDIBLE

Masseters, temporales and medial pterygoidei.

Test

0 Elevation is impossible.
In myasthenia, the mandible drops and the subject is unable to raise it.
No contraction can be felt in the masseters at the lateral side of the face.
No contraction can be palpated endo-buccally at the medial angle formed by the mandible and maxilla.

1 The movement should be repeated five times. It is weak and range of movement is incomplete. There is no contact between upper and lower jaws. The distance between the two jaws is measured between central upper and lower incisors with callipers.

2 The movement should be repeated five times in full.

3 The movement should be repeated ten times with teeth remaining clenched for two seconds.

4 The movement is performed fully. Teeth clench forcefully. This is controlled with a tongue blade upon which the examiner tugs. Right and left sides are tested first separately, then together.

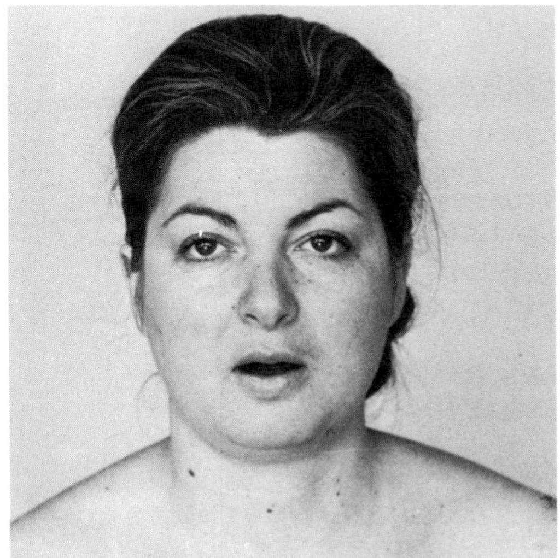

Dropped jaw: 'Open your mouth, say "ah"'.

LOWERING THE MANDIBLE

The movement may be active or passive. Bilaterally it involves:

—pterygoideus lateralis, anterior bellies of supra-hyoids,
—mylohyoideus lateralis (after lifting hyoid bone and tongue), and
—geniohyoideus (whose function is the same as that of the mylohyoideus).

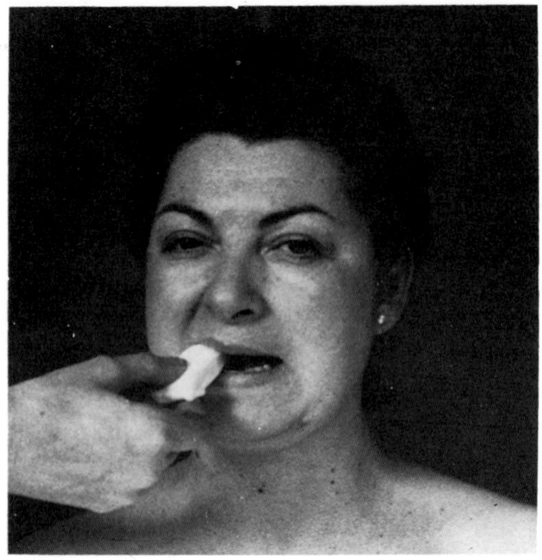

Testing bite. Tubes of different diameters.

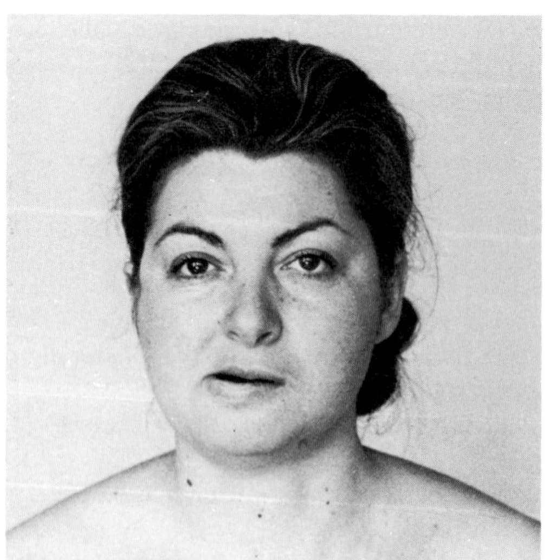

Mandibular deviation.

Test

The examiner places resistance beneath the mandible, then asks the subject to lower it. Depending upon the resistance overcome, the force of the movement is graded weak, fair or strong. The opening between the two jaws is measured between the incisors in millimetres, and should be about 40 mm.

MANDIBULAR PROTRUSION

Simultaneous, bilateral contraction of pterygoideus lateralis and masseter, superficial portion.

Test Possibility or impossibility of protruding the mandible forward. The amount varies, generally 3 to 4 mm (no bony lesions present), measured between upper and lower incisives.

MANDIBULAR RETRACTION

Simultaneous, bilateral contraction of temporalis, posterior fibres, medial pterygoideus, supra- and infra-hyoids.

Test Possibility or impossibility of pushing the examiner's fingers, placed behind the ramus of the mandible. Extent is slight, 1 to 2 mm.

MANDIBULAR DEVIATION (sideward movement)

Unilateral contraction of pterygoideus lateralis on side opposite the direction of the movement.

Test Possibility or impossibility of moving the mandible from side to side. Compare right and left sides. Extent varies from 8 to 11 mm.

MUSCLES OF MASTICATION: EXPRESSION

With the jaws forcefully closed, and participation of the platysma, these muscles express rage.

NOTES

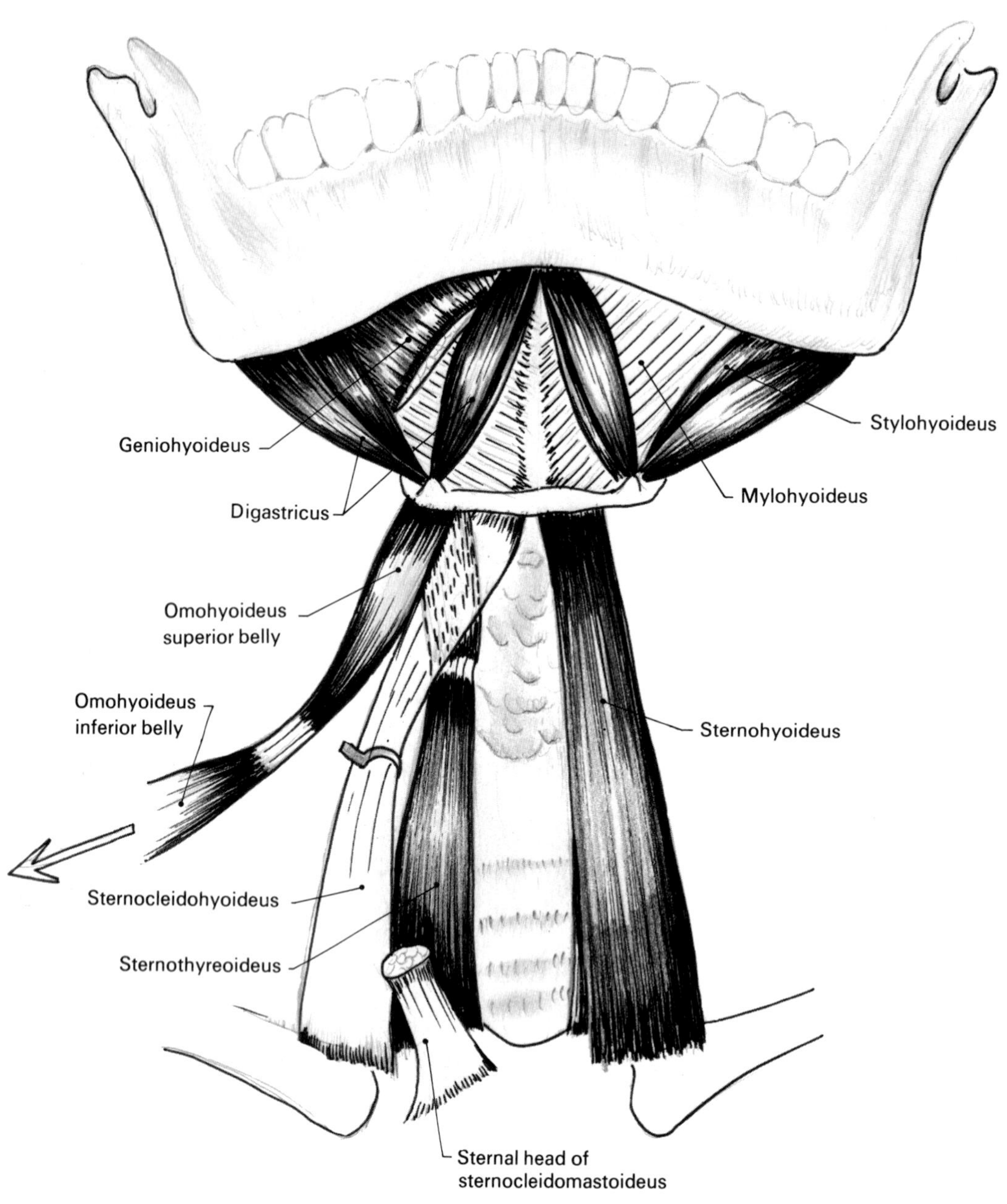
Stylohyoideus
Geniohyoideus
Mylohyoideus
Digastricus
Omohyoideus superior belly
Omohyoideus inferior belly
Sternohyoideus
Sternocleidohyoideus
Sternothyreoideus
Sternal head of sternocleidomastoideus

HYOID MUSCLES

These muscles form the anterior junction between head and thorax. They are located above and below the hyoid bone. The hyoid bone is very mobile. It is suspended by muscles and tendons from the tip of the styloid process of the temporal bone at the C4 level. These muscles are listed individually but a pathological state is necessary in order to test them analytically. Here, their presence is considered with the active mobility of the hyoid bone. In testing, the examiner should note their presence or absence, and the active mobility of the hyoid bone.

SUPRA-HYOID MUSCLES

There are four pairs located anteriorly and subcutaneously in the neck above the hyoid bone.
—mylohyoideus
—geniohyoideus
—digastricus
—stylohyoideus.

MYLOHYOIDEUS

Flat, triangular muscle; with its contralateral partner, it forms a muscular floor for the oral cavity.

Origin Entire length of the mylohyoid ridge of the mandible.

Insertion Into the body of the hyoid bone and median raphe from chin to hyoid bone.

Innervation Mylohyoid branch of the inferior alveolar division, trigeminal nerve (V).

Function Raises floor of mouth and acts in swallowing. When the mandible is fixed, it elevates the hyoid bone and the base of the tongue. When the hyoid bone is fixed, it lowers the mandible.

GENIOHYOIDEUS

Short, narrow muscle situated immediately above the inner border of the mylohyoideus. It participates in forming the muscular floor of the mouth.

Origin Inferior, genial tubercle on surface of mandibular symphysis.

Insertion Anterior surface of the body of the hyoid bone.

Innervation Branch of C1, C2 through the hypoglossal (XII).

Function Acts in swallowing. When the mandible is fixed, it elevates the hyoid bone and the larynx.

DIGASTRICUS

A muscle formed of two fleshy bellies, an anterior and a posterior. They are separated by a tendon which is attached to the hyoid bone by a fibrous loop.

DIGASTRICUS, ANTERIOR BELLY

Origin A depression on the inner side of the lower border of the mandible, close to the symphysis.

Insertion Into the central tendon with an expansion into the side of the body and the greater cornu of the hyoid bone.

Innervation Myohyoid branch of the inferior alveolar division of the trigeminal (V).

DIGASTRICUS, POSTERIOR BELLY

Origin Digastric groove on the inner side of the mastoid process of the temporal bone.

Insertion Into the central tendon with an expansion into the body and the greater cornu of the hyoid bone.

Innervation Branches of the facial (VII).

Function

—Simultaneous contraction of both bellies raises the hyoid bone and the larynx.
—Contraction of anterior belly: raises the hyoid bone forward.
—Contraction of posterior belly: raises the hyoid bone backwards.

STYLOHYOIDEUS

Small, spindle-shaped muscle.

Origin Back and outer surface of the styloid process near its base, anteriorly and medially to the mastoid.

Insertion Into the body of the hyoid bone at the junction of the greater cornu, just above the omohyoideus.

Innervation Stylohyoid branch from the posterior trunk of the facial (VII).

Function Same as that of the posterior belly of the digastricus, it raises the hyoid bone backwards.

INFRA-HYOID MUSCLES

Flat, thin muscles located in front of the trachea and the larynx in two layers:
—Superficial, composed of the omohyoideus and the sternocleidohyoideus.
—Deeper layer, composed of the thyrohyoideus and the sternothyroideus.

Function When the thorax is fixed, they depress the hyoid bone and thus indirectly depress the mandible.

OMOHYOIDEUS (scapulohyoideus)

Composed of two fleshy bellies united by a central tendon.

OMOHYOIDEUS, INFERIOR (POSTERIOR) BELLY

Origin Upper border of the scapula medially to the coracoid notch and occasionally the transverse ligament which crosses the suprascapular notch.

Insertion Central tendon.

OMOHYOIDEUS, SUPERIOR (ANTERIOR) BELLY

Origin Lower border of the body of the hyoid bone.

Insertion Central tendon. The central tendon is held in place by a process of the deep cervical aponeurosis. This process extends down to attach to the clavicle and the first rib.

Global innervation Ansa cervicalis formed by medial branches of C2 and C3, and a descending ramus of the hypoglossal from C1.

Global function Depresses the hyoid bone.
Contracts violently in sobbing; forcefully raises the skin of the supra-clavicular fossa. Approximates scapula towards hyoid bone. Accessory inspiratory muscle. May be palpated only in very thin people, such as some elderly patients.

STERNOCLEIDOHYOIDEUS

A thin, ribbon-like muscle situated just laterally to the median axis, covering the thyroid gland.

Origin Medial portion of the lower border of the hyoid bone.

Insertion Posterior surface of manubrium sterni, posterior sterno-clavicular ligament, and medial end of clavicle.

Innervation Ansa cervicalis C2 and C3, and ramus of hypoglossal (XII).

Functions Depresses the hyoid bone and larynx.

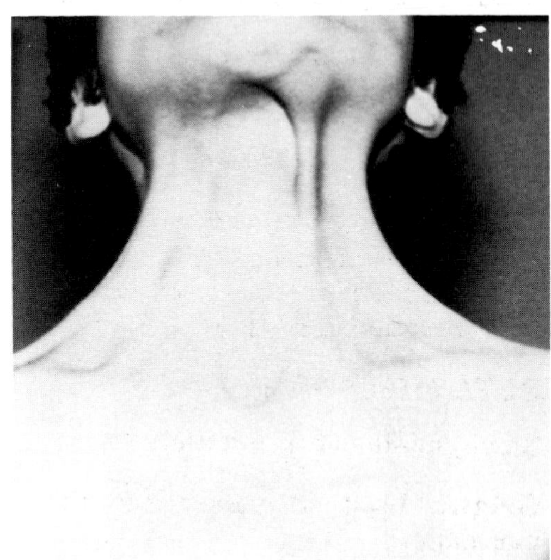

Paralysis of right supra- and infra-hyoid muscles.

THYROHYOIDEUS

A small, quadrilateral, deep layer muscle.

Origin Oblique line on the side of the thyroid cartilage.

Insertion Lower border of the body and greater cornu of the hyoid bone.

Innervation Thyroid branch of C1 through descendens hypoglossi C1 and C2.

Function Elevates thyroid cartilage, depresses hyoid bone and larynx.

STERNOTHYROIDEUS

Deep muscle situated beneath the sternohyoideus extending the thyrohyoideus.

Origin Posterior medial surface of manubrium sterni below and deep to origin of sternohyoideus and edge of the first costal cartilage.

Insertion Oblique line on the side of the ala of the thyroid cartilage.

Innervation Ansa cervicalis C1, C2, and C3, and ramus of hypoglossal (XII).

Function Depresses thyroid cartilage and larynx.

QUICK TESTS

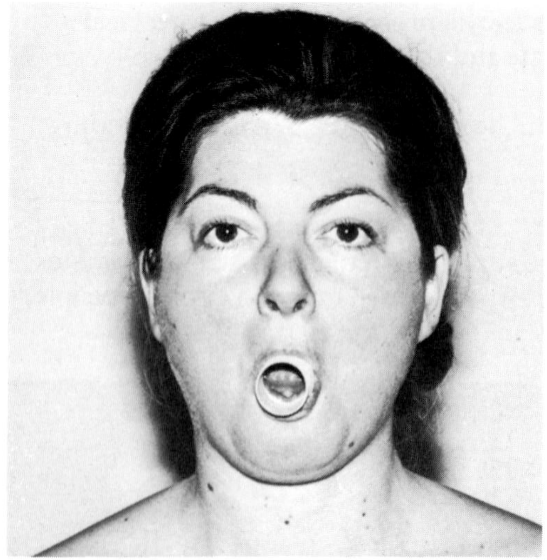

Lips tightened around a large tube.

—Orbicularis oris, superior and inferior portions in lengthened position, bilateral contraction.
—Depressor septi.
—Buccinator.

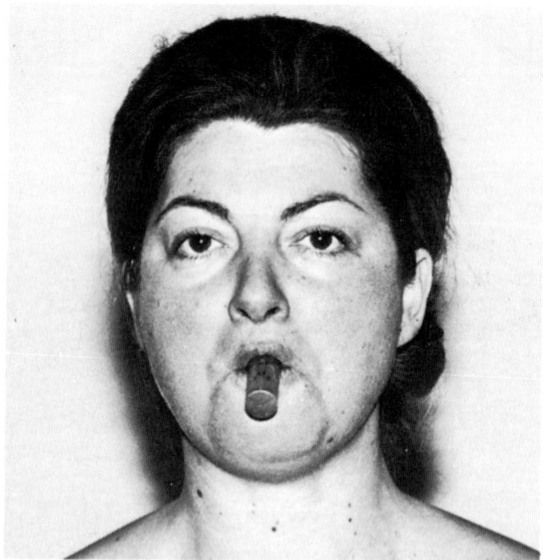

Lips tightened around an intermediate tube.

—Orbicularis oris, superior and inferior portions in middle position, bilateral contraction.
—Depressor labii inferioris.
—Mentalis.

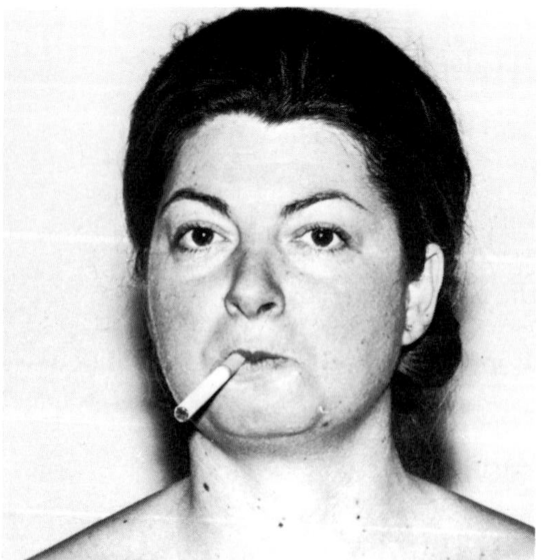

Lips tightened around a small tube.

—Orbicularis oris, supra- and infra-orbital portions in shortened position, unilateral contraction.

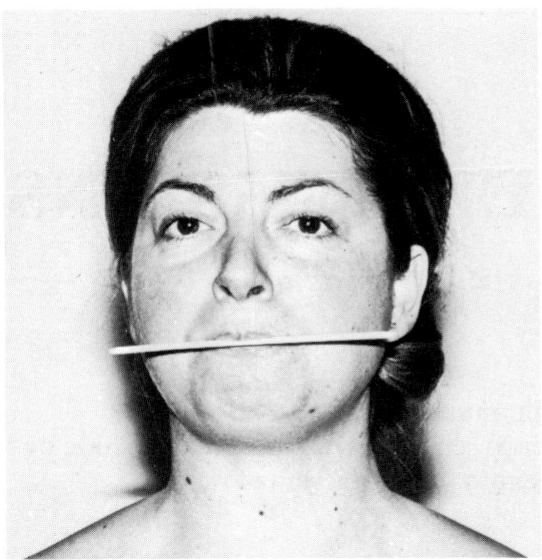

Stretched lips pressed together.

—Buccinator.
—Risorius.
—Mentalis.

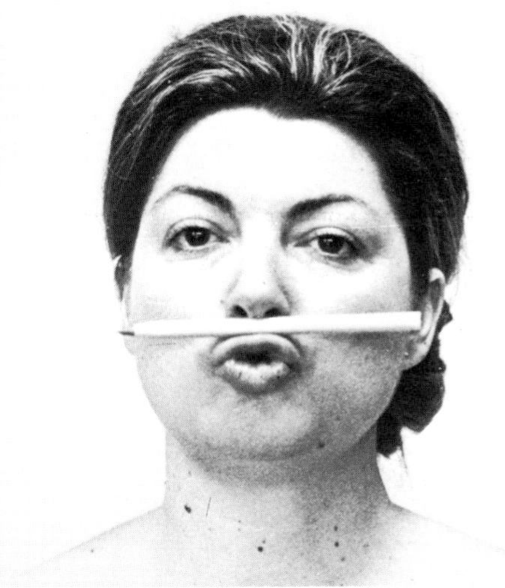

Raising the upper lip up.

—Levator labii superioris.
—Depressor labii inferioris.
—Mentalis.

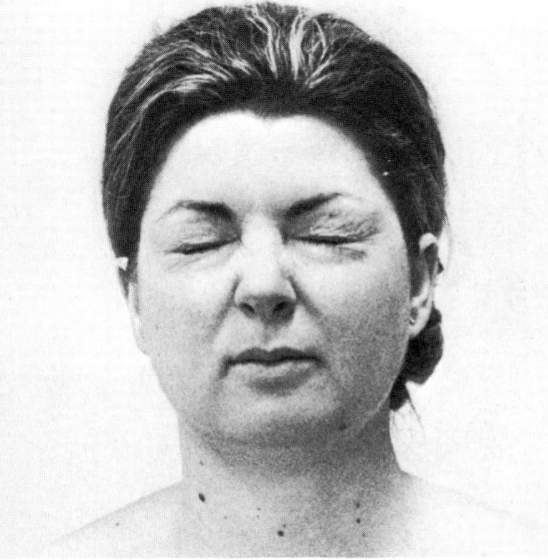

Eyes closed tightly.

—Orbicularis oculi, supra- and infra-orbital portions in shortened position, bilateral contraction.
—Zygomaticus major in maximally shortened position.

PRINCIPAL FACIAL EXPRESSIONS

Primary expressions involve the simple contraction of one main muscle.
Secondary expressions are more intense. Different muscle groups reinforce the simple contraction by degrees, giving scope to the expression of feeling and refinement to the expression of mood.

PRINCIPAL PRIMARY EXPRESSIONS

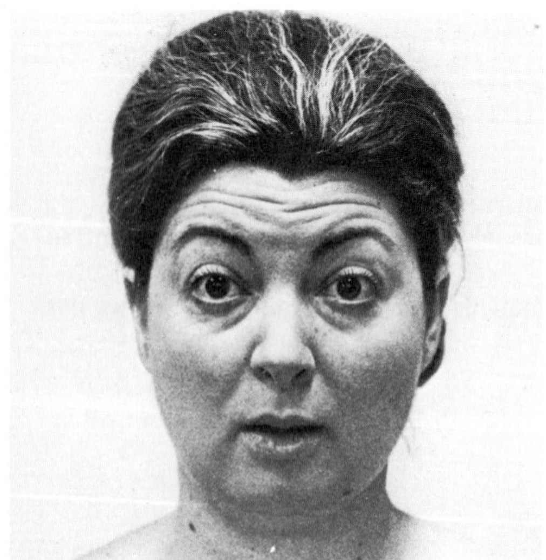

Astonishment: frontalis.

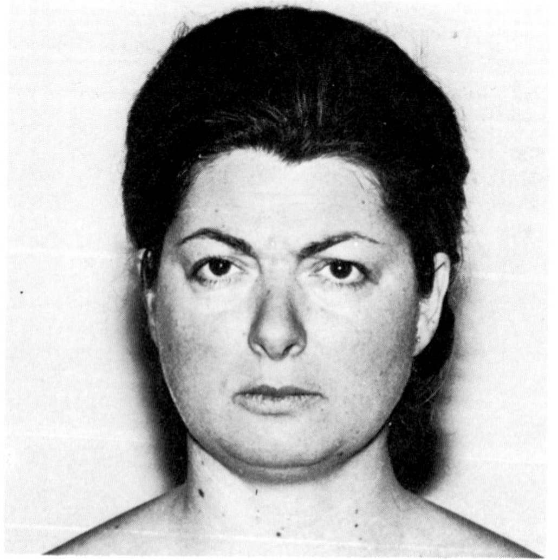

Sternness, anger: corrugator supercilii, horizontal fibres.

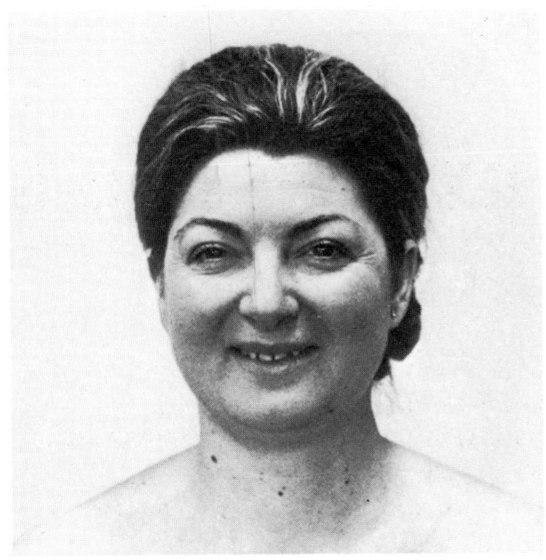

Smile : zygomaticus major.

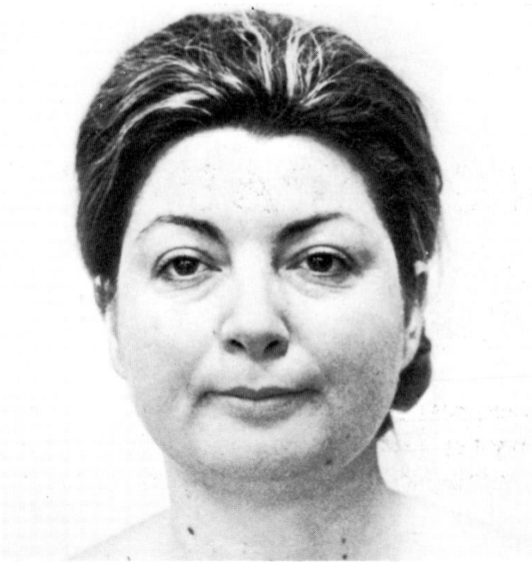

Disapproval : buccinator and risorius.

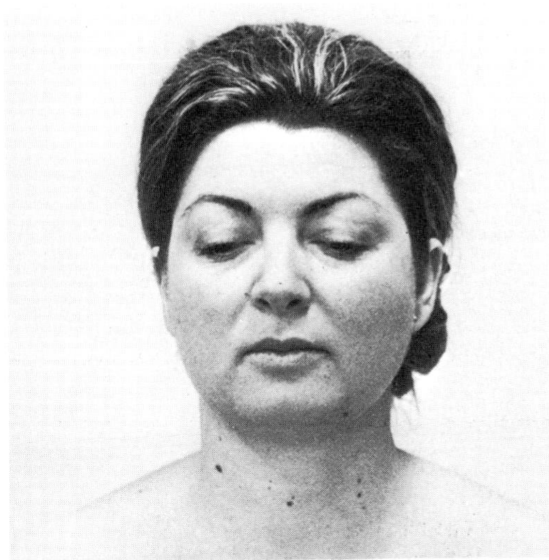

Meditation : balance between levator palpebrae superioris and orbicularis oculi.

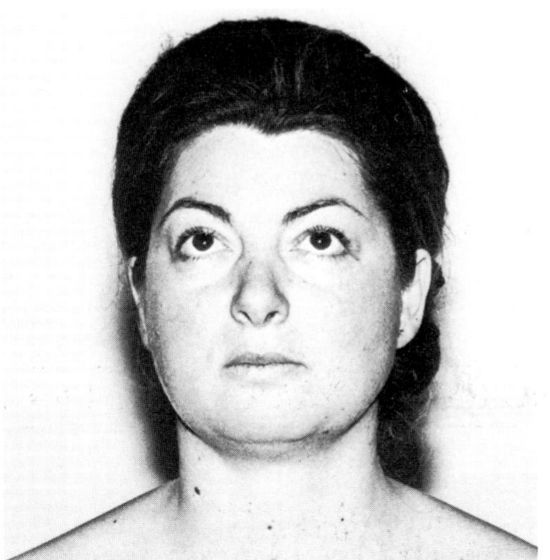

Adoration : Levator palpebrae superioris.

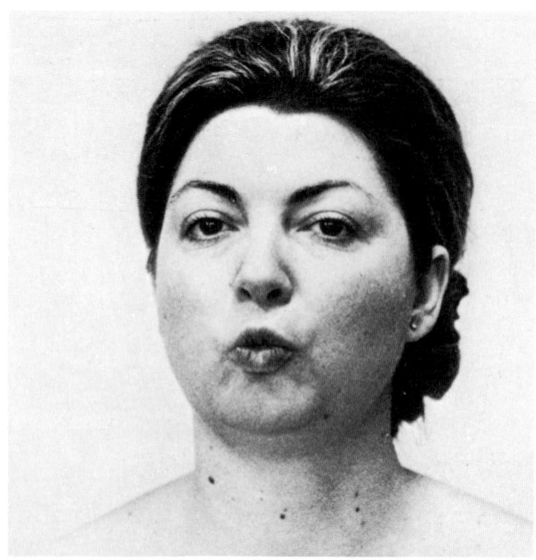
Reserve: orbicularis oris.

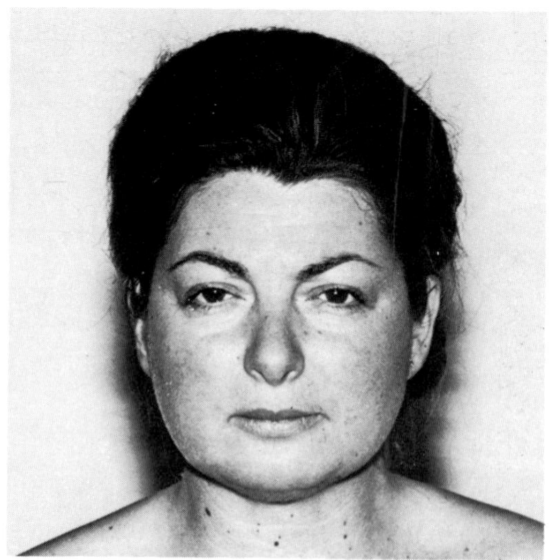
Searching: corrugator supercilii.

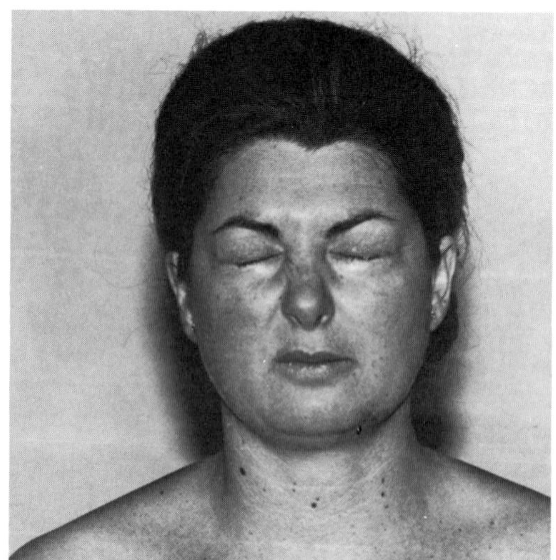
Protection: orbicularis oculi, superior and inferior fibres.

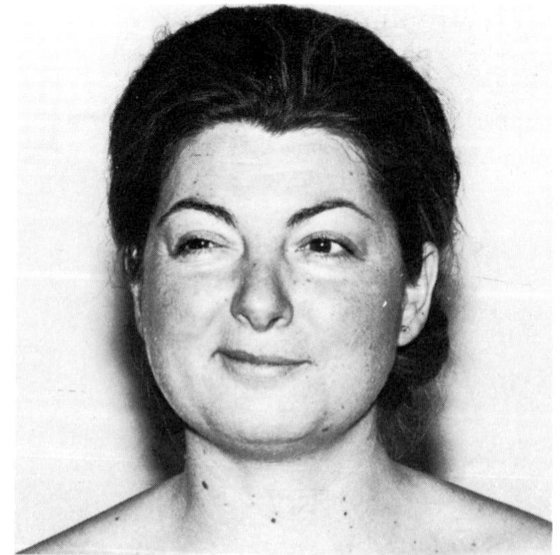
Complicity (wink): levator anguli oris and zygomaticus major.

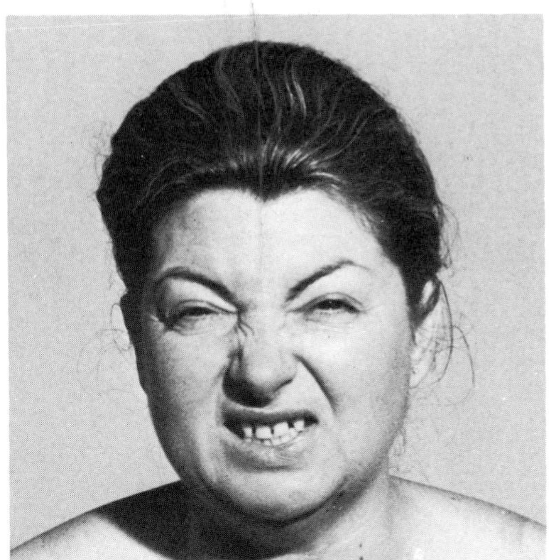

Rage.

Bilateral contraction of:
—Procerus.
—Nasalis dilating fibres.
—Dilatator naris.
—Levator anguli oris.
—Levator labii superioris, superficial layer.
—Zygomaticus minor.
—Orbicularis oculi, supra- and infra-orbital portions.
—Masticators.

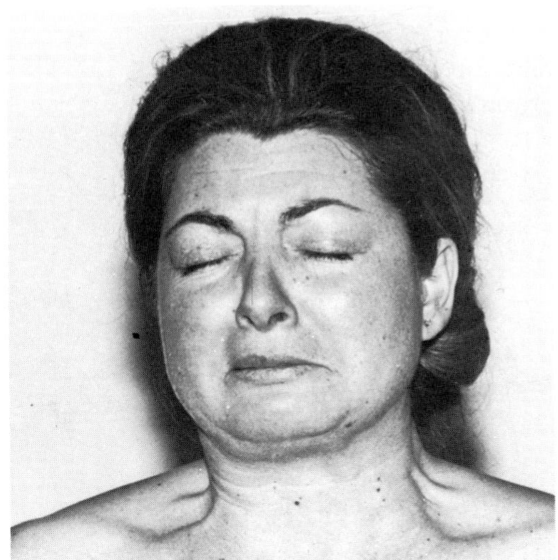

Overt sobbing, grief.

Bilateral contraction of:
—Nasalis, constricting fibres.
—Orbicularis oculi.
—Depressor anguli oris.
—Platysma.
—Mentalis.

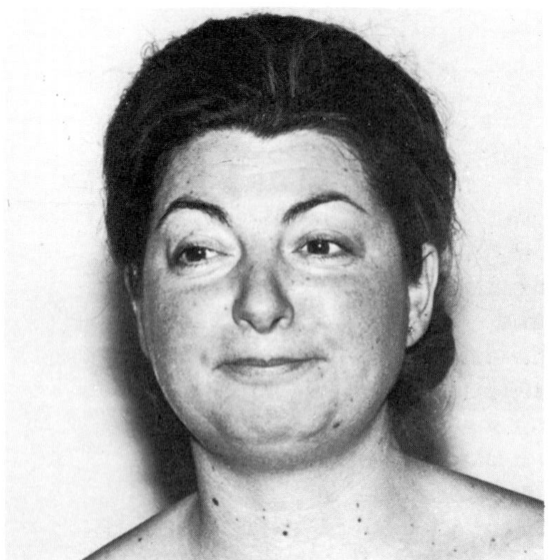

Irony.

Bilateral contraction of:
—Frontalis.
—Zygomaticus major.
—Buccinator.
—Orbicularis oculi, infra-orbital portion.
—Risorius.

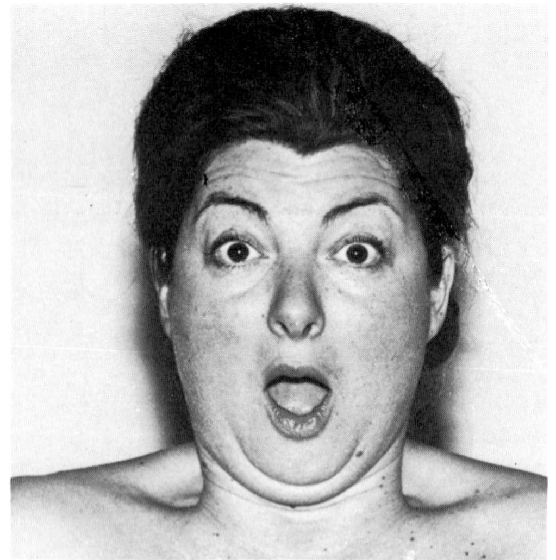

Fright.

Bilateral contraction of:
—Frontalis, especially lateral fibres.
—Infra-hyoids.
—Upper trapezius.
—Neck extensors.

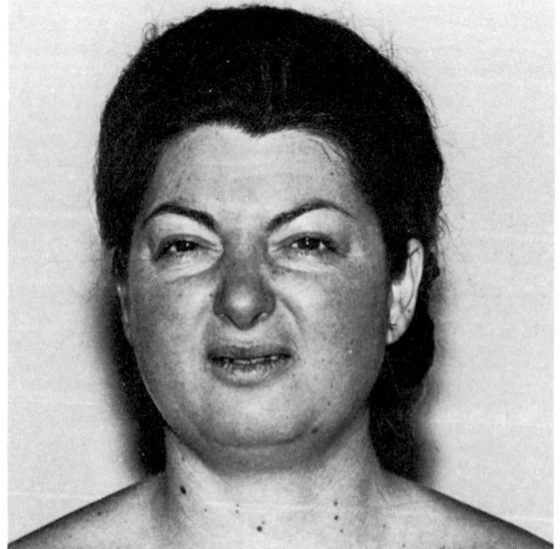

Disgust.

Bilateral contraction of:
—Nasalis.
—Procerus.
—Zygomaticus minor.
—Levator labii superioris.

Sniggering or embarrassed laughter.

Unilateral contraction of:
—Nasalis.
—Procerus.
—Levator anguli oris.
—Dilatator naris.
—Zygomaticus major.
—Zygomaticus minor.
—Buccinator.
—Risorius.

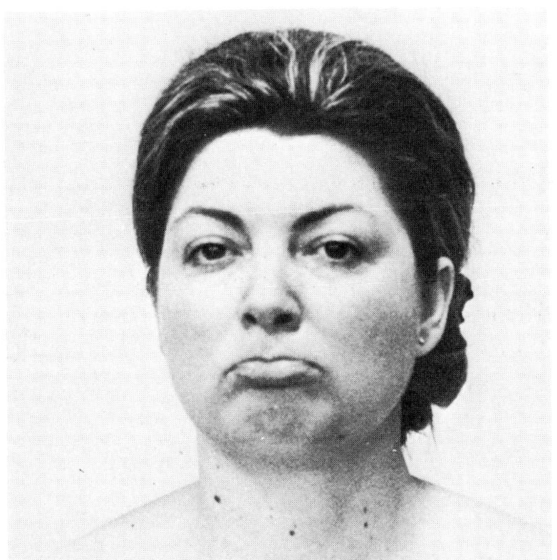

Pouting.

Bilateral contraction of:
—Mentalis.
—Depressor labii inferioris.
—Depressor anguli oris.
—Platysma, anterior fibres.

Mockery.

Bilateral contraction of:
—Orbicularis oculi, infra-orbital portion.
—Zygomaticus major.
—Risorius.

BIBLIOGRAPHY

Basmajian J V 1977 Anatomie. Maloine. Sombabec iteé

Chouard C, Charanchon R, Morgon A, Cathala H P 1972 Le Nerf Facial (anatomie pathologie en chirurgie). Masson, Paris

Chusid J G 1970 Correlative Neuroanatomy and Functional Neurology, 14th edn. Lange Medical Publications, Los Altos, Calif., p 94

Cuyer E 1896 Anatomie artistique du corps humain. 3rd edn. Baillière, Paris

Cuyer E 1906 La mimica. Version in Spanish (Jorro D (ed)). Madrid

Daniels L, Williams M, Worthingham C 1958 Le testing. 2nd edn. Maloine, Paris

Duchenne G B (de Boulogne) 1876 Mécanisme de la Physionomie humaine ou analyse électro-physiologique de l'expression des passions. 2nd edn.

Duchenne G B (de Boulogne) 1867 Physiologie des mouvements démontrée àl'aide de l'expérimentation électrique. Paris

Ermiane R 1949 Jeux musculaires et expressions du visage. Librairie Le François

Kendall H O, Kendall F P, Wadsworth G E 1974 Les Muscles. Bilan et étude fonctionnelle. Maloine, Paris

Mantegazza P 1889 La Physionomie et l'expression des sentiments. 2nd edn. Ancienne librairie Germer. Alcan F (ed). Baillière, Paris. Bibliothèque scientifique internationale.

Moreaux A 19-- Anatomie artistique de l'homme. Maloine, Paris

Perlemuter L, Waligora 19-- Nerfs crâniens et organes correspondants. Masson, Paris

Rouvière H 1962 Anatomie humaine par Cordier Delmas. Vol 1 : Tête et cou. Masson, Paris

SECTION TWO

Evaluating Motor Function in Trunk and Limbs of Patients with a Peripheral Nerve Lesion

A. Miranda and Collaborators

Translator's note In this section Mr Miranda and collaborators have asked that expressions such as 'flexion of the forearm on the arm' and 'extension of first phalanx on metacarpal' be used in place of 'elbow flexion' or 'metacarpophalangeal joint extension' in order to convey the idea of mobilizing a distal segment from a fixed origin.

D. Thomas

PART 1

Introduction, Testing Protocol and Material

INTRODUCTION

Manual testing of individual muscles is designed to determine peripheral lesions. Manual testing allows the establishment of qualitative and quantitative grading of muscle contraction and its strength. It enables us to follow the progression of the disorder and to foresee certain imbalances and contractures. Lastly, it permits us to establish and monitor an appropriate and progressive treatment programme.
There are important matters to be considered with respect to the patient, the examiner and the tests.

PROTOCOL FOR THE MANUAL TESTING OF INDIVIDUAL MUSCLES IN PERIPHERAL LESIONS OF THE TRUNK AND LIMBS

THE SUBJECT

To be considered:
—Age, sex, and morphology
—Weight (relation between body weight and muscle force)
—Pain
—Bone and ligament fragility
—Variation in range of motion from one individual to another
—Muscle force varies depending upon:
• Age (maximum force between 20 and 30 years)
• Difference between tonic and phasic muscles
• Angle of joint
• Normal use (i.e. occupation, sport etc.)
• Dominant limb
—Fatigue
—Degree of comprehension and cooperation.

THE EXAMINER

The different evaluations should be done by the same person, with possible confirmation by another examiner, personal subjectivity being taken into account. The test is objective up to grade 3, from then on it becomes a measure of relative force.

EVALUATIONS

Testing should be done:
—at the same time of day (mornings if possible).
—in the same order (so that reactions and fatigue can be compared).
Testing is compared with the sound side when possible.
Testing is done at regular intervals (which may vary depending upon the pathology).

MATERIAL NECESSARY FOR MANUAL TESTING OF INDIVIDUAL MUSCLES

Little material is needed:
—hard-surfaced treatment table
—stool
—tilt-table

—sliding board and talc (if this is too difficult to use, the examiner may support the limb)
—cushions
—baton
—toothbrush
—metric tape-measure
—goniometer
—portable lamp which will direct light across the patient's skin in such a way that the slightest movement of the skin is visible to the examiner
—wood block (used as a lift under foot).

POSITIONING THE SUBJECT

The subject should wear as little clothing as possible, and lie comfortably as relaxed as possible, in a calm, well-lit, correctly heated room.

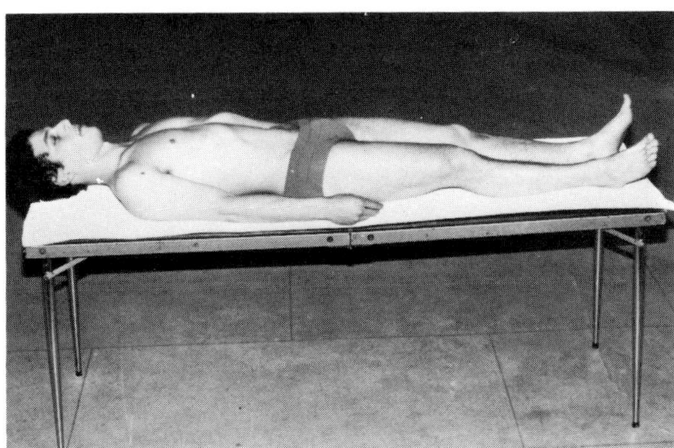

Supine

Spinal column straight, girdles aligned, upper limbs along sides of body (small cushion beneath head if necessary).

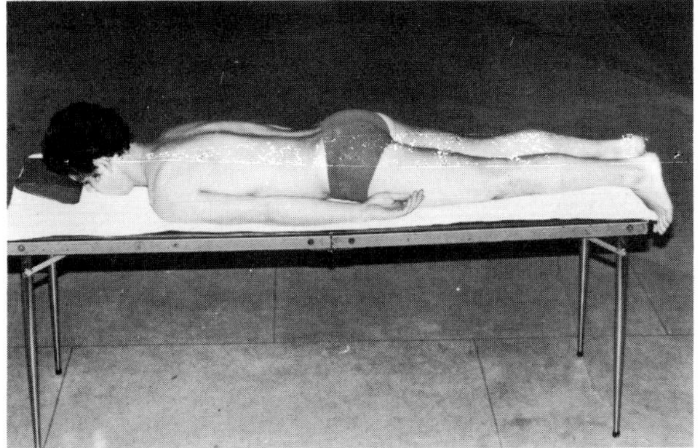

Prone

Spinal column straight, girdles aligned.
Cushion under forehead, feet hanging over end of table, cushion under shins (if necessary).
Avoid abdominal cushion which may alter the test.

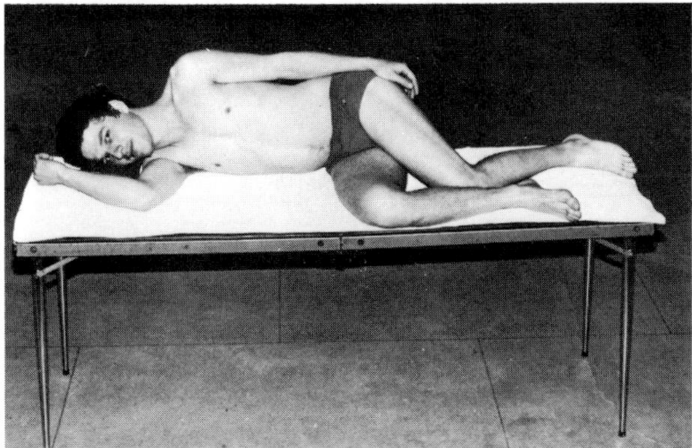

Side-lying

Cushion shoulder-height under head.
Seek maximum stability.

Side prone lying (not shown)

Halfway between prone and sidelying, cushion shoulder-height under head, stabilized by flexed lower extremity.

Supine on tilt-table at 45°

Subject clothed to permit sliding.

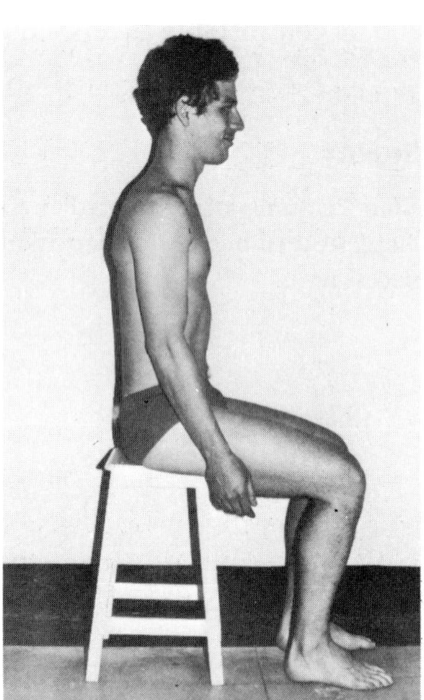

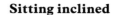

Sitting

All lower extremity joints at right angles.
Spinal column erect.
Head straight.

Sitting inclined

Sitting with trunk reclining at about 45°.

On all fours

Kneeling on hands and knees.

Standing (not shown)

Standing 'at ease'.

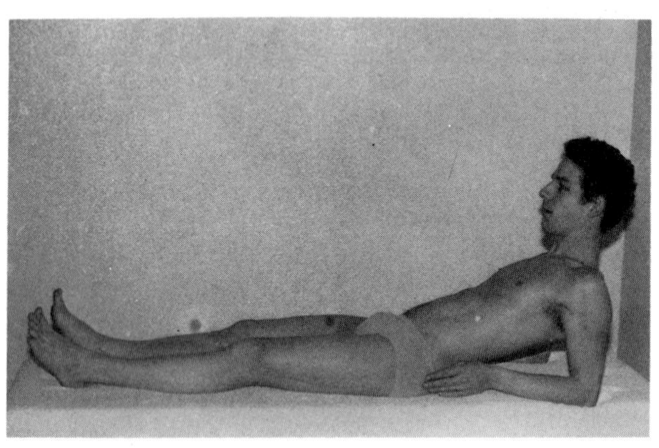

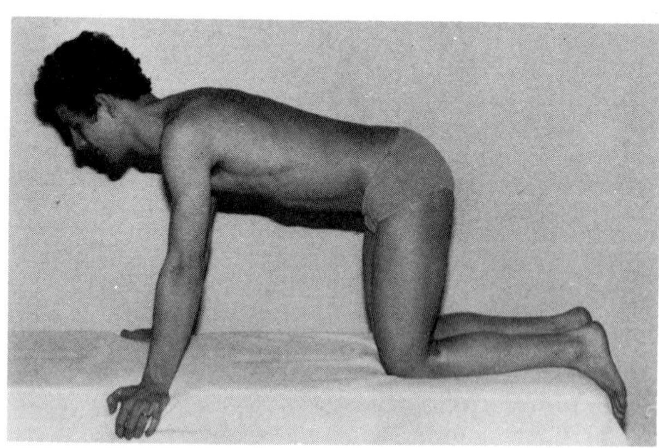

POSITION OF EXAMINER

—should be stable.
—is as close as possible to the subject and the muscle to be tested. (For photographic purposes, illustrations in this book may not always reflect this position.)
—allows a constant, overall view of the patient.
—wear nothing on upper extremities, neither watch nor jewellery, to avoid injuring the patient.

The test begins with a **qualitative evaluation.**
Before grading a muscle, an attentive **visual assessment** of the muscle is essential.
This is always done under bright, glancing light.
The examiner notes:
—absence of or decreased skin wrinkles,
—whether or not muscular tendons protrude,
—volume variation in the fleshy part of the muscle:

- Either flattening: muscular atrophy (decrease in muscle relief) which can be localized or generalized, not to be confused with thinness.
- Or protrusion: muscular hypertrophy (increase in muscle volume) is found for example in myopathies, myositis or pseudohypertrophia, and in compensating adiposis by fatty deposits or oedema.

This volume variation will be shown objectively with precise measurements. They are taken on a supine, relaxed subject, at fixed distances from subcutaneous bony landmarks, after preliminary markings traced at different segmental levels depending on the pathology and the muscle group to be evaluated.
—Upper limb, from the olecranon:
at + 6 cm, + 9 cm, + 12 cm for the arm,
at − 3 cm, − 6 cm, − 9 cm for the forearm.
—Lower limb, from superior border of patella:
at + 12 cm, + 15 cm + 20 cm for the thigh,
at − 15 cm, − 20 cm for the leg.
Use a pliable tape-measure and compare with homologous points on the sound side whenever possible. Verify positions of tape to avoid errors. For measurements of perimeter, encircle limb with tape at a tangent to marking and perpendicular to the segment. Tape-measure is taut but does not indent skin.
Measurements are registered on the evaluation form and are particularly useful in noting recovery speed in muscular atrophy cases.
The **palpatory assessment** completes the qualitative evaluation.
Palpation is done on the belly of the muscle when possible, on the tendon when not possible.
—The belly or fleshy part of the muscle is easily depressed; it is supple and extensible on the relaxed muscle, hard and firm on the contracted muscle.
—The tendon rolls beneath the finger. It has a firm consistency and is inextensible.
Testing continues with a **quantitative evaluation** which should cover norms to be respected and possible causes for error.

Norms to be respected

—Ensure that subject is perfectly relaxed.
—Avoid frequent position changes to postpone onset of fatigue (see chart page 151).
—Examine articular integrity:

- Capsulo-ligamentary tightness is verified by passive mobilization of the segment. (Proximal and distal joint positions are taken into account considering that certain two-joint muscles can limit movement.)
- Evaluation is done on a relaxed subject, with respect to the zero position of joint range of motion which corresponds to joint alignment in the anatomical position.

- Several methods are possible:
  —Objective:
  either in relation to measuring devices (goniometer, tape-measure).
  or in relation to the organization of body segments in space (opposition of two segments to each other).
  —Subjective:
  Most frequently used because it is fast (the use of a goniometer is not always possible and often requires presence of two examiners). Practised judgement is often comparable to a goniometer estimate.
- When articular range of motion is limited, it should be taken into account in muscle testing, and the usual technique modified.

—Examine range of antagonists.
—Fix firmly proximal or distal segment (by hand or straps) to avoid compensation.
—Do not compress muscle masses with hand holds.
—When applying resistance, the lever must be as long as possible but avoid including another joint.
—Address the subject in a language he understands. Then demonstrate the movement or perform it on the contralateral sound side. If testing a very young child or a patient who is mentally deficient, use cutaneous stimulation (scratching, tooth-brush . . .), and passive movements of the limb to be tested to awaken its motor image.
—Use verbal stimulation: speak in a loud voice to incite maximal muscular response.

Possible causes for error

—Inspection and palpatation may be misleading on an adipose or oedematous subject.
—Presence of contractures.
Examine for this systematically before muscle testing, on a relaxed subject in the correct anatomical alignment, fixing proximal and distal segments to avoid compensation. Through a slow, passive movement, bring the muscle to its maximally stretched position in all its components (watch for poly-articular muscles) while recognizing pain, and bone and muscle fragility.

The degree and range of these contractures may alter performance and must be noted on the evaluation form.

—Weakness of stabilizing muscles in the correct performance of the movement may lead to error (a manual hold may avoid this).
—Substitution of an agonist muscle (always check by palpation of muscle belly or tendon).
—At times dissociation of two synergistic muscles is impossible (then the examiner knows he evaluates the function).
—Misinterpretation of the command by the subject may lead to confusion and instructions must be unambiguous.

GRADING SYSTEM FOR MANUAL TESTING INDIVIDUAL MUSCLES

HISTORY

Grading systems were established for individual muscle testing.
—In 1917, R. W. Lovett gave the following grades: Trace, Mediocre, Fair, Good, and Normal.
—In 1922, C. L. Lowman opted for numerical grading from 0 to 9.
—In 1936, H. O. and F. P. Kendall used a system of grading with percentages 0%, 5%, 20%, 30%, 80%, and 100%.
—In 1940, S. Brunnstrom and M. Dennen introduced the notions of + and −.
—In 1946, M. Williams, L. Daniels, and C. Worthingham, while working for the National Foundation for Infantile Paralysis, established the 0 to 5 system internationally (introduced in France by Pol Lecoeur).

GRADING FROM 0 TO 5

0 (zero)	There is no evidence of contraction.
1 (trace)	Presence of minimal contraction but no movement without gravity.
2 (mediocre)	Full range of movement without gravity.
3 (fair)	Full range of movement against gravity.
4 (good)	Full range of movement against gravity, with partial resistance or notion of fatigability.
5 (normal)	Full range of movement against gravity with normal resistance. Muscle is sound.

This grading system may be refined by the addition of + or − signs.

1+	Trace of movement.
2−	Range of movement incomplete without gravity.
2+	Beginning of movement against gravity.
3−	Range of movement incomplete against gravity.
3+	Range of movement complete against gravity and light resistance.

This international grading of 0 to 5 will therefore be used in this book in order to preserve a common approach and test interpretations understandable by all.
It is useful to repeat that this assessment remains subjective, and therefore subject to inter observer variation.
Grades given from 0 to 5 in order to clarify the explanation of tests are not commonly used in that order. When testing, first look for grade 3 (against gravity), then depending on what is encountered, find grade 2 or 4 and 5.
Palpation is always added to grades 0, 1, 2 and 3 (not always illustrated) in order to avoid any substitution by another muscle.
The 0, 1, 2, 3, 4, 5 system cannot be used in certain cases (for example intrinsic muscles of the hand, of the foot). Gravity has little effect on these small muscles. It is also the case for very powerful muscles (for example triceps surae, trunk muscles).
For patients who cannot maintain the vertical, positions must be adapted (a light resistance is furnished to replace gravity).
Finally, when articular range of motion is limited, limb displacement is not always possible. When the range of movement is limited:
—Either tests the muscle in one part of its range of motion, lengthened position, middle, or shortened position,
—Or test muscle function isometrically.

OBSERVATIONS CONCERNING GRADING

at 0 *no contraction is detected under bright glancing light.*

—This remains a gross approximation.
—Palpation is difficult on certain muscle bellies and tendons.

at 1 *presence of minimal contraction but no movement.*

—There may be the risk of confusion with adjacent agonists.

at 2 *full range of movement without gravity.*

—Gravity is not to be considered for muscles mobilizing a small segment, nor for very powerful muscles.

at 3 *full range of movement against gravity.*

—Repeat the movement four to five times.
—Applying light resistance corresponding to the weight of the mobilized segment in cases where range of movement exceeds 90° or for a segment upon which gravity affords no resistance.

at 4 *full range of movement against gravity with partial resistance.*

—Repeat the movement ten times.
—The use of partial resistance.
—Determine whether the subject shows fatigue in comparing affected side with the sound. Force is subnormal.
—Resistance is applied as perpendicularly as possible to the distal portion of the segment (the photographs in this book do not always include this condition in order to let the photographer work). The position is sometimes different so as not to risk damaging the tendon.

at 5 *muscle normal.*

Note:

—From grade 4 on, the question of subjectivity of single muscle response.
—As soon as resistance intervenes, contraction irradiates in order to stabilize the articular pivot.
—The degree of resistance remains very variable from one examiner to another.
—Finally, insufficiency of manual tests of certain individual muscles.
Possibility of using the break-test (isometric in muscle's most shortened position) or added weights (M. R. dynamometers, etc., not used in this book).

Grading of contracture or tightness

This is done by means of one of the three following codes, and noted on the evaluation form.

1.	C	2C	3C
2.	C	CC	CCC
3.	C+	C++	C+++

EVALUATION FORM

In following the evolution of peripheral neurological disorders, it is necessary to fill out an evaluation form upon which are also noted testing dates and the name(s) of the examiner(s).
The form proposed on the following pages is organized so that at a glance one may read the level(s) and the side(s) of muscular deficiencies (see page 129).
Only the muscular functions tested in this book appear on this form.

LIST OF ABBREVIATIONS USED IN EVALUATION FORMS

abd.	abduction/abductor
acc.	accessory
add.	adduction/adductor
ant.	anterior
ax.	axillary
brac. p.	brachial plexus
brev.	brevis
c	cervical nerve root, secondary importance
C	cervical nerve root, primary importance
CPM	Carpometacarpal joint
cut.	cutaneous
DIP	distal interphalangeal joint
dep.	depression
dev.	deviation
dir.	direction
elev.	elevation
ever.	eversion
expir.	expiration
ext.	extension/external
fib.	fibres
flex.	flexion
IP	interphalangeal joint
inf.	inferior
inspir.	inspiration
int.	internal, medial
inver.	inversion
l	lumbar nerve root, secondary importance
L	lumbar nerve root, primary importance
lat.	lateral (external)
long.	longus
MCP	metacarpophalangeal joint
MTP	metatarsophalangeal joint
med.	median/medial (internal)
mus. cut.	musculocutaneous
opp.	opposition
PIP	proximal interphalangeal joint
pl.	plantar
post.	posterior
rad. dev.	radial deviation
rot.	rotation
s	sacral nerve root, secondary importance
S	sacral nerve root, primary importance
sac. pl.	sacral plexus
sc.	scapular
sp.	spinal
sup.	supra/superior
supin.	supination
t	thoracic nerve root, secondary importance
T	thoracic nerve root, primary importance
uln. dev.	ulnar deviation
upr.	upper

Region	RIGHT SIDE					Nerve / Segment	Date and examiner	Action	LEFT SIDE					Region
HEAD & NECK						(Sp. acc. XI) C2 C3	Sternocleidomastoideus	flex. dev. rot.						HEAD & NECK
						Cl . . . C8	Anterior vertebral head and neck flexors	flex.						
						C1 . . . T1 C3 C4 + XI	Neck extensors Upper trapezius	ext.						
SCAPULA						Sp acc. XI c2 C3 C4	Middle trapezius sup. fibres	elev.						SCAPULA
							Middle trapezius inf. fibres	add.						
							Lower trapezius	dep.						
						C4 C5	Levator scapulae	elev.						
						C4 C5	Rhomboids	downward rotation						
						brac. p. med. cord. c7 C8 T1	Pectoralis minor	ant. dep.						
						C5 C6 C7	Serratus anterior	abd. upward rotation						
SHOULDER						axillary C5 C6	Anterior deltoid	flex.						SHOULDER
						mus. cut. C6 C7	Coracobrachialis							
						axillary C5 C6	Middle deltoid	abd.						
						sup. sc. C5 C6	Supra spinatus							
						axillary C5 C6	Posterior deltoid	ext. abd.						

Region	Nerve / roots	Muscle	Action
SHOULDER	upr. trunk post. cord. axillary C5 C6	Subscapularis	int. rot.
	sup. sc. C5 C6	Infraspinatus	ext. rot.
	axillary C5 C6	Teres minor	
	lat. cord.	Pectoralis major upper portion	add.
	med. cord.	Pectoralis major middle portion	
	C5 . . . T1	Pectoralis major lower portion	
	post. cord. C5 C6 c7	Teres major	add.
	post. cord. C6 C7 C8	Latissimus dorsi	ext. add. int. rot.
ELBOW	mus. cut.	Biceps brachii	flex.
	C5 C6	Brachialis	
	radial c5 C6	Brachioradialis	
	radial c6 C7 c8	Triceps brachii and anconeus	ext.
	radial C6 c7	Supinator	supin.
	Med C6 C7 / c7 C8	Pronator teres Pronator quadratus	pron.

Region						Nerve	Muscle	Action								
WRIST						ulnar c7 C8 T1	Flexor carpi ulnaris	flex. uln. dev.								
						median C6 C7 c8	Flexor carpi radialis Palmaris longus	flex.								
						radial C6 C7	Extensor carpi radialis longus	ext. radial dev.								
						radial c6 C7 c8	Extensor carpi radialis brevis	ext.								
						radial C7 C8	Extensor carpi ulnaris	ext. uln. dev.								
HAND						median c7 C8 T1	2 3 flexor digitorum 4 superficialis 5	2 3 4 5 PIP flex.								
						median + ulnar c7 C8 T1	2 3 flexor digitorum 4 profundus 5	2 3 4 5 DIP flex.								
						radial c6 C7 c8	2 Extensor digitorum & indicis propius 3 Extensor digitorum 4 5 Extensor digitorum & digiti quinti propius	2 3 4 5 MCP ext								
						median c7 C8 T1	Flex. pollicis longus	IP flex.								
						radial c6 C7 c8	Abd. pollicis longus	thumb anteposition								
						radial c6 C7 c8	Ext. pollicis brevis	thumb abduction								

WRIST
HAND

HAND

					radial c6 C7 c8	Ext, pollicis longus	IP ext. thumb retroposition										
					median C8 tl	Abd. pollicis brevis	thumb anteposition										
					median C8 tl	Opponens pollicis	opp. of thumb										
					median C8 T1	Flex. pollicis brevis superficial head	MCP flex. & anteposition										
					ulnar C8 T1	Flex. pollicis brevis deep head	MCP flex.										
					ulnar C8 T1	Add. pollicis	thumb add.										
					median + ulnar C7 C8 T1	1st 2nd 3rd 4th Lumbricales 1st 2nd 3rd 4th	MCP flex.										
					ulnar C8 T1	1st 2nd 3rd 4th Interossei palmares 1st 2nd 3rd 4th	finger add.										
					ulnar C8 T1	1st 2nd 3rd 4th Interossei dorsales 1st 2nd 3rd 4th	finger abd.										
					ulnar c8 T1	Flexor digiti minimi	MCP flex. of 5th										
					ulnar c8 T1	Abd. digiti minimi	abd. of 5th										
					ulnar c8 T1	Opponnens digiti minimi	opp. of 5th										

HAND

Region						Nerve	Muscle	Function										
TRUNK						iliohypogas-tricus, ilio-inguinalis	Rectus abdominis	flex										
TRUNK							Obliquus externus abdominis	trunk flex.rot										
TRUNK						T7 . . . T12	Obliquus internus abdominis	pelvic flex. rot										
TRUNK							Transversus abdominis	expir.										
TRUNK						phrenic c3 C4 c5	Diaphragm	inspir.										
TRUNK						intercostales T1 . . . T11	Intercostales externi	inspir.										
TRUNK						T1 . . . S3	Thoracic spine extensors	ext.										
TRUNK							Lumbar spine extensors											
TRUNK						T12 L1 l2	Quadratus lumborum	pelvic hitching										
HIP						lumb. plex. L1 . . . L4	Psoas	flex.										
HIP						femoral L2 . . . L4	Iliacus											
HIP						femoral L1 L2 L3	Sartorius	flex. abd. ext. rot.										
HIP						superior gluteal L4 L5 S1	Tensor fasciae latae	flex. abd. int. rot.										
HIP							Gluteus medius	abd.										
HIP							Gluteus minimus	int. rot.										
HIP						sacral plexus L3 . . . S2	External hip rotators	ext. rot.										
HIP						obturator + femoral L2 . . . S1	Hip adductors	add.										

						Nerve	Muscle	Action										
						inferior gluteal l4 l5 S1	Gluteus maximus	ext.										
KNEE						femoral L2 L3 L4	Quadriceps femoris (hip straight)	ext.										
							Vastus lateralis Vastus intermedius Vastus medialis											
						sciatic	Biceps femoris	flex. ext. rot.										
						L4 . . . S2 l5 S1	Semi-membranosus Semitendinosis Popliteus	flex. int. rot.										
ANKLE						peroneal L4 l5	Tibialis anterior	dorsiflex add. sup.										
						peroneal l4 L5 sl	Exten. hallucis longus	big toe dorsiflex										
							Exten. digitorum long Exten. digitorum brev	toe dorsiflex										
							Exten. digitorum peroneus tertius	foot dorsiflex. abd. pron.										
						mus. cut. l4 L5 sl	Peroneus brevis	foot abd. pron.										
							Personeus longus	dep, M1 abd. pron.										
						int. popliteal tibial l5 S1 S2	Triceps surae	pl. flex.										
						tibial l5 S1 s2	Soleus											
						tibial l5 S1	Tibialis posterior	add. supin										

Region						Nerve / segments	Muscle	Action											Region
						tibial l5 S1	Flexor hallucis long.	IP flex.											
							Flexor digitorum & quadratus plantae	DIP flex.											
FOOT						med. plantar tibial l5 S1	Abductor hallucis	big toe abd.											FOOT
						med. plantar l5 S1	Flexor hallucis brevis	MCP flex.											
						lat. plantar tibial S1 s2	Adductor hallucis	big toe add.											
						med. plantar l5 S1	Flexor digitorum brevis	PIP flex.											
						med. & lat. plantar S1 s2 s3	Lumbricales + Interossei	MCP flex.											
						lat. plantar S1 s2 s3	Interossei plantares	toe add.											
							Interossei dorsales	toe abd.											
						lat. plantar	Flexor digiti minimi brevis	MCP flex. of 5th											
							Abductor digiti minimi	Abd. of 5th											
						S1 S2	Opponens digiti minimi	Opp. of 5th											
						sacral plexus	muscles of the perineum	rectal sphincter control											

QUICK TESTS

Quick tests enable instant, gross strength examination by hand of synergistic muscle groups. The examiner, having thus a quick clinical notion of a possible deficiency in a muscle group, may then proceed into detailed individual muscle testing.

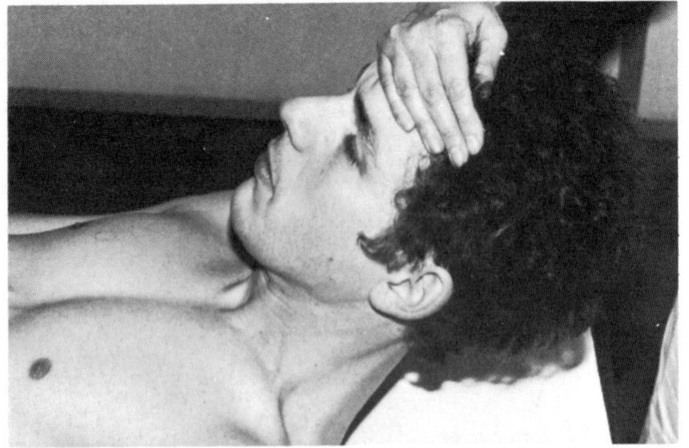

Anterior head and neck flexors.

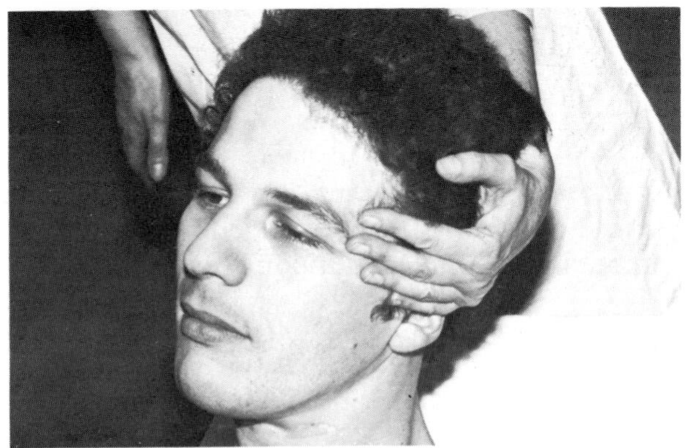

Lateral head and neck flexors.

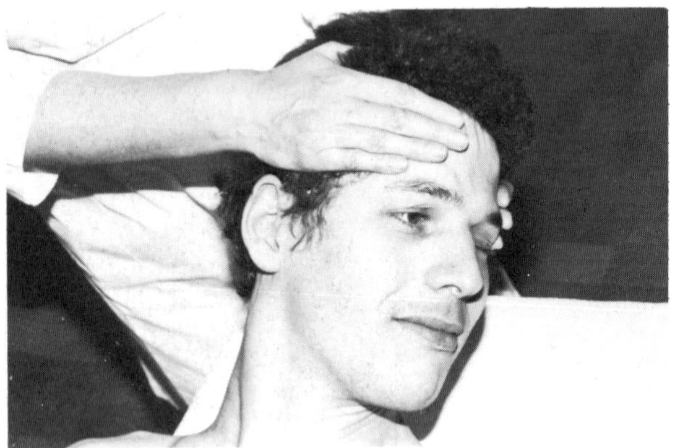

Head and neck rotators.

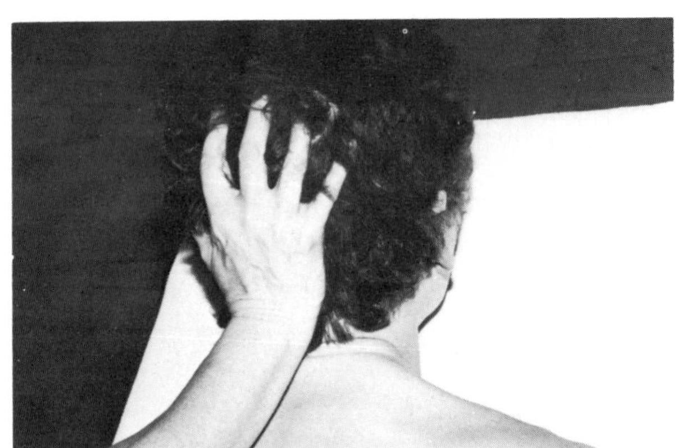

Head and neck extensors.

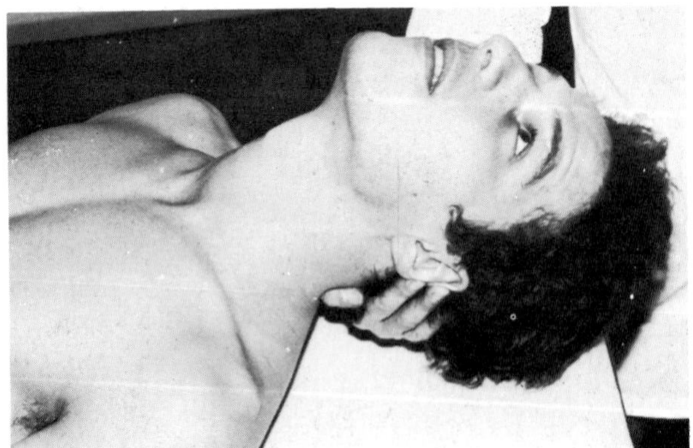

Head and neck extensors. Alternative.

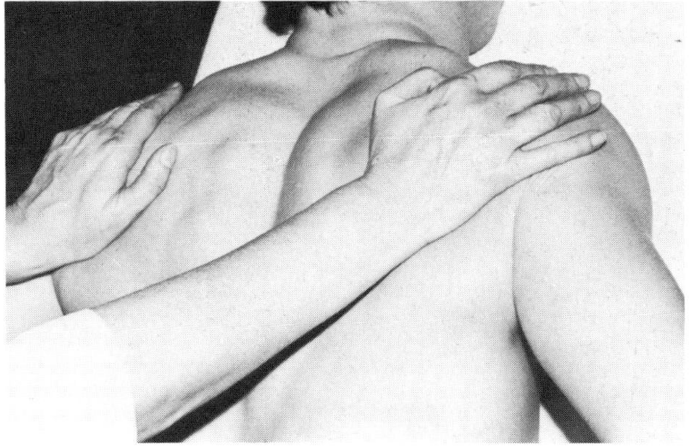
Scapular adductors.

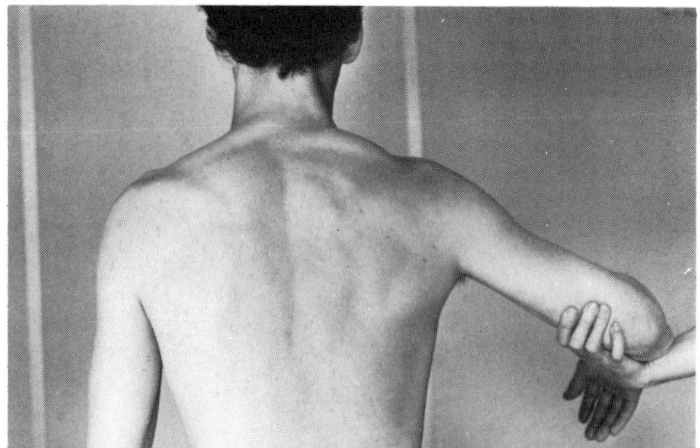
Arm and scapular adductors.

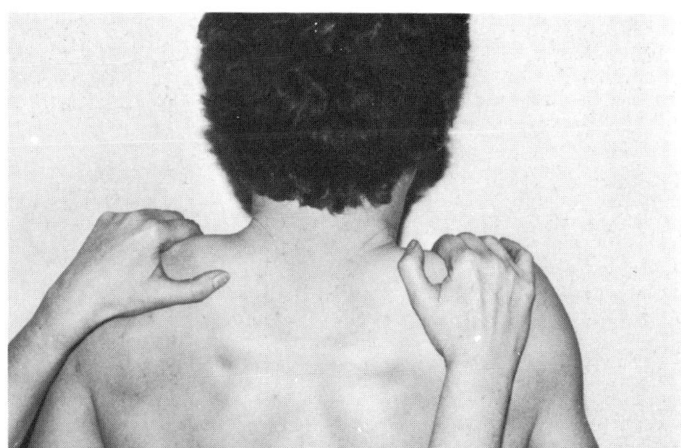
Scapular elevators.

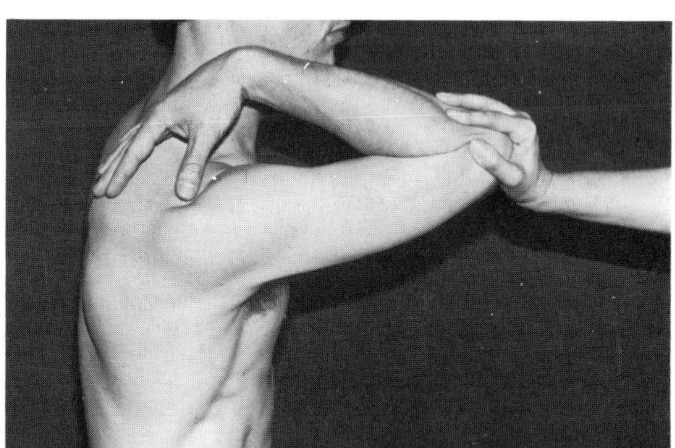
Scapular abductors.

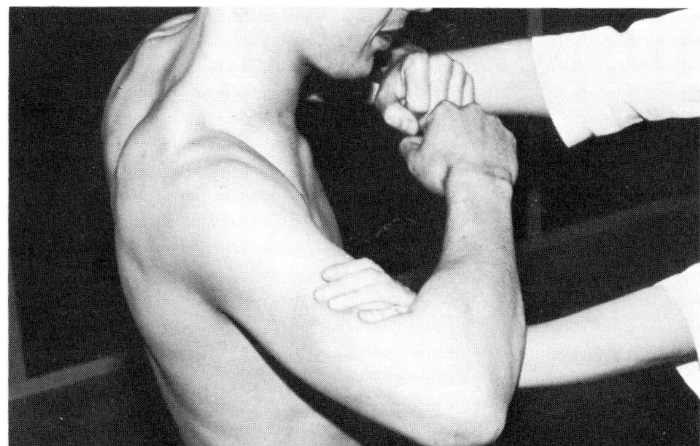
Shoulder joint flexors.

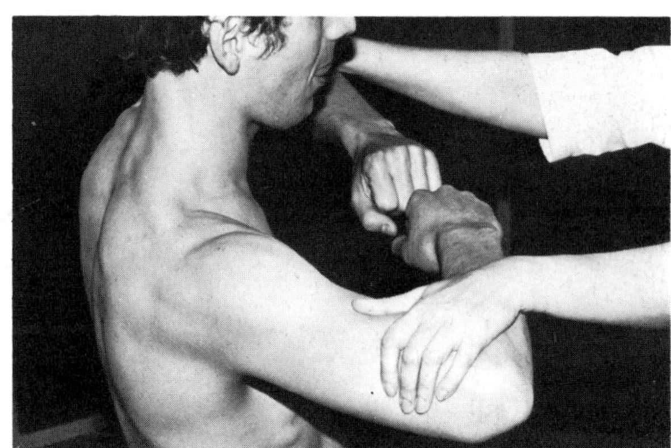
Shoulder joint abductors.

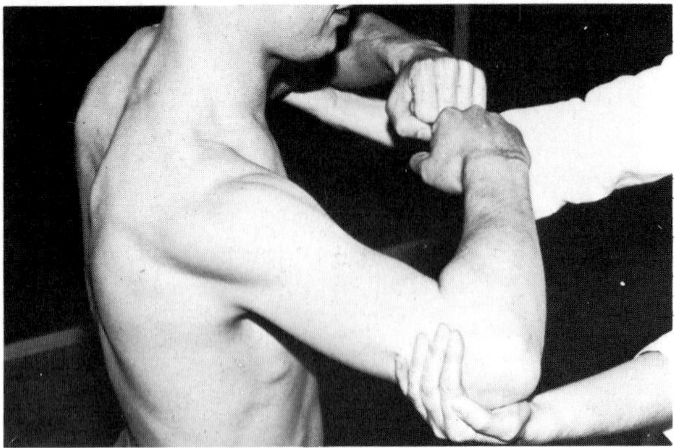

Shoulder joint extensors.

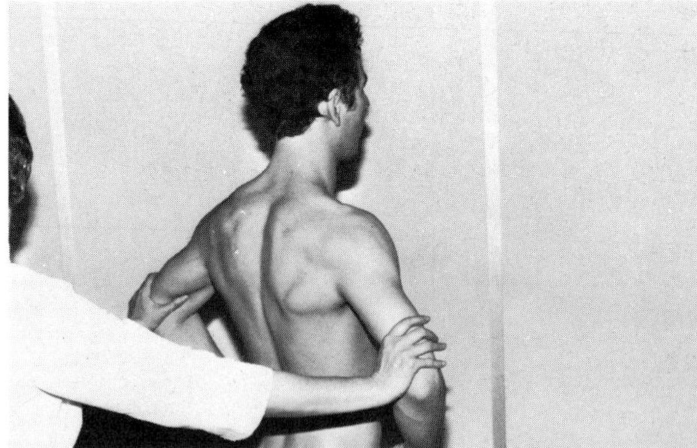

Shoulder joint extensors. Specifically for posterior deltoid (hands on hips).

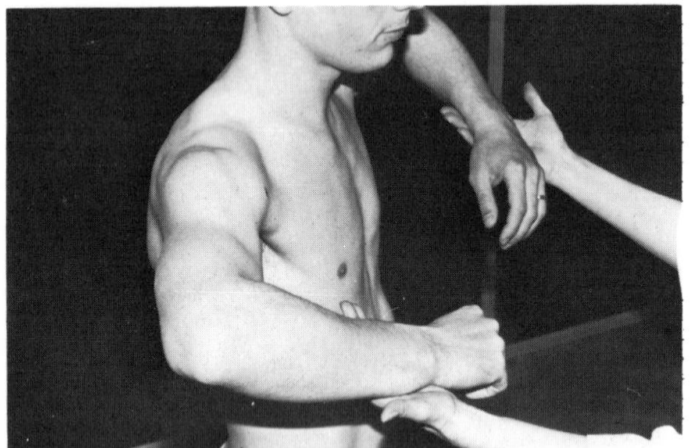

Internal rotators of the shoulder joint.

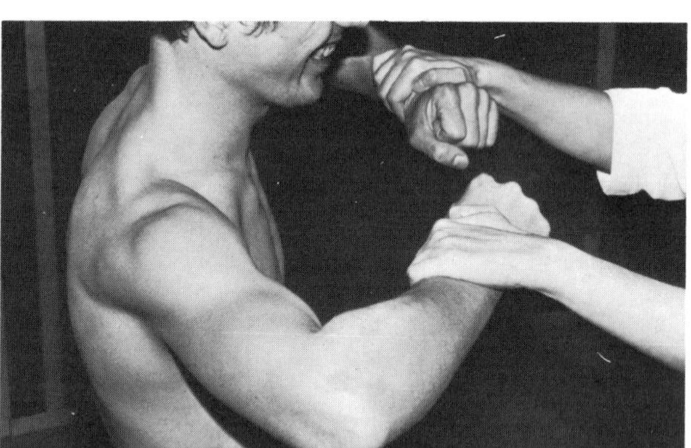

External rotators of the shoulder joint.

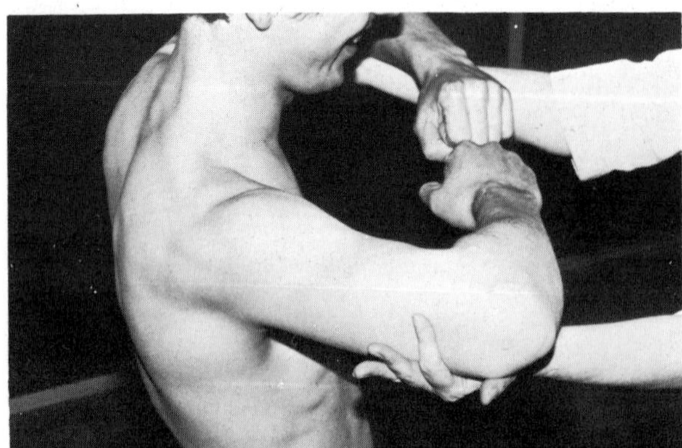

Shoulder joint adductors and depressors.

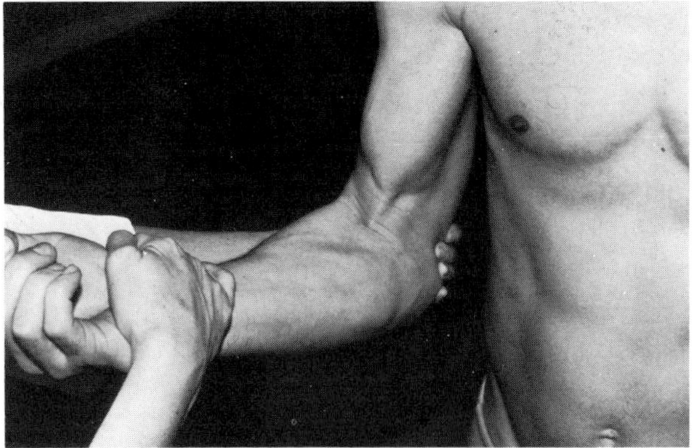
Elbow flexors.

'Pull' muscles.

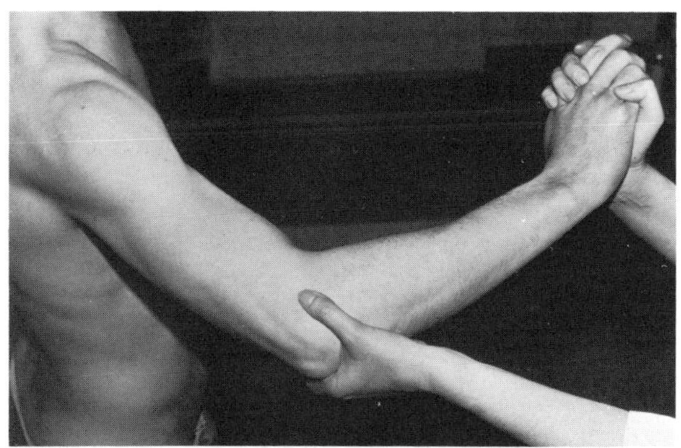
Elbow extensors and 'push' muscles.

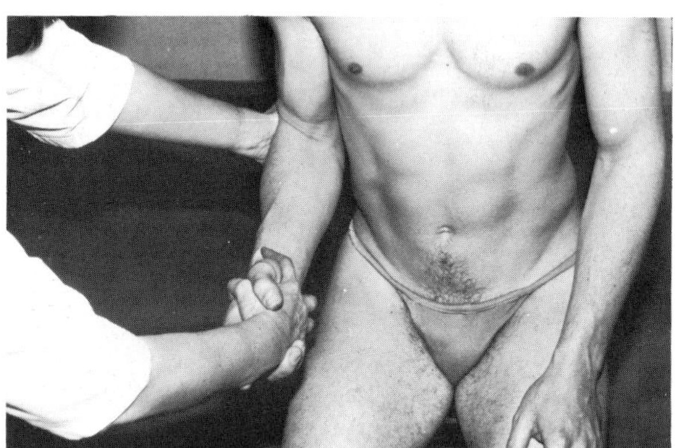
Pronators, supinators.

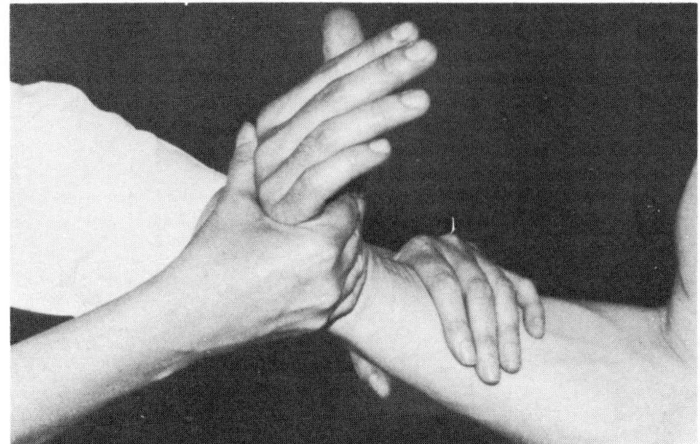
Wrist flexors.

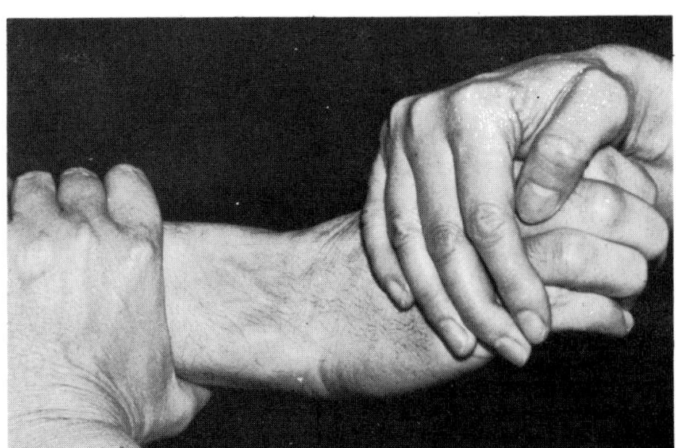
Wrist extensors.

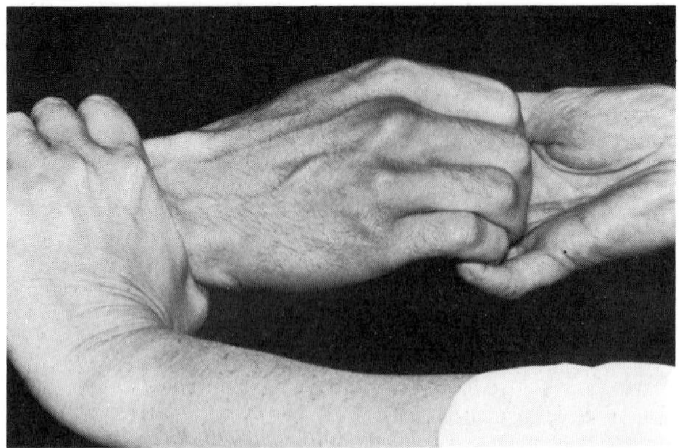
Finger flexors.

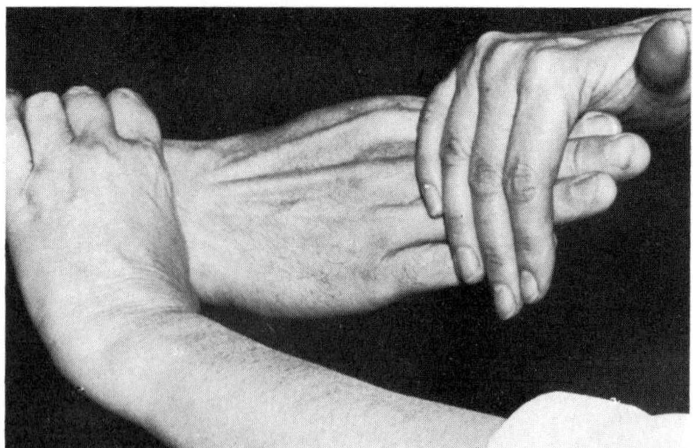
Finger extensors.

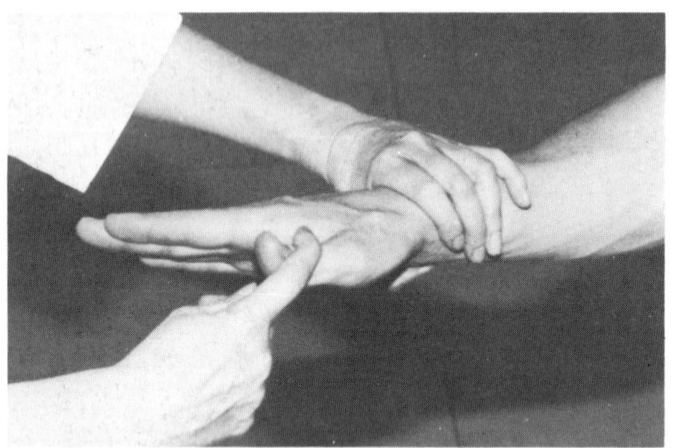
Thumb extensors and abductors.

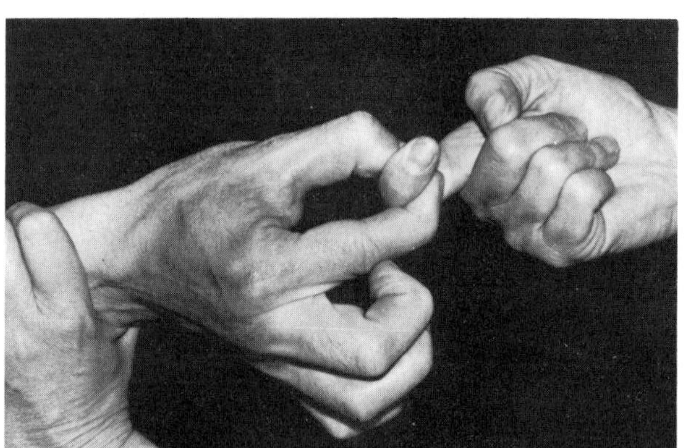
Intrinsic muscles of the thumb.

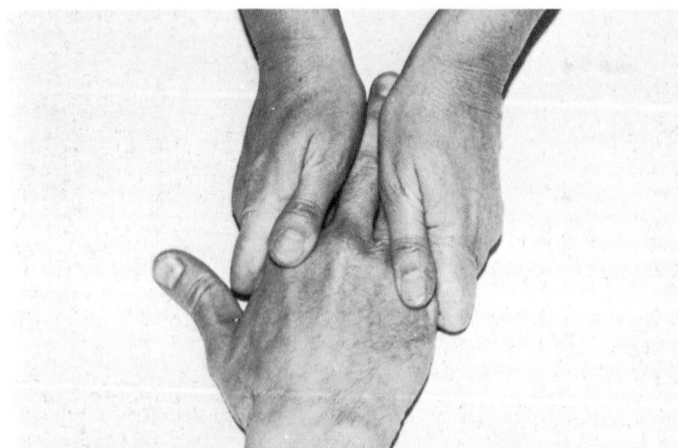
Finger abductors.

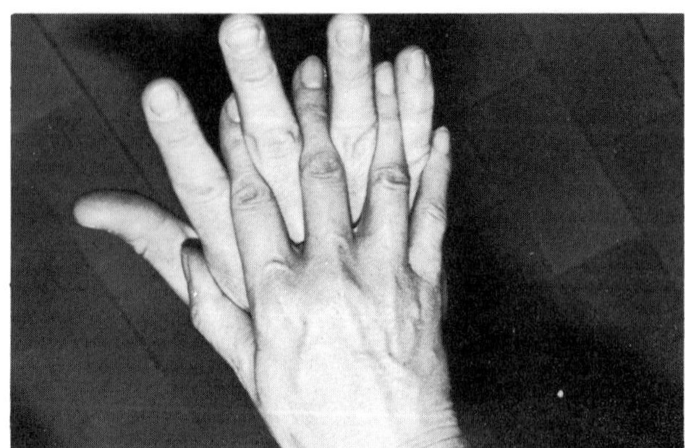
Finger adductors.

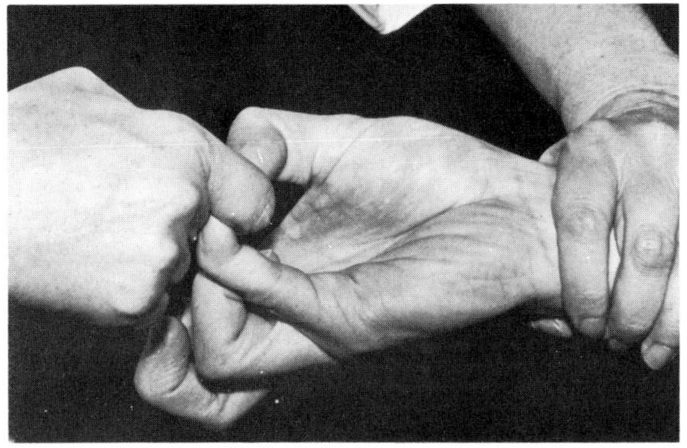

Intrinsic muscles of thumb and little finger.

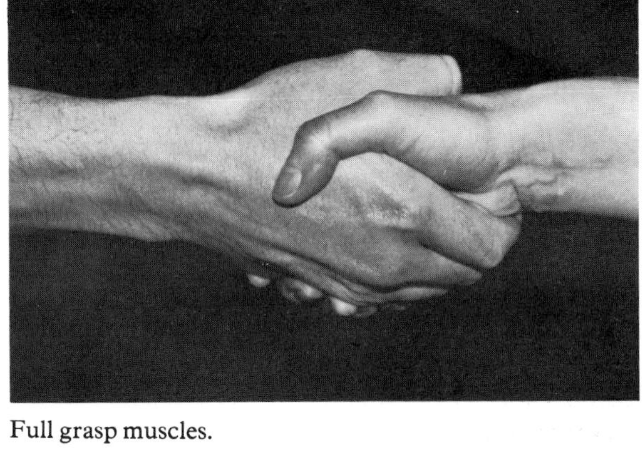

Full grasp muscles.

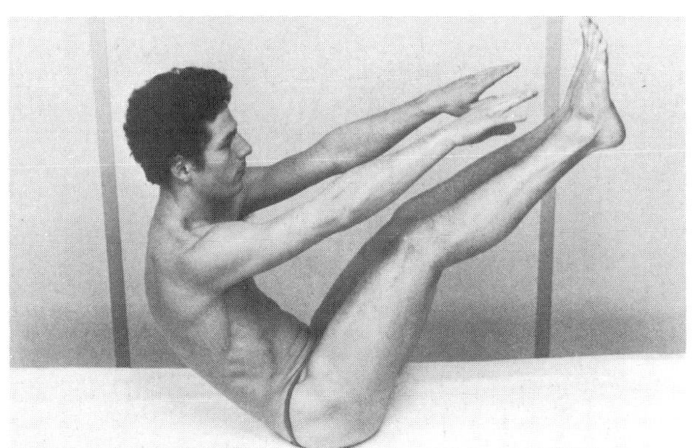

Trunk flexors.

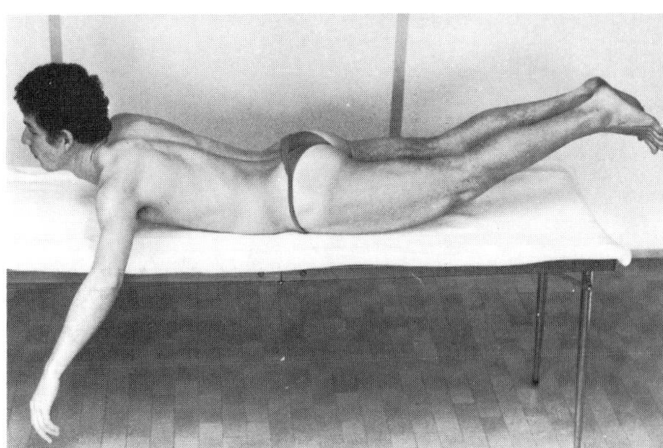

Extensors of trunk and lower extremities.

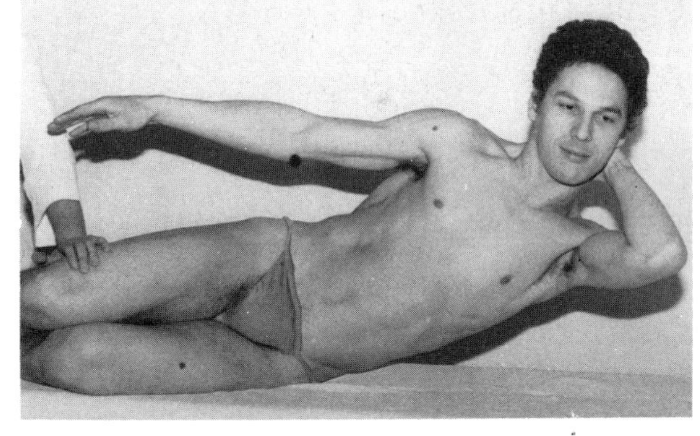

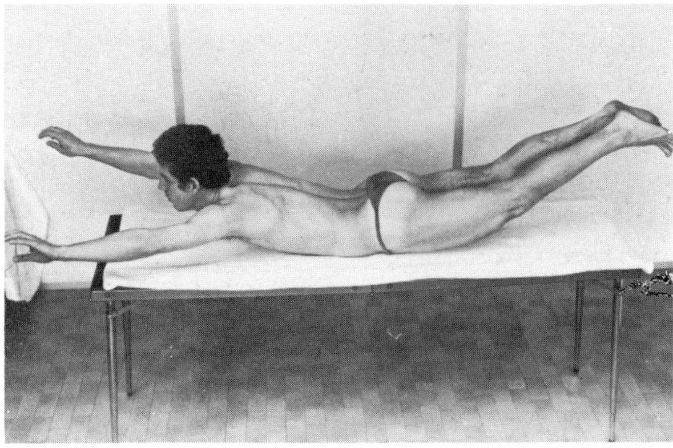

Extensors of trunk and lower extremities with elevation of upper extremities.

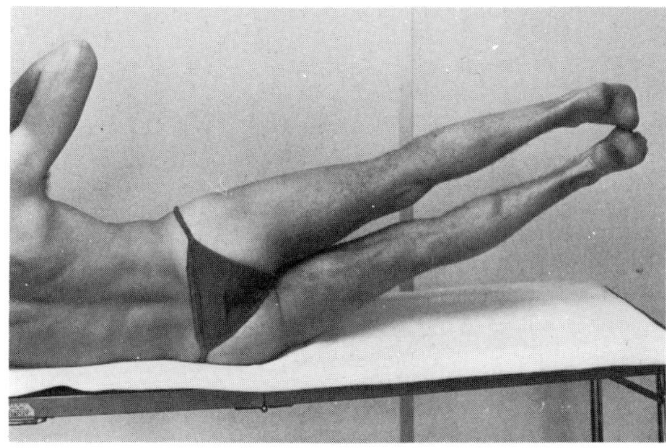

Lower trunk lateral flexors.

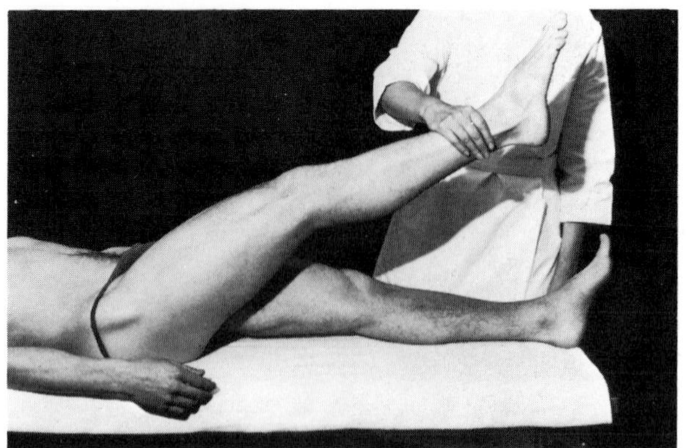

Hip flexors and knee extensors.

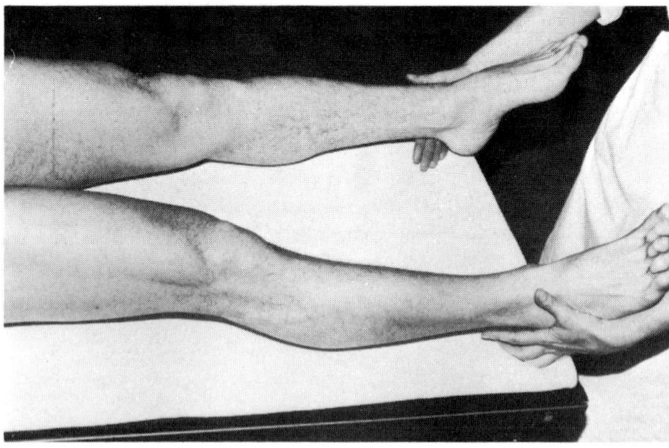

Hip abductors.

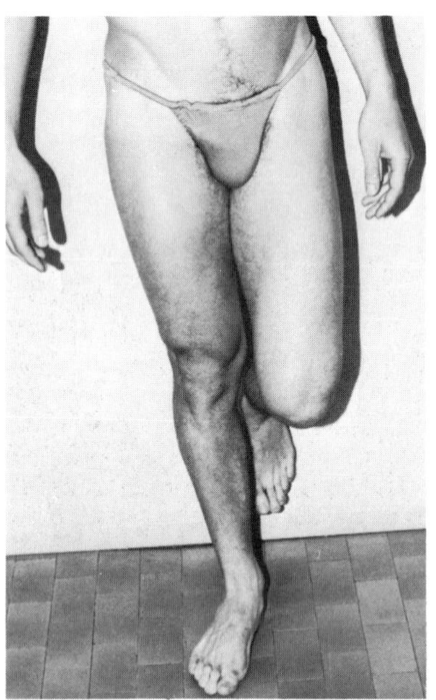

Hip abductors, unilateral stance.

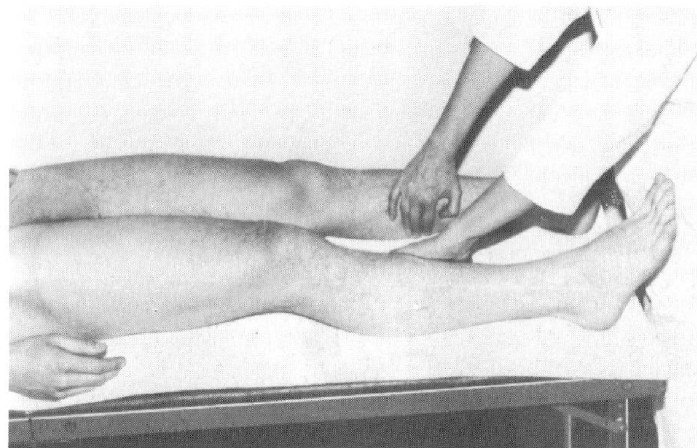

Hip adductors.

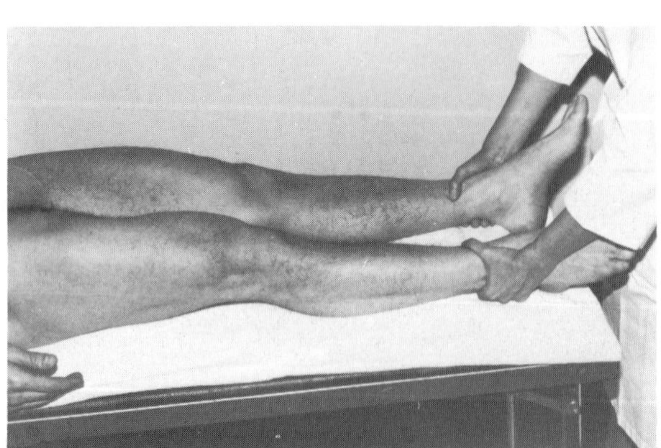

Hip external rotators.

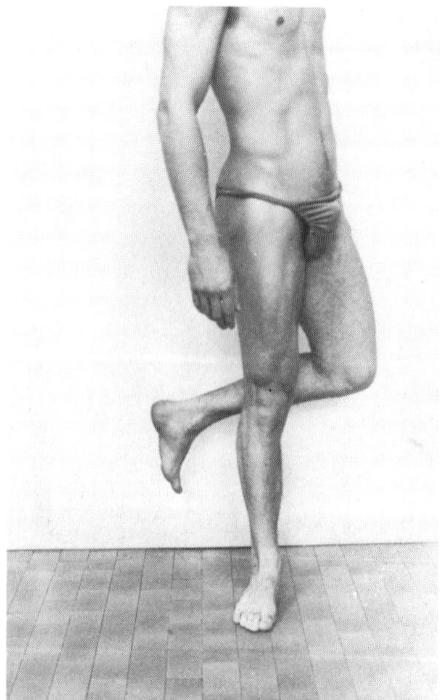

Hip external rotators, unilateral stance.

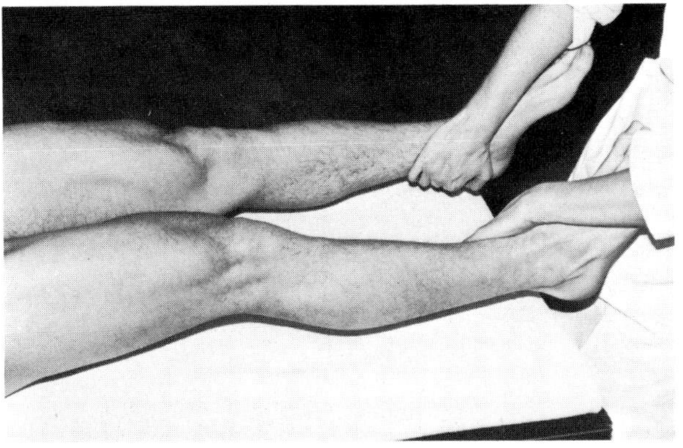

Hip internal rotators.

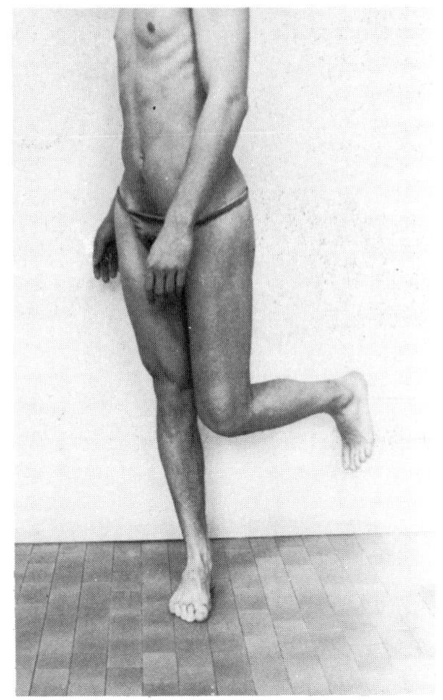

Hip internal rotators, unilateral stance.

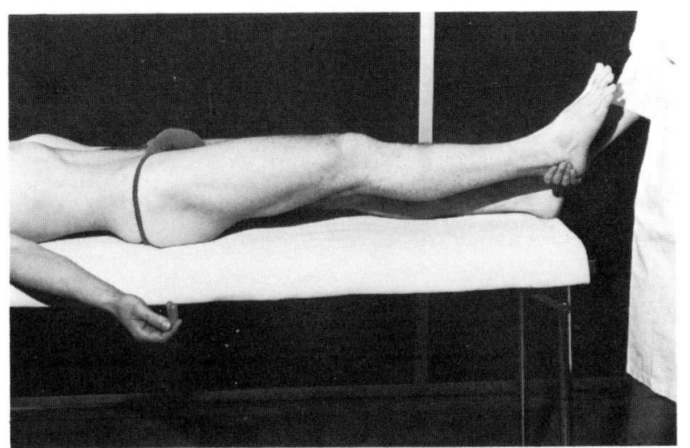

Hip extensors.

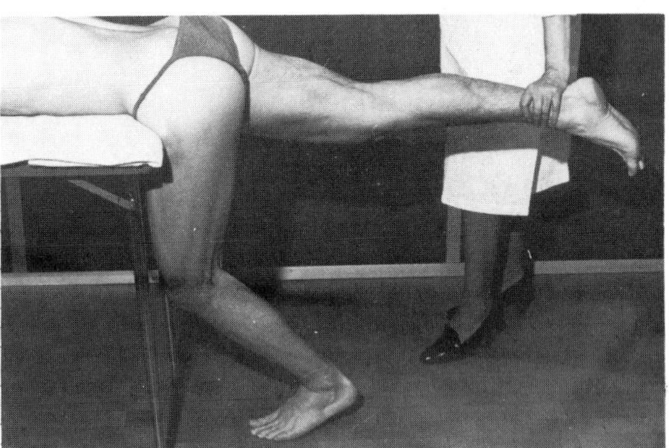

Hip extensors. Alternative.

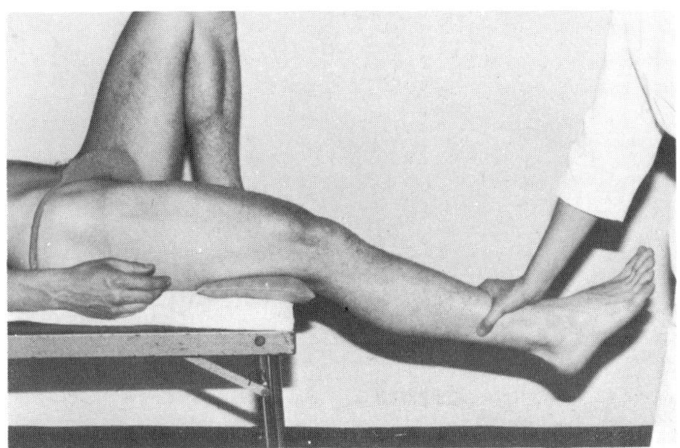

Knee extensors.

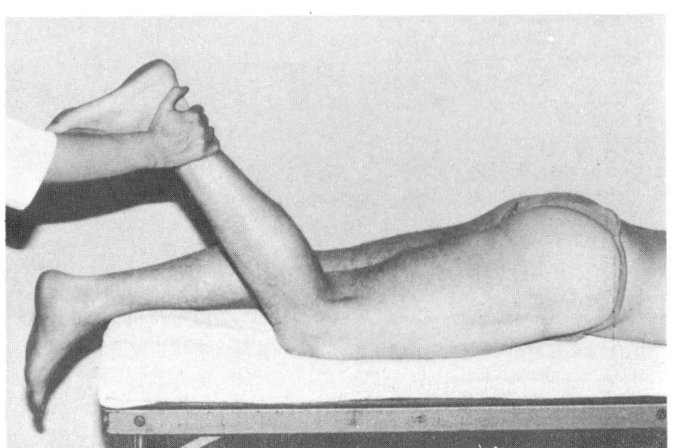

Knee flexors.

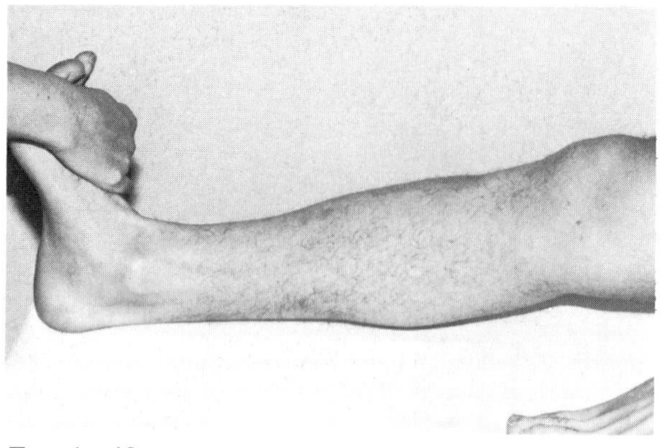
Foot dorsiflexors.

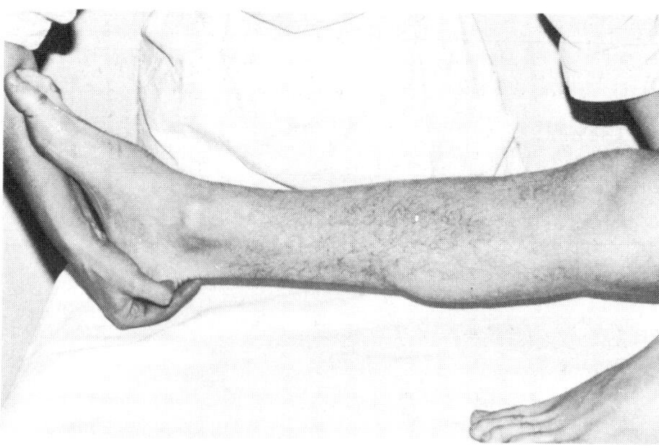
Foot plantar flexors.

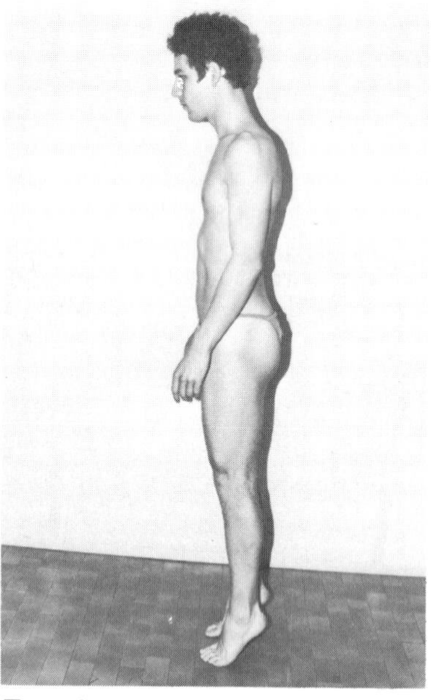
Foot plantar flexors, bilateral stance.

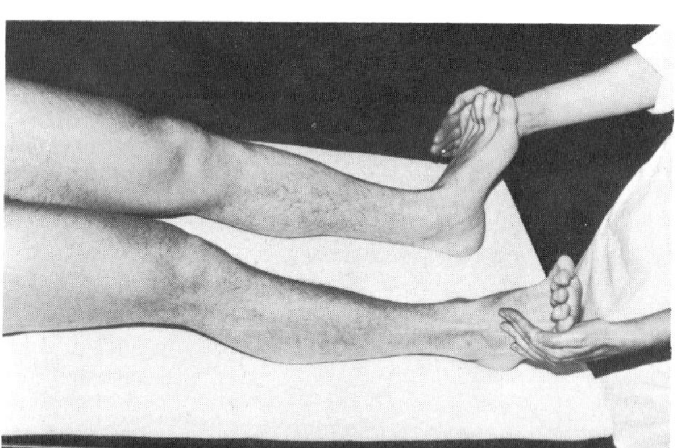
Foot abductors and evertors.

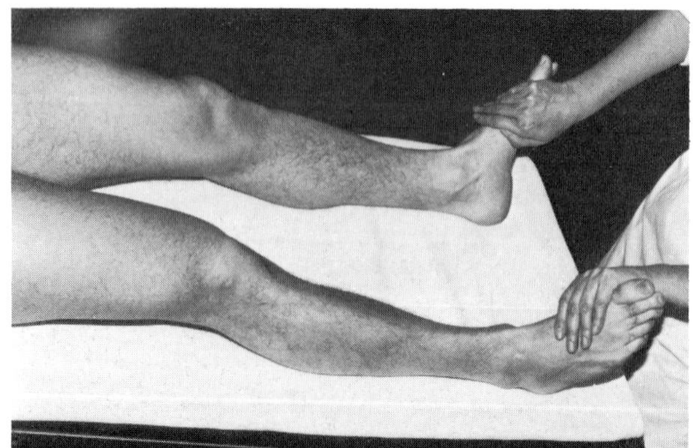
Foot adductors and invertors.

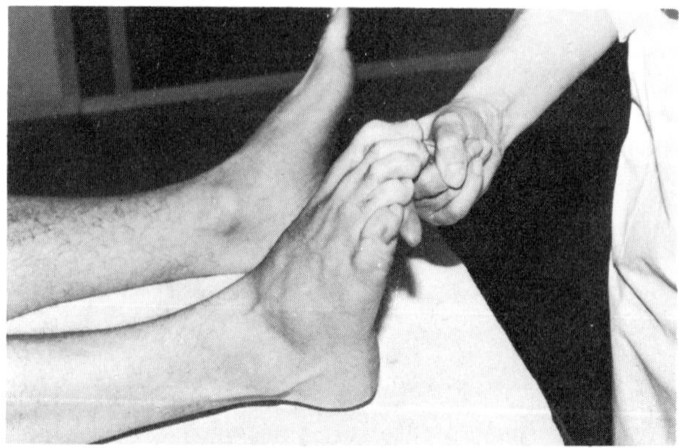
Plantar flexors of toes.

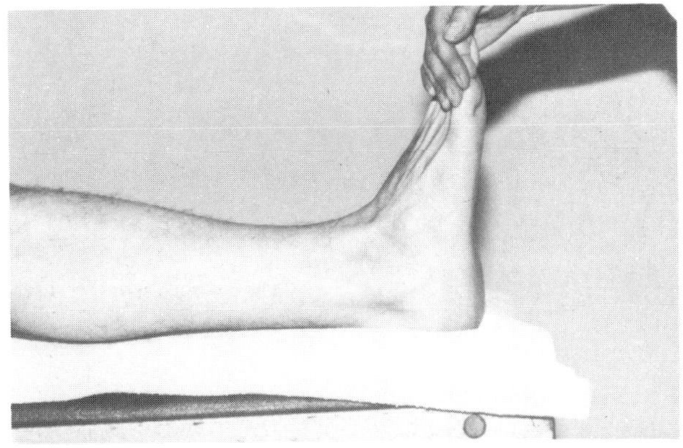
Dorsiflexors of toes.

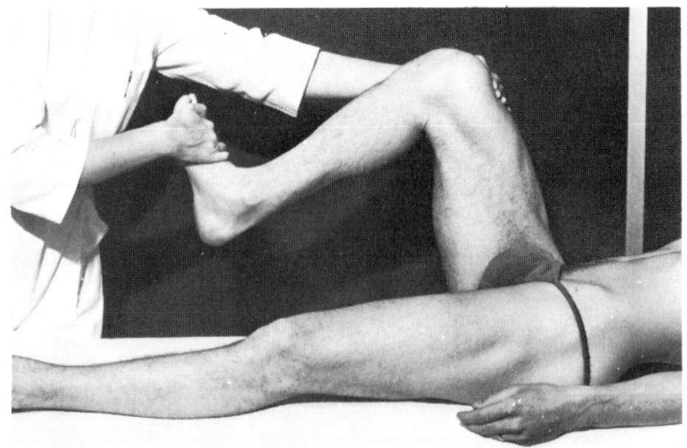
Triple flexion, or withdrawal muscles.

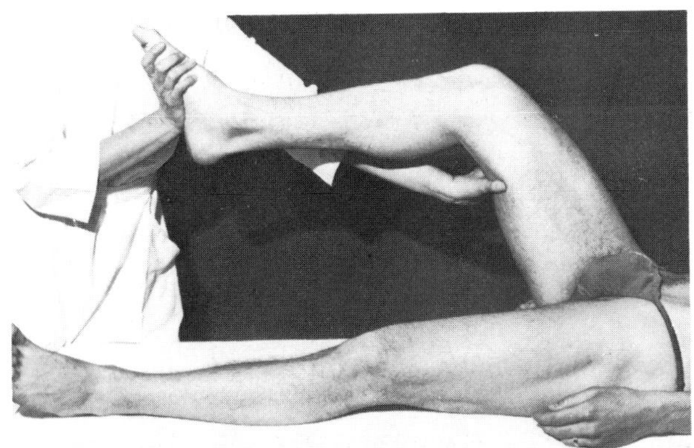
Triple extension muscles.

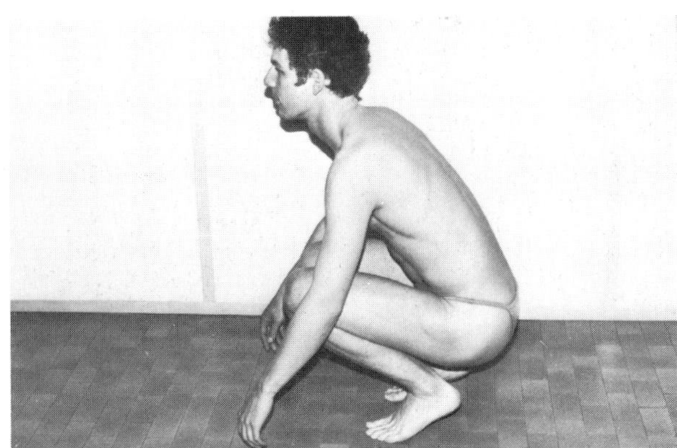
Crouching muscles.

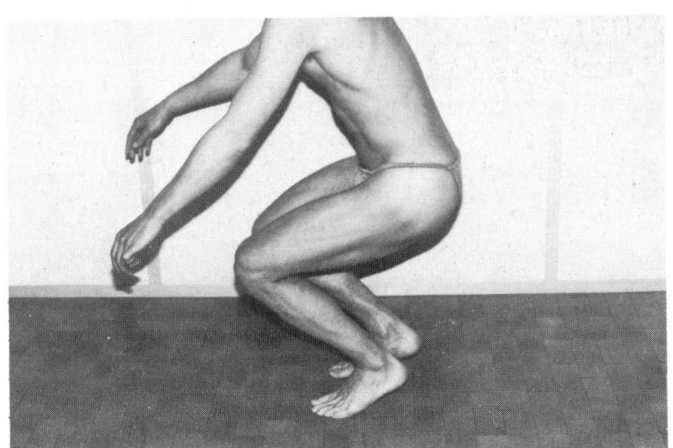
Rising muscles.

PART 2

Individual Muscle Testing

POSITION CHART FOR TESTING DIFFERENT GRADES OF INDIVIDUAL MUSCLES

	I—Supine	Grades							Page
		0	1	2	3	4	5	F*	
NECK	Sternocleidomastoideus	0	1		3	4	5		161
	Anterior vertebral head and neck flexors				3	4	5		165
UPPER EXTREMITY	Middle trapezius. Superior fibres			2					175
	Levator scapulae			2					183
	Pectoralis minor				3	4	5		187
	Serratus anterior				3	4	5		189
	Anterior deltoid	0	1						193
	Middle deltoid			2					195
	Supraspinatus			2					201
	Pectoralis major	0	1		3	4	5		203
	Biceps brachii	0	1						213
	Brachialis	0	1						215
	Brachioradialis	0	1						217
	Pronator teres				3	4	5		225
	Pronator quadratus				3	4	5		225

* F = functional evaluation.

I—Supine		Grades							Page
		0	1	2	3	4	5	F*	
TRUNK	Rectus abdominis	0	1	2	3	4	5		278–80
	Obliquus externus abdominis	0	1	2	3	4	5		283–4
	Obliquus internus abdominis	0	1	2	3	4	5		287–8
	Transversus abdominis	0	1	2		4	5		291–3
	Diaphragm	0	1		3	4	5		295–6
	Quadratus lumborum			2		4	5		311–2
	Muscles of perineum	0	1	2	3	4	5		316
LOWER EXTREMITY	Iliopsoas	0	1		3	4	5		319–21
	Sartorius	0	1	2	3	4	5		323–4
	Tensor fasciae latae	0	1						329
	Gluteus medius	0	1	2					332
	Gluteus minimus	0	1	2	3	4	5		335–7
	Hip external rotators			2	3	4	5		340–1
	Adductors	0	1	2	3				344–5
	Quadriceps femoris	0	1		3	4	5		350–1
	Tibialis anterior	0	1						359
	Extensor hallucis long.	0	1	2	3	4	5		361
	Extensor digitorum long.	0	1	2	3	4	5		366–7
	Extensor digitorum brev.	0	1	2	3	4	5		365
	Peroneus tertius	0	1	2					366–7
	Peroneus brevis	0	1	2					369
	Peroneus longus			2					372
	Tibialis posterior	0	1	2		4	5		381
	Flexor hallucis long.	0	1	2	3	4	5		383
	Flexor digitorum long.	0	1	2	3	4	5		385

* F = functional evaluation.

	II—Prone	Grades							Page
		0	1	2	3	4	5	F*	
NECK	Neck extensors. Upper trapezius	0	1		3	4	5		170–1
UPPER EXTREMITY	Middle trapezius. Superior fibres	0	1						175
	Middle trapezius. Inferior fibres	0	1	2	3	4	5		176
	Lower trapezius	0	1	2	3	4	5	F	179–80
	Rhomboids	0	1		3	4	5		183
	Posterior deltoid				3	4	5		197
	Subscapularis	0	1	2	3	4	5		199
	Infraspinatus	0	1	2	3	4	5		201
	Teres minor	0	1	2	3	4	5		201
	Teres major	0	1	2					207
	Latissimus dorsi	0	1	2	3	4	5		209–10
	Triceps brachii & anconeus	0	1		3	4	5		219–20
	Supinator			2					223
	Pronator teres			2					225
	Pronator quadratus			2					225
TRUNK	Thoracic spine extensors	0	1	2	3	4	5		307
	Lumbar spine extensors	0	1	2	3	4	5	F	308–9
	Quadratus lumborum				3	4	5		311–2
LOWER EXTREMITY	Gluteus maximus	0	1		3	4	5		347
	Knee flexors (hamstrings)	0	1		3	4	5		355–6
	Triceps surae	0	1	2					376
	Soleus	0	1						378–9

* F = functional evaluation.

	III—Side-lying	Grades							Page
		0	1	2	3	4	5	F*	
NECK	Anterior vertebral head and neck flexors			2					165
NECK	Neck extensors. Upper trapezius			2					170
UPPER EXTREMITY	Anterior deltoid			2					193
UPPER EXTREMITY	Coracobrachialis			2					193
UPPER EXTREMITY	Latissimus dorsi			2					209
UPPER EXTREMITY	Triceps brachii			2					219
TRUNK	Lumbar spine extensors			2					308
LOWER EXTREMITY	Iliopsoas			2					319
LOWER EXTREMITY	Tensor fasciae latae				3	4	5		329
LOWER EXTREMITY	Gluteus medius				3	4	5		332
LOWER EXTREMITY	Gluteus minimus				3				336
LOWER EXTREMITY	Hip external rotators				3				341
LOWER EXTREMITY	Adductors				3	4	5		345
LOWER EXTREMITY	Gluteus maximus			2					347
LOWER EXTREMITY	Quadriceps femoris			2					350
LOWER EXTREMITY	Knee flexors (hamstrings)			2					355
LOWER EXTREMITY	Tibialis anterior			2					359
LOWER EXTREMITY	Extensor digitorum long. & peroneus tertius			2					366
LOWER EXTREMITY	Peroneus brevis				3	4	5		369
LOWER EXTREMITY	Peroneus longus	0	1		3	4	5		372–3
LOWER EXTREMITY	Tibialis posterior				3				381
	—Side-lying prone								
UPPER EXTREMITY	Teres major				3	4	5		207

* F = functional evaluation.

	IV—SITTING	Grades							Page
		0	1	2	3	4	5	F*	
NECK	Sternocleidomastoideus			2					161
	Anterior vertebral head and neck flexors-scaleni	0	1						165
	Head and neck extensors	0	1						170
UPPER EXTREMITY	Middle trapezius. Superior fibres				3	4	5		175
	Levator scapulae				3	4	5		183
	Rhomboids			2					183
	Pectoralis minor	0	1	2					187
	Serratus anterior	0	1	2					189
	Anterior deltoid				3	4	5		193
	Coracobrachialis	0	1		3	4	5		193
	Middle deltoid	0	1		3	4	5		195
	Supraspinatus	0	1		3	4	5		195
	Posterior deltoid	0	1	2					197
	Subscapularis	0	1						199
	Pectoralis major			2					203
	Latissimus dorsi							F	209
	Biceps brachii			2	3	4	5		213
	Brachialis			2	3	4	5		215
	Brachoradialis			2	3	4	5		217
	Triceps brachii & anconeus			2				F	219
	Supinator				3	4	5		223
	Pronator teres	0	1		3	4	5		225
	Pronator quadratus				3	4	5		225
	Flexor carpi ulnaris	0	1	2	3	4	5		227
	Flexor carpi radialis	0	1	2	3	4	5		229
	Palmaris longus	0	1	2	3	4	5		230
	Extensor carpi radialis longus	0	1	2	3	4	5		233
	Extensor carpi radialis brevis	0	1	2	3	4	5		235
	Extensor carpi ulnaris	0	1	2	3	4	5		237
	Flexor digitorum superficialis	0	1	2	3	4	5		239

* F = functional evaluation.

	IV—SITTING	Grades							Page
		0	1	2	3	4	5	F*	
UPPER EXTREMITY	Flexor digitorum profundus	0	1	2	3	4	5		241
	Extensor digitorum	0	1	2	3	4	5		243
	Extensor indicis	0	1	2	3	4	5		243
	Extensor digiti minimi	0	1		3	4	5		243
	Flexor pollicis long.	0	1	2	3	4	5		245
	Abductor pollicis long.	0	1	2	3	4	5		247
	Extensor pollicis brev.	0	1	2	3	4	5		248–9
	Extensor pollicis long.	0	1	2	3	4	5		250–1
	Abductor pollicis brev.	0	1	2	3	4	5		253
	Opponens pollicis			2	3	4	5		255
	Flexor pollicis brev. Superficial head	0	1	2	3	4	5		257
	Flexor pollicis brev. Deep head			2	3	4	5		257
	Adductor pollicis	0	1	2	3	4	5		259
	Lumbricales			2	3	4	5		261
	Interossei palmares			2	3	4	5		263
	Interossei dorsales	0	1	2	3	4	5		265
	Abductor digiti minimi	0	1	2	3	4	5		267
	Flexor digiti minimi brevis	0	1	2	3	4	5		269
	Opponens digiti minimi			2	3	4	5		271
TRUNK	Transversus abdominis				3		5		291, 293
	Diaphragm			2					295
	Intercostales externi	0	1	2	3	4	5		303
LOWER EXTREMITY	Iliopsoas				3	4	5		320, 321
	Quadriceps femoris				3	4	5		350
	Popliteus	0	1	2	3	4	5		357
	Tibialis anterior				3	4	5		359
	Extensor digitorum longus				3	4	5		366–7
	Peroneus tertius				3	4	5		366–7

* F = functional evaluation.

V—SITTING INCLINED		Grades							Page
		0	1	2	3	4	5	F*	
LOWER EXTREMITY	Tensor fascia latae			2					329
	Abductor hallucis	0	1	2	3	4	5		387
	Flexor hallucis brev.	0	1	2	3	4	5		389
	Adductor hallucis	0	1	2	3	4	5		391
	Flexor digitorum brev.	0	1	2	3	4	5		393
	Lumbricales pedis			2	3	4	5		394
	Interossei plantares			2	3	4	5		395
	Interossei dorsales	0	1	2	3	4	5		396
	Flexor digiti minimi brevis	0	1	2	3	4	5		397
	Abductor digiti minimi	0	1	2	3	4	5		398
	Opponens digiti minimi			2	3	4	5		399

VI—ON ALL FOURS		Grades							Page
		0	1	2	3	4	5	F*	
TRUNK	Transversus abdominis				3		5		292

VII—SUPINE (AT 45°) ON A TILT TABLE		Grades							Page
		0	1	2	3	4	5	F*	
LOWER EXTREMITY	Iliopsoas				3	4	5		320, 321
	Sartorius				3	4	5		323–4
	Triceps surae			2					376
	Soleus			2					378

* F = functional evaluation.

	VIII—STANDING	Grades							Page
		0	1	2	3	4	5	F*	
UPPER EXTREMITY	Serratus anterior							F	190
TRUNK	Transversus abdominis						5		293
	Quadratus lumborum				3				311
LOWER EXTREMITY	Iliopsoas					4	5		321
	Gluteus medius							F	333
	Gluteus minimus			2					335
	Hip external rotators			2					340
	Triceps surae				3	4	5		377
	Soleus				3	4	5		378–9

* F = functional evaluation.

HEAD AND NECK MUSCLES

STERNOCLEIDOMASTOIDEUS

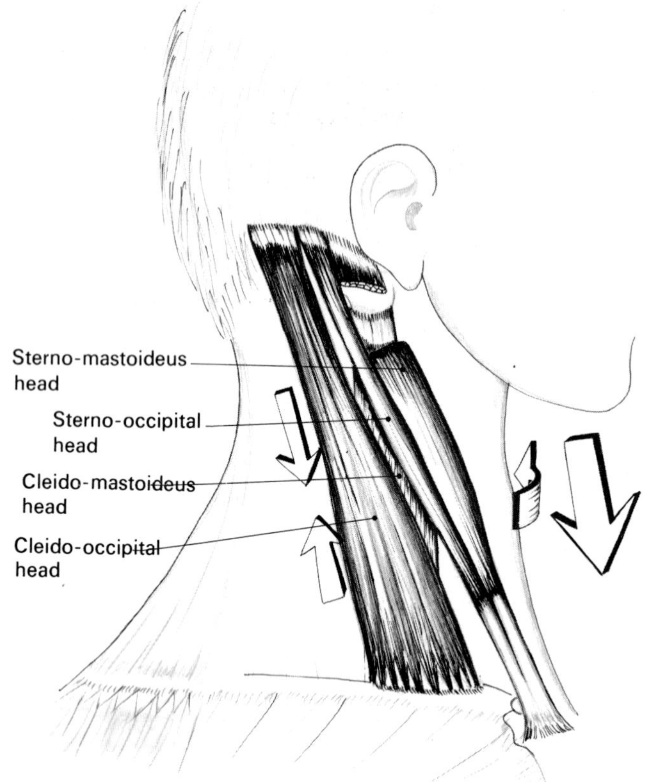

Composed of four heads

Origin and insertion

—Sterno-mastoideus head: Runs from anterior surface of sternal manubrium to the lateral surface of mastoid process.
—Sterno-occipital head: Common origin with sterno-mastoideus head. Inserts into lateral part of superior nuchal line of occipital bone.
—Cleido-mastoideus head: Runs from upper surface of medial third of clavicle, behind the cleido-occipital head, to the mastoid process.
—Cleido-occipital head: Runs from upper surface of medial third of clavicle into lateral two-thirds of superior nuchal line of occipital bone.

Innervation

—Cervical (C2, C3).
—Lateral branch of spinal (XI).

Function

Unilateral contraction

—Rotation of head and neck to side opposite contracting muscle (sternal portion).
—Ipsilateral latero-flexion of head and neck (clavicular portion).
—Flexion of head and neck.

Bilateral contraction

Straight head and neck flexion. Anterior vertebral head and neck flexors, particularly longus colli, must stabilize the neck. If they do not simultaneously contract, the sternocleidomastoideus can become an extensor.
Note: In a supine position, head flexion is made difficult or even impossible by the absence of abdominal muscles.

Contracture or tightness

Spasmodic, congenital torticollis.
It is also seen in cervico-thoracic scoliosis.
Wry or stiff neck, the head is fixed in the position of the shortened muscle. One of the muscle heads may dominate.

Deficiency

Subject cannot bring chin towards sternum. However, he is able to straighten neck lordosis by using anterior vertebral neck flexors.
In upright or sitting position head and neck flexor paralysis causes a forward projection of the head with chin protrusion and cervical column hyperlordosis.
Unilateral affection of sternocleidomastoideus: the head is rotated towards the affected side.

Verify strength of abdominal muscles before testing sternocleidomastoideus.

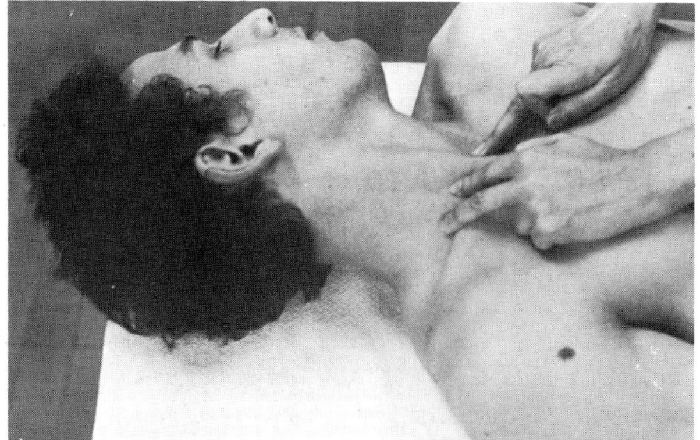

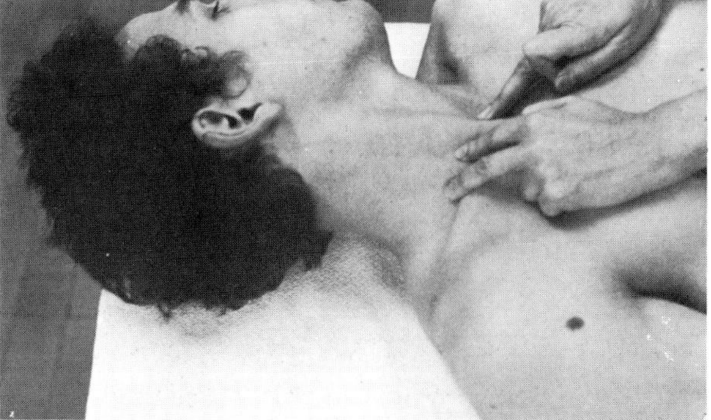

0 and 1 Subject supine.

Upper limbs in external rotation along sides of trunk so as to fix shoulder girdle, shoulders depressed.
Ask subject to flex his head and neck, turning head to opposite side, leaning it ipsilaterally.
Muscular fibres are palpable on the antero-lateral portion of the neck.
Clavicular and sternal heads can be differentiated by palpating their insertions.

Grade 0: No contraction is felt.
Grade 1: Contraction is felt but there is no joint movement.

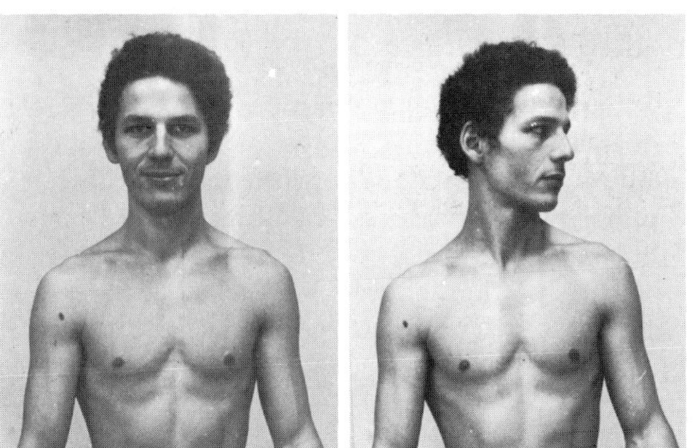

2 Difficult to grade.

Grade 2 is determined only after failure to achieve grade 3.
Subject sitting or supine.
Ask for opposite rotation of head and neck.
Palpate muscle to verify its contraction and to assess its volume.

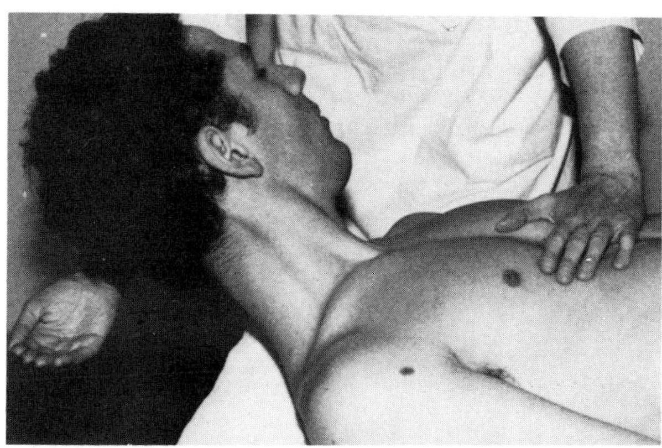

3 Subject supine.

Head hanging over end of table and supported by examiner.
Upper limbs along sides of trunk in external rotation.
Ask for head and neck flexion with ipsilateral latero-flexion and contralateral rotation.
If abdominals are insufficient, examiner holds thoracic segment in place with light pressure.
Range of movement should be complete.

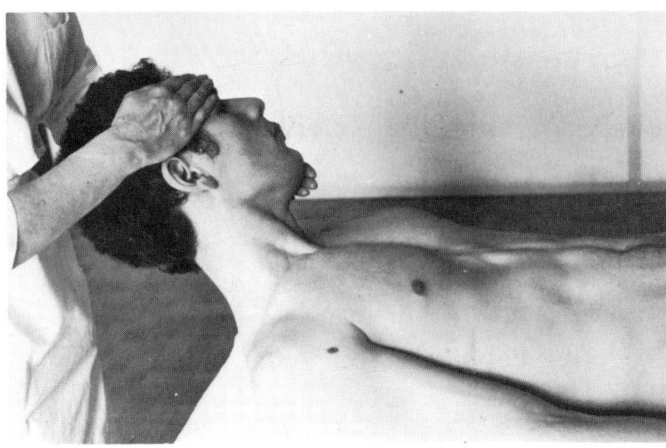

4 and 5 Same position.

Same precautions and movement.
Resistance to the components of the movement is placed on the temporal, supra-auricular region and laterally to the chin on the opposite side.

Grade 4: Power is sub-normal.
Grade 5: Power is normal.

ANTERIOR VERTEBRAL HEAD AND NECK MUSCLES

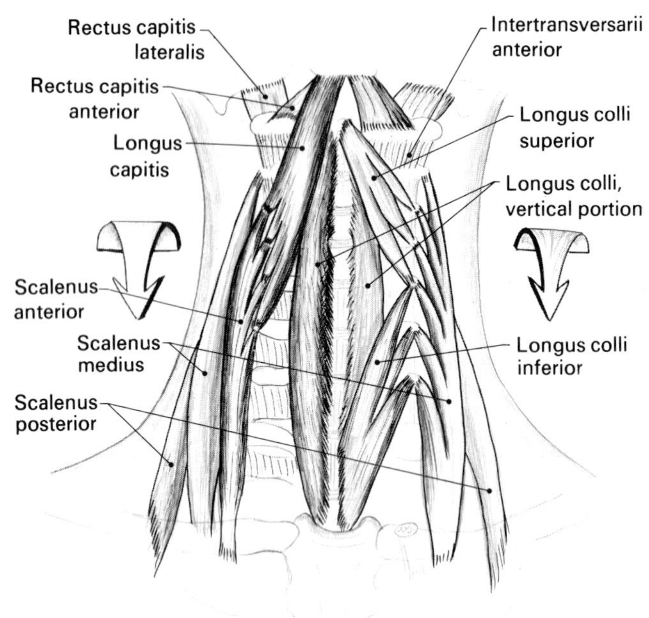

LONGUS COLLI

Origin

—Superior lateral oblique portion: from anterior tubercle of transverse processes of third to fifth cervical vertebrae.
—Inferior lateral oblique portion: from lateral surface of bodies of second and third thoracic vertebrae.
—Vertical portion: anterior surface of bodies of last three cervical vertebrae and first three thoracic vertebrae.

Insertion

—Superior lateral oblique portion: anterior tubercle of atlas.
—Inferior lateral oblique portion: anterior tubercle of transverse processes of last three cervical vertebrae.
—Vertical portion: anterior surface of bodies of second, third, and fourth cervical vertebrae.

Innervation. Cervical plexus (C2 to C7).

Function

—Unilateral: flexion and ipsilateral latero-flexion of cervical column.
—Bilateral: straightens cervical lordosis then flexes the cervical column. Plays an important role in statics of neck which it stabilizes strongly.

LONGUS CAPITIS

Origin Anterior tubercle of transverse processes of third to sixth cervical vertebrae.

Insertion Inferior surface of basilar process of occipital bone, anterior to foramen magnum.

Innervation Cervical plexus (C1 to C3).

Function

—Unilateral:
Flexion and ipsilateral latero-flexion of the head and cervical column.
—Bilateral:
Flexion of upper part of cervical column.

RECTUS CAPITIS ANTERIOR

Origin Anterior surface of lateral mass of atlas and anterior margin of transverse process of atlas.

Insertion Inferior surface of basilar process of occipital bone.

Innervation Cervical plexus (C1, C2).

Function

—Unilateral: head flexion with ipsilateral latero-flexion and rotation.
—Bilateral: flexion at the atlanto-occipital level.

RECTUS CAPITIS LATERALIS (FIRST INTERTRANSVERSARIUS)

Origin Anterior part of transverse process of atlas.

Insertion Inferior surface of basilar process of occipital bone.

Innervation Cervical plexus (C1, C2).

Function

—Unilateral: ipsilateral latero-flexion at the atlanto-occipital level.
—Bilateral: head flexion at the atlanto-occipital level.

INTERTRANSVERSARII

Short muscles running from one transverse process to the next.

Innervation Cervical of same level.

Function Ipsilateral latero-flexion of one vertebra on the next.

SCALENI (ANTERIOR, MEDIUS, POSTERIOR)

Origin

—Scalenus anterior: anterior tubercles of transverse processes of third to sixth cervical vertebrae.
—Scalenus medius: transverse processes of last six cervical vertebrae.
—Scalenus posterior: posterior tubercles of transverse processes of fourth to sixth cervical vertebrae.

Insertion

—Scalenus anterior: Lisfranc's scalene tubercle on upper surface of first rib.
—Scalenus medius: upper surface of first rib posterior to the groove.
—Scalenus posterior: Superior border and lateral surface of second rib.

Innervation

—Scalenus anterior: cervical plexus (C5 to C7).
—Scalenus medius: cervical plexus (C4 to C8).
—Scalenus posterior: cervical plexus (C7, C8).

Function

—Unilateral: ipsilateral latero-flexion and contralateral rotation.
—Bilateral: flexion of cervical column.
—Accessory inspiration muscle, when cervical column fixed.

ACCESSORY MUSCLES

Mylohyoideus, thyrohyoideus, sternocleidohyoideus, sternohyoideus, omohyoideus, stylohyoideus, digastricus and platysma. These muscles act from a distance upon the cervical column: a stabilizing role. They help in flexing the cervical column when the mandible is locked against the maxilla (see pages 104 and 105).

COMMON FUNCTION OF ANTERIOR VERTEBRAL HEAD AND NECK MUSCLES

These muscles start head and neck flexion and straighten cervical lordosis.

Deficiency When these muscles are paralysed, and even if the sternocleidomastoidei are present, the patient cannot pull his chin towards the sternum. His head projects forward, the chin sticking out.

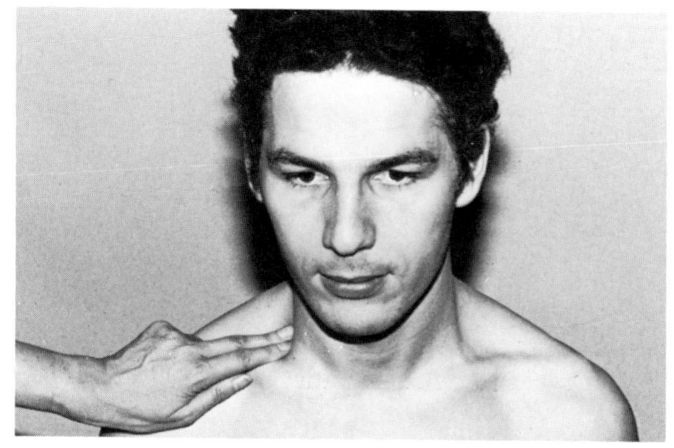

0 and 1

Of the head and neck muscles, only the scalenes are palpable. They are found laterally to and behind the clavicular portion of the sternocleidomastoideus.

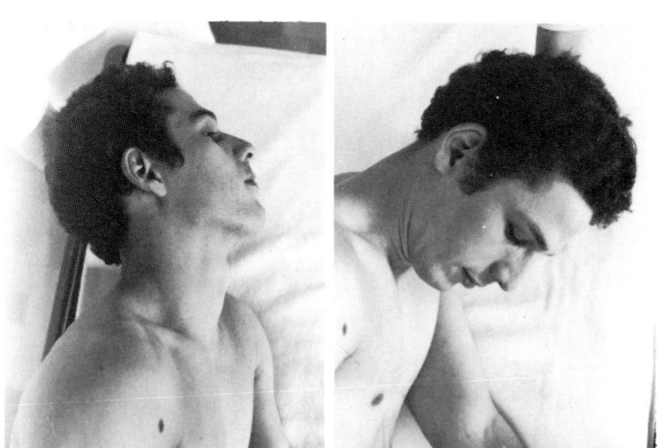

2 Subject side-lying.

Examiner supports head.
From a position of head and neck extension, ask subject to flex his head and neck in such a manner that he curves the cervical region, pulling his chin in as close to the sternum as possible.
Range of movement should be complete.

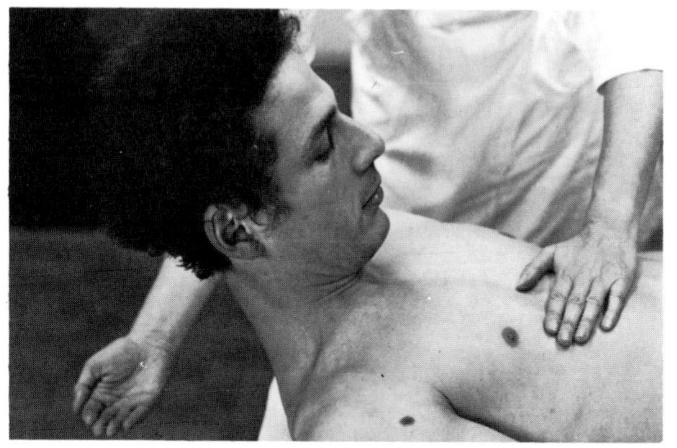

3 Subject supine.

Subject's head hangs over the end of table and is supported by examiner.
Upper limbs in external rotation along sides of trunk so as to fix the shoulder girdle.
Hold the thorax in place.
From a position of head and neck extension, ask for flexion.
Watch curving of the cervical region.
The chin comes as close as possible to the sternum.
Range of movement should be complete.

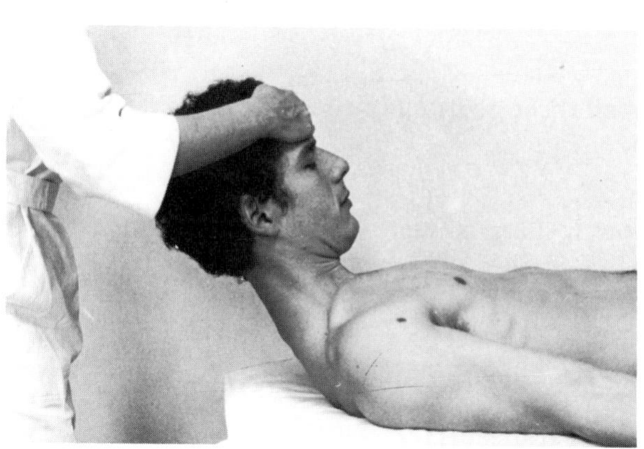

4 and 5 Same position.

Ask for the same movement.
Resistance to the movement is placed on the forehead.

Grade 4: Power is sub-normal.
Grade 5: Power is normal.

TRAPEZIUS SUPERIOR AND HEAD AND NECK EXTENSORS

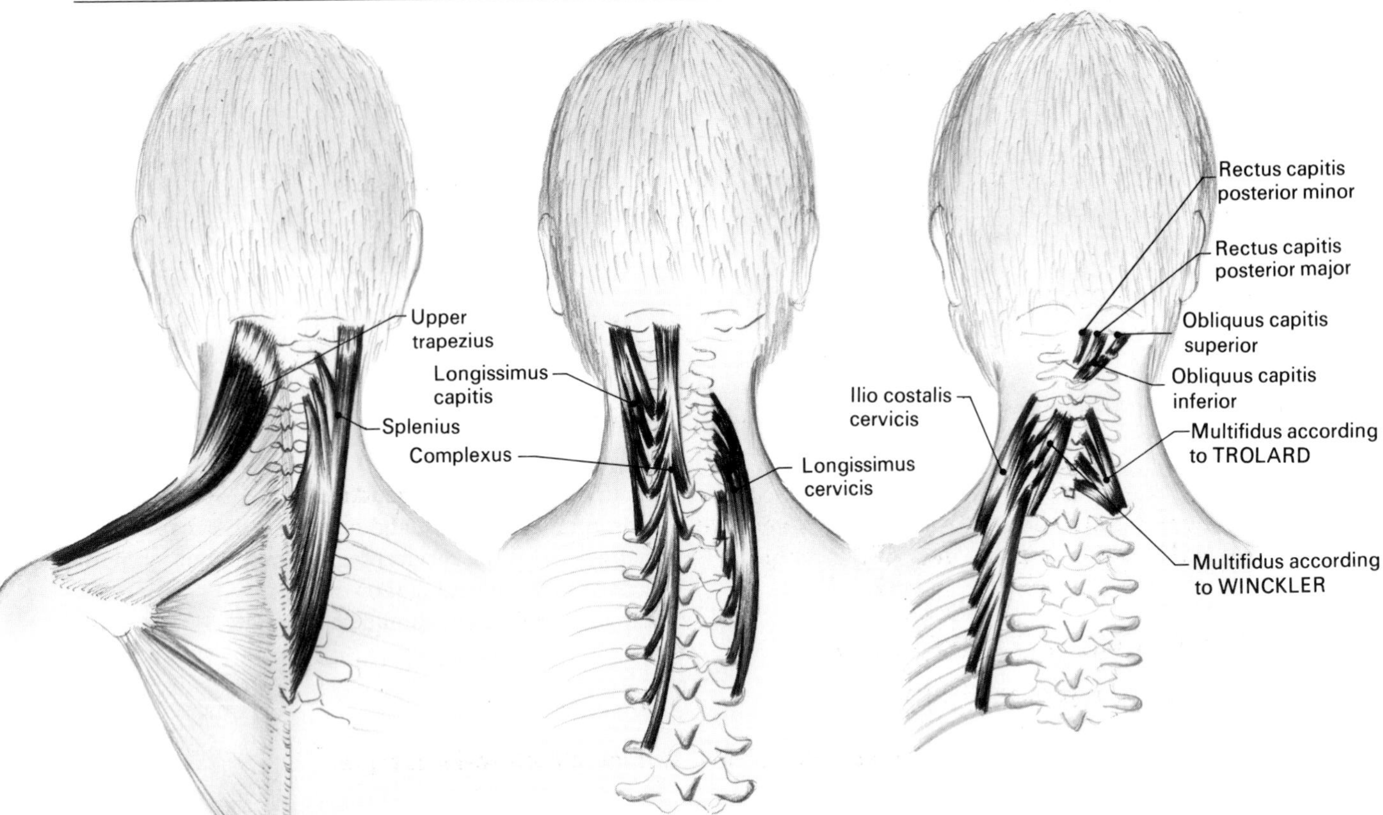

TRAPEZIUS SUPERIOR

Origin

—External occipital protuberance.
—Medial third of superior nuchal line.
—Posterior cervical ligament.
—Spinous processes of first two cervical vertebrae.

Insertion

—Lateral third of superior surface and posterior margin of clavicle.
—Acromioclavicular joint.

Innervation

—Cervical (C3, C4),
—Lateral branch of spinal (XI).
—The upper trapezius is the most richly innervated of all three portions of the trapezius.

Function

—Unilateral: head and upper cervical column extension, ipsilateral latero-flexion, and contralateral rotation.
—Bilateral: head and upper cervical column extension.

Contracture or tightness Locks the head in the muscle's shortened position.

Deficiency Causes predominance of contralateral upper trapezius.

RECTUS CAPITIS POSTERIOR MAJOR

Origin Spinous process of axis.

Insertion Inferior nuchal line of occipital bone.

Function

—Unilateral: ipsilateral latero-flexion of head. Ipsilateral rotation at atlanto-axial level.
—Bilateral: head extension on cervical column.

RECTUS CAPITIS POSTERIOR MINOR

Origin Posterior tubercle of the atlas.

Insertion Medial third of inferior nuchal line of occipital bone.

Function

—Unilateral: mild ipsilateral latero-flexion at atlanto-occipital level.
—Bilateral: head extension on cervical column.

OBLIQUUS CAPITIS INFERIOR

Origin Spinous process of axis.

Insertion Transverse process of atlas on inferior surface and posterior margin.

Function

—Unilateral: extension, ipsilateral latero-flexion and rotation at atlanto-axial level.
—Bilateral: extension at atlanto-axial level. Important static stabilizing function.

OBLIQUUS CAPITIS SUPERIOR

Origin Superior surface of transverse process of atlas.

Insertion Lateral third of inferior nuchal line of occipital bone.

Function

—Unilateral: contralateral rotation at occipito-atlantal level, weak latero-flexion.
—Bilateral: Extension at occipito-atlantal level.

MULTIFIDUS CERVICIS

Origin and insertion (see page 304)

—According to Trolard: Fibres originate in one transverse process to insert into laminae and spinous processes of the four vertebrae above the vertebra of origin.
—According to Winckler: Fibres originate in transverse processes of four adjacent vertebrae to insert into lamina and spinous process of next vertebra up.

Function

—Unilateral: ipsilateral latero-flexion and contralateral rotation.
—Bilateral: cervical column extension.

INTERSPINALES CERVICIS

Origin and insertion Located on either side of spinous processes, connects spines of vertebrae, originating in one to insert into the next, up to the axis.

Function Extends one vertebra in relation to the next.

SPINALIS CERVICIS

Spinalis thoracis goes up to the spinous process of the seventh cervical vertebra. Spinalis cervicis goes up to the third cervical vertebra.

Function Extends one cervical vertebra in relation to the next.

LONGISSIMUS CAPITIS

Origin Base of transverse processes of last four cervical vertebrae and first thoracic vertebra.

Insertion Summit and posterior margin of mastoid process.

Function

—Unilateral: ipsilateral latero-flexion and slight rotation of head and neck.
—Bilateral: extends head and neck.

COMPLEXUS

Origin

—Transverse processes of first six thoracic and last four cervical vertebrae.
—Spinous processes of first thoracic vertebra and of seventh cervical vertebra.

Insertion Between two curved lines of the occipital bone, laterally to the lateral crest.

Function

—Unilateral: slight ipsilateral latero-flexion of head and neck.
—Bilateral: head and neck extension.

LONGISSIMUS CERVICIS

Origin Summit of transverse processes of the first five thoracic vertebrae.

Insertion Summit of transverse processes of the last five cervical vertebrae.

Function

—Unilateral: ipsilateral latero-flexion of lower cervical column.
—Bilateral: extension of lower cervical column.

ILIOCOSTALIS LUMBORUM

Iliocostalis cervicis: see page 305.

Origin The lumbar region: it attaches to the superior borders of the first six ribs.

Insertion Posterior tubercles of transverse processes of last five cervical vertebrae.

Function

—Unilateral: ipsilateral latero-flexion of lower cervical column.
—Bilateral: extension of lower cervical column.
—Accessory inspiration muscle.

SPLENIUS CAPITIS AND CERVICIS

Origin

—Spinous process of seventh cervical vertebra.
—Posterior cervical ligament.
—Spinous processes of first four thoracic vertebrae.
—Inter-spinous ligament.
It divides into two heads.

Insertion

—Splenius capitis: below insertion of sternocleidomastoideus, lateral half of superior nuchal line, and into mastoid process.
—Splenius cervicis: into posterior tubercle of transverse processes of first three cervical vertebrae.

Function

—Unilateral: ipsilateral latero-flexion and rotation of head and neck.
—Bilateral: extension of head and neck.

INNERVATION

All are innervated by posterior rami of cervical nerves.

COMMON FUNCTION OF HEAD AND NECK EXTENSORS

All nuchal muscles produce head and neck extension at different levels, depending upon their insertions, with participation of the levator scapulae.
Two particular levels should be noted:
—Upper neck or infra-occipital level, muscles acting on the head (occipito-atlantal level and atlanto-axial level).
—Cervical level per se: cervical muscles and cervical level muscles reaching the head.
They act at the cervical level and at the head level depending upon their insertions.
The sternocleidomastoideus can become a head and neck extensor depending upon the head position.
All posterior vertebral muscles of the head and neck take part in head posture through a synergistic balance with flexors. They play an important role in head and cervical stability.

Contracture or tightness

Splenius and complexus can be shortened and cause head and neck posture of extension and latero-flexion towards the contracted side.

Deficiency

Deficiency of the cervical extensors is serious since stability of the head and neck is no longer assured. Some balance is achieved with the flexors, particularly with the sternocleidomastoideus, which may compensate for extension. But a sudden change in position can throw the head off balance. When paralysis is severe and the flexors are affected, the subject should wear a supportive occipital collar, and eventually an occipital collar with a chin piece.

Global testing of trapezius superior and head and neck extensors

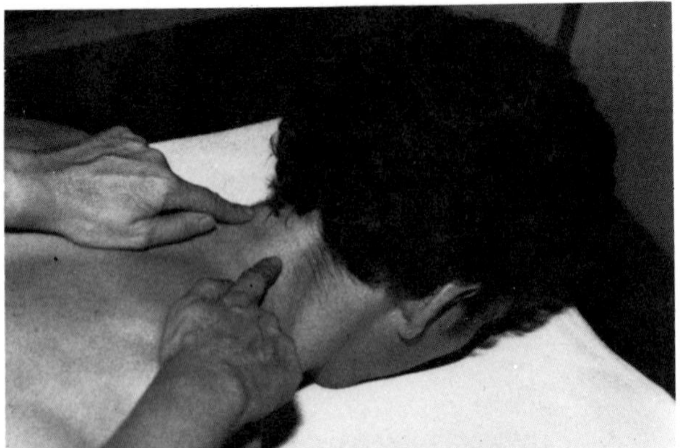

0 and 1 Subject prone.
Upper limbs along sides of body.
Ask subject for head and neck extension.
Distinguishing the different muscles through palpation is very difficult. Global palpation is possible on either side of the median nuchal line.

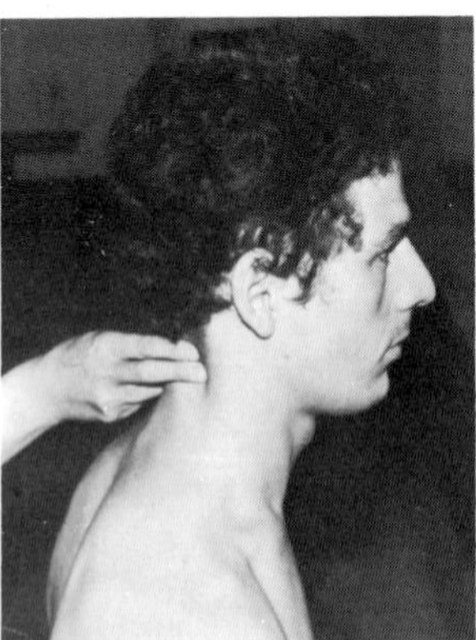

Contraction of the splenius can be discerned between the upper trapezius and the sternocleidomastoideus.

Grade 0: No contraction is felt.
Grade 1: Contraction is felt but there is no joint movement.

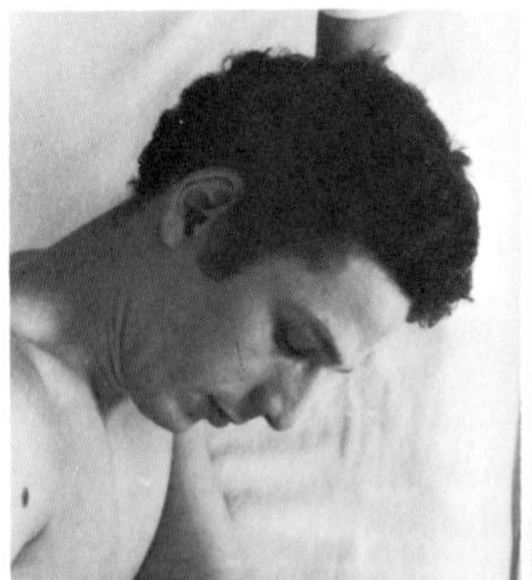

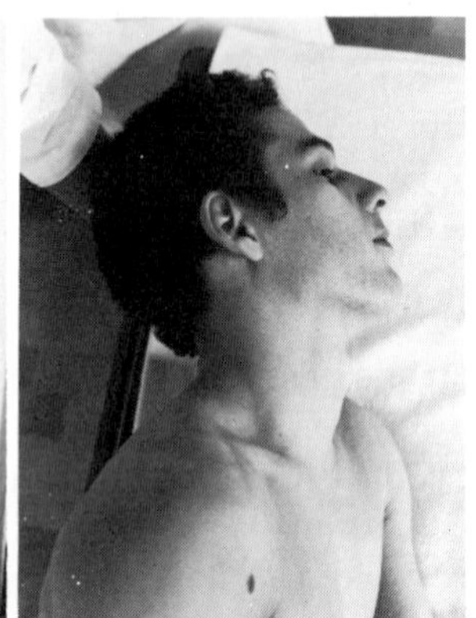

2 Subject side-lying.

Support head.
From a position of flexion, ask subject to extend his head and neck.
Range of movement should be complete.

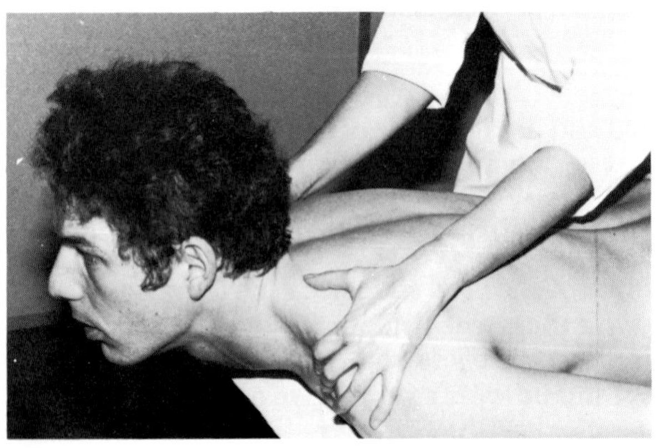

3 Subject prone.

Head hangs over end of table, neck in flexion.
Upper limbs along sides of body.
Ask subject to extend his head and neck.
Range of movement should be complete.

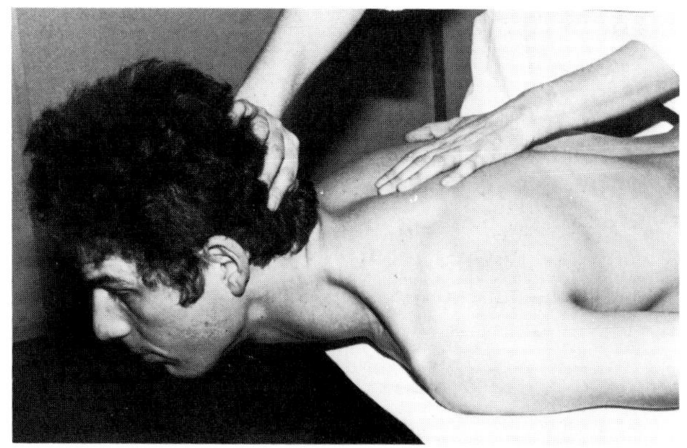

4 and 5 Same position.

Same movement.
Resistance to the movement is placed on the occipital region.

Grade 4: Power is sub-normal.
Grade 5: Power is normal.

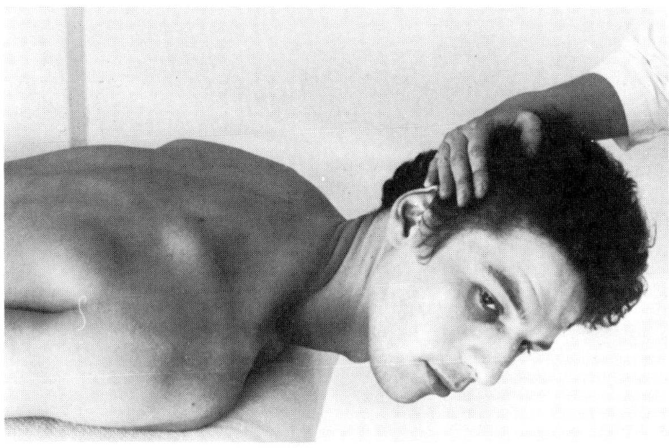

Testing extensors on side of the contraction (especially the splenius)

Subject prone.
Head hangs over end of table, neck in flexion.
Upper limbs along sides of body.
Ask for an extension and ipsilateral latero-flexion of head and neck.

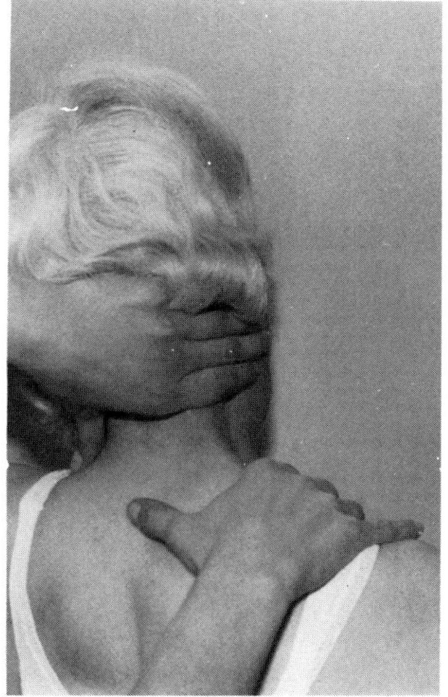

Tightness of trapezius superior.

UPPER EXTREMITY MUSCLES

TRAPEZIUS MEDIUS

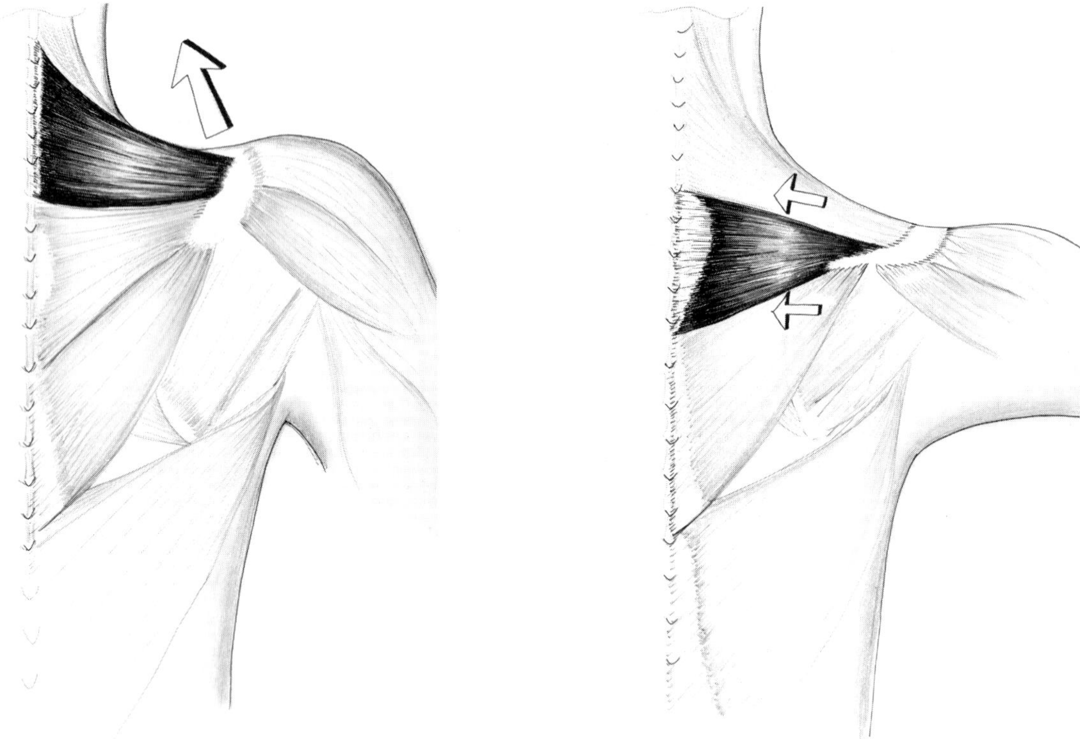

Upper portion (C2, C6)   Lower portion (C7,T4)

Origin

—Spinous processes from axis to the fourth thoracic vertebra.
—Posterior cervical and supra-spinous ligaments.

Insertion

—Superior lip of posterior border of spine of scapula.
—Medial margin of acromion.

Innervation

—Cervical (c2, C3, C4).
—Lateral branch of spinal (XI).

Function

Upper portion:
—From fixed cervical spine: elevates shoulder girdle up and backwards.
—From fixed shoulder: extends cervical column.
Lower portion:
—Adducts scapula.
—Stabilizes scapula during shoulder movements.

Contracture or tightness of upper portion

—Unilateral: lateral deviation of cervical column and scapular elevation.
—Bilateral: elevation of shoulder.

Deficiency Weakness of the lower portion causes a winging of the scapulae, a rounding of shoulders and functional impairment in the use of the upper extremities.

Testing upper portion of middle trapezius

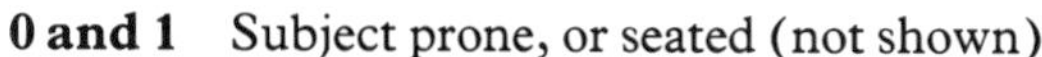

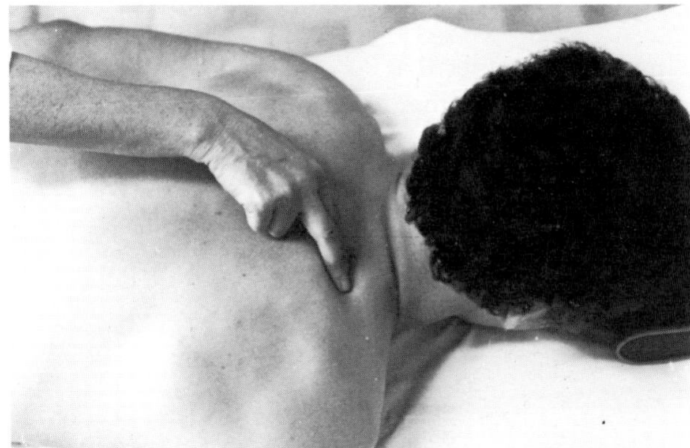

0 and 1 Subject prone, or seated (not shown).

Ask subject to raise his shoulder with a slight posterior component.
Examiner palpates fibres of upper portion over the supra-spinous fossa.

Grade 0: No contraction is felt.
Grade 1: Contraction is felt but there is no joint movement.

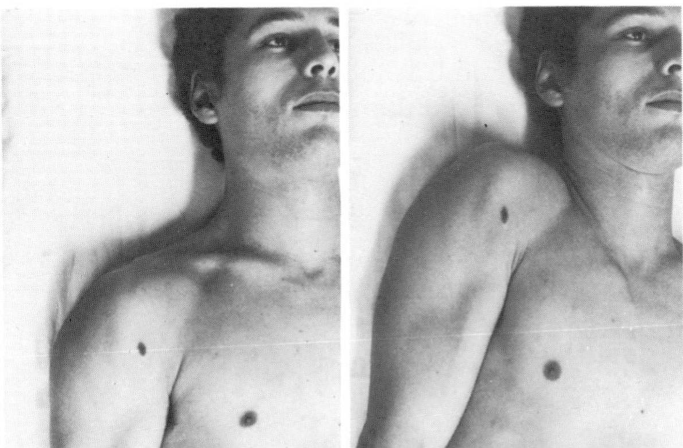

2 Subject supine.

Upper limbs along sides of body.
Ask subject for the same movement.
Watch for possible compensation by shoulder girdle protraction.
Range of movement should be complete.

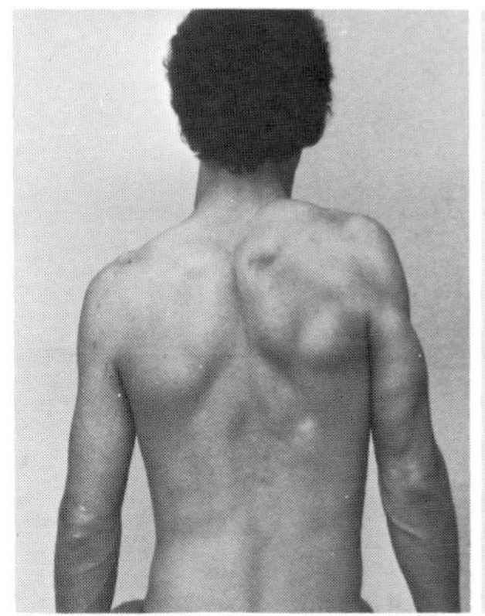

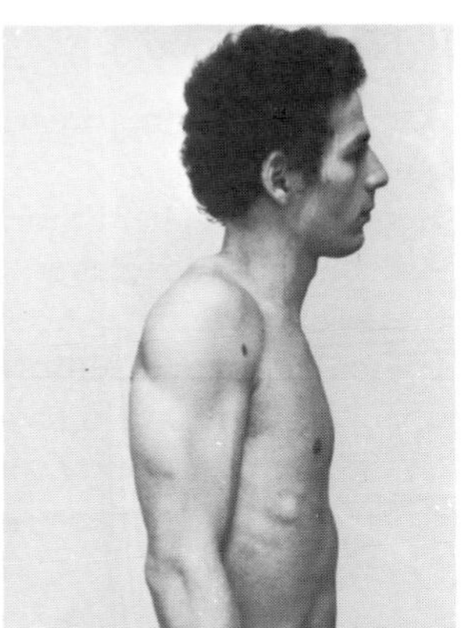

3 Subject seated on a stool.

Upper limbs along sides of body.
Same risk of compensation.
Ask subject for same movement.
Range of movement should be complete.

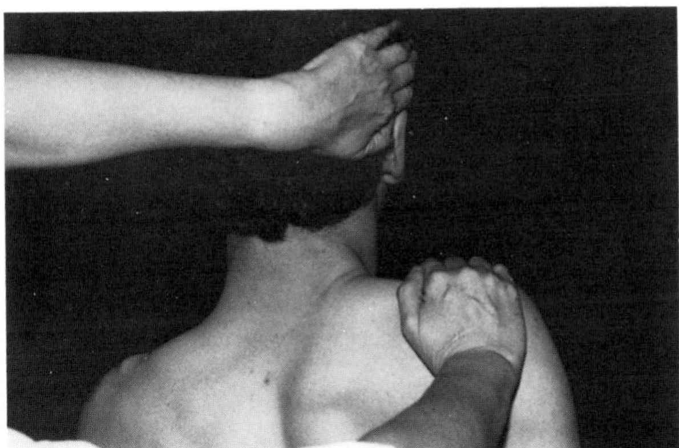

4 and 5 Same position.

Hold occipital region and upper portion of the cervical spine.
Resistance to the movement is placed in the middle of supra-spinous fossa.

Grade 4: Power is sub-normal.
Grade 5: Power is normal.

Testing lower portion of middle trapezius

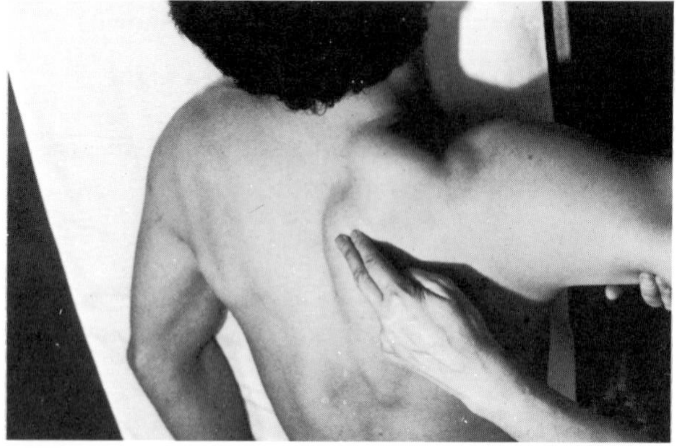

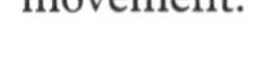

0 and 1 Subject prone.

Upper limb abducted at 90°, rotation neutral and supported by examiner.
Ask subject for direct scapular adduction.
Palpate between spinal border of scapula and spinal column.

Grade 0: No contraction is felt.
Grade 1: Contraction is felt but there is no joint movement.

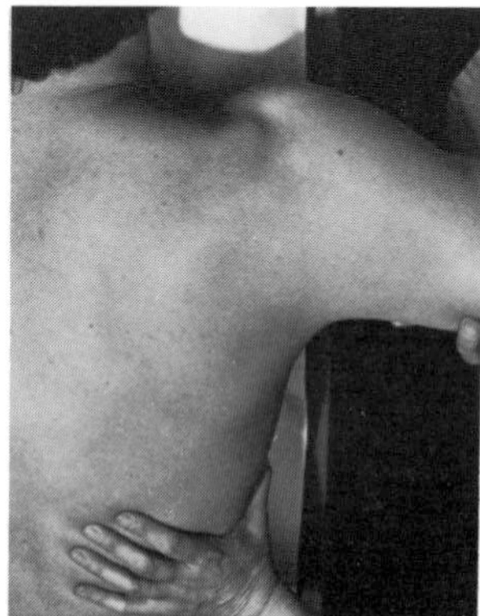

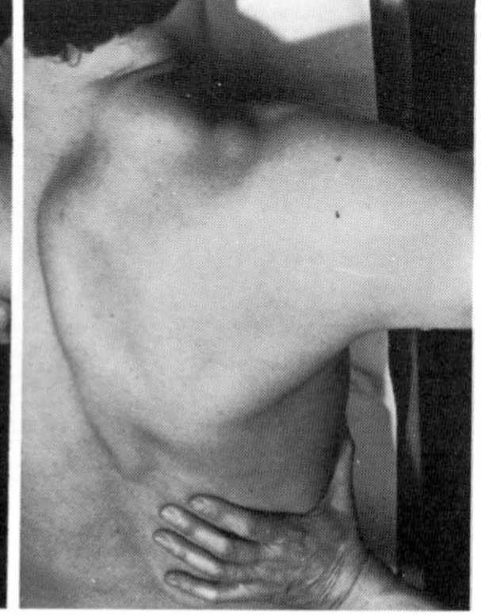

2 Subject prone.

Support upper limb in abduction, rotation neutral.
Hold ipsilateral hemithorax in place so as to avoid its elevation.
Ask subject for the same movement.
Spinal border of scapula should remain parallel to spinous processes.
Range of movement should be complete.

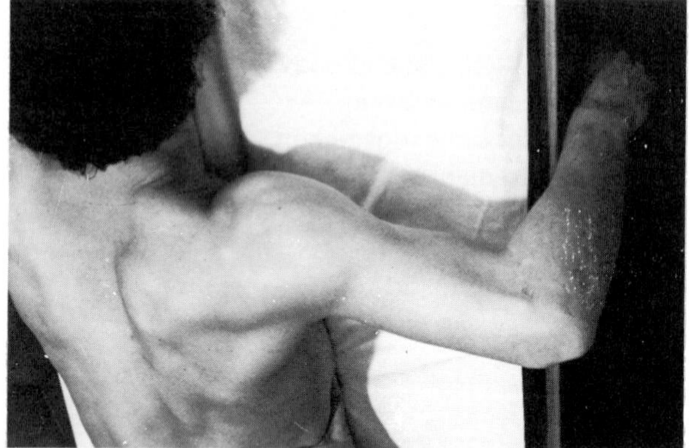

3 Same position.

Arm in abduction, forearm hanging over edge of table.
Hold ipsilateral hemithorax in place.
Ask for same movement.
Range of movement should be complete.
If shoulder joint extensors are paralysed, support the upper limb placing light resistance to the movement on the lateral border of the scapula (not shown).

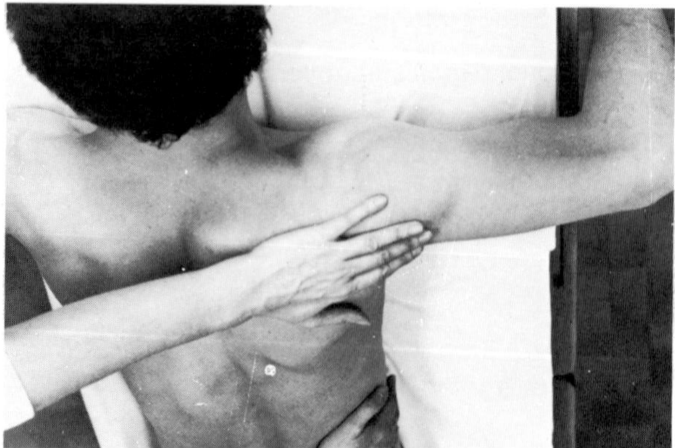

4 and 5 Same position.

Same hold and same movement.
Resistance to the movement is placed on lateral border of scapula. If shoulder joint is sound, resistance may be placed on inferior third of upper arm.

Grade 4: Power is sub-normal.
Grade 5: Power is normal.

NOTES

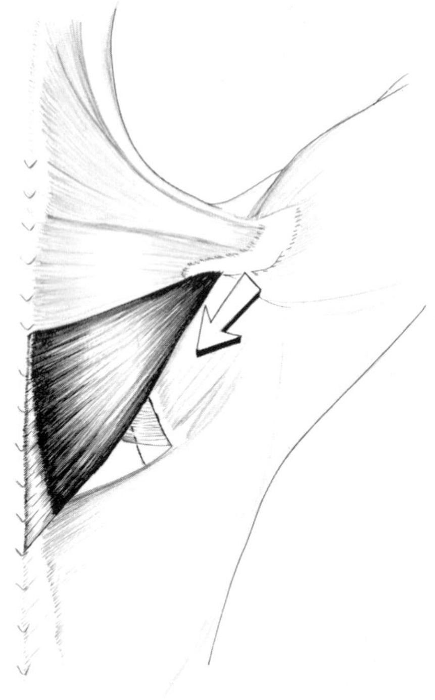

Origin

—Spinous processes of fifth to tenth thoracic vertebrae.
—Supra-spinous ligament.

Insertion

—Trapezius tubercle of posterior border of scapular spine.

Innervation

—Cervical (C2, C3, C4).
—Lateral branch of spinal (XI).

Function

Depression and adduction of scapula.
Pulls scapula down and inwards, rotating it slightly simultaneously so that the inferior angle moves a little away from the median axis.

Contracture or tightness Fairly rare. However the lower trapezius can be retracted on the concave side of severe scoliosis.

Deficiency Inferior angle of scapula wings away from spine and ribcage even when rhomboids and levator scapulae are sound.

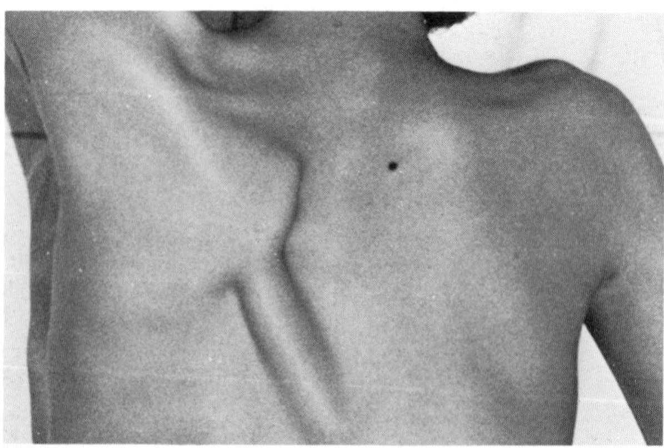

Subject whose trapezius is partially paralysed, only the lowest fibres of the trapezius inferior are preserved, forming a cord.

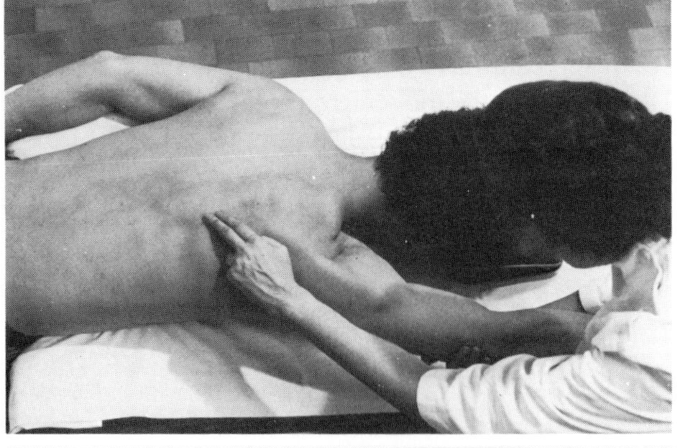

0 and 1 Subject prone.

Arm placed in flexion in direction of the fibres and supported by examiner.
Ask subject to lower and adduct his scapula.
Palpate between scapula and lower thoracic region.

Grade 0: No contraction is felt.
Grade 1: Contraction is felt but there is no joint movement.

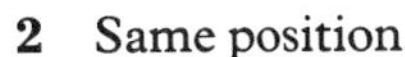

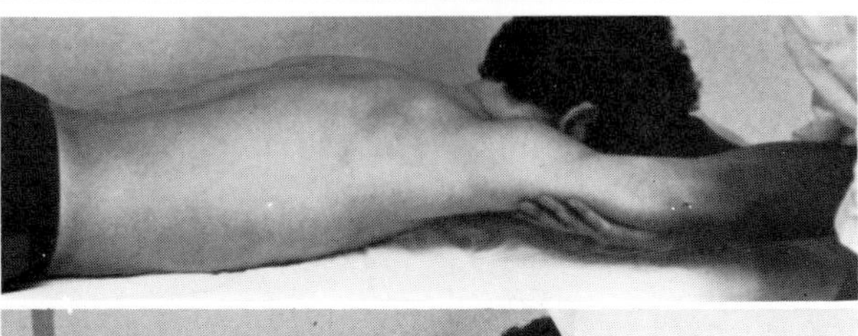

2 Same position.

Examiner supports the upper limb placed in flexion.
Ask subject for the same movement: lowering and adduction of scapula bringing inferior angle slightly outwards.
Range of movement should be complete.

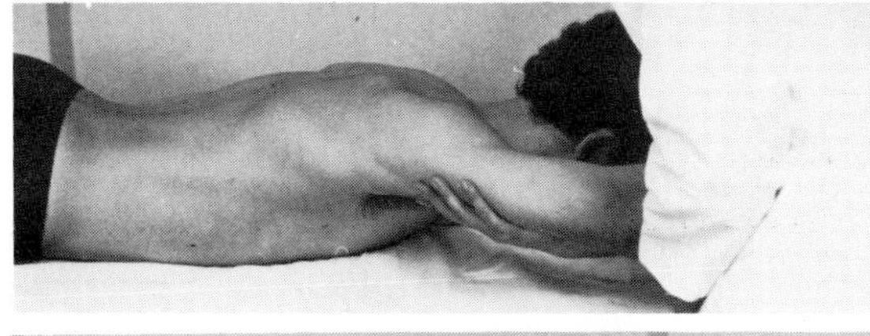

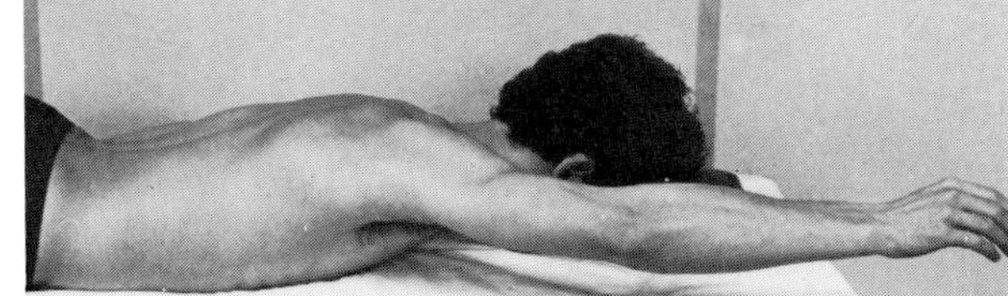

3 Same position.

Same movement.
Resistance is provided by weight of upper limb.
Avoid elevation of ipsilateral hemithorax.
Range of movement should be complete.
Note: When muscles of upper limb are paralysed, it should be supported and light resistance applied to the superior, lateral portion of the scapula (not shown).

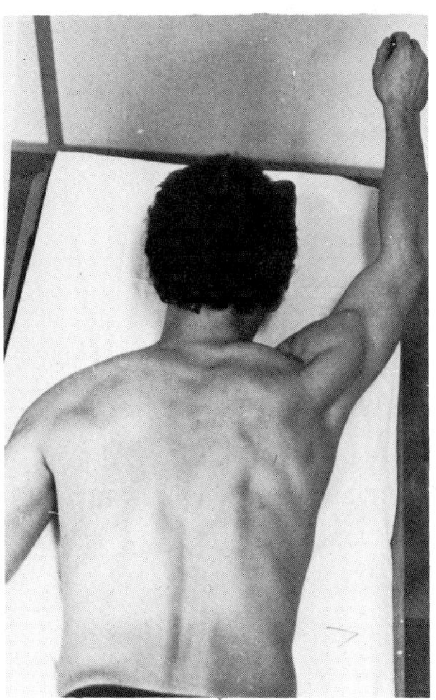

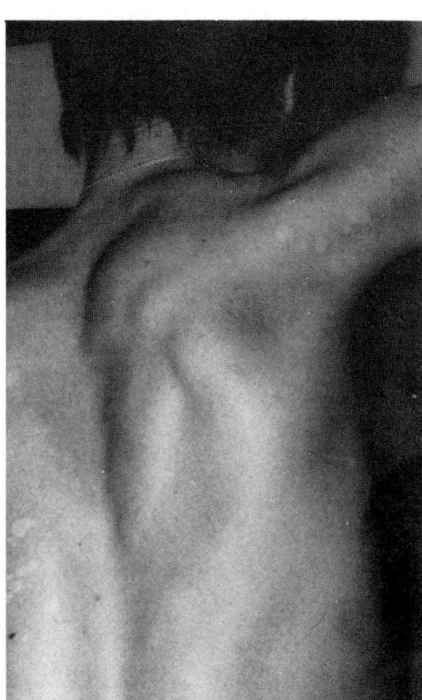

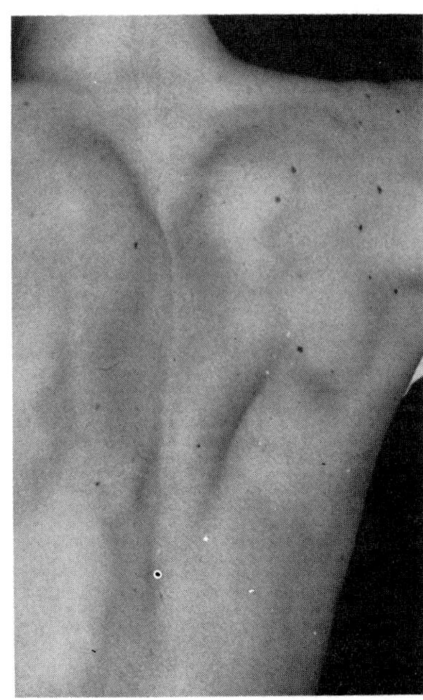

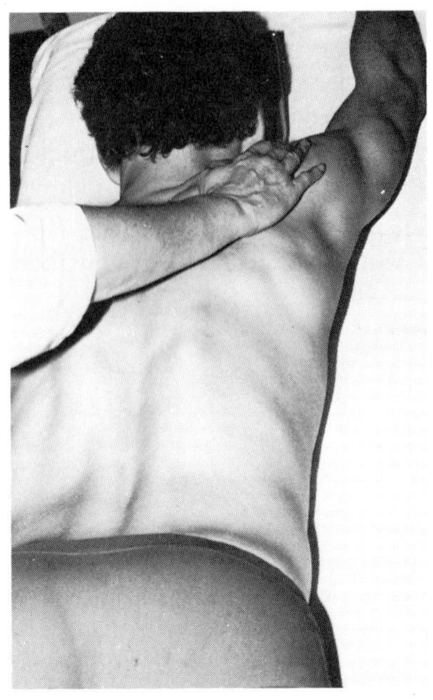

4 and 5 Same position.

Same precaution and movement.
Resistance to the movement is placed on the superior lateral angle of the scapula or on the inferior third of the arm (not shown).

Grade 4: Power is sub-normal.
Grade 5: Power is normal.

Functional muscle testing in traction

The muscle is in its most shortened position.
Ask subject to maintain position while examiner pulls on upper portion of arm.

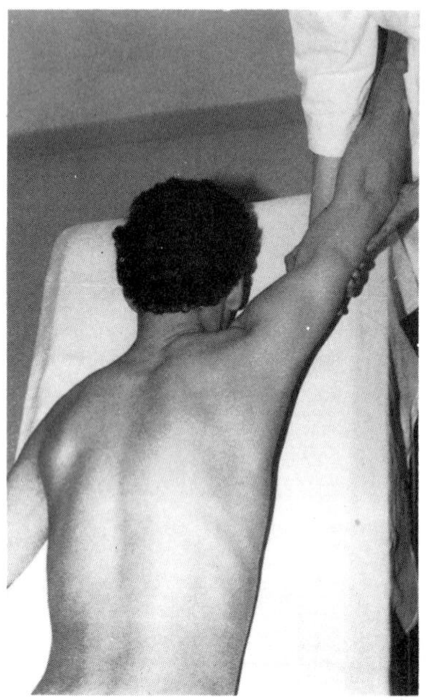

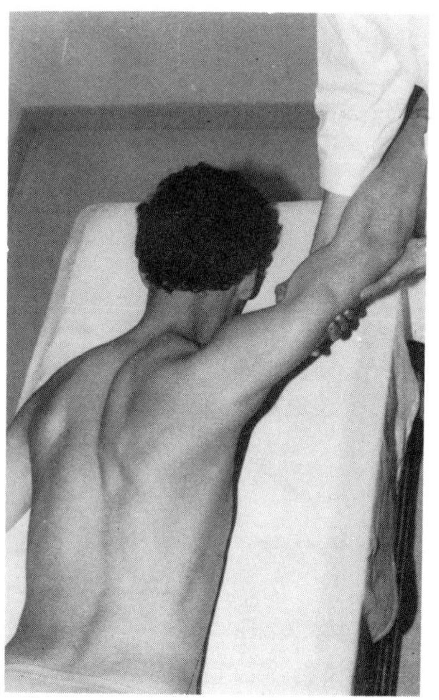

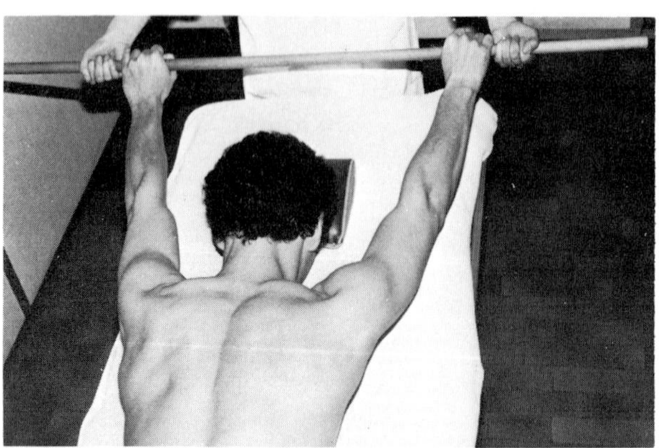

Functional muscle testing with baton

Bilateral, comparative assessment.
Same as test in traction, baton is held with both hands.

NOTES

RHOMBOIDEUS MINOR, RHOMBOIDEUS MAJOR, and LEVATOR SCAPULAE

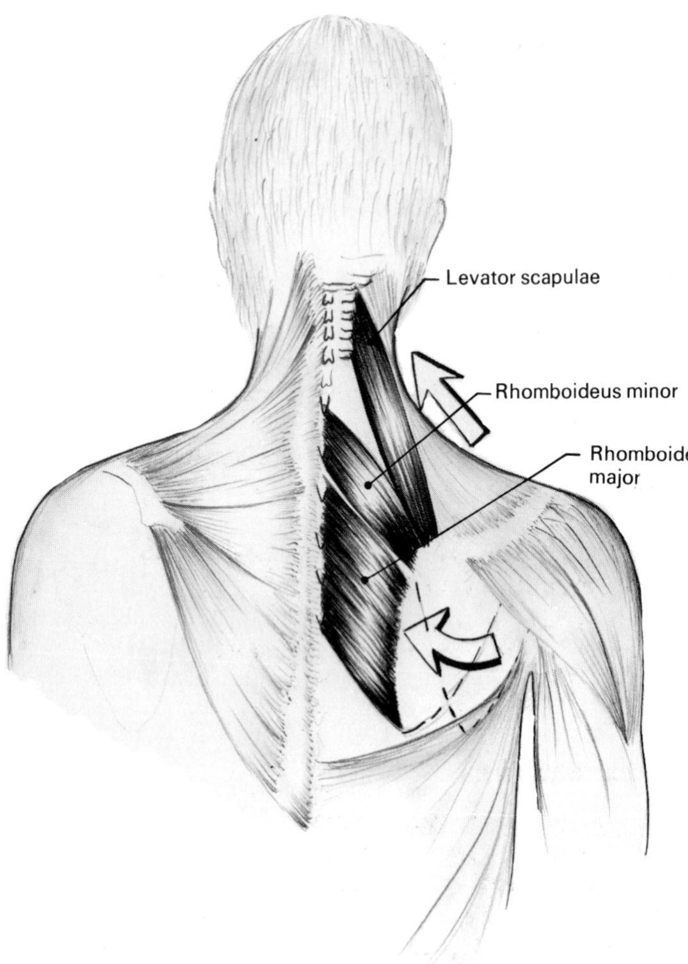

RHOMBOIDS

Origin

—Rhomboideus minor: spinal processes of first thoracic and of seventh cervical vertebrae, and posterior cervical ligament.
—Rhomboideus major: spinous processes of second to fifth thoracic vertebrae.

Insertion

—Rhomboideus minor: spinal border of scapula just above the root of the spine.
—Rhomboideus major: spinal border of scapula between spine and inferior angle.

Innervation

Nerve to levator scapulae and rhomboids (C4, C5).

LEVATOR SCAPULAE

Origin Transverse processes of first four cervical vertebrae.

Insertion Superior medial angle of scapula.

Innervation Nerve to levator scapulae and rhomboids (C4, C5).

Function

Several of their functions are similar, although with some variations.

Rhomboids Elevation of the medial angle of the scapula and adduction of the scapula upwards while rotating it so that the inferior angle approaches the vertebral column more so than the medial angle. The scapula rotates around the lateral angle which also elevates: the glenoid cavity is oriented downwards and outwards. Thus the entire scapula elevates slightly.
The rhomboids are synergistic stabilizers of shoulder movements as well, especially in:
—external rotation of the arm (infraspinatus and teres minor).
—adduction (synergistic pair: teres major and rhomboids).
—extension (latissimus dorsi, teres major, posterior deltoid).
When the scapula is fixed, the rhomboids participate in extension of the upper thoracic spine as well.

Levator scapulae Rotation of scapula around the lateral angle. Both medial and inferior angles elevate, the inferior angle approaches the median axis of the body, protruding slightly under the skin. It raises the entire scapula.
When the scapula is fixed, the levator scapulae participates in extension and ipsilateral latero-flexion of the upper part of the cervical column.

Deficiency

Paralysis of levator scapulae and rhomboids causes an abduction and global drop of the scapula, its inferior angle pulling away from the median axis. Adduction is partially assured by the trapezius and latissimus dorsi.
When rhomboids are paralysed, the scapula is drawn outwards and forwards by the dominating action of serratus anterior. Paralysis of levator scapulae diminishes the elevating power of the scapula.

Testing rhomboids and levator scapulae

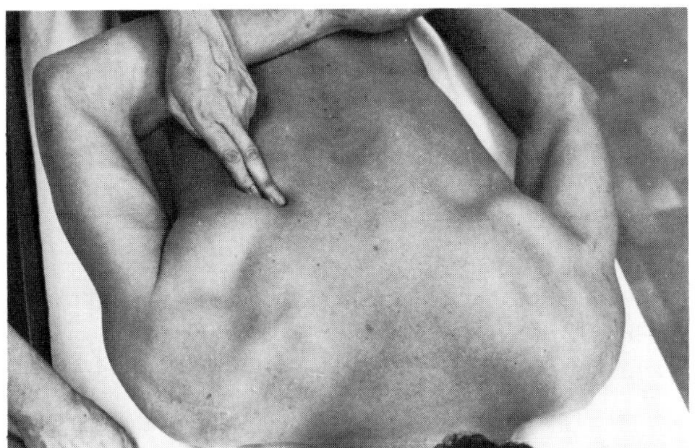

0 and 1 Subject prone, or seated (not shown).

Arms along sides of body, elbow flexed, forearm behind back.
Ask subject for medial rotation of scapula by drawing his arm back and adducting it.
Palpate the rhomboids in the angle formed by the trapezius and the spinal border of the scapula at eighth rib level.
Rhomboids are not to be confused with trapezius fibres which are more superficial.

Grade 0: No contraction is felt.
Grade 1: Contraction is felt but there is no joint movement.

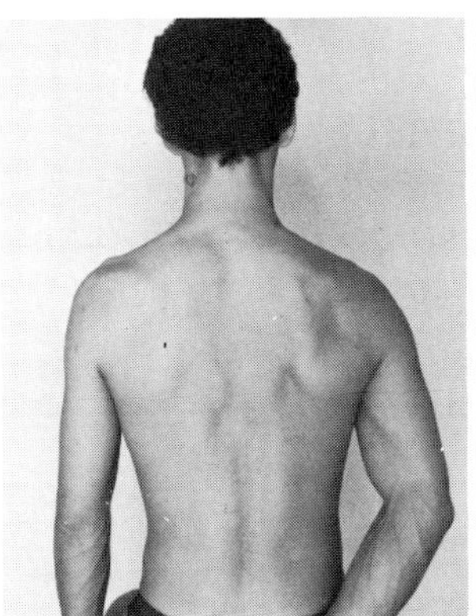

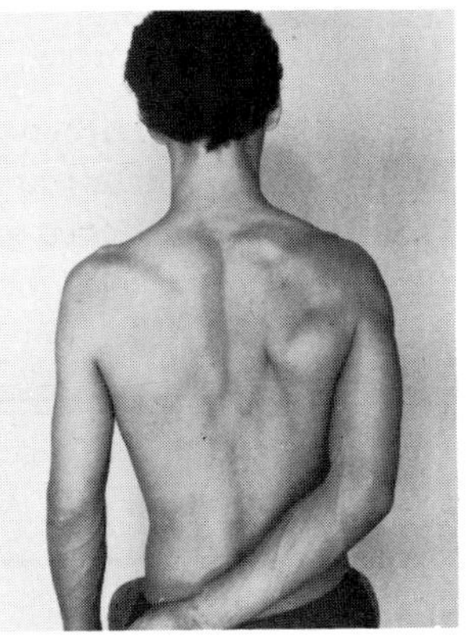

2 Subject sitting.

Arm and forearm in same position.
Ask subject for same movement.
Scapula should elevate, its inferior angle pulling closer to the median axis, tracing a movement of medial rotation. If spinal border of scapula remains parallel to spinal column, middle trapezius is dominating.
Range of movement should be complete.

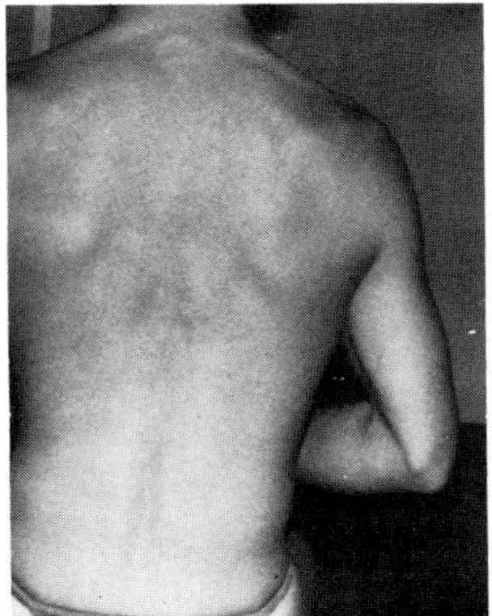

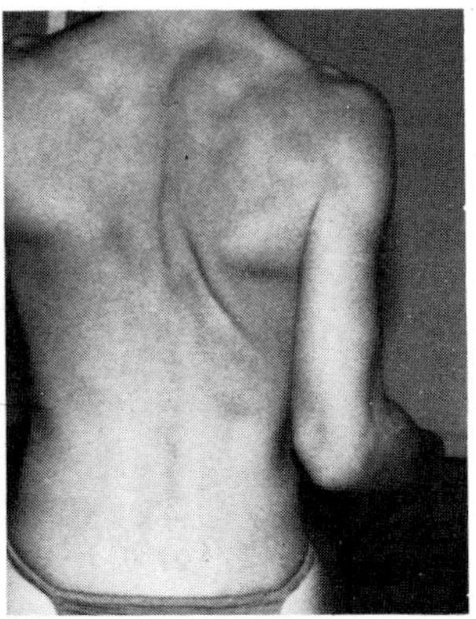

2 *Alternative.*

Subject sitting.
Arm along side of body, elbow flexed.
Ask subject for medial rotation of scapula by drawing arm backwards and adducting it.
Watch the scapula which should elevate, its inferior angle approaching the spinal column.

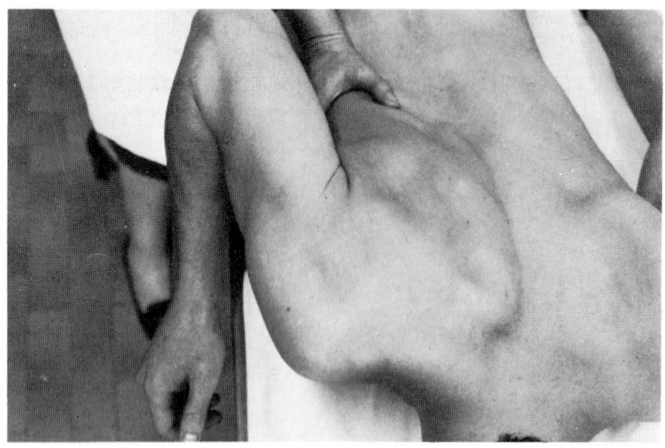

3 Subject prone.

Hold inferior portion of ipsilateral hemithorax in place.
Arm is beside body, elbow flexed.
Ask subject to rotate the scapula medially by drawing arm backwards and adducting it.
Range of movement should be complete.

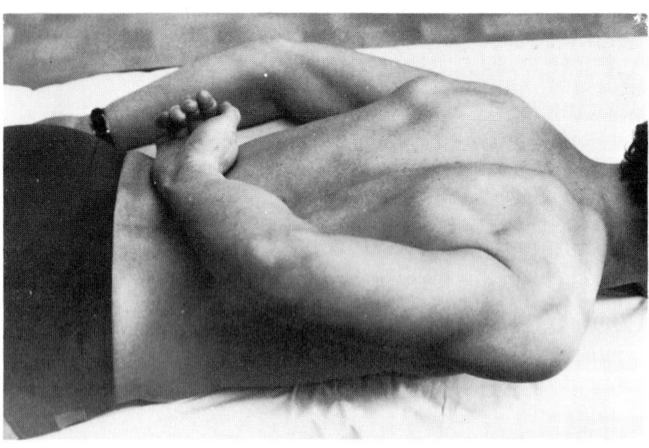

3 *Alternative.*

Subject prone.
Arm beside body, elbow flexed, forearm behind the back.
Ask subject to rotate the scapula medially by drawing his arm backwards and adducting it so that his hand moves towards the opposite buttock.
Range of movement should be complete.

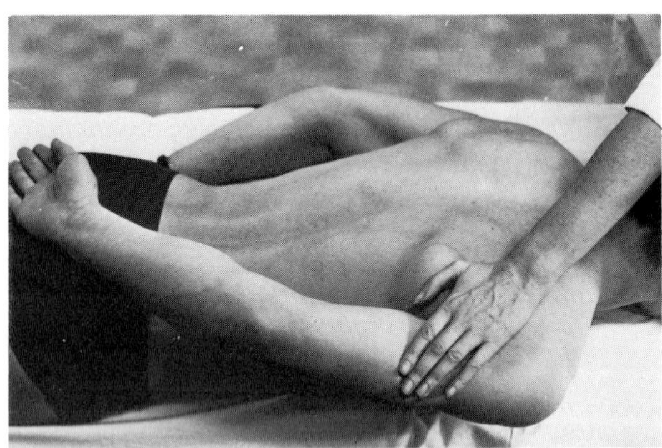

4 and 5 Subject prone.

Examiner can use either testing technique.
Ask subject to rotate scapula medially.
In both cases resistance to the movement is placed on the inferior portion of the lateral border of the scapula.

Grade 4: Power is sub-normal.
Grade 5: Power is normal.

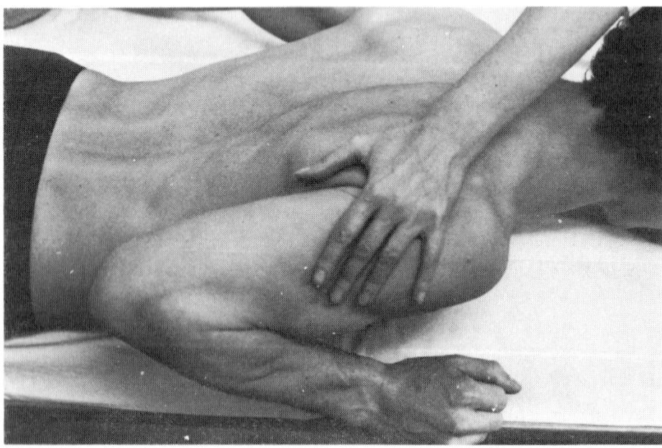

Suggested test for separating levator scapulae from upper portion of middle trapezius

This muscle is tested in its function of elevation and protraction (not retraction) of shoulder girdle.

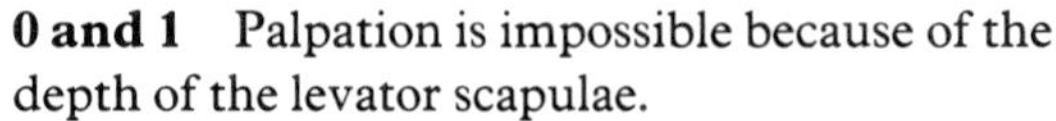

0 and 1 Palpation is impossible because of the depth of the levator scapulae.

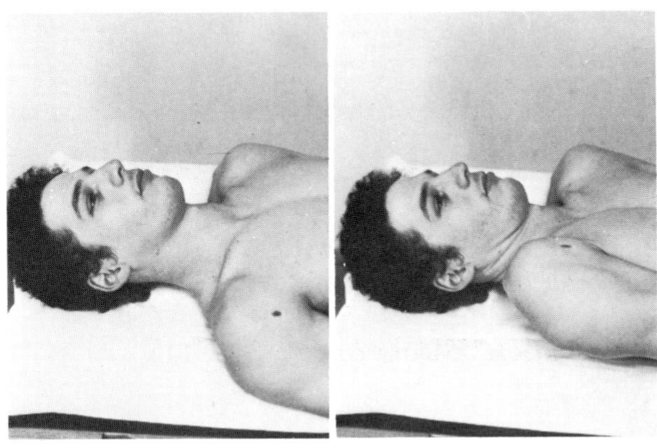

2 Subject supine.

Upper limbs relaxed along sides of body.
Ask subject to elevate his shoulder, raising it slightly up and forward towards his chin.
Range of movement should be complete.

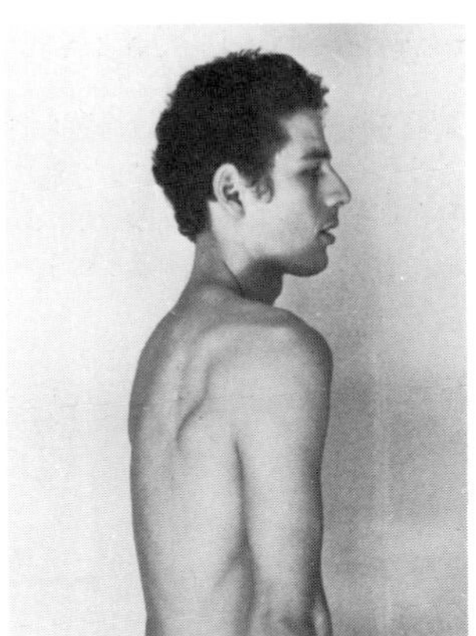

3 Subject sitting.

Upper limb relaxed along side of body.
Ask for same movement.
Range of movement should be complete.

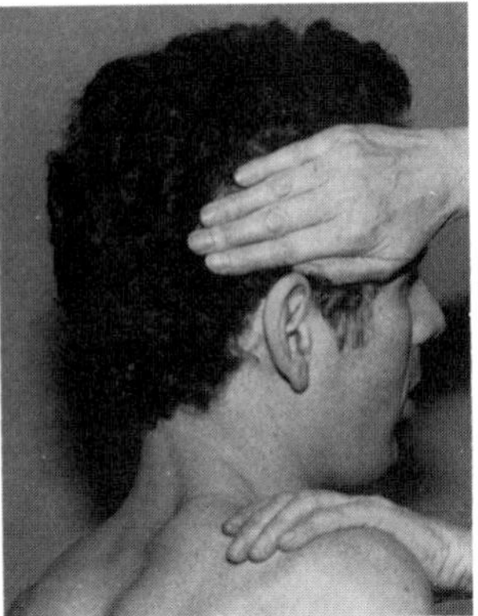

4 and 5 Same position.

Hold subject's head above the ear.
Ask subject for the same movement.
Resistance to the movement is placed over the antero-superior aspect of shoulder girdle.

Grade 4: Power is sub-normal.
Grade 5: Power is normal.

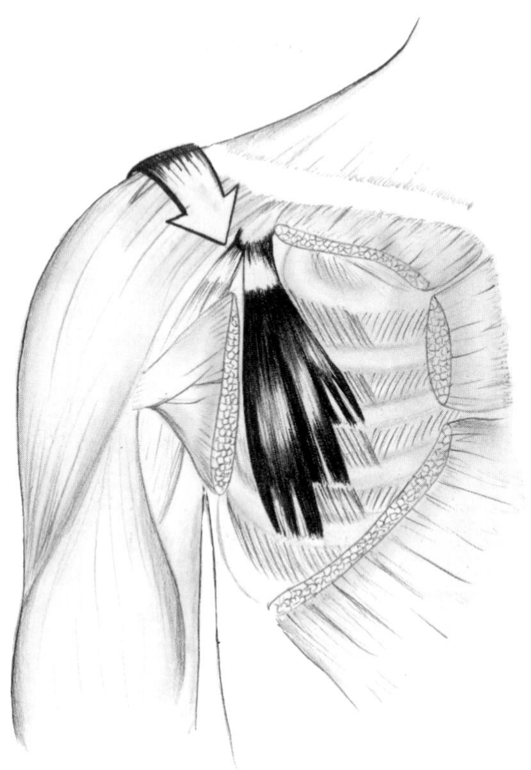

Origin

—Anterior surface of third to fifth ribs near costal cartilage.
—Aponeurosis over intercostal muscles.

Insertion Superior surface and medial border of coracoid process of scapula.

Innervation Medial cord (c7, C8, T1).

Function

Depresses shoulder girdle and projects it forward tilting the scapula anteriorly and slightly outward.
With scapula fixed, pectoralis minor is an accessory muscle of inspiration.
It stabilizes the shoulder girdle when walking with crutches.

Contracture or tightness

Shoulder girdle is protracted forward and inward.
Scapular movements are limited.

Deficiency

Decrease in shoulder girdle forward tilting possibilities.
Shoulder instability during movements of depression and arm extension secondary to poor scapular fixation.

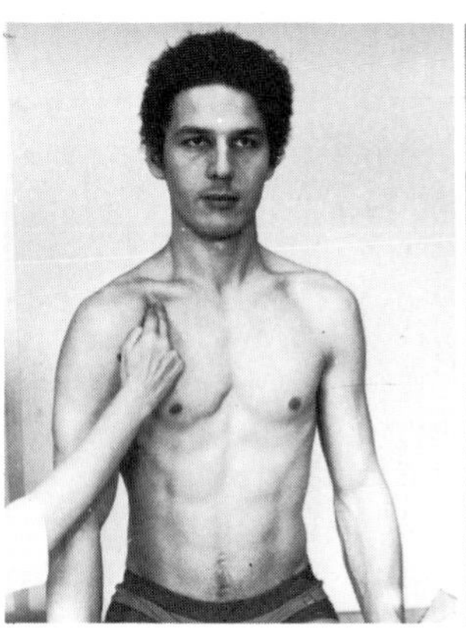

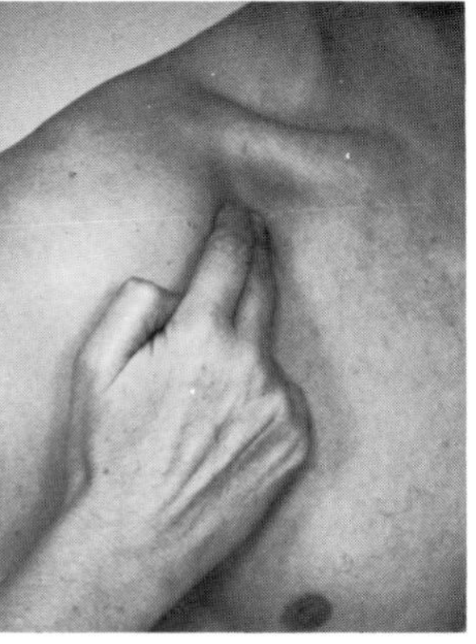

0 and 1 Subject sitting, or supine (not shown).

Due to its anatomical location, palpation of pectoralis minor is difficult.
However, it is possible above the delto-pectoral groove just medially to the apex of the coracoid process.
Ask subject to bring shoulder girdle forward and down.
Note: Some examiners slide their fingers beneath the pectoralis major palpating pectoralis minor at the third, fourth, and fifth rib level.

Grade 0: No contraction is felt.
Grade 1: Contraction is felt but there is no joint movement.

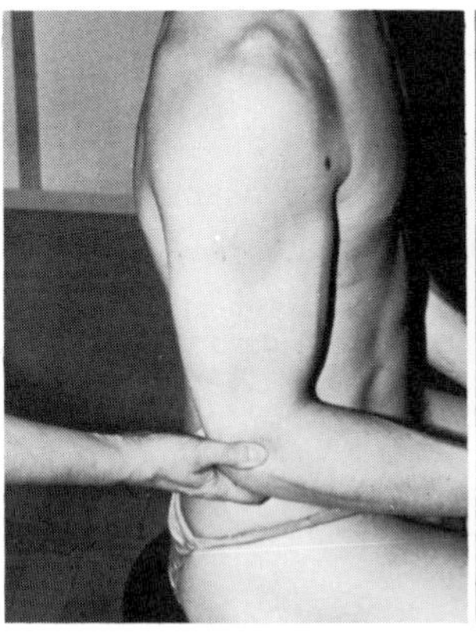

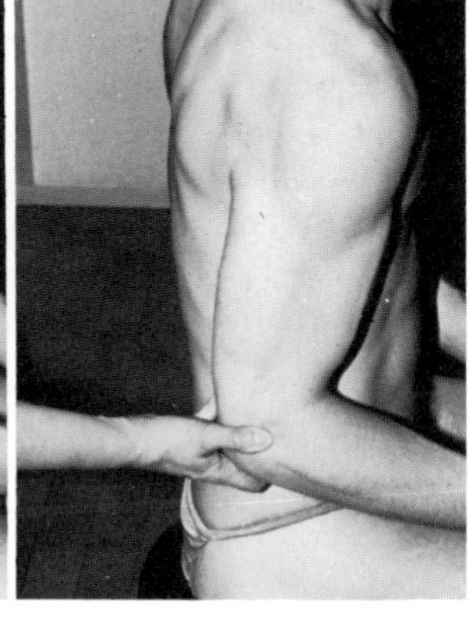

2 Subject sitting on a stool.

Hold posterior part of elbow.
Ask subject for same movement.
Subject should not push on his elbow as he moves his shoulder girdle forward and down.
Thorax must remain immobile, its contralateral rotation might give the illusion of movement.
Range of movement should be complete.

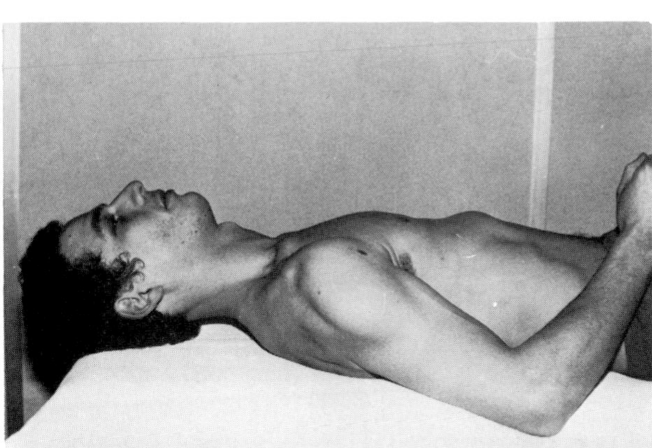

3 Subject supine.

Arms beside body, elbow rests on the table.
Avoid the same compensatory movement.
Ask subject for same movement.
Range of movement should be complete.

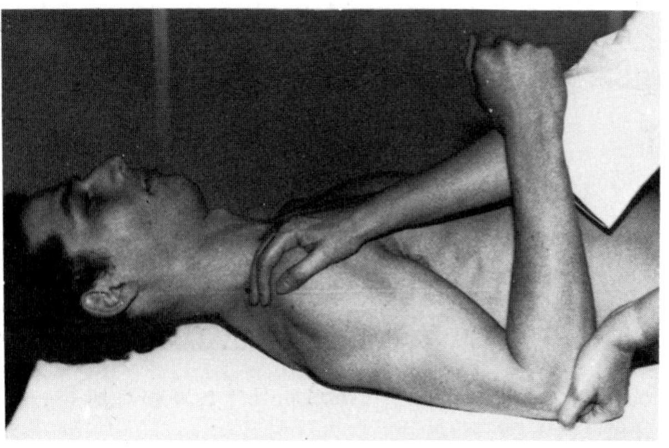

4 and 5 Same position.

Same precautions and movement.
Resistance to the movement is placed on anterior surface of shoulder girdle.

Grade 4: Power is sub-normal.
Grade 5: Power is normal.

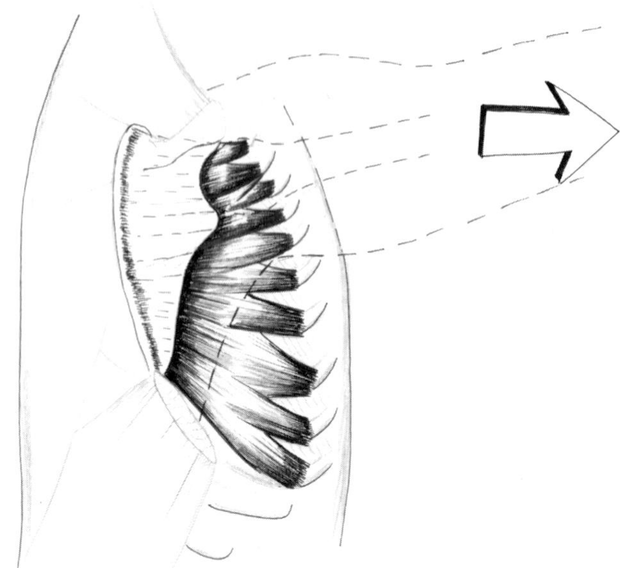

Origin

In fleshy digitations.
It is divided into three portions:
—Superior portion: from first and second ribs.
—Middle portion: from second to fourth ribs.
—Inferior portion: from fifth to tenth ribs.
Digitations originate in the lateral surface of ribs and intercostal aponeurosis.

Insertion

—Superior angle of scapula.
—Costal surface of vertebral border of scapula.
—Inferior angle of scapula.

Innervation Long thoracic (Charles Bell's) (C5, C6, and C7).

Function

It draws scapula forward, out, and slightly up, rotating it laterally (upwards).
It pulls the spinal border of the scapula against the ribcage and permits upper extremity and shoulder elevation above the horizontal line.
With insertion fixed by scapular stabilizers, it is an accessory muscle of inspiration.

Contracture or tightness Exceptional.

Deficiency

Wing scapula (scapular alata).
Vertebral border of scapula wings out away from ribcage.
Upper extremity flexion and abduction above the horizontal line are hampered though superior fibres of middle trapezius may partially substitute.

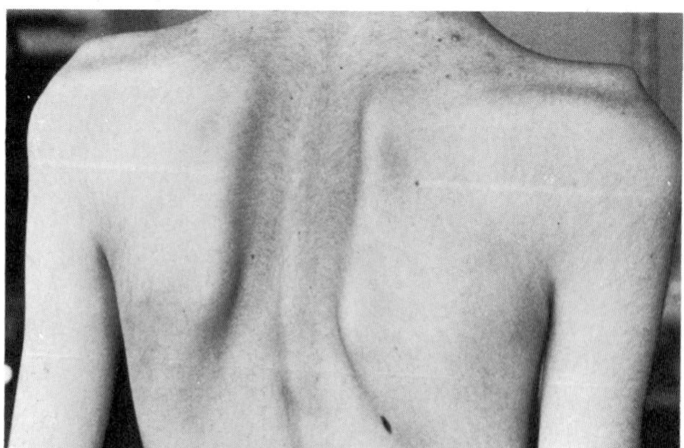

Patient with paralysis of serratus anterior. Note winging of spinal border of scapula.

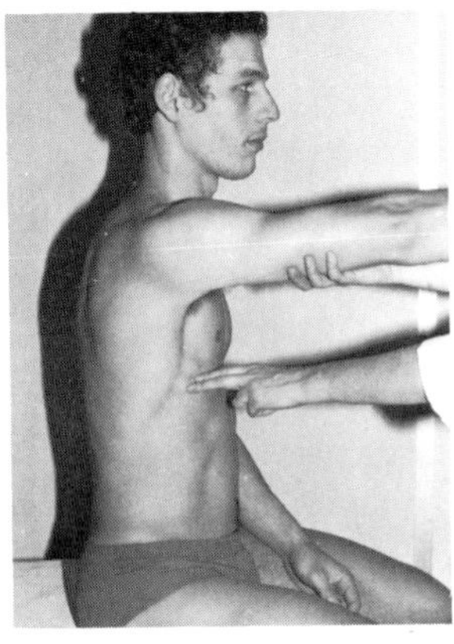
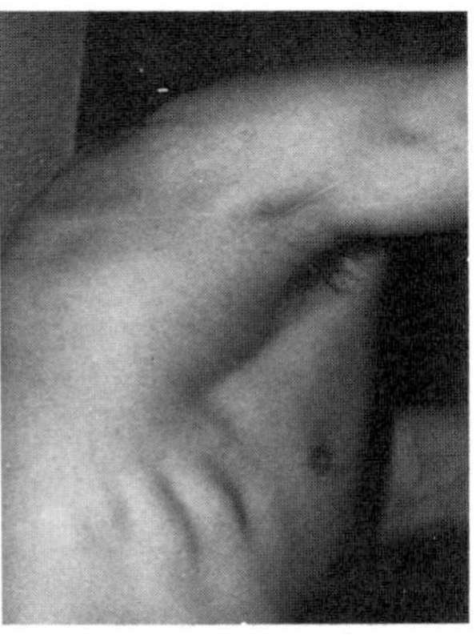

0 and 1 Subject sitting on a stool.

Hold upper limb in forward horizontal position.
Ask subject to push his shoulder girdle forward bringing his upper extremity forward.
Inferior digitations are palpable from the seventh to tenth ribs on anterior lateral surface of thorax (palpation is difficult on under-muscled or overweight subjects).
Serratus anterior is not to be confused with latissimus dorsi, the fibres of which are slightly posterior.
Watch spinal border of scapula wing out in case of deficiency.

Grade 0: No contraction is felt.
Grade 1: Contraction is felt but there is no joint movement.

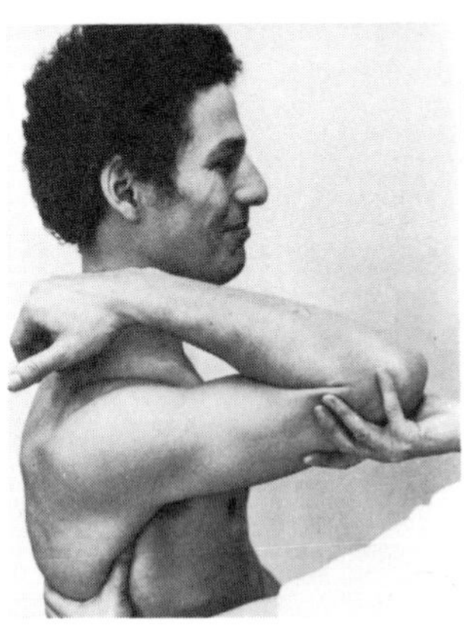
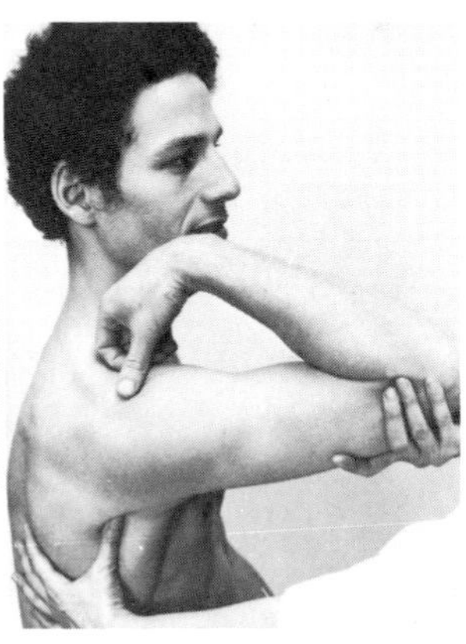

2 Same position.

Support upper limb in forward horizontal position, elbow flexed.
Examiner holds scapula with palm of hand to control lateral (upward) rotation.
Ask for same movement.
Spinal border of scapula should be pulled in against ribcage, since shoulder girdle forward projection can be partially compensated for by pectoral muscles.
Avoid contralateral trunk rotation in flexion.
Range of movement should be complete.
Note: This test may also be done with arm in forward horizontal position on a horizontal sliding board.

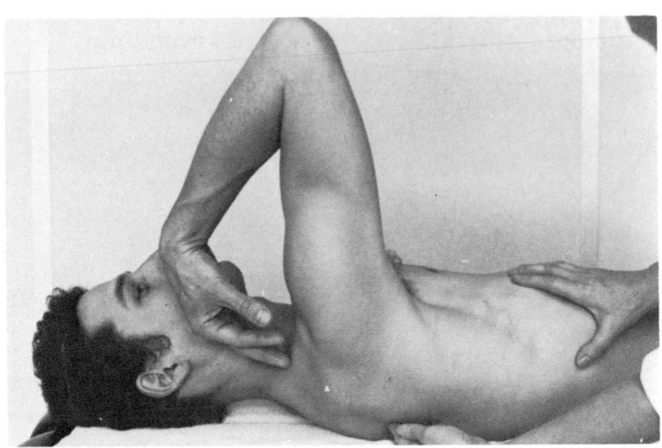

3 Subject supine.

Arm in flexion, elbow flexed.
Hold ipsilateral hemithorax in place.
Control scapular movement with palpation.
Ask subject for same movement.
Range of movement should be complete.

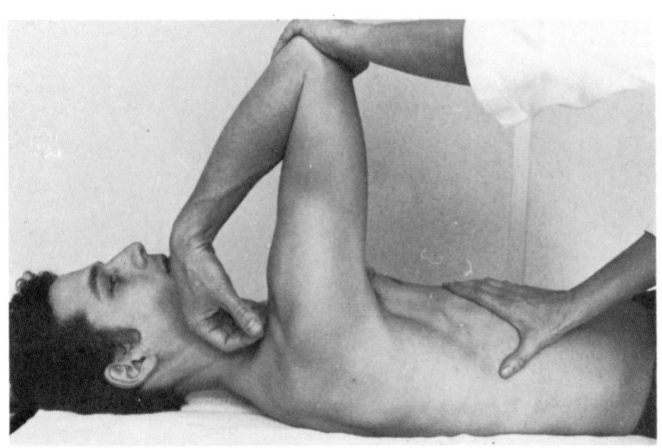

4 and 5 Same position.

Same precaution, same movement.
Resistance to the movement is placed on the elbow.

Grade 4: Power is sub-normal.
Grade 5: Power is normal.

Subject stands, leaning slightly forward, supporting himself, arms horizontal against a wall.

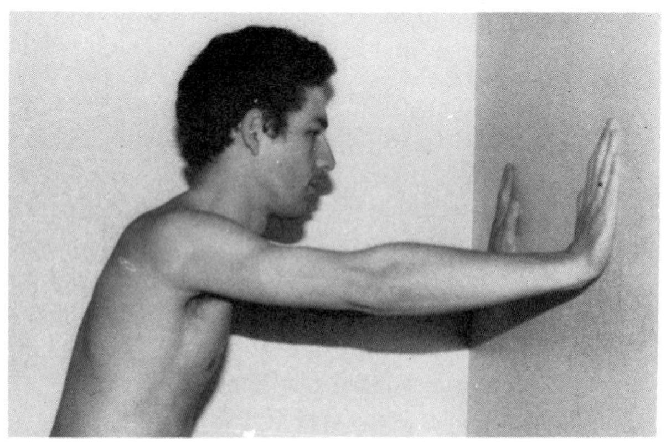

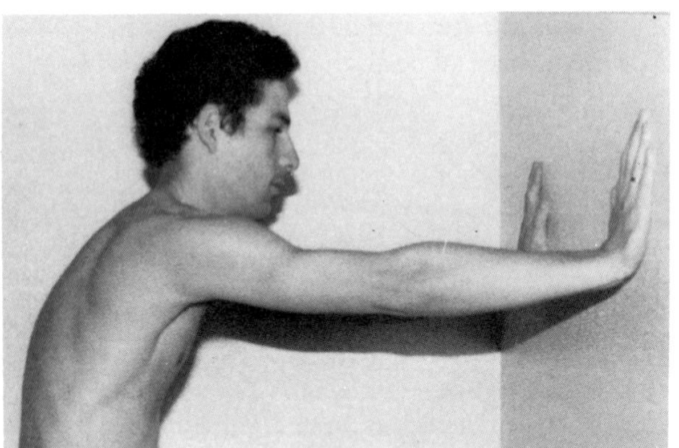

Ask subject to pull trunk back while keeping shoulder girdle and upper extremity immobile: contraction of serratus anterior in closed loop.

NOTES

DELTOIDEUS, ANTERIOR PORTION and CORACO-BRACHIALIS

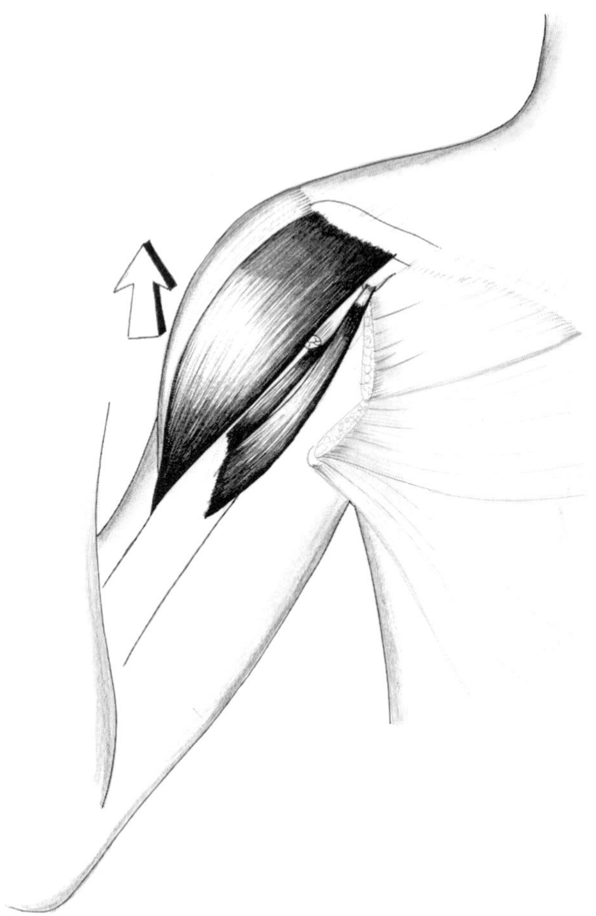

ANTERIOR DELTOID

Origin Lateral third anterior border and superior surface of clavicle.

Insertion Anterior ridge of deltoid tuberosity on anterior border of humerus.

Innervation Axillary (C5, C6).

CORACO-BRACHIALIS

Origin Medial part of tip of coracoid process of scapula by conjoined tendon with short head of biceps brachii.

Insertion Medial surface of humeral shaft above its middle part.

Innervation Musculocutaneous (C6, C7).

Function Both muscles flex the arm.

Anterior deltoid Flexes and slightly abducts the arm. It participates in direct abduction with the middle deltoid, in horizontal adduction with the clavicular portion of pectoralis major, and slightly in internal rotation of the arm.

Coraco-brachialis Suspends and stabilizes the humeral head during adduction.
It flexes and slightly adducts the arm.

Deficiency

A lesion of the anterior deltoid is a serious functional handicap. It severely diminishes shoulder flexion capacities. All functional gestures involving flexion are impaired, such as grooming, eating, etc. Coraco-brachialis, biceps, and the clavicular portion of pectoralis major partly make up for the deficiency.
A lesion of coraco-brachialis causes shoulder pain during adduction.

Anterior deltoid is tested with slight abduction.
If the coraco-brachialis is more specifically being tested, shoulder flexion is done with slight adduction.

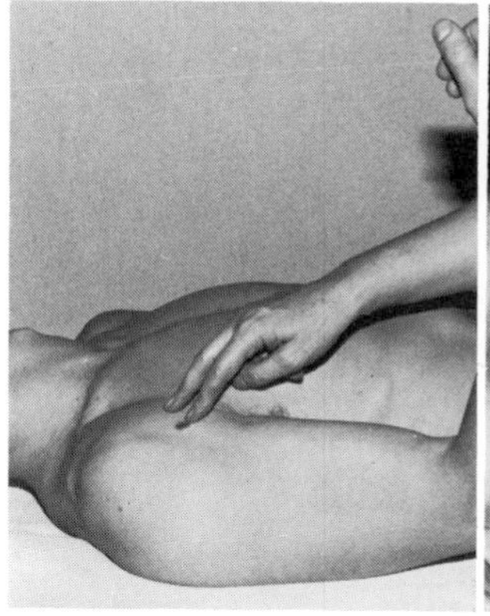
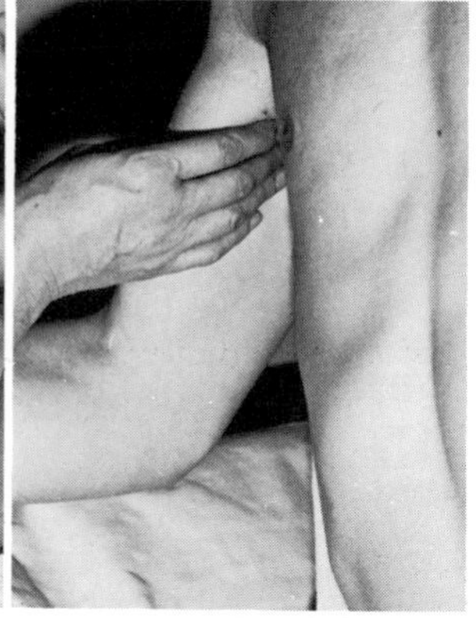

0 and 1 Subject supine or sitting.

Ask subject for arm flexion, elbow flexed, forearm supported by examiner.
Palpate anterior deltoid on the anterior aspect of the shoulder girdle.
Palpate coraco-brachialis on the superior, medial surface of the arm behind short head of biceps. It is more easily found when the arm is separated from the trunk (. . . crucifixion muscle).

Grade 0: No contraction is felt.
Grade 1: Contraction is felt but there is no joint movement.

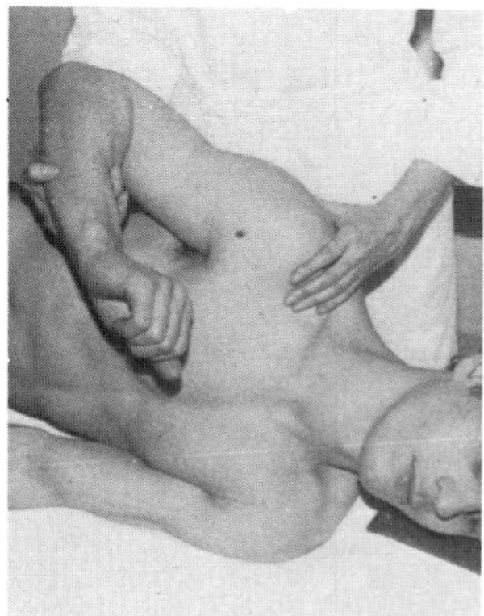
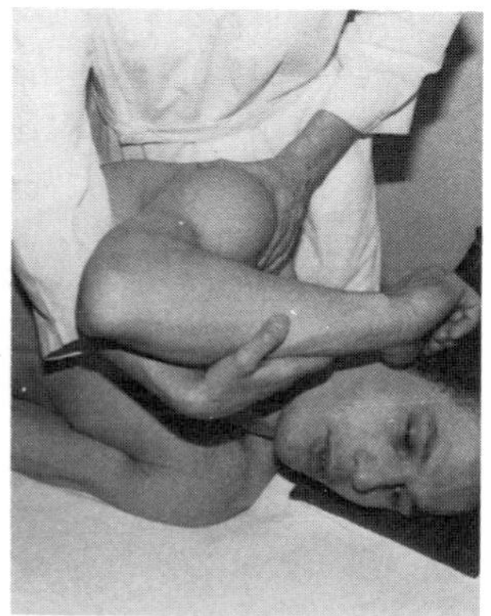

2 Subject side-lying.

Hold upper aspect of shoulder girdle.
Support upper limb, elbow flexed, in order to diminish the biceps' action on shoulder.
Ask subject to flex the arm, considering possibilities of substitution by clavicular portion of pectoralis major.
Range of movement should be complete.

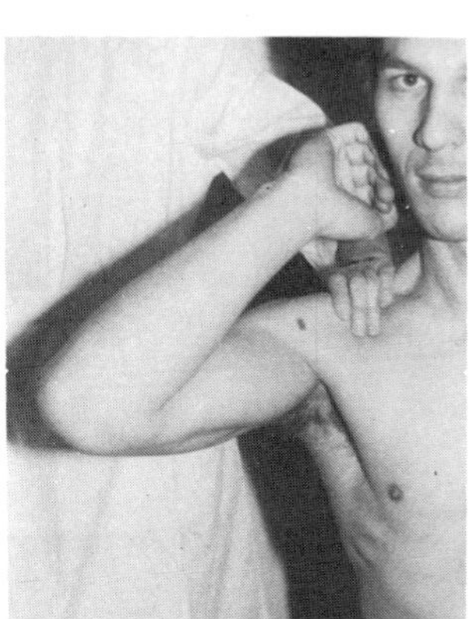

3 Subject sitting.

Hold acromio-clavicular angle in place.
Ask subject for same movement, elbow flexed, considering same possibility of substitution.
Range of movement should be complete.

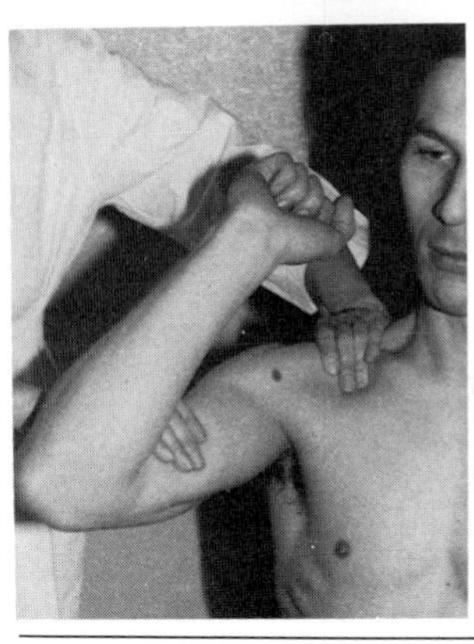
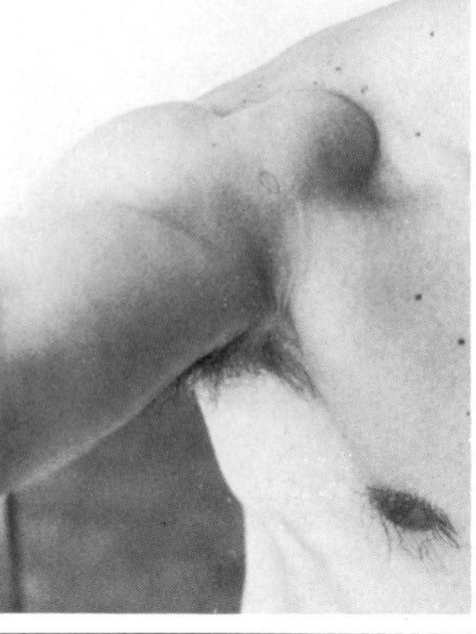

4 and 5 Same position.

Same hold and same movement.
Resistance to the movement is placed on the inferior third of anterior aspect of arm.

Grade 4: Power is sub-normal.
Grade 5: Power is normal.

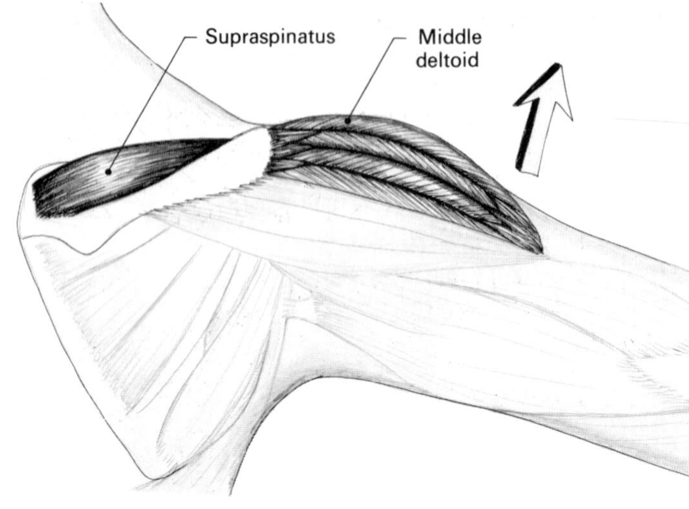

MIDDLE DELTOID

Origin Lateral margin and apex of acromion.

Insertion Angle of deltoid tuberosity on lateral surface of humerus.

Innervation Axillary (C5, C6).

SUPRASPINATUS

Origin Medial two-thirds of supra-spinous fossa of scapula, and superior surface of spine of scapula.

Insertion Superior facet of greater tubercle of humerus adhering to shoulder joint capsule.

Innervation Suprascapular (C5, C6).

Function Both abduct the arm.
Fibres of anterior and posterior deltoids are abductors as well.
The principal function of supraspinatus is on the humeral head which it prevents from dislocating downwards.
Supraspinatus is a true active ligament of the shoulder joint.

Deficiency Leads to wasting of the shoulder form.

In presence of paralysis of middle deltoid and even of the whole deltoid, the supraspinatus can abduct the shoulder with the serratus anterior and trapezius; however, there is a significant loss of strength.
Wasting of supraspinatus (paralysis, rupture, . . .) leads to joint instability and may provoke pain.
Wasting of both deltoid and supraspinatus makes shoulder joint abduction impossible.

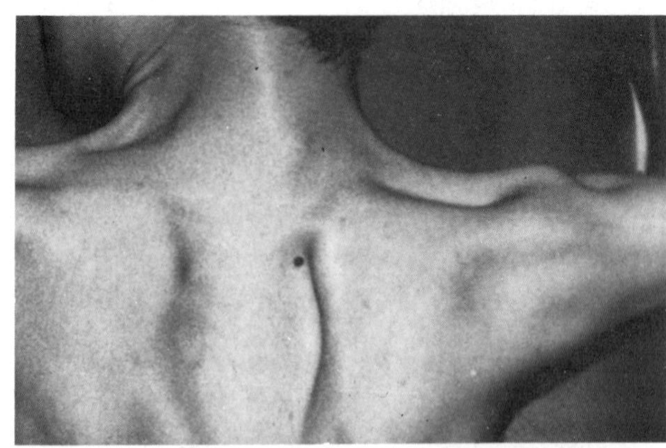

Wasting of shoulder relief and of the supra-spinous fossa on a subject with an important deficiency of the scapular girdle.

Middle deltoid and supraspinatus are tested together since it is too difficult to separate them in their common function of shoulder abduction.

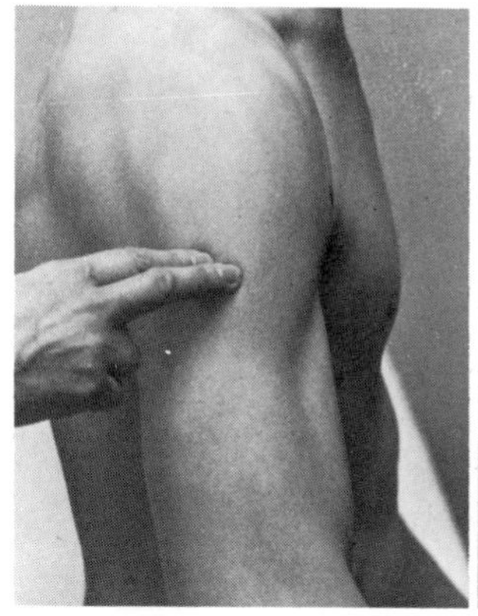

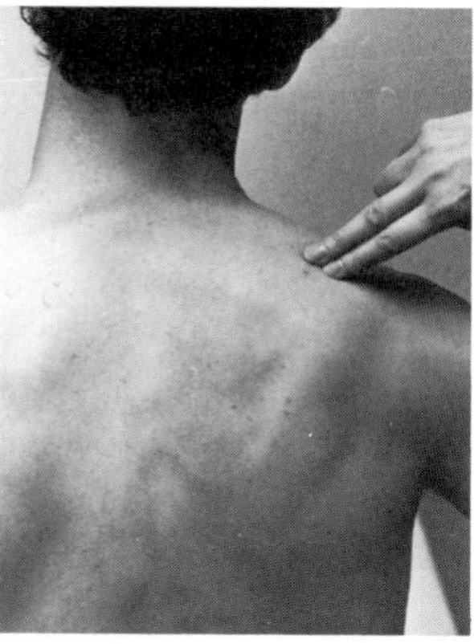

0 and 1 Subject sitting, or supine (not shown).

Upper limb along side of body.
Ask subject for arm abduction in indifferent rotation.
The middle deltoid is palpated on the lateral portion of the shoulder girdle.
In order to palpate the supraspinatus, the subject's head should be inclined on the same side so as to relax the trapezius which is still relaxed at beginning of movement. Fibres are palpated in the medial two-thirds of the supra-spinous fossa.

Grade 0: No contraction is felt.
Grade 1: Contraction is felt but there is no joint movement.

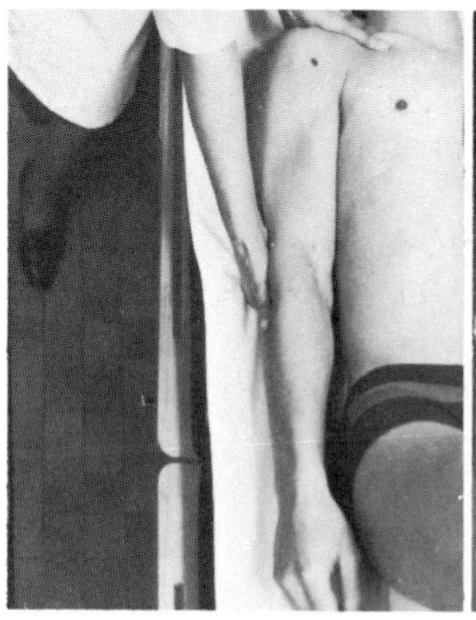

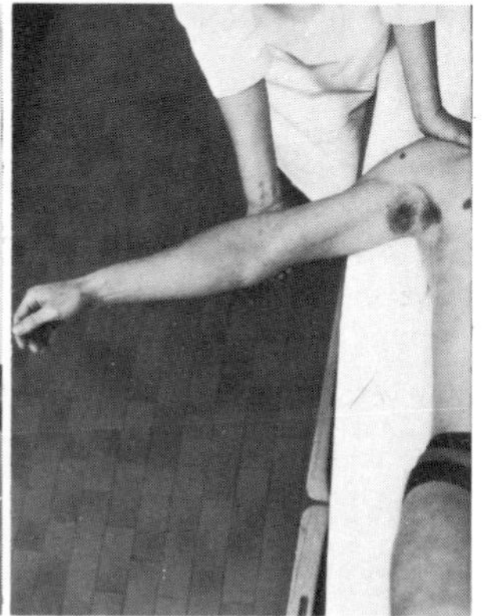

2 Subject supine.

Upper limb along side of body in indifferent rotation.
Hold acromio-clavicular angle in place. Examiner supports the elbow.
Ask subject to abduct his arm to 80° in indifferent rotation.
Watch for trick movement, especially during flexion with external rotation.
Range of movement should be complete.

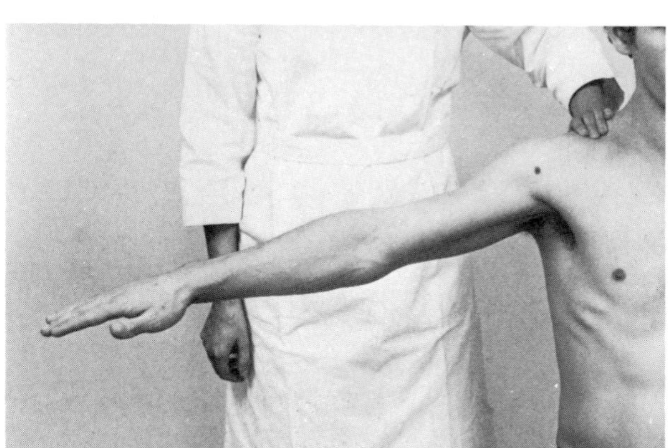

3 Subject sitting on a stool.

Upper limb along side of body in indifferent rotation.
Hold acromio-clavicular angle in place.
Ask subject for arm abduction avoiding possible trick movement.
Range of movement should be complete.

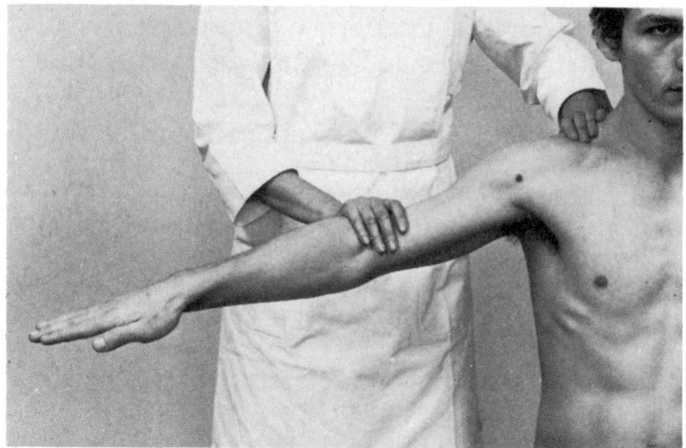

4 and 5 Same position.

Same hold, precautions and movement.
Resistance to the movement is placed above the elbow.

Grade 4: Power is sub-normal.
Grade 5: Power is normal.

DELTOIDEUS, POSTERIOR PORTION

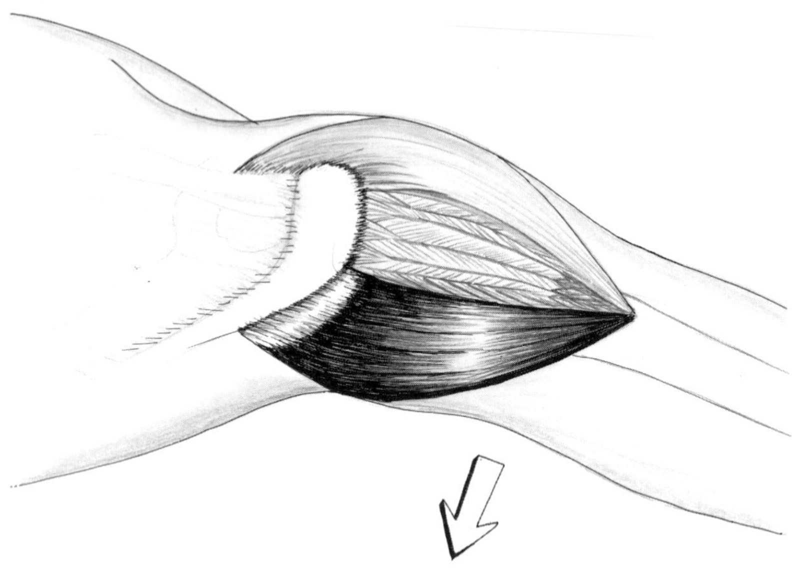

Origin Inferior aspect of posterior border of spine of scapula.

Insertion Posterior ridge of deltoid tuberosity on lateral surface of humerus.

Innervation Axillary (C5, C6).

Function

Extension of the arm: it is one of the main shoulder joint extensors. It functions synergistically with latissimus dorsi and teres major. But its role is important since it extends and abducts the arms while the other extensors extend and adduct it.
Posterior abduction of the arm: posterior deltoid participates in abduction. But, above about 50°, it may become an adductor, thus reversing its function.
It contributes slightly to external rotation of the arm.

Deficiency

Decrease in capacity for arm extension with abduction.
Difficulty in completing gestures requiring extension when arm is in abduction: hindrance in placing hands on hips, in pockets or behind back.
Teres major and latissimus dorsi may substitute, in which case the arm goes markedly into adduction.

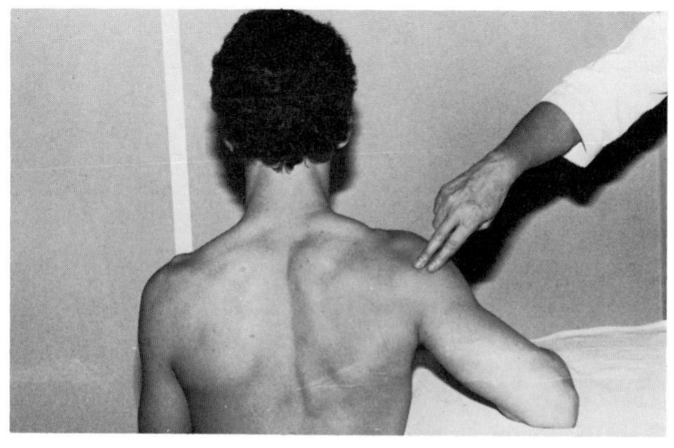

0 and 1 Subject seated on a stool.

Arm abducted at 50°, elbow flexed, forearm resting on table.
Ask subject to extend while posteriorly abducting his arm.
Examiner palpates posterior deltoid on posterior aspect of shoulder joint between spine of scapula and posterior ridge of deltoid tuberosity.

Grade 0: No contraction is felt.
Grade 1: Contraction is felt but there is no joint movement.

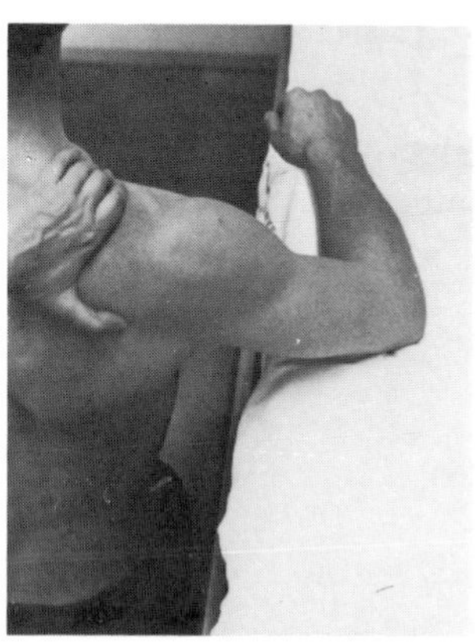

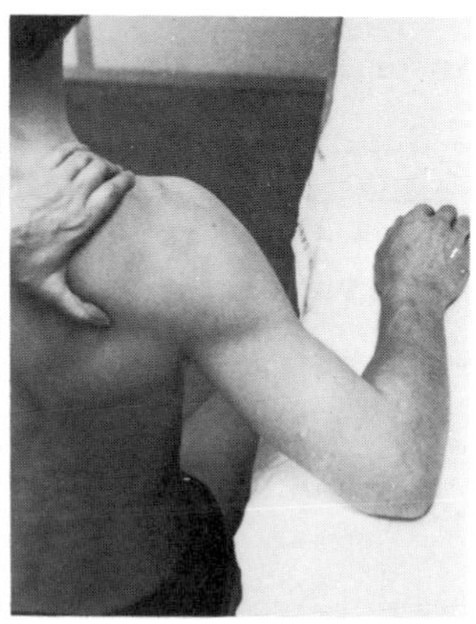

2 Same position.

Arm is placed in flexion.
Hold acromio-clavicular angle in place.
Ask for same movement.
Avoid posterior rotation of ipsilateral hemithorax.
Range of movement should be complete.

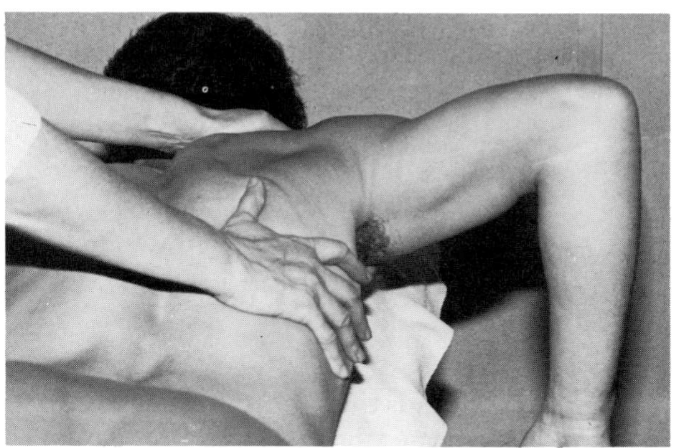

3 Subject prone at edge of table.

Upper limb hangs in flexion.
Examiner holds acromio-clavicular angle and ipsilateral hemithorax in place to avoid its elevation.
Ask subject to extend while abducting his arm.
Range of movement should be complete.

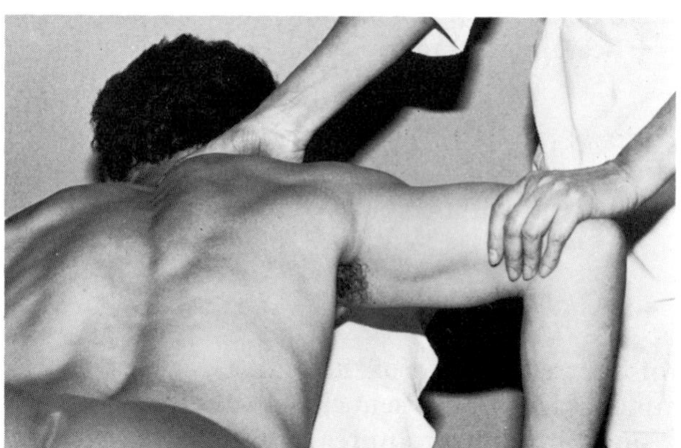

4 and 5 Same position.

Hold acromio-clavicular angle in place.
Same precautions and movement.
Resistance to the movement is placed on the inferior third of posterior aspect of arm.

Grade 4: Power is sub-normal.
Grade 5: Power is normal.

SUBSCAPULARIS

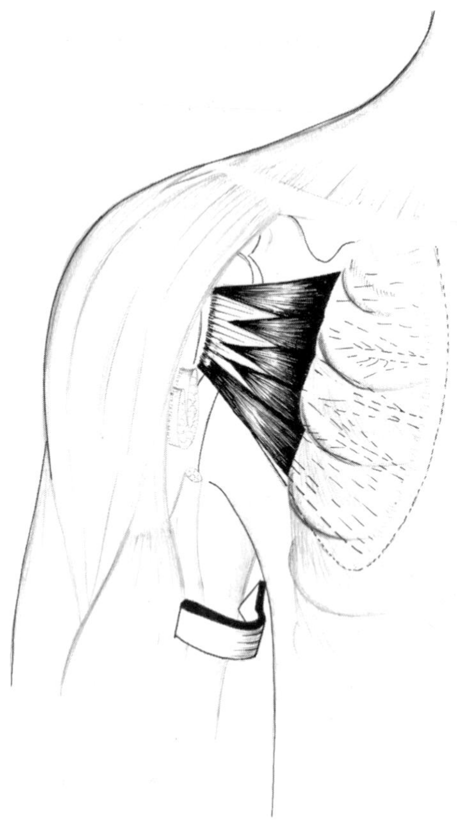

Origin Subscapular fossa of anterior surface of scapula.

Insertion

—Superior aspect of lesser tuberosity of humerus.
—Adheres to capsule of shoulder joint.

Innervation

—Upper trunk.
—Posterior cord.
—Branch of axillary (C5, C6).

Function

Internal rotation of the arm: in spite of a slight adduction component, it is the purest internal rotator. It acts synergistically with forearm pronators when pronation is performed against strong resistance: functional pronation.
Anatomical location of subscapularis makes it a true active ligament of the shoulder joint: it plays a large role in shoulder joint stability by retaining humeral head in glenoid cavity.

Contracture or tightness

This muscle is often contracted.
Its contracture, with the other internal rotators, leads to a posture of internal rotation and adduction, and a diminution of external rotation.

Deficiency

Isolated deficiency of this muscle diminishes strength of internal rotation of the arm.
Associated with deficiency of other internal rotators: functional hindrance in all movements bringing the hand towards the body results, specifically, in precision grip which requires functional pronation.

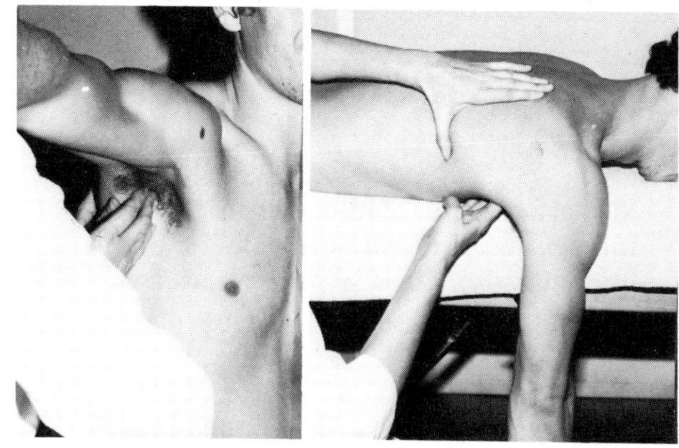

0 and 1 Subject sitting or prone.

Upper limb in abduction, resting on examiner's shoulder (subject sitting).
Upper limb hanging over edge of table (subject prone).
The scapula is thus in abduction, its anterior surface partially accessible.
Ask subject for internal rotation of arm.
Examiner palpates subscapularis fibres from anterior surface of scapula in axilla.
Palpation may be difficult if all other muscles are normal.

Grade 0: No contraction is felt.
Grade 1: Contraction is felt but there is no joint movement.

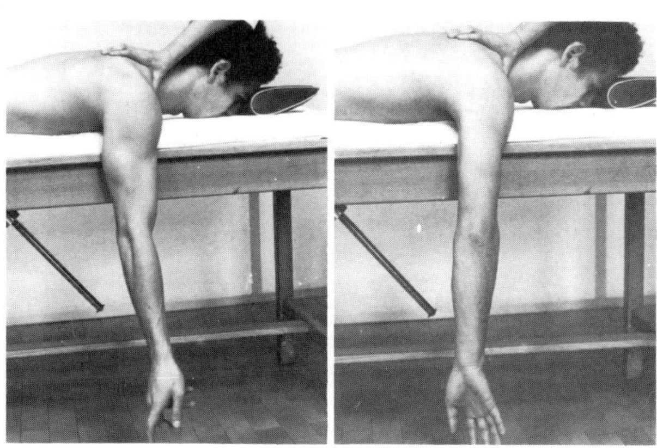

2 Subject prone.

Upper limb hanging over edge of table in external rotation.
Hold acromio-clavicular angle in place.
Ask subject for internal rotation of arm, bringing its anterior aspect inward from the outward position.
The arm must rotate internally and not the forearm which may pronate giving an illusion of rotation.
Range of movement should be complete.
Note: If pull on shoulder is painful, a horizontal sliding board is placed beneath ulnar aspect of forearm.
Note: It is possible to use this test with subject supine, upper extremity resting along side of body.

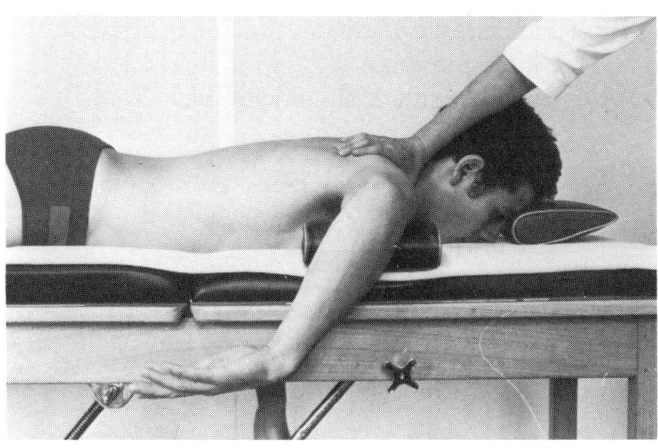

3 Subject prone.

Arm in abduction, cushion under elbow, forearm hanging over edge of table.
Hold acromio-clavicular angle in place.
Ask subject for internal rotation of the arm, bringing his hand up and backwards.
This test is only valid for the most shortened position, the middle and lengthened positions being aided by gravity.
Range of movement should be complete.

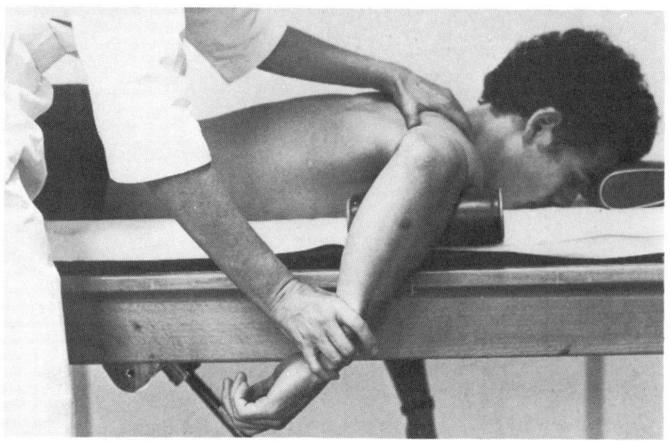

4 and 5 Same position.

Hold acromio-clavicular angle in place.
Ask for same movement.
Resistance to the movement is placed on inferior third of anterior aspect of forearm.

Grade 4: Power is sub-normal.
Grade 5: Power is normal.

INFRASPINATUS and TERES MINOR

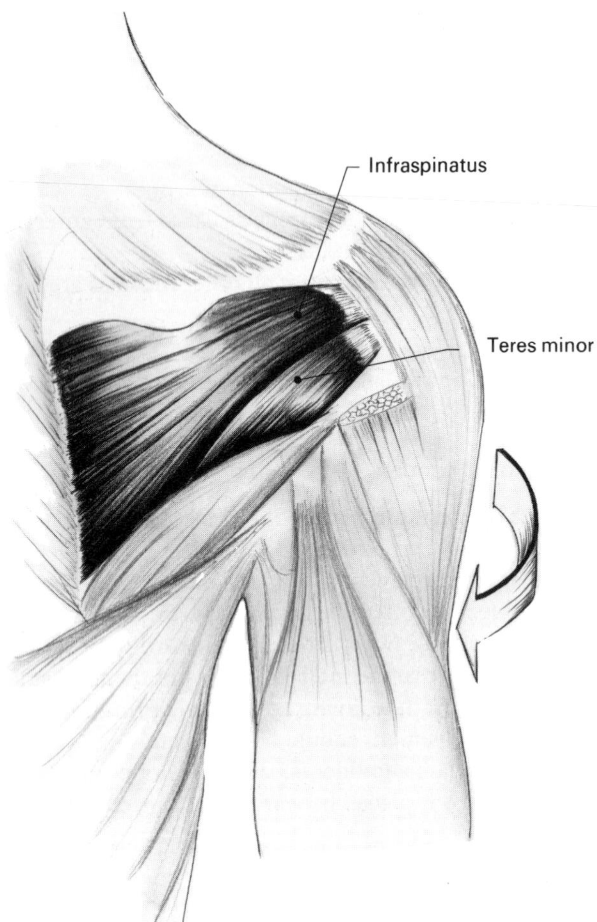

INFRASPINATUS

Origin

—Medial two-thirds of infra-spinous fossa.
—Inferior surface of spine of scapula.
—Intermuscular fascia separating it from other muscles.

Insertion

—Middle facet, greater tuberosity of humerus between supraspinatus and teres minor.
—Adheres to capsule of shoulder joint.

Innervation Suprascapular (C5, C6).

TERES MINOR

Origin

—Lateral portion of upper part of infra-spinous fossa of scapula.
—Intermuscular septa separating it from other muscles.

Insertion

—Posterior surface of greater tuberosity of humerus below infraspinatus.
—Adheres to capsule of shoulder joint.

Innervation Axillary (C5, C6).

Function

Their functions are identical: external rotation of the arm.
They act in synergy with forearm supinators as soon as the movement is performed against strong resistance: functional supination.
They play an important role in writing and all movements bringing the arm from an inward to an outward position.
Contributing to the stability of the shoulder joint by retaining humeral head in glenoid cavity, they function as true active ligaments of the shoulder joint.
Note: External rotators of the shoulder are less powerful than its internal rotators.

Deficiency

Active, external rotation of shoulder joint becomes impossible; these muscles have no agonists.
Functional impairment in activities of daily living.

Subject presenting atrophy of supra- and infra-spinous fossa.

Infraspinatus and teres minor are tested together. They cannot be separated in their common function.

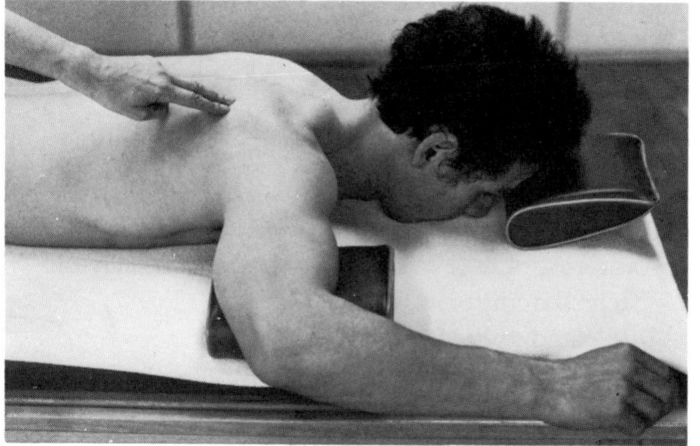

0 and 1 Subject prone.

Arm in abduction, cushion under elbow, forearm hanging over edge of table.
Ask subject for external rotation of arm, elbow held at 90°.
Infraspinatus is palpated in the infra-spinous fossa.
Teres minor is palpated at upper part of axillary border of scapula above teres major (not shown).
Scapular movement by trapezius and rhomboids may give an illusion of external rotation.

Grade 0: No contraction is felt.
Grade 1: Contraction is felt but there is no joint movement.

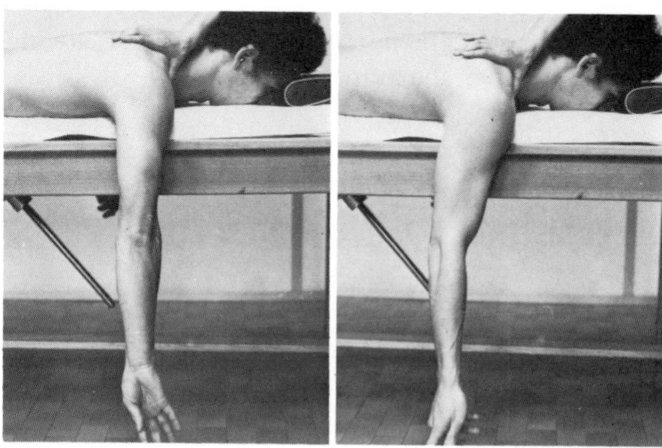

2 Subject prone.

Entire arm hangs over edge of table in internal rotation.
Hold acromio-clavicular angle in place.
Ask subject for external rotation of the arm, bringing its anterior aspect out from an inward position.
External rotation takes place in the arm and not in the forearm which may supinate giving an illusion of rotation.
Range of movement should be complete.
Note: If pull on shoulder is painful, place ulnar aspect of forearm on a horizontal sliding board.
Note: This test is also possible with subject in supine position, upper limb along side of body.

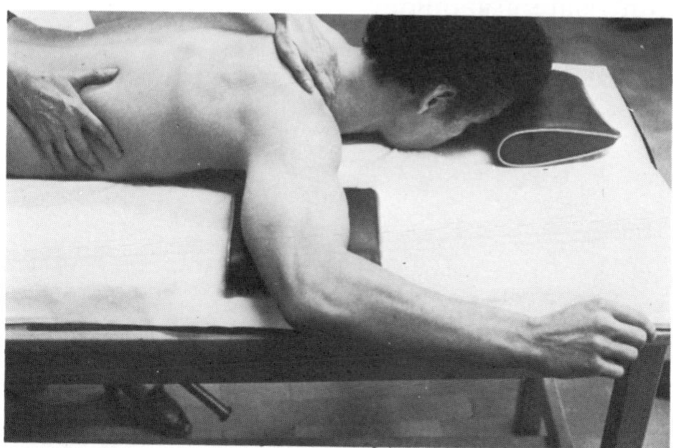

3 Subject prone.

Arm in abduction, cushion under elbow.
Forearm hangs over edge of table.
Hold acromio-clavicular angle in place.
Ask subject for external rotation of arm, elbow flexed at 90°, bringing hand forward and up.
Avoid elevation and rotation of ipsilateral hemi-thorax by holding it in place.
Range of movement should be complete.

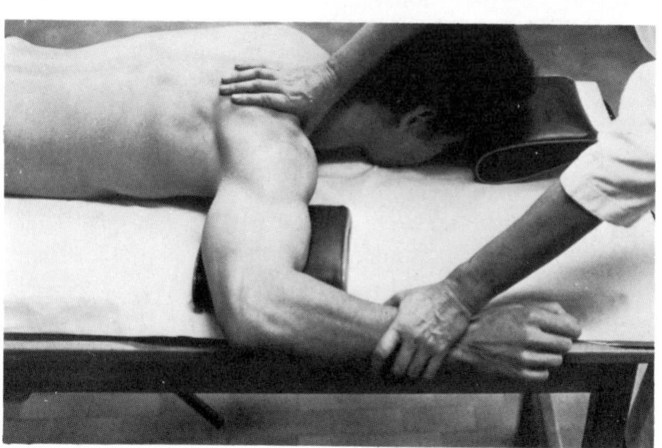

4 and 5 Same position.

Avoid same risks of trick movement.
Hold acromio-clavicular angle in place.
Ask for same movement.
Resistance to the movement is placed on the inferior third of the forearm.

Grade 4: Power is sub-normal.
Grade 5: Power is normal.

PECTORALIS MAJOR

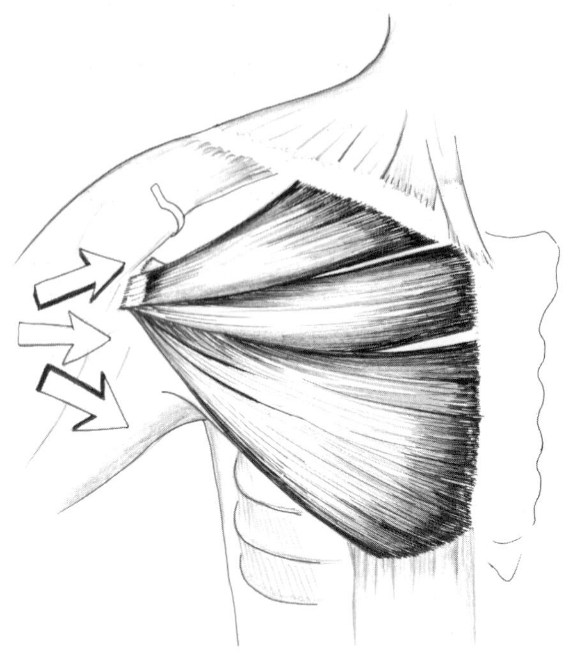

Composed of three portions.

Origin

Upper or clavicular portion:
—Anterior border of medial two-thirds of clavicle.
—Sternal manubrium.
Middle and lower portions:
—Anterior lateral surface of sternum.
—Anterior surface of first five or six costal cartilages.
—Aponeurosis of rectus abdominis.

Insertion Common insertion into anterior lateral bicipital ridge of the humerus.

Innervation

—Medial cord.
—Lateral cord.
—Ansa pectoralis (C5, C6, C7, C8, T1).

Function

With thorax fixed.
—Arm adduction:
Clavicular portion adducts and flexes arm towards contralateral shoulder.
Middle portion adducts arm horizontally.
Lower portion adducts and lowers arm towards contralateral hip. It acts in synergy with pectoralis minor.
—Arm internal rotation:
With humerus fixed.
- Pulls the trunk up. It is a climbing muscle.
- Accessory muscle of inspiration.

Contracture or tightness

Limits abduction and external rotation capacity of arm.
Locks shoulder girdle in forward depression.

Deficiency

Global wasting is rare, selective wasting is frequent.
Deficiency is partially compensated for by other adductors and internal rotators.
Shoulder girdle depression strength is decreased: affects walking with crutches.
Difficulty in hugging loads to front of body.

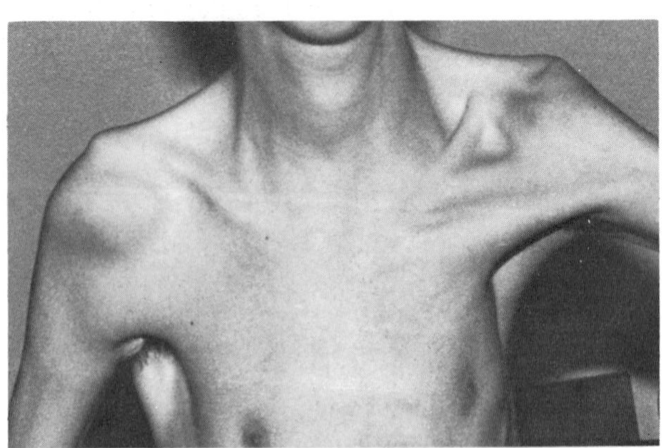

Subject presenting a contracture of middle portion, a partial deficiency of clavicular portion and a wasting of inferior portion.

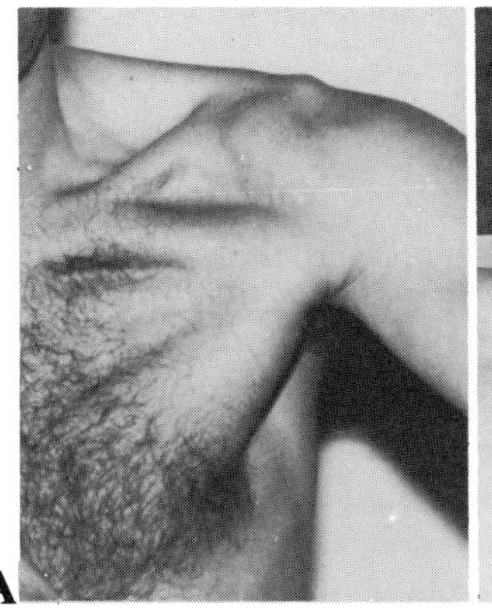

A

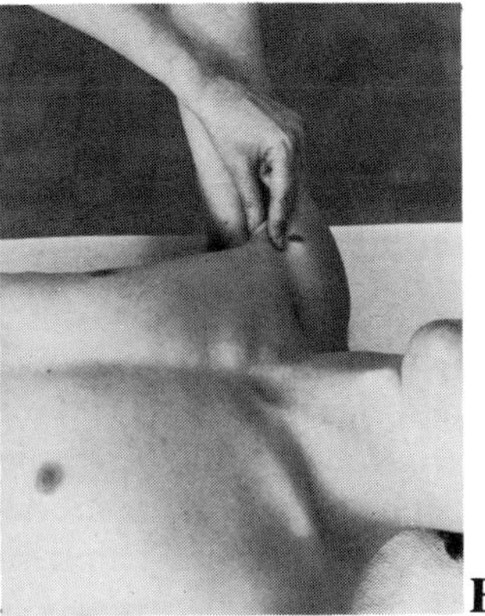

B

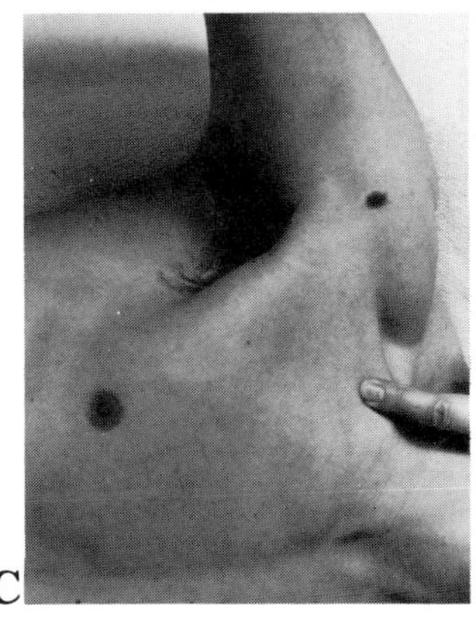

C

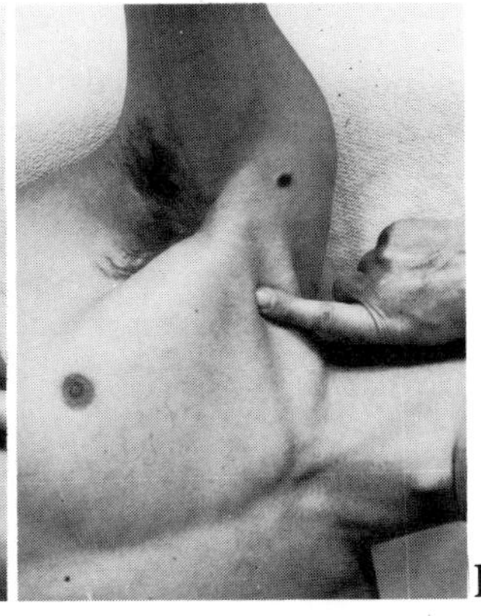

D

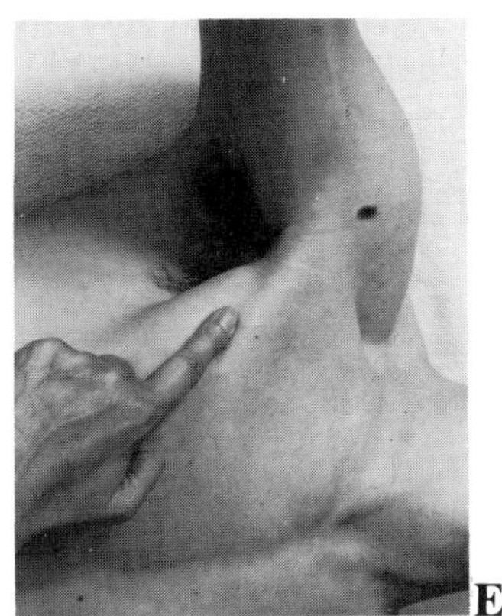

E

0 and 1 Subject supine

Upper limb in abduction.
Ask subject for adduction and internal rotation of upper limb.
Examiner palpates common tendon near the delto-pectoral groove.
Each portion is palpable:
—Clavicular portion: beneath middle of clavicle. Ask subject to flex and adduct his upper limb towards opposite shoulder.
—Middle portion: between sternal manubrium and common terminal tendon. Ask subject for direct arm adduction.
—Lower portion: between sternum and common terminal tendon.
Ask subject to adduct his upper limb towards opposite hip.

Grade 0: No contraction is felt.
Grade 1: Contraction is felt but there is no joint movement.

A—The three portions of pectoralis major.
B—Palpating common terminal tendon.
C—Palpating clavicular portion.
D—Palpating middle portion.
E—Palpating lower portion.

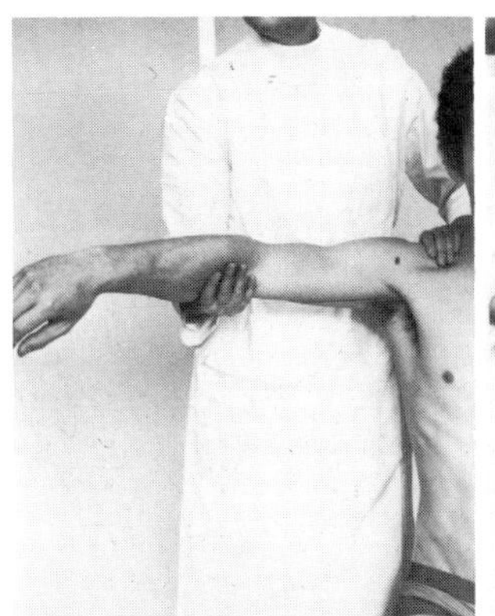

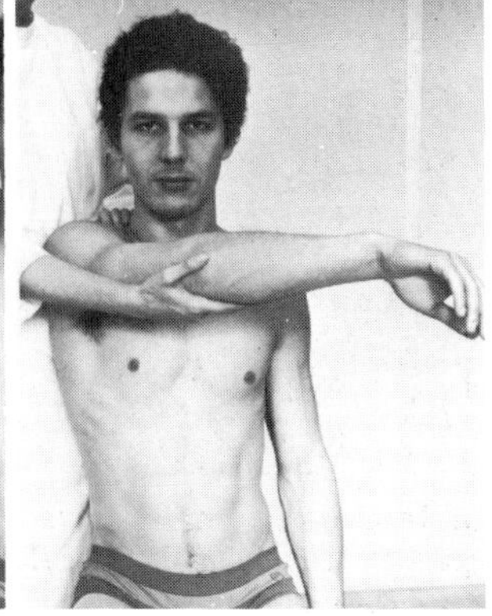

2 Three portions tested together

Subject sitting on a stool.
Hold upper limb in abduction at elbow level.
Hold acromio-clavicular angle in place.
Ask subject for adduction and internal rotation of the upper limb.
Avoid ipsilateral trunk rotation.
Range of movement should be complete.

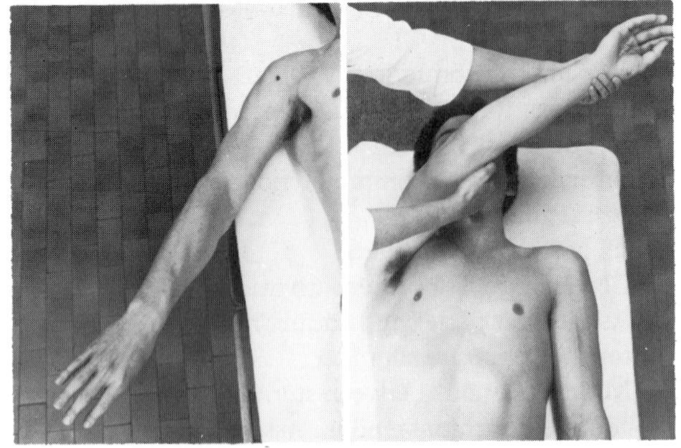

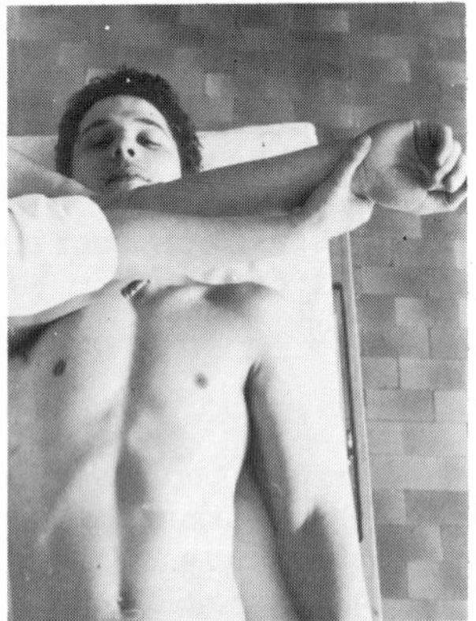

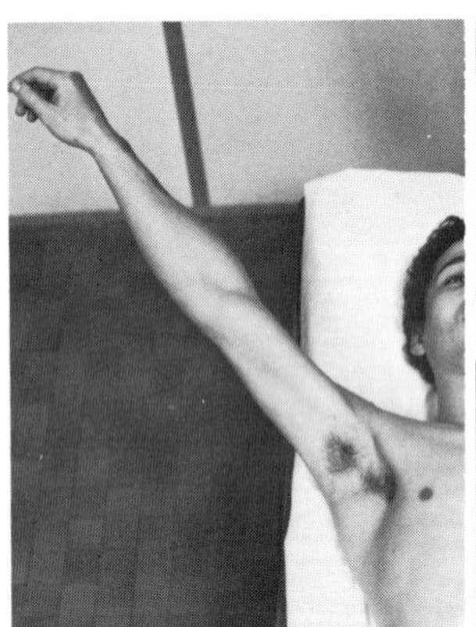

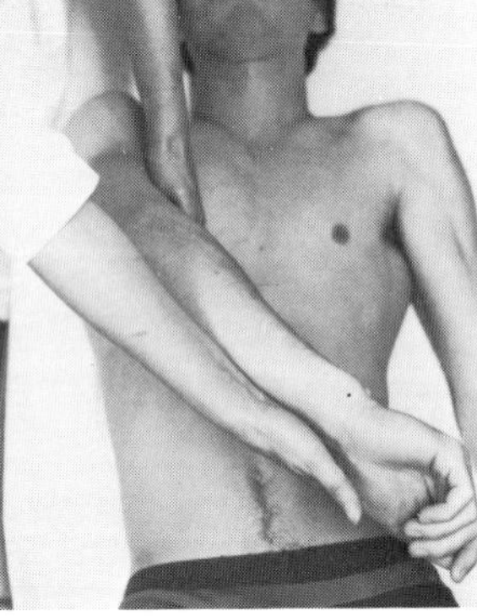

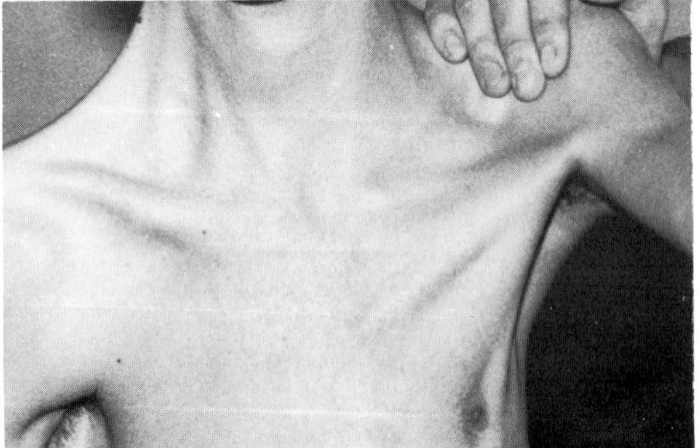

Subject presenting partial deficiency of clavicular portion and contracture of middle and lower portions.

3 Three portions tested separately

Subject supine.

Clavicular portion:

Upper limb abducted at 70° in internal rotation.
Ask subject to adduct upper limb towards opposite shoulder.
From above, support forearm while applying light resistance to anterior aspect of arm since gravity aids movement in most shortened range.
Avoid ipsilateral trunk elevation.
Range of movement should be complete.

Middle portion:

Upper limb abducted at 90° in internal rotation.
Ask subject for straight adduction.
From above, support forearm and apply light resistance to anterior aspect of arm since gravity aids movement in most shortened range.
Avoid same trick movement.
Range of movement should be complete.

Lower portion:

Upper limb abducted at 120° in internal rotation.
Ask subject to adduct upper limb towards opposite hip.
From above, support forearm and apply light resistance to anterior aspect of arm since gravity aids movement in most shortened range.
Avoid same trick movement.
Range of movement should be complete.

4 and 5 Three portions tested separately

Same positions, precautions and movements. Resistance to the movements of the different portions is placed on the inferior third of anterior aspect of the arm.

Grade 4: Power is sub-normal.
Grade 5: Power is normal.

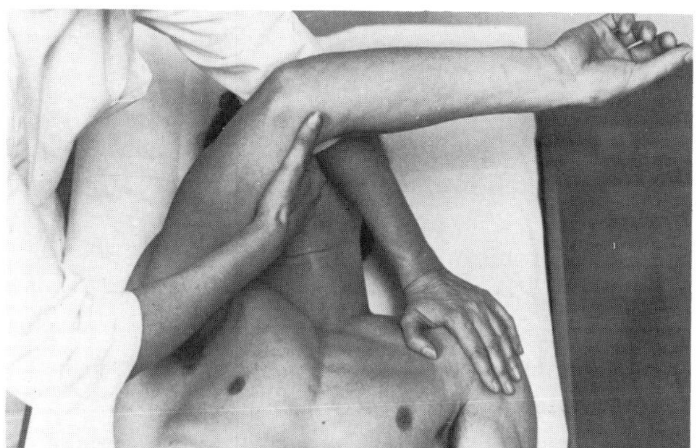

Clavicular portion: hold opposite shoulder in place.

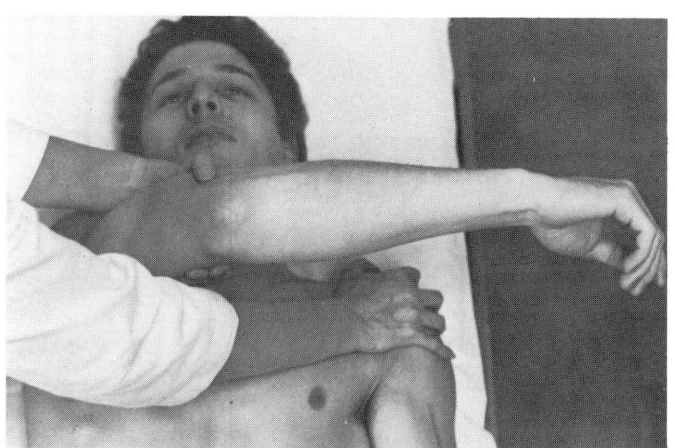

Middle portion: hold opposite shoulder in place.

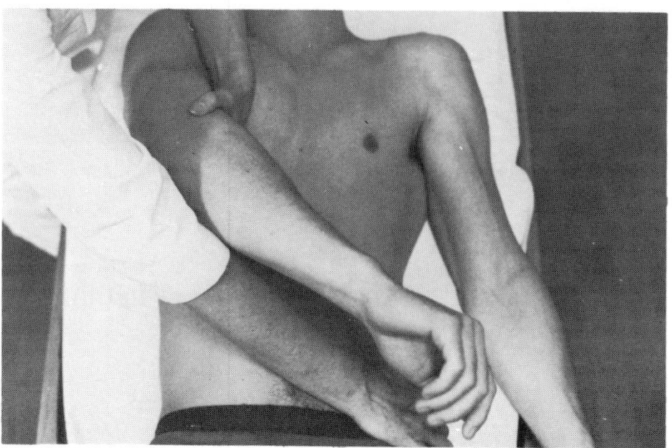

Lower portion: hold opposite iliac crest in place.

TERES MAJOR

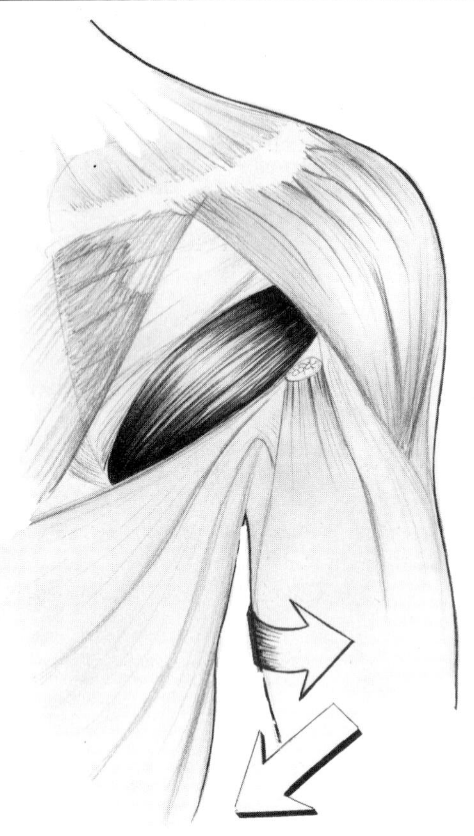

Origin

—Inferior lateral part of infra-spinous fossa of dorsal surface of scapula.
—Intermuscular septa separating it from infraspinatus and teres minor.

Insertion Medial lip of bicipital groove of humerus.

Innervation Posterior cord (C5, C6, c7).

Function

With scapula fixed.
—Arm adduction: this function requires good scapular stabilizers to prevent upward rotation of scapula.
—Internal rotation of arm.
—Arm extension, movement which is limited according to Duchenne de Boulogne.
Teres major participates in arm extension with pectoralis major and latissimus dorsi.
With humerus fixed: it draws the inferior angle of scapula outward and forward.

Contracture or tightness

Limitation of arm abduction and flexion.
Scapula is tilted outward towards the humerus, shoulder girdle remains elevated.

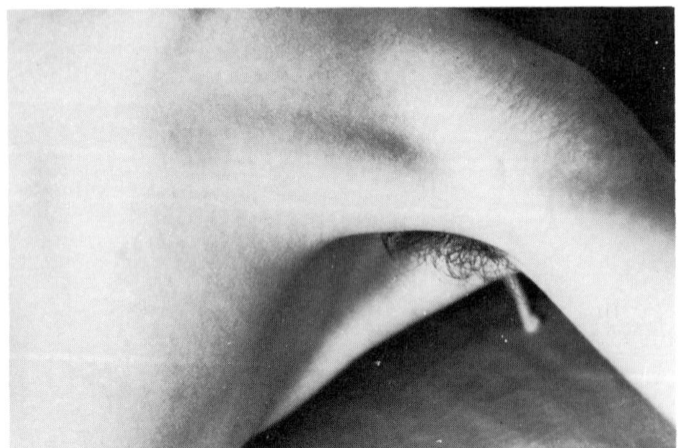

Subject presenting contracture of teres major.

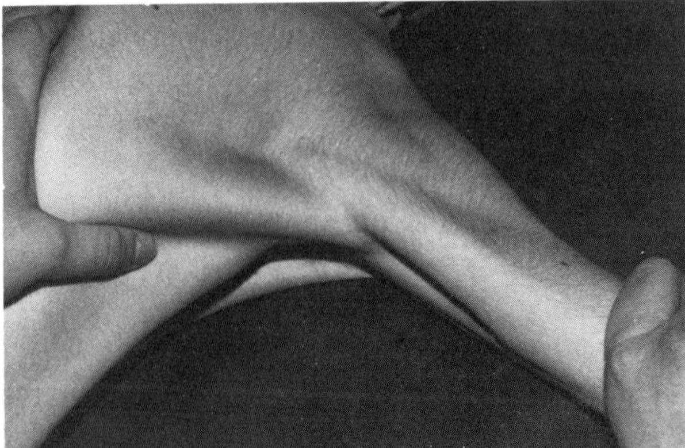

Examination of contracture.

Deficiency

Isolated deficiency is rare. The other depressors, internal rotators and adductors are so powerful that paralysis of teres major does not induce a marked functional impairment.
Gravity makes up for the deficiency.
Its paralysis, combined with that of all the above mentioned muscles causes severe functional hindrance; shoulder depression becomes impossible.

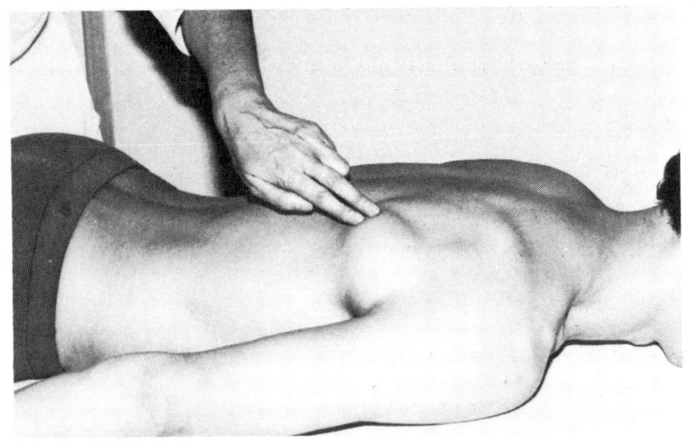

0 and 1 Subject prone.

Upper limb in slight abduction and internal rotation.
Ask subject to adduct upper limb, moving hand towards ipsilateral buttock.
Palpate teres major at lower lateral aspect of posterior surface of scapula, not to be confused with fibres of latissimus dorsi.

Grade 0: No contraction is felt.
Grade 1: Contraction is felt but there is no joint movement.

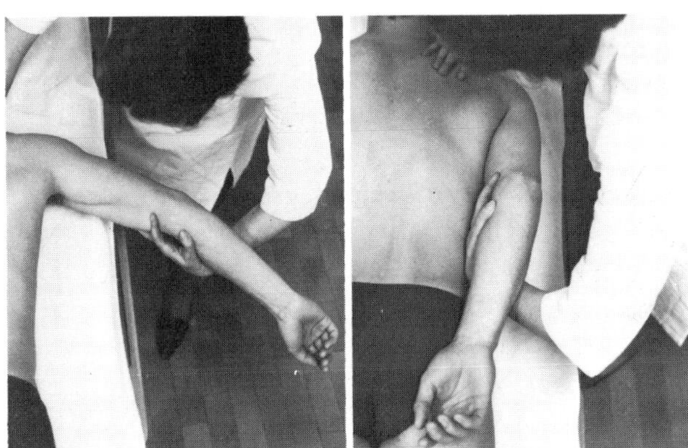

2 Same position.

Hold acromio-clavicular angle in place.
Upper limb abducted at 80°, supported by examiner at elbow level.
Ask for same movement.
Watch for possible trick movements, especially by rhomboids.
Range of movement should be complete.

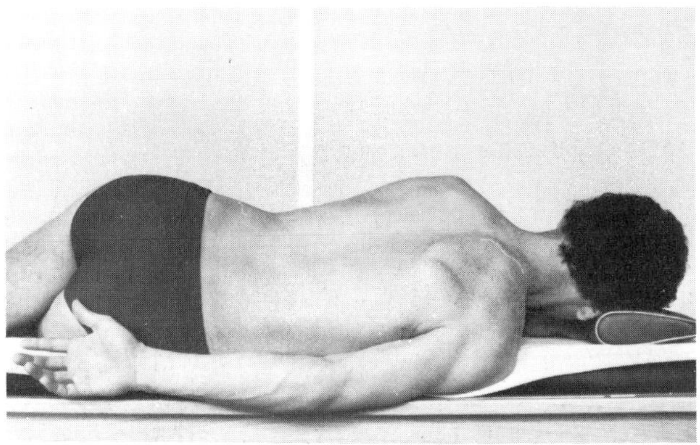

3 Subject side-lying prone.

Upper limb hanging in internal rotation.
Ask subject to bring his hand up to ipsilateral buttock.
Range of movement should be complete.

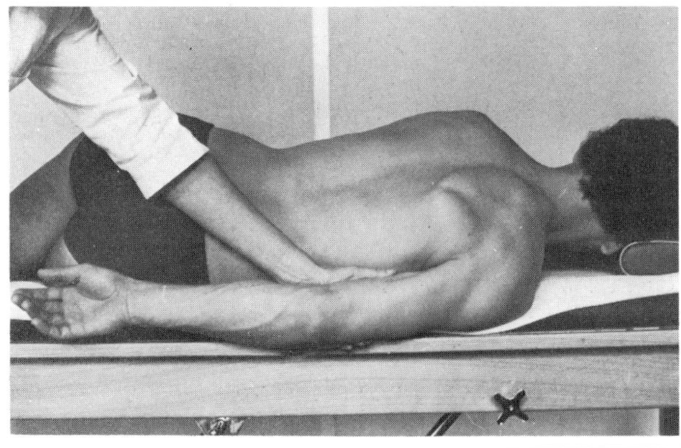

4 and 5 Same position.

Same movement.
Resistance to the movement is placed just over the elbow flexion crease.

Grade 4: Power is sub-normal.
Grade 5: Power is normal.

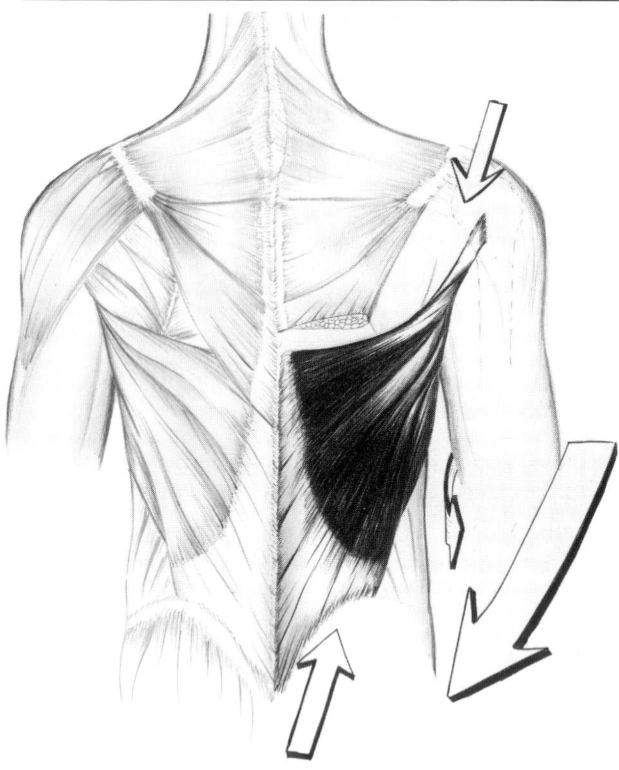

Origin

Through lumbo-sacral aponeurosis from:
—Spinous processes of last six thoracic vertebrae, five lumbar vertebrae, and sacrum.
—Lateral lip of posterior part of ilium.
In muscular fibres from:
—Last three or four ribs.
—Inferior angle of scapula.

Insertion Floor of bicipital groove of humerus.

Innervation Posterior trunk (C6, C7, C8).

Function

With pelvis fixed.
—Superior fibres:
- Extension, adduction and internal rotation of the arm.
- Depresses shoulder, opening the chest forward.

—Inferior and lateral fibres:
- Depression of shoulder girdle,
- Latero-flexion of trunk.

With humerus fixed.
Ipsilateral elevation of pelvis, closing the inter costoiliac space.
Bilateral function.
—Extension of spine ('standing-at-attention' muscle).
—Expands chest forward facilitating inspiration.
Latissimus dorsi, connecting the two shoulder girdles, is essential for walking with crutches (paraplegic's blessing), and for depressing upper extremity (downward stroke, climbing, . . .).
It participates in approximating scapula to spine.

Contracture or tightness Limits arm flexion and abduction, with shoulder girdle depression.

Deficiency Difficulty in bringing the hand towards contralateral buttock.
Difficulty in lifting buttocks from testing table by pushing on hands when subject is sitting.
Severe functional hindrance in walking with crutches, even if other depressors are sound.

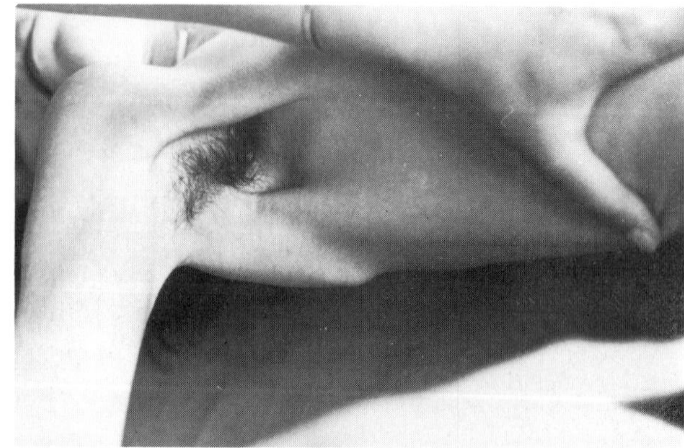

Subject presenting a contracture of latissimus dorsi.

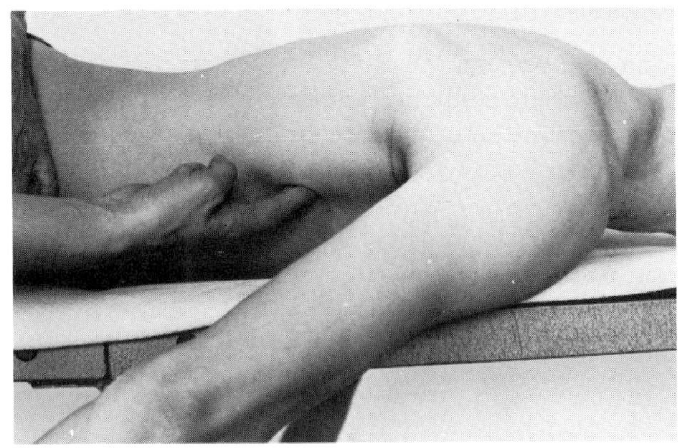

0 and 1 Subject prone.

Upper limb separated from trunk in internal rotation, supported by examiner.
Ask subject to bring his hand towards his contralateral buttock, extending and adducting his arm and depressing the shoulder girdle.
Palpate on the postero-inferior aspect of the shoulder and the lateral portion of the thorax.

Grade 0: No contraction is felt.
Grade 1: Contraction is felt but there is no joint movement.

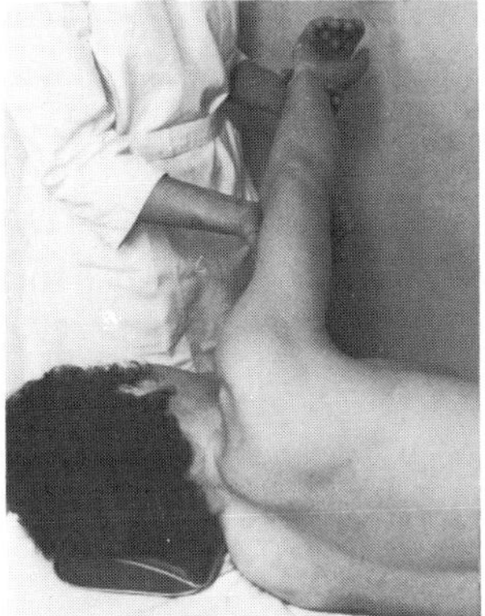

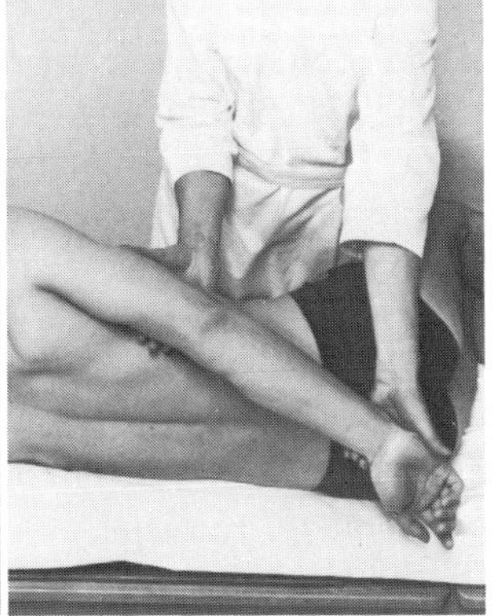

2 *First method.*

Subject side-lying.
Support upper limb in internal rotation.
From a position of flexion and abduction, ask subject to extend and adduct his arm, depressing the shoulder girdle, bringing his hand towards the contralateral buttock.
This test is specifically directed towards the extension component.
Range of movement should be complete.

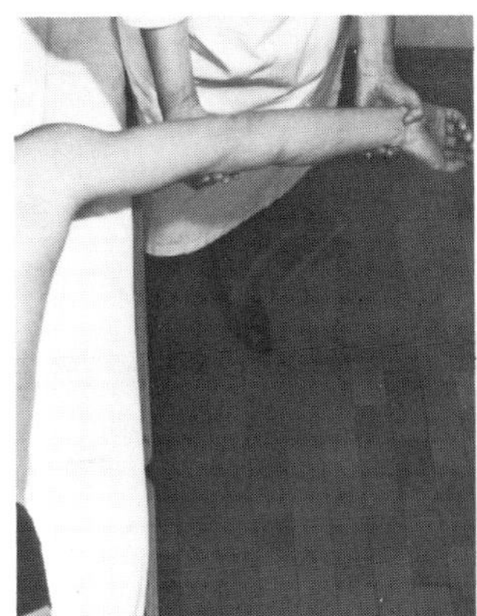

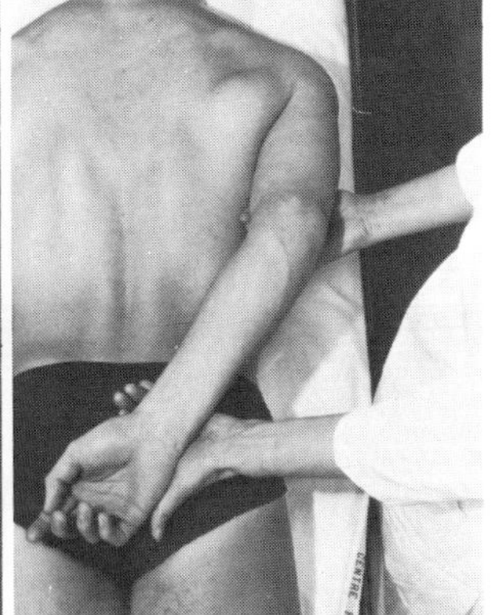

2 *Second method.*

Subject prone.
Support upper limb in internal rotation.
From a position of abduction and slight flexion, ask subject to adduct and extend his upper limb while depressing shoulder girdle and latero-flexing trunk ipsilaterally.
This test is specifically directed towards the adduction component, as the subject's hand moves towards the contralateral buttock.
Range of movement should be complete.

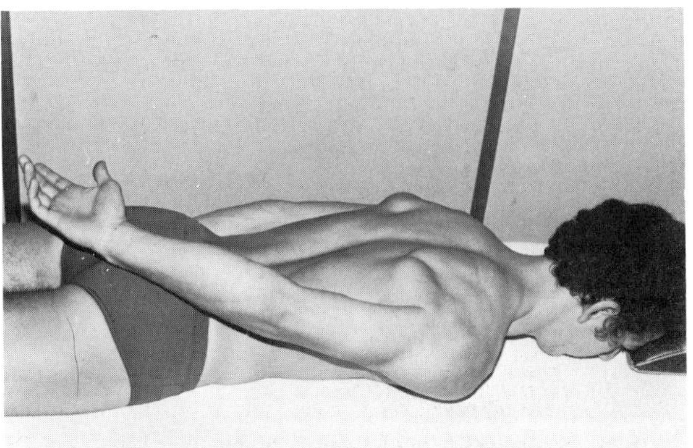

3 Subject prone.

Same starting position, upper limb no longer supported by examiner.
Subject brings his hand to contralateral buttock, shoulder girdle depresses and trunk flexes laterally.
Range of movement should be complete.

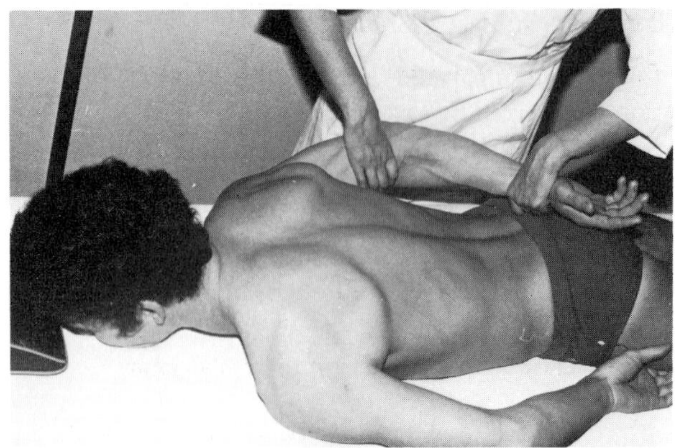

4 and 5 Same position.

Same movement.
Resistance to adduction and extension is placed on inferior third, medial aspect of the arm.
Examiner holds wrist in order to prevent internal rotation.

Grade 4: Power is sub-normal.
Grade 5: Power is normal.

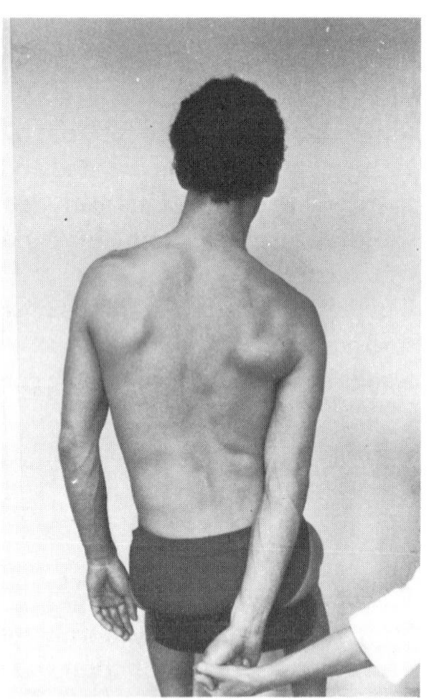

Functional evaluation

Hold shoulder girdle depressed, upper limb stretched in adduction, extension and internal rotation. Trunk is flexed laterally.
Subject clenches his fist.
Against clenched fist, examiner applies resistance to the movement.

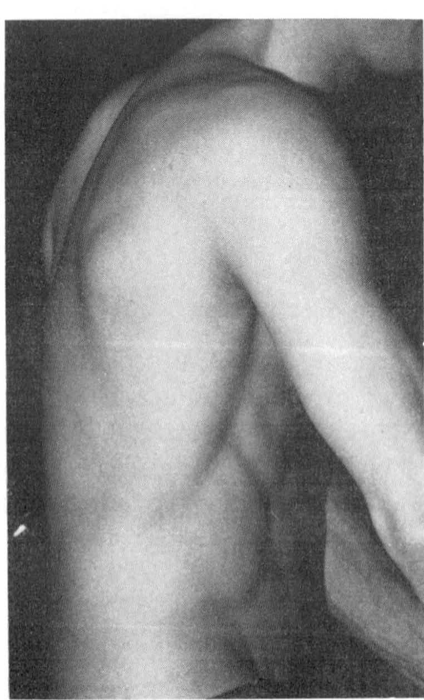

Latissimus dorsi in contraction.

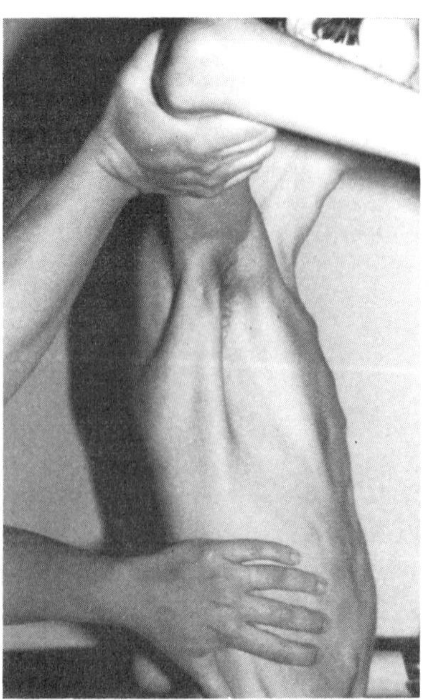

Showing contracture of latissimus dorsi

NOTES

BICEPS BRACHII

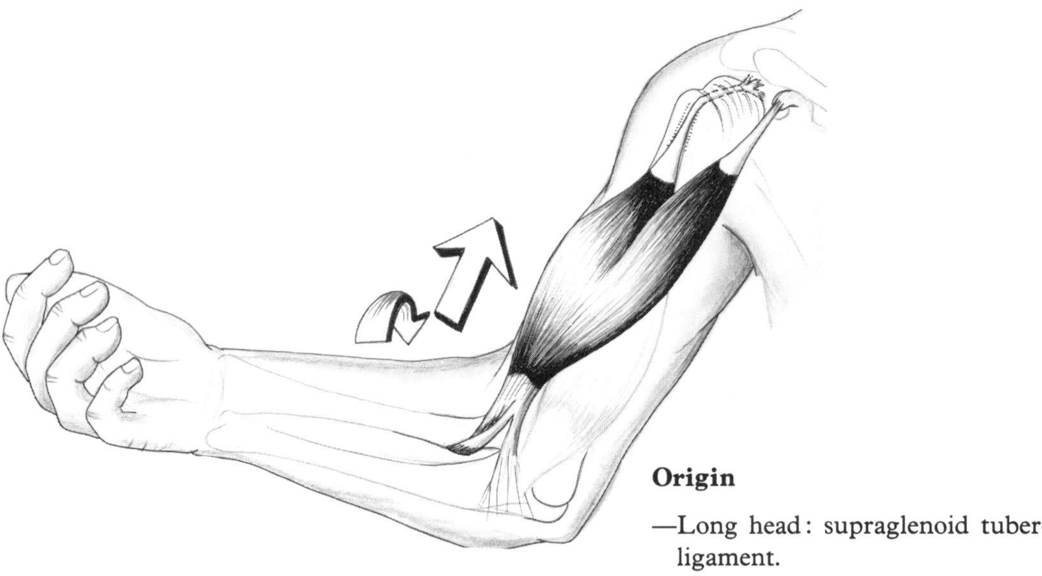

Origin

—Long head: supraglenoid tuberosity and glenoid ligament.
—Short head: apex of coracoid process of scapula by tendon common with coracobrachialis.

Insertion Common on bicipital tuberosity of radius and through the lacertus fibrosus to medial part of antebrachial aponeurosis.

Innervation Musculocutaneous (C5, C6).

Function

Flexion of the forearm on the arm.
Forearm supination: when the supinator is paralysed, supination is possible because of the biceps; its action is more important when the elbow is flexed at 90°.
Long head stabilizes shoulder joint by retaining humeral head in glenoid cavity.
Maximal efficiency of elbow flexion is at about 90°; the force of the biceps at that point coincides with the tangential force.
With forearm fixed, elbow flexors can approximate the arm and the entire body towards the forearm (pull-ups).

Contracture or tightness Flexum deformity of the elbow.

Deficiency

Often associated with that of the brachialis.
When biceps is paralysed the humeral head separates itself from the glenoid cavity during resisted exercises.
Functional impairment is severe since elbow flexors participate in all activities of daily living. Wasting of biceps seriously disrupts all combined patterns of movement: pulling, screwing, opening a door, . . .
Carrying and lifting can trigger pain.

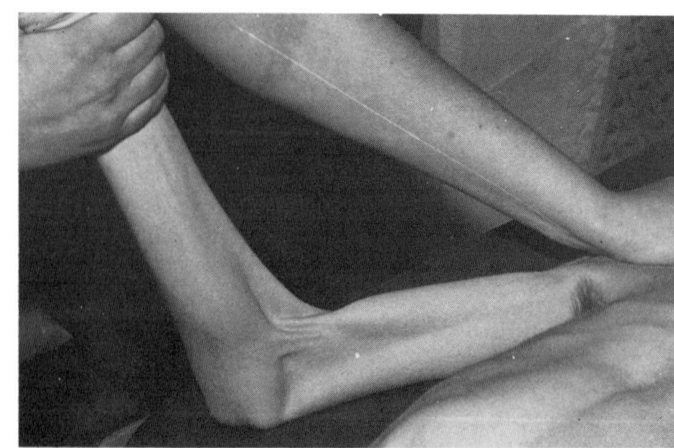

Subject presenting contracture of biceps brachii and brachio radialis.

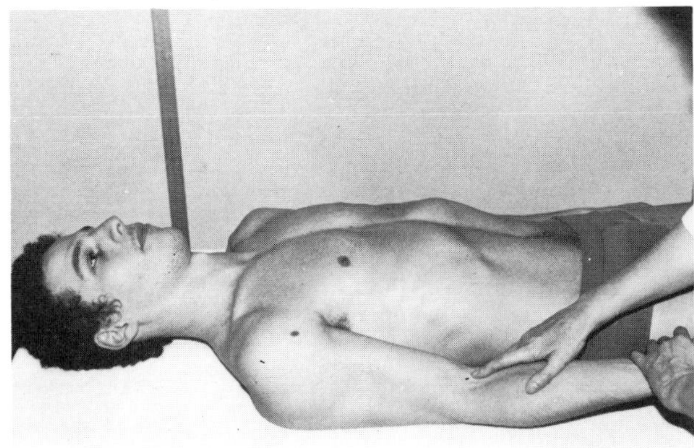

0 and 1 Subject supine, or sitting (not shown).

Arm along side of body.
Hold forearm in slight flexion.
Ask subject to flex his forearm on his arm, supinating the forearm.
Palpate tendon in elbow flexion crease, muscle belly on anterior surface of arm (not shown).

Grade 0: No contraction is felt. No tendon excursion is detected.

Grade 1: Contraction is felt but there is no joint movement.

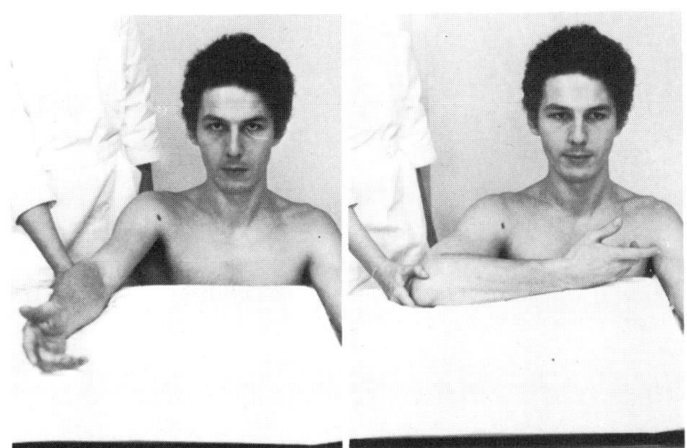

2 Subject sitting.

Arm resting on a sliding board in flexion and slight abduction.
Forearm in supination.
Hold lower extremity of humerus in order to avoid all shoulder movement, particularly extension which could give an illusion of flexion.
Wrist should remain relaxed.
Ask subject to flex his forearm on his arm while maintaining supination.
Range of movement should be complete.

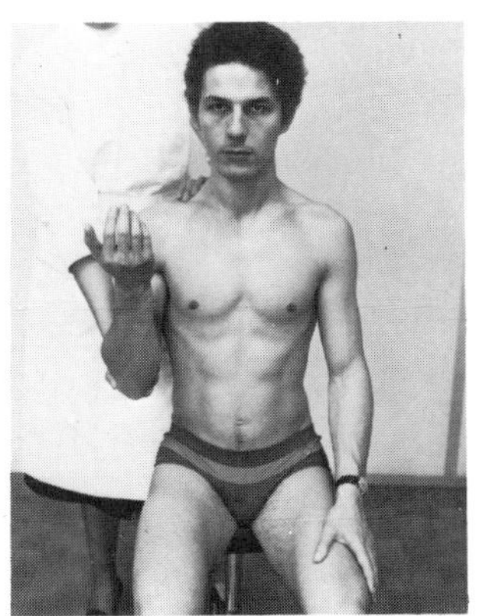

3 Subject sitting on a stool.

Arm along side of body.
Hold acromio-clavicular angle and lower extremity of humerus in place.
Wrist should remain relaxed throughout the movement.
Ask subject to flex his forearm on his arm while supinating.
Range of movement should be complete.

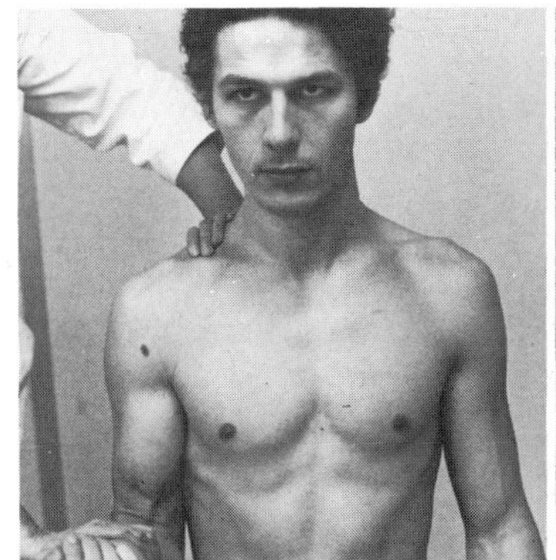

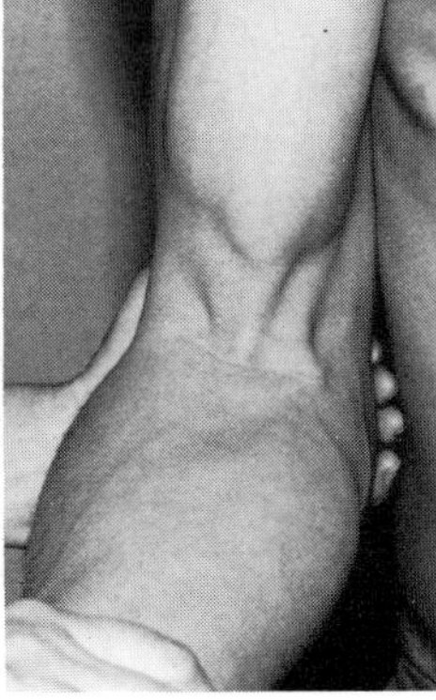

4 and 5 Same position.

Hold acromio-clavicular angle in place.
Same precaution and movement.
Resistance to flexion and supination is placed on the lower third, anterior aspect of the forearm.

Grade 4: Power is sub-normal.
Grade 5: Power is normal.

Watch for protrusion of biceps' tendon and its expansion oriented towards the medial portion of antebrachial aponeurosis.

BRACHIALIS

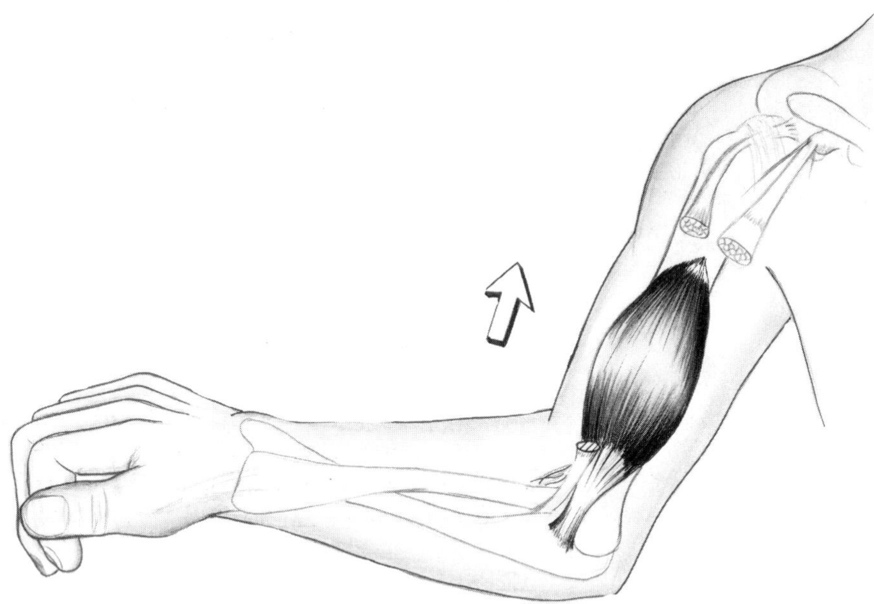

Origin

—Lower third, anterior and lateral surfaces of humerus.
—Medial and lateral intermuscular septa.

Insertion

—Anterior inferior surface of the coronoid process of the ulna.
—Expansion into lateral aspect of forearm aponeurosis.

Innervation

—Musculocutaneous (C5, C6).
—Sometimes a branch of the radial.

Function

Flexion of the forearm on the arm independently of forearm position.
It has no action in pronation or supination.
Maximal efficiency of combined elbow flexors is at about 90°.
Brachialis is maximally efficient at about 135° of elbow flexion.

Deficiency Decreases the power of elbow flexion.

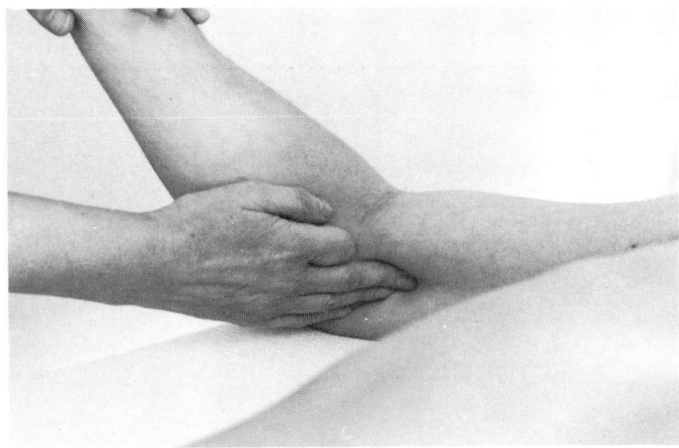

0 and 1 Subject supine, or sitting (not shown).

Place forearm in pronation.
Ask subject to flex his forearm on his arm, avoiding supination. Palpate brachialis from internal aspect of arm medially to and under the biceps' tendon.

Grade 0: No contraction is felt.
Grade 1: Contraction is felt but there is no joint movement.
Watch for supination.

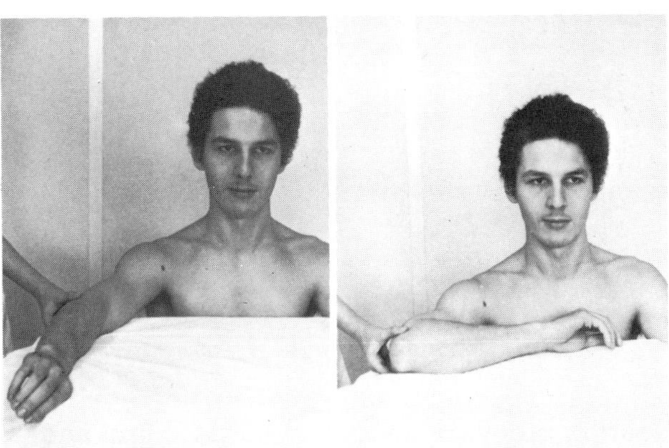

2 Subject sitting.

Upper extremity to be examined rests on a sliding board, arm in flexion and slight abduction, elbow extended, forearm in pronation.
Hold lower extremity of humerus so as to avoid all shoulder movement, particularly extension which could give an impression of elbow flexion.
The wrist must remain relaxed.
Ask subject to flex his forearm on his arm, avoiding supination.
Range of movement should be complete.

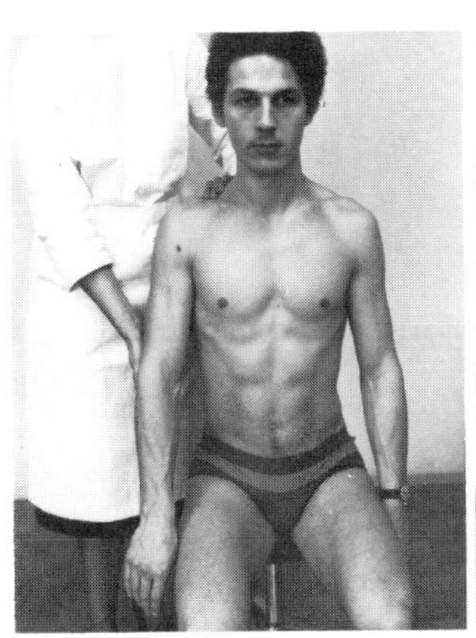

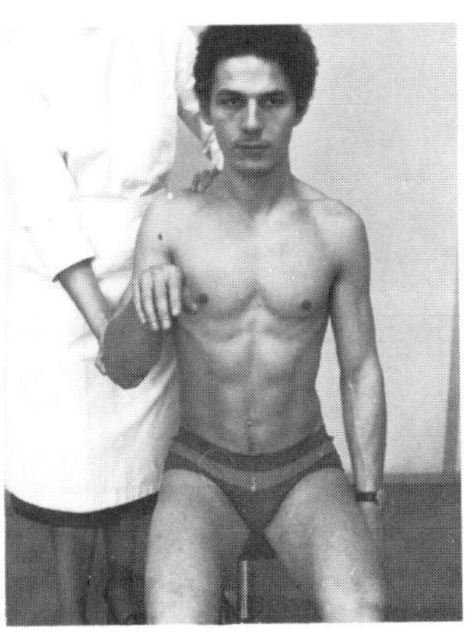

3 Subject sitting.

Arm along side of body, elbow extended, forearm in pronation.
Hold acromio-clavicular angle in order to avoid all shoulder movement.
Hold lower extremity of humerus.
Subject flexes his forearm on his arm, the forearm in pronation to eliminate supination component (to be watched for throughout the movement).
Range of movement should be complete.

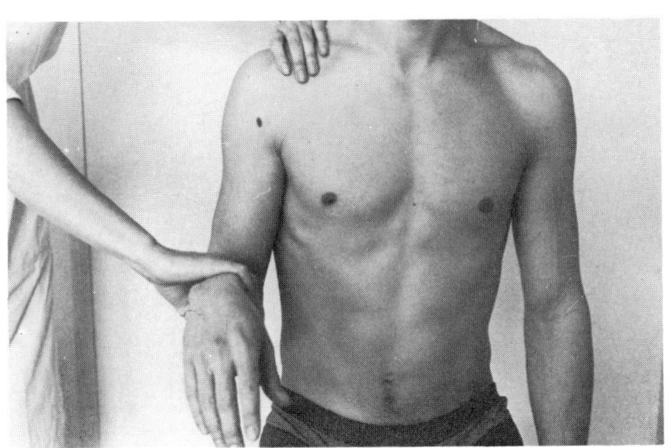

4 and 5 Same position.

Hold lower extremity of humerus.
Same precautions and movement.
Resistance to the movement is placed on inferior third of the forearm.

Grade 4: Power is sub-normal.
Grade 5: Power is normal.

BRACHIO RADIALIS

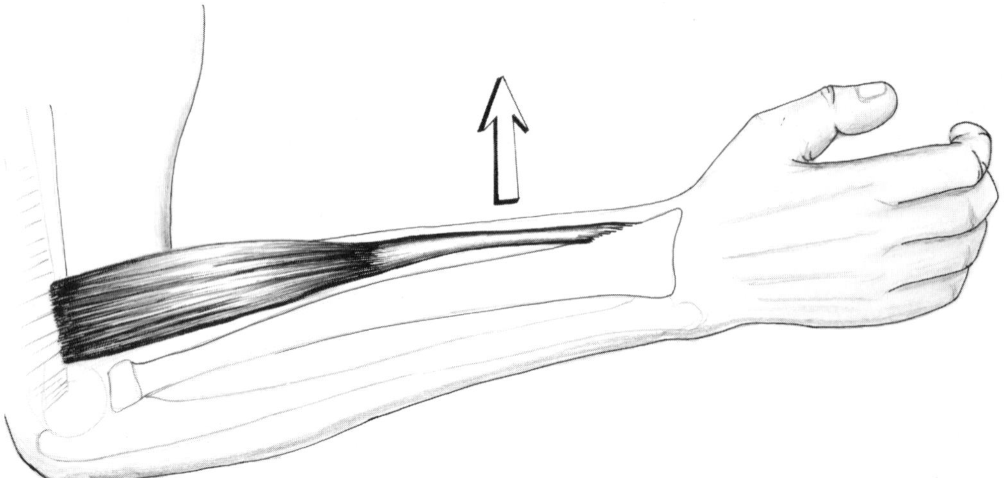

Origin

—Lower third of lateral border of humerus above extensor carpi radialis longus,
—Lateral intermuscular septum.

Insertion Base of styloid process of radius, at upper part of groove of abductor pollicis longus and extensor pollicis brevis.
Brachio radialis tendon forms the lateral margin of the radial artery groove (pulse groove).

Innervation Radial (C5, C6).

Function

Flexion of the forearm on the arm in mid prono-supination.
According to Kapandji, its maximal flexion force is at 100° to 110° of elbow flexion, when its force coincides with the tangential force.
When forearm is in full pronation, it becomes supinator.
When forearm is in full supination, it becomes pronator.
When forearm is fixed, brachio radialis plays an important role in flexing the forearm on the arm. It stabilizes the elbow joint: static role.

Contracture or tightness

Rare, nearly always associated with that of the biceps. It participates in locking the elbow into a flexum deformity.

Deficiency Decreases strength of elbow flexion.

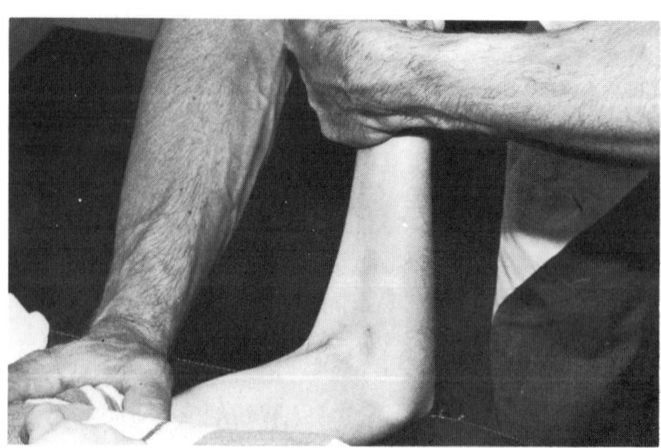

Subject presenting a contracture of brachio radialis.

Because of its important stabilizing role, brachio radialis may be tested dynamically or statically.

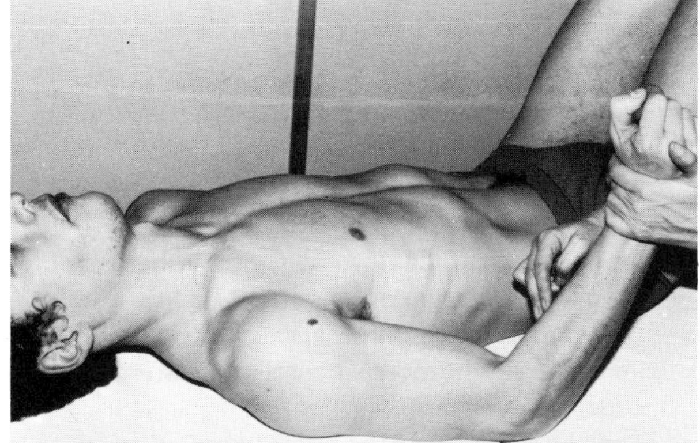

0 and 1 Subject supine, or seated (not shown).

Arm along side of body, forearm in mid prono-supination, elbow in intermediate flexion.
Ask subject to flex his forearm on his arm.
Palpate on anterior lateral portion of the forearm.
The movement may be made by biceps brachii and brachialis in spite of deficiency of brachio radialis.

Grade 0: No contraction is felt.
Grade 1: Contraction is felt but there is no joint movement.

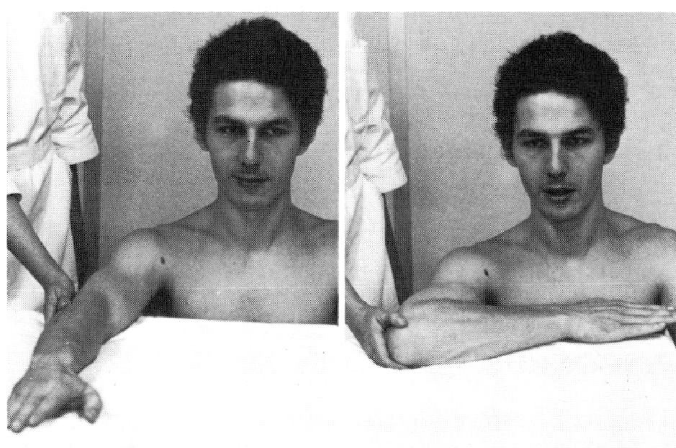

2 Subject sitting.

Arm to be tested rests on a sliding board, elbow extended, forearm in mid prono-supination.
Hold lower extremity of humerus in place.
Ask subject to flex forearm on arm, keeping forearm in mid prono-supination.
Verify contraction of brachio radialis with palpation.
Range of movement should be complete.

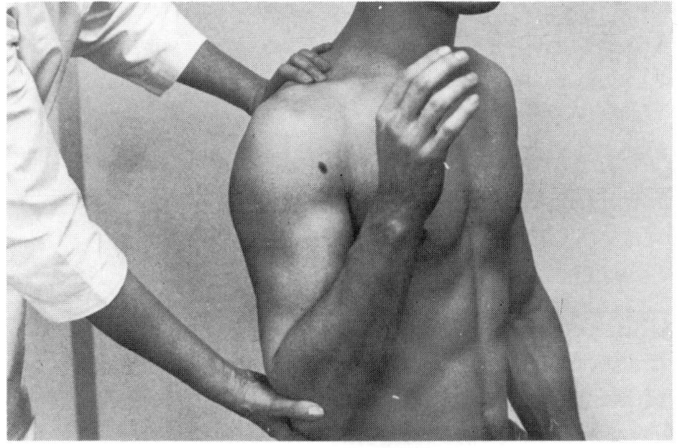

3 Subject sitting.

Arm along side of body, elbow against trunk, forearm hanging in mid prono-supination.
Hold acromio-clavicular angle and lower extremity of humerus in place.
Ask subject to flex his forearm on arm, keeping forearm in mid prono-supination.
Range of movement should be complete.

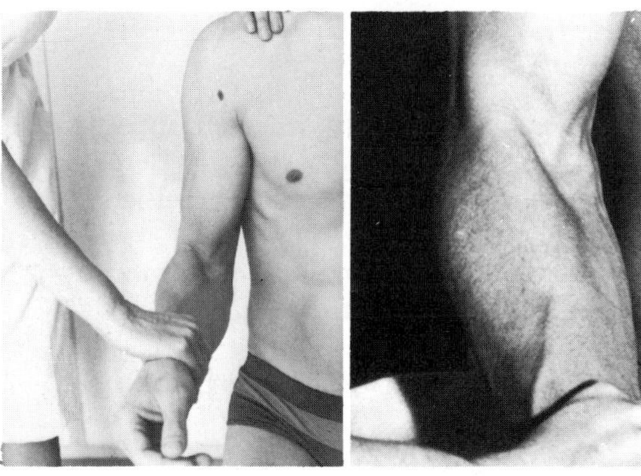

4 and 5 Same position.

Hold lower extremity of humerus.
Ask for same movement.
Resistance to the movement is placed on the lower third, lateral portion of forearm.
Note: This muscle can be tested with dynamic or static tests (at about 100° of elbow flexion).

Grade 4: Power is sub-normal.
Grade 5: Power is normal.

TRICEPS BRACHII AND ANCONEUS

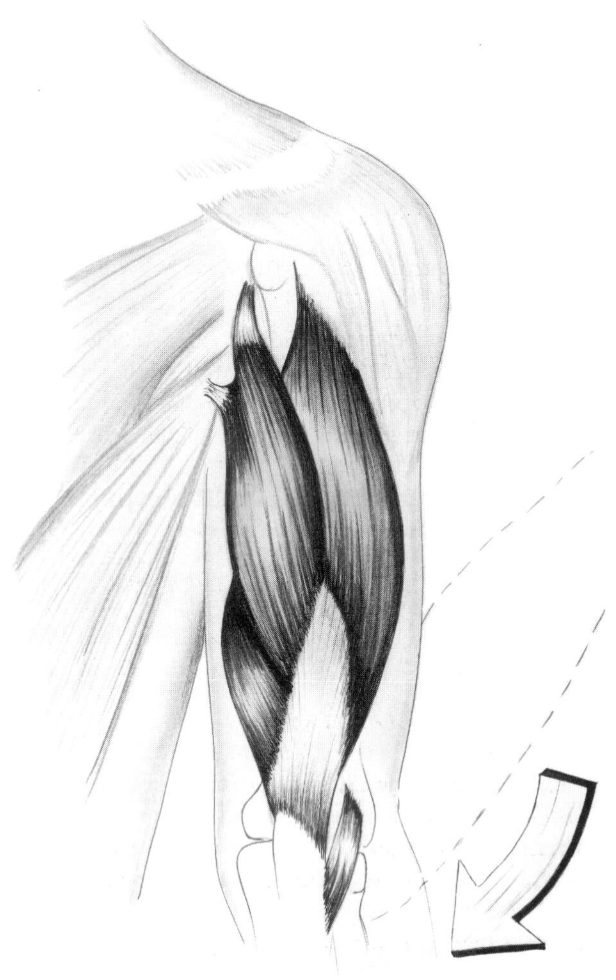

TRICEPS BRACHII

Composed of three heads.

Origin

—Long head (scapular): infraglenoid tuberosity of scapula, and tendon of latissimus dorsi.
—Lateral head (long humeral): upper half of posterior surface of humeral shaft close to lateral border.
—Medial head (short humeral): lower posterior surface of humerus between the radial nerve groove and the medial and lateral intermuscular septa.

Insertion Common through tricipital tendon into posterior part of upper surface of olecranon and its lateral and medial surfaces, and antebrachial aponeurosis.

ANCONEUS

Origin Posterior surface of lateral epicondyle of humerus.

Insertion Upper fourth of posterior surface of ulna and lateral surface of olecranon.

Innervation

Both are innervated by the radial (c6, C7, c8).

Function

Both extend the forearm on the arm. Each head independently of the others can extend the elbow. However, lateral and medial heads are more efficient extensors than the long head.

The long head takes part in retaining humeral head in glenoid cavity, especially when simultaneously lowering arm and extending elbow strongly (hammering . . .).
It also participates in arm extension and stabilizes the arm synergistically with coraco-brachialis during adduction.
The long head is thus an active ligament preventing forward and outward humeral head dislocation.
Flexion of the arm improves triceps function during extension of forearm on the arm.
Triceps is essential in walking with crutches as well as for all pushing activities.
Its efficiency varies with elbow flexion, maximal force being at about 20° to 30°. At this point the muscular pull coincides with the line of tangential force.
Anconeus tightens the elbow joint capsule.

Deficiency

Functional impairment is important even for gestures requiring little strength. However, in certain arm positions gravity may help.
Precision movements are hindered by the absence of triceps' moderating action, particularly fast movements: the hand cannot stop with precision when reaching for or manipulating objects.
Walking with crutches and all pushing activities are impossible.

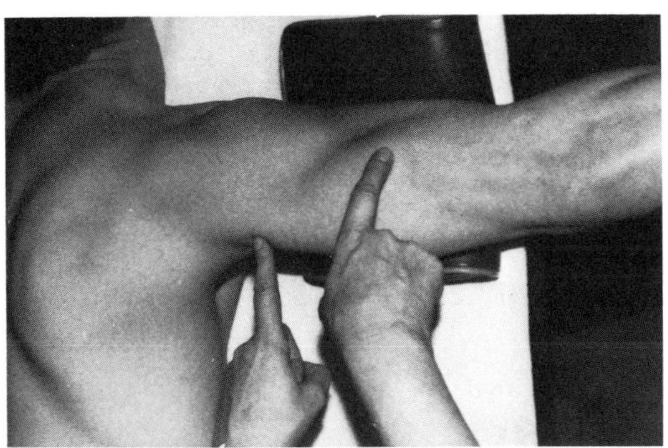

0 and 1 Subject prone, or sitting (not shown).

Arm in abduction, forearm hanging over edge of table.
Cushion under lower portion of arm.
Ask subject to extend the forearm on the arm.
Examiner can palpate tricipital tendon.
The different heads may also be palpated (photograph illustrates palpation of long and lateral heads).
Anconeus is palpated laterally to the inferior portion of tricipital tendon.

Grade 0: No contraction is felt.
Grade 1: Contraction is felt but there is no joint movement.

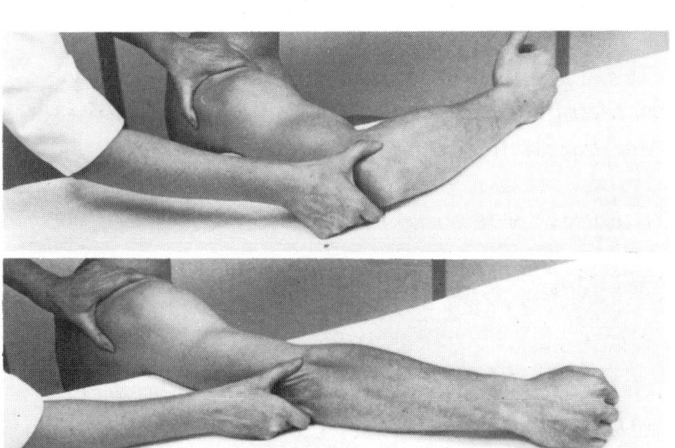

2 Subject sitting.

Arm resting in abduction on table.
Elbow flexed.
Hold acromio-clavicular angle and lower extremity of humerus in place.
Ask subject to extend his forearm on the arm.
During movement, forearm should remain in mid prono-supination.
Range of movement should be complete.

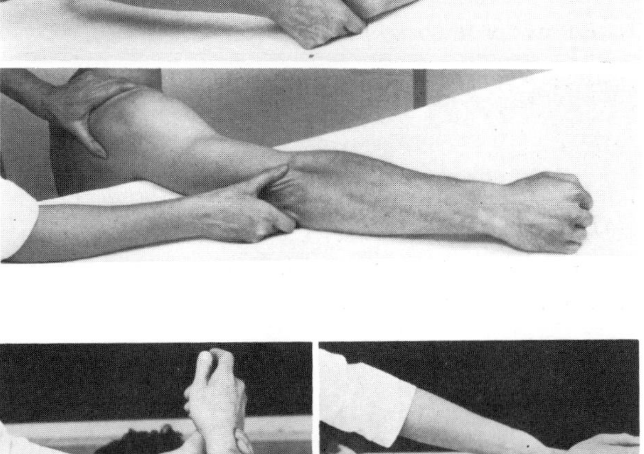

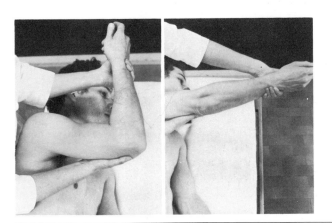

2 *Alternative.*

Subject side-lying.
Arm in flexion: this position facilitates extension of forearm on the arm.
Support arm, holding it in this position throughout the test.
Support forearm and ask subject to extend his forearm on his arm.
Range of movement should be complete.

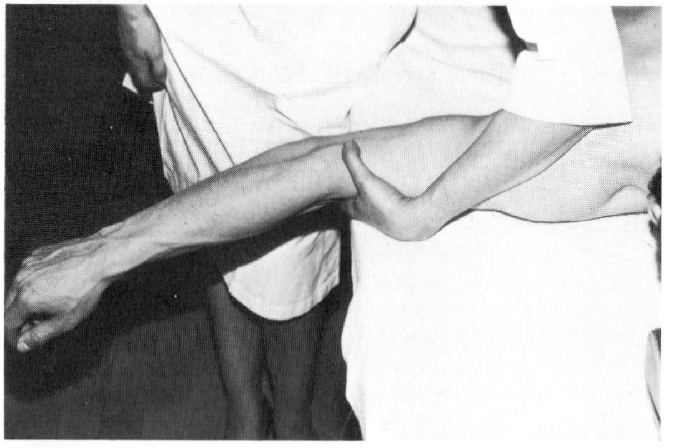

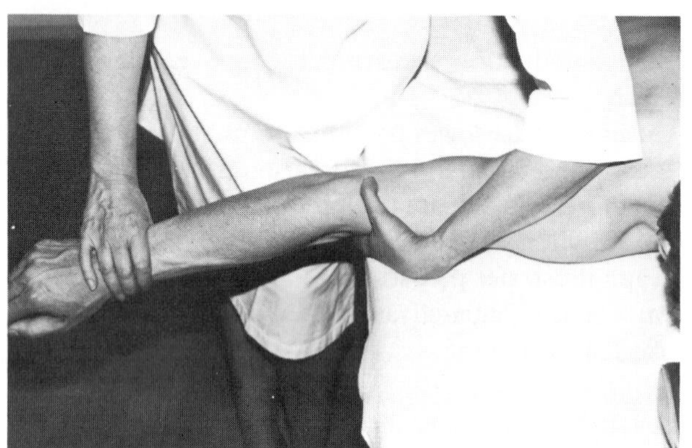

3 Subject prone.

Arm in abduction, forearm hanging over edge of table.
Examiner cradles lower third of arm holding shoulder in place with his own forearm.
Ask subject to extend his forearm on his arm.
Forearm should stay in mid prono-supination during the movement.
Range of movement should be complete.

4 and 5 Same position.

Same hold and movement.
Resistance to the movement is placed on the lower third of the forearm.

Grade 4: Power is sub-normal.
Grade 5: Power is normal.

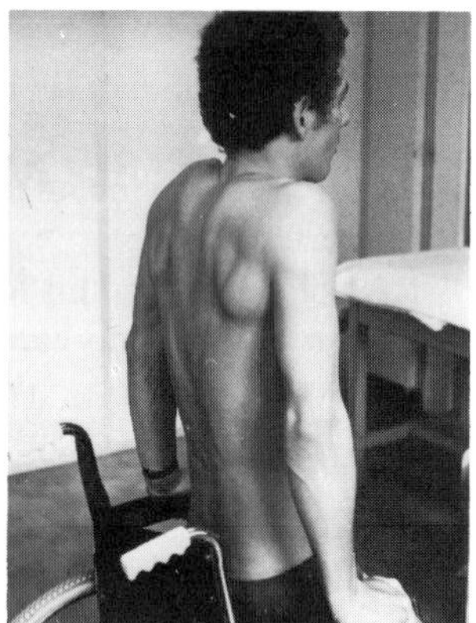

Functional evaluation

Global force of the triceps may be tested in an instant by asking the subject to lift himself from his seat by pushing on the arm rests. He must extend his elbows, his shoulders remaining relaxed in order not to use his depressors (shoulder girdle rises).

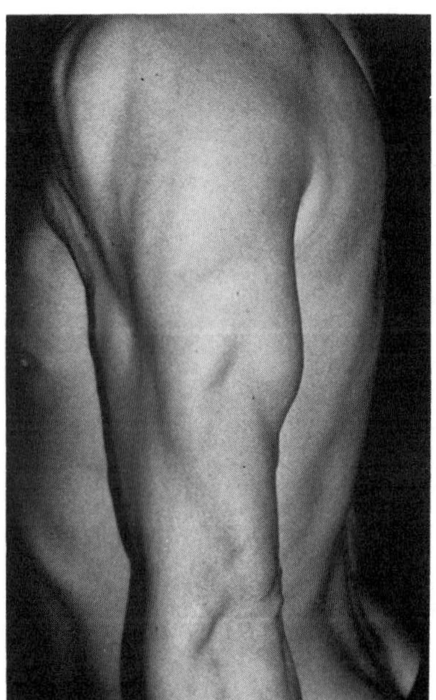

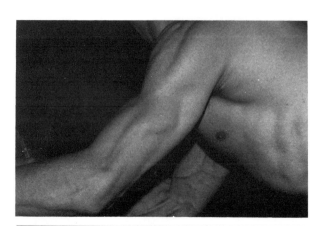

Triceps in contraction.

NOTES

SUPINATOR

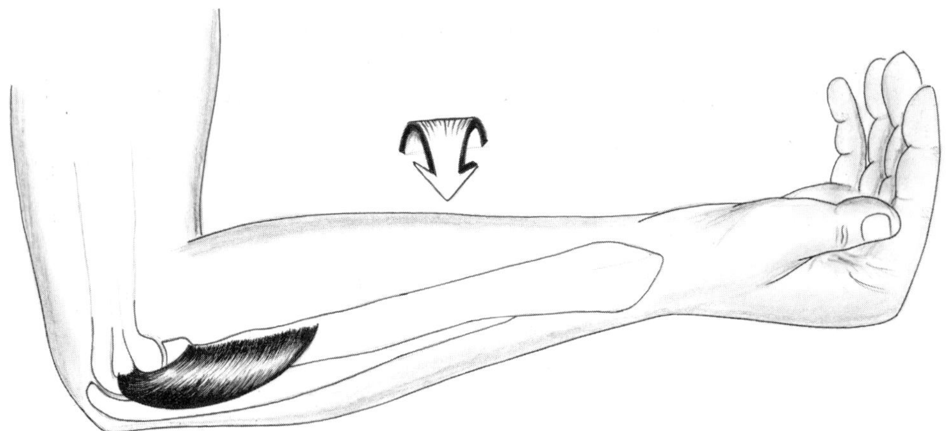

Composed of two heads.

Origin

Superficial head:
—Common extensor tendon on anterior surface of lateral epicondyle of humerus.
—Supinator crest of ulna.
Deep head:
—Bicipital fossa of ulna.

Insertion

Superficial head:
—Oblique line of anterior border of the radius.
Deep head:
—Posterior, lateral, and anterior surfaces of neck of radius.

Innervation Radial (C5, C6, c7).

Function

Rotation of forearm from inward to outward, supinating independently of elbow position.
It is assisted by biceps brachii.
When supination does not require important muscular effort, supinator contracts with biceps.
Against heavy resistance, shoulder joint adductors and external rotators assist supination, contracting synergistically.
The supinator is the only muscle able to produce last few degrees of supination.

Deficiency

Severe functional impairment due to loss of supination, partially made up for by biceps brachii and the shoulder muscles.

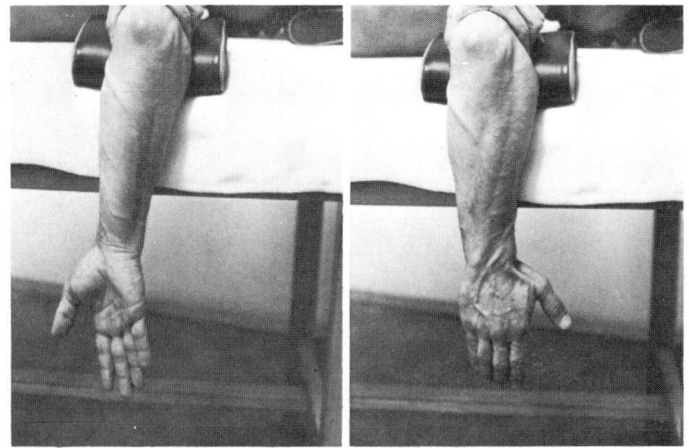

0 and 1

The supinator is a deep muscle and is impalpable except when the muscles covering it have wasted.

2 Subject prone.

Arm in abduction.
Forearm hanging over edge of table.
Cushion under elbow.
Hold inferior portion of posterior aspect of arm in place.
From a position of forearm pronation, ask subject to supinate.
Range of movement should be complete.

2 *Alternative* (not shown).

Subject supine.
Elbow against body.
Forearm vertical.
Ask subject to supinate forearm from a position of pronation (hand-puppet-like movements).
Range of movement should be complete.

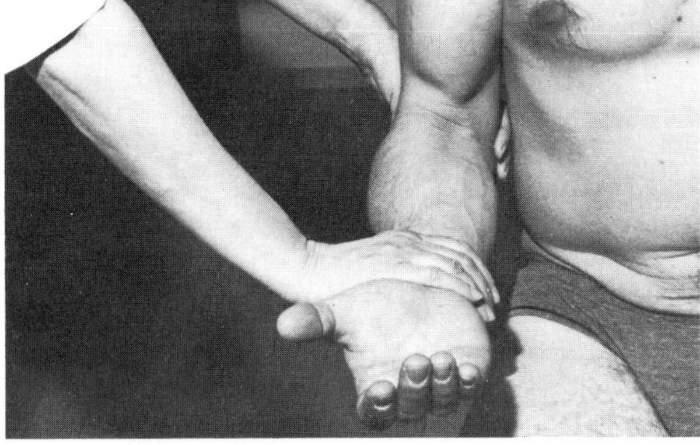

3, 4 and 5 Subject sitting on a stool.

Arm held against body.
Hold lower extremity of humerus in place.
Elbow flexed at 90°.
Resistance to the movement is placed on lower third of forearm.
Resistance is minimal for grade 3 and increases for grades 4 and 5.

Grade 4: Power is sub-normal.
Grade 5: Power is normal.

Alternatives.

For the various grades, it is possible to avoid the supinating component of biceps by flexing or extending the elbow completely.

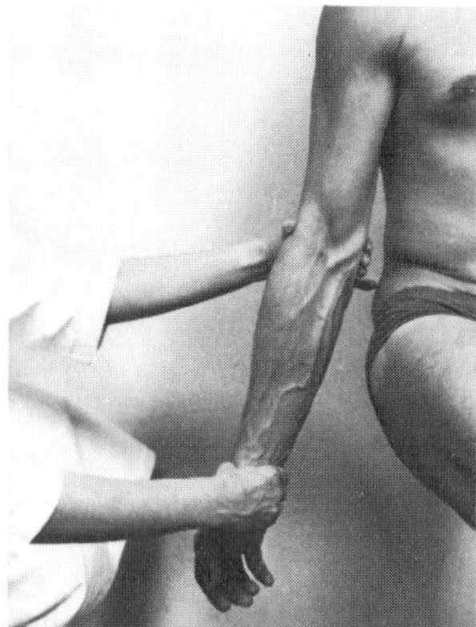

Elbow extended: Grades 3, 4 and 5.

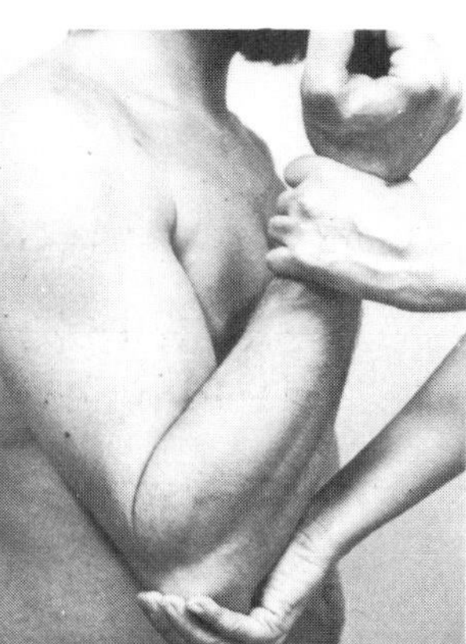

Elbow flexed: Grades 3, 4, 5.

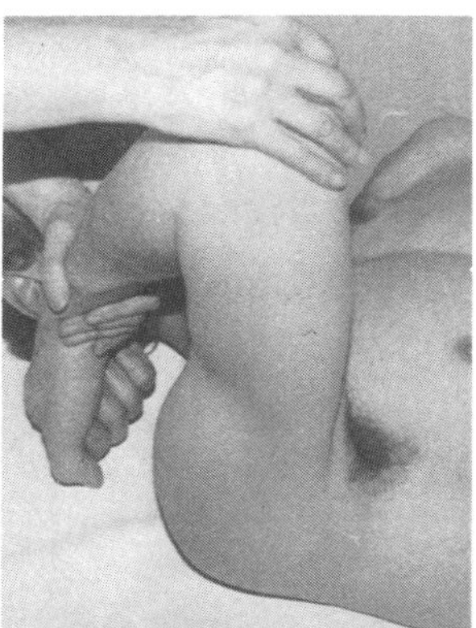

Elbow flexed: Grades 3, 4 and 5.

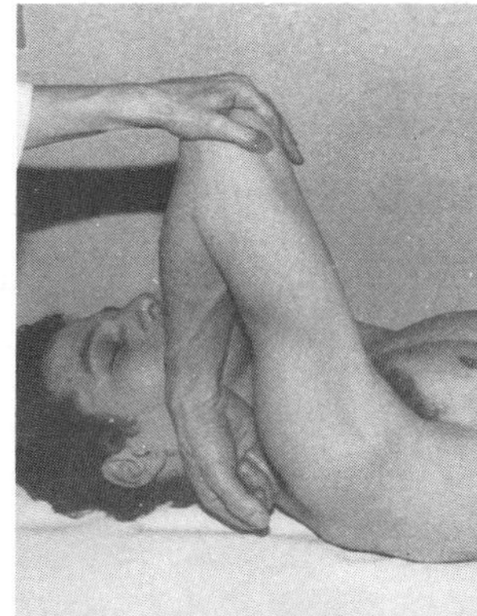

Elbow flexed: Grade 2.

PRONATOR TERES and PRONATOR QUADRATUS

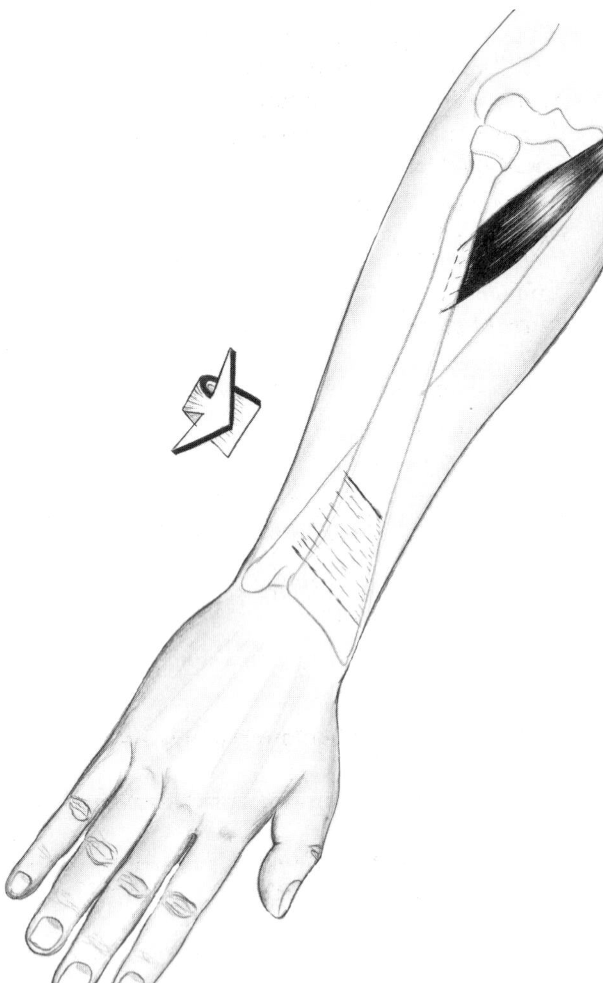

PRONATOR TERES

Origin In two heads:

Humeral head:
—Common flexor tendon on anterior surface of medial condyle.
—Antebrachial aponeurosis.
Ulnar head:
—Anterior surface of coronoid process of ulna.
—Tendon of brachialis.

Insertion Middle third of lateral surface of radius.

Innervation Median (C6, C7).

PRONATOR QUADRATUS

Origin Lower fourth of anterior surface of ulna.

Insertion Lower fourth of anterior and medial surfaces, and lateral border of radius.

Innervation Median (c7, C8, T1).

Function

Common function: forearm pronation (rotation from outward to inward).
Pronator quadratus pronates forearm independently of other movements.
Pronator teres flexes the forearm on the arm.
Isolated pronation is rare in activities of daily living.
The shoulder, with its abduction and internal rotation, participates in all movements requiring pronation.

Deficiency Functional impairment is considerable since it compromises pronation which is necessary for so many movements: screwing, eating, cutting, writing, dressing . . .

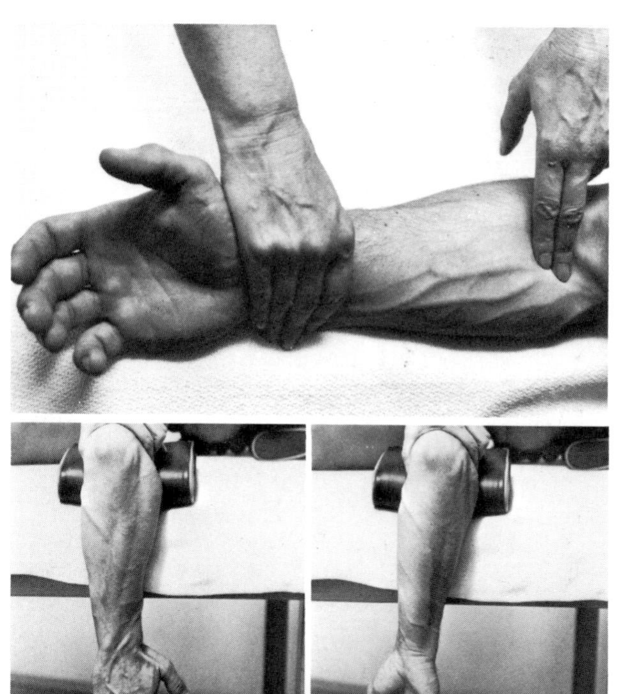

0 and 1 Subject sitting.

Forearm and hand resting in supination on table.
Elbow flexed at 90°.
Ask subject for forearm pronation.
Pronator teres is palpated on the proximal antero-medial portion of the forearm.
Pronator quadratus cannot be palpated.
For pronator teres:

Grade 0: No contraction is felt.
Grade 1: Contraction is felt but there is no joint movement.

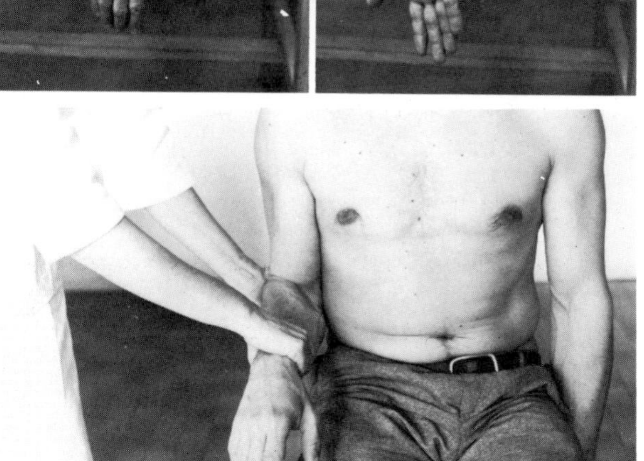

2 Subject prone.

Arm in abduction.
Cushion under elbow.
Forearm hanging over edge of table.
Hold arm in place just over olecranon.
From a position of supination, ask subject to pronate forearm.
Be sure that shoulder remains neutral.
Range of movement should be complete.

3, 4 and 5 Subject sitting on a stool.

Forearm in supination, elbow flexed at 90° and held tightly against body.
Hold lower extremity of humerus in place.
Resistance to the movement is placed on lower third of forearm.
Resistance is minimal for grade 3 and increases for grades 4 and 5.

Grade 4: Power is sub-normal.
Grade 5: Power is normal.

Photograph A represents an alternative for grades 4 and 5.
More specific testing of the pronator quadratus is possible from a position of full elbow flexion (photograph B) or extension (not shown).
Pronator teres can be tested when elbow is slightly flexed (photograph C).

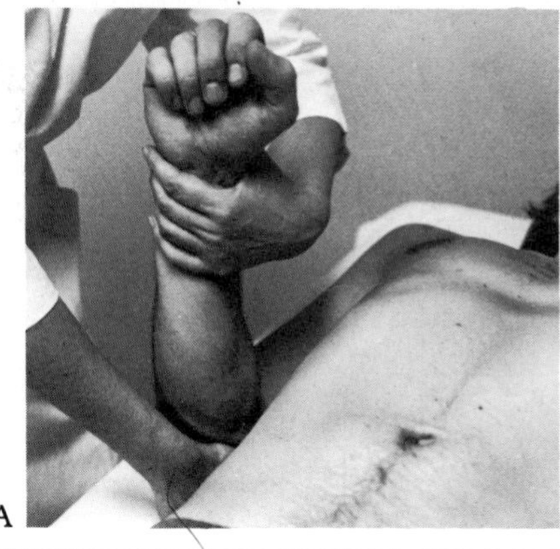

A

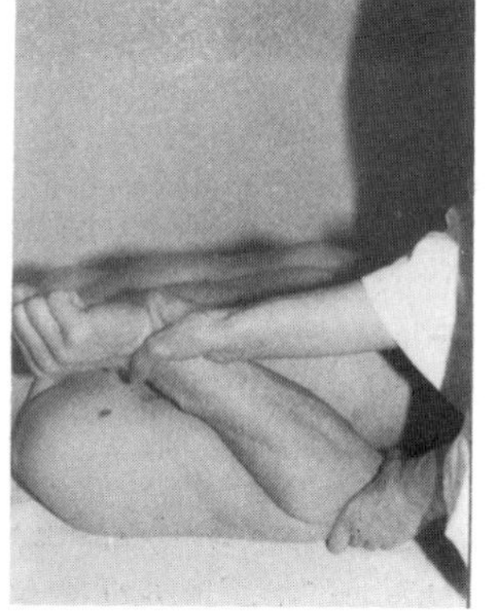

B

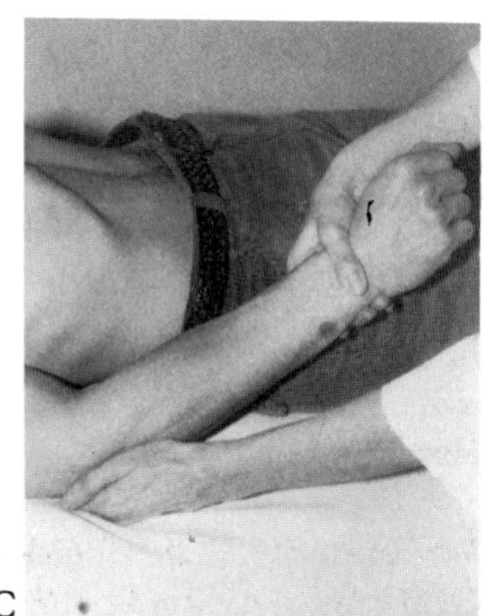

C

FLEXOR CARPI ULNARIS

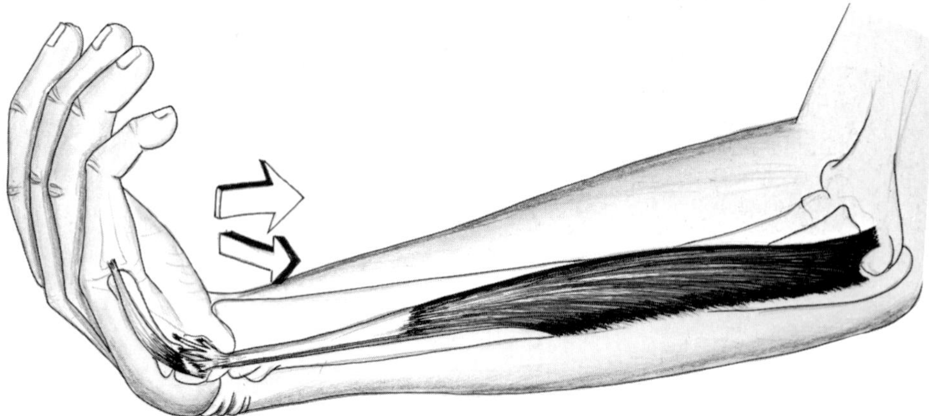

Origin In two heads:

Humeral head:
—By common flexor tendon from anterior surface of medial condyle of humerus.
Ulnar head:
—Medial border of olecranon.
—Upper two thirds of medial lip of ulnar ridge.
—Deep surface of antebrachial aponeurosis.

Insertion

—Primary: pisiform.
—Expansions: hamate, fourth and fifth metacarpals, and fascia of abductor digiti minimi.

Innervation Ulnar (c7, C8, T1).

Function

Flexion of hand on forearm. Flexor carpi ulnaris draws the medial side of hand so that palm faces outward; the fifth metacarpal is then placed anteriorly in relation to the metacarpus.

It participates in hand adduction slightly, since adduction is limited by wrist flexion.

Through its expansion, it flexes the fifth metacarpal on the carpus.

It stabilizes the wrist joint especially during finger extension, acting synergistically with hypothenar muscles.

It participates in stabilizing elbow joint when the latter is flexed.

Contracture or tightness

Positions the wrist and fifth metacarpal in flexion and ulnar deviation so that the palm of the hand faces outward.

Deficiency Leads to predominance of the abducting component of flexor carpi radialis.

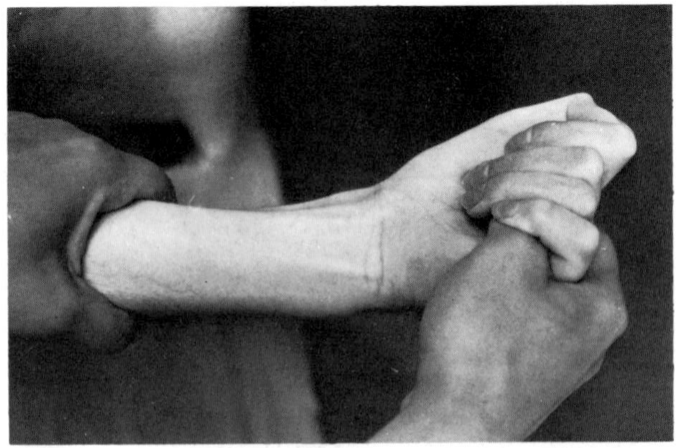

Subject presenting contracture of flexor carpi ulnaris. Examination of the contracture.

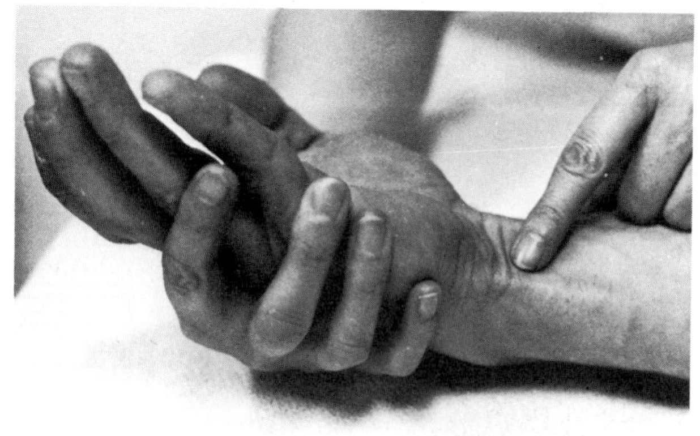

0 and 1 Subject sitting.

Forearm resting on table in supination, hand held by examiner.
Ask subject to flex his hand on forearm in slight ulnar deviation drawing the fifth metacarpal anteriorly to the metacarpus.
Palpate the tendon towards the medial aspect of the wrist above the pisiform.

Grade 0: No tendon movement is felt.
Grade 1: Tendon movement is felt but there is no joint movement.

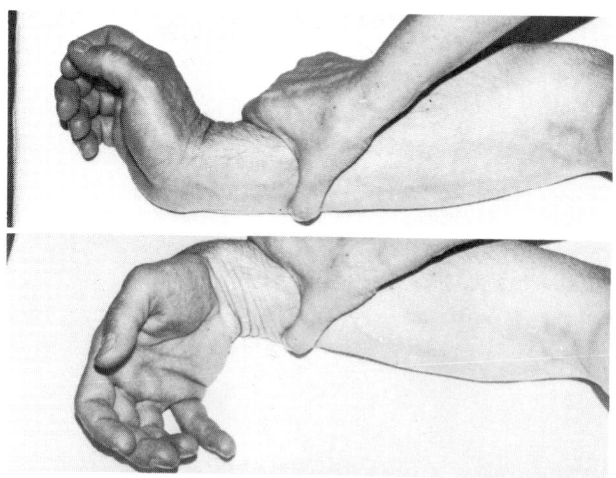

2 Subject sitting.

Forearm in about 45° supination resting on table.
Hand on table, fingers relaxed.
Hold forearm in place just above the styloids.
Ask subject to flex his hand on his forearm, drawing the hand towards the forearm in a gathering movement.
Range of movement should be complete.

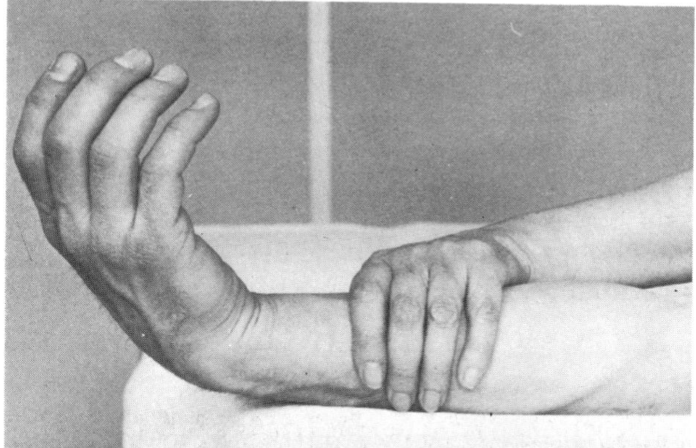

3 Subject sitting.

Forearm in supination resting on table.
Hand hangs over edge of table.
Hold forearm in place just over the styloids.
From a position of extension and radial deviation, ask the subject to flex and adduct the hand on the forearm in such a way that the palm faces outward.
Range of movement should be complete.

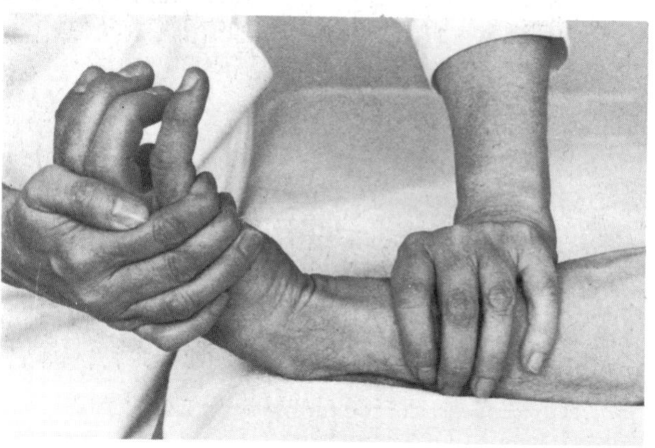

4 and 5 Same position.

Same hold and movement.
Resistance to the components of the movement is placed on the palmar aspect of head of fifth metacarpal.

Grade 4: Power is sub-normal.
Grade 5: Power is normal.

FLEXOR CARPI RADIALIS

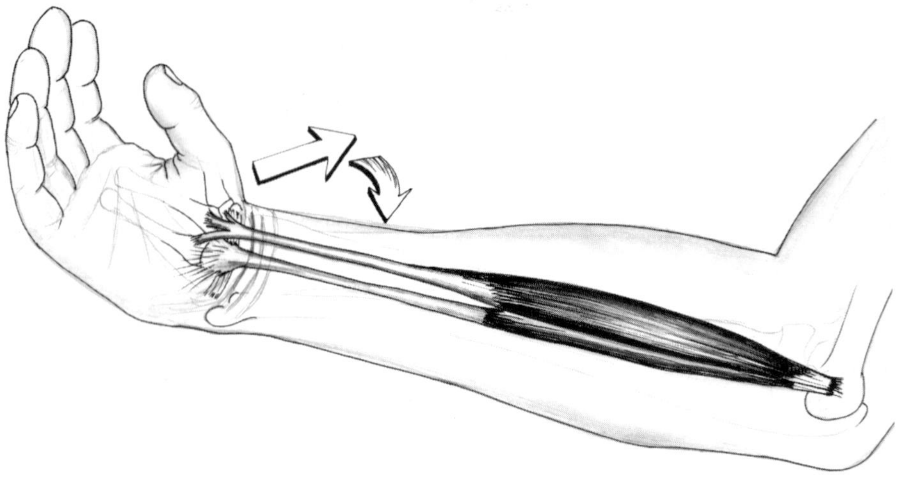

Origin

—Common flexor tendon from anterior surface of medial condyle.
—Antebrachial aponeurosis.

Insertion

—Anterior surface of base of second metacarpal bone.
—Expansion into base of third metacarpal and trapezium.
—Its tendon forms the medial margin of the radial artery groove (pulse groove).

Innervation Median (C6, C7, c8).

Function

Flexion of hand on forearm.
Forearm pronation: palm of hand faces slightly inward.
Participates in radial deviation of hand.
Flexor carpi radialis is a synergistic stabilizer of the wrist during finger extension (Steindler's effect).
It partially stabilizes the elbow joint during flexion.

Contracture or tightness

Locks wrist in a vicious position of flexion, radial-deviation deformity.

Deficiency

Decrease in force of hand flexion as well as of radial deviation. Hand goes into ulnar deviation.

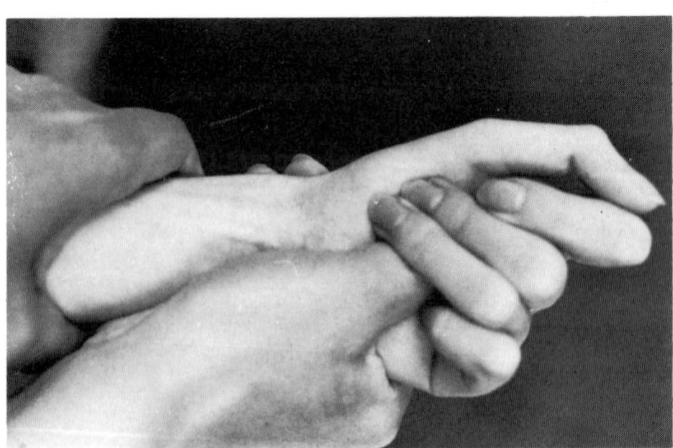

Subject presenting a contracture of flexor carpi radialis and palmaris longus.

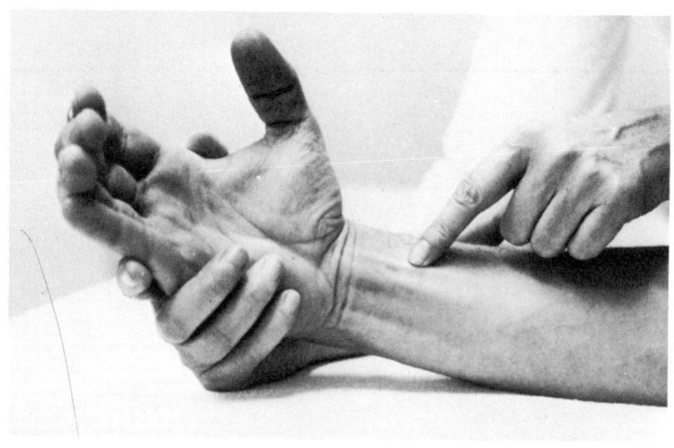

0 and 1 Subject sitting.

Forearm resting on ulnar side, hand supported by examiner.
Ask subject to flex hand on forearm in radial deviation, pronating forearm slightly.
Fingers should remain relaxed.
Flexor carpi radialis tendon is easy to palpate on the anterior aspect of the wrist towards the second metacarpal.

Grade 0: No tendon movement felt.
Grade 1: Tendon movement is felt but there is no joint movement.

2 Subject sitting.

Forearm and hand resting on ulnar side.
Hold forearm in place.
Fingers should remain relaxed.
Ask subject to flex his hand on forearm in slight radial deviation, pronating forearm.
Range of movement should be complete.

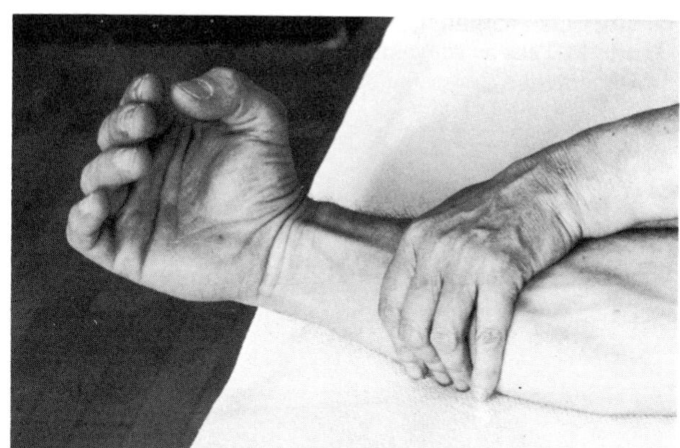

3 Subject sitting.

Forearm in supination, resting on table.
Hand hanging over edge of table.
Fingers relaxed.
Hold forearm in place.
Ask subject for same movement.
Range of movement should be complete.

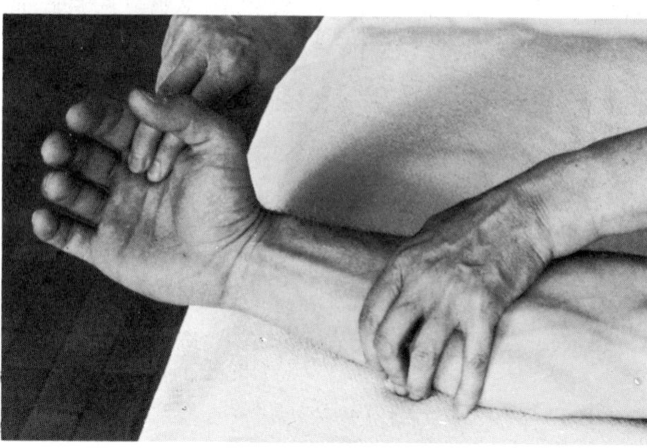

4 and 5 Same position.

Same hold and same movement.
Resistance to the components of the movement is placed on the palmar surface of the head of the second metacarpal.

Grade 4: Power is sub-normal.
Grade 5: Power is normal.

PALMARIS LONGUS

Origin

—Common flexor tendon on anterior surface of medial condyle.
—Deep surface of antebrachial aponeurosis.

Insertion

—Middle fibres: transverse carpal ligament and middle palmar aponeurosis.
—Lateral and medial fibres: thenar and hypothenar eminences.

Innervation

Median (C6, C7, c8).

Function

Direct hand flexion on forearm, and tightening of palmar aponeurosis.

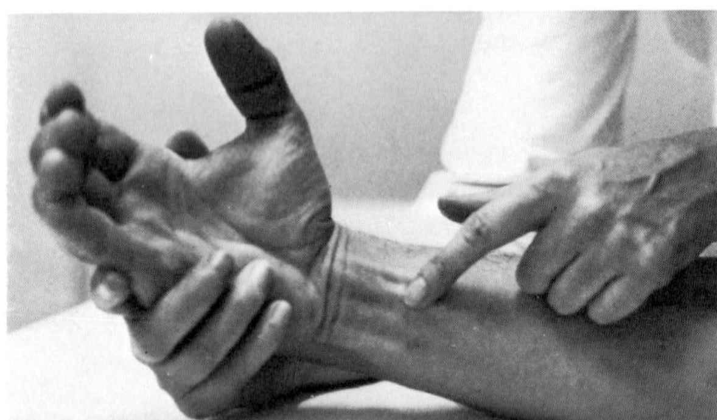

0 and 1 Subject sitting.

Forearm resting on ulnar side.
Examiner supports hand.
Ask subject for straight hand flexion on forearm.
The tendon is palpable on the anterior, median aspect of the wrist medially to the flexor carpi radialis tendon.

Grade 0: No tendon movement is felt.
Grade 1: Tendon movement is felt but there is no joint movement.

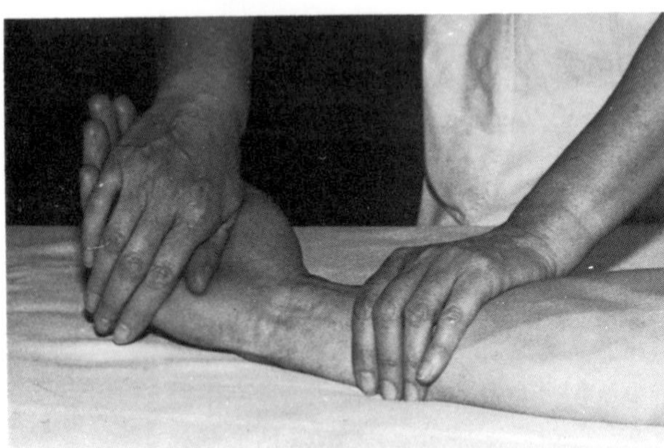

Test palmaris longus with straight hand flexion on forearm.
Use tests for flexor carpi radialis asking for straight wrist flexion without deviation (grades 4 and 5 are shown).

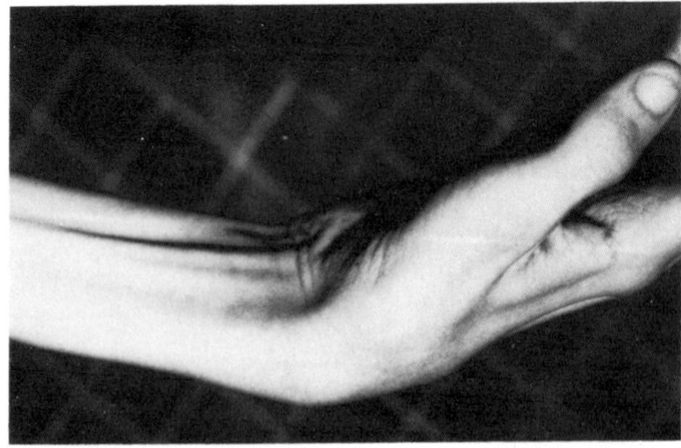

Contracture of palmaris longus.

NOTES

EXTENSOR CARPI RADIALIS LONGUS

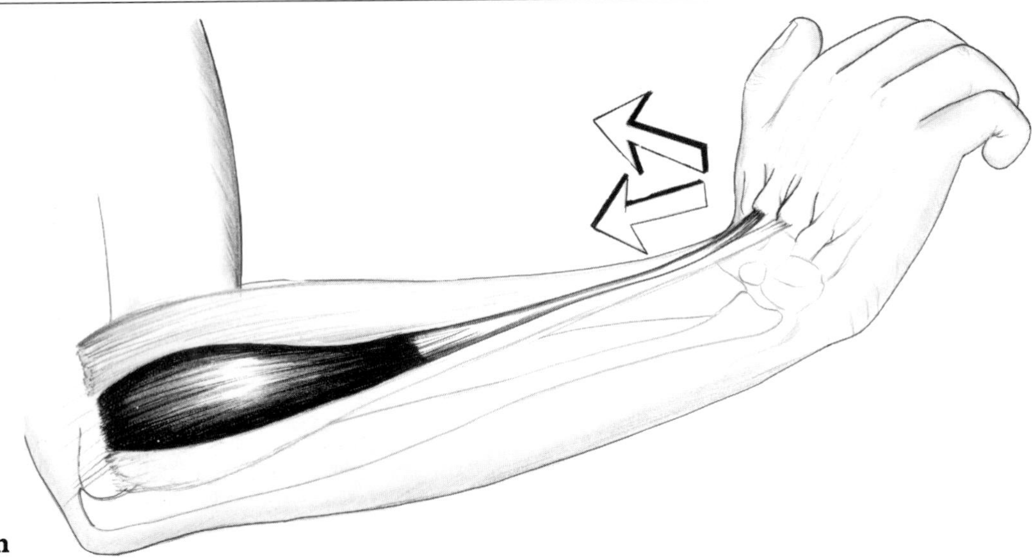

Origin

—Lateral border of humerus over lateral epicondyle.
—Lateral intermuscular septum.
—Common extensor tendon.

Insertion Lateral tubercle of base of second metacarpal bone.

Innervation Radial (C6, C7).

Function

Extension and abduction of hand on forearm.
Wrist stabilization: extensor carpi radialis longus is synergist of finger flexors. It is also a fixator and stabilizer of the wrist in the various prehensile activities. It participates accessorily in elbow flexion (Steindler's effect).

Deficiency

When the extensor carpi radialis longus is affected, the hand has a tendency to go into extension and adduction.
The extensor carpi ulnaris is then predominant.
Approximating the hand towards the body is more difficult (hand to mouth).
Paralysis of wrist extensors hampers function since they act synergistically with the flexors. Flexion force diminishes from 5 to 1 (see photograph page 234).

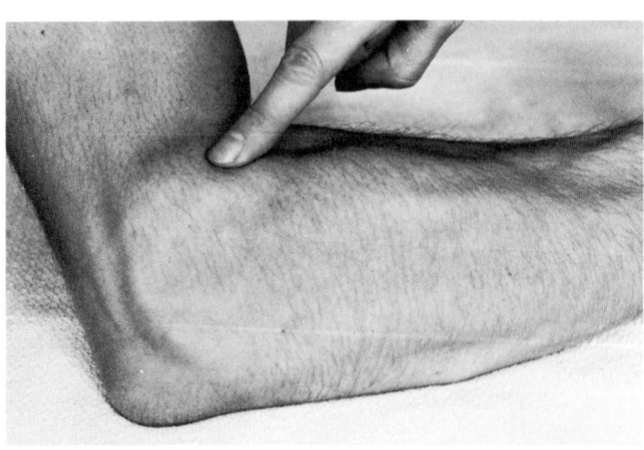

0 and 1 Subject sitting.

Forearm and hand resting in pronation on table, elbow flexed at 90°.
Ask subject to extend and abduct hand on forearm.
Fingers should remain relaxed.
Palpation of fibres is possible on lateral aspect of elbow above extensor carpi radialis brevis.

Grade 0: No contraction is felt.
Grade 1: Contraction is felt but there is no joint movement.

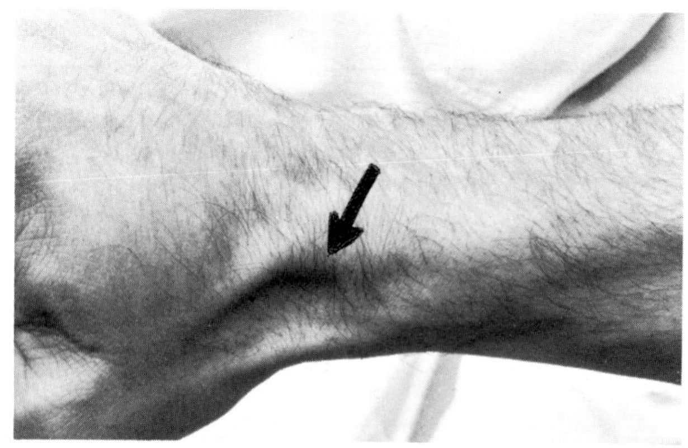

0 and 1

Palpation of tendon is possible but difficult above base of second metacarpal between extensor pollicis longus and extensor digitorum tendon to index.

Grade 0: No tendon movement is felt.
Grade 1: Tendon movement is felt but there is no joint movement.

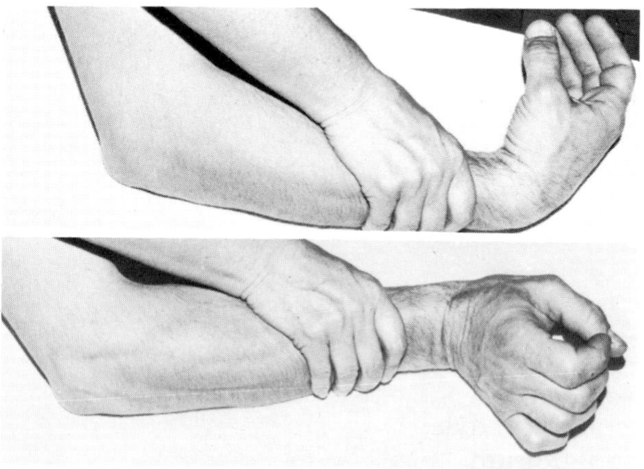

2 Subject sitting.

Forearm placed on table in about 45° supination.
Hand resting on table, fingers relaxed.
Hold forearm in place just above the styloids.
Ask subject to extend and abduct his hand on forearm.
Range of movement should be complete.

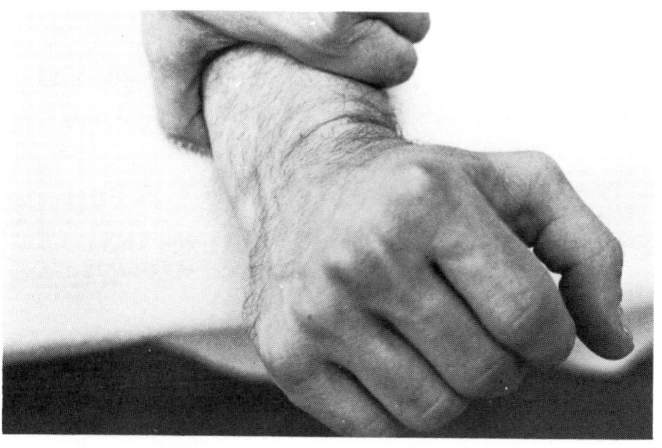

3 Subject sitting.

Forearm in pronation on table.
Hand hangs over edge of table.
Hold forearm in place just over styloids.
Ask subject to extend and abduct his hand on forearm.
Fingers should remain relaxed.
Range of movement should be complete.

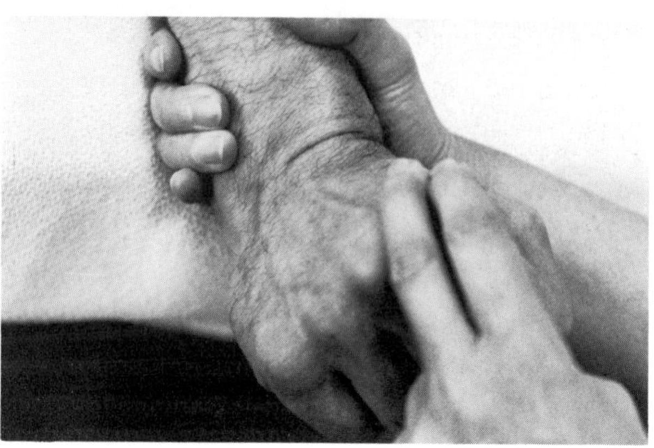

4 and 5 Same position.

Hold volar aspect of wrist.
Ask subject for the same movement.
Resistance to the components of the movement is placed on the dorsal aspect of metacarpus at the second metacarpal level.

Grade 4: Power is sub-normal.
Grade 5: Power is normal.

EXTENSOR CARPI RADIALIS BREVIS

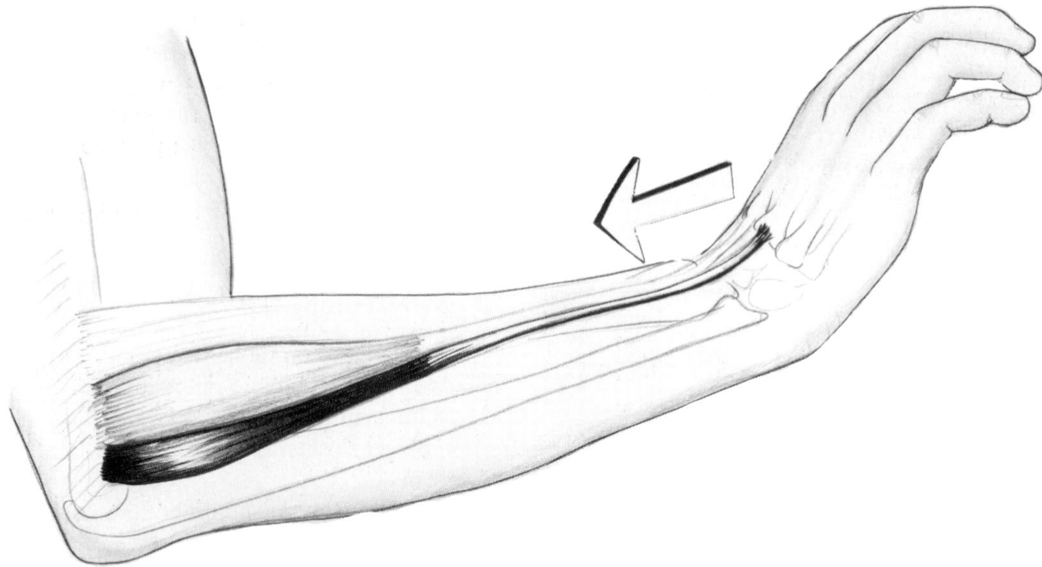

Origin

—Common extensor tendon on anterior surface of lateral epicondyle.
—Intermuscular aponeurosis separating from extensor digitorum.

Insertion Dorsal surface of base of styloid process of third metacarpal bone.

Innervation Radial (c6, C7, c8).

Function

Direct extension of hand on forearm.
Wrist stabilization: extensor carpi radialis brevis is a synergist of finger flexors. It also fixes and stabilizes the wrist in the various prehensile activities.
It participates accessorily in elbow flexion (Steindler's effect).

Deficiency

Isolated wasting is rare.
However, direct wrist extension is still possible due to the combined action of extensor carpi ulnaris and extensor carpi radialis longus.

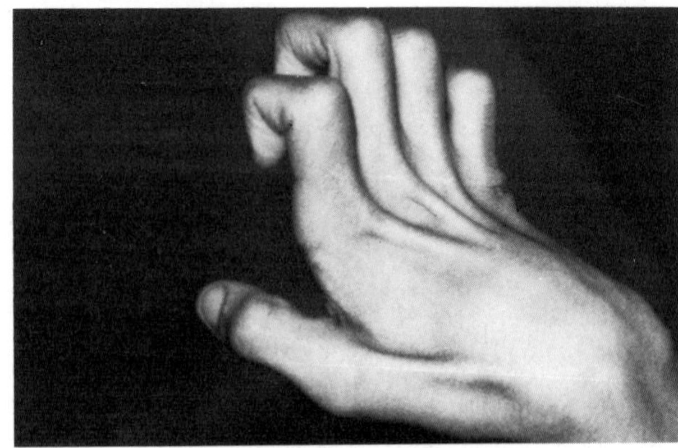

Paralysis of wrist extensors. Inefficient substitution of extensor digitorum.

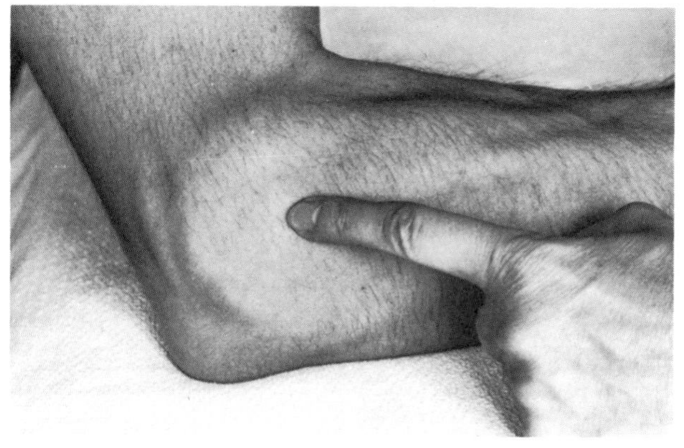

0 and 1 Subject sitting.

Forearm and hand in pronation, resting on table, elbow flexed at 90°.
Ask subject for straight hand extension on forearm.
Fingers should remain relaxed.
Palpation of tendon at wrist medially to extensor carpi radialis longus is difficult.
Palpation of muscle fibres is possible on lateral aspect of elbow beneath extensor carpi radialis longus.

Grade 0: No contraction is felt.
Grade 1: Contraction is felt but there is no joint movement.

2 Subject sitting.

Forearm and hand resting on table on ulnar side.
Hold lower third of forearm in place.
From a position of hand flexion, ask subject for straight extension.
Fingers should remain relaxed throughout the movement.
If hand goes into abduction, extensor carpi radialis longus is predominant.
Range of movement should be complete.

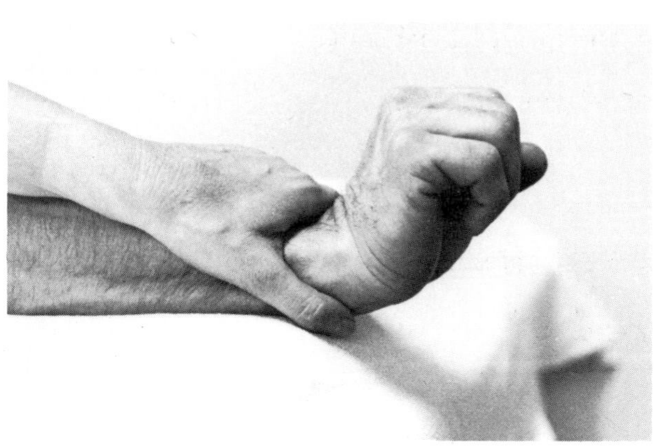

3 Subject sitting.

Forearm in pronation on table.
Hand hangs over edge of table.
Hold forearm in place just over styloids.
Ask for straight extension of hand on forearm.
Range of movement should be complete.

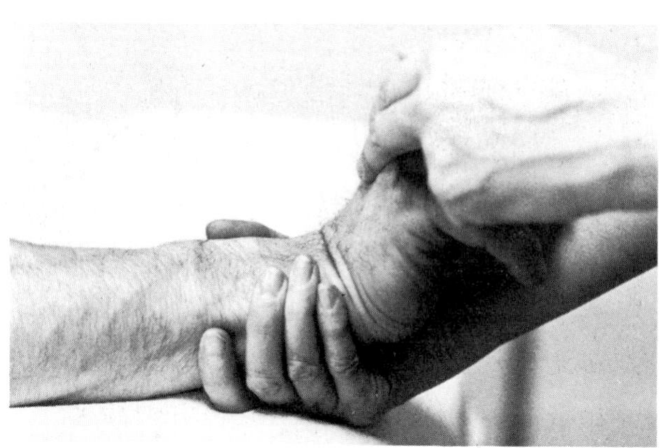

4 and 5 Same position.

Hold volar aspect of wrist in place.
Ask subject to complete the same movement.
Resistance to the movement is placed on dorsal aspect of hand at third metacarpal level.

Grade 4: Power is sub-normal.
Grade 5: Power is normal.

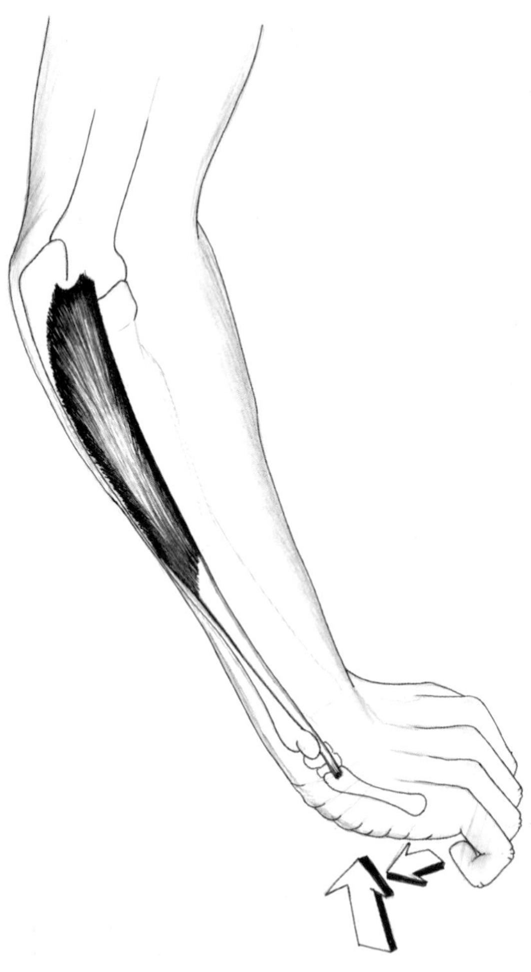

Origin

—Common extensor tendon on anterior surface of lateral epicondyle.
—Upper two-thirds of posterior border of ulna.
—Antebrachial aponeurosis.

Insertion Posterior medial tubercle of base of fifth metacarpal.

Innervation Radial (C7, C8).

Function

Extension and adduction of hand on forearm.
It is a synergistic stabilizer of finger flexors.
Its synergistic stabilizing action is necessary during thumb abduction.
It participates in extension of forearm on the arm.

Deficiency

Decreases strength of wrist extension and adduction.
Hand extension is drawn into radial deviation.
Wrist is not stabilized during thumb abduction.
Paralysis of extensor carpi ulnaris impairs hand function in all prehensile activities.

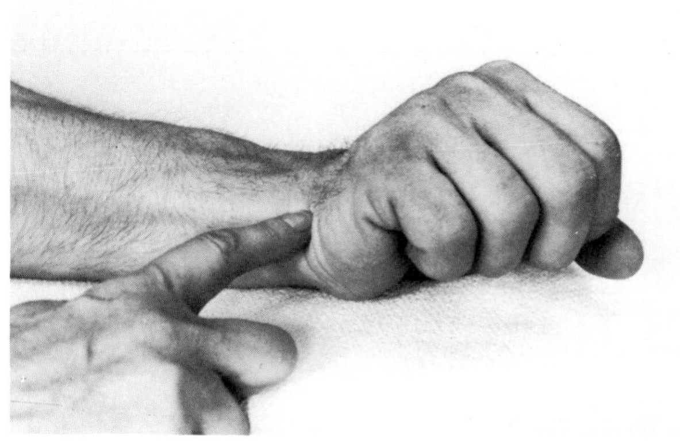

0 and 1 Subject sitting.

Forearm and hand in pronation on table.
Ask subject to adduct and extend his hand on his forearm, fist half closed, fingers relaxed.
The tendon is palpable under the styloid process of the ulna, above base of fifth metacarpal.
Avoid confusing with tendons of extensor digitorum to fifth, and extensor digiti minimi, which are located laterally to extensor carpi ulnaris tendon.

Grade 0: No tendon movement is felt.
Grade 1: Tendon movement is felt but there is no joint movement.

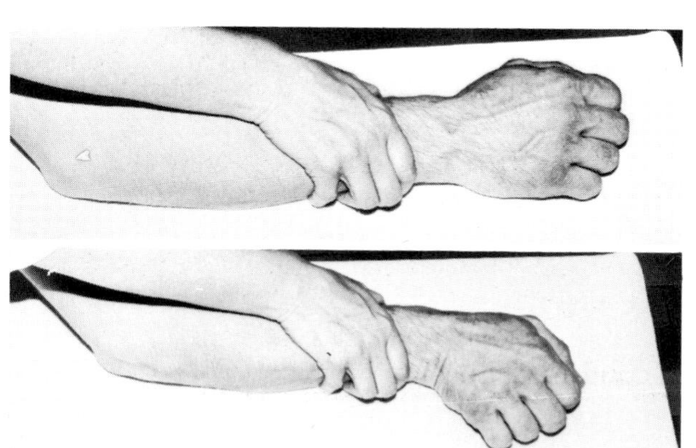

2 Subject sitting.

Forearm and hand in about 45° pronation, resting on table.
Hold lower third of forearm in place.
Ask subject for ulnar deviation of hand, fingers remain relaxed.
Range of movement should be complete.

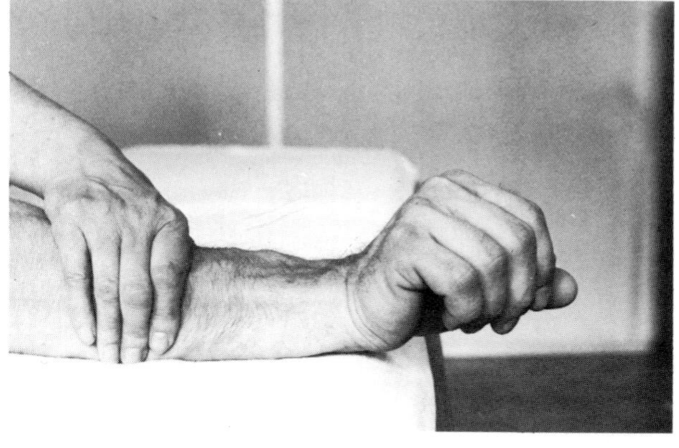

3 Subject sitting.

Forearm in pronation on table. Hand hangs over edge of table.
Hold lower third of forearm in place.
From a position of wrist flexion and radial deviation, ask subject for extension and ulnar deviation of hand on forearm.
Range of movement should be complete.

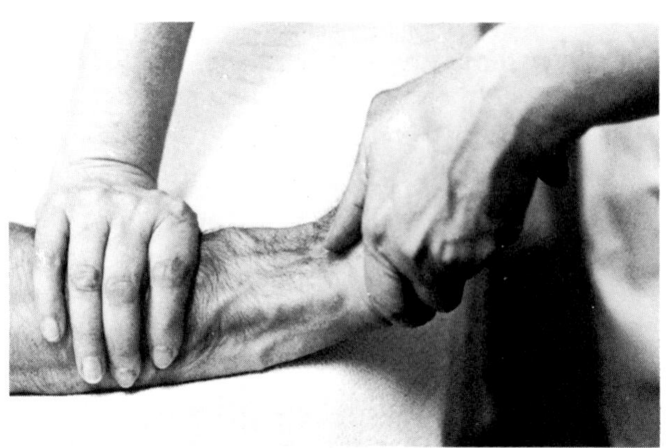

4 and 5 Same position.

Same hold.
Ask subject for same movement.
Resistance to the components of the movement is placed on ulnar side of fifth metacarpal.

Grade 4: Power is sub-normal.
Grade 5: Power is normal.

FLEXOR DIGITORUM SUPERFICIALIS

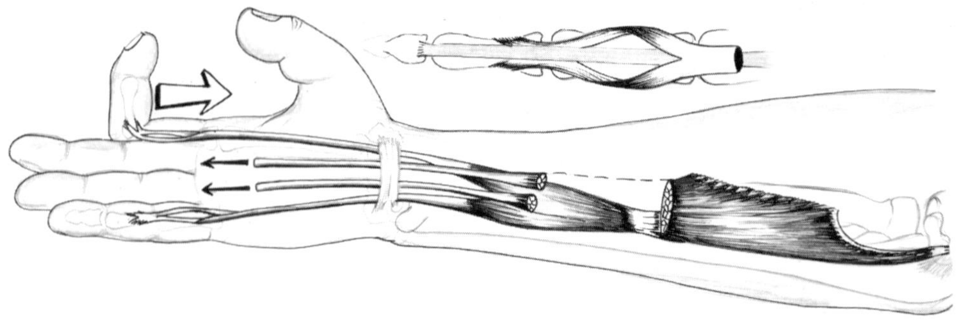

Origin In two heads:

Humero-ulnar head:
—Common flexor tendon on anterior surface of medial condyle.
—Medial lateral ligament of elbow joint.
—Anterior surface of coronoid process of ulna.
Radial head:
—Anterior border of radius.
A fibrous arch joins the two heads, the median nerve passes under it.

Insertion At proximal phalanx level the tendon divides into two slips to allow passage of flexor profundus tendon (perforans).
The two slips insert into the sides of the middle phalanx.

Innervation Median (c7, C8, T1).

Function

Flexion of middle phalanx of last four fingers on proximal phalanx.
It weakly participates in flexion of proximal phalanx.
Its efficiency varies greatly in function to wrist and metacarpophalangeal joint position (tenodesis effect).
It is less efficient when these joints are flexed.
The synergistic stabilizing action of wrist extensors is necessary for the function of flexor digitorum superficialis.
It plays an important role in subterminal opposition, full palmar prehension, and digito-palmar prehension.
It partially stabilizes the elbow when this joint is flexed.

Contracture or tightness Participates in claw-hand deformity.

Deficiency Difficulty in flexing the middle phalanx of last four fingers; substitution by the flexor profundus is weak.
Hand function is hindered by its paralysis: drawing, typing, playing the piano, the violin . . .
Difficulty in power grips.

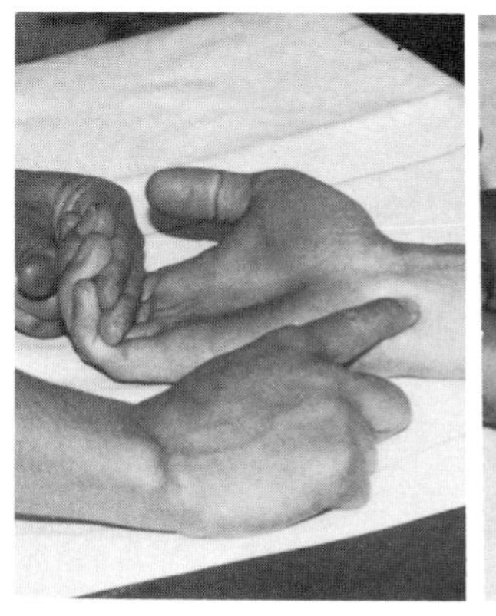
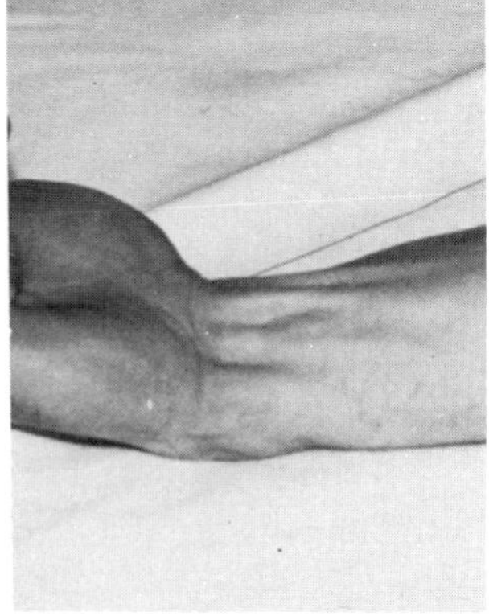

0 and 1 Subject sitting.

Forearm and hand in supination on the table. Hand supported by examiner.
Ask subject to flex his middle phalanx on the proximal phalanx while the distal phalanx remains in extension.
Palpation of tendons on palmar aspect of fingers is impossible.
It is possible on anterior aspect of wrist medially to flexor carpi radialis and palmaris longus.

Grade 0: No tendon movement is felt.
Grade 1: Tendon movement is felt but there is no joint movement.

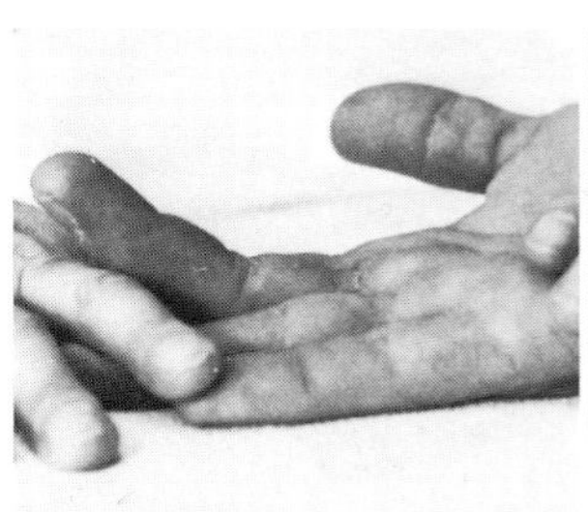
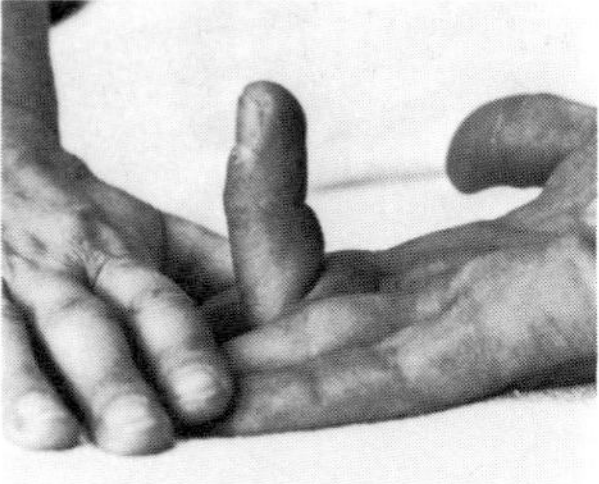
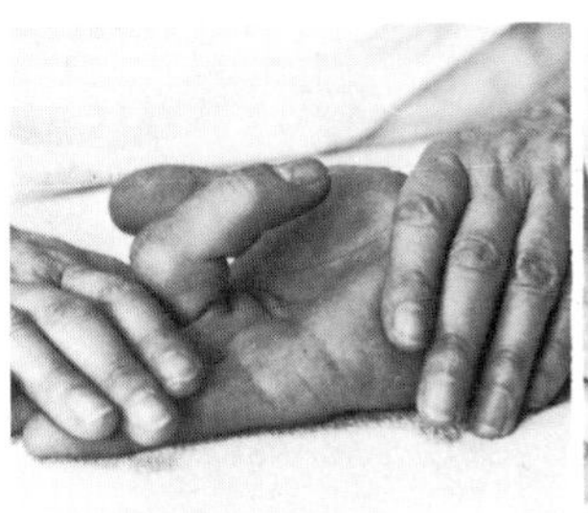
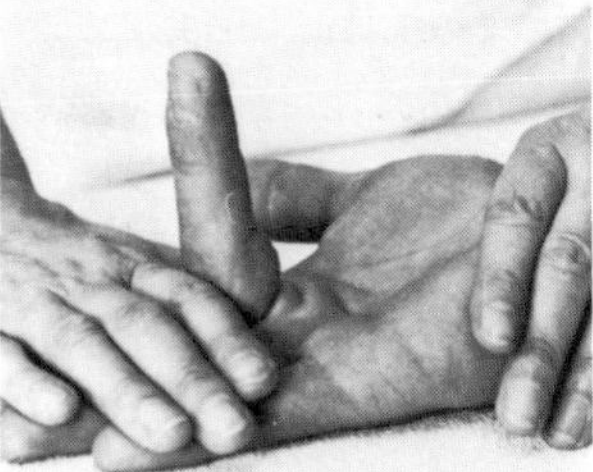

2 Subject sitting.

Forearm in supination.
Hold anterior aspect of wrist straight, including part of metacarpals.
Two tests are possible:
1. Hold fingers adjacent to the one being tested flat on the table. Metacarpophalangeal joint of the tested finger is straight. Ask for flexion of middle phalanx on the proximal phalanx.
Range of movement should be complete.
2. Hold fingers adjacent to the one being tested flat on the table in such a way that the metacarpophalangeal joint is placed in flexion by the examiner's hand. Ask for same movement.
Range of movement should be complete.

These two tests avoid the involvement of flexor digitorum profundus.
The first test is the most efficient.

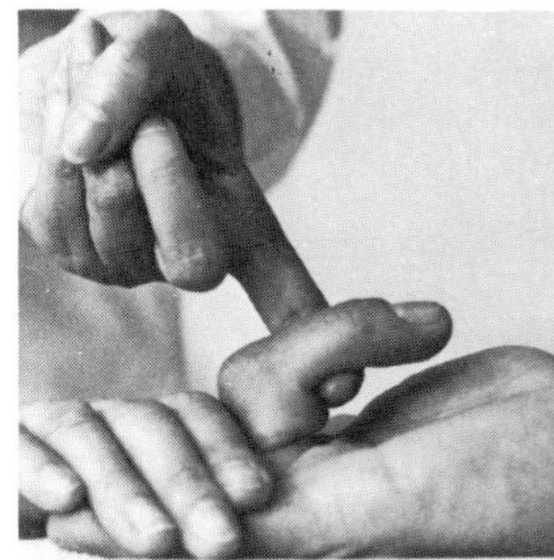
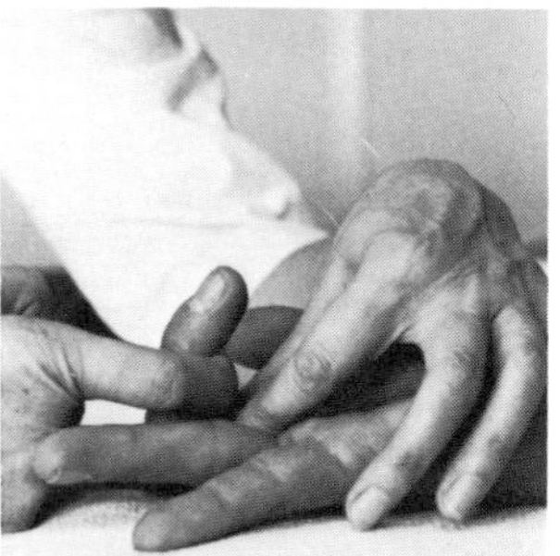

3, 4, and 5

Test 1: Stabilize proximal phalanx laterally.
Test 2: Hold adjacent fingers.
In both cases resistance to the movement is placed on the palmar aspect of the middle phalanx. Resistance is minimal for grade 3, and increases for grades 4 and 5.

Grade 4: Power is sub-normal.
Grade 5: Power is normal.

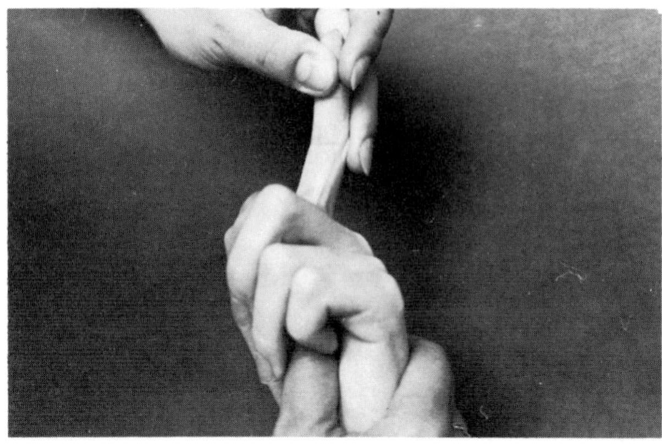

Strong position in finger flexion.
Examination of contracture of flexor digitorum superficialis.

FLEXOR DIGITORUM PROFUNDUS

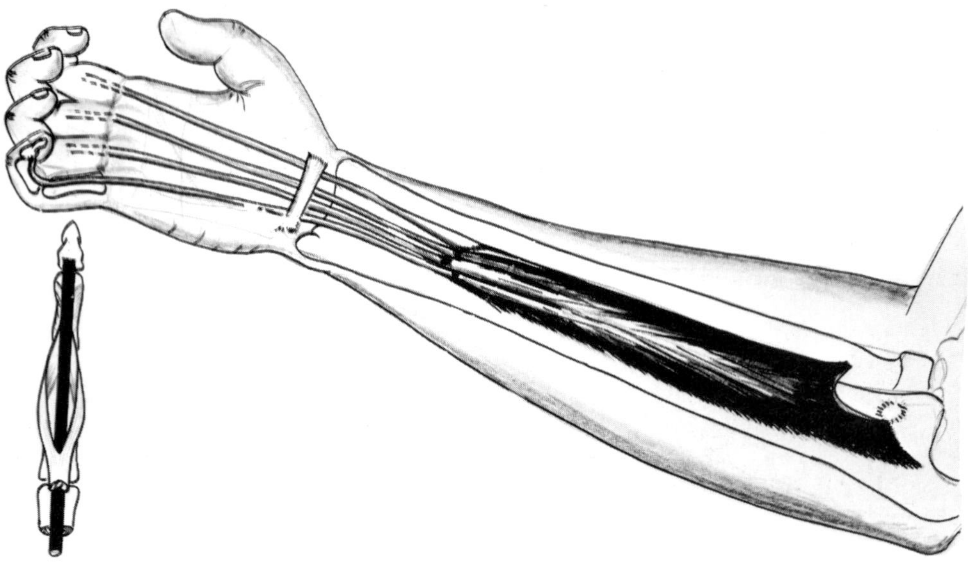

Origin

—Upper three quarters of the anterior and medial surfaces of the ulna.
—Anterior surface of interosseus membrane.
—Antebrachial aponeurosis.
—Medial and anterior surfaces of the coronoid process.

Insertion Situated posteriorly to the superficialis, it passes through the two divisions of its tendons at the level of the middle phalanges and inserts into the base of the distal phalanges of the last four fingers.

Innervation

—Two medial bellies: ulnar (C8, T1).
—Two lateral bellies: median (c7, C8, T1).

Function

Flexion of distal phalanges of last four digits on middle phalanges.
It participates a little in flexion of middle phalanges on proximal phalanges, proximal phalanges on metacarpals, and in wrist flexion.
When wrist and metacarpophalangeal joints are flexed, the strength of flexor profundus decreases markedly.
Its maximal efficiency is when these joints are in extension. During activities of daily living, these two joints place themselves automatically in extension.
Extensors are synergistic stabilizers of finger flexors.
Flexor profundus digitorum is essential in gripping, it participates in prehension by distal opposition and full palmar prehension.

Contracture or tightness Participates in claw-hand deformity.

Deficiency Loss of flexion capacities of the distal phalanx.
Paralysis of flexor digitorum profundus impairs hand function, e.g. playing piano, and different prehensile activities, especially prehension by distal opposition.
The distal phalanx of last four digits may pass into hyperextension.

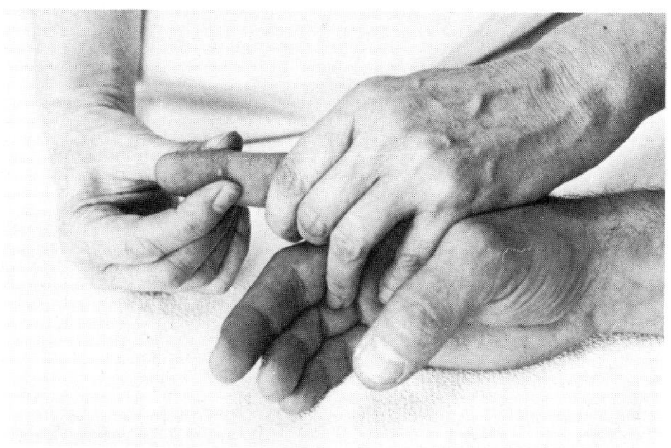

0 and 1 Subject sitting.

Forearm and hand resting on ulnar side.
Stabilize first two phalanges.
Ask subject for flexion of distal phalanx on the middle.
The tendon is palpable at palmar aspect of distal interphalangeal joint.

Grade 0: No tendon movement is felt.
Grade 1: Tendon movement is felt but there is no joint movement.

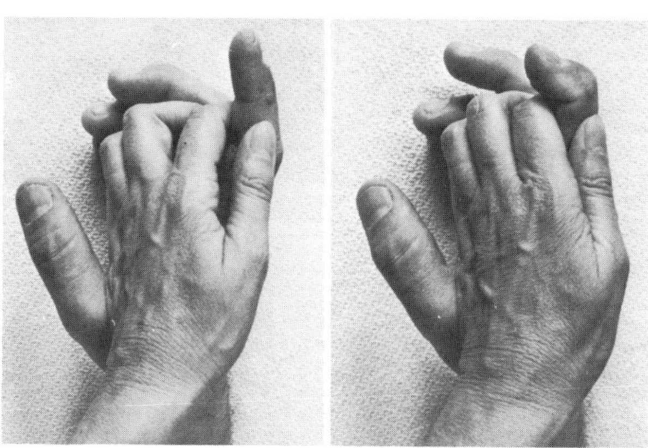

2 Same position.

Same hold and movement.
Fingers may be tested separately or together (not shown).
Range of movement should be complete.

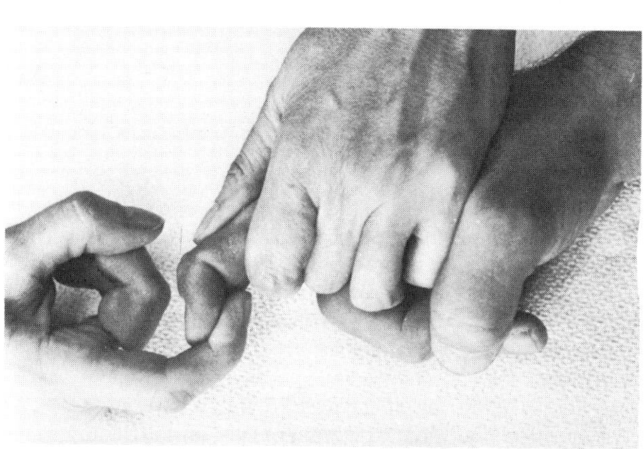

3, 4, and 5 Same position.

Same hold and movement.
Resistance to the movement is placed on the palmar aspect of the distal phalanx.
Resistance is minimal for grade 3 and increases for grades 4 and 5.

Grade 4: Power is sub-normal.
Grade 5: Power is normal.

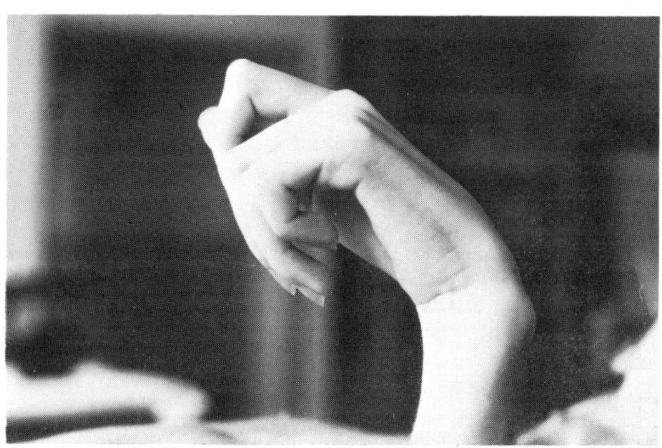

Subject demonstrating a contracture of all wrist and finger flexors.

EXTENSOR DIGITORUM, EXTENSOR INDICIS, and EXTENSOR DIGITI MINIMI

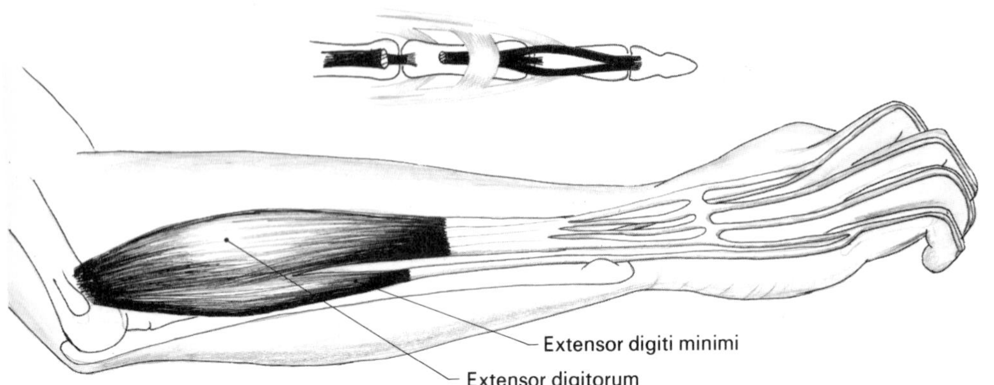

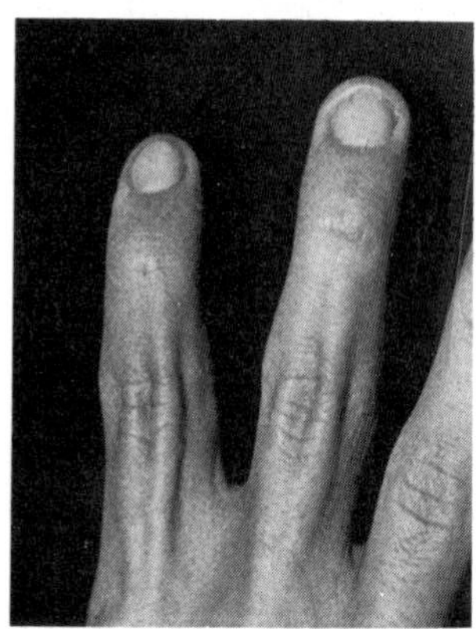

Division of extensor digitorum.

EXTENSOR DIGITORUM

Origin—Common extensor tendon on anterior surface of lateral epicondyle.
—Intermuscular septa.
—Antebrachial aponeurosis.

Insertion Dorsal surface of the three phalanges of the last four fingers.

EXTENSOR INDICIS

Origin—Posterior surface of ulna.
—Interosseus membrane.

Insertion It blends with the extensor digitorum tendon to the index on its medial side at metacarpophalangeal joint level.

EXTENSOR DIGITI MINIMI

Origin—Common extensor tendon on anterior surface of lateral epicondyle.
—Intermuscular septa.
—Antebrachial aponeurosis.

Insertion It interlaces with the tendon of extensor digitorum to the little finger on its medial side at metacarpophalangeal joint level.

Innervation All three muscles are innervated by the radial (c6, C7, c8).

Function
The extensor digitorum extends all three phalanges, acting more selectively on the proximal phalanx. This muscle is assisted by the extensors of the index and little fingers.
If each digit is extended individually, one notices:
—Extension and radial deviation of the index towards thumb.
—Straight extension of the middle finger.
—Extension and ulnar deviation of ring and little fingers.
According to Duchenne de Boulogne, electrical stimulation applied to extensor indicis and to extensor digiti minimi causes the index finger to extend and deviate away from the middle finger, and the little finger to extend and deviate away from the middle finger.
Extensor digitorum and the extensors of index and little fingers participate in extension of hand on forearm.

Deficiency
Loss of extension of proximal phalanges.
Proximal phalanges may assume a flexed position functionally, impairing the hand since proximal phalanges are no longer stabilized when middle and distal phalanges are in use.
Decrease in hand extension strength.
Extensors of index and little fingers may be spared during certain metabolic or toxic affections of the radial nerve afflicting the extensor digitorum; the patient may still be able to point with the index and little finger together.

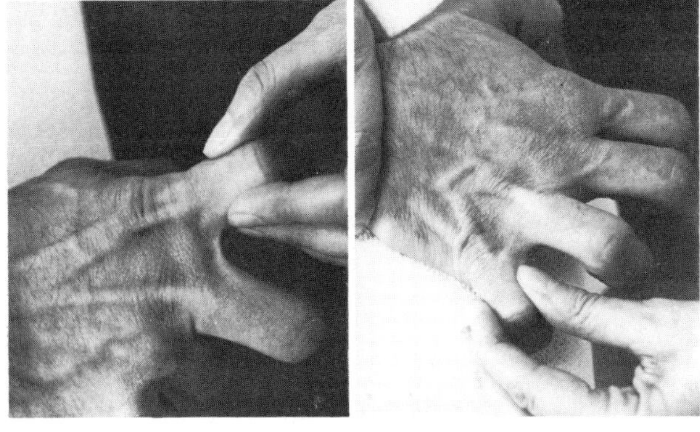

Showing extensor indicis and digiti minimi.

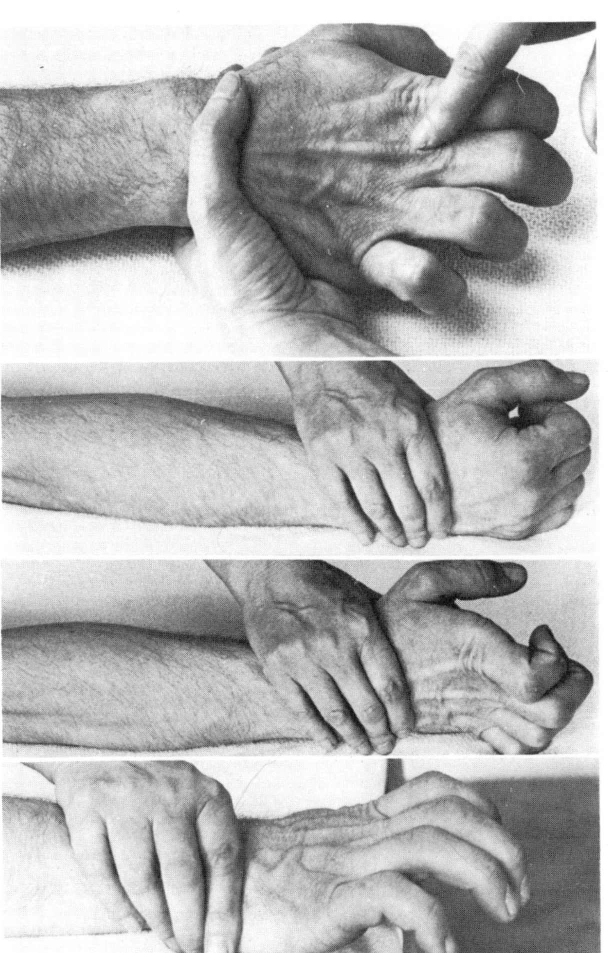

0 and 1 Subject sitting.

Forearm in pronation on table, hand supported by examiner.
Ask subject to extend his proximal phalanges, middle and distal phalanges remaining relaxed.
Palpate tendons of extensor digitorum over each metacarpal or at metacarpophalangeal joint level.
Tendons of extensor indicis and extensor digiti minimi are found medially to tendons of extensor digitorum of index and little fingers.

Grade 0: No tendon movement is felt.
Grade 1: Tendon movement is felt but there is no joint movement.

2 Subject sitting.

Forearm and hand resting on ulnar side.
Hold wrist and metacarpus straight.
Examiner places his hand over radial aspect of wrist keeping volar aspect of metacarpals in place.
Ask subject to extend his proximal phalanges on metacarpus, middle and distal phalanges remaining relaxed.
Range of movement should be complete.

3 Subject sitting.

Forearm in pronation on table.
Fingers hanging over edge of table.
Hold dorsal aspect of wrist including part of metacarpals.
Ask for same movement.
Range of movement should be complete.

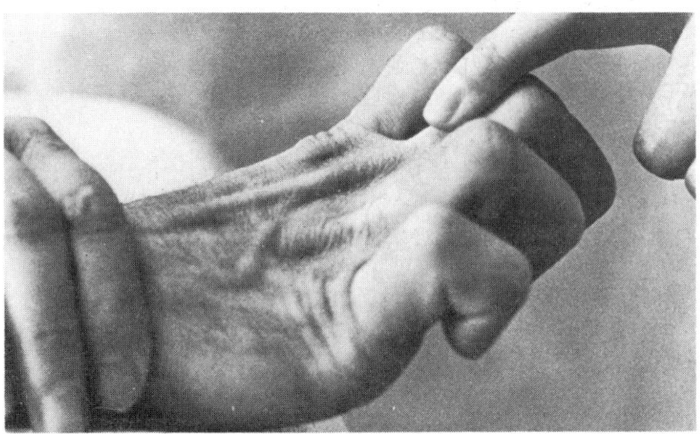

4 and 5 Same position.

Same hold.
Ask for same movement.
Resistance to the movement is placed on dorsal aspect of proximal phalanx.
Fingers may be tested separately or together (not shown).

Grade 4: Power is sub-normal.
Grade 5: Power is normal.

FLEXOR POLLICIS LONGUS

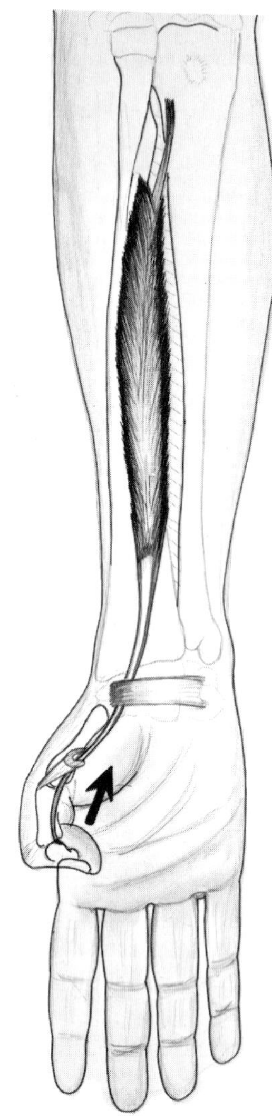

Origin

—Upper three-quarters of anterior surface of radius.
—Anterior surface of interosseus membrane.
—Expansion to coronoid process of ulna.

Insertion Base of distal phalanx of thumb on palmar surface.

Innervation Median (c7, C8, T1).

Function

Flexion of middle phalanx on the proximal.
Accessory flexion of proximal phalanx on first metacarpal.
Essential role in various types of prehensile activities, particularly in terminal opposition.
A synergistic stabilizing contraction of extensor pollicis brevis is necessary to flex the middle phalanx on the proximal.
Patients deprived of thenar muscles use the flexor pollicis longus in prehensile activities; the long fingers flex towards the thumb.

Contracture or tightness Marked flexion position of middle phalanx.

Deficiency Severe functional impairment for precision grips.

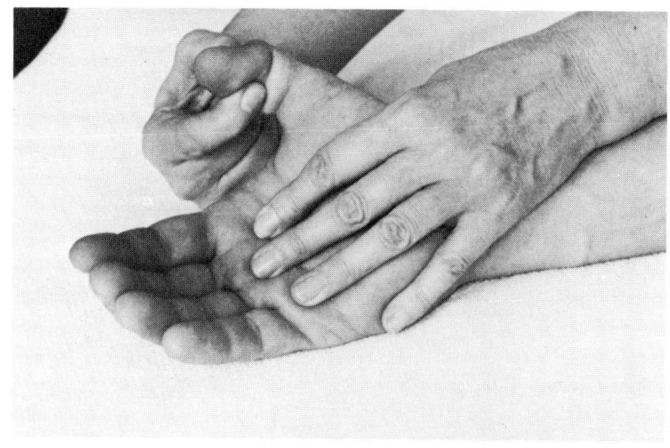

0 and 1 Subject sitting.

Forearm and hand in supination on table.
Ask subject to flex his middle phalanx on the proximal.
The tendon can be palpated at thumb interphalangeal joint level.

Grade 0: No tendon movement is felt.
Grade 1: Tendon movement is felt but there is no joint movement.

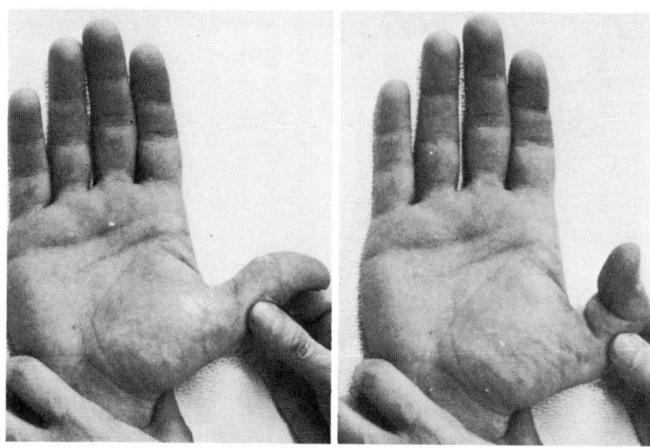

2 Same position.

Hold proximal phalanx laterally and the anterior aspect of the wrist in place.
Ask for the same movement.
Range of movement should be complete.

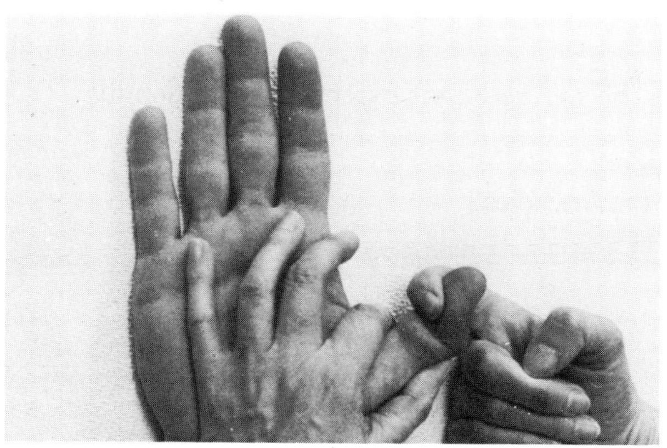

3, 4 and 5 Same position.

Fixate thenar eminence and hold proximal phalanx laterally.
Ask for the same movement.
Resistance to the movement is placed on the palmar aspect of the middle phalanx.
Resistance is minimal for grade 3 and increases for grades 4 and 5.

Grade 4: Power is sub-normal.
Grade 5: Power is normal.

ABDUCTOR POLLICIS LONGUS

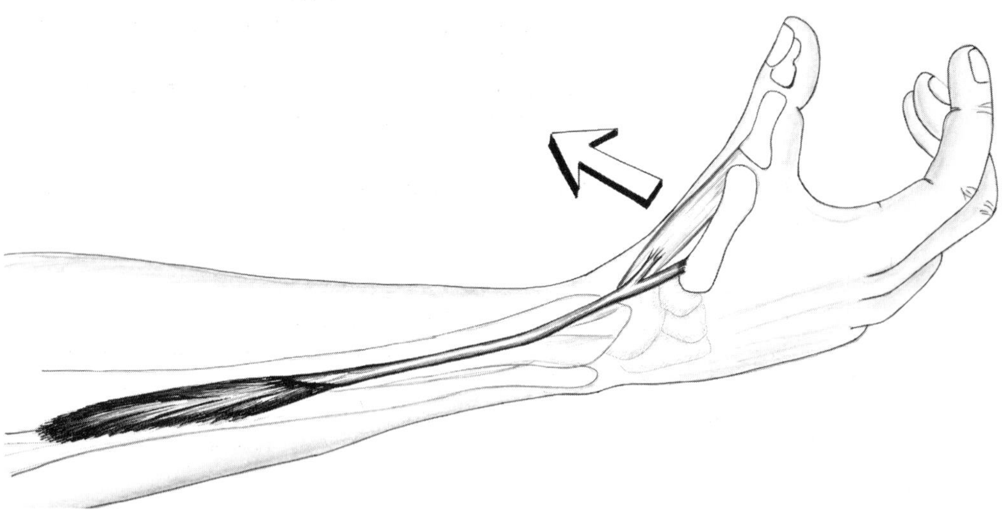

Origin

—Lateral margin of posterior surface of ulna between the supinator and extensor pollicis longus.
—Middle third of posterior surface of radius.
—Adjacent posterior surface of interosseus membrane.
—Intermuscular aponeurosis separating it from the other muscles.
Outlines the anatomical snuff box laterally and anteriorly.

Insertion

—Lateral tubercle of base of first metacarpal bone.
—Expansions into abductor pollicis brevis and thenar eminence aponeurosis.

Innervation

Radial (c6, C7, c8).

Function

Brings the first metacarpal bone into anterior position and radial abduction on carpus.
It participates in flexion of hand on forearm.
It starts thumb opposition acting with the most lateral thenar muscles.
Note: The synergistic stabilizing contraction of the extensor carpi ulnaris is necessary during the action of abductor pollicis longus.

Deficiency

The thumb 'drops' into the hand.
Spherical grasp is difficult to perform.

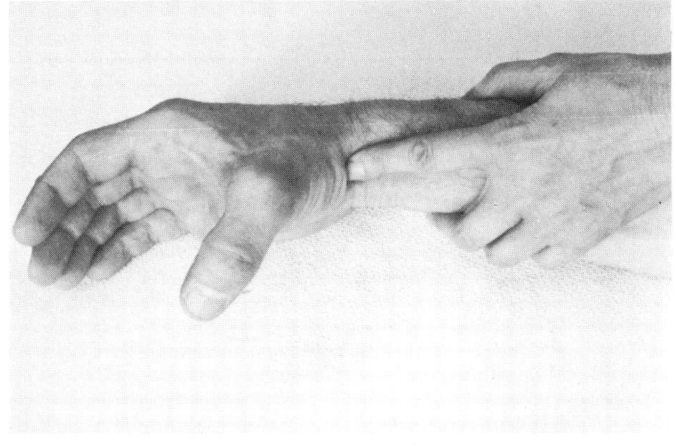

0 and 1 Subject sitting.

Forearm and hand resting on ulnar side.
Ask subject for movement anteriorally with radial abduction of the first metacarpal bone.
The tendon can be palpated at the anterior lateral part of the base of the first metacarpal.
This is the most anterior muscle of the anatomical snuff box.
It is easy to confuse contraction of this muscle with that of the extensor pollicis brevis.

Grade 0: No tendon movement is felt.
Grade 1: Tendon movement is felt but there is no joint movement.

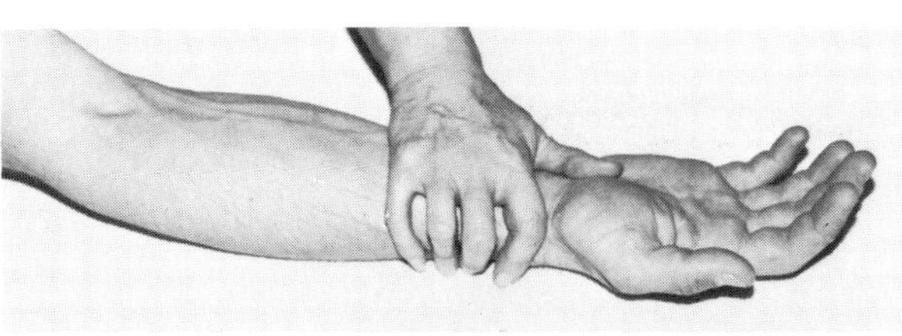

2

Forearm and hand in supination on table.
Hold wrist straight, with thumb in neutral position.
Ask subject for anterior movement with radial abduction of the first metacarpal bone.
Range of movement should be complete.

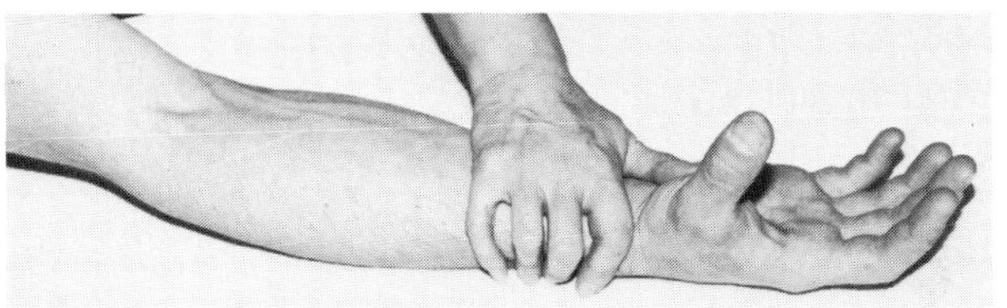

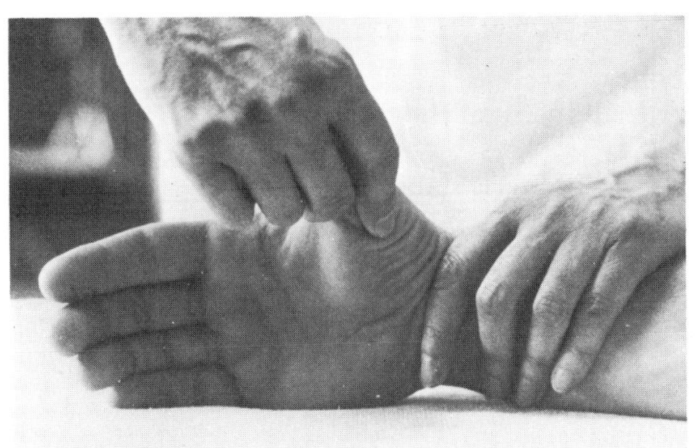

3, 4, and 5

Forearm and hand resting on ulnar side.
Same hold and movement.
Resistance to the movement is placed on the anterior lateral part of the head of the first metacarpal.
Resistance is minimal for grade 3 and increases for grades 4 and 5.

Grade 4: Power is sub-normal.
Grade 5: power is normal.

EXTENSOR POLLICIS BREVIS

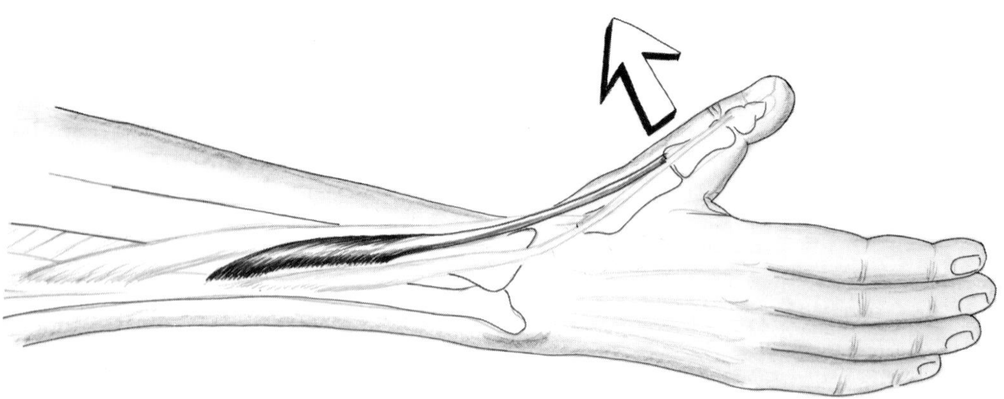

Origin

—Posterior surface of radius distally to abductor pollicis longus.
—Posterior surface of interosseus membrane.
—Intermuscular aponeurosis separating it from the other muscles.
It outlines the anatomical snuff box anteriorly and laterally.

Insertion Dorsal surface of proximal phalanx of thumb.

Innervation Radial (c6, C7, c8).

Function

—Extension of proximal phalanx on first metacarpal bone.
—Radial abduction of first metacarpal bone; it is the true abductor of the thumb.
It participates in radial deviation of the hand.
During contraction of extensor pollicis brevis, extensor carpi ulnaris stabilizes the wrist.

Deficiency

Thumb 'drops' into hand, pulled there by the thenar muscles. Its paralysis hinders grip of large objects and use of scissors.

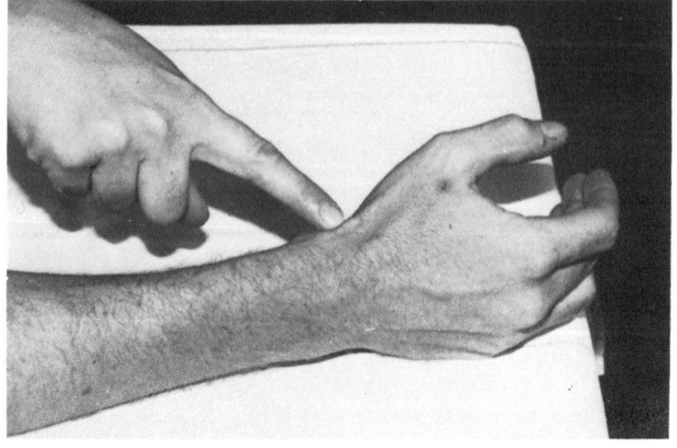

0 and 1 Subject sitting.

Forearm and hand resting on ulnar side.
Ask subject for extension of the proximal phalanx on the first metacarpal with radial abduction of thumb.
The tendon is palpable at the lateral margin of the anatomical snuff box posteriorly and medially to abductor pollicis longus.
Confusion with extensor pollicis longus is easy.

Grade 0: No tendon movement is felt.
Grade 1: Tendon movement is felt but there is no joint movement.

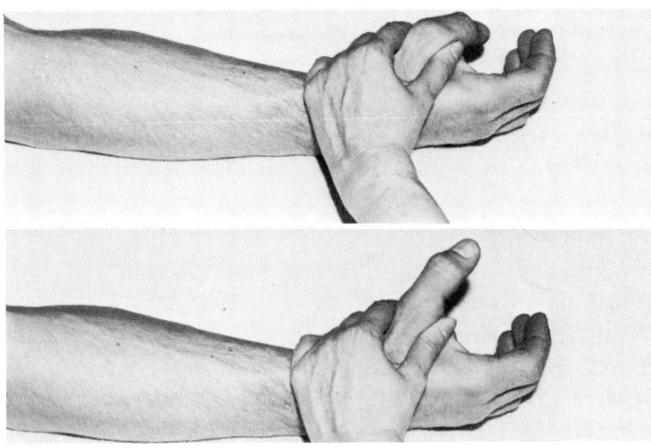

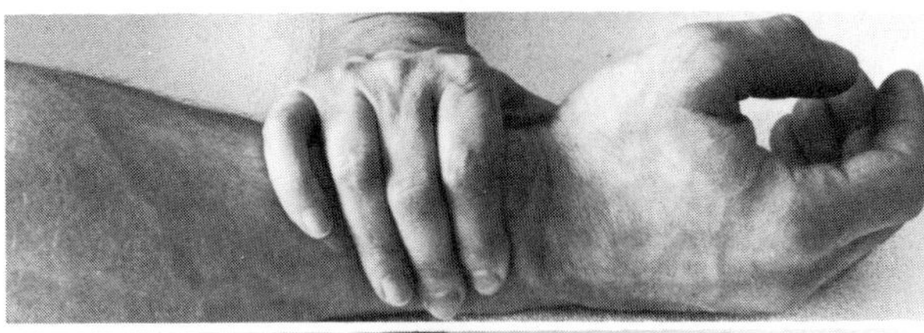

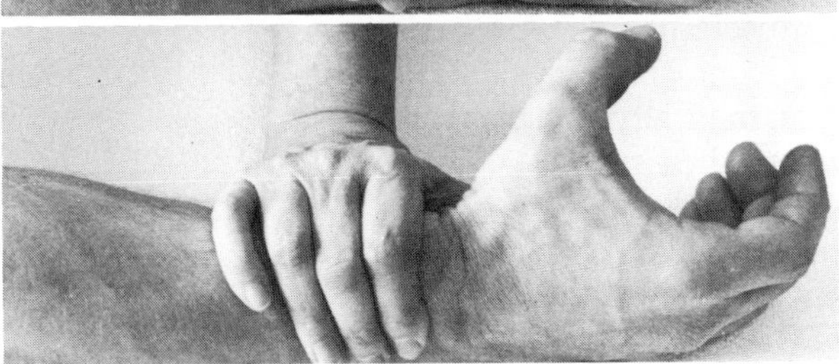

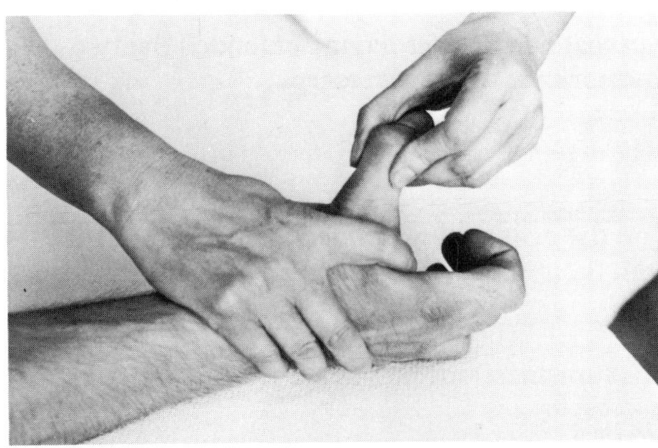

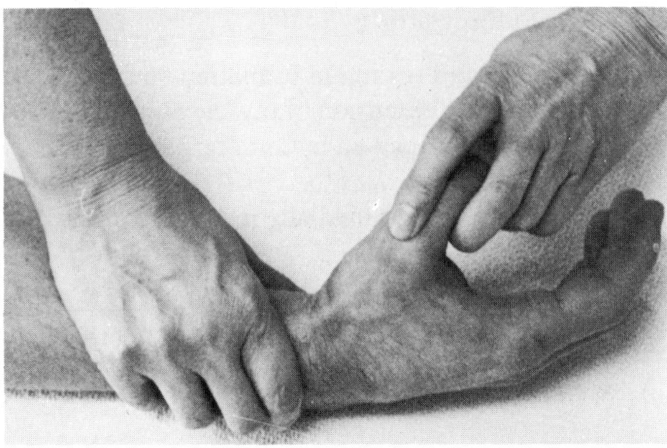

2 *Specific testing of extension of proximal phalanx.*

Same position.
Hold first metacarpal bone.
Ask subject for extension of proximal phalanx on first metacarpal while distal phalanx remains relaxed.
Range of movement should be complete.

2 *Global testing of proximal phalanx extension and of thumb abduction.*

Same position.
Hold wrist straight.
Ask subject for extension of proximal phalanx on first metacarpal and radial abduction of the thumb.
The thumb should be drawn neither backward (substitution by extensor pollicis longus), nor forward (substitution by abductor pollicis longus).
Range of movement should be complete.

3, 4, and 5 *Extension of proximal phalanx.*

Same position.
Hold first metacarpal in place.
Ask subject for extension of proximal phalanx on first metacarpal.
The distal phalanx should remain relaxed.
Resistance is placed on both sides of proximal phalanx.
Resistance is minimal for grade 3 and increases for grades 4 and 5.

Grade 4: Power is sub-normal.
Grade 5: Power is normal.

3, 4, and 5 *Extension of proximal phalanx with thumb radial abduction.*
Same position.
Hold wrist straight.
Possible substitutions by abductor pollicis longus and extensor pollicis longus should be avoided.
Ask subject for extension of proximal phalanx on metacarpal bone and radial abduction of the thumb.
Resistance to the two components of movement is placed on dorsal aspect of first phalanx.
Resistance is minimal for grade 3 and increases for grades 4 and 5.

Grade 4: Power is sub-normal.
Grade 5: Power is normal.

EXTENSOR POLLICIS LONGUS

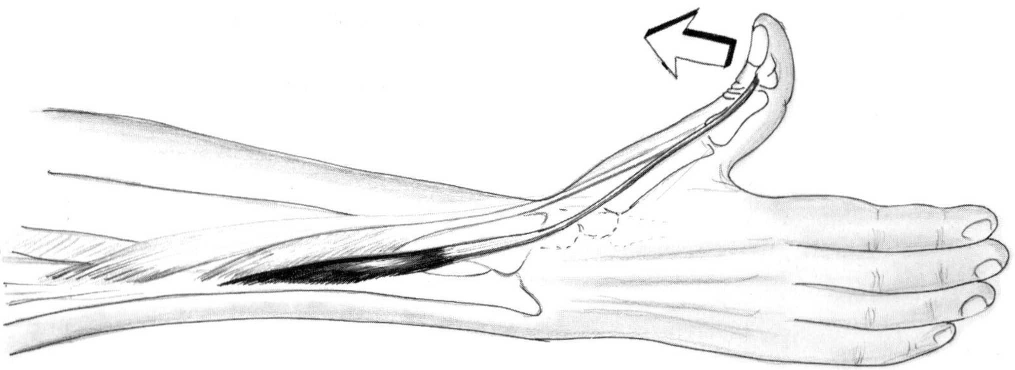

Origin

—Middle third of posterior surface of ulna below abductor pollicis longus, above extensor indicis.
—Posterior surface of interosseus membrane.
—Intermuscular aponeurosis separating it from other muscles.

Insertion Its tendon outlines the anatomical snuff box medially and posteriorly.
It inserts into the dorsal surface of base of distal phalanx of the thumb.

Innervation Radial (c6, C7, c8).

Function

Extension of distal phalanx on the proximal.
It then extends the proximal phalanx on the first metacarpal bone. After having extended the two phalanges it extends the first metacarpal and draws it posteriorly to the metacarpus.
Accessorily it participates in hand radial deviation.

Deficiency

Decreases the capacity for extension of distal phalanx on the proximal which is partly accomplished by the thenar muscles, except for the opponens, through their expansions into the tendon of extensor pollicis longus.
Loss of extension and posterior position of the thumb.
Paralysis of the extensor pollicis longus causes a certain clumsiness but no serious functional hindrance.

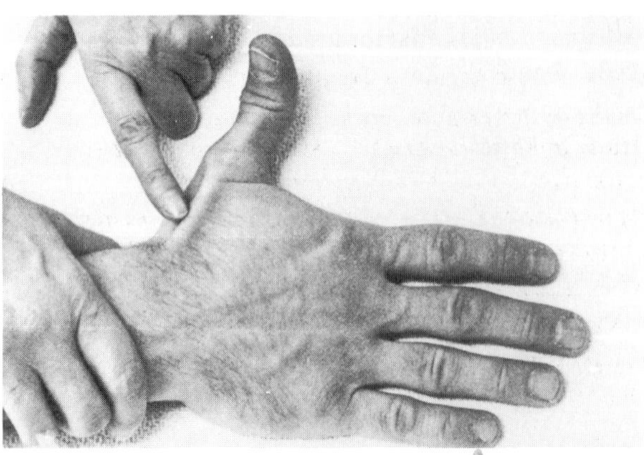

0 and 1 Subject sitting.

Forearm and hand resting in pronation on table.
Ask subject for an extension of middle phalanx on the proximal accompanied by metacarpal posterior position.
The tendon is palpated medially in the anatomical snuff box.

Grade 0: No tendon movement is felt.
Grade 1: Tendon movement is felt but there is no joint movement.

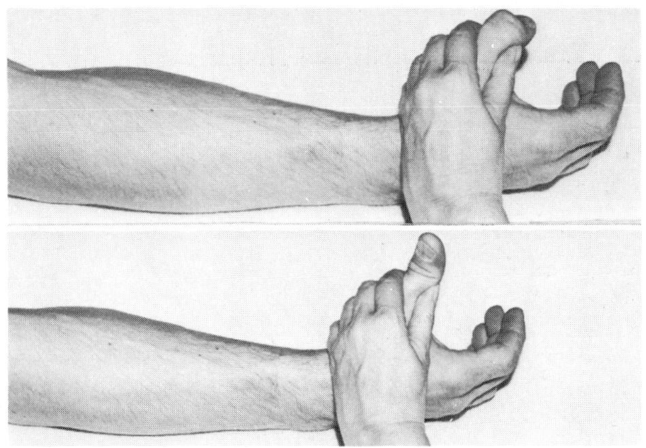

2 *Specific testing of distal phalanx extension.*

Forearm and hand resting on ulnar side.
Hold both sides of proximal phalanx.
Ask subject for extension of distal phalanx on the proximal.
Range of movement should be complete.

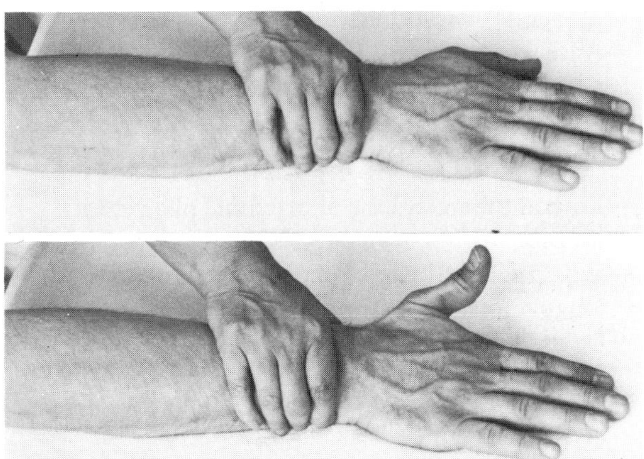

2 *Global testing of distal phalanx extension and extension of thumb.*

Forearm and hand resting in pronation.
Hold wrist straight.
Ask subject to extend his distal phalanx on the proximal and to position his thumb in such a way that it rises from the table.
Range of movement should be complete.

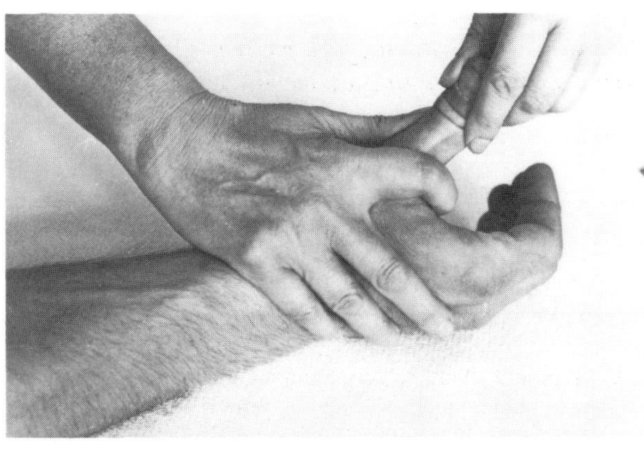

3, 4, and 5 *Distal phalanx extension.*

Forearm and hand resting on ulnar side.
Hold both sides of proximal phalanx.
Ask subject for extension of distal phalanx on the proximal.
Resistance to the movement is placed on both sides of distal phalanx.
Resistance is minimal for grade 3 and increases for grades 4 and 5.

Grade 4: Power is sub-normal.
Grade 5: Power is normal.

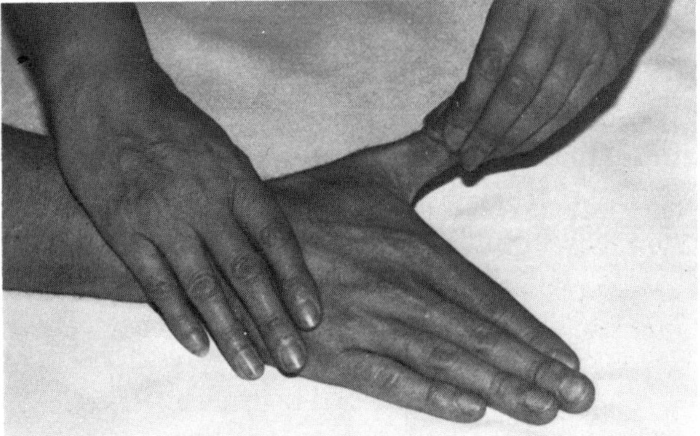

3, 4, and 5 *Extension of distal phalanx and extension of thumb.*

Forearm and hand in pronation on table.
Hold wrist straight.
Ask subject to extend his distal phalanx on the proximal and to extend his thumb.
Resistance to the two components of the movement is placed on both sides of the distal phalanx.
Resistance is minimal for grade 3 and increases for grades 4 and 5.

Grade 4: Power is sub-normal.
Grade 5: Power is normal.

ABDUCTOR POLLICIS BREVIS

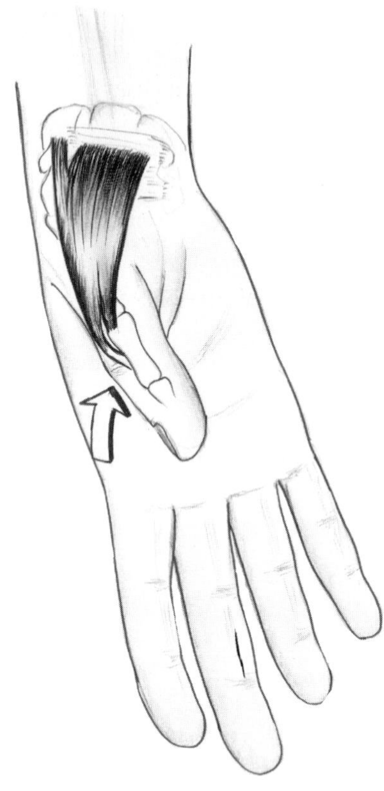

Origin

—Tubercle of scaphoid.
—Anterior transverse carpal ligament.
It receives an expansion of abductor pollicis longus.

Insertion

—Lateral tubercle, base of proximal phalanx of thumb.
—Expansion to the tendon of extensor pollicis longus, forming part of the extensor expansion (lamina dorsales).

Innervation Median (C8, t1).

This muscle is situated anteriorly to the first metacarpal, and does not deserve its name; it does not lead the thumb into abduction but into palmar abduction.

Function

Thumb abduction in palmar plane.
Internal rotation of proximal phalanx completing that of first metacarpal induced by the opponens. It also leads the first metacarpal into adduction.
It partially flexes the proximal phalanx over the first metacarpal and extends the distal phalanx on the proximal phalanx.
Its isolated contraction most specifically opposes the thumb to the index and middle fingers; it is an essential motor of the tridigital grip (three jaw chuck grip).
It takes part in prehensile activities and especially subterminal opposition and full palmar prehension.

Deficiency

Isolated deficiency is rare.
Its paralysis, even associated with that of the opponens, is partially compensated by the superficial head of the flexor pollicis brevis.

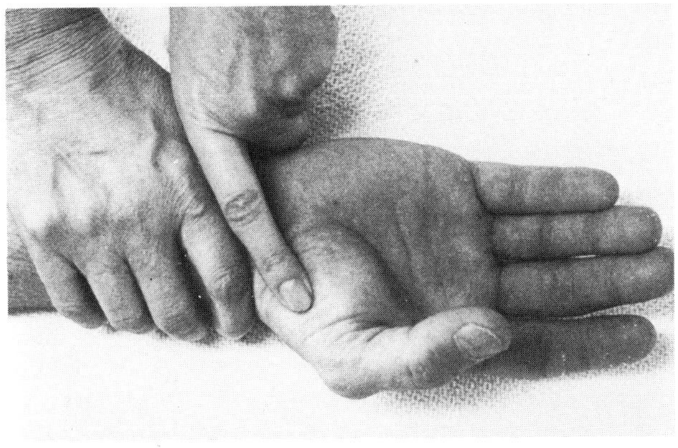

0 and 1 Subject sitting.

Forearm and hand in supination resting on table. Ask subject for palmar abduction of thumb, proximal phalanx slightly flexed, so that the pulp of the thumb looks towards the index and middle fingers. Abductor pollicis brevis, being the most superficial of the thenar muscles, is easily palpated on the lateral part of the thenar eminence at first metacarpal level.

Grade 0: No contraction is felt.

Grade 1: Contraction is felt but there is no joint movement.

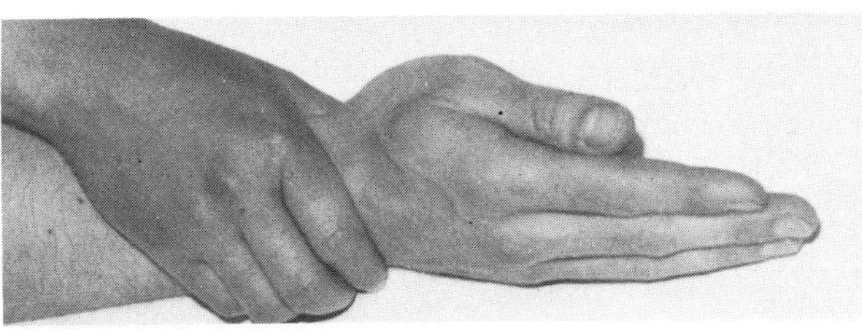

2 Forearm and hand resting on ulnar side.

Hold wrist straight.
Ask for the same movement with slight flexion of proximal phalanx on first metacarpal and extension of middle phalanx on the first.

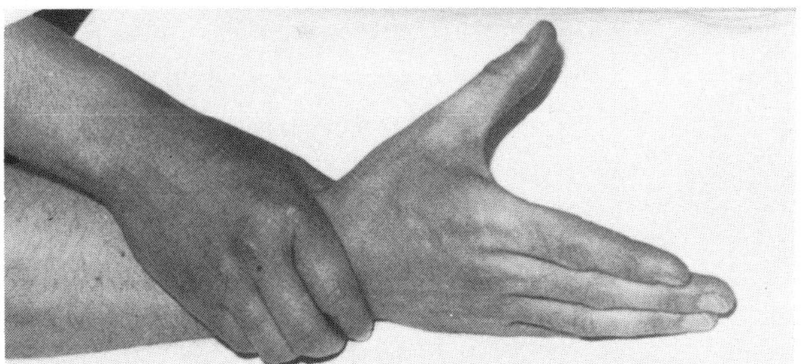

Range of movement should be complete.

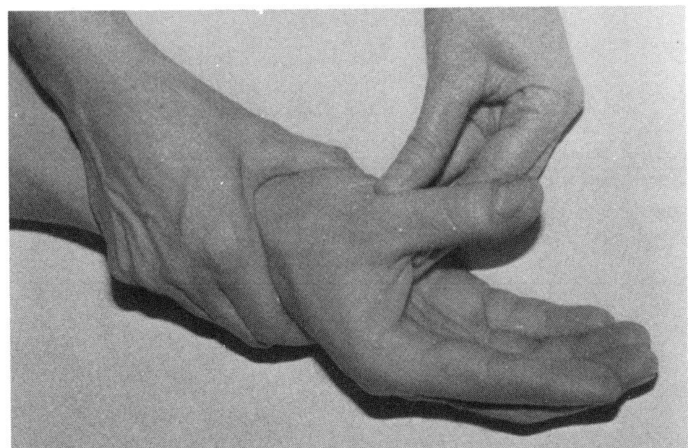

3, 4, and 5 Same position.

Same hold and movement.
Resistance to the movement is placed at metacarpophalangeal joint level.
Resistance is minimal for grade 3 and increases for grades 4 and 5.

Grade 4: Power is sub-normal.
Grade 5: Power is normal.

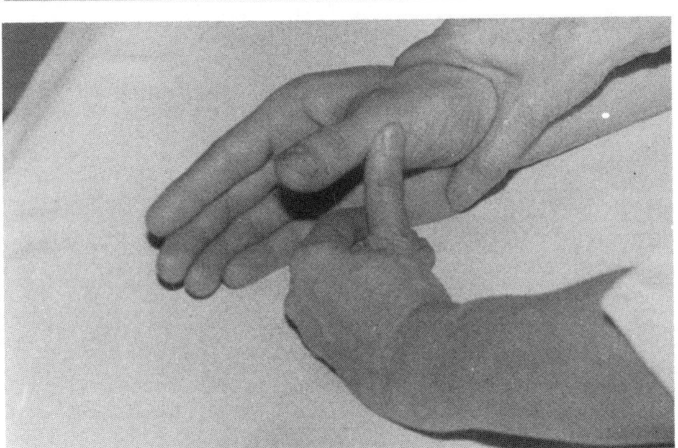

OPPONENS POLLICIS

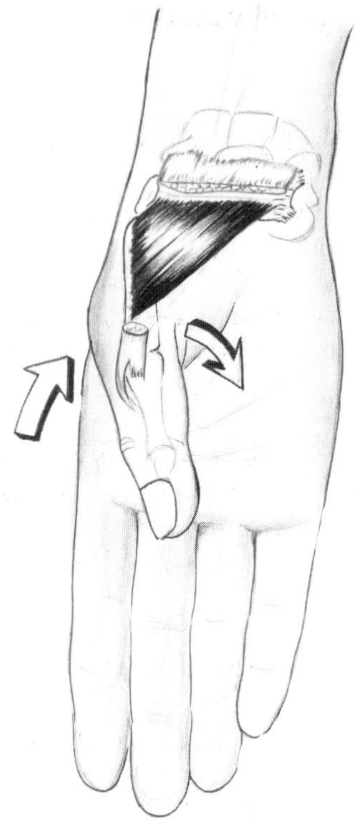

Origin

—Ridge of trapezium bone.
—Anterior transverse carpal ligament.

Insertion entire length of lateral border of first metacarpal bone.

Innervation Median (C8, t1).

Function

Internal axial rotation of first metacarpal bone.
Anterior position and adduction of first metacarpal.
These movements place the thumb anteriorly to the second metacarpal so that its palmar surface is facing inward.
The opponens is active in practically all prehensile activities requiring internal rotation of the thumb.
Note: The opponens pollicis has no action on the phalanges.

Deficiency Difficulty in picking up small objects.

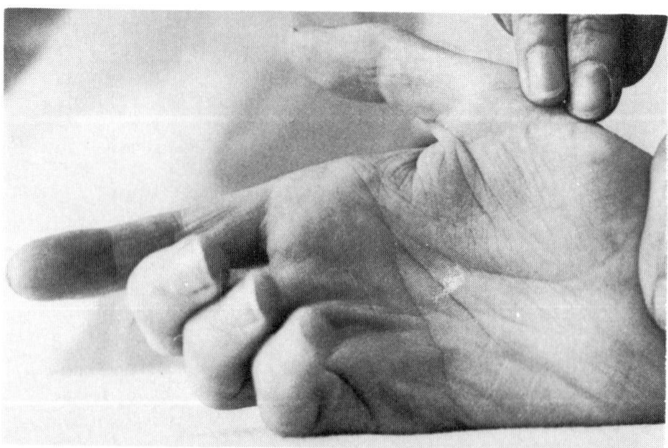

Subject presenting a paralysis of intrinsic thumb muscles. Note importance of muscular wasting.

0 and 1

A deep layer muscle, the opponens pollicis is palpable at the lateral border of the first metacarpal only if the other thenar muscles have atrophied.

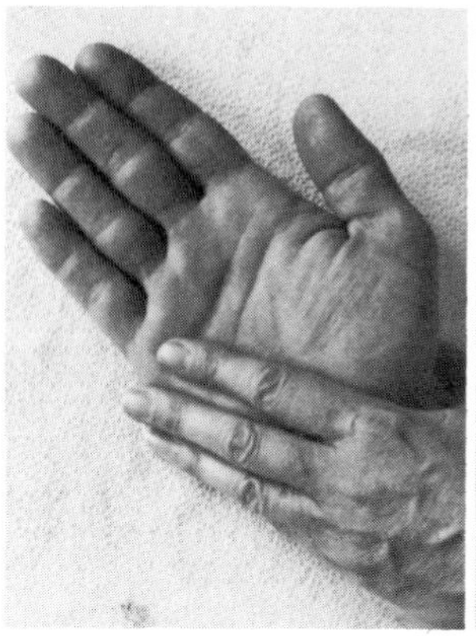

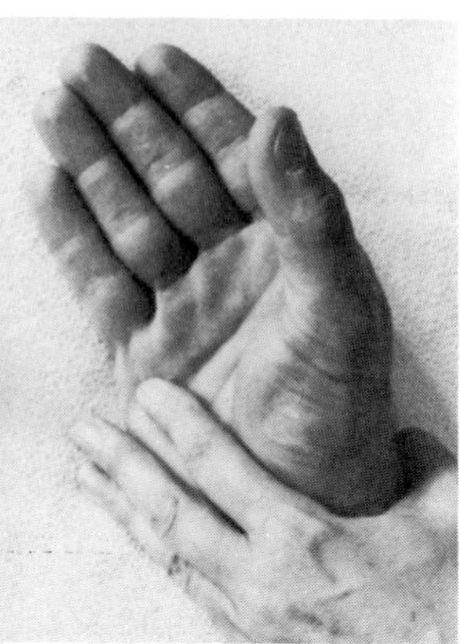

2 Subject sitting.

Forearm and hand in supination resting on table.
Hold medial aspect of carpus.
Phalanges should remain relaxed throughout the movement.
Ask subject to bring his thumb into opposition: internal rotation with slight anterior movement so that the thumb is oriented towards the palmar aspect of the fingers.
The first metacarpal is at the level of the second.
If the thumb skims the palm of the hand with proximal phalanx in important flexion, the deep head of flexor pollicis brevis dominates.
If anterior position is emphasized, abductor pollicis brevis dominates.
If the thumb goes into adduction against the second metacarpal, with a slight external rotation, the adductor dominates.
Range of movement should be complete.

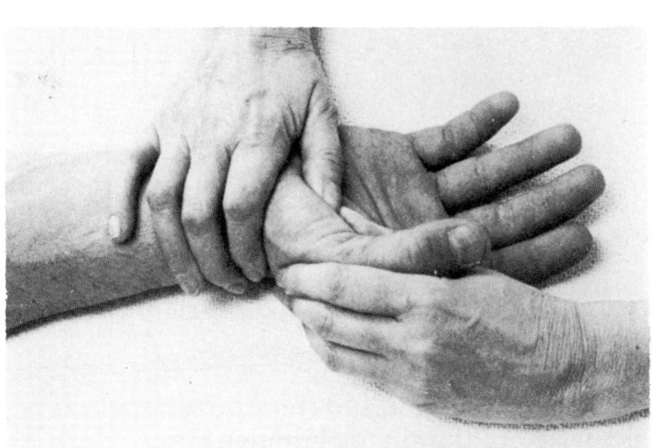

3, 4, and 5 Same position.

Ask for the same movement.
The same compensatory movements are possible.
Resistance to the movement is placed along the entire length of medial part of first metacarpal.
Resistance is minimal for grade 3 and increases for grades 4 and 5.

Grade 4: Power is sub-normal.
Grade 5: Power is normal.

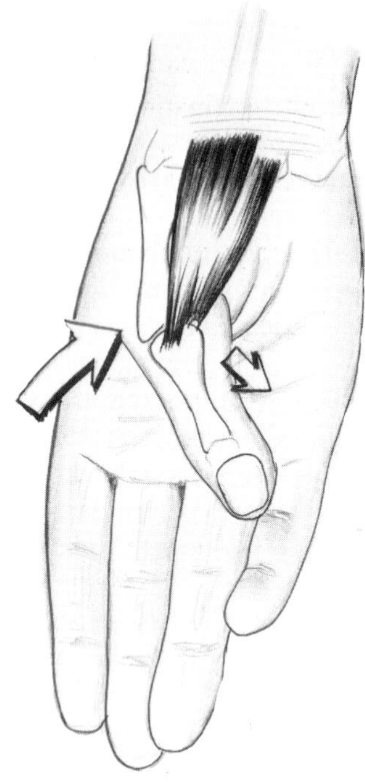

Composed of two heads.

Origin

Superficial head:
—Ridge of trapezium.
—Anterior transverse carpal ligament.
Deep head:
—Trapezoid.
—Capitate.

Insertion

—Lateral sesamoid of thumb metacarpophalangeal joint.
—Lateral tubercle of base of proximal phalanx of thumb.

Innervation

—Superficial head: median (C8, T1).
—Deep head: ulnar (C8, T1).

Function

Common for both heads:

Flexion of proximal phalanx on the first metacarpal.
They participate in extension of distal phalanx on the proximal.

Deep head:

Adduction of first metacarpal bone so that it is placed anteriorly to second and third metacarpals. It is this movement which causes the ulnar side of the distal phalanx of thumb to glide along the metacarpal heads of the long fingers.
It participates in thumb opposition by rotating it internally.

Superficial head:

Anterior position of thumb (this movement is less than that produced by abductor pollicis brevis) with important internal axial rotation, thus it participates very actively in opposition. This series of movements opposes the thumb with the last two digits, flexing the metacarpophalangeal joint. The flexor pollicis brevis participates in subterminal and subtermino-lateral opposition, and full palmar prehension.
It allows contact of thumb with little finger.

Deficiency

Isolated loss is rare.
Decrease in flexion capacity of proximal phalanx on first metacarpal.
Difficulty in power grip between thumb and fingers.

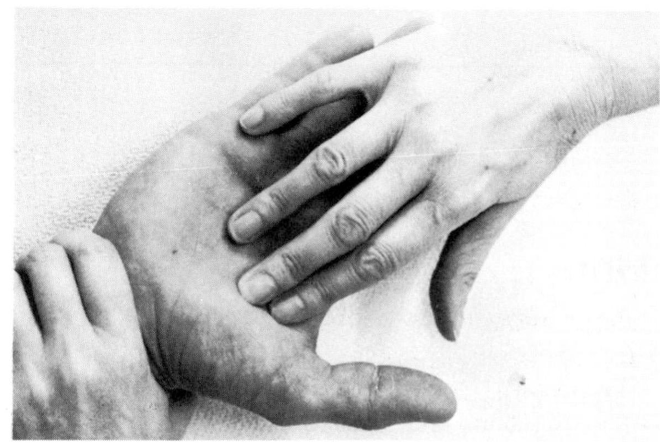

0 and 1 Subject sitting.

Forearm and hand in supination resting on table. Ask subject to flex the proximal phalanx on the first metacarpal, bringing his thumb towards the palmar aspect of his hand.
Only the superficial head can be palpated. It is found on the thenar eminence medially to abductor pollicis brevis.

Grade 0: No contraction is felt.
Grade 1: Contraction is felt but there is no joint movement.

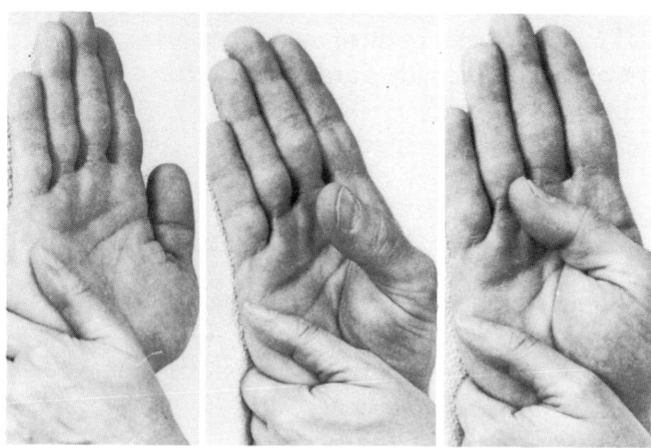

2 Same position.

Hold medial aspect of carpus, the thumb being in anatomical position.

Superficial head:

Ask for flexion of proximal phalanx on first metacarpal bringing the first metacarpal into anterior position and opposition so that the pulp of the thumb is oriented towards the proximal phalanges of the last two fingers.

Deep head:

Ask for flexion of proximal phalanx on first metacarpal which is drawn in front of second and third metacarpals. This is the movement which causes the side of the distal phalanx to glide over the heads of the metacarpals.
Range of movement should be complete.

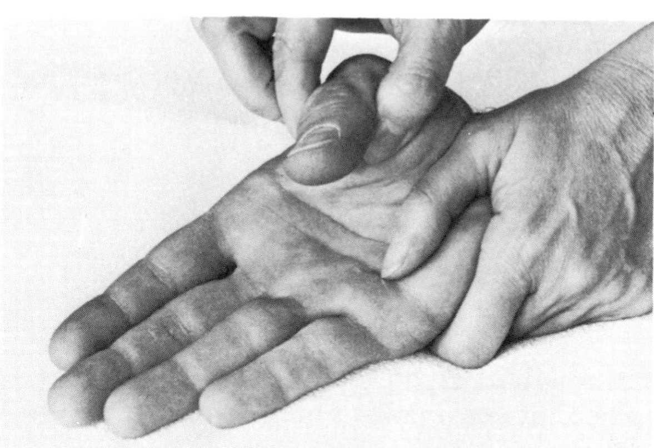

3, 4, and 5 Superficial head.

Same position, hold, and movement.
Resistance to the movement is placed on the sides of the proximal phalanx. Flexion of proximal phalanx is resisted by pulling the thumb back towards its initial position.
Resistance is minimal for grade 3 and increases for grades 4 and 5.

Grade 4: Power is sub-normal.
Grade 5: Power is normal.

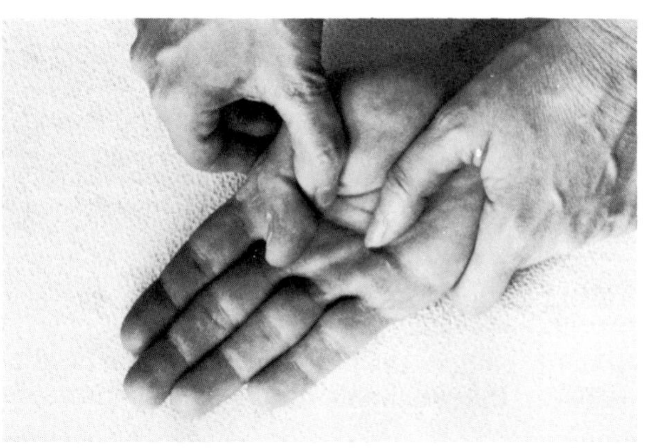

3, 4, and 5 Deep head.

Same position, hold and movement.
Resistance to the movement is placed on the palmar aspect of the proximal phalanx. Flexion of proximal phalanx is resisted by pulling the thumb back towards its initial position.
Resistance is minimal for grade 3 and increases for grades 4 and 5.

Grade 4: Power is sub-normal.
Grade 5: Power is normal.

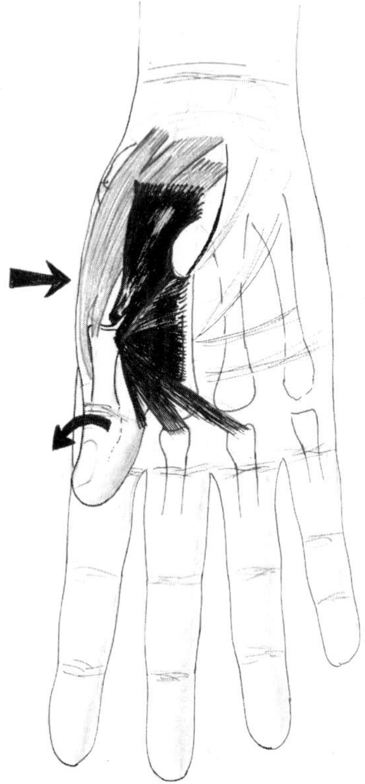

Origin

Oblique head (adductor obliquus pollicis):
—Capitate.
—Trapezoid.
—Base of second and third metacarpal bones.
—Interosseous carpal ligament.
Transverse head (adductor transverses pollicis):
—Anterior border of third metacarpal bone.
—Capsule of second, third, and fourth metacarpophalangeal joints.
—Deep palmar aponeurosis.

Insertion

By a common tendon into:
—Medial sesamoid of metacarpophalangeal joint of thumb.
—Medial tuberosity of base of proximal phalanx of thumb.

Innervation Ulnar (C8, T1).

Function

Adduction of first metacarpal bone: the adductor pollicis draws the first metacarpal towards the second and places it slightly anteriorly to it.
Adductor pollicis extends the distal phalanx on the first through its expansion. It also provokes a movement of external rotation. It is assisted by the deep head of flexor pollicis brevis with which it is sometimes closely related.
Adductor pollicis brings the thumb back into resting position, whatever the starting position.
It participates in all power grips of the thumb.
Note: In all movements it is assisted by the first palmar interosseus.

Factor limiting movement Movement is partially limited by contact with the second metacarpal bone.

Contracture or tightness Retracts the first web space.

Deficiency Froment's sign: loss of subtermino-lateral opposition.
Flexor pollicis longus may substitute in this deficiency.
Note: If the thenar muscles are paralysed, except for adductor pollicis, the isolated contraction of this muscle is quite useful since it enables gripping objects between the proximal phalanx and the palm of the hand.

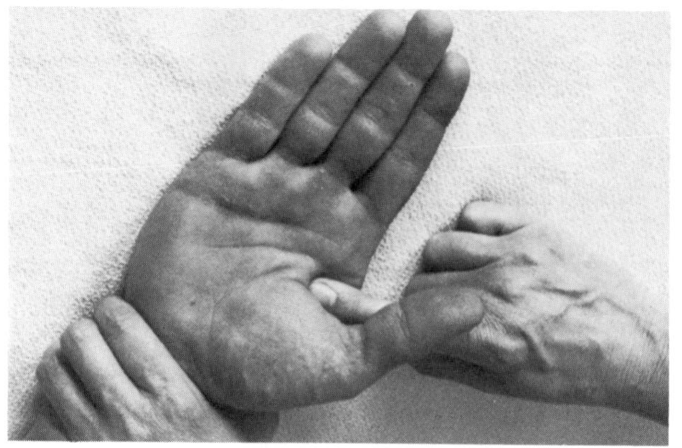

0 and 1 Subject sitting.

Forearm and hand in supination resting on table.
Ask subject to bring his thumb inward, slightly anteriorly to the second metacarpal.
Fibres can be palpated at the inferior part of the thenar eminence, at second metacarpal level.

Grade 0: No contraction is felt.
Grade 1: Contraction is felt but there is no joint movement.

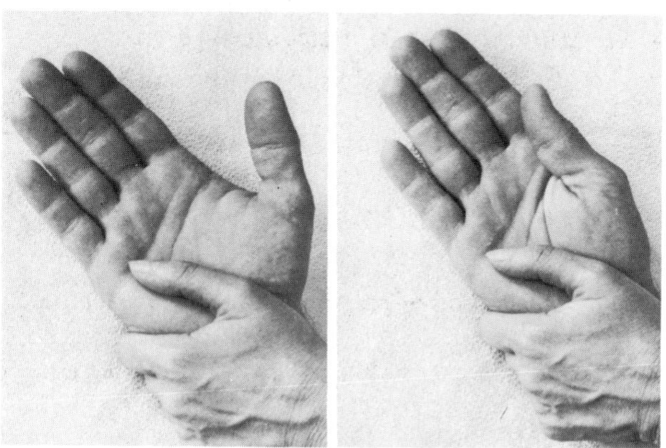

2 Same position.

Hold medial aspect of carpus and metacarpus.
Ask subject for thumb adduction. The thumb should be moved towards the second metacarpal with slight external rotation. The first metacarpal is in front of the second but should not go into full anterior position, movement produced by abductor pollicis brevis.
Range of movement should be complete.

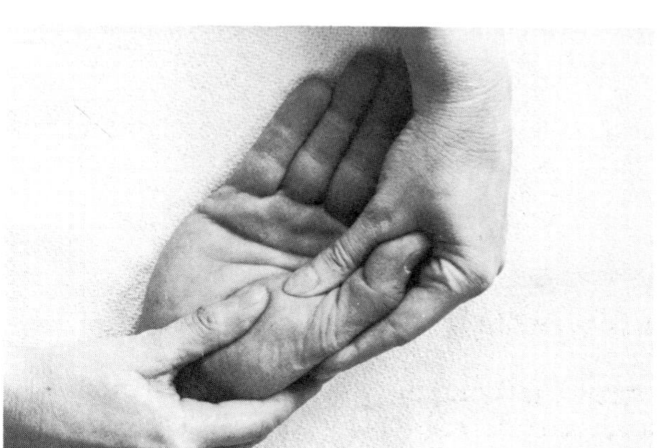

3, 4, and 5 Same position.

Ask subject for same movement.
Resistence to the movement is placed along the entire length of the first metacarpal.
Resistance is minimal for grade 3 and increases for grades 4 and 5.

Grade 4: Power is sub-normal.
Grade 5: Power is normal.

LUMBRICALES

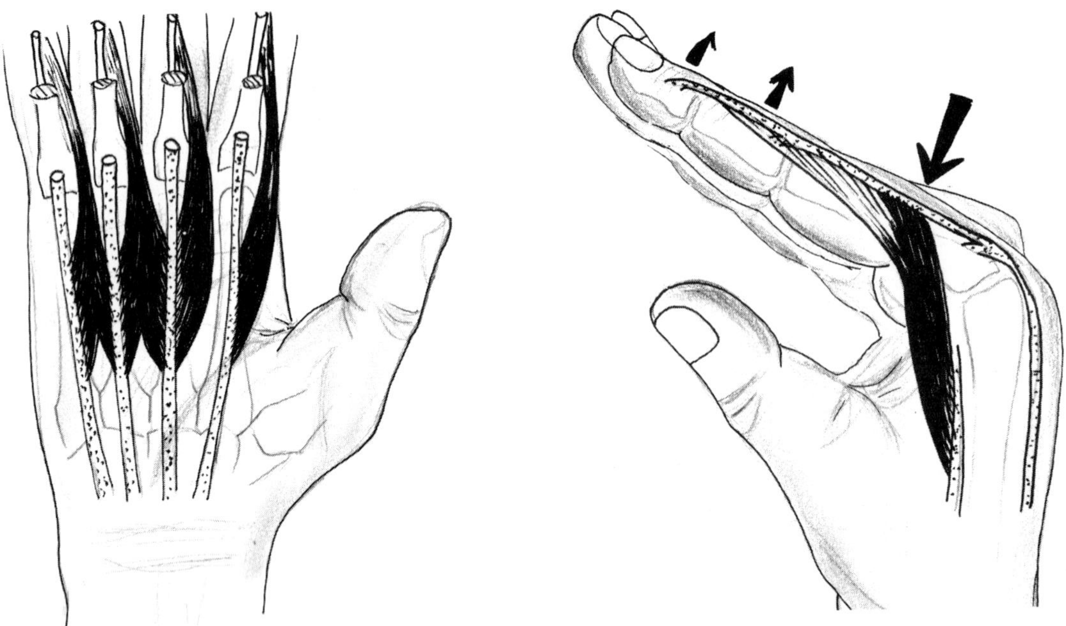

Origin

—First and second lumbricales:
Radial side and anterior aspect of flexor profundus tendon of index and middle fingers.
—Third and fourth lumbricales:
Sides and anterior aspect of flexor profundus tendons of ring and little fingers between which they are located.

Insertion Radial side of corresponding extensor tendon at level of proximal, middle and distal phalanges.

Innervation

—First and second lumbricales: median (C7, C8, T1).
—Third and fourth lumbricales: ulnar (C8, T1).

Function

Flexion of proximal phalanges on metacarpals.
Extension of middle and distal phalanges independently of metacarpophalangeal joint position.
Lumbricales act synergistically with dorsal and palmar interossei.
It is difficult to distinguish the activity of each in their common function.
The first lumbrical participates a little in index finger abduction.

Contracture or tightness Exceptional.

Deficiency

Decrease of flexion capacity of proximal phalanx on metacarpal, and of extension of middle and distal phalanges.
Claw-hand deformity: extension of proximal phalanx associated with flexion of middle and distal phalanges.
Functional impairment in writing, holding cards, squeezing objects (power grips, digito-palmar grips) . . .

Single muscle testing of lumbricales is difficult since their action is shared with that of dorsal and palmar interossei.

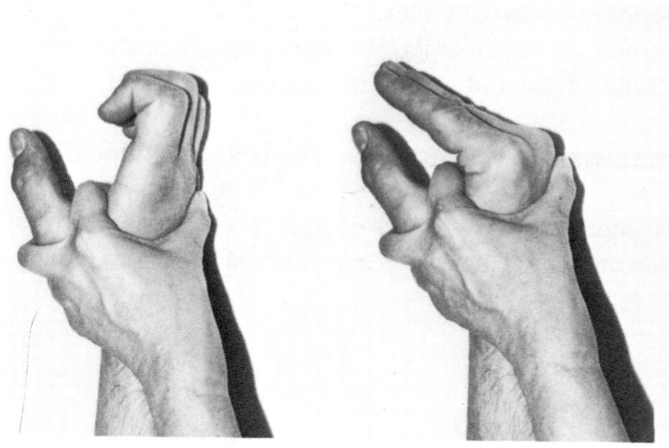

0 and 1

Lumbricales are deep muscles and cannot be palpated.

2 Subject sitting.

Wrist and hand resting on ulnar side.
Wrist straight.
Hold metacarpus in place.
Proximal phalanges in extension, middle and distal phalanges in flexion.
Ask subject for flexion of proximal phalanges and extension of middle and distal phalanges.
Range of movement should be complete.

3, 4, and 5 Same position.

Wrist is held straight so as not to interfere with manual resistance.
Resistance to the movement is placed on palmar aspect of proximal phalanx.
Resistance is minimal for grade 3 and increases for grades 4 and 5.
Note: Global testing of lumbricales is possible.

Grade 4: Power is sub-normal.
Grade 5: Power is normal.

Lumbricales may be tested in their function as extensors of middle and distal phalanges when the metacarpophalangeal joints are in flexion.
Hold both sides of proximal phalanx.
Resistance to the movement is placed on the dorsal aspect of the middle and distal phalanges.

INTEROSSEI PALMARES

They are four in number, the middle finger has none. The first palmar interosseus is often closely associated with the adductor pollicis.

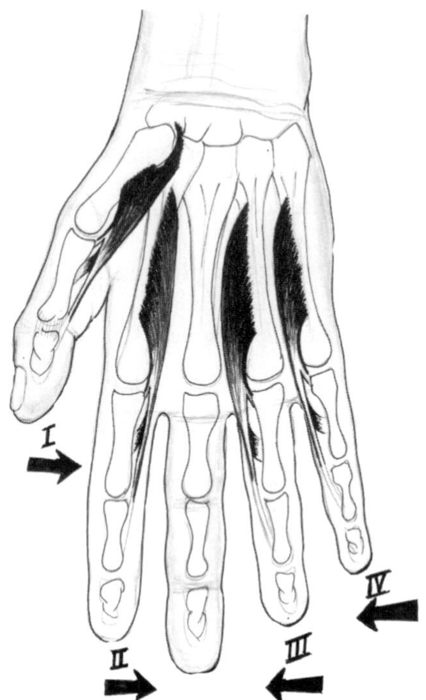

Origin

Anterior part of sides of metacarpal bones.

—First palmare interosseus: superior part of ulnar surface of shaft of first metacarpal. Superior part of second metacarpal bone and trapezium bone.

—Second palmar interosseus: ulnar surface of shaft of second metacarpal.

—Third palmar interosseus: radial surface of shaft of fourth metacarpal.

—Fourth palmar interosseus: radial surface of shaft of fifth metacarpal.

Course

—The first runs towards the thumb.

—The second runs towards the index.

—The third runs towards the ring finger.

—The fourth runs towards the little finger.

Insertion

—Deep head: lateral tubercle of base of proximal phalanx.

—Superficial head: corresponding extensor tendon at level of proximal, middle and distal phalanges.
Over the proximal phalanx, they form the extensor expansion.

Innervation Ulnar (C8, T1).

Function

Adduction of thumb, index, ring and little fingers.
The first palmar interosseus draws the thumb close to the index, assisted by the adductor pollicis and deep head of flexor pollicis brevis.
The second draws the index close to the middle finger.
The third draws the ring finger close to the middle fingers.
The fourth draws the little finger close to the ring finger.
Palmar interossei flex the proximal phalanx on the metacarpus and extend the middle and distal phalanges. With the dorsal interossei and lumbricales, they are the true extensors of the middle and distal phalanges.

Contracture or tightness Exceptional.

Deficiency

Their isolated paralysis is rare. Associated with paralysis of dorsal interossei, it leads to atrophy of interossei spaces: the so-called skeleton hand.
Paralysis of interossei and lumbricales makes flexion of the proximal phalanges impossible and markedly decreases extension of middle and distal phalanges.
It is impossible to grasp an object with force.
There is considerable functional impairment in any kind of prehensile movement: e.g. writing, . . .

Testing is done with the fingers straight. Convergence into flexion is thus eliminated.

Testing interossei palmares in their function of adduction

0 and 1

Since the interossei palmares are located deep in the intermetacarpal spaces, their palpation is extremely difficult.

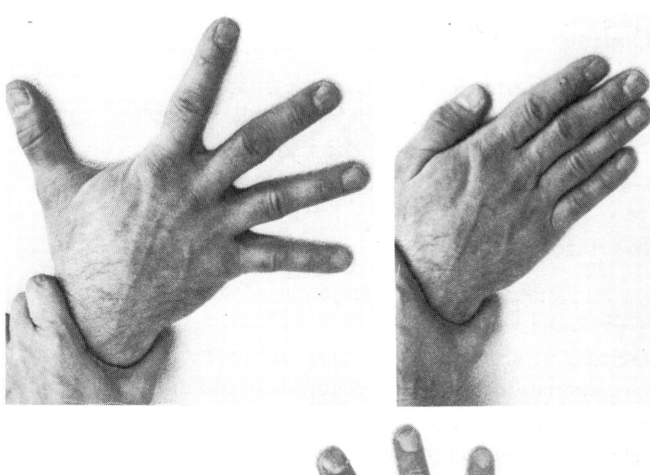

2 Subject sitting.

Forearm and hand in pronation resting on table.
Hold dorsal aspect of wrist.
Spread fingers apart.
Ask subject to approximate his thumb, his index, ring finger and little finger to the middle axis of his hand.
Range of movement should be complete.

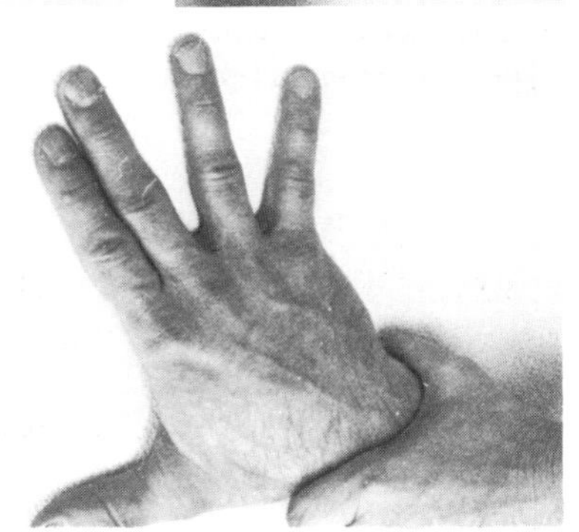

Testing second interosseus palmaris. Global testing

3, 4, and 5 Same position.

Same hold and movement.
Resistance to the movement is placed on the sides of the phalanges, or of the distal phalanx only.
Resistance is minimal for grade 3 and increases for grades 4 and 5.

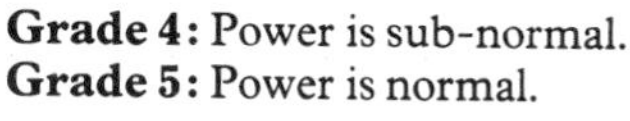

Grade 4: Power is sub-normal.
Grade 5: Power is normal.

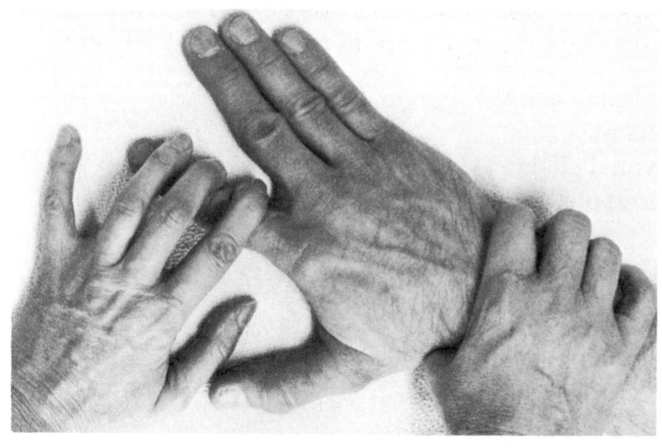

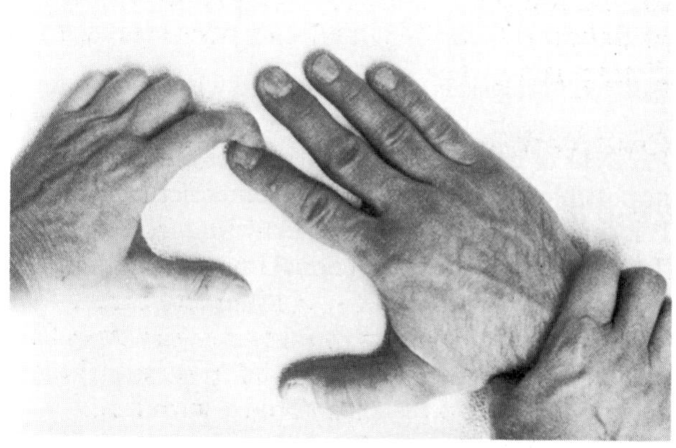

INTEROSSEI DORSALES

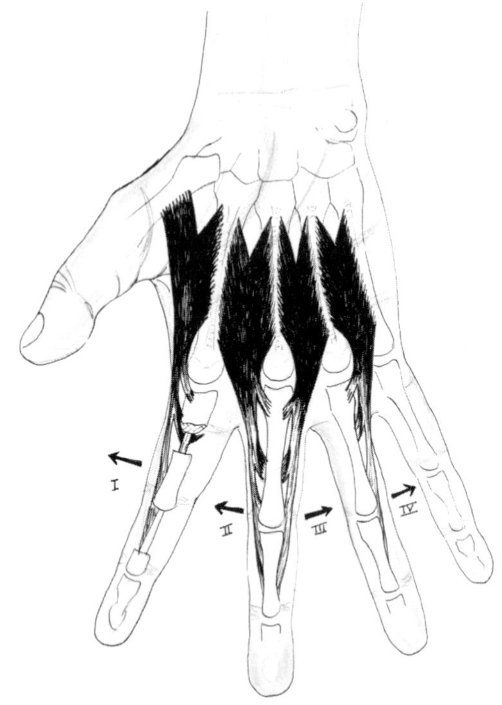

They are four in number.

Origin Posterior part of lateral surfaces of metacarpal shafts in each interosseus space.

Course

—The first runs towards the index finger.
—The second runs towards the middle finger.
—The third runs towards the middle finger.
—The fourth runs towards the ring finger.

Insertion

—Deep insertion: lateral tubercle of base of proximal phalanx.
—Superficial insertion: corresponding extensor tendon at level of proximal middle and distal phalanges.

Over the proximal phalanx, they form the extensor expansion.

Innervation Ulnar (C8, T1).

Function

Abduction of index, middle and ring fingers.
The first dorsal interosseus draws the index finger towards the thumb.
The second draws the middle finger towards the index.
The third draws the middle finger towards the ring finger.
The fourth draws the ring finger towards the little finger.
They flex the proximal phalanx on the metacarpus and extend the middle and distal phalanges acting with the palmar interossei and lumbricales.
The first dorsal interosseus participates in thumb adduction when the index is fixed in neutral (0°) position of abduction-adduction.

Contracture or tightness Exceptional.

Deficiency

Isolated paralysis is rare. Associated with that of the palmar interossei, it leads to atrophy of interossei spaces: the so-called skeleton hand.
Paralysis of interossei and lumbricales makes flexion of proximal phalanx impossible and greatly decreases extension of middle and distal phalanges.
It is impossible to grasp an object with force.
There is considerable functional impairment in any kind of prehensile activity: e.g. writing, . . .

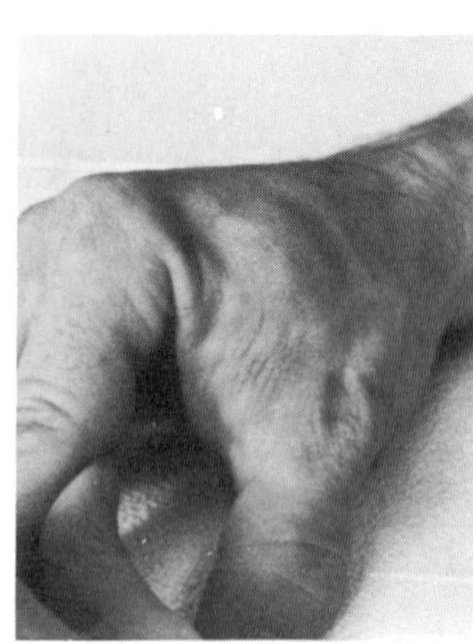

Showing first dorsal interosseus in the first web space.

Testing interossei dorsales in their function of abduction

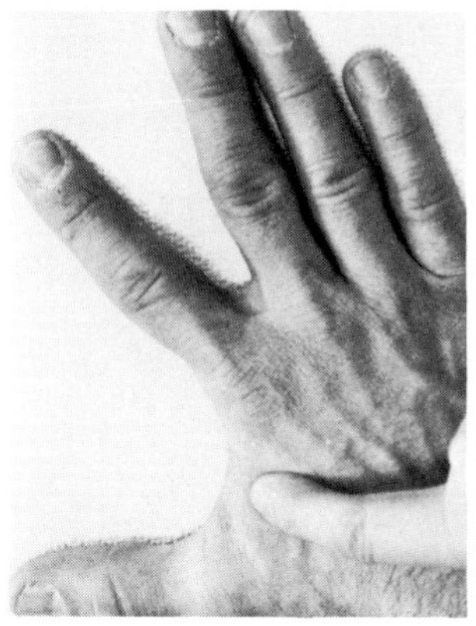
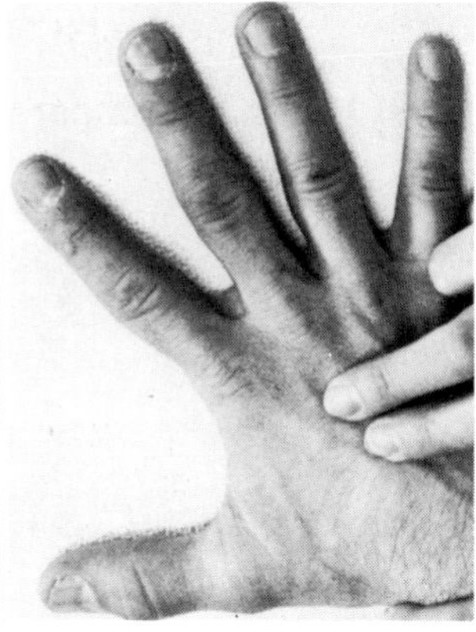

0 and 1 Subject sitting.

Forearm and hand in pronation resting on the table.
Wrist straight.
Ask subject to spread each finger.
Palpate first interosseus dorsalis in the first web space.
Differentiate its palpation from that of the adductor; the latter is more proximal and is located deeper.
Other interossei dorsales are palpated in their corresponding spaces.

Grade 0: No contraction is felt.
Grade 1: Contraction is felt but there is no joint movement.

2 Same position.

Hold dorsal aspect of wrist.
Separate the untested fingers so as to test each interosseus.
First interosseus: index deviates towards thumb.
Second interosseus: middle finger deviates towards index.
Third interosseus: middle finger deviates towards ring finger.
Fourth interosseus: ring finger deviates towards little finger.
Watch out for substitution by the extensor digitorum.
Range of movement should be complete.

Testing second and third interossei

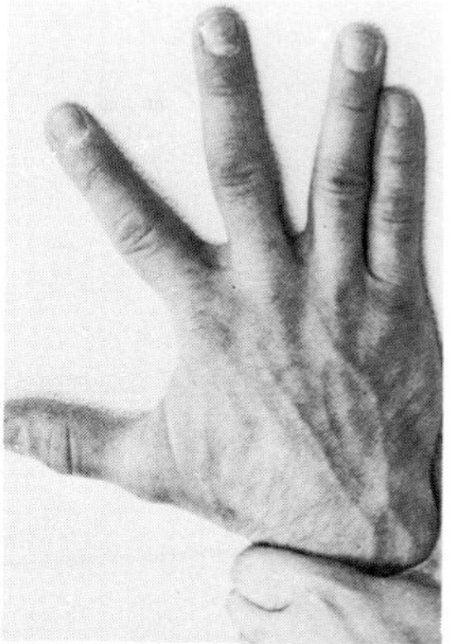
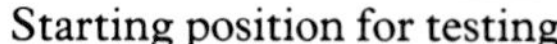

Starting position for testing

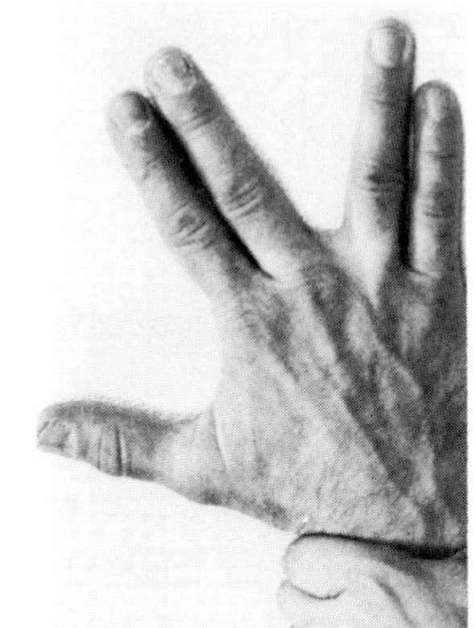

second

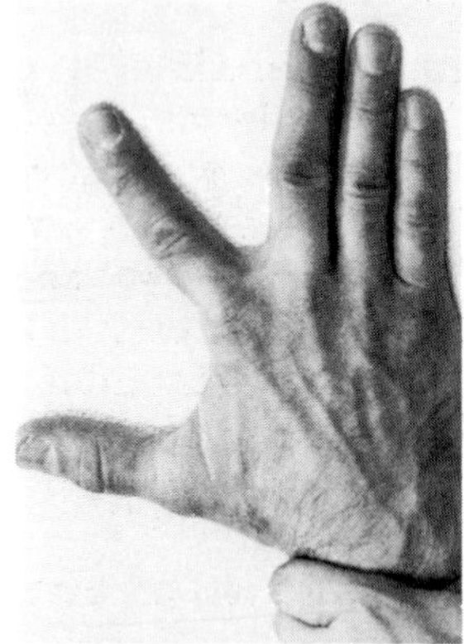

and third interossei.

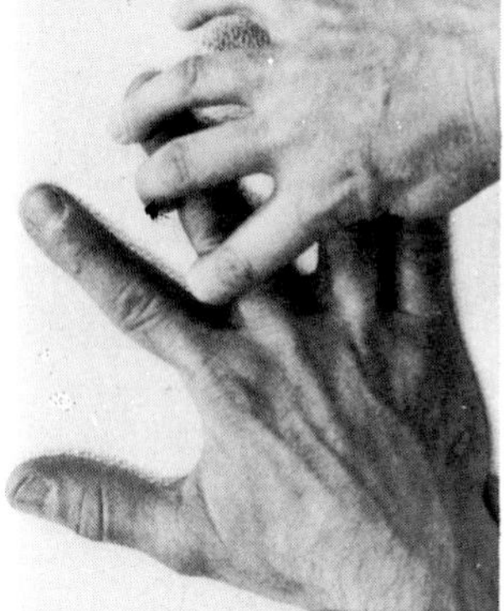
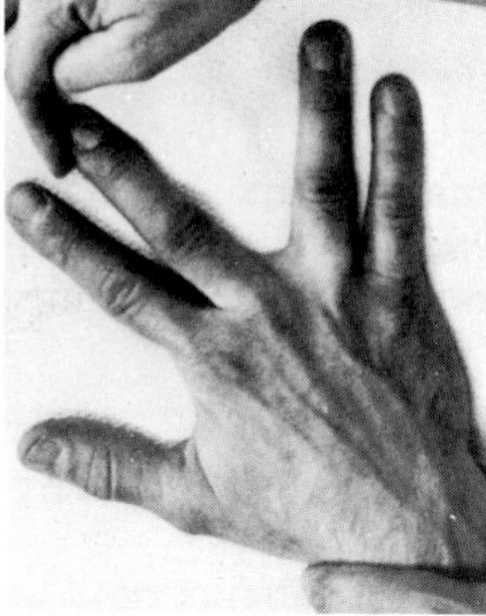

3, 4, and 5 Same position.

Same hold and movement.
Resistance to the movement is placed on the sides of the phalanges or on the lateral side of the distal phalanx.
Resistance is minimal for grade 3 and increases for grades 4 and 5.

Grade 4: Power is sub-normal.
Grade 5: Power is normal.

ABDUCTOR DIGITI MINIMI

Sometimes called adductor (in relation to mid-line of body). Can be compared to a dorsal interosseus.

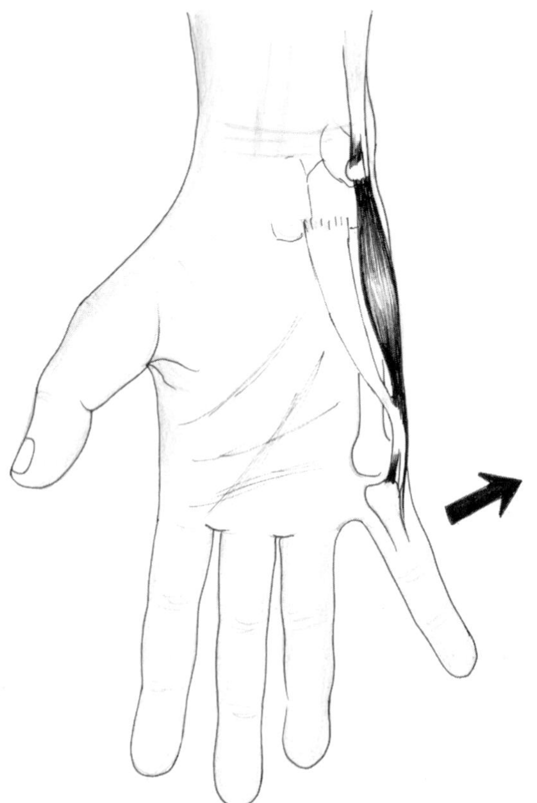

Origin

—Distal part of pisiform bone.
—Tendon of flexor carpi ulnaris.

Insertion

—Medial tubercle of base of proximal phalanx of little finger.
—Anterior (glenoid) ligament of metacarpophalangeal joint and sesamoid bone.
—Extensor tendon of the little finger. It participates in formation of the extensor expansion.

Innervation Ulnar (c8, T1).

Function

Abduction of little finger in relation to mid-line of hand.
Abductor digiti minimi participates in flexion of proximal phalanx on metacarpus.
It also participates in extension of middle and distal phalanges through its expansions.
Thus its function is similar to that of the interossei among which it is included by certain anatomists.

Deficiency

Marked decrease in abduction capacity of little finger, partly compensated by the extensor digitorum and extensor digiti minimi.
The extensor carpi ulnaris also draws the fifth metacarpal into abduction.

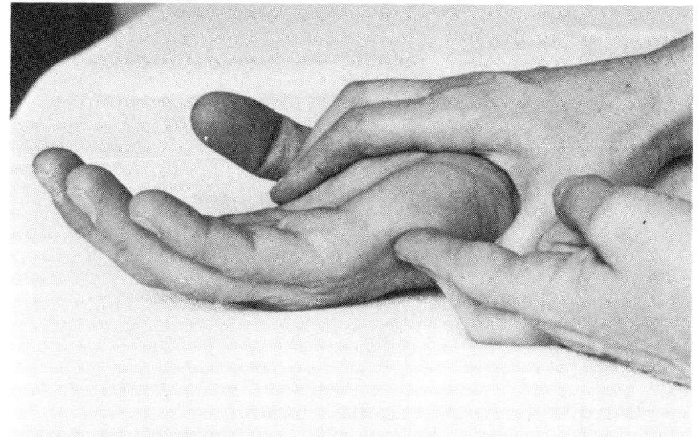

0 and 1 Subject sitting.

Forearm and hand in supination resting on table.
Ask subject to abduct his little finger.
Palpation is easy on the ulnar side of the fifth metacarpal.

Grade 0: No contraction is felt.
Grade 1: Contraction is felt but there is no joint movement.

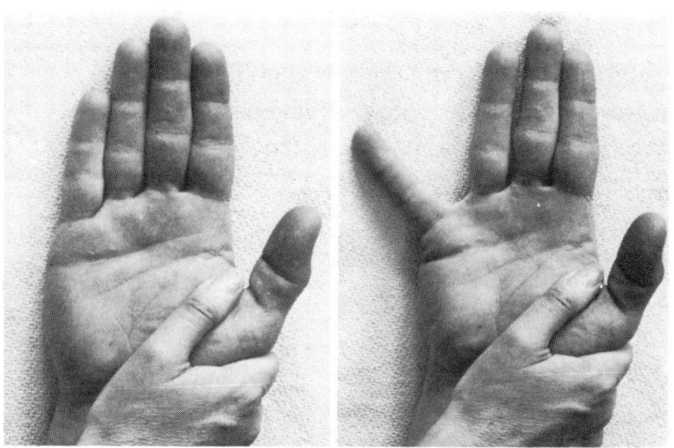

2 Same position.

Hold carpus leaving fifth metacarpal free.
Ask subject to abduct his little finger.
Range of movement should be complete.

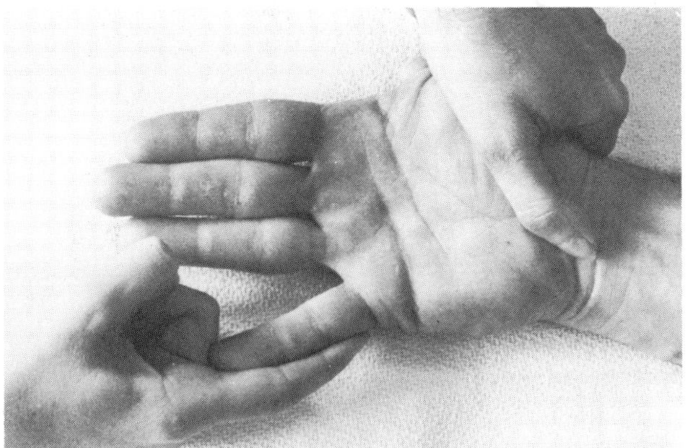

3, 4, and 5 Same position.

Same hold and movement.
Resistance to the movement is placed on the medial aspect of the proximal phalanx or on the medial aspect of the distal phalanx (not shown).
Resistance is minimal for grade 3 and increases for grades 4 and 5.

Grade 4: Power is sub-normal.
Grade 5: Power is normal.

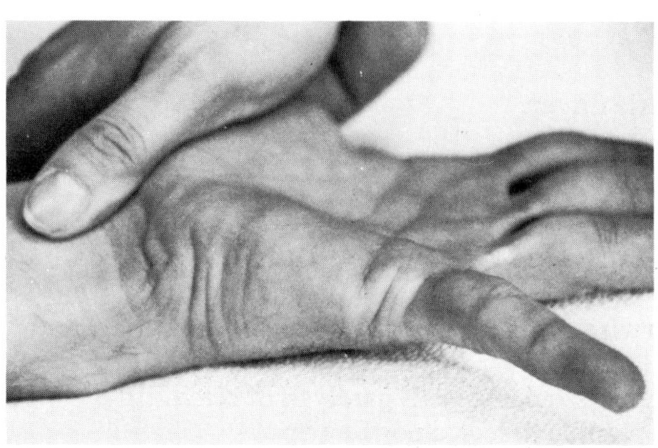

Note contraction of palmaris brevis shown by flexion crease running obliquely upwards and backwards.

FLEXOR DIGITI MINIMI BREVIS

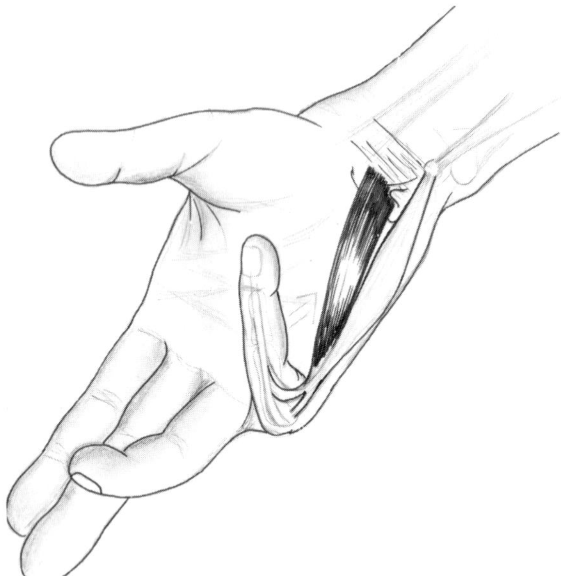

Origin

—Hamulus (hook) of hamate bone.
—Anterior transverse carpal ligament.

Insertion

It interlaces with the tendon of abductor digiti minimi and inserts into medial tubercle of base of proximal phalanx, anterior (glenoid) ligament, and sesamoid bone.

Innervation

Ulnar (c8, T1).

Function

Flexion of proximal phalanx on metacarpal bone of little finger.
It abducts the little finger slightly and participates in opposition.

Contracture or tightness Flexed position of proximal phalanx.

Deficiency

Decrease in flexor capacity of proximal phalanx, otherwise compensated by the fourth lumbrical and the fourth palmar interosseus.
Slight decrease in little finger abduction and opposition. Associated with paralysis of remaining hypothenar muscles, its deficiency makes it nearly impossible to lock a power grip.

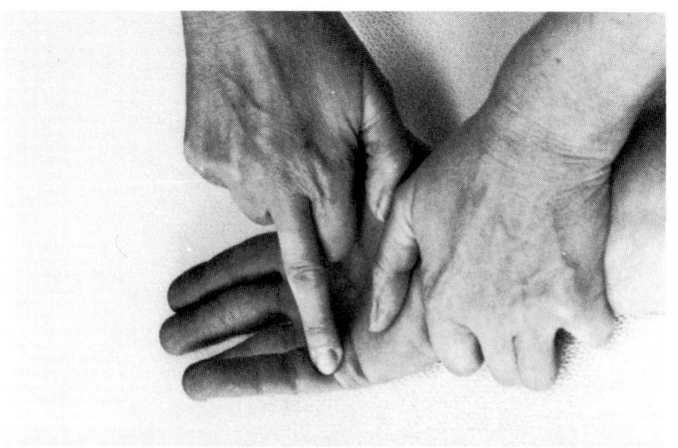

0 and 1 Subject sitting.

Forearm and hand in supination resting on table.
Ask subject to flex the proximal phalanx on the fifth metacarpal bone.
The tendon is palpated at the latero-medial part of the head of the fifth metacarpal.

Grade 0: No contraction is felt.
Grade 1: Contraction is felt but there is no joint movement.

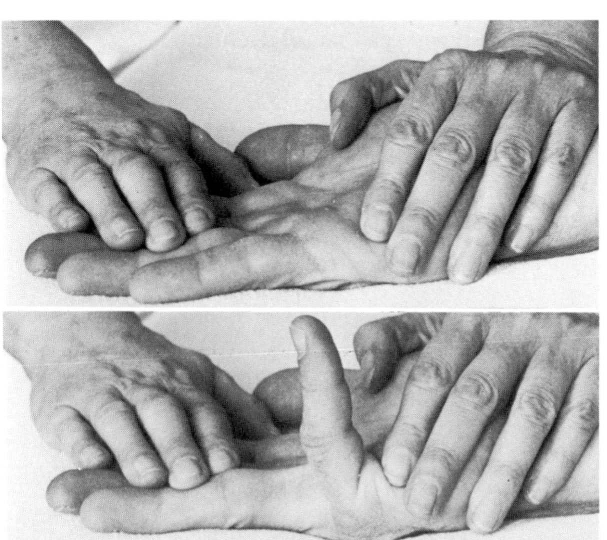

2 Same position.

Hold the medial aspect of the metacarpus and the first four fingers.
Ask subject to flex the proximal phalanx on the fifth metacarpal.
Range of movement should be complete.

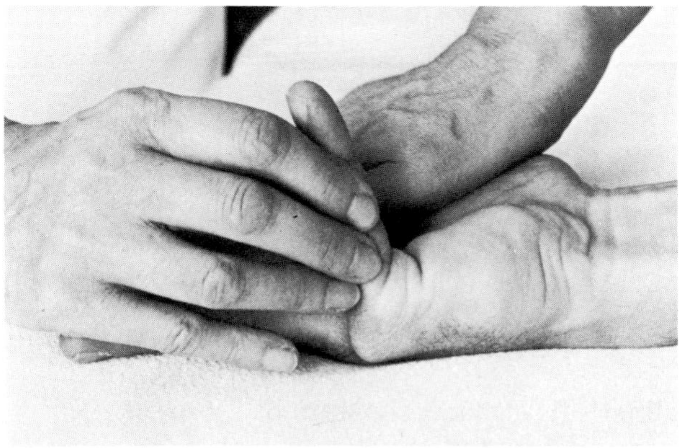

3, 4, and 5 Same position.

Hold the metacarpus.
Ask for the same movement.
Resistance to the movement is placed on the sides of the proximal phalanx.
Resistance is minimal for grade 3 and increases for grades 4 and 5.

Grade 4: Power is sub-normal.
Grade 5: Power is normal.

OPPONENS DIGITI MINIMI

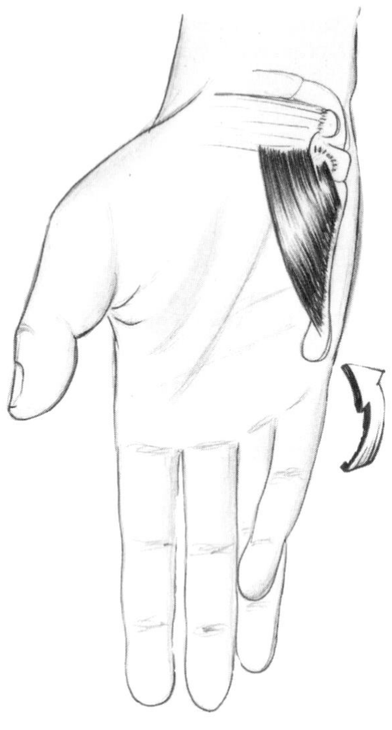

Origin

—Hamulus (hook) of hamate bone.
—Anterior transverse carpal ligament.

Insertion Medial border and medial surface of entire length of fifth metacarpal.

Innervation Ulnar (c8, T1).

Function

External rotation of the fifth metacarpal so that the palmar surface faces the thumb.
Flexion of the fifth metacarpal bone, bringing it forward and outward.
Thus it participates in the hollowing of the palm.
Note: Acting with the other hypothenar muscles, it permits gripping with the palm of the hand. This grip is used by subjects whose fingers have been amputated.

Deficiency

Decreases the power of opposition grip.
Locking a power grip becomes nearly impossible.

0 and 1

Palpation at the ulnar side of the fifth metacarpal is only possible if the other hypothenar muscles have atrophied.

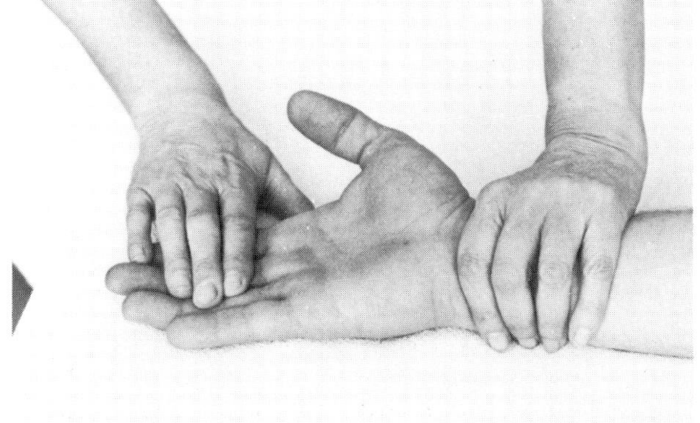

2 Subject sitting.

Forearm and hand in supination resting on table.
Hold anterior aspect of wrist and the first four fingers.
Ask subject to bring his little finger into opposition.
Range of movement should be complete.

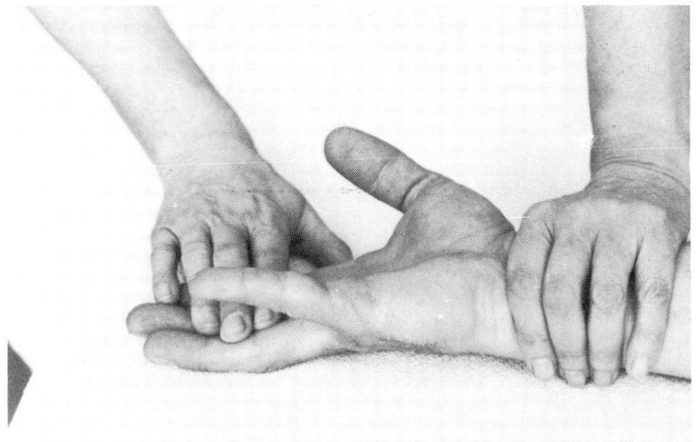

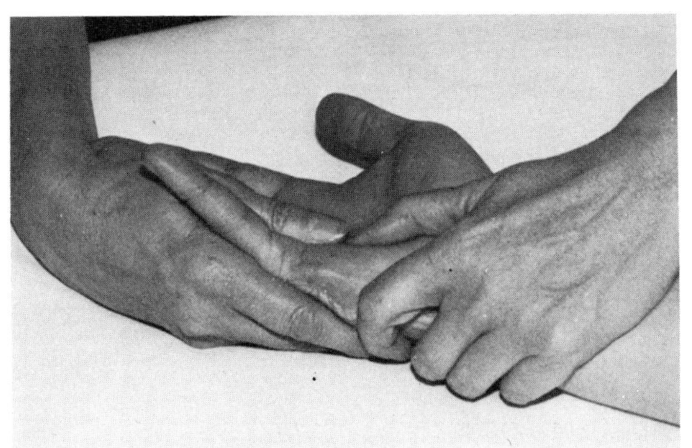

3, 4, and 5 Same position.

Resistance to the movement is placed along the entire length of the radial side of the palmar surface of the fifth metacarpal.
Resistance is minimal for grade 3 and increases for grades 4 and 5.

Grade 4: Power is sub-normal.
Grade 5: Power is normal.

TRUNK MUSCLES

MUSCLES OF THE ABDOMEN

GROSSIORD'S STAR

Observe displacement of the navel (umbilicus) during abdominal muscle contraction. It is attracted by dominating muscles.
Subject supine.
Ask him to cough.

1. Upper portion of left and right recti abdominis dominates.
2. Left obliquus abdominis externus dominates.
3. Left obliqui dominate.
4. Left obliquus abdominis internus dominates.
5. Lower portion of left and right recti abdominis dominates.
6. Right obliquus abdominis internus dominates.
7. Right obliqui dominate.
8. Right obliquus abdominis externus dominates.

This test reveals dominant muscles at a glance.

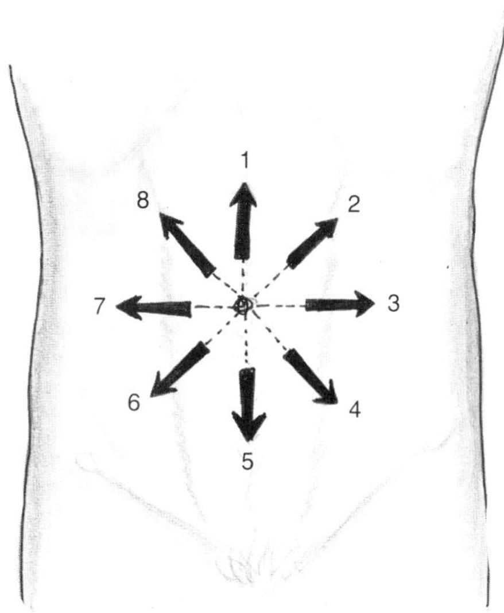

RECTUS ABDOMINIS

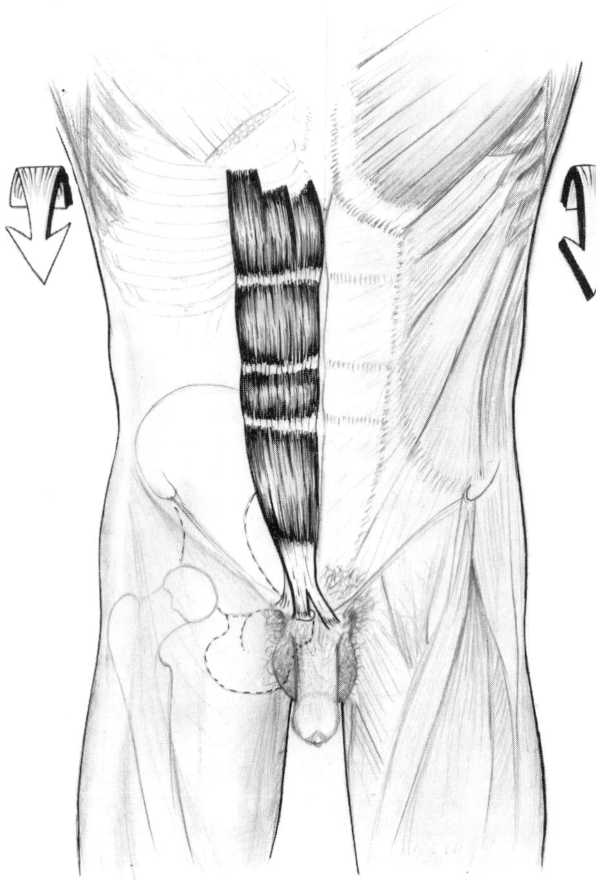

Origin

—Fifth, sixth, and seventh costal cartilages.
—Most lateral digitation spreads over the fifth rib.
—Most medial digitation spreads over the anterior chondroxyphoid ligament and anterior surface of xyphoid process.

Insertion Through two tendons:

—Lateral tendon: upper border and anterior surface of pubis from the pubic spine to the symphysis.
—Medial tendon: anterior surface of symphysis, some fibres crossing to contralateral symphysis.

Innervation

—Last six intercostal (T7 to T12)
—Iliohypogastric and ilio-inguinal (L1).

Note: The pyramidalis abdominis, located close to the inferior portion of the recti, runs from symphysis and pubis to anterior and lateral surface of inferior part of linea alba.

Function

With pelvis fixed

Subject supine, trunk flexion with increase of thoracic kyphosis and straightening of lumbar lordosis. Thorax is drawn toward pubis. Standing, trunk flexion is affected by gravity. The recti contract only in the last range of this movement.

With thorax fixed

Tilts pelvis anteriorly (symphysis up) and approximates it to the thorax, flattening lumbar lordosis.
Recti abdominis contract intensely on a supine subject raising and lowering the lower extremities (counteracting the lordotic action of iliopsoas).
They stabilize the thorax during head and neck flexion.
Discrete role in spinal column statics.
Recti abdominis contract during heavy exercise and body functions as do the other abdominal muscles.

Deficiency

Marked decrease in capacity for trunk flexion and pelvic anterior tilt (otherwise compensated for by the obliqui).
Tendency towards posterior pelvic tilt inducing lumbar hyperlordosis.
Impairment of body functions.

Supra-umbilical portion of recti abdominis is tested in trunk-raising, i.e. dynamic trunk flexion on lower extremities.
The infra-umbilical portion is tested in the lowering of the lower extremities with static holds at different levels.

TESTING RECTUS ABDOMINIS, SUPRA-UMBILICAL PORTION

Certain examiners prefer testing the supra-umbilical portion of recti abdominis with the lower extremities extended on the table, other examiners prefer them flexed. In either case it is unlikely that the action of hip flexors, particularly iliopsoas, will be eliminated. Placing the hips in flexion, among other things, avoids lumbar lordosis and tilts the pelvis anteriorly. It is therefore impossible, as the trunk curls, to watch the performance of the lumbar spine which the supra-umbilical portion of rectus abdominis should straighten.
Holding the subject's feet during anterior trunk-raising favours the action of hip flexors greatly. It is therefore preferable not to hold them during this test. However, holding them at the end of the movement helps to stabilize the subject, particularly when the centre of gravity is high.

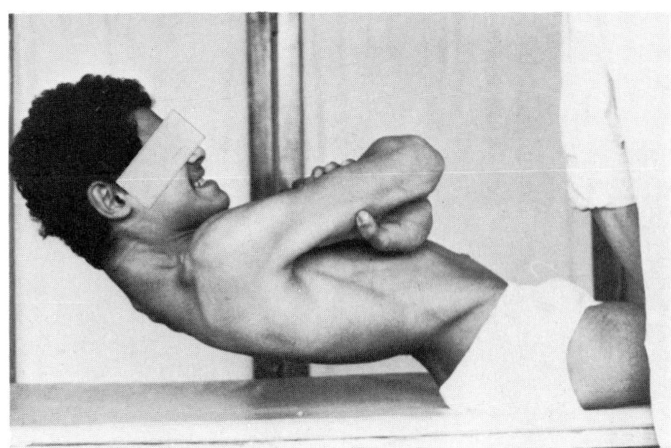

When lower limbs are extended during the test, and if the lumbar spine remains in lordosis, hip flexors (particularly iliopsoas) dominate.

Before testing the recti abdominis, test the mobility of the spinal column, the extensibility of the erector spinae and the strength of the head and neck flexors.
The proposed test is global. The difference between right and left sides is assessed through palpation and eventual navel displacement (Grossiord's star).

SUPRA-UMBILICAL PORTION OF RECTI ABDOMINIS

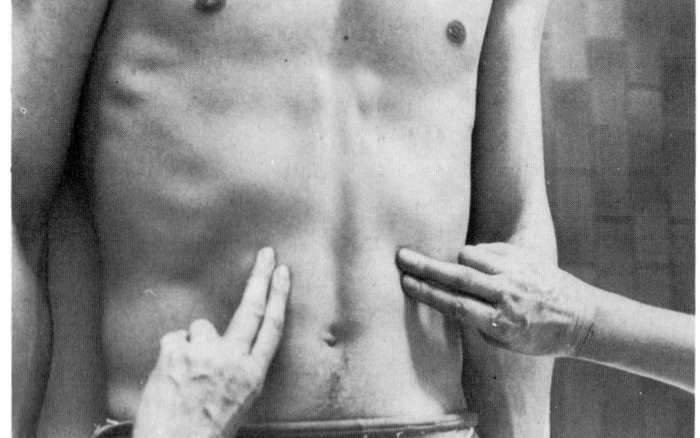

0 and 1 Subject supine.

Upper limbs along sides of body.
Hips flexed, feet resting on table in order to relax abdomen and stabilize lumbar spine.
Ask subject to cough, breathe out, or to flex his head and neck.
Palpate on each side of the linea alba between the navel and xiphoid process.
Note navel displacement as it is drawn towards the dominating side.

Grade 0: No contraction is felt
Grade 1: Contraction is felt but there is no joint movement.

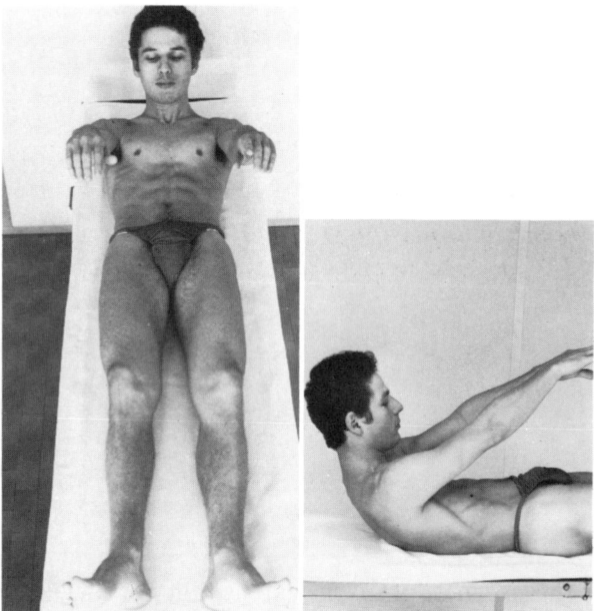

2 Subject supine.

Lower extremities extended but not fixed.
Ask subject to flex trunk gradually, his arms held straight out before him.
The subject should not propel himself but must curl his trunk progressively in such a way that the scapular spines leave the table while the inferior angles of the scapula remain in contact with it.
During the test:
—Be sure recti abdominis and not the obliqui are performing the movement.
—Assess muscle volume by palpation, note any asymmetry.

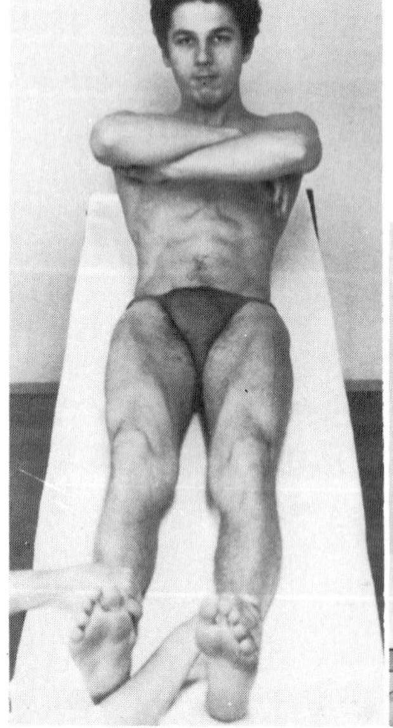

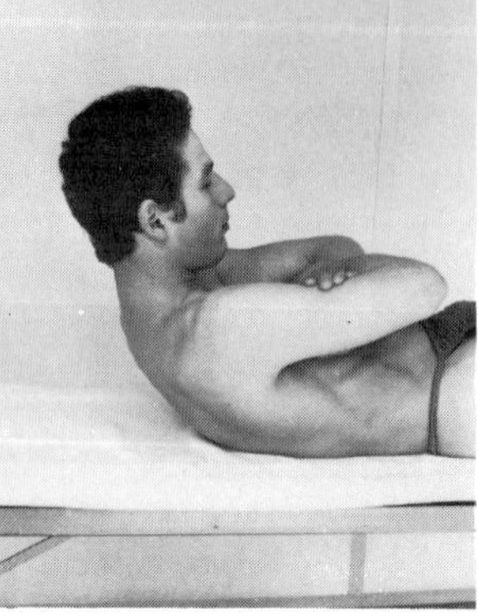

3 Subject supine.

Lower limbs extended.
Arms crossed over the chest.
Ask subject for gradual trunk flexion without propulsion until scapulae have cleared the table.
Stabilize subject by holding his feet at the end of the movement.
Observe lumbar spine, lordosis of which should straighten.
Assess muscle volume by palpation.
Hold thighs if hip flexors are deficient.

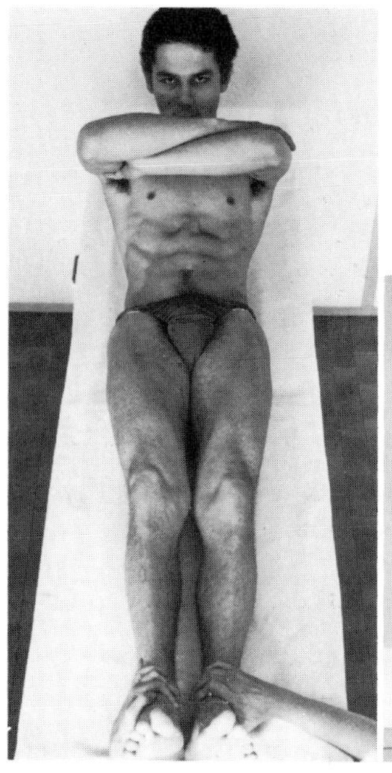
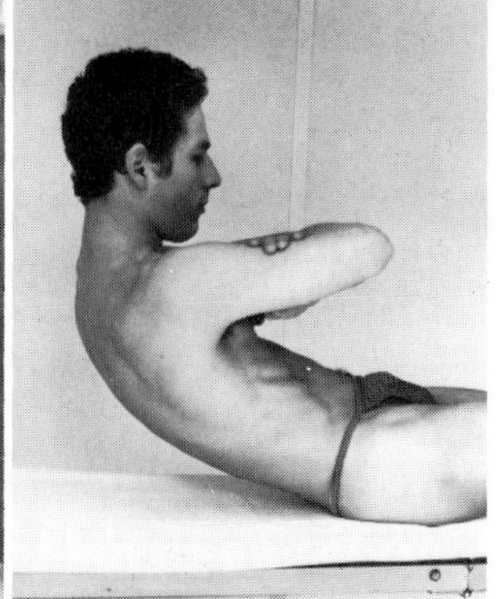

4 Subject supine.

Lower limbs extended.
Arms crossed over chest.
Ask subject for gradual trunk flexion without propulsion.
The entire thoracic and lumbar region should roll up from the table until the subject is sitting.
Observe lumbar region of the spine, lordosis of which should straighten.
Stabilize subject by holding his feet at end of movement.

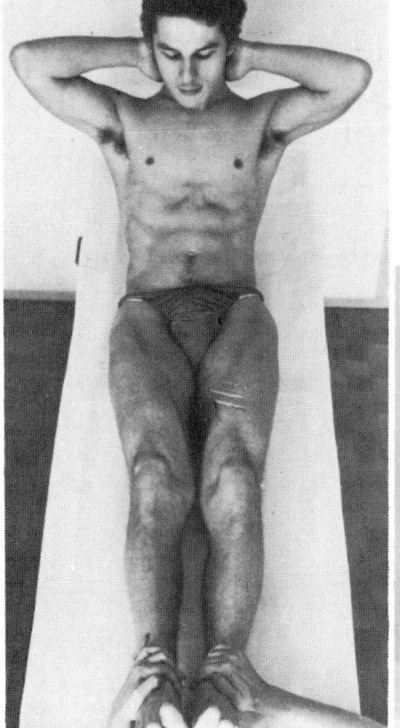
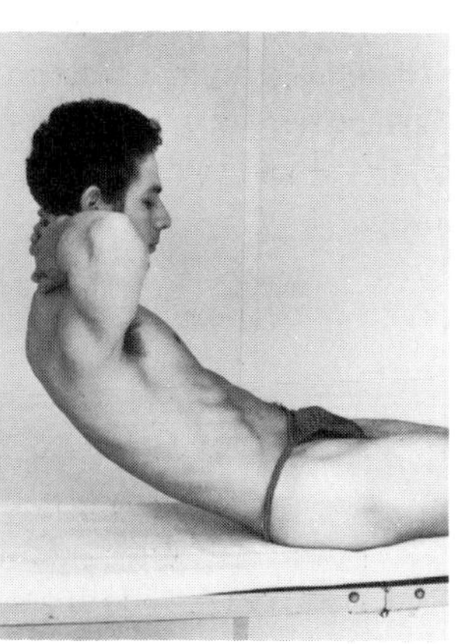

5 Subject supine.

Arms back, elbows flexed, hands behind neck.
Ask for the same movement without propulsion.

TESTING RECTUS ABDOMINIS, INFRA-UMBILICAL PORTION

The test proposed is global. Watch for difference between right and left sides of infra-umbilical portions of recti.

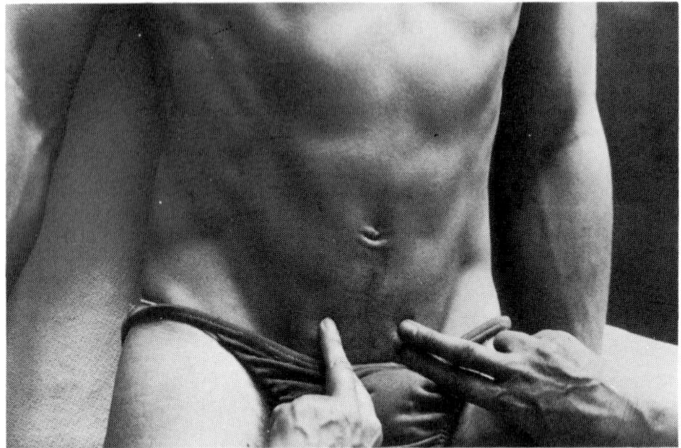

0 and 1 Subject supine.

Lower limbs flexed.
Ask subject to lift his feet.
Palpate on each side of the linea alba between the infra-umbilical and supra-pubic regions.

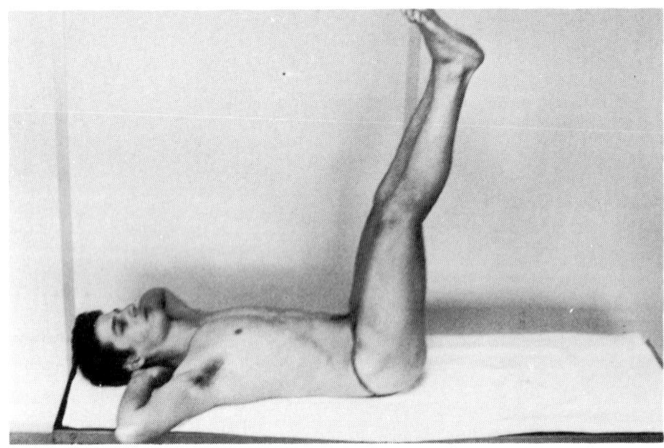

2 Subject supine.

Lower limbs are raised passively to 90°, lumbar region of spine flattened on table.
Examiner supports lower limbs (not shown).
Ask subject to hold this position, observe his lumbar region.

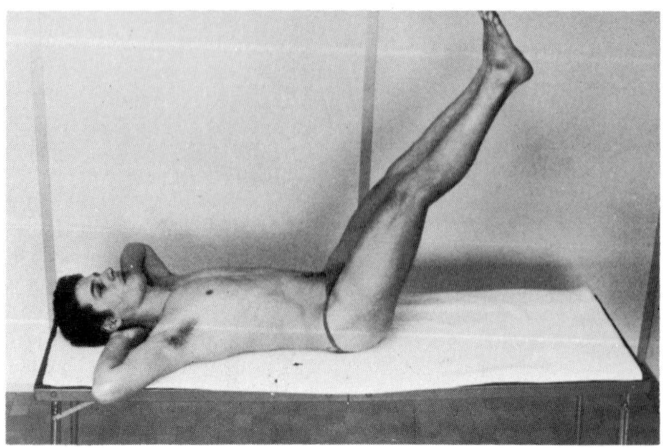

3

Same test, lower limbs held at 60°.

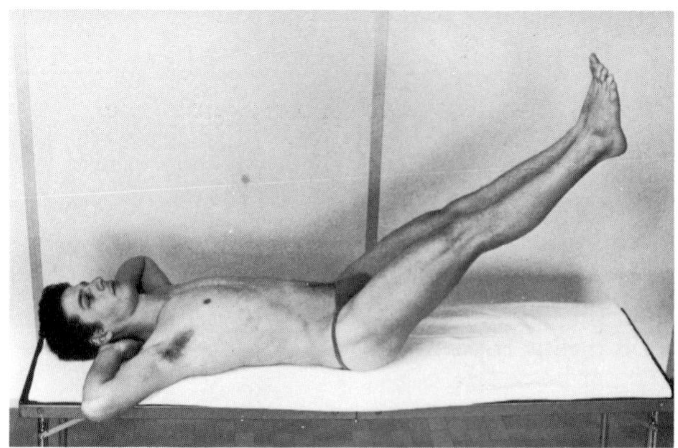

4

Same test, lower limbs held at 45°.

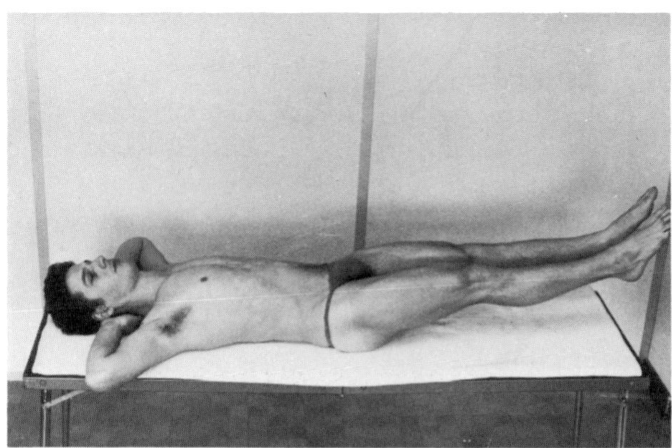

5

Same test, lower limbs held just above the treatment table.

OBLIQUUS EXTERNUS ABDOMINIS

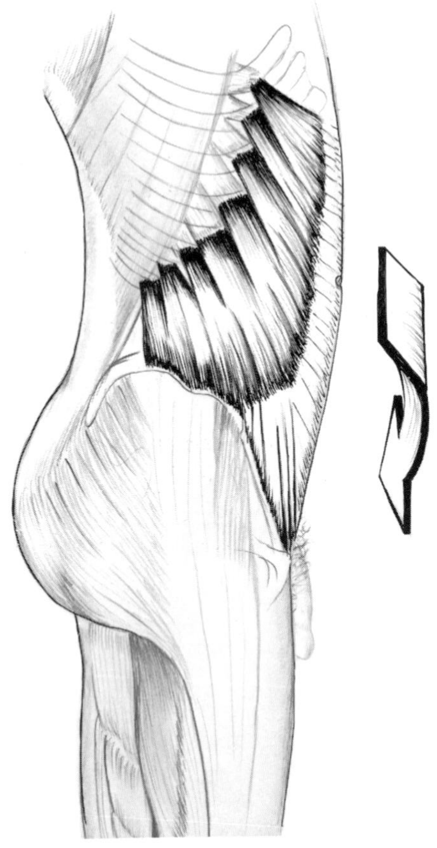

Origin

—Lateral surface and lower margin of lower seven or eight ribs.
—Digitations interlace with those of serratus anterior in the upper part and with those of latissimus dorsi in the lower part.

Insertion Widely spread:

—Superior fibres: into wide aponeurosis of obliquus externus located anteriorly to rectus abdominis. The aponeurosis contributes to forming the linea alba.
—Middle fibres: form the pillars which insert into:
- pubic spine.
- pectineal crest.

Some fibres cross over to contralateral side to insert into inguinal surface, pubic spine, and pectineal crest.
—Inferior fibres:
- anterior two-thirds of lateral lip of iliac crest.
- lateral third of inguinal ligament.

Innervation

—Last six intercostal (T7 to T12).
—Iliohypogastric and ilio-inguinal (L1).

Function

Unilateral contraction

Unilateral contraction of right obliquus externus causes trunk flexion and ipsilateral latero-flexion to the right and rotation of upper trunk to the left (when subject is hanging by his arms, obliquus externus abdominis causes an ipsilateral rotation).
The right obliquus externus contracts synergistically with the left obliquus internus. Their fibres are oriented in the same direction.

Bilateral contraction

The obliqui externi abdominis participate in trunk flexion acting in synergy with the recti. They form a real muscular support system, accentuate thoracic kyphosis, and flatten lumbar lordosis by tilting the pelvis anteriorly.
They contract strongly at raising and lowering of lower extremities.
They play a discrete role in spinal column stability and posture.
They contract during body functions but to a lesser degree than the transversus abdominis.

Tightness

Rare. However, in certain subjects it induces a depression of the lower thoracic wall. The thorax tends to flatten and the thoracic kyphosis to accentuate.

Deficiency

Decrease in ability of the trunk to flex. Possible diastasis of the linea alba by overstretching the aponeurosis of the anterior abdominal wall.
Pelvis tends to tilt posteriorly, accentuating lumbar lordosis.

TESTING RIGHT OBLIQUUS EXTERNUS ABDOMINIS

Ensure suppleness of the vertebral column and spinal muscle length before testing. During test, lower extremities are not held for same reasons cited for recti abdominis.

0 and 1 Subject supine.

Left lower limb flexed.
Ask subject to cough, raise his head, or breathe out.
If deficiency is significant a bulge may appear.
Ask subject then to flex and rotate his trunk so that the right shoulder is directed towards the left iliac crest.
Palpate anterior lateral and superior portion of abdomen.
The left obliquus internus abdominis will contract at the same time.
Watch for eventual navel displacement as it is drawn towards the dominating side.

Grade 0: No contraction is felt.
Grade 1: Contraction is felt but there is no joint movement.

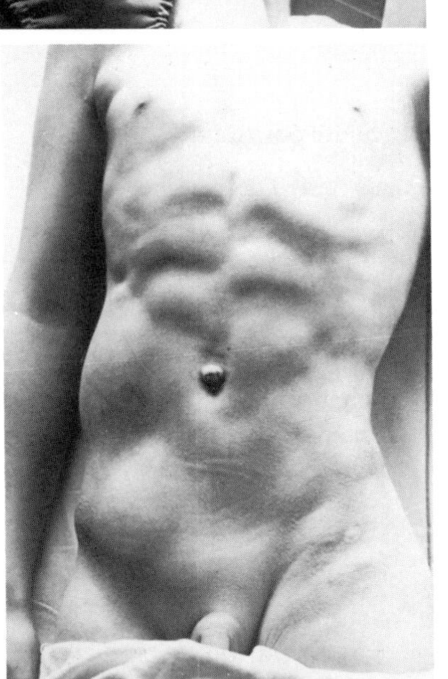

Subject presenting paralysis of right obliquus externus abdominis
Note the bulge.
Also note deficiency in right obliquus internus abdominis.

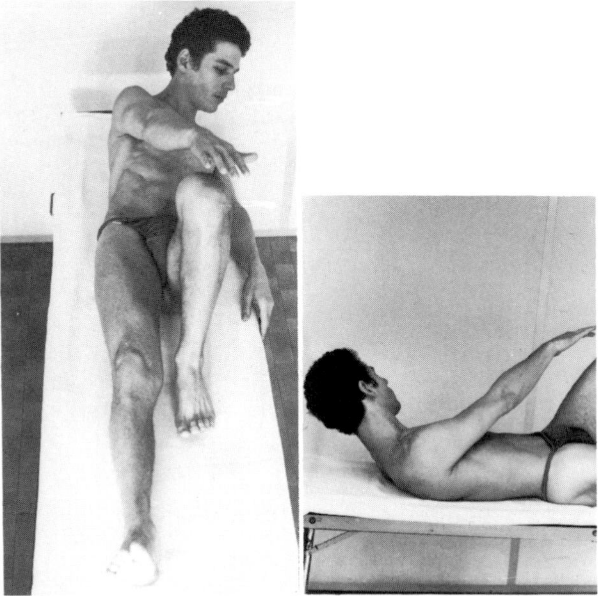

2 Same position.

Ask subject to flex and rotate his trunk so that with arm straight his right hand moves towards his bent left knee.
Subject should flex his trunk until the scapular spine is lifted from the table while the inferior angle remains in contact with it.
The movement must be performed without momentum.
Palpate to be sure that the obliquus externus is functioning and not the ipsilateral rectus: the subject would then flex his trunk to the same side but without rotation.

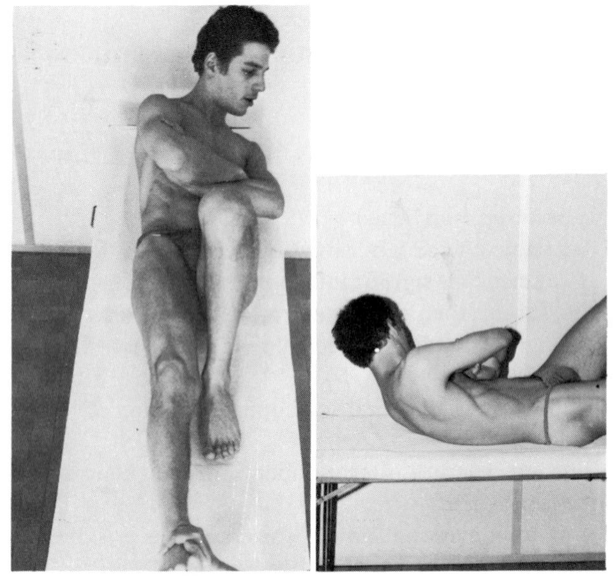

3 Same position.

Arms folded across chest.
Ask for same movement.
Scapula should completely clear the table.
Lower thoracic region remains in contact with the table.
The movement is performed without momentum.
Stabilize subject by holding his feet at end of movement.

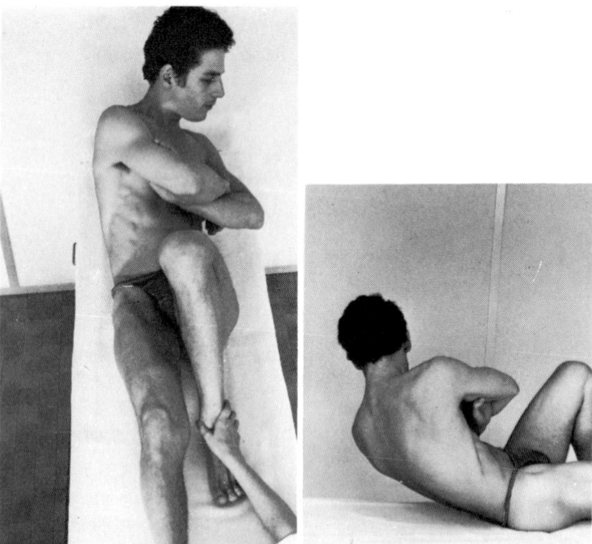

4 Same position.

Arms folded across chest.
Same precaution.
Entire vertebral column should rise from the table.
Movement should be performed without momentum.
Subject arrives at sitting position gradually.
Watch lumbar spine which should remain flat.
Stabilize subject by holding his feet at end of movement.

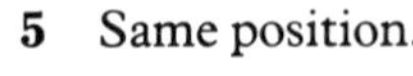

5 Same position.

Same test as above, with hands behind neck.

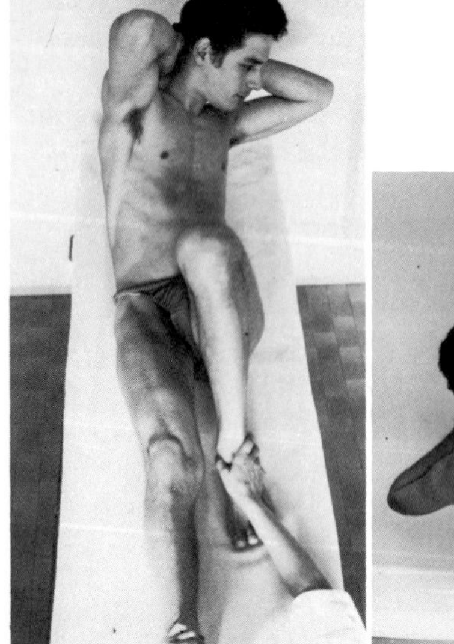

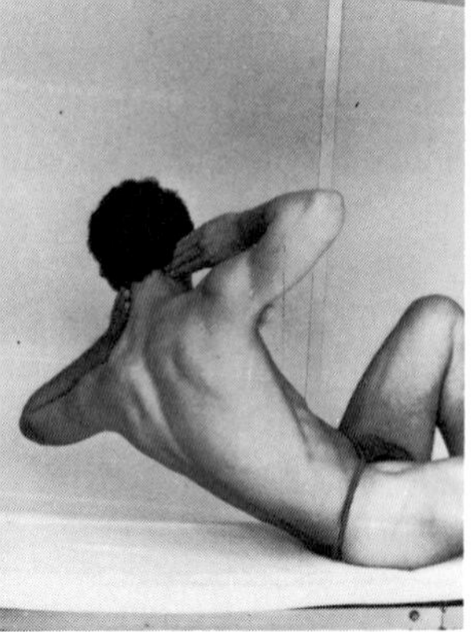

NOTES

OBLIQUUS INTERNUS ABDOMINIS

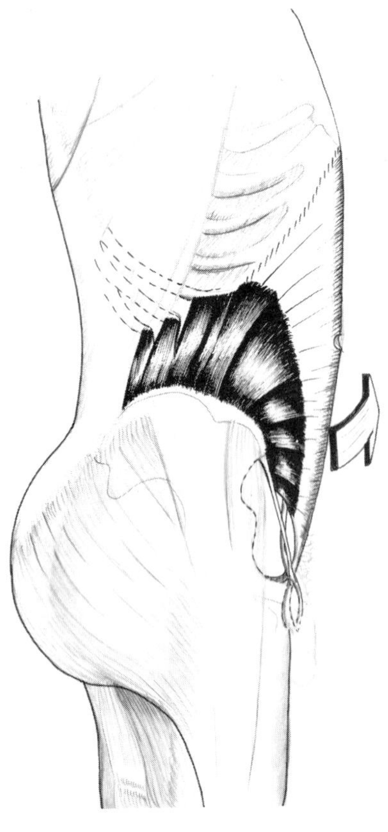

Origin

—Anterior two-thirds of iliac crest.
—Lateral third of inguinal ligament.
—Posterior third of lateral lip of iliac crest and spinous process of fifth lumbar vertebra through lumbar aponeurosis.

Insertion

—Posterior fibres: inferior border of last three or four ribs.
—Middle fibres: into a wide aponeurosis which in the upper two-thirds of the abdominal wall divides into two lamellae, one passing in front of, the other behind rectus abdominis. The lower third passes in front of this muscle. This aponeurosis inserts into the linea alba, contributing to its formation.
—Inferior fibres: onto pubis and pubic symphysis forming, with the transversus abdominis and the cord, the conjoined tendon.
The most inferior fibres form the cremaster muscle.

Innervation

—Last four intercostal (T9, T10, T11, T12).
—Iliohypogastric and ilio-inguinal (L1).

Function

Unilateral contraction.

—Ipsilateral pelvic hitching and rotation towards the contralateral side (left obliquus internus rotates pelvis to the right).
—With pelvis fixed, it participates in ipsilateral trunk latero-flexion acting with the quadratus lumborum.
During flexion and rotation of the trunk, it contracts synergistically with the contralateral obliquus externus whose fibres are oriented in the same direction.

Bilateral contraction

The obliqui interni abdominis flex the trunk, acting with obliqui externi and the recti. Together they accentuate thoracic kyphosis and flatten lumbar lordosis by tilting the pelvis anteriorly.
They contract strongly in supine position during leg raising or lowering.
They play a discrete role in vertebral column stability, and they contract during forward rotation of the pelvis during gait.
The obliqui interni abdominis contract during body functions though to a lesser degree than transversus abdominis.

Deficiency

Decrease in trunk flexion capacity, in pelvic hitching and rotation.
Consequences of their deficiency are identical to those of the other abdominal muscles.

TESTING LEFT OBLIQUUS INTERNUS ABDOMINIS

0 and 1 Subject supine.

Left lower limb flexed, foot resting on table.
Ask subject to raise left hemipelvis, rotating it towards right shoulder.
Palpate the obliquus internus abdominis above and medially to the anterior superior iliac spine in lower, lateral part of abdomen.
Contraction may be witnessed during cough or expiration.

Grade 0: No contraction is felt.
Grade 1: Contraction is felt but there is no joint movement.

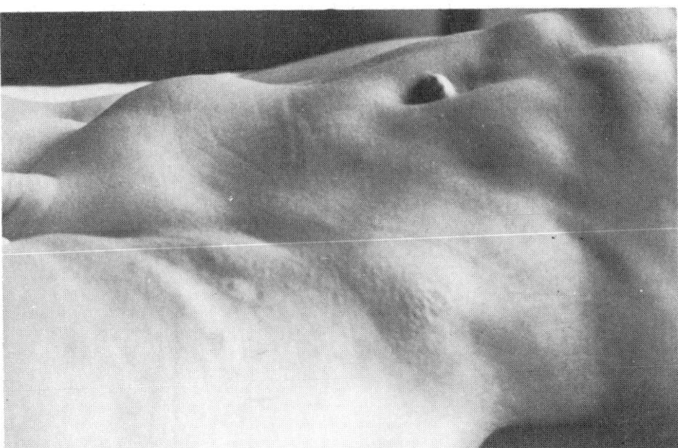

Subject presenting a paralysis of right obliquus internus abdominis. Note the bulge.

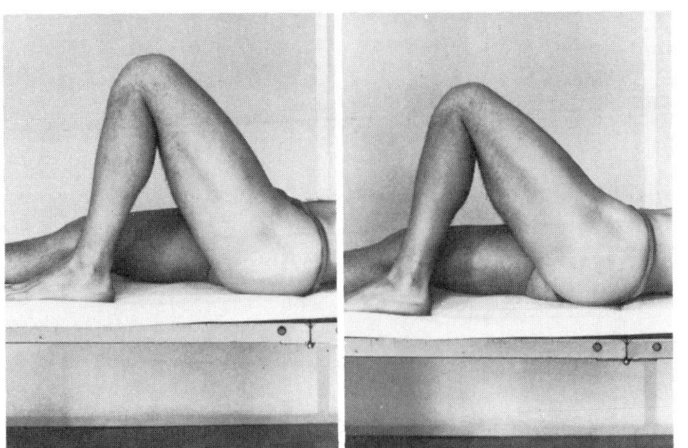

2 Same position.

Ask subject to raise and rotate his left hemipelvis towards his right shoulder.
During the movement the left lower extremity should remain relaxed. The subject should not push on his heel.
Palpate to verify contraction.
Grade 2 is given if subject lifts his hemipelvis in incomplete range of movement.

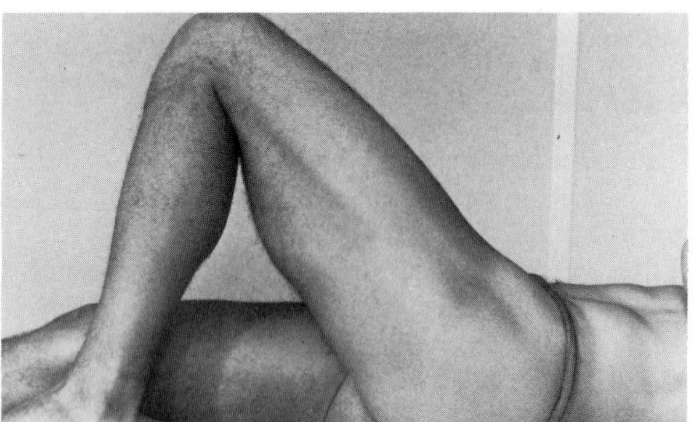

3

Same test performed in full range of movement.

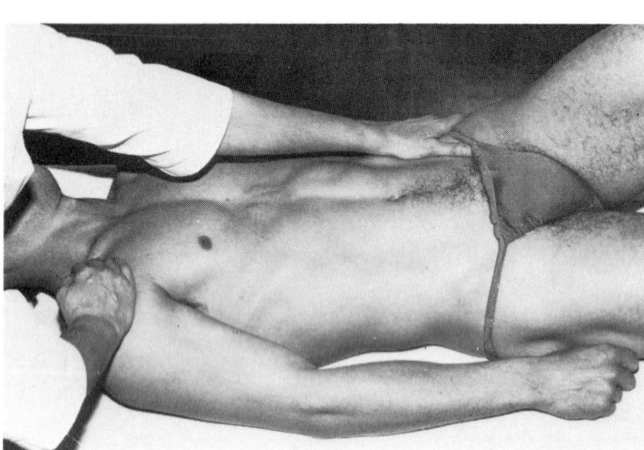

4 and 5

Same test against resistance.
Hold right shoulder in place.
Resistance to the movement is placed on the anterior aspect of the iliac crest.

Grade 4: Power is sub-normal.
Grade 5: Power is normal.

NOTES

TRANSVERSUS ABDOMINIS

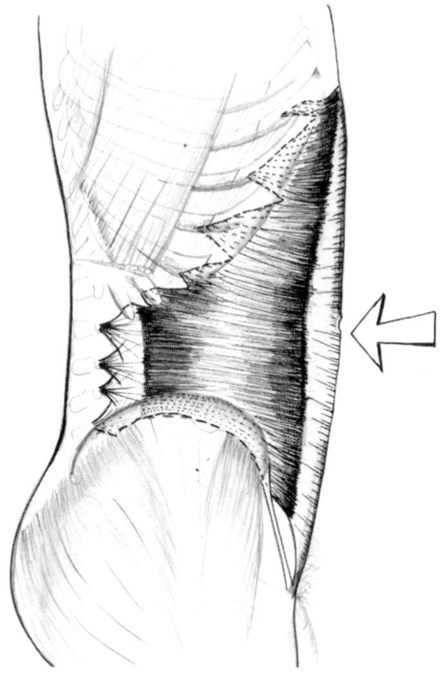

Origin

—Inner surface of lower six ribs.
—Transverse processes of first four or five lumbar vertebrae.
—Anterior two-thirds of medial lip of iliac crest.
—Lateral third of inguinal ligament.

Insertion

—Anterior aponeurosis of transversus passes posteriorly to rectus abdominis in its upper two-thirds and anteriorly to it in its lower third. It inserts into the linea alba contributing to its formation.
—Inferior fibres, with the obliquus internus, form the conjoined tendon.
The most inferior fibres form the cremaster muscle.

Innervation

—Last six intercostal (T7 or T9 to T12).
—Iliohypogastric and ilio-inguinal (L1).

Function

By itself, it constitutes a true girdle especially in its inferior portion. It plays an important role in supporting the viscera.
Its contraction depresses the abdominal wall much more than does that of the other abdominal muscles.
It contracts strongly during forced expiration. It increases intra-abdominal pressure, driving the diaphragm back and upward. It is a direct antagonist of the diaphragm.
Note: The diaphragm needs the visceral support against which its central tendon pushes. Thus the transversus is a synergist of the diaphragm.
It has no action on trunk latero-flexion but it facilitates the action of the other anterior lateral abdominal muscles by retaining the viscera.
The transversus participates in body functions: defaecation, vomiting, delivery (in childbirth), micturition, coughing, expectoration, phonation . . .
It also contracts during intense efforts such as lifting, pulling, pushing . . .

Deficiency

Vital capacity is decreased because of reduced capacity for forced expiration and lack of visceral support during diaphragm contraction.
Decreased visceral support indirectly tends to increase lumbar lordosis.
Impairment of body functions.

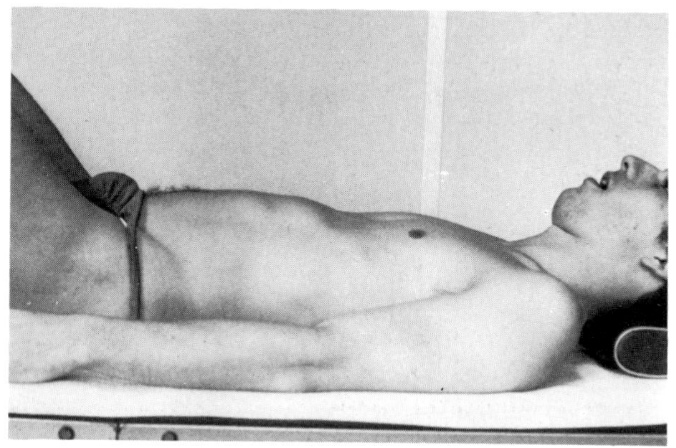

0 Subject supine.

Upper limbs along sides of body.
Cushion under knees.
Ask subject to cough.
If coughing is impossible and abdomen protrudes, examiner may give grade 0.

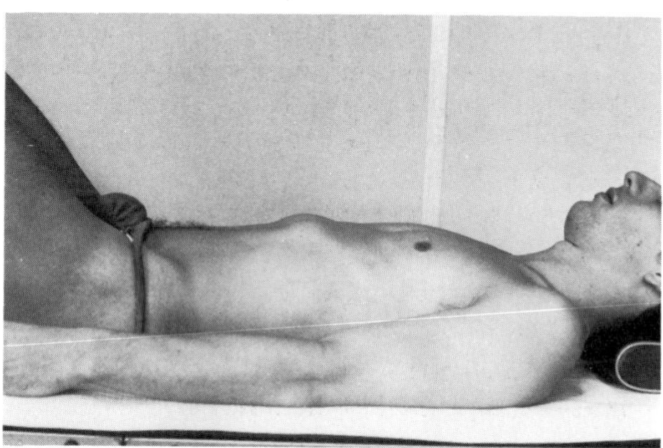

1 Same position.

Ask subject to breathe out.
Grade 1 is given if the abdominal wall remains motionless during expiration.

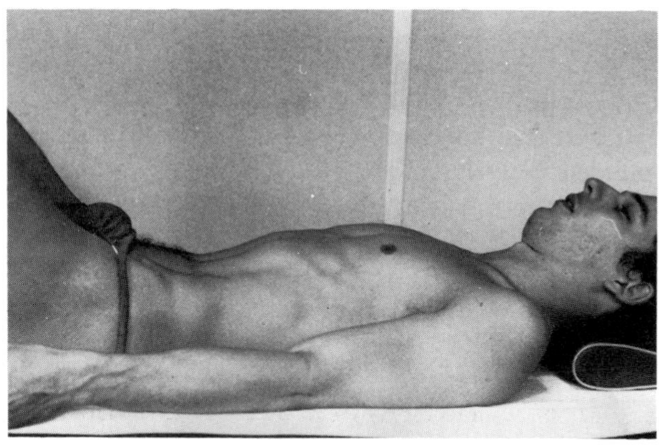

2 Same position.

Ask subject to breathe out.
Grade 2 is given if examiner observes a depression of the abdominal wall.

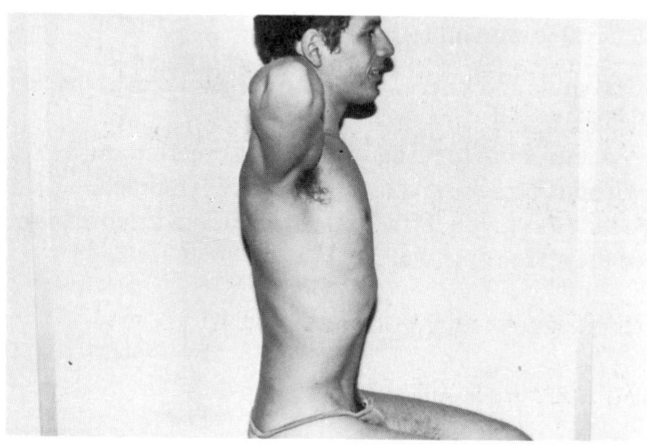

3 *Alternative.*

Subject sitting on a stool.
Arms in abduction, hands behind the neck.
Ask subject to breathe out.
Grade 3 is given if examiner observes a depression of the abdominal wall.

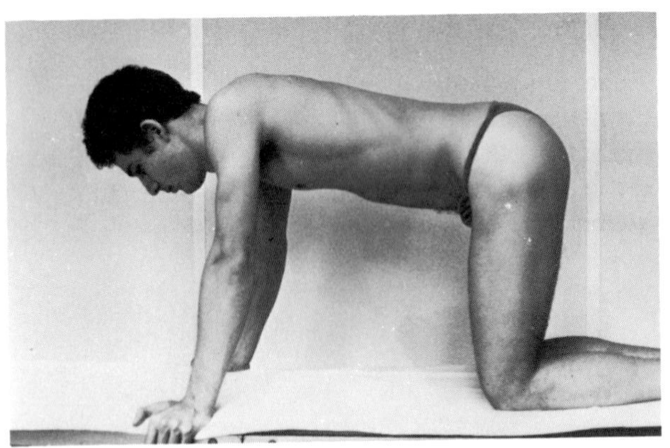

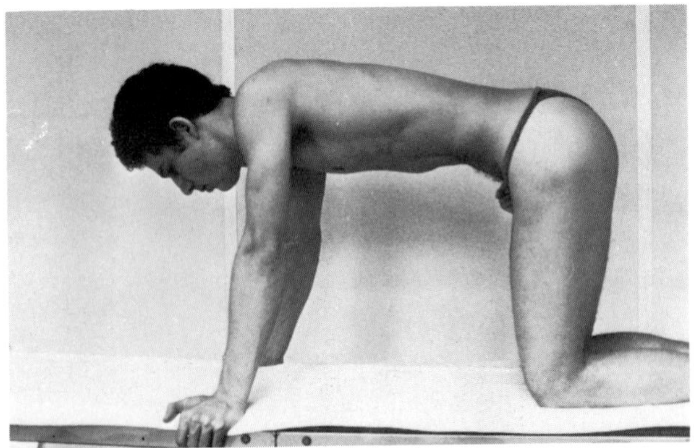

3 Subject on all fours, on hands and knees, arms and thighs are vertical, back is flat.
Ask subject to breathe out.
In this position the transversus works against visceral weight.
Grade 3 is given if the examiner notices a depression in the abdominal wall during expiration.
Precaution: the back must remain immobile.

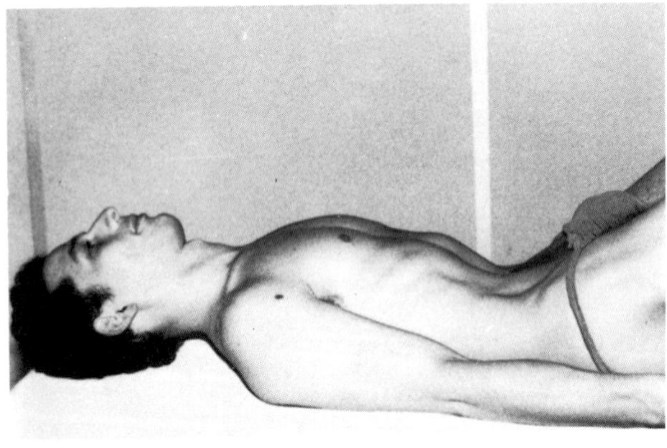

4 Subject supine.

Upper limbs along sides of body.
Cushion under knees.
Ask subject to breathe in, hold his breath then without breathing out, to suck in his stomach.
Grade 4 is given if the examiner notices a depression of the abdominal wall.

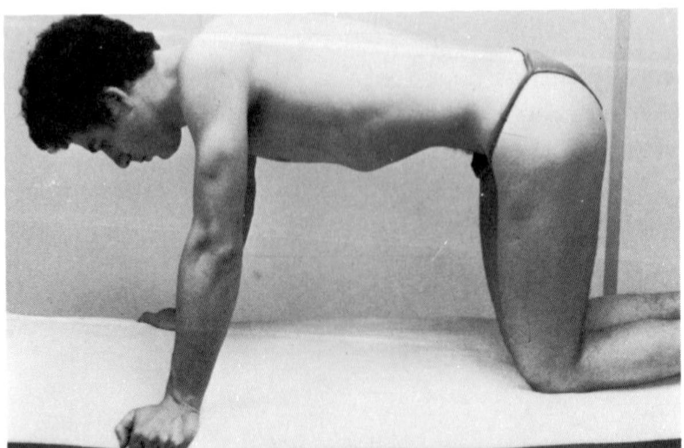

5 Subject on all fours.

On hands and knees, arms and thighs vertical, back is flat.
Ask subject to breathe in, hold his breath then without breathing out, to suck in his stomach.
Grade 5 is given if the examiner notices a depression of the abdominal wall.

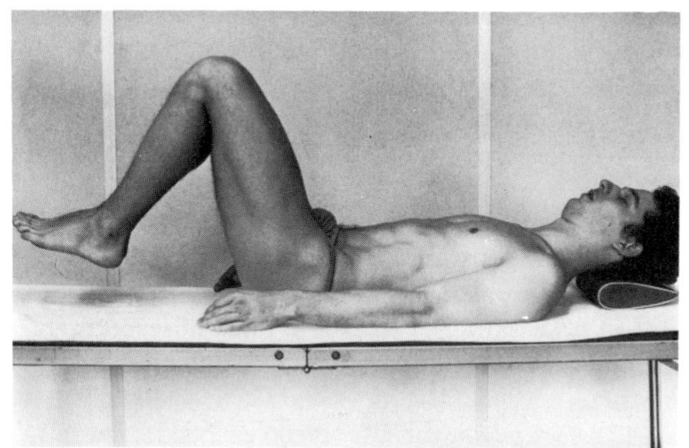

5 *Alternative.*

Subject supine.
Upper limbs along sides of body.
Ask subject to breathe in, hold his breath then without breathing out, to suck in his stomach.
Subject must be able to lift his feet from the table.

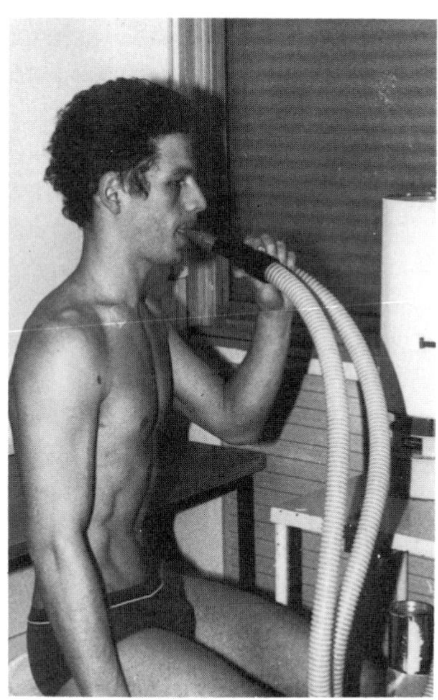

5 *Method used by certain examiners.*

Subject sitting or standing (not shown).
Ask subject to blow into a spirometer, this provides a certain resistance.
Grade 5 is given if the examiner notices a depression of the abdominal wall.

DIAPHRAGM

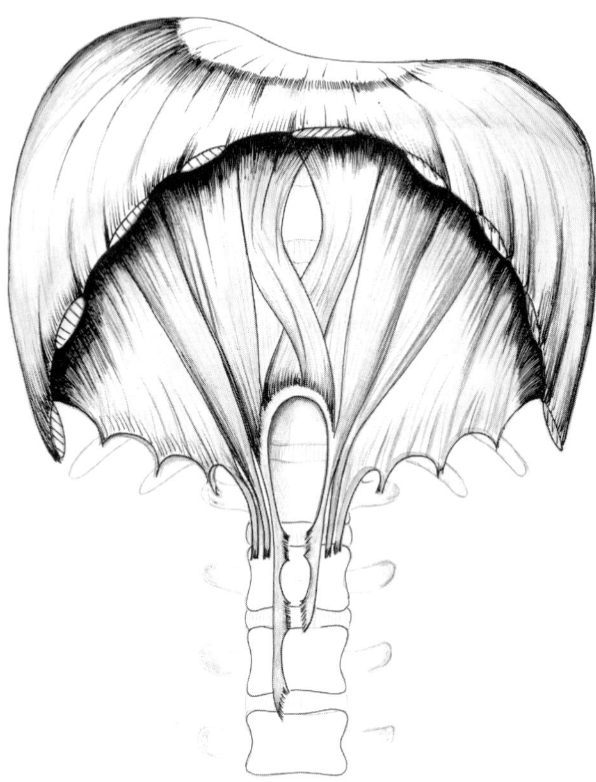

Origin

On spine

—*Right pillar:* anterior surface of bodies of second, third, and fourth lumbar vertebrae and adjacent intervertebral discs.

—*Left pillar:* anterior surface of body of second (and sometimes first) lumbar vertebra and adjacent discs.

—*Accessory pillar:* there is one on each side, laterally to the main pillars, originating on anterior surface of second lumbar vertebra.

—*Lumbocostal arch (medial arcuate ligament):* arches from the anterior surface of second lumbar vertebra to the costal process of the first lumbar vertebra.

On ribs

—Medial surface of last six costal cartilages and last three or four ribs, by digitations intermingling with those of the transversus abdominis.

—Some fibres originate in fibrotendinous arches:

- Lateral arcuate ligament: from costal process of first lumbar vertebra to twelfth rib.
- Senac's fibrous arches running from the twelfth to eleventh and from the eleventh to the tenth ribs.

On sternum Posterior surface of xiphoid process.

Insertion The different portions converge towards the sides of the central tendon.

Innervation Phrenic (c3, C4, C5).

Function

—Increases vertical diameter by lowering central tendon.
—Increases transversal diameter by raising lower ribs.
—Increases anterior posterior diameter by raising upper ribs, acting through the sternum.
In a first phase it lowers the central tendon which confronts the tension of the structures of mediastinum and the abdominal viscera contained by the abdominal wall.
The action of the diaphragm requires good abdominal muscle tone.
When the thorax is fixed, the pillars of the diaphragm increase lumbar lordosis, according to F. Mezieres.
Some portions of the diaphragm are more actively contracted than others, depending upon body position, since they must contain the viscera.
Supine: posterior portion,
Prone: anterior portion,
Side-lying: ipsilateral portion.
The diaphragm contracts during defecation, micturation, vomiting, delivery, hiccups, yawning, laughing . . .

Deficiency

Vital capacity is greatly diminished.
Subject breathes in with upper thorax. Accessory inspiration muscles develop (scalene hypertrophia in certain polio patients).
During inspiration the abdominal wall depresses.
The efficiency of a good diaphragm is much impaired if abdominals are deficient.
Abdominal spasm blocks the excursion of the diaphragm.

Note: The proposed hand testing is controversial. Special radiographic techniques are necessary aids for evaluating the diaphragm.

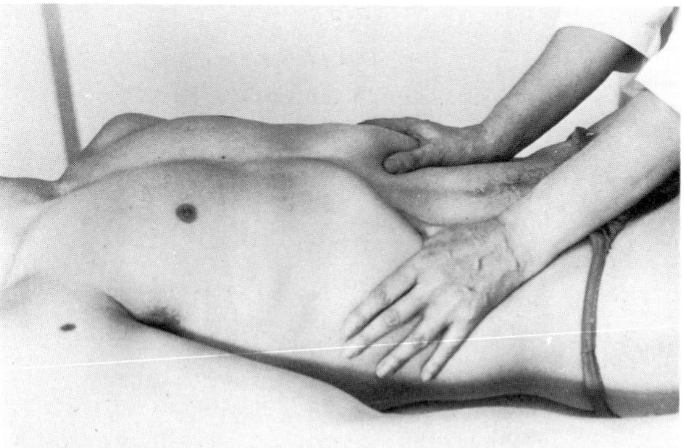

0 and 1 Subject supine.

Upper limbs along sides of body.
Ask subject to sniff several times consecutively, inducing a series of depression-inflations of the thorax.
Examiner places his hands at the base of the thorax, his fingers spread over the lower ribs, his thumbs under the lower edge of the ribcage.
Deficient domes rise creating a depression in the abdominal wall.
Grade 0 is given if the domes rise distinctly.
Grade 1 is given if the domes rise only a little.
Watch for possible asymmetry.

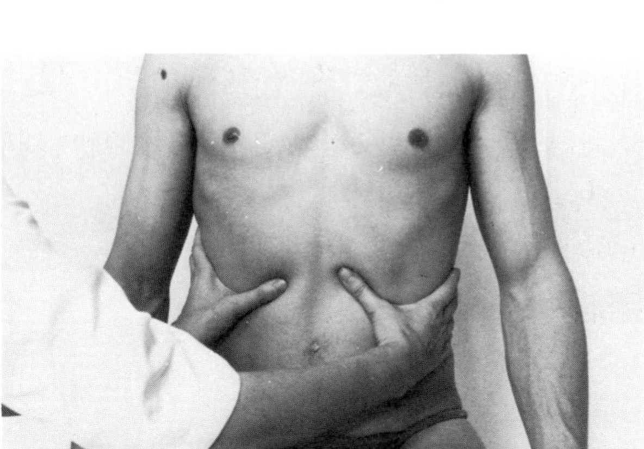

2 Subject sitting.

Upper limbs relaxed along sides of body.
Same test.
Same placement of examiner's hands.
Grade 2 is given if the domes remain immobile.
Ventilation is inadequate.

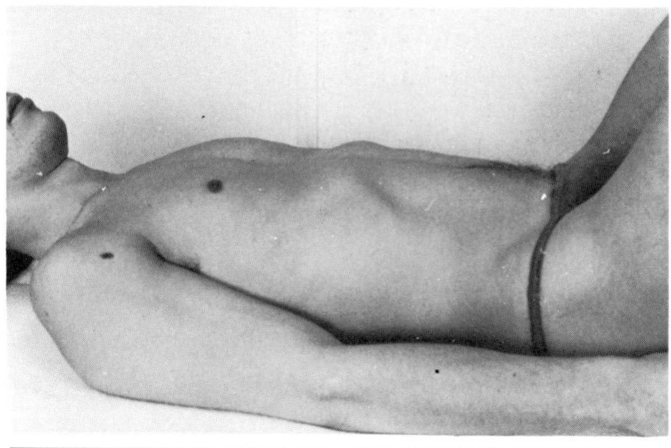

3 Subject supine.

Ask subject to breathe in deeply.
Grade 3 is given if the lower ribs spread and the abdominal wall lifts. The diaphragm confronts visceral resistance.
Lumbar spine should remain immobile throughout the test; its hyperlordosis can give the illusion of diaphragm contraction.

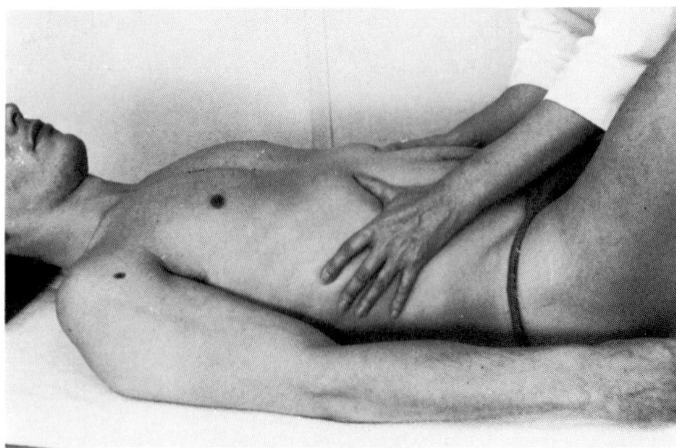

4 and 5 Subject supine.

Upper limbs along sides of body.
Examiner places hands at base of ribcage.
Ask subject to breathe in.
At each maximal inspiration the examiner resists the push of the diaphragm, and may notice an eventual asymmetry.

Grade 4: Power is sub-normal.
Grade 5: Power is normal.

NOTES

INTERCOSTALES EXTERNI

There are eleven on each side, composed of parallel bundles oriented obliquely downward and forward. The intercostales externi are assisted by the levatores costarum, subclavius and the serratus posterior superior.

Origin Lateral lip of subcostal groove of lower border of the upper rib.

Insertion Upper border and adjacent lateral surface of rib below the rib of origin.
The intercostales do not fill the entire intercostal space, they run from the posterior part of the costotransverse joint to the costochondral joint.

Innervation Intercostal (T1 to T12).

LEVATORES COSTARUM

There are twelve pairs.

Origin Transverse processes of seventh cervical vertebra and upper eleven thoracic vertebrae.

Insertion Between tubercle and angle on lateral surface of rib below vertebra of origin.

Innervation Intercostal.

SUBCLAVIUS

Origin Subclavius groove on lower surface of clavicle.

Insertion Upper surface of first cartilage and adjacent part of first rib.

Innervation Branch of upper trunk (C5, C6).

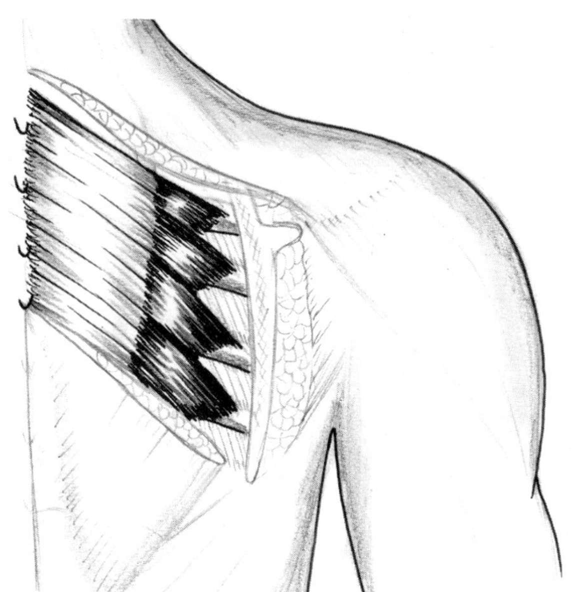

SERRATUS POSTERIOR SUPERIOR

Origin Spinous processes of seventh cervical and upper three thoracic vertebrae.

Insertion Digitation on first rib and lateral surface of posterior part of second to fifth thoracic vertebrae.

Innervation Intercostal (T1 to T4).

Function

Intercostales externi are muscles of inspiration.
They give cohesion to the thorax and prevent its distension by connecting the ribs.
Intercostales externi are assisted by levatores costarum, subclavius and serratus posterior superior.

Contracture or tightness

Closes one or several intercostal spaces.

INTERCOSTALES INTIMI and ACCESSORY EXPIRATION MUSCLES

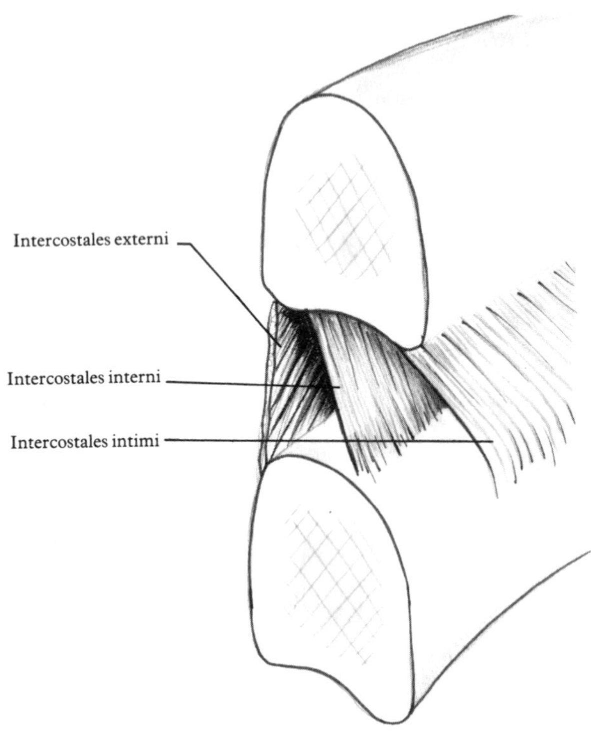

Cross view of intercostales.

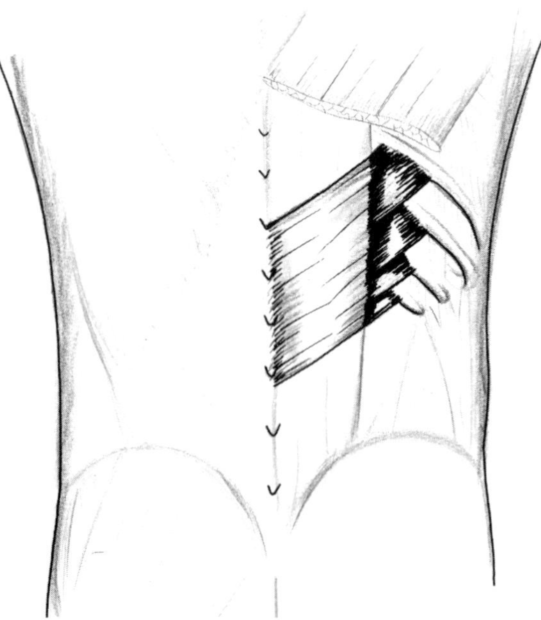

INTERCOSTALES INTIMI

They occupy the innermost intercostal spaces extending from the posterior angle of the ribs to about five or six centimetres from the sternum. The most posterior fibres of intercostales intimi are sometimes called infracostales.

Origin Medial lip of subcostal groove of rib above.

Insertion Upper border and adjacent medial surface of rib below the rib of origin.

Innervation Intercostal (T1 to T12).

INTERCOSTALES INTERNI

They are situated medially to intercostales externi. They occupy the intercostal space from the axillary line to the edge of the sternum.

Origin Entire height of lateral lip of subcostal groove.

Insertion Fibres run obliquely downward and backward and insert onto superior border of rib below the rib of origin.

Innervation Intercostal (T1 to T12).

SERRATUS POSTERIOR INFERIOR

Origin Lumbar fascia at level of twelfth thoracic and first three lumbar vertebrae.

Insertion Posterior part of lateral surface of last four ribs.

Innervation Intercostal (T9 to T12).

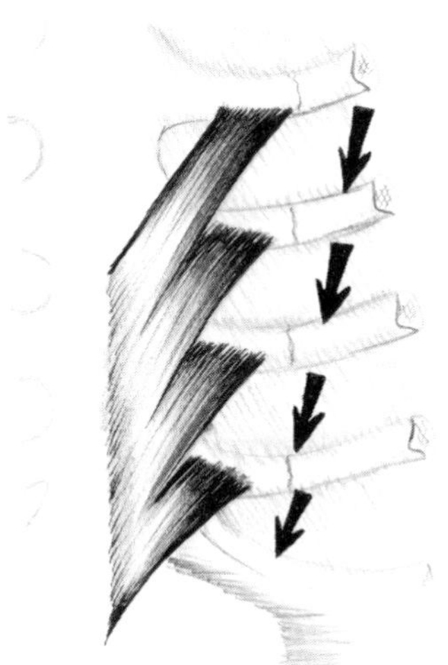

Posterior view of sternum and adjacent ribs.

TRANSVERSUS THORACIS

Origin Lateral margin of lower portion of posterior surface of sternum.
Border and posterior surface of xyphoid process.

Insertion Fibres are directed upward and outward; they insert into lower borders and posterior surfaces of third to sixth costal cartilages.

Innervation Intercostal (T1 to T8).

Function

All these are muscles of expiration. As do the intercostales externi, the intercostales intimi and interni give cohesion to the ribcage and prevent its distension.
It is impossible to evaluate them separately.

Contracture or tightness

Closes one or more of the intercostal spaces.

NOTES

Testing intercostales externi

These are difficult to test.
The examiner can, however, assess their contraction with palpation and by observing rib movement.

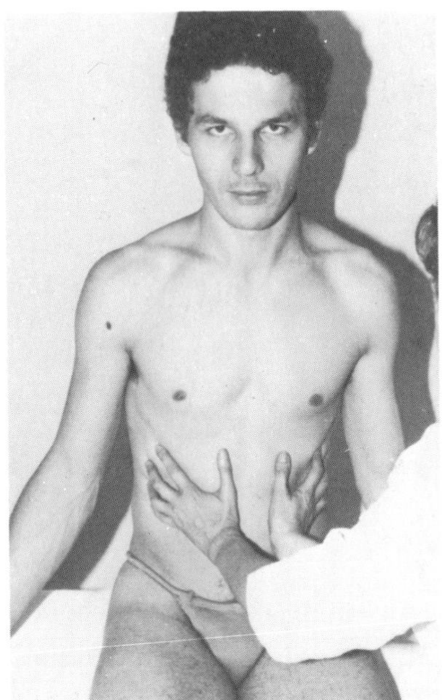

Subject sitting.

Upper limbs slightly separated from trunk.
Examiner places his hands on lateral portion of ribcage, his fingers placed in the intercostal spaces.
Ask subject to breathe in deeply and slowly.
Upper intercostales externi are difficult to palpate because they are covered by the scapular girdle muscles.
In the lower thorax, dissociating them is difficult because of diaphragm activity.

Suggested grading

0 Depression appears at level of intercostal space.
1 There is no depression.
2 Partial rib separation.
3 Normal rib separation.
4 Rib separation is performed against slight resistance.
5 Subject is capable of overcoming resistance several times successively.

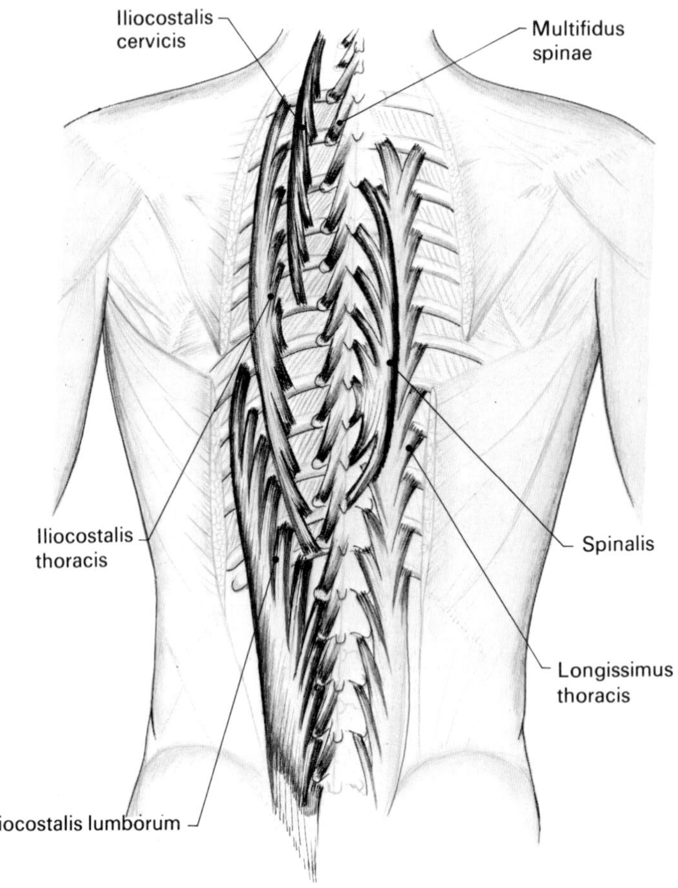

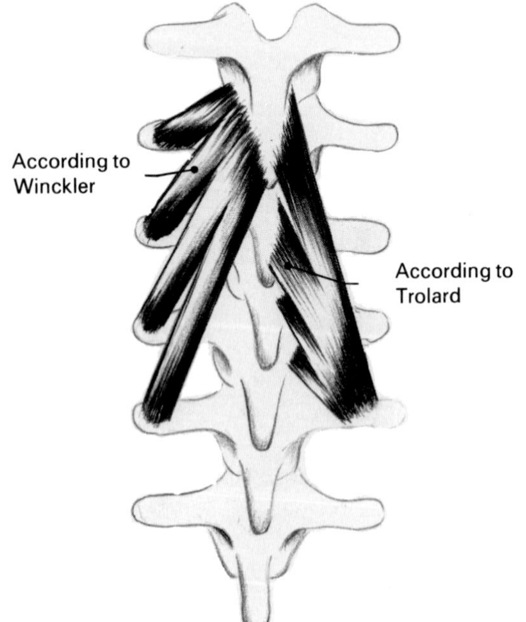

MULTIFIDUS SPINAE

Composed of laminaris brevis, spinalis brevis, laminaris longus and spinalis longus, the multifidus spinae runs from the fleshy part of the common muscle mass to the axis.

According to Trolard

Fibres originate from one transverse process to insert into laminae and spines of the four vertebrae situated above the one of origin.

According to Winckler

Fibres originate from transverse processes of four adjacent vertebrae and insert into lamina and spine of the vertebra immediately above the last one of origin.

LONGISSIMUS THORACIS

Origin Tendinous portion of common muscle mass.

Insertion By two sorts of expansions from lumbar region to the second rib.
—Lateral or costal insertions: inferior border of costal processes in lumbar region.
Inferior border of ribs medially to the posterior angle in the thoracic region.
—Medial or transverse insertions: accessory tubercles of costal processes in lumbar region.
Transverse processes in the thoracic region.

ILIOCOSTALIS

Composed of iliocostalis lumborum (sacrolumbalis), iliocostalis thoracis and iliocostalis cervicis.

SACROLUMBALIS

Origin Tendinous portion of common muscle mass and medial lip of posterior third of iliac crest.

Insertion

—Costal processes of lumbar vertebrae.
—Posterior angle of last six ribs.

ILIOCOSTALIS THORACIS

Origin Superior border of last six ribs.

Insertion Posterior angle of first six ribs.

ILIOCOSTALIS CERVICIS

Origin Superior border of first six ribs.

Insertion Posterior tubercle of transverse processes of last five cervical vertebrae.

SPINALIS THORACIS

Origin Spinous processes of eleventh thoracic to second lumbar vertebrae.

Insertion Spinous processes of first ten thoracic vertebrae.

INTERSPINALIS

Short muscular fasciculi connecting spines of vertebrae from lumbar region to second cervical vertebra (two in each space), not always present in the middle thoracic region.

INNERVATION OF SPINAL MUSCLES

Posterior primary rami of spinal nerves (T1 to S3).

GLOBAL FUNCTION OF SPINAL MUSCLES

They all extend the spine. The deep spinal muscles are erectors.
Sacrolumbalis or iliocostalis and longissimus thoracis support the spine in the sagittal plane. They flex the spine ipsilaterally, and the contralateral group straighten it when the column is flexed laterally. They rotate the spine ipsilaterally.
Spinalis forms part of the support system in the sagittal plane. It flexes the spine laterally to the ipsilateral side, but only in the thoracic region.
Multifidus spinae stabilizes and locks one vertebral level, and rotates the spine contralaterally.
Interspinalis locks and extends one vertebral level.

The activity of spinal muscles depends on the gravity line of each individual. If this line is anterior, muscle activity is increased.
During gait, spinal muscles contract on the side of the swinging limb.

Accessory role of spinal muscles in respiration

—Expiration: lower portion of sacrolumbalis and longissimus thoracis to which serratus posterior inferior must also be added.
—Inspiration: upper fibres of iliocostalis to which the serratus posterior superior is added.

Contracture or tightness

—Bilateral: accentuates lumbar lordosis and diminishes thoracic kyphosis.
—Unilateral: leads to ipsilateral latero-flexion of the spinal column.

Deficiency

When bilateral, the spinal column collapses anteriorly leading to a total kyphosis.

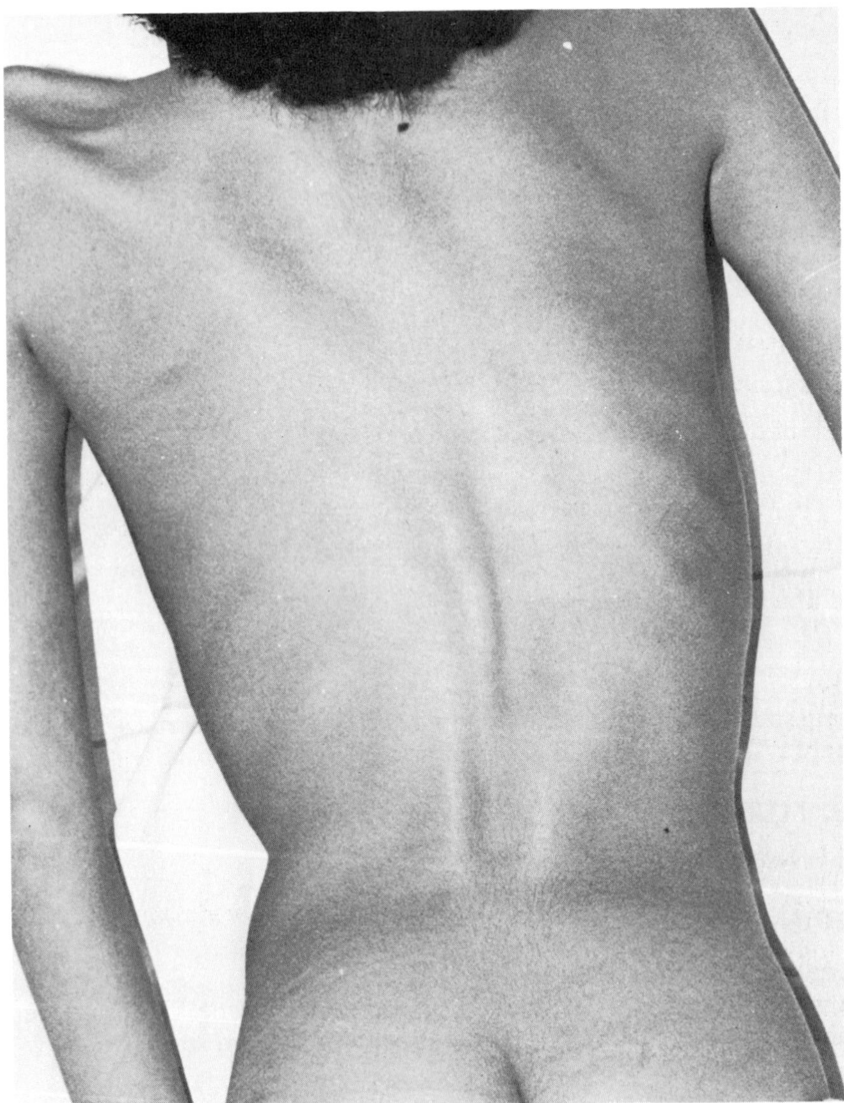

Subject presenting a contracture of the left lumbar and lumbothoracic spinal muscles. Note left latero-flexion of lumbo-thoracic spine.

Erector spinae are evaluated in their function of voluntary extension of the spine. They are evaluated globally since it is difficult to separate the individual action of each muscle. Thoracic and lumbar levels are tested separately.

TESTING THORACIC SPINAL MUSCLES

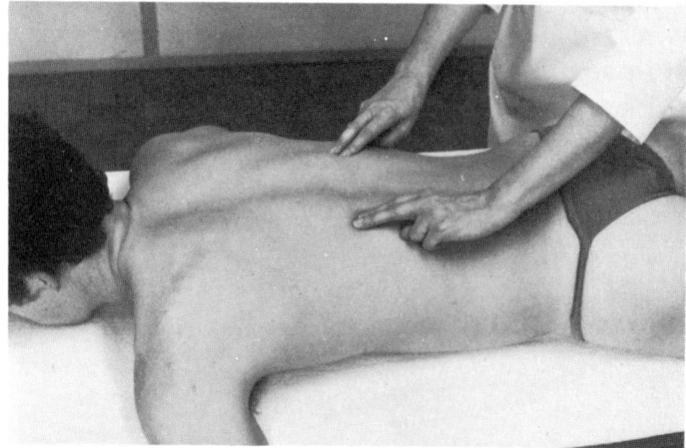

0 and 1 Subject prone.

Arms hanging down on each side of the table.
Cushion under abdomen.
Ask subject to lift his thorax from the table.
Palpate on either side of spinous processes in the upper and lower thoracic region.
At the same time, watch for speckling of the skin, visible under bright glancing light.
Precaution: Subject might cheat by drawing scapulae together.
Contraction of superficial thoracic muscles can give the illusion of spinal muscle contraction.

Grade 0: No contraction is felt.
Grade 1: Contraction is felt but there is no joint movement.

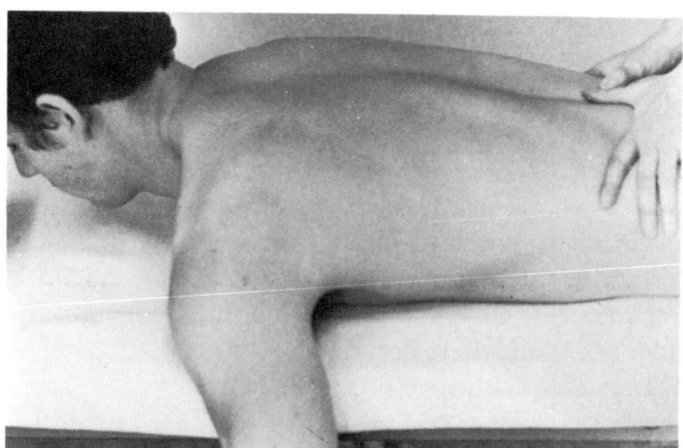

2 Same position.

Same precautions.
Stabilize thorax in place at lumbo-thoracic junction level.
Subject performs the movement, scapulae remain relaxed.
Watch for speckling of skin and eventual latero-flexion of the spine.
Grade 2 is given if the subject manages to lift his upper trunk, part of the thorax remaining on the table.

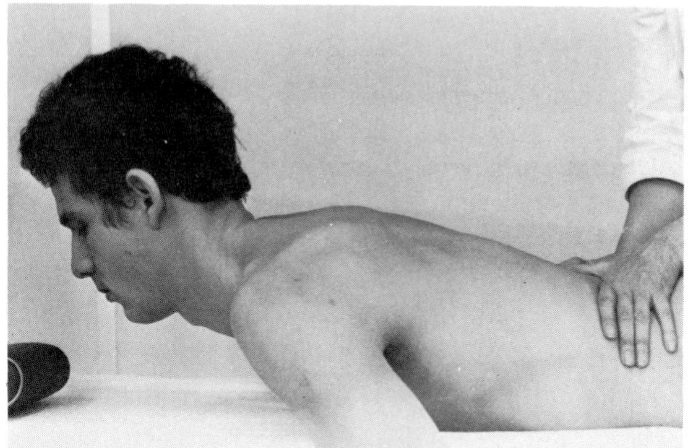

3 Same test, performed with same precautions, in complete range of movements.

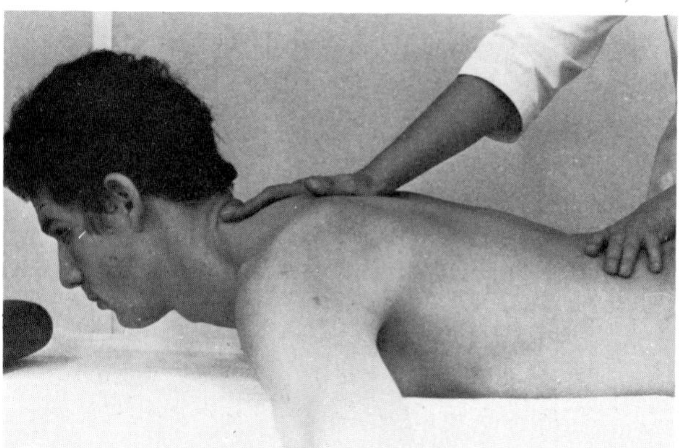

4 and 5 Same position.

Same precautions and movement.
Resistance to the movement is placed at the level of first thoracic vertebrae.

Grade 4: Power is sub-normal.
Grade 5: Power is normal.

TESTING LUMBAR SPINAL MUSCLES

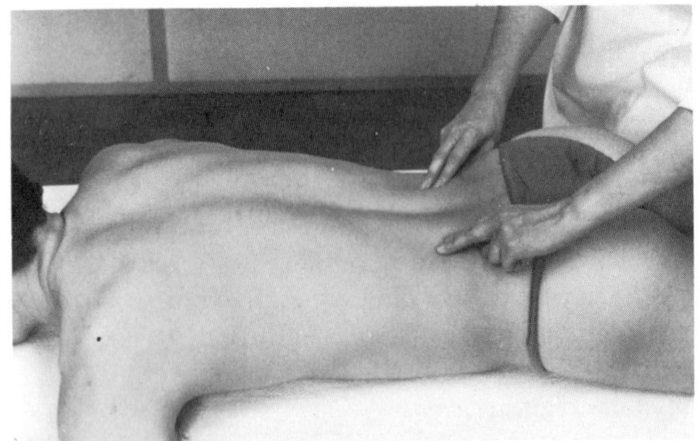

0 and 1 Subject prone.

Arms hanging down on either side of the table to avoid confusion with the latissimus dorsi.
Ask subject to lift his trunk.
Palpate at lumbar level on each side of the spinous processes.
Watch for speckling of skin.

Grade 0: No contraction is felt.
Grade 1: Contraction is felt but there is no joint movement.

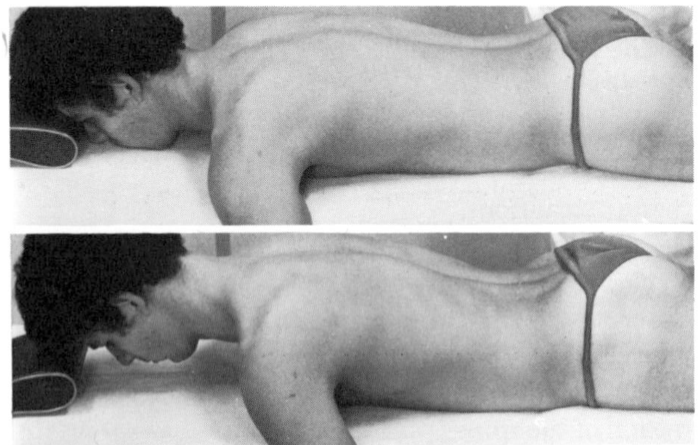

2 Same position.

Stabilize lower limbs down by placing hands on posterior aspect of thighs.
Ask for trunk extension.
If hip extensors are deficient the buttocks rise, in which case the sacrum should be held down.
In case of asymmetry: lumbar spine extends laterally toward the dominant side.
Palpate to estimate muscle volume, compare both sides and grade them accordingly.
Watch for speckling of skin.
Grade 2 is given if part of the thoracic cage remains in contact with the table.

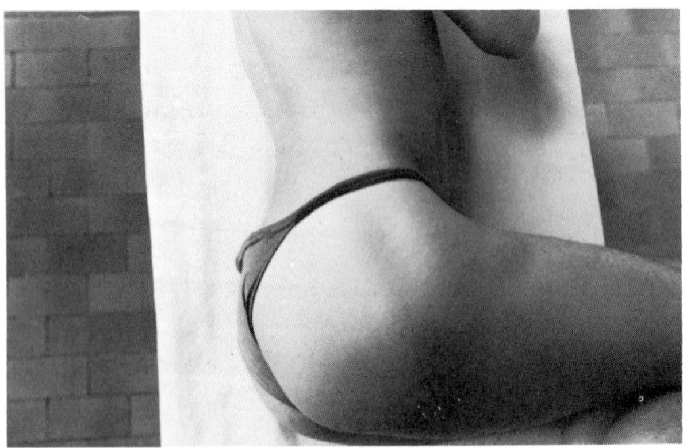

2 *Alternative.*

Subject side-lying.
Hips flexed at 90°.
Ask subject to extend lumbar spine while hips and knees remain flexed.
Range of movement should be complete.

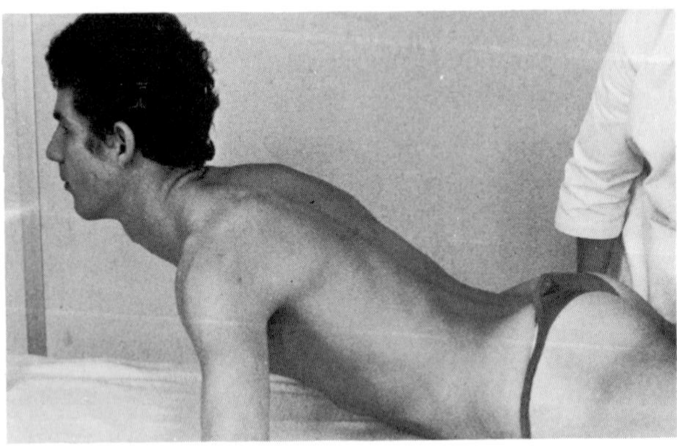

3 Subject prone.

Stabilize lower limbs in place on posterior aspect of thighs.
Ask subject to extend trunk.
Upper limbs should remain relaxed throughout the movement.
As lumbar spine extends, thorax and upper abdomen clear the table.
Watch for any asymmetry.

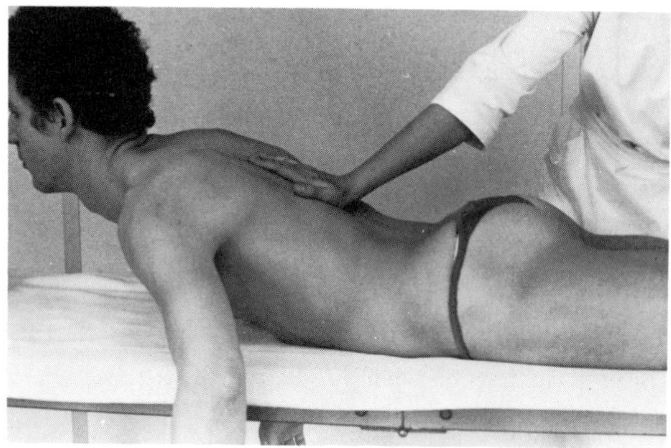

4 and 5 Same position.

Stabilization and movement.
Resistance to the movement is placed over the thoracic region.

Grade 4: Power is sub-normal.
Grade 5: Power is normal.

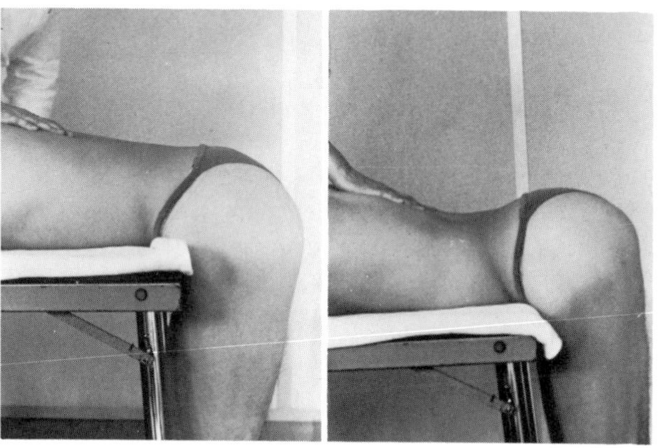

3, 4, and 5 *Alternative.*

Subject prone, back straight, lower limbs hanging over end of table.
Ask subject to raise his buttocks while keeping lower extremities relaxed.
Grade 3 is given if the movement is executed in complete range of movement.
For grades 4 and 5, resistance to the movement is placed on the sacrum.

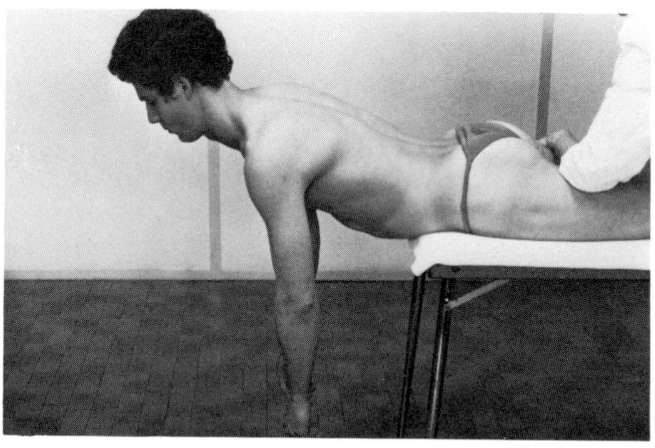

Functional evaluation

Subject prone, lower extremities extended on table top, trunk hanging over the end of the table.
Stabilize thighs. Test quickly since the position is uncomfortable.
From a position of complete flexion, ask subject for trunk extension.
Grade 3 is given if trunk rises above the horizontal level.
For grades 4 and 5, resistance to the movement is placed over the thoracic region.

QUADRATUS LUMBORUM

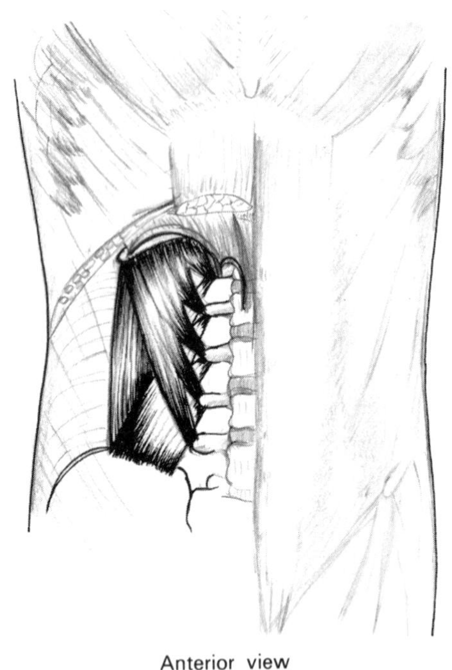

Anterior view

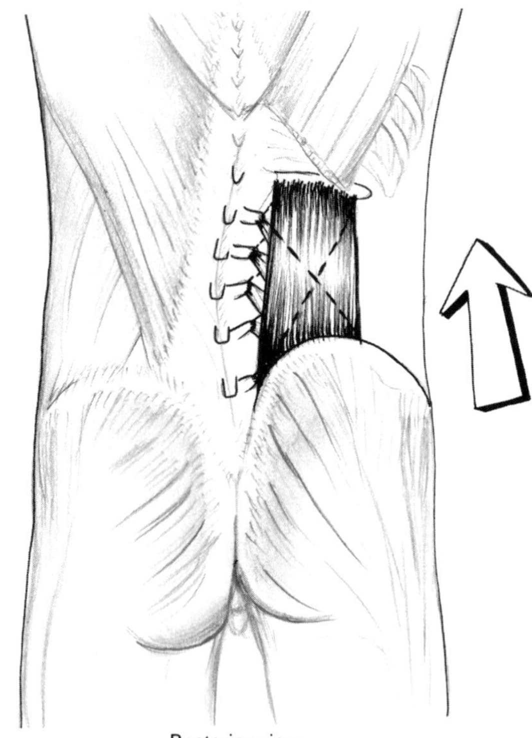

Posterior view

Composed of three portions.

ILIO-COSTAL AND ILIO-TRANSVERSARY PORTIONS

Origin Common on the posterior third of iliac crest.

Insertion

—Ilio-costal belly : inferior border of twelfth rib.
—Ilio-transverse belly : tips of transverse processes of first four lumbar vertebrae.

COSTO-TRANSVERSARY PORTION

Origin Inferior border of twelfth rib.

Insertion Tips of transverse processes of five lumbar vertebrae.

Innervation Lumbar plexus (T12, L1, 12).

Function

—With pelvis fixed : ipsilateral latero-flexion of trunk.
—With thorax fixed : ipsilateral pelvic hitching. It laterally flexes the lumbar spine for which it constitutes a rigging system of support.
—Accessory function : it lowers the last rib during forced expiration. During gait it contracts during ipsilateral pelvic forward rotation.

Contracture or tightness Closes the ipsilateral, lateral infracostal region. Gives an illusion of lower extremity shortening.

Deficiency Hindrance during swing phase of gait.

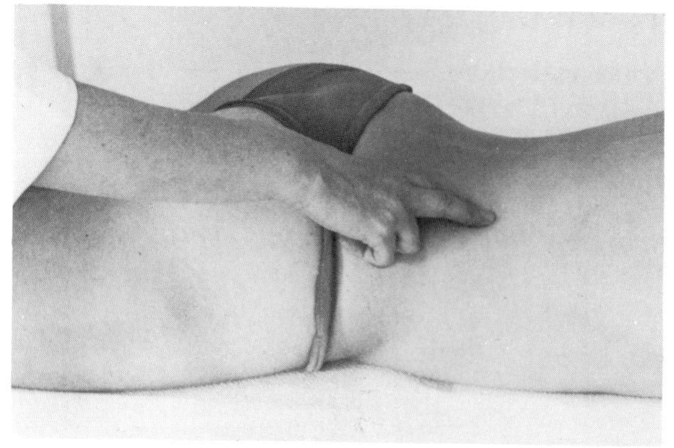

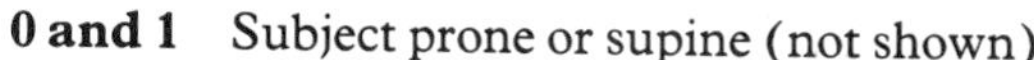

0 and 1 Subject prone or supine (not shown).

Feet hanging over end of table.
Place lower extremity in slight abduction to give the direction of the movement.
Ask subject to hitch ipsilateral pelvis.
Palpate in the costo-iliac space laterally to erector spinae muscle mass.
Palpation is difficult because the quadratus lumborum is deep and partially covered by the spine extensor common muscle mass, and postero-lateral flank muscles.
Palpation becomes possible only after atrophy of these muscles.
Palpate postero-lateral aspect of thoracic cage to be certain there is no confusion with latissimus dorsi.

Grade 0: No contraction is felt.
Grade 1: Contraction is felt but there is no movement.

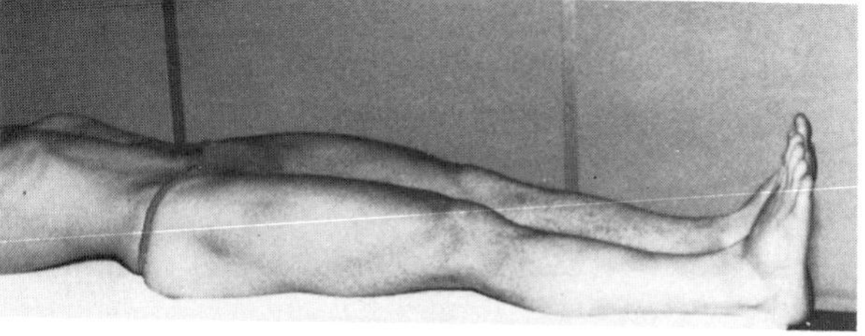

2 Subject supine (or prone, not shown).

Same precautions.
Side to be tested is at table's edge, upper extremity hanging in order to relax latissimus dorsi.
Ask subject to hitch ipsilateral pelvis.
Range of movement should be complete.

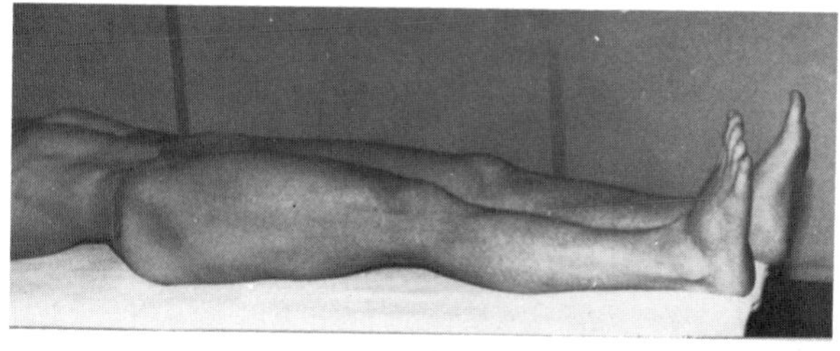

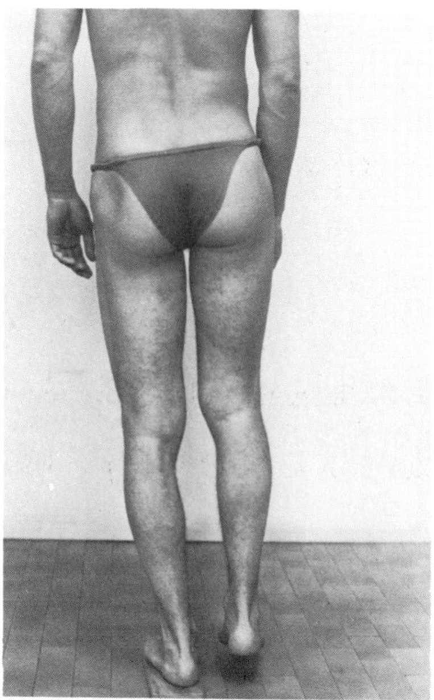

3 Subject standing.

Foot on non-tested side is on a block.
Ask subject to hitch ipsilateral pelvis.
Watch for same possible trick movements.
Subject should close the costo-iliac space.
Range of movement should be complete.

3 *Alternative.*

Subject prone.
Upper limb hanging over edge of table.
Same possible substitutions.
Ask subject to hitch ipsilateral pelvis.
Light resistance to the movement is placed on the iliac crest and lower third of the thigh.
Range of movement should be complete.

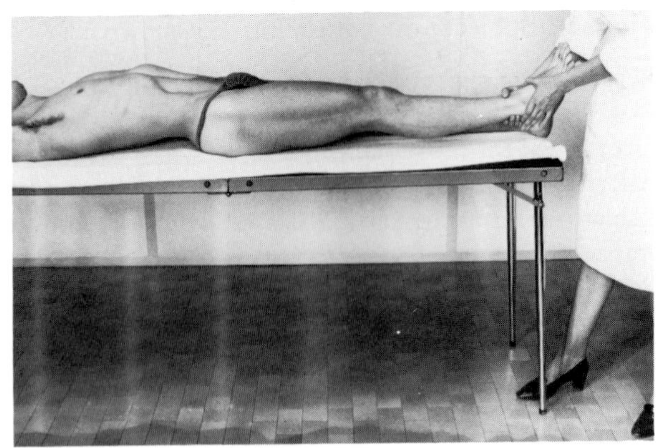

4 and 5 Subject supine or prone.

Same precautions.
Ask subject to perform same movement.
Grasp ankle at malleoli level and exert resistance to the movement.

Grade 4: Power is sub-normal.
Grade 5: Power is normal.

NOTES

MUSCLES OF THE PERINEUM

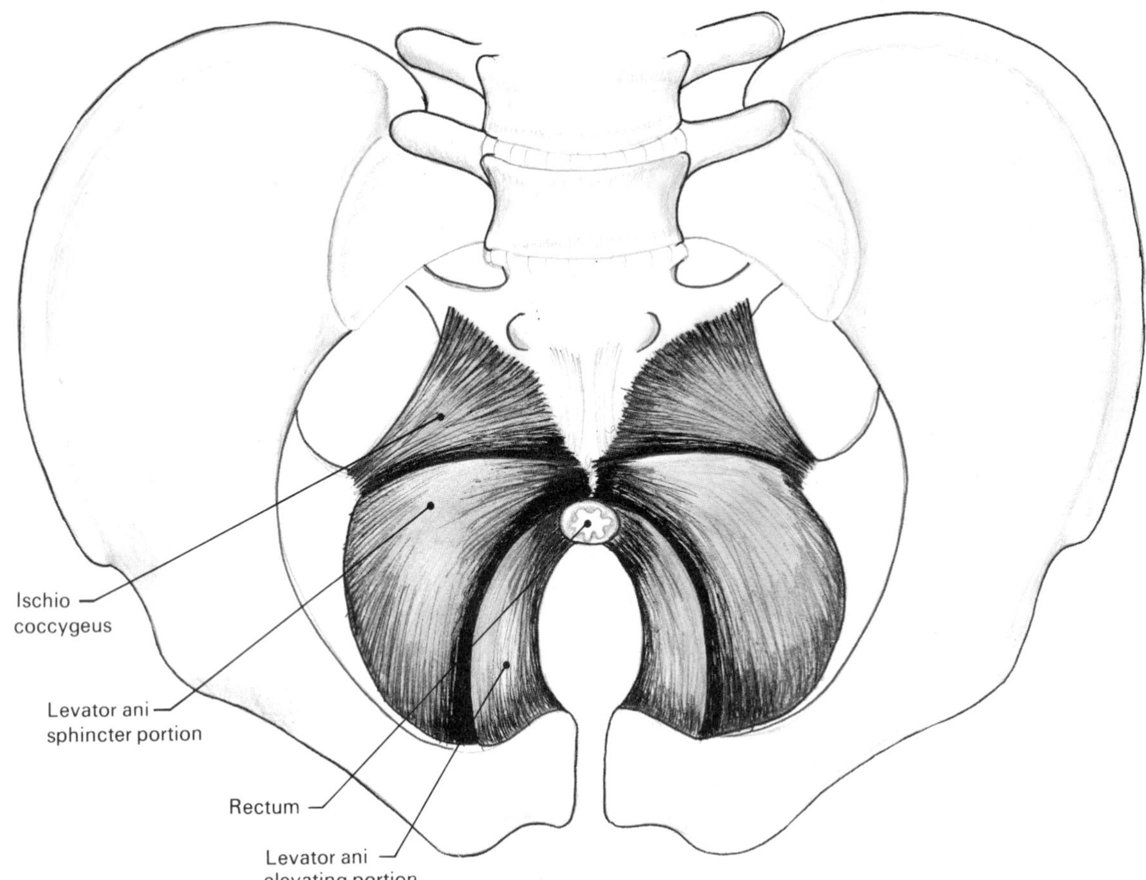

These muscles form the floor of the pelvic cavity. They support the intra pelvic organs and assist in occluding the lower end of the rectum. They are organized in several layers, only the muscles relating to the rectum will be reviewed.

LEVATOR ANI

Insertions are similar in male and female. The only difference concerns genital organs. A deep layer muscle, it is composed of two portions.

LATERAL OR SPHINCTER PORTION

Origin

—Posterior surface of body of os pubis.
—Obturator aponeurosis.
—Medial surface of spine of ischium.

Insertion

—Median fibrous anococcygeal raphe.
—Lateral side of inferior part of coccyx.

MEDIAL OR ELEVATING PORTION

Origin Posterior surface of body of os pubis medially, below, and above the most medial fibres of the sphincter portion.

Insertion

—Front and lateral sides of rectum.
—Inferior fibres descend in the wall of the anal canal to the skin of the anus.

Innervation Branch of sacral plexus.

Function

Synergist of abdominal muscles, it contributes to maintaining intra-abdominal pressure during effort or pushing.

Sphincter portion

—Compresses the rectum from outside to inside,
—Approximates anterior and posterior walls, constricts the rectum,
—Fibres inserting into coccyx counteract its movement of backward deviation.

Elevating portion

—Draws the anal canal up and forward.
—Dilates the canal through traction exerted on anterior and lateral walls.

ISCHIOCOCCYGEUS

Situated behind and parallel to levator ani, its origins, insertions, and relations are similar in male and female.

Origin

—Medial surface of spine of the ischium, posteriorly to levator ani.
—Occasionally, from adjacent portion of anterior margin of great sacrosciatic notch.

Insertion Margin and anterior surface of last two sacral and first three coccygeal vertebrae.

Innervation Branch of sacral plexus.

Function Supports intra-pelvic organs and counteracts the backward deviation movement of the coccyx.

TRANSVERSUS PERINEI PROFUNDUS

Situated in the deep layer.
Wider in female than in male, its relations with genital organs are different.

Origin Ischium and medial surface of the inferior ischial ramus.

Insertion Central tendinous point of the perineum posteriorly to urethra and anteriorly to sphincter ani externus.

Innervation Branch of the medial pudendum.

Function Support of urogenital floor.
Supports prostate in male, vagina in female.

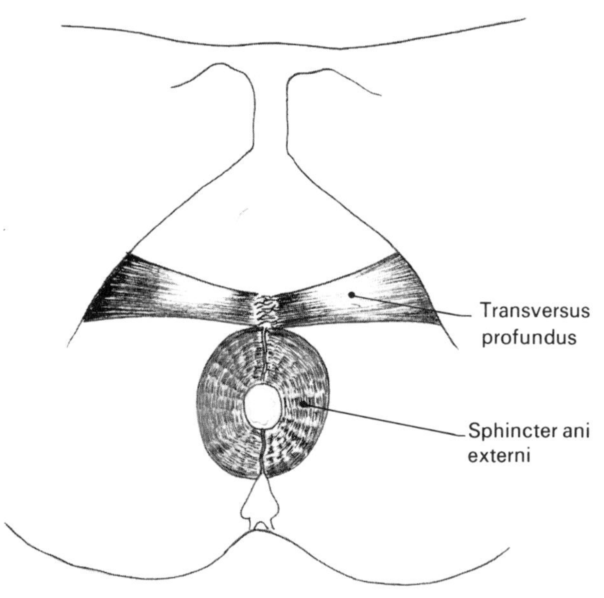

TRANSVERSUS PERINEI SUPERFICIALIS

Occasional muscle situated in the superficial layer.
Site and relations are similar in male and female.

Origin

—Medial surface of ischium.
—Inferior ischial ramus.

Insertion Central tendinous point of perineum.

Innervation Branches of medial pudendum,

Function

Constricts anterior part of anal canal.
Stabilizes the perineal body and thus permits contraction of bulbocavernosus.

SPHINCTER ANI EXTERNUS

Situated in the superficial layer around the anal portion of the rectum.
Site and relations are similar in male and female.

Origin

—Apex of coccyx.
—Anococcygeal raphe.

Insertion Fibres decussate around anus and meet anteriorly in central point of perineum and deep surface of skin.

Innervation Branch of sacral plexus.

Function Closes anal orifice.

Suggested test for muscles of the perineum

Perineal muscles contract synergistically with abdominal muscles during violent efforts or when withholding abdominal pushing. For all grades, ask subject to cough.

0 and 1 During cough, the perineal floor yields to pressure forming an important prolapsus (palpation is possible with a finger glove).

2 During the cough, the perineal floor yields partially forming a discrete prolapsus.

3 The muscles do not yield to the pressure.

4 and 5 Muscles of the perineal floor contract efficiently. No matter or gas escape during effort.

LOWER EXTREMITY MUSCLES

Translator's note

Terminology of the walking cycle used in the French language edition of this book considers both limbs at the same time. Its direct translation into English is difficult to interpret for the English-speaking student who is accustomed to descriptions of gait expressed in percentages and based on observation of a single limb. Gait has therefore been translated using Dr Perry's terminology which divides it into two phases:

—The stance phase (60% of the cycle), subdivided into

- initial contact or heel strike (a single moment),
- loading response (0% to about 15%),
- mid stance (about 15% to 40%),
- terminal stance (about 40% to 55%), and
- pre-swing (about 55% to 60%).

—The swing phase of gait (40% of the cycle) is subdivided into

- initial swing (about 60% to 70%),
- mid swing (about 70% to 80%), and
- terminal swing (about 80% to 100%).

These two phases with their subdivisions are used as headings in this section for description of muscle activity during the gait cycle by J. B. Piera and A. Grossiord.

D. Thomas

PSOAS ILIACUS (ILIOPSOAS)

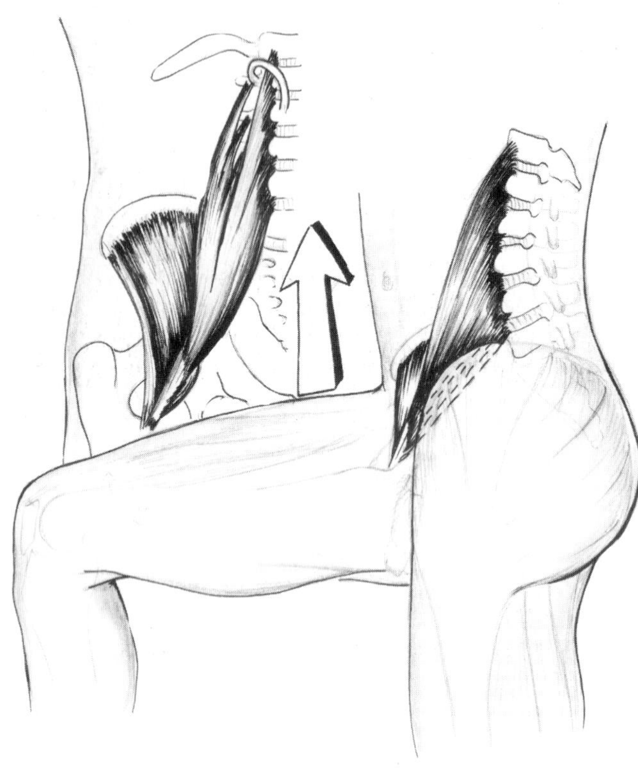

PSOAS

Psoas minor is associated with it.

Origin

—Lateral surfaces of bodies of twelfth thoracic to fifth lumbar vertebrae close to the intervertebral discs.
—Lateral surfaces of adjacent intervertebral discs.
—Tendinous arches connecting the vertebral origins.
—Costal processes of first four lumbar vertebrae and inferior border of twelfth rib.

Insertion Summit aspect of lesser trochanter.

ILIACUS

Iliacus minor is associated with it.

Origin

—Upper two-thirds of iliac fossa.
—Upper half of sacroiliac symphysis.
—Superior surface of ala of sacrum.

Insertion Common with the psoas, or independent close to it on the lesser trochanter.

Innervation

—Psoas: lumbar plexus (L1 . . . L4).
—Iliacus: femoral (L2, L3, L4).

Function

With trunk fixed

Iliopsoas flexes the thigh on the pelvis and rotates it slightly externally. Certain authors describe it as an adductor.
Iliopsoas flexes the pelvis on the lumbar spine, tilting the pelvis anteriorly (symphysis up).

With femur fixed

The psoas draws lumbar vertebrae forward, flexing the spine ipsilaterally and rotating it to the opposite side (especially the third lumbar vertebra).
The iliacus tilts the pelvis and rotates it.
Note: According to F. Meziere, in the standing position, when both spine and femur are semi-fixed, the iliopsoas draws the femur into internal rotation.

Iliopsoas muscle activity during gait cycle

(J. B. Piera and A. Grossiord)
Stance phase. Iliopsoas starts firing at the end of the stance phase. By an eccentric contraction it eliminates the tendency towards excessive hip extension.
Swing phase. Iliopsoas keeps firing but in concentric contraction, flexing the hip, bringing the thigh forward. It seems, even though electromyographic analysis is quite difficult, that the psoas does not fire in normal gait. The activity of the iliacus stops before the terminal swing. Psoas seems to participate very little in gait (difficult to demonstrate).

Contracture or tightness

—Flexion deformity of the hip.
—Lumbar hyperlordosis, drawing the pelvis into posterior tilt.
Lumbar hyperlordosis and pelvic tilt persist in standing position.

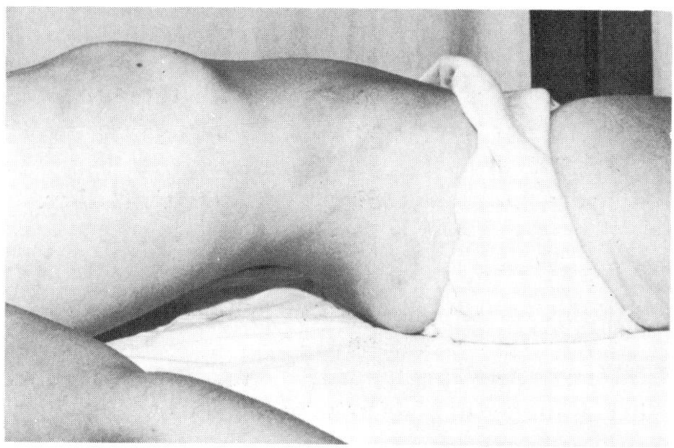

Contracture of iliopsoas.
Note lumbar hyperlordosis and posterior pelvic tilt.

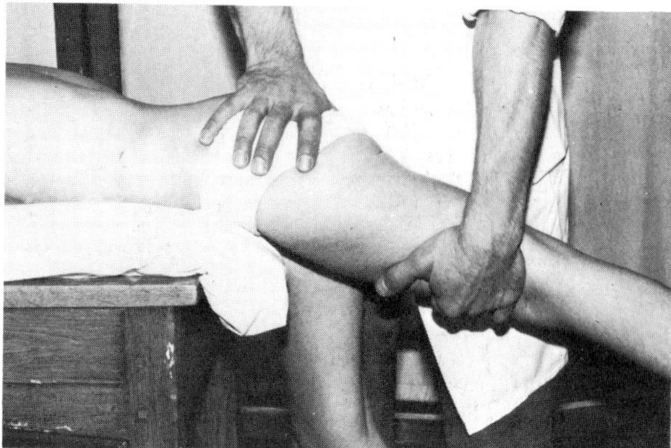

Examining contracture of iliopsoas.

Deficiency

—Marked decrease of hip flexion capacity.
—Functional hindrance during gait cycle, partly compensated by remaining hip flexors and abdominal muscles. During gait cycle, subject limps, swinging his limb by raising his pelvis.
—Difficulty in rapid gait, running, walking up a slope, and passing from sitting or supine position to standing.
—Flattening of lumbar lordosis in upright position.

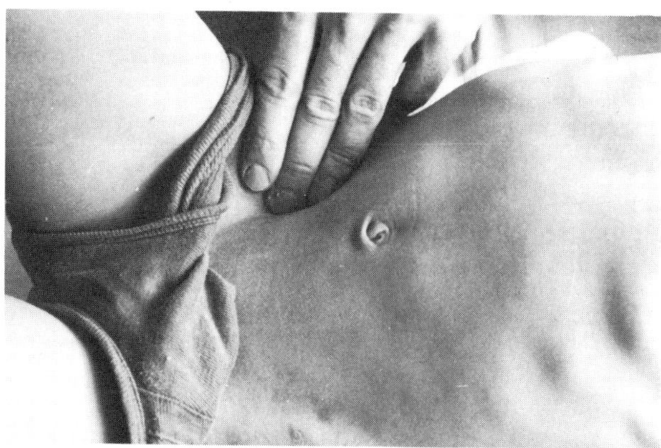

0 and 1 Subject supine.

Iliopsoas, being a deep layer muscle, is difficult to palpate. It can be palpated, however, especially in children, in the inferolateral portion of the abdomen laterally to the rectus abdominis. Ask subject to flex his thigh on his pelvis. To palpate the muscle at abdominal level, the examiner must exert pressure. Palpation of the terminal tendon is possible in the medial aspect of the inguinal crease when surrounding muscles are atrophied.

Grade 0: No contraction is felt.
Grade 1: Contraction is felt, but there is no joint movement.

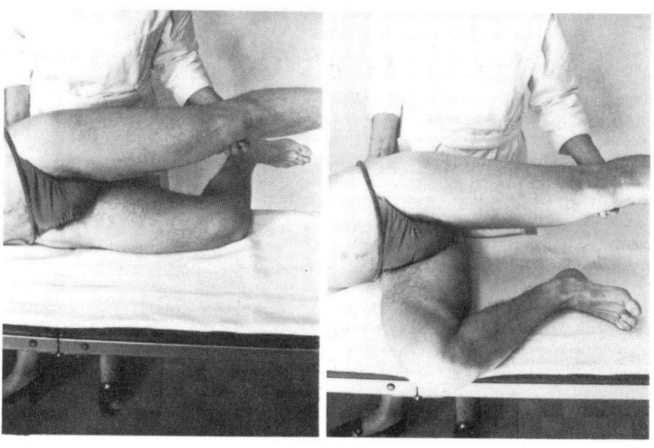

2 Subject side-lying.

Limb to be tested is on the table, hip extended, knee flexed.
Support contralateral limb.
Stabilize sacro-lumbar region to prevent lordosis flattening.
Ask subject to flex thigh on pelvis.
Range of movement should be complete.

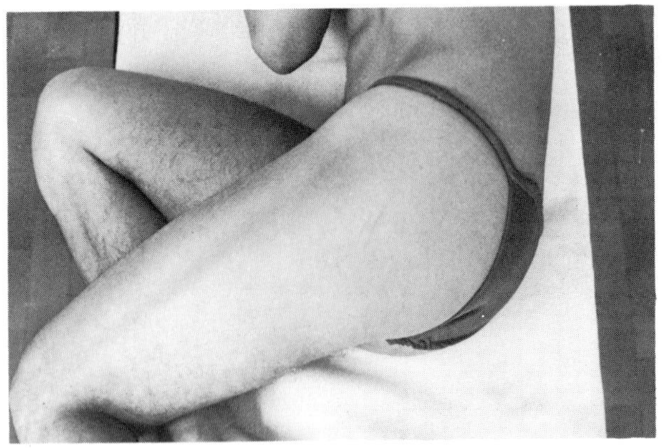

Compensation by pelvic anterior tilt and flattening of lumbar lordosis.
Compensation is characteristic of grade 2 but can appear in all grades if the pelvis is not held, and if examiner does not watch the lumbar spine.

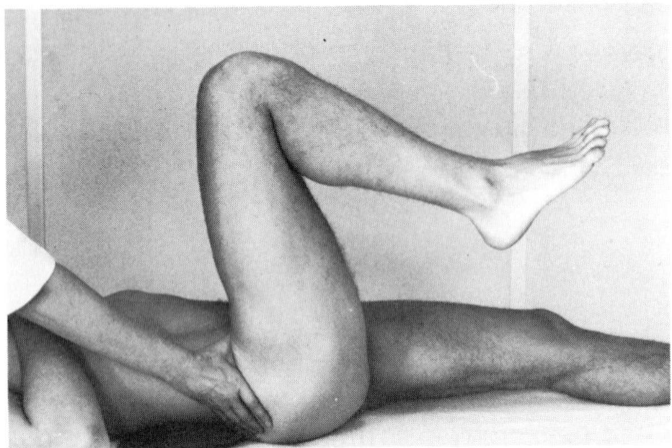

3 Subject supine.

Lower extremities extended.
Cushion under lumbar spine preserves lordosis.
Stabilize iliac crest.
Ask subject to flex thigh on pelvis, keeping his knee flexed throughout the movement.
After 90° flexion, minimal resistance to the movement is placed on the anterior aspect, lower third of thigh.
Flexion with abduction and external rotation demonstrates activity of sartorius. Flexion with abduction and internal rotation demonstrates predominance of tensor fasciae latae.
Range of movement should be complete.

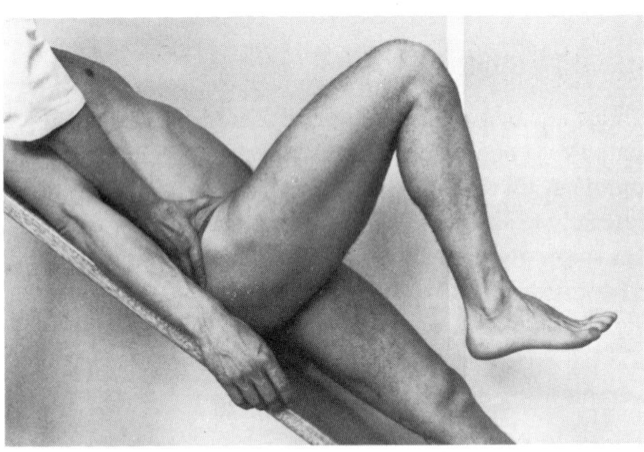

3 *Alternative.*

Subject supine on tilt table.
Cushion under lumbar region.
Same stabilization.
Same precautions.
Ask subject for same movement.
Range of movement should be complete.

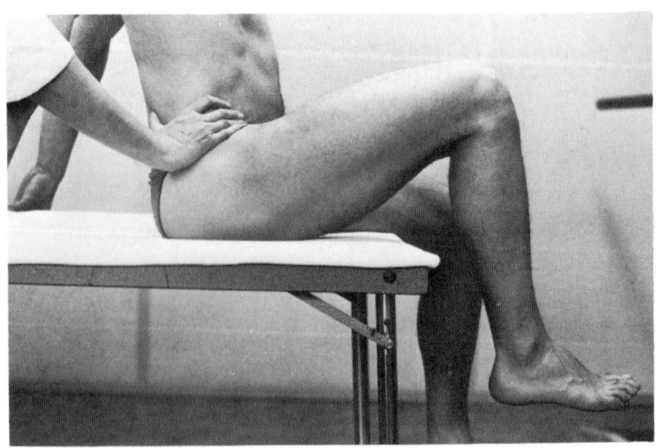

3 *Alternative.*

Subject sitting, legs hanging over edge of table.
Stabilize iliac crest in place.
Ask subject for same movement.
Watch for same compensations; ensure that pelvis and lumbar spine remain neutral.
Range of movement should be complete.
Note: Tests only concern iliopsoas in its most shortened position.

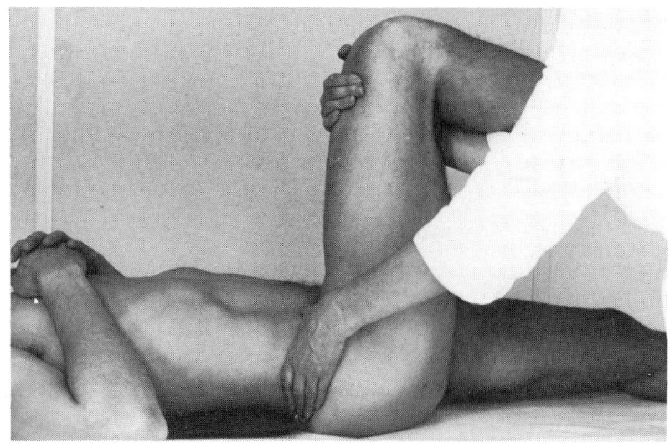

4 and 5 Subject supine.

Cushion under lumbar region.
Same stabilization and precautions.
Ask subject for same movement.
Resistance to the movement is placed on the lower third, anterior aspect of the thigh.

Grade 4: Power is sub-normal.
Grade 5: Power is normal.

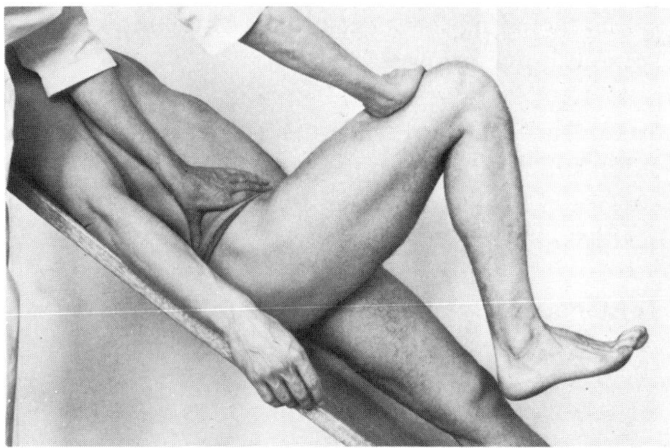

4 and 5 *Alternative.*

Subject supine on tilt table.
Same stabilization, precautions and movement.
Place resistance in same place.

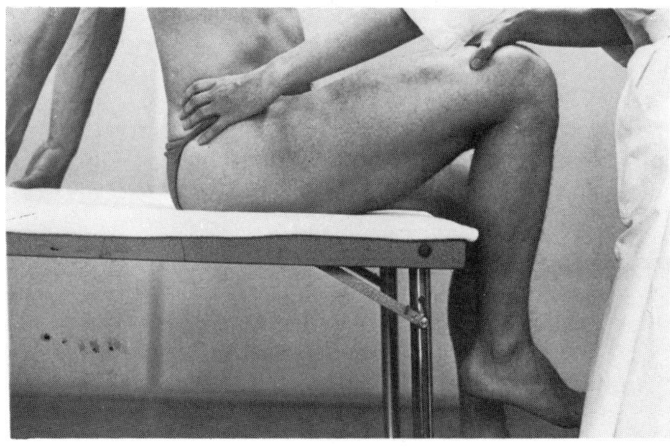

4 and 5 *Alternative.*

Subject sitting, legs hanging over edge of table.
Same stabilization, precautions and movement.
Place resistance in same place.

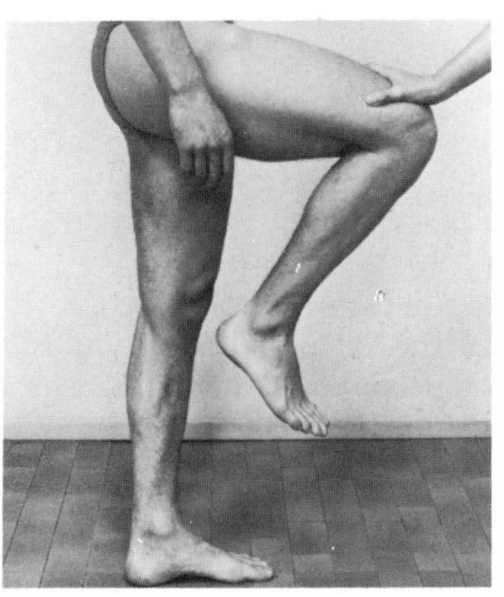

4 and 5 *Alternative.*

Subject standing.
Same precautions and movement.
Place resistance in same place.
Note: Examiner should stand beside patient (not shown for photographic purposes).
Watch for compensating pelvic anterior tilt and flattening of lumbar lordosis.

SARTORIUS

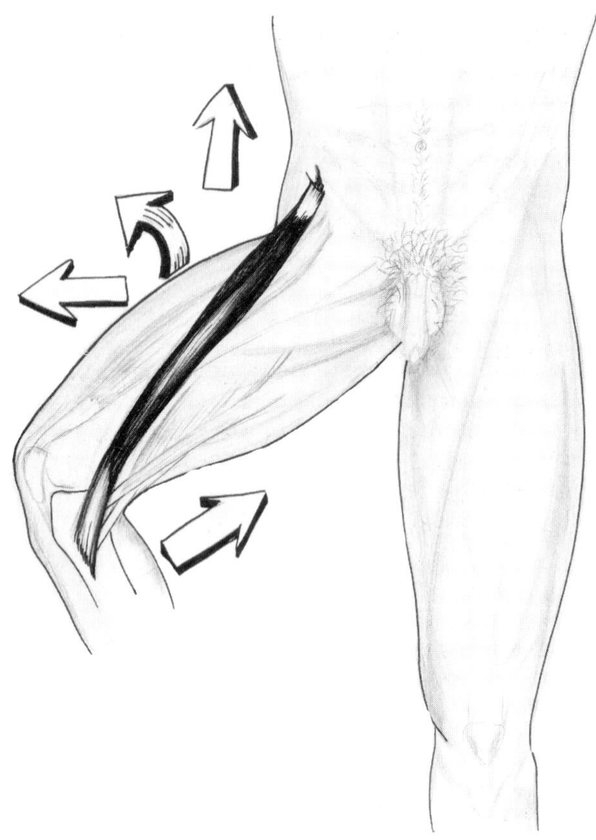

Origin Lateral surface of anterior superior iliac spine, including the superior part of the inter-spinous notch below it.

Insertion The muscle crosses anterior surface of thigh and inserts into upper quarter of medial surface of tibia on pes anserinus muscle insertion surface.

Innervation Femoral (L1, L2, L3).

Function

On hip. External rotation of thigh on pelvis in flexion and abduction. It stabilizes the pelvis in the sagittal plane and tilts it posteriorly.
On knee. Flexion of leg on thigh. Internal rotation of leg when knee is flexed. Stabilizes knee, acting with pes anserinus muscles.

Sartorius muscle activity during gait cycle

(J. B. Piera and A. Grossiord)
Stance phase. With the pes anserinus muscles, gracilis and semitendinosus, the sartorius, a biarticular muscle, wraps around the medial aspect of the knee joint and opposes, during weight bearing, the increase of the physiological valgus angle. Thus it stabilizes this joint and plays the role of an active ligament.
Swing phase. It activates at terminal swing. At the same time as a hip and knee flexor, it participates as a braking force in controlling knee extension and assists the iliacus in hip flexion. Furthermore, it is an external rotator of the lower limb swinging from the pelvis; this rotation is at its maximum at heel strike. Finally, it prepares to play its role of an active medial ligament of the knee.

Contracture or tightness Extreme hip position of external rotation, flexion and abduction.

Deficiency Isolated involvement is rare and has little functional significance.

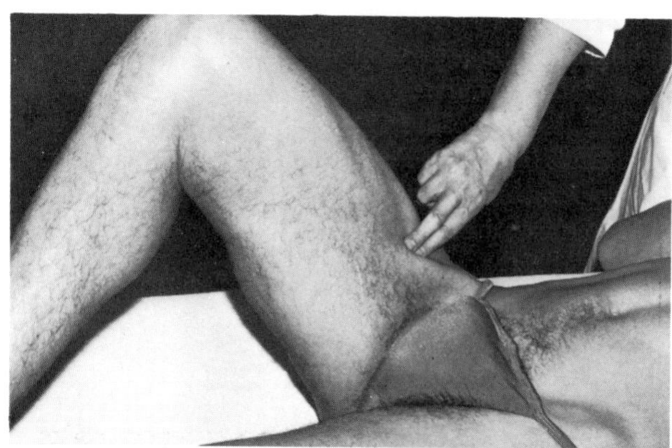

0 and 1 Subject supine.

Lower extremities extended.
Ask subject to flex thigh on pelvis in external rotation and abduction.
Sartorius fibres are palpable under anterior superior iliac spine, on anterior aspect of thigh, or on medial aspect of knee.

Grade 0: No contraction is felt.
Grade 1: Contraction is felt but there is no joint movement.

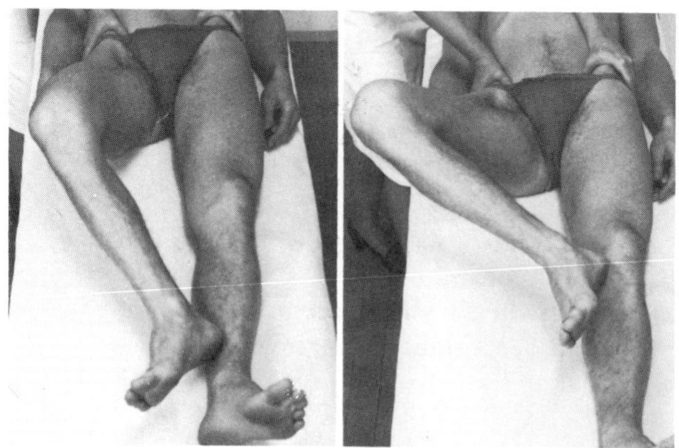

2 Same position.

Stabilize iliac crest.
Ask subject to slide his heel up the opposite shin in such a way that he flexes, abducts and externally rotates his hip.
Range of movement should be complete.

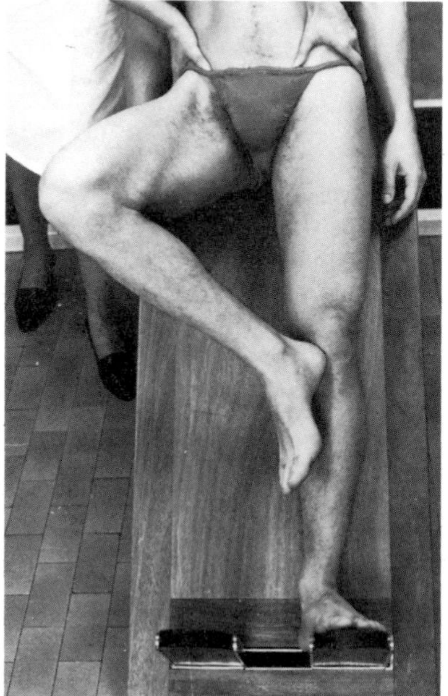

3 Subject supine on tilt table.

Stabilize iliac crest.
Ask for same movement.
Range of movement should be complete.

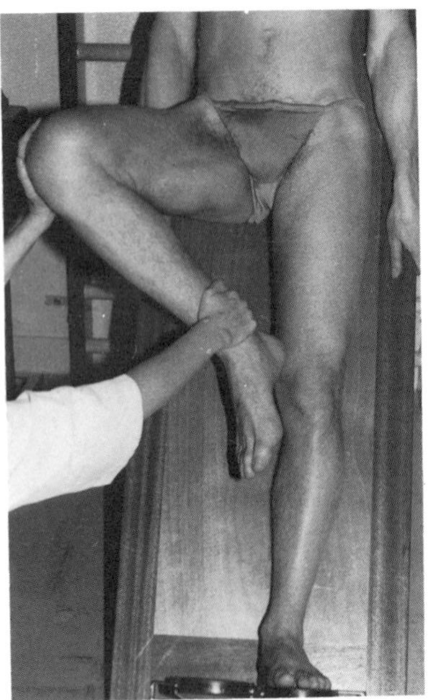

4 and 5 Same position.

Same stabilization and movement.
Resistance to the different components of movement is placed on the lower third, lateral aspect of the thigh, and lower third, medial aspect of the leg.

Grade 4: Power is sub-normal.
Grade 5: Power is normal.

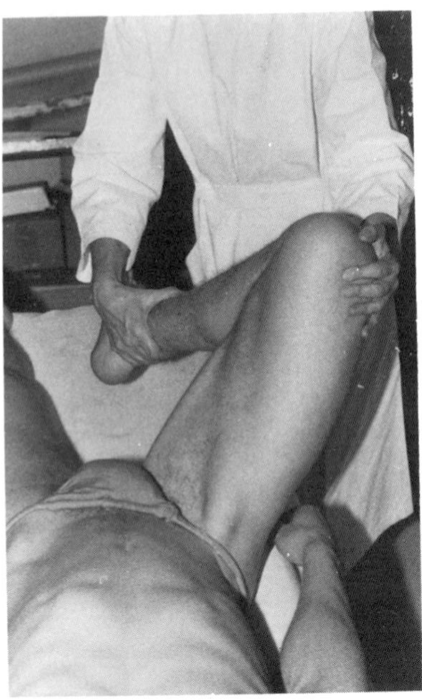

3, 4, and 5 *Alternative.*

Subject supine.
Place resistance in same places.
For grade 3, resistance is minimal.

NOTES

TENSOR FASCIAE LATAE

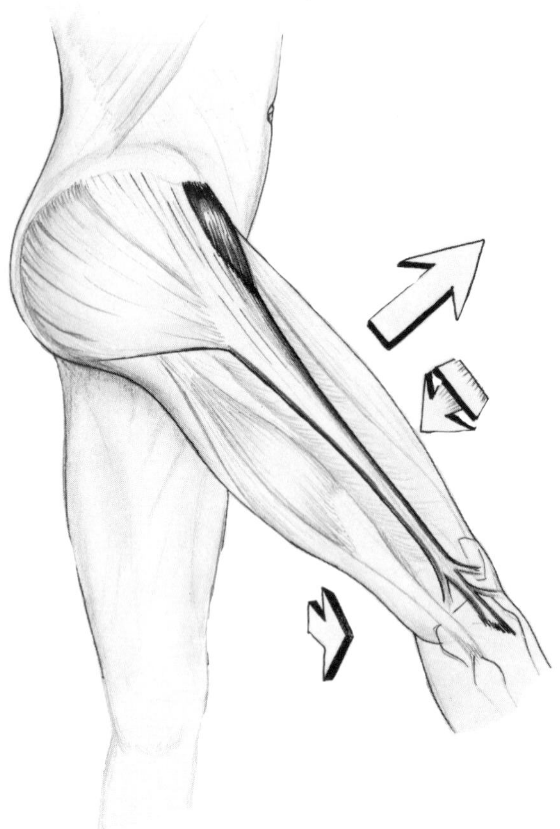

Origin

—Lateral surface of anterior superior iliac spine.
—Anterior part of lateral lip of iliac crest.

Insertion Muscle inserts into the iliotibial band (Maissiat's band) or fascia latae, which inserts into Gerdy's tubercle on the lateral tuberosity of the tibia.

Innervation Superior gluteal (L4, L5, S1).

Function

On hip. Flexion with abduction of thigh on pelvis. It is at maximum efficiency when pelvis is tilted anteriorly. When the thigh is in internal rotation, it balances the external rotary action of sartorius. When tibia is fixed, it participates in horizontal stability of the pelvis.
On knee. It participates in extension of leg on thigh, and in external rotation when the knee is flexed. It constitutes a true active ligament of the knee joint, assuring transverse knee stability, and balancing its action with that of the pes anserinus muscles.
It prevents subluxation of iliotibial band at level of greater trochanter tuberosity.

Tensor fasciae latae muscle activity during gait cycle

(J. B. Piera and A. Grossiord)

Stance phase

—Initial contact, loading response: It contracts as does gluteus medius throughout this period of double support, and participates in pelvic stability, this time not in relation to the femur but to the tibia since it is a biarticular muscle. At knee level it plays an active, lateral ligament role, its valgus component balancing the action of the pes anserinus muscle.
—Mid stance: It gives lateral stability to the pelvis, with gluteus minimus and gluteus medius.

Swing phase

It contracts to assure the lateral balance of the thigh in opposition to the adductors. It keeps the thigh in balance; otherwise the external rotators would predominate at heel strike.

Contracture or tightness

Very frequent in infantile neuro-orthopaedics.
Since this is a two-joint muscle, it leads to several deformities:
—Tendency for hip abduction.
—Ipsilateral posterior pelvic tilt.
—Valgus of the knee and leg lateral rotation.

Contracture of tensor fasciae latae is often associated with a disability of the lower extremities, pelvis, and trunk muscles. Associated with contracture of contralateral adductors and contralateral flank muscles, it favours lateral pelvic tilt and lumbar spine latero-flexion.

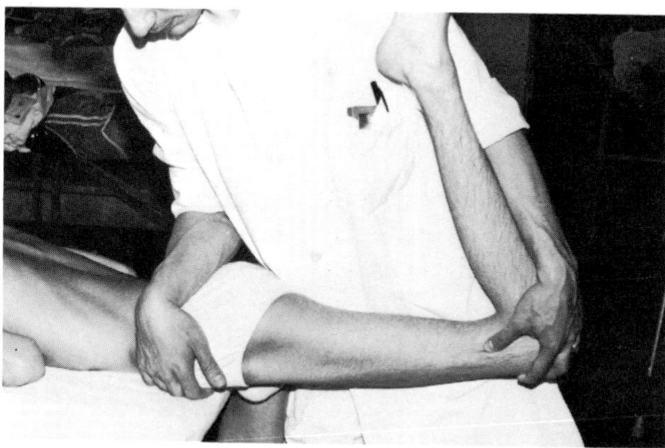

Examining the contracture.

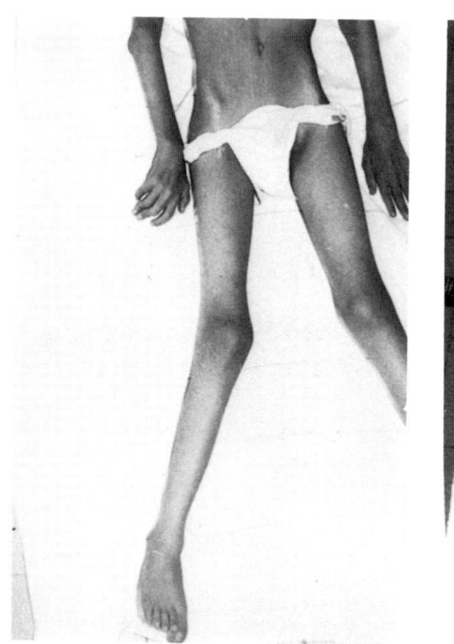

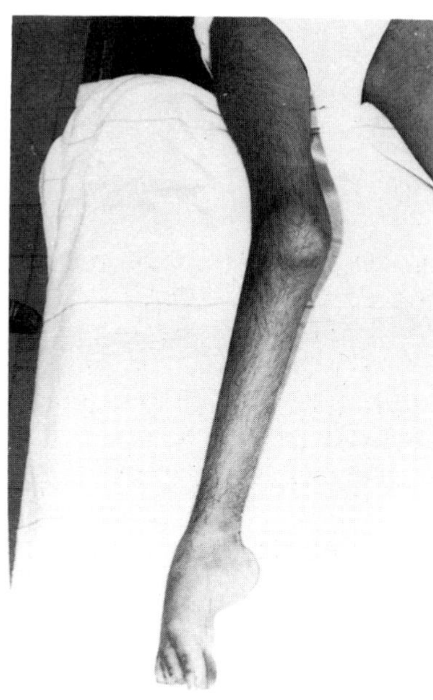

Contracture of tensor fasciae latae. Note genuvalgum and external rotation of leg.

Deficiency

During gait the lower extremity goes into external rotation.
Lateral instability of pelvis and knee.
Paralysis of tensor fasciae latae is often associated with that of gluteus medius and minimus. It leads to contralateral pelvic drop in the stance phase of the gait cycle or when standing on the affected limb (Trendelenburg's and Duchenne de Boulogne's signs).

NOTES

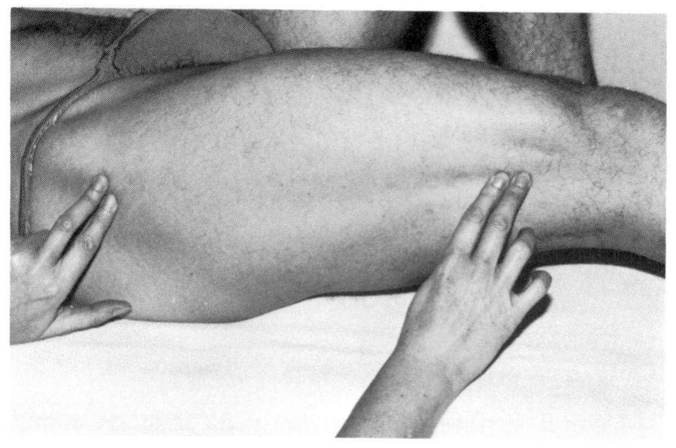

0 and 1 Subject sitting inclined.

Lower extremities extended.
Ask subject for abduction and internal rotation of the thigh on the pelvis.
Tensor fasciae latae is palpable at its origin beneath and slightly posteriorly to the anterior superior iliac spine.
Examiner can feel the band along its course at the lateral aspect of thigh or at knee level, between the biceps tendon and the patella.

Grade 0: No contraction is felt.
Grade 1: Contraction is felt but there is no joint movement.

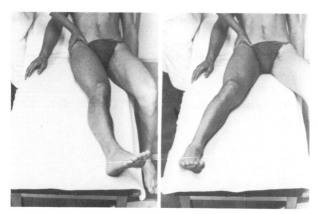

2 Same position.

Limb to be tested is in adduction.
Contralateral limb abducted or flexed, braced on a stool.
Stabilize iliac crest firmly.
Ask subject for abduction and internal rotation of the thigh on the pelvis.
Range of movement should be complete.

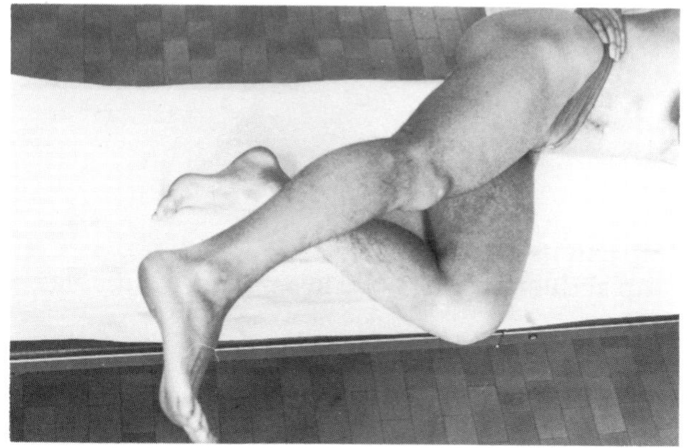

3 Subject side-lying.

Contralateral limb flexed, stabilizing the pelvis.
Stabilize iliac crest firmly.
Ask subject for abduction to 30°, flexion and internal rotation of the thigh on the pelvis, knee extended.
Range of movement should be complete.

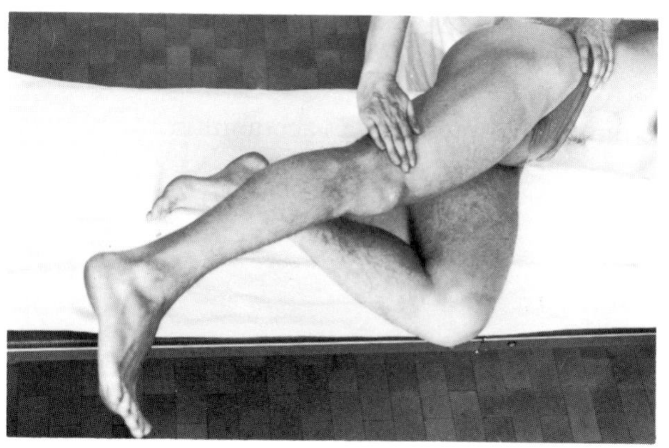

4 and 5 Same position.

Same stabilization and movement.
Resistance to abduction and flexion is placed on the inferior third, lateral aspect of the thigh.

Grade 4: Power is sub-normal.
Grade 5: Power is normal.

4 and 5 *Alternative* (not shown).

Resistance can be placed on the upper third of lateral aspect of leg when dealing with subjects presenting a stable knee.

GLUTEUS MEDIUS

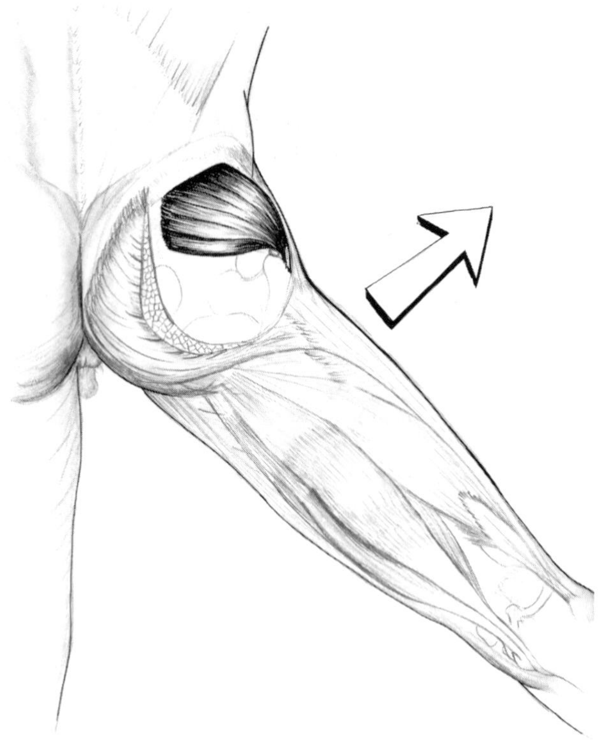

Origin

—Lateral surface of ilium between iliac crest and anterior and posterior gluteal lines.
—Lateral lip of iliac crest.
—Deep surface of gluteal aponeurosis.

Insertion Lateral surface of the greater trochanter.

Innervation Superior gluteal (L4, L5, S1).

Function

With pelvis fixed

Abduction of thigh on pelvis:
—Internal rotation of thigh on pelvis by anterior fibres.
—External rotation of thigh on pelvis by posterior fibres.
Anterior fibres assist in hip flexion, posterior fibres in hip extension.
Gluteus medius reaches maximal efficiency at 30°–35° hip abduction when the lever is perpendicular to the pull of the muscle.
Power of gluteus maximus also depends on pelvic position in the sagittal plane: it is strongest when position is neutral.

With femur fixed

—It stabilizes the pelvis horizontally, acting with the other hip abductors, and specially during single limb stance.
—It tilts the pelvis ipsilaterally.
Note: Gluteus medius participates in hip stability by retaining femoral head in acetabulum.

Gluteus medius muscle activity during gait cycle

(J. B. Piera and A. Grossiord)

Stance phase

—Initial contact: Its role, obviously, is to stabilize the pelvis horizontally. In relation to the greater trochanter which remains relatively stable, it prevents contralateral pelvic drop and limits lateral pelvic movement by its eccentric contraction.

—Mid stance: It maintains horizontal pelvic balance with the tensor fasciae latae. Its deficiency is marked by an increase in contralateral pelvic drop (Trendelenburg's sign), leading to a shortening of the stance phase. The subject shifts trunk laterally over stance limb (Duchenne de Boulogne's sign).

Contracture or tightness

Hip assumes a strongly abducted position.
Adduction is limited. The imbalance is more important when the contracture is associated with that of other muscles originating in the lateral surface of the ilium and contralateral hip adductors.

Deficiency

Marked decrease of abduction capacity, partly compensated for by remaining abductors.
Contralateral pelvic drop in standing position. The subject shifts his trunk laterally over the affected limb to bring his centre of gravity inside the base of support.
During gait, Trendelenburg and Duchenne's sign.

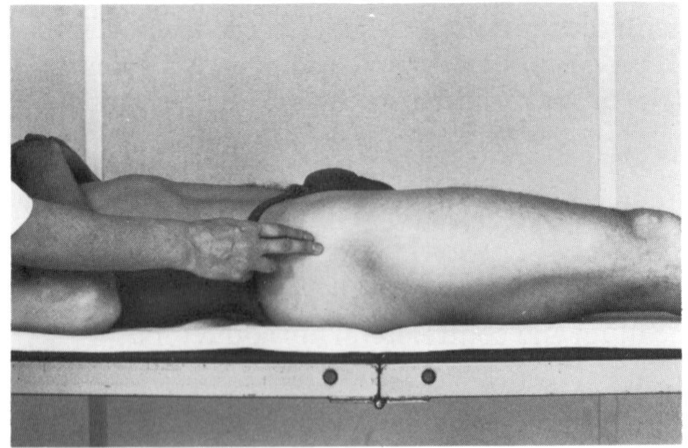

0 and 1 Subject supine.

Lower limbs extended.
Ask subject to abduct thigh on pelvis along table top without rotation.
Palpate on lateral surface of ilium over the greater trochanter.
Do not confuse with gluteus minimus or tensor fasciae latae which are more anterior.

Grade 0: No contraction is felt.
Grade 1: Contraction is felt but there is no joint movement.

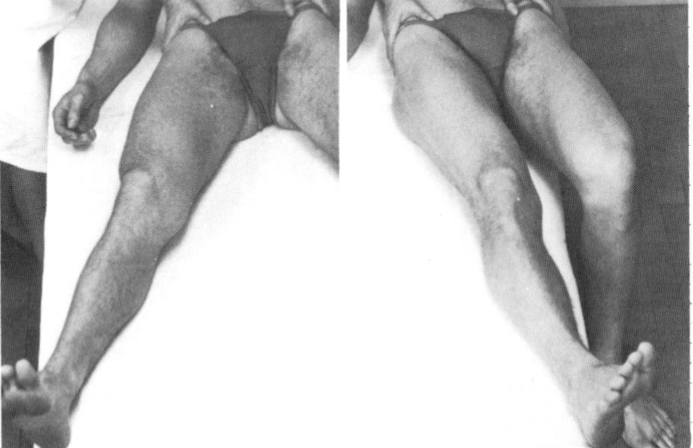

2 Subject supine.

Trunk deviated towards tested side in order to diminish lateral trunk muscle action. Limb to be tested in adduction. Contralateral limb in abduction, flexed and braced on a stool.
Stabilize iliac crest firmly to prevent its tilting.
Ask subject to abduct thigh on pelvis, remaining strictly on table top, with neither rotation nor flexion.
Range of movement should be complete.

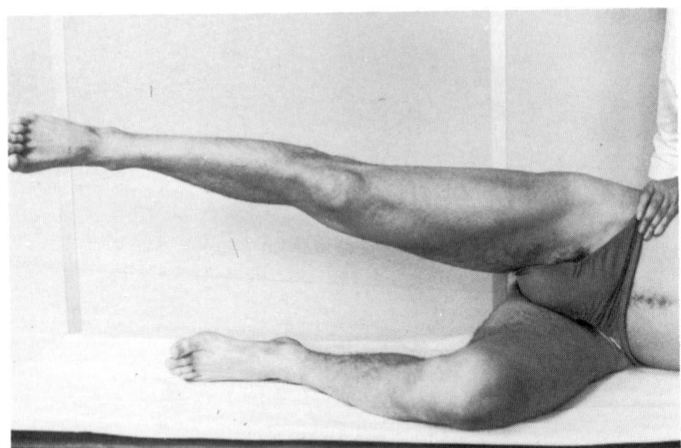

3 Subject side-lying.

Contralateral limb flexed, stabilizing pelvis and trunk.
Stabilize iliac crest.
Ask subject to abduct thigh on pelvis.
Avoid hip flexion and rotation.
Range of movement should be complete.

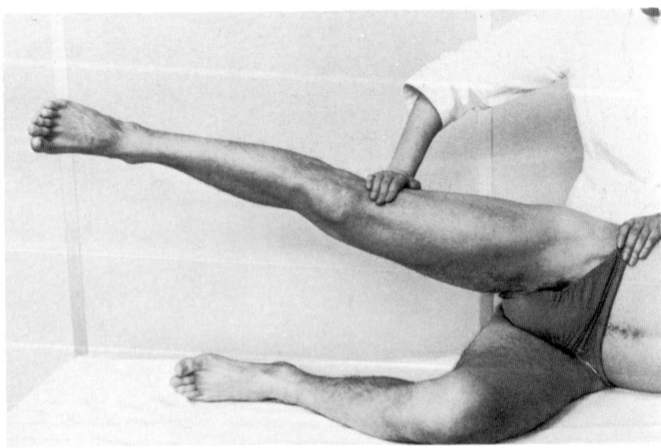

4 and 5 Same position.

Same stabilization, precautions, and movement.
Resistance to the movement is placed on the lower third, lateral aspect of the thigh.

Grade 4: Power is sub-normal.
Grade 5: Power is normal.

3, 4, and 5 *Alternative* (not shown).

Test can be performed with flexed knee in order to decrease the action of tensor fasciae latae.

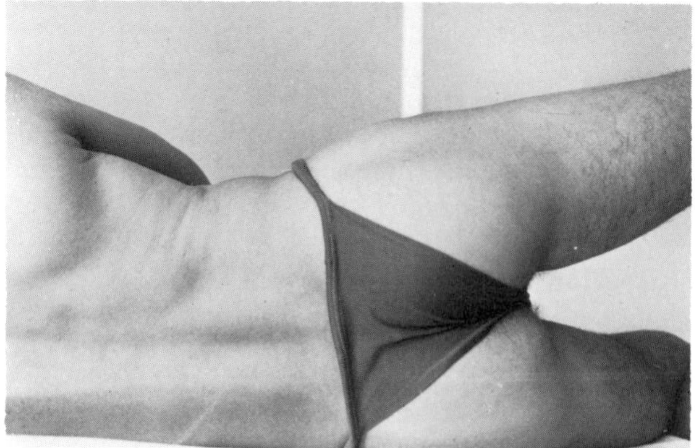

Possible error: Subject contracts lateral trunk muscles thus raising the pelvis.

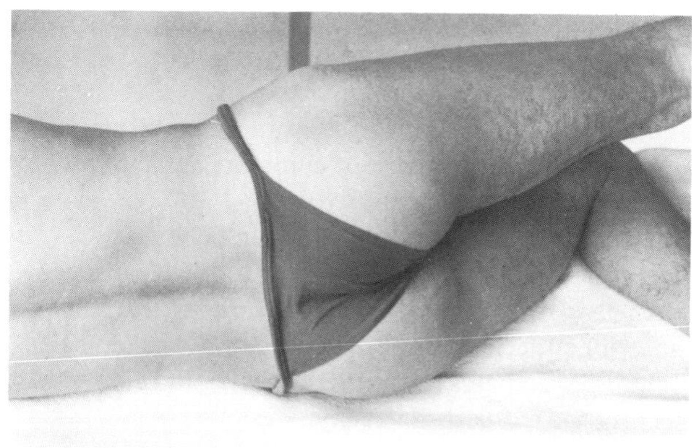

Another possible error: Subject lifts lower extremity in flexion thus leading to action of tensor fasciae latae and gluteus minimus.

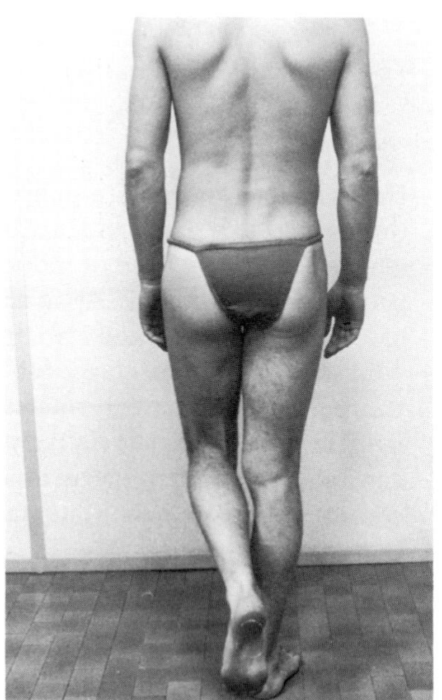

Normal, single limb stance.

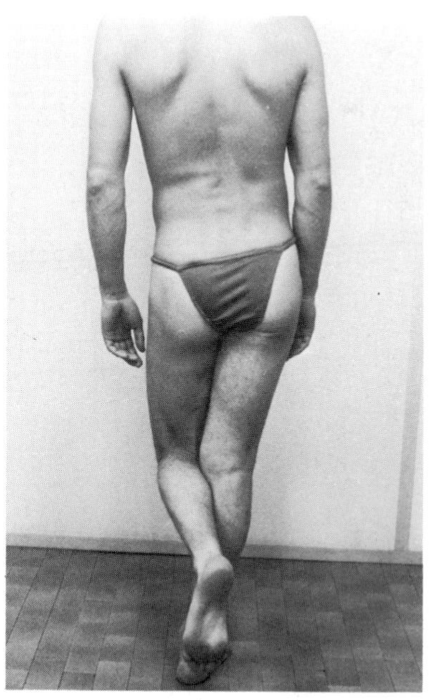

Disability of gluteus medius causes contralateral pelvic drop.

PETIT FESSIER *(Gluteus minimus)*

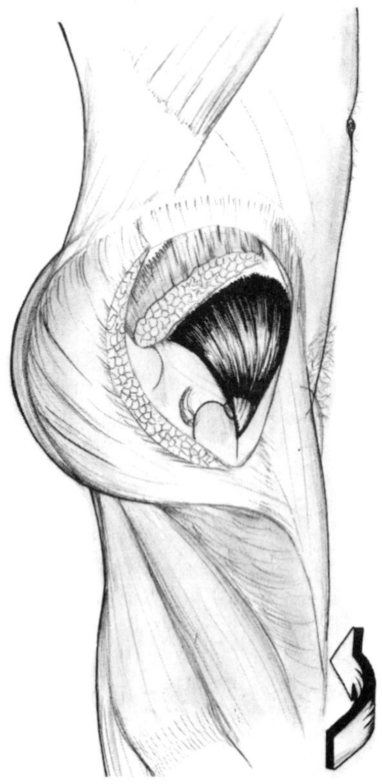

Origin Lateral surface of ilium between the anterior gluteal line, the supra acetabulum depression and the greater sciatic notch.

Insertion

—Anterior surface of greater trochanter.
—Expansion into iliofemoral ligament and hip joint capsule.

Innervation Superior gluteal (L4, L5, S1).

Function

—Internal rotation of thigh on pelvis.
—Abduction of thigh on pelvis, more particularly when pelvis is in slight or complete anterior tilt.
—Horizontal pelvic stability.
It participates in flexion of thigh on pelvis.
In single limb stance when femur is fixed, it rotates the pelvis ipsilaterally, bringing the contralateral anterior superior iliac spine ahead of the tested side. It also participates in tilting the pelvis and side flexing the trunk ipsilaterally.

Gluteus minimus muscle activity during gait cycle

(J. B. Piera and A. Grossiord)

Stance phase

—Initial contact, loading response: It contracts throughout this phase and, acting from the femur which is relatively fixed, it induces the start of pelvic external rotation. As an accessory, it participates in pelvic horizontal stability.
—Mid stance, terminal stance: It contracts throughout this phase. It rotates the pelvis externally in relation to the relatively fixed femur: contralateral forward pelvic rotation. It stops contracting at pre-swing; its stabilizing action is then useless.

Contracture or tightness

Often associated with that of tensor fasciae latae.
Position of thigh in slight abduction and internal rotation.
Ipsilateral lateral pelvic tilt.

Deficiency

Marked decrease of internal rotation otherwise compensated for by tensor fasciae latae and anterior fibres of gluteus medius. When flexing thigh on pelvis, femur tends to go into external rotation.
Decrease in abduction strength.
Slight instability in gait.

TESTING GLUTEUS MINIMUS AS INTERNAL ROTATOR OF THIGH ON PELVIS

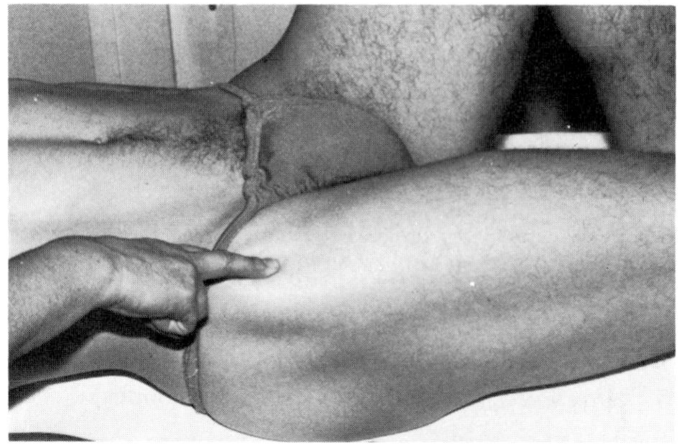

0 and 1 Subject supine.

Lower limb extended.
Ask subject for internal rotation of thigh on pelvis.
Gluteus minimus, covered by tensor fasciae latae and anterior fibres of gluteus medius, is nearly impalpable. However, if the latter two are atrophied, it is palpable anteriorly to and above the greater trochanter.

Grade 0: No contraction is felt.
Grade 1: Contraction is felt but there is no joint movement.

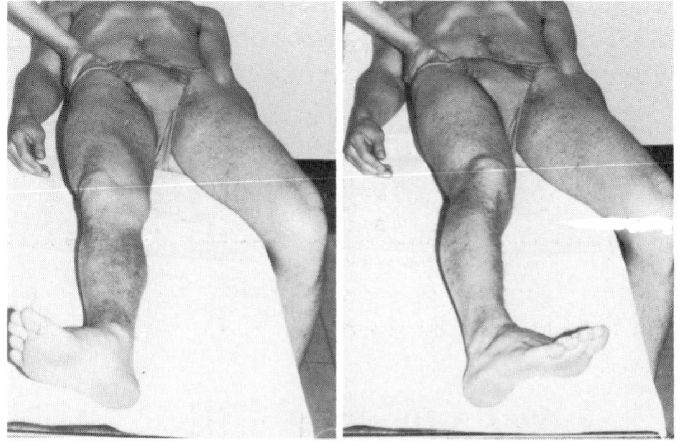

2 Same position.

Contralateral lower limb flexed, and braced on stool to stabilize pelvis.
Stabilize iliac crest to prevent pelvis from tilting.
From a position of external thigh rotation on the pelvis, ask subject for internal rotation.
Ensure thigh movement occurs, not leg or foot; inversion of the latter could give an impression of internal rotation.
Range of movement should be complete.

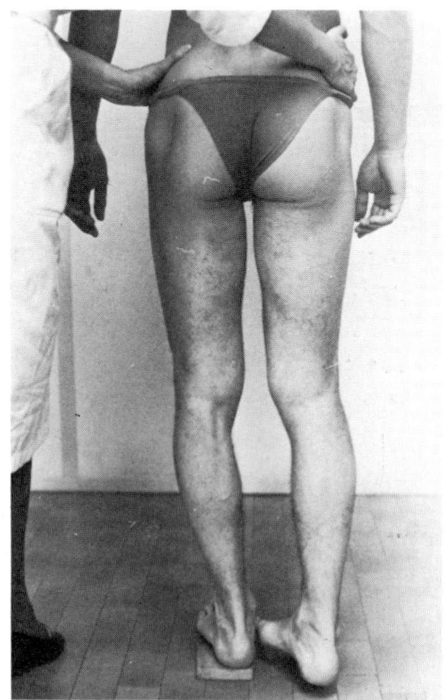

2 *Alternative.*

Subject standing.
Block under contralateral foot.
Stabilize pelvis to prevent rotation.
Ask subject for internal rotation of thigh on pelvis.
Range of movement should be complete.

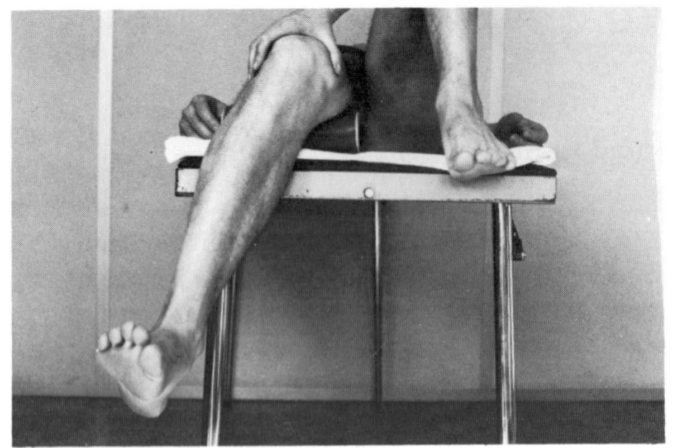

3 Subject supine.

Leg hanging on tested side.
Contralateral lower limb flexed, braced on table to stabilize pelvis.
Stabilize lower third, anterior aspect of the thigh.
Ask subject for internal rotation of thigh on pelvis, drawing leg and foot outwards.
Range of movement should be complete.

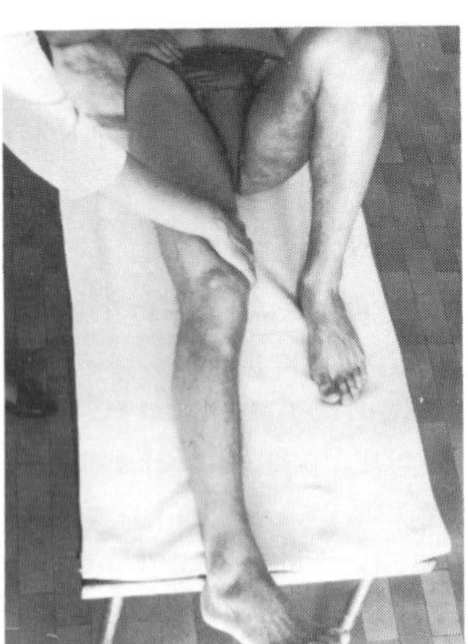

3 *Alternative* (for subjects with fragile bones).

Subject supine. Contralateral lower limb flexed, braced on table or stool to stabilize pelvis.
Stabilize iliac crest to prevent pelvis from tilting.
Ask subject for internal rotation of thigh on pelvis.
Place slight resistance to the movement on lower third, medial aspect of thigh.
Range of movement should be complete.
Note: This alternative may be used for grades 4 and 5.

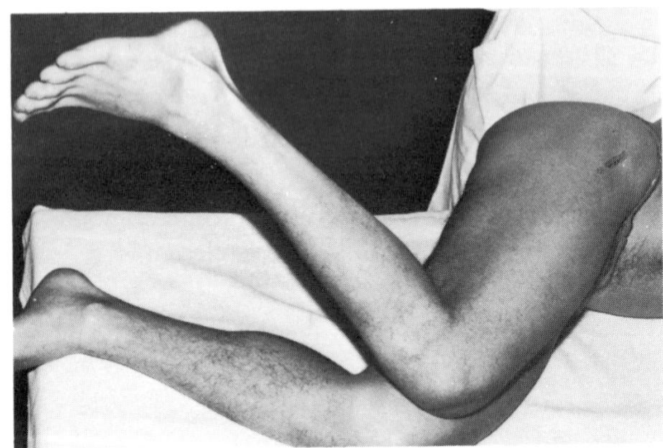

3 *Alternative* (gravity acting upon entire length of limb).

Subject side-lying.
Lower limbs flexed.
Stabilize iliac crest.
Ask subject for internal rotation of thigh on pelvis, drawing leg and foot upwards.
Range of movement should be complete.

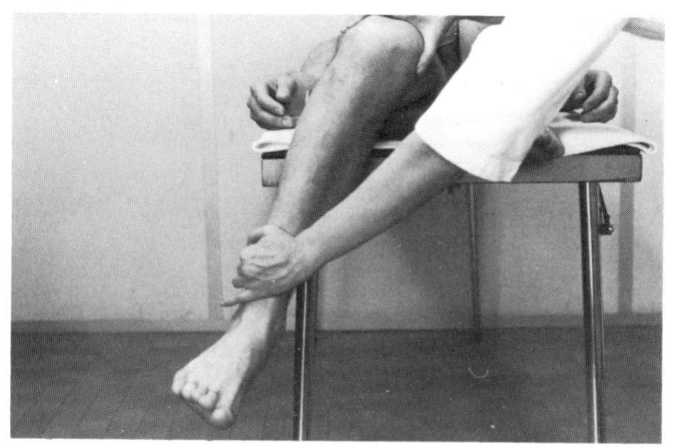

4 and 5 Subject supine.

Leg hanging on tested side.
Contralateral lower limb flexed, braced on table in order to stabilize pelvis.
Stabilize lower third, anterior aspect of thigh.
Ask subject for internal rotation of thigh on pelvis, drawing leg and foot outwards.
Resistance to the movement is placed on lower third of leg.

Grade 4: Power is sub-normal.
Grade 5: Power is normal.

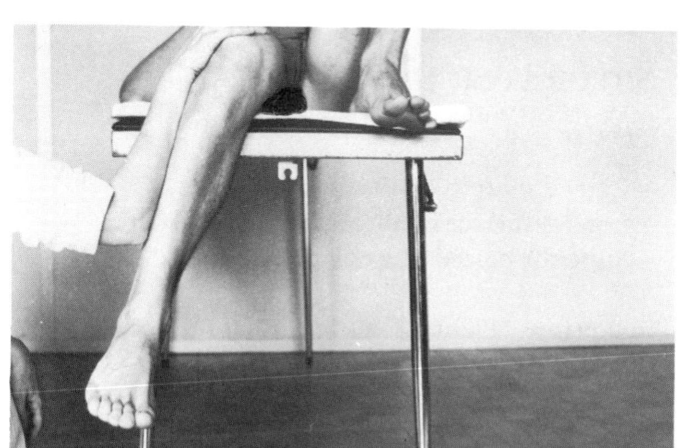

Note: On subjects with bone fragility, examiner uses his forearm to place resistance along entire length of the leg.

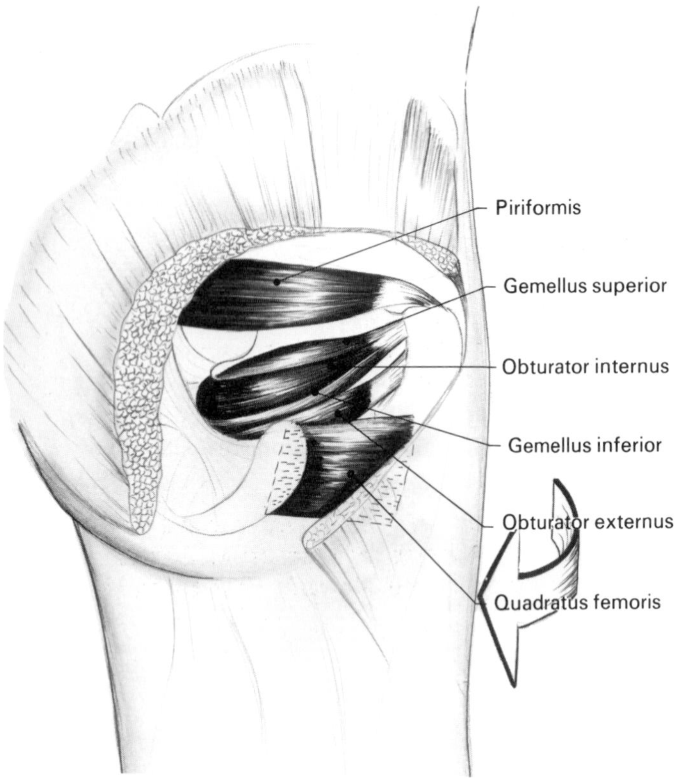

PIRIFORMIS

Origin Anterior surface of sacrum on second, third and fourth sacral vertebrae.

Insertion Upper surface of greater trochanter.

Innervation Sacral plexus (S1, S2).

OBTURATOR EXTERNUS

Origin

—Lateral margin of obturator foramen.
—Lateral surface of obturator membrane.
—Superior border of ischiopubic ramus.

Insertion Digital fossa of medial surface of greater trochanter.

Innervation Obturator (L3, L4).

OBTURATOR INTERNUS

Origin Medial margin of obturator foramen and of obturator membrane.

Insertion Medial surface of greater trochanter anteriorly to digital fossa.

Innervation Nerve to obturator internus and gemellus superior (L5, S1, S2).

GEMELLUS SUPERIOR

Origin Lateral surface of ischial spine.

Insertion Blends with obturator internus tendon and inserts with it into medial surface of greater trochanter.

Innervation Nerve to obturator internus and gemellus superior (L5, S1, S2).

GEMELLUS INFERIOR

Origin Upper part of ischial tuberosity.

Insertion Blends with obturator internus tendon.

Innervation Nerve to gemellus inferior and quadratus femoris (L4, L5, S1).

QUADRATUS FEMORIS

Origin Anterior to ischial tuberosity.

Insertion Quadrate tubercle of femur on posterior inferior angle of greater trochanter stretching over linea quadrata.

Innervation Nerve to gemellus inferior and quadratus femoris (L4, L5, S1).

Function

All are external rotators of the thigh on the pelvis.
—The piriformis abducts the hip and stabilizes it when the femur is fixed.
—The quadratus femoris adducts the hip.
They participate in stabilizing the hip joint by retaining femoral head in acetabulum.
Certain authors maintain that these muscles are external rotators only when the hip joint is in extension. If the hip is flexed their action changes and they become horizontal abductors.

External rotators of the hip, muscle activity during gait cycle

(B. J. Dolto)
In the stance phase, during forward movement, the weight-bearing surface is displaced forward since pelvic functional posterior tilt increases as the stance limb passes from flexion into extension. At heel strike the hip is in flexion, pelvic posterior tilt decreases (or disappears). Contralateral pelvic forward rotation begins which is effected by hip external rotators.

Contracture Hip is held in marked external rotation.

Deficiency

Decrease in the ability to rotate the hip joint externally, partly compensated for by remaining external rotators (gluteus maximus, hamstrings, posterior fibres of gluteus medius, pectineus, adductor brevis, adductor longus, superior fibres of adductor magnus and sartorius).
Internal rotation of thigh at rest and during gait cycle.
Decrease in push-off in gait cycle and alteration in angle of foot.
Paralysis of hip external rotators affects the entire lower extremity, it can lead to genuvalgum (knock knees) and foot pronation.

0 and 1

Palpating external hip rotators is impossible if the gluteus maximus is normal. If it is atrophied, examination of the subject should be in supine position.
Lower limbs extended.
Ask subject for external rotation of the thigh on the pelvis.
Hip external rotators are palpable posteriorly to greater trochanter.

Grade 0: No contraction is felt.
Grade 1: Contraction is felt but there is no joint movement.

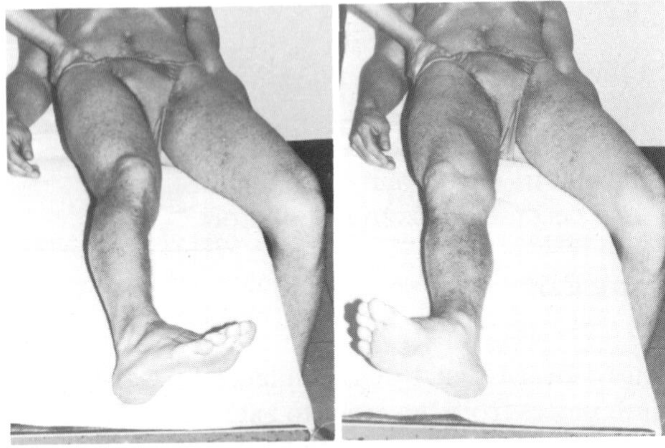

2 Same position.

Contralateral lower limb flexed and braced on a stool to stabilize pelvis.
Stabilize iliac crest to avoid pelvic girdle movement.
Limb to be tested in internal rotation.
Ask subject for external rotation of thigh on pelvis.
Ensure external rotation of thigh; foot eversion may give the impression of lower limb external rotation.
Range of movement should be complete.

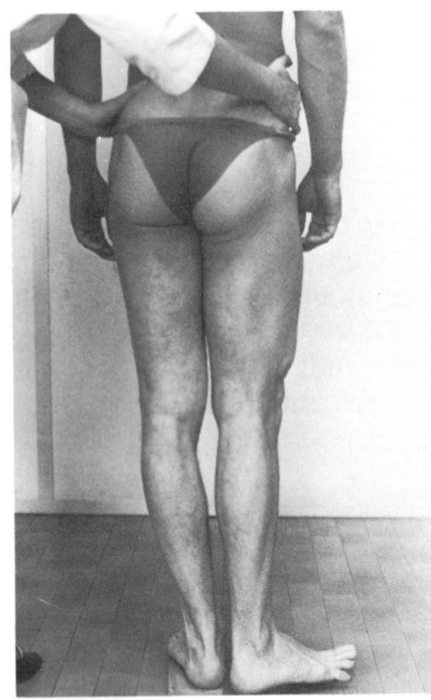

2 *Alternative.*

Subject standing.
Block under contralateral foot.
Stabilize pelvis to avoid its rotation.
Ask subject for same movement.
Range of movement should be complete.

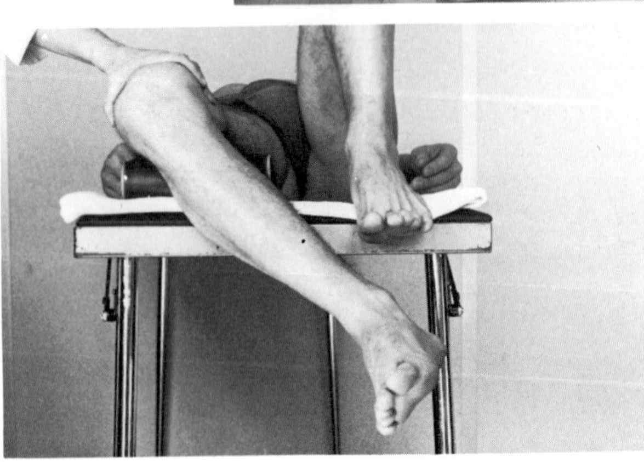

3 Subject supine.

Leg hanging on tested side.
Contralateral lower limb flexed and braced on table.
Stabilize lower third, anterior aspect of thigh.
Ask subject for external rotation of thigh on pelvis, drawing leg and foot inwards.
Range of movement should be complete.

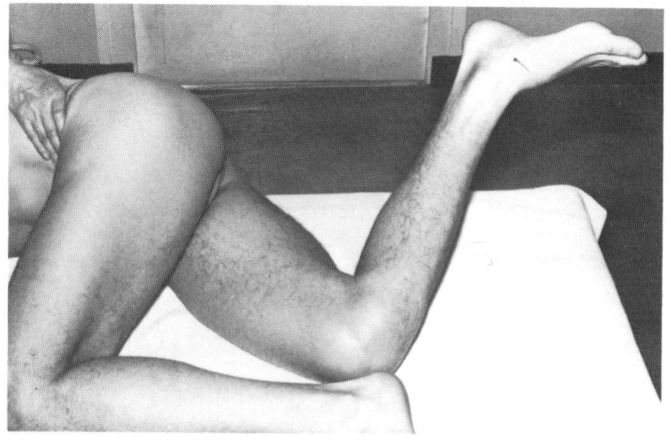

3 *Alternative* (gravity acting on entire length of limb).

Subject side-lying.
Limb to be tested on table, knee flexed.
Contralateral lower limb flexed, braced on table stabilizing pelvis.
Stabilize pelvis.
Ask subject for external rotation of thigh on pelvis, drawing leg and foot upwards.
Range of movement should be full.
Note: This test may be used for grades 4 and 5, placing resistance on lower third of leg.

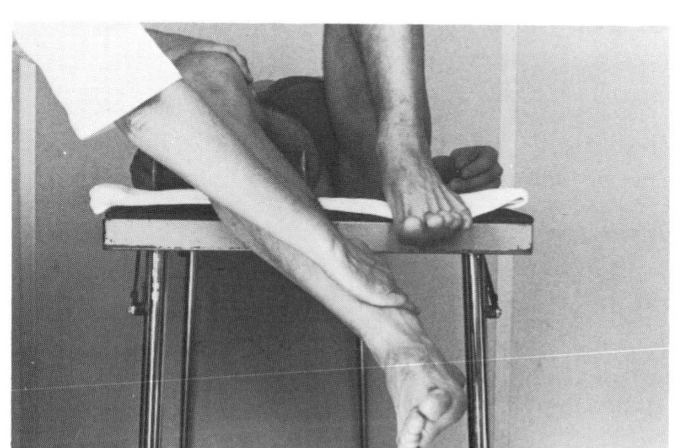

4 and 5 Subject supine.

Leg hanging on tested side.
Contralateral lower limb flexed and braced on table.
Stabilize lower third, anterior aspect of thigh.
Ask subject for external rotation of thigh on pelvis, drawing leg and foot inwards.
Resistance to the movement is placed on lower third of leg.

Grade 4: Power is sub-normal.
Grade 5: Power is normal.

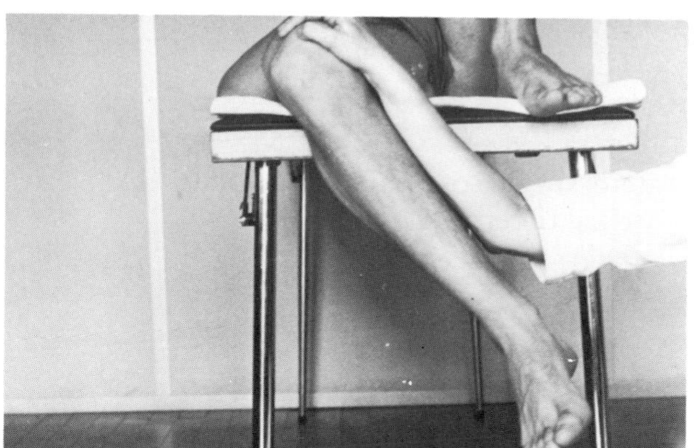

Note: For subjects with fragile bones, examiner uses his forearm to place resistance along entire length of leg.

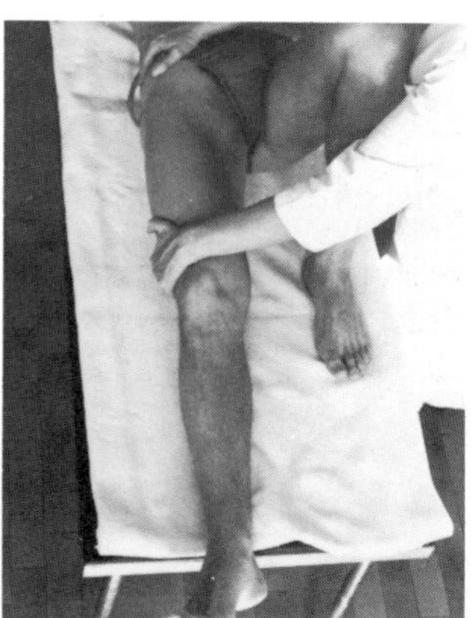

3, 4, and 5 *Alternative.* For subjects with bone fragility.

Subject supine.
Contralateral lower limb flexed, braced on table or on a stool to stabilize pelvis.
Stabilize iliac crest to prevent pelvic movement.
Ask subject for external rotation of thigh on pelvis.
Resistance to the movement is placed on lower third, lateral aspect of thigh.
Resistance is minimal for grade 3 and increases for grades 4 and 5.

Grade 4: Power is sub-normal.
Grade 5: Power is normal.

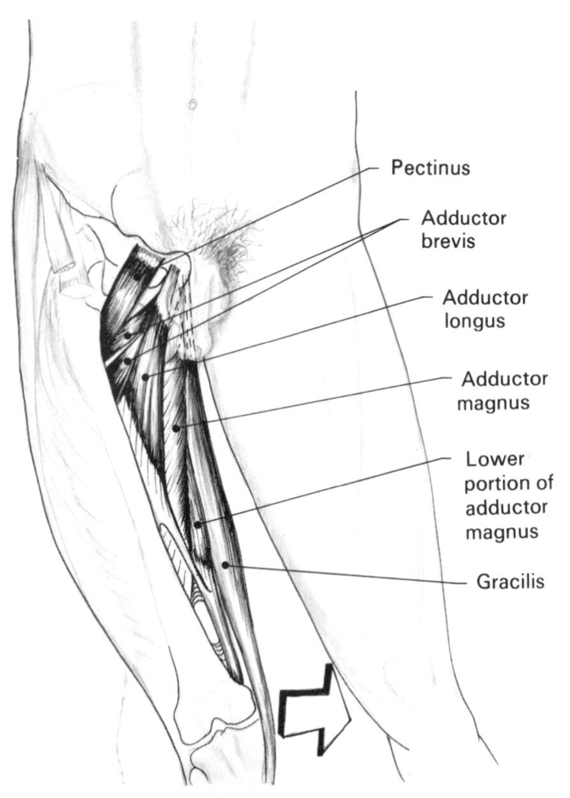

PECTINEUS

Origin Pectineal line from iliopectineal eminence to pubic spine. Anterior lip of infra-pubic groove.

Insertion Pectineal line of femur from lesser trochanter to linea aspera.

Innervation Femoral and obturator (L2, L3, L4).

ADDUCTOR LONGUS

Origin Anterior surface of pubis in angle between crest and symphysis.
Inferior surface of pubic spine.

Insertion Middle half of medial lip of linea aspera. Expansion into medial portion of adductor magnus.

Innervation Femoral and obturator (L2, L3, 14).

ADDUCTOR BREVIS

Origin Superior surface of ischiopubic ramus.

Insertion By two portions, superior and inferior, between lateral and middle lips of line extending from lesser trochanter to linea aspera.

Innervation Obturator (L2, L3, 14).

ADDUCTOR MAGNUS

Origin

—Upper and middle portions: posterior two-thirds of lateral surface of ischiopubic ramus.
—Lower or posterior portion: posterior inferior part of ischial tuberosity.

Insertion

—Upper portion: rough line leading from the greater trochanter to linea aspera.
—Middle portion: lower three-quarters of lateral margin of medial ridge of linea aspera.
—Lower portion: adductor tubercle on the medial condyle of femur.

Innervation Obturator and sciatic (12, L3, L4, 15, s1).

GRACILIS

Origin Anterior surface of pubis laterally to symphysis, extending over pubic descending ramus.

Insertion Upper quarter of medial surface of tibia with pes anserinus muscles.

Innervation Obturator (L2, L3, 14).

Function

—Adduction of thigh on pelvis.
—External rotation of thigh on pelvis, except for lower portion which participates in internal rotation.

The adductors, except for lower portion of adductor magnus, participate in flexion of thigh on pelvis. From a certain degree of flexion (about 50°), they can act as extensors.
Lower portion of adductor magnus participates in thigh extension on pelvis.
They play a stabilizing role on pelvis, balancing their action with that of the abductors.
Gracilis participates in flexion of leg on thigh and internal rotation of leg.
It is an adductor and internal rotator of the hip joint.

Hip adductors, muscle activity during gait cycle (J. B. Piera and A. Grossiord)

Stance phase

—Terminal stance: They start firing here, at the end of hip abductor activity. At this point, pelvic lateral shift towards the weight-bearing limb starts reversing towards the contralateral side, i.e. the swinging limb; the adductors control this lateral displacement from the stance side femur, a relatively fixed point, by a braking contraction.
—Pre-swing: They contract with gracilis and, because of the external rotation position of the pelvis (internal rotation of femur in relation to pelvis), their contraction will not only lead to internal rotation of the pelvis, but also start hip flexion.

Swing phase

The adductors and especially adductor magnus maintain their activity, the importance of which we have seen on lateral shift of centre of gravity. They become horizontal hip stabilizers in opposition to tensor fasciae latae. At terminal swing they adjust the foot angle before heel strike and prevent excessive and abrupt external rotation of the limb which could otherwise occur.

Gracilis muscle activity during gait cycle (J. B. Piera and A. Grossiord)

Stance phase

—Initial contact: The gracilis, a two-joint muscle, with the semitendinosus and the sartorius, wraps round the medial aspect of the knee joint, counteracting the tendency towards valgus angle increase during weight-bearing. Hence it stabilizes this joint, performing an active ligament role.
—Terminal stance: It contracts and, because of the external rotation position of the pelvis (internal rotation of femur in relation to pelvis), the contraction will not only lead to pelvic internal rotation, but will start hip flexion as well.

Swing phase

It contracts with the sartorius.

Deficiency

Marked decrease of adduction capacity, weakly compensated by hamstrings, quadratus femoris, obturator externus, obturator internus, and gemelli (gravity also partially compensates for their paralysis).
Transversal instability of the pelvis.
Functional disability during gait causes limping.
During swing phase the lower extremity is thrust forward and outward by the predominant abductors.
Contracture or tightness (see next page 344)

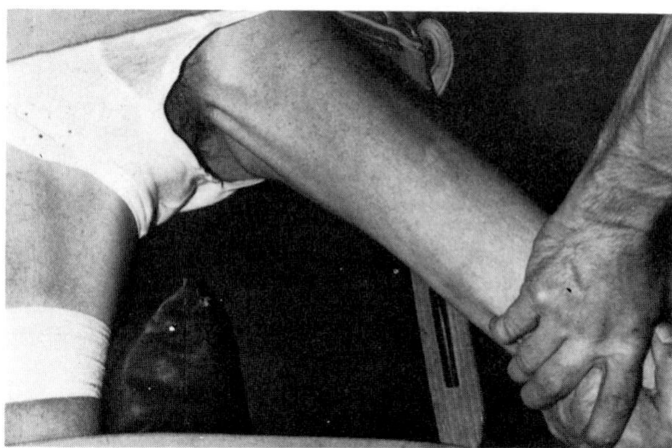

Showing contracture of adductors.

Contracture or tightness

Limits abduction.
Tendency to dislocate femoral head aggravated by simultaneous abductor deficiency.
Pelvic lateral tilt and elevation towards contractured side: during gait the subject must lengthen his limb by placing his foot in plantar flexion.
Associated with contracture of contralateral tensor fasciae latae, the imbalance is even greater: oblique pelvis.

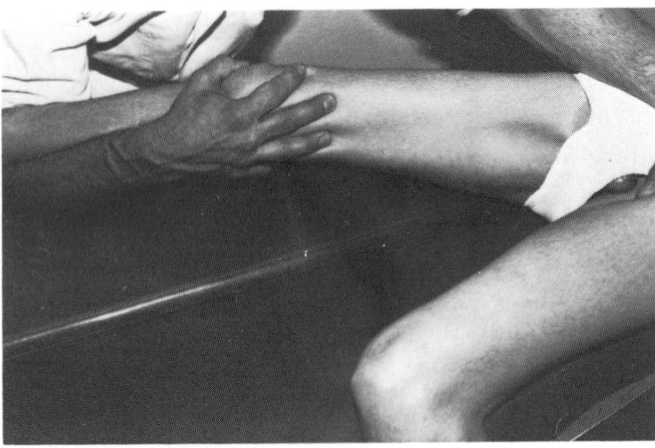

Showing contracture of adductors.

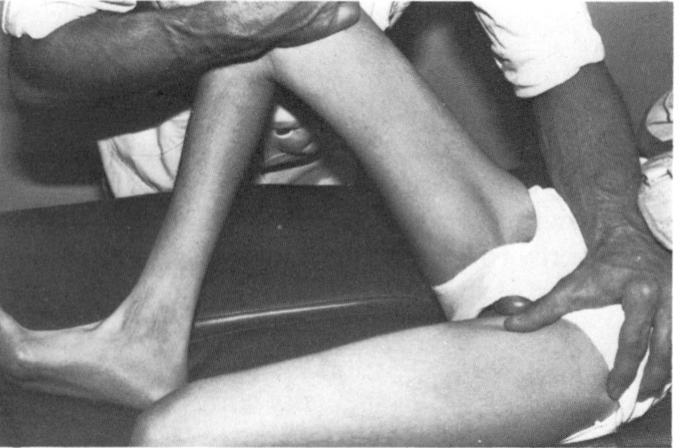

Examining contracture of upper portion of adductors, especially adductor brevis and pectineus. This position decreases identification of gracilis tension considerably.

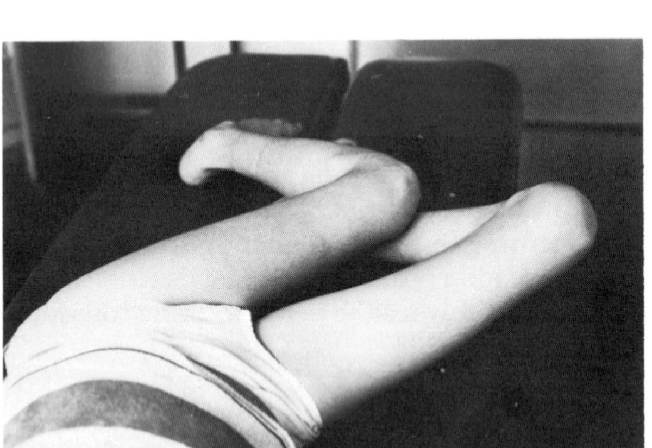

Subject in 'windscreen wiper' position. This position aggravates contracture of left adductors.

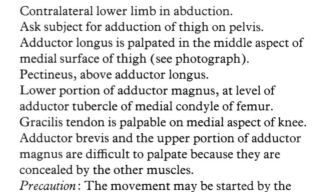

0 and 1 Subject supine.

Contralateral lower limb in abduction.
Ask subject for adduction of thigh on pelvis.
Adductor longus is palpated in the middle aspect of medial surface of thigh (see photograph).
Pectineus, above adductor longus.
Lower portion of adductor magnus, at level of adductor tubercle of medial condyle of femur.
Gracilis tendon is palpable on medial aspect of knee.
Adductor brevis and the upper portion of adductor magnus are difficult to palpate because they are concealed by the other muscles.
Precaution: The movement may be started by the hamstrings.

Grade 0: No contraction is felt.
Grade 1: Contraction is felt but there is no joint movement.

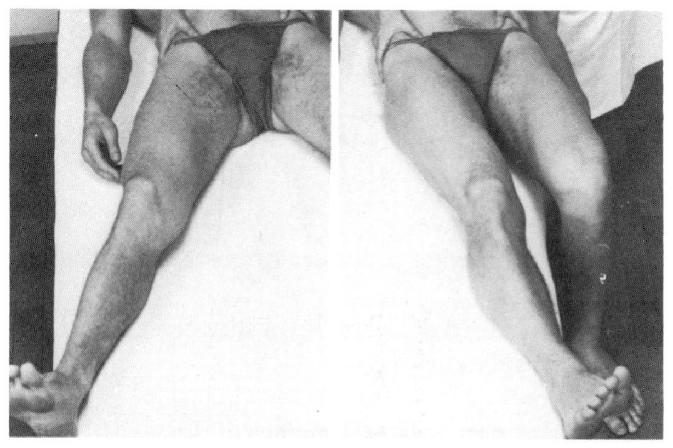

2 Subject supine.

Limb to be tested in abduction.
Contralateral limb abducted or flexed, braced on table or stool.
Stabilize pelvis, subject may compensate by raising it.
Ask subject for adduction of thigh on pelvis, knee extended and patella facing upward to avoid compensatory rotation.
Range of movement should be complete.

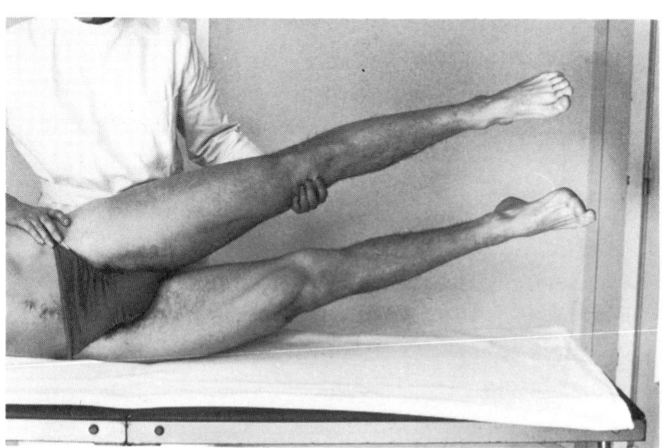

3 Subject side-lying.

Limb to be tested on the table.
Contralateral limb abducted and supported by examiner.
Stabilize iliac crest.
Ask subject for adduction of thigh on pelvis, avoiding same compensations.
Range of movement should be complete.

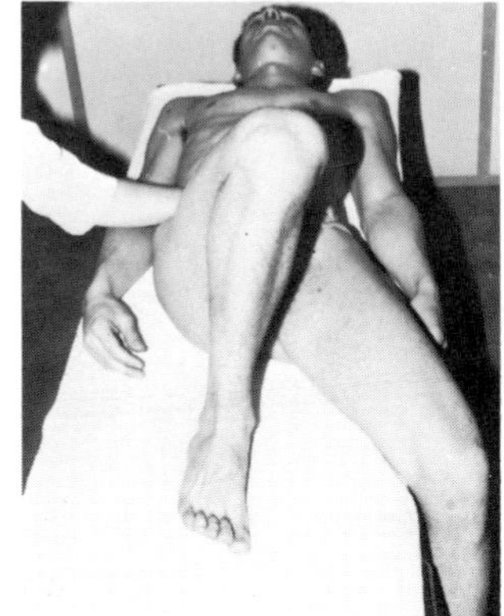

3 *Alternative* (psoas and long adductors neutralized).

Tests specifically the upper adductors (adductor brevis and pectineus).
Subject supine.
Limb to be tested in flexion and abduction.
Contralateral limb abducted, flexed over edge of table, and foot resting on stool.
Stabilize iliac crest.
Ask subject for adduction of thigh on pelvis.
From above, slight resistance is placed on upper third, medial aspect of thigh.
Range of movement should be complete.

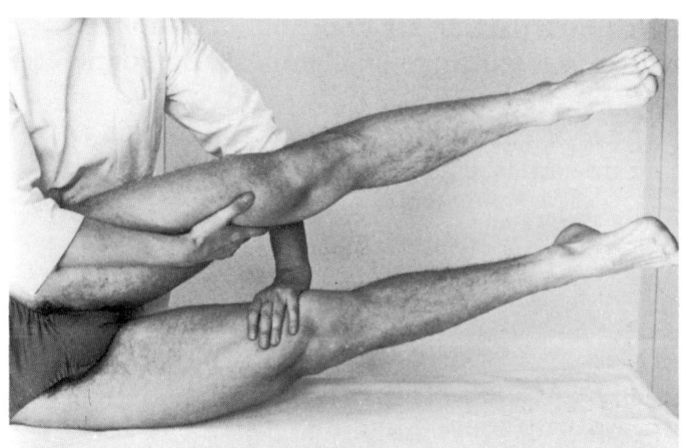

4 and 5 Same position.

Same movement and precautions.
Resistance to the movement is placed on lower third, medial aspect of thigh.

Grade 4: Power is sub-normal.
Grade 5: Power is normal.

GLUTEUS MAXIMUS

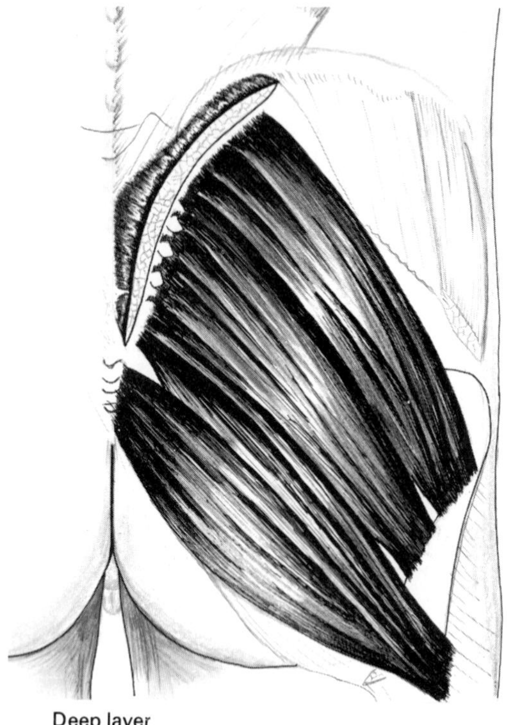

Deep layer

Superficial layer

Origin

Superficial layer:
—Crest and medial processes of posterior surface of sacrum.
—Posterior third of lateral lip of iliac crest.
—Gluteal aponeurosis.
Deep layer:
—Posterior part of lateral surface of ilium.
—Lateral margin of sacrum and sacral posterior lateral process.
—Sacroiliac ligaments.
—Posterior surface of coccyx.
—Posterior surface of great sacro-sciatic ligament.

Insertion

Superficial layer:
—Posterior margin of iliotibial tract of fasciae latae.
Deep layer:
—Gluteal tuberosity of femur and lateral intermuscular fascia.

Innervation Inferior gluteal (l4, l5, S1, S2).

Functions

—Extension of thigh on pelvis.
—External rotation of thigh on pelvis.
—Extension of pelvis on thigh when acting from a fixed femur.
—Stabilizes the pelvis in the sagittal plane.
—Tilts pelvis anteriorly.
In weight-bearing, it balances and stabilizes the trunk on the lower extremity.
Certain authors say that the superior fibres of gluteus maximus participate in hip abduction (Faraboeuf's gluteal deltoid), the inferior in hip adduction. According to Duchenne de Boulogne, it plays no role either in hip abduction or in hip adduction.
The gluteus maximus does not contract during normal gait, or in standing. However, its action is very clear in jumping, running, climbing stairs, walking up slopes, carrying loads . . .

Contracture or tightness

Rare, it limits flexion of thigh on pelvis.

Deficiency

With paralysis of glutei maximi, the subject can stand, and walk practically without limping on flat terrain. However, he cannot rise from a sitting position, walk up stairs or inclines, carry loads, jump or run . . .

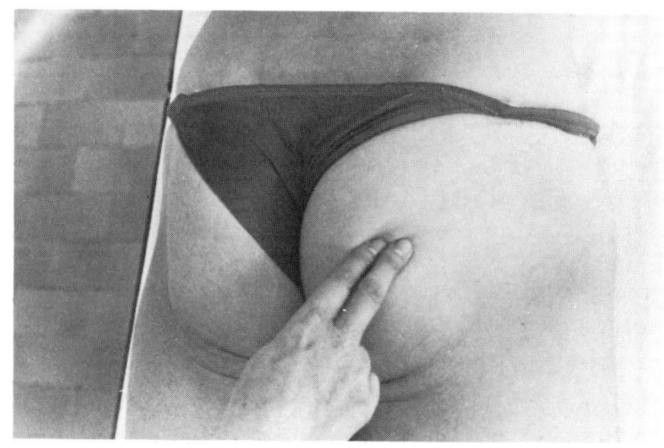

0 and 1 Subject prone.

Ask subject to tighten his buttocks, or to extend thigh on pelvis.
Fibres are palpable across entire buttock surface. The gluteal skin crease tends to be horizontal and loses its form. Contraction of levator ani may give the impression of a gluteus maximus contraction.

Grade 0: No contraction is felt.
Grade 1: Contraction is felt but there is no joint movement.

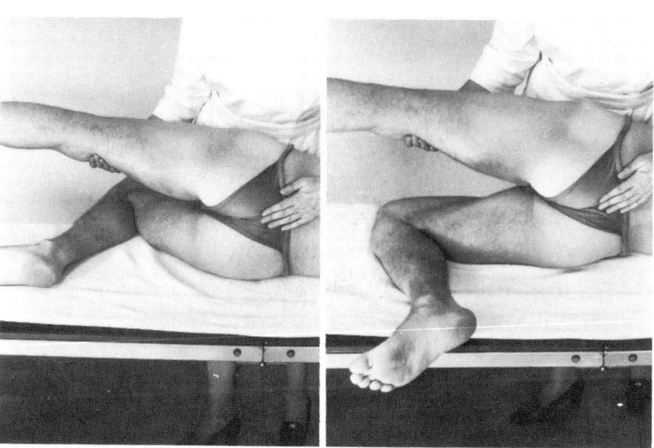

2 Subject side-lying.

Limb to be tested on table, hip and knee flexed.
Stabilize pelvis at sacrum to avoid posterior tilt and lumbar hyperlordosis (possible compensation by the erector spinae).
Ask subject to extend thigh on pelvis, knee flexed to eliminate possible hamstring extensor activity.
Range of movement should be complete.
Note: Check muscle length of hip flexors.

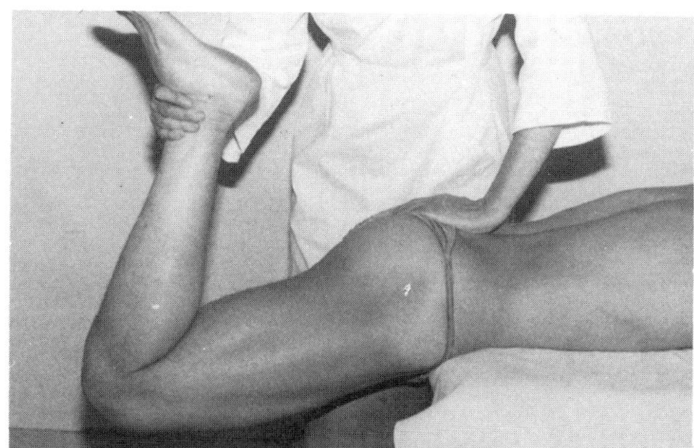

3 Subject prone, lower limbs flexed over the end of the table to avoid lordosis.

Stabilize sacrum.
Stabilize leg to avoid excessive hamstring participation.
Ask subject to extend thigh on pelvis, keeping knee flexed.
Range of movement should be complete.
Note: When lumbar spine extensors are very deficient, examiner should stabilize pelvis firmly.

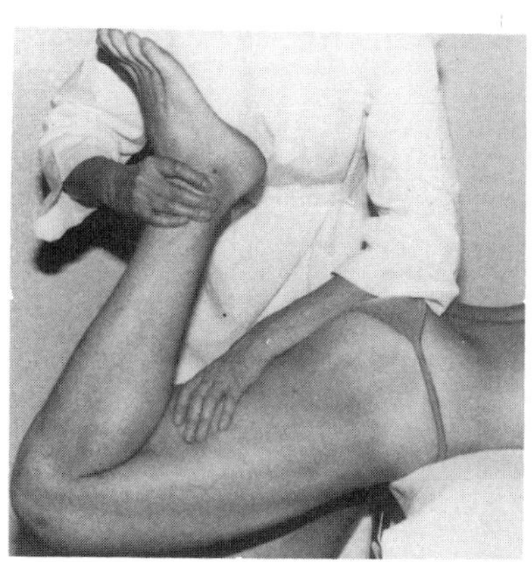

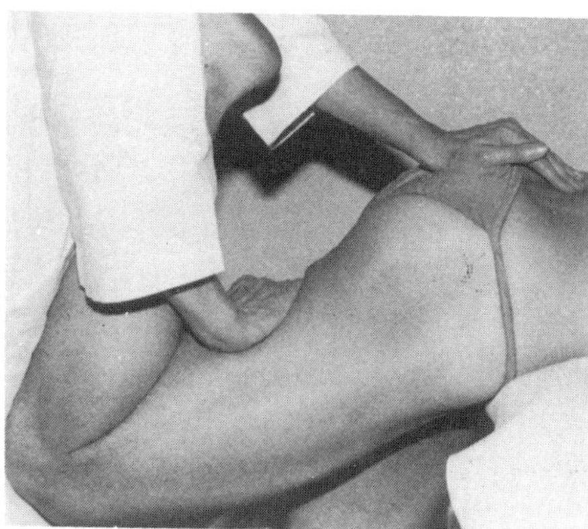

4 and 5 Same position.

Same stabilization and movement.
Place resistance to movement on inferior third of posterior surface of thigh.

Grade 4: Power is subnormal.
Grade 5: Power is normal.

QUADRICEPS FEMORIS

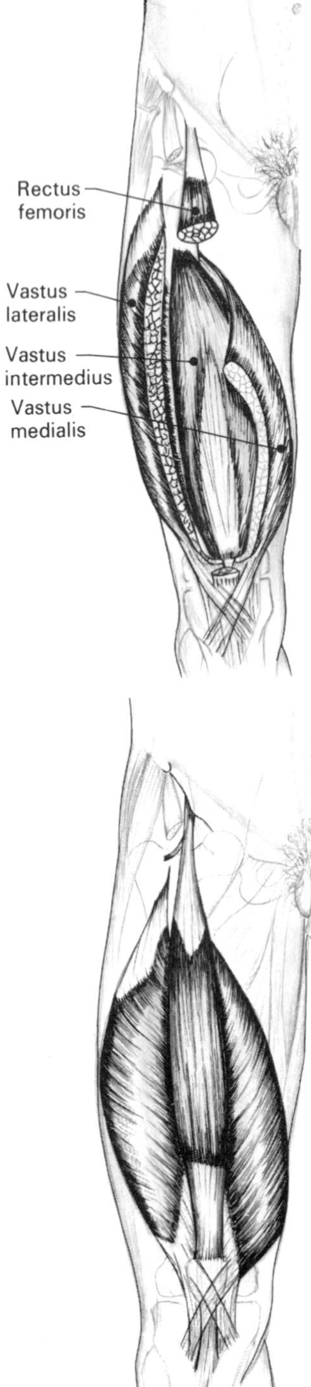

Composed of four heads.

RECTUS FEMORIS

Origin

—Direct tendinous head: lateral surface of anterior inferior iliac spine.
—Reflected head: groove on upper rim of acetabulum.
—Recurrent head: from junction of superior and inferior border of greater trochanter.

Insertion Upper border of patella and patellar ligament anteriorly to insertion of vasti.

VASTUS MEDIALIS

Origin

—Lower half of intertrochanteric line.
—Medial lip of linea aspera.
—Upper part of medial supracondylar line.
—Medial intermuscular septum.
—Tendon of lower portion of adductor magnus.

Insertion

—Upper border of patella.
—Direct and crossed expansions into ridge of the vasti on anterior surface of upper extremity of tibia.

VASTUS LATERALIS

Origin

—Lateral margin of intertrochanteric line, reaching over crest outlining anteriorly and inferiorly the greater trochanter, up to the medial border of its anterior surface.
—Lateral intermuscular septum.

Insertion

—Upper border of patella.
—Sends expansions in same manner as vastus medialis.

VASTUS INTERMEDIUS (CRUREUS)

Origin

—Upper three-quarters of anterior and lateral surfaces of femur.
—Lateral intermuscular septum.

Insertion Upper border of patella posteriorly to lateral and medial vasti.

SUBCRUREUS

Origin Anterior surface of lower part of femur below the vastus intermedius.

Insertion Upper part of the cul de sac of the capsular ligament and bursa.

Innervation Common to all.
Femoral (L2, L3, L4).

Function Extension of lower leg on thigh: The rectus and vastus intermedius play a dynamic role, lateral and medial vasti, a stabilizing one.
The direct and crossed fibres of the lateral and medial vasti take part in transverse knee stability by controlling contralateral gaping of the joint.

Dominance of one vastus (especially the vastus lateralis) tends to subluxate the patella.
Rectus femoris, being a two-joint muscle, has a special function:
—Extension of lower leg on thigh.
—Flexion of thigh on pelvis.
Thus its dynamic role is important. The more the knee is in flexion and the more the hip is in extension (lengthened position), the more it will flex the hip. It contracts in all movements requiring knee extension and hip flexion. It is a true active ligament of the hip.
Quadriceps femoris is necessary for running, jumping, walking up stairs (concentric contraction), walking down stairs (eccentric contraction), rising from a sitting or supine position . . .
Note: Subcrureus draws the cul-de-sac of the capsular ligament upwards.

Quadriceps femoris muscle activity during gait cycle

(J. B. Piera and A. Grossiord)

Stance phase

—Initial contact, loading response: At heel strike, the thrust of body weight tends to flex the knee. The quadriceps contracts to stop this flexion. The vastus intermedius, the vastus medialis, and the vastus lateralis are more active than the rectus femoris; being single-joint muscles, the former supply a dynamic concentric contraction, whereas the rectus femoris, a two-joint muscle, is used more as a strap since, because of its origin, insertion, and course, its role in limiting knee flexion becomes more important as hip flexion decreases. Quadriceps contraction prevents knee flexion under body weight.
—Mid stance: Quadriceps contracts only at the beginning of stance. Contrary to Scherb's work, more recent electromyographic studies (Perry, Rainaut, and Lotteau) have shown that, among the various heads of the quadriceps, only the vastus intermedius and the vastus lateralis are acting. In other words, during loading response when the weight-bearing knee is flexed, the quadriceps is only partly contracted, knee stability being essentially assured by triceps surae which draws it backwards.
—Terminal stance, pre-swing: the rectus femoris and the vastus intermedius contract to control knee flexion.

Swing phase

The rectus femoris participates as an accessory muscle in flexion of thigh on pelvis.

Contracture or tightness

Decrease of the ability to flex the knee: functional impedance during gait. The subject has a tendency to walk with the knee straight, producing a limp. Contracture of rectus femoris is quite common: it limits knee flexion especially when hip is in extension. During examination of thigh extension on pelvis, if knee is placed in flexion, contracture of rectus femoris provokes a posterior tilt of pelvis and lumbar hyperlordosis.

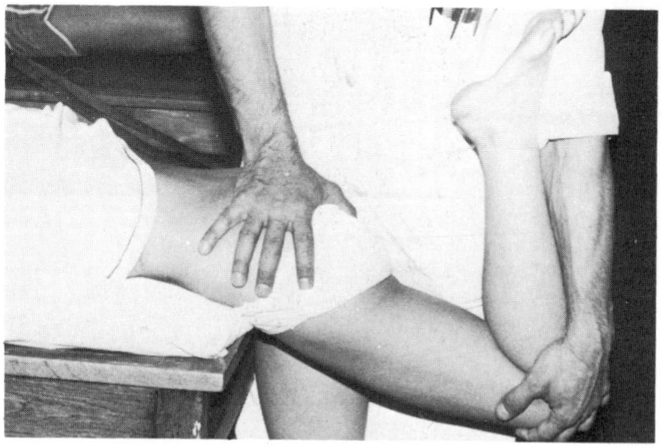

Showing contracture of rectus femoris.

Deficiency

Inability of extending leg on thigh.
Inability of walking up or down stairs, or rising from sitting or lying positions (patients with muscular dystrophy have difficulty in rising from a low seat). The subject can lock his knee with his hand.
Gait disturbance: The patient thrusts his leg forward during swing and locks his knee in hyperextension to prevent falling, hence he limps because of an exaggerated elevation of his centre of gravity. The patient can also lock his knee with his hand.

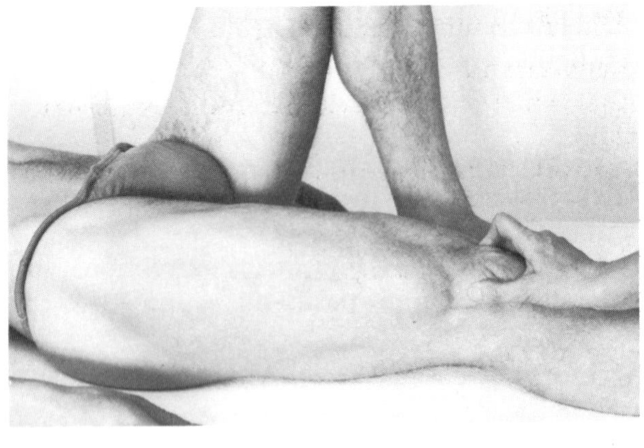

0 and 1 Subject supine.

Limb to be tested extended, patella straight up.
Contralateral limb flexed, and fixed on table.
Hold patella down. Ask subject to move it proximally. This movement may be obtained by asking subject for foot dorsiflexion. Quadriceps femoris tendon is palpated above the patella.
Tensor fasciae latae contraction may give impression of quadriceps activity.
Palpation of vasti is possible on lateral and medial aspects of thigh (not shown).
Note: On certain patients rectus femoris tendon is found between tendons at origins of sartorius and tensor fasciae latae.

Grade 0: No contraction is felt.
Grade 1: Contraction is felt. There is weak patellar movement but no knee joint motion.

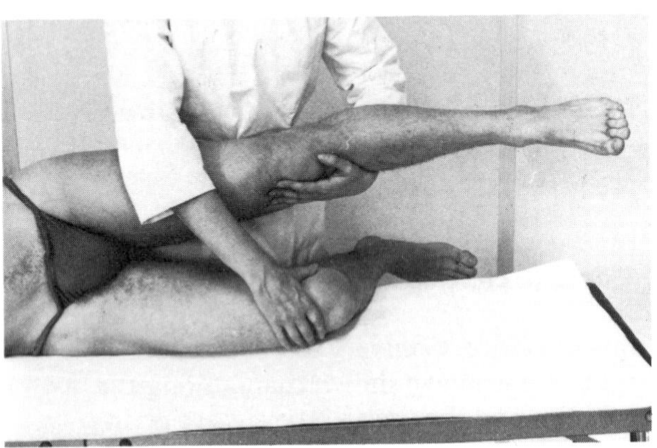

2 Subject side-lying.

Limb to be tested on table, hip straight, knee flexed.
Contralateral limb supported by examiner.
Stabilize lower third, anterior aspect of the thigh.
Ask subject to extend lower leg on thigh, avoiding all rotational components.
Range of movement should be complete.

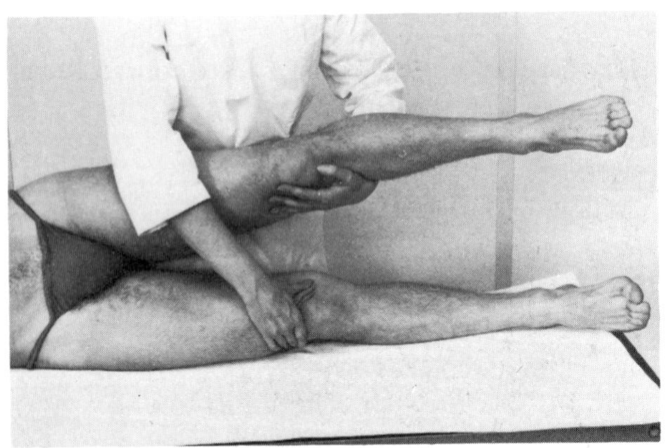

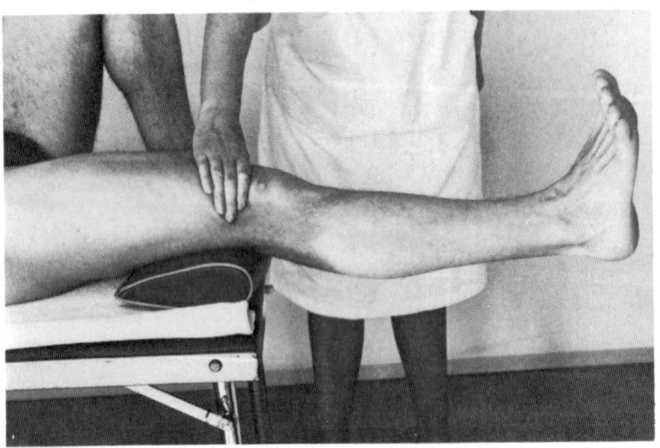

3 Subject supine.

Knee at edge of table, leg hanging, cushion under popliteal space.
Contralateral limb flexed, braced on table.
Stabilize lower third, anterior aspect of thigh without squeezing muscular bulk.
Ask subject to extend lower leg on thigh, avoiding all rotation components.
Range of movement should be complete.

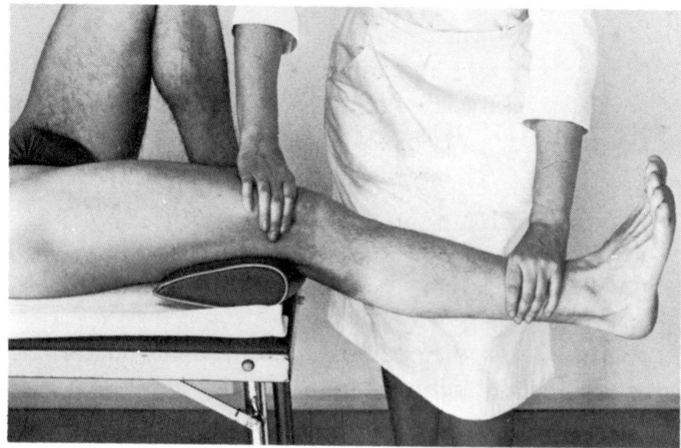

4 and 5 Same position.

Same stabilization
Ask for same movement, avoiding same compensations.
Resistance to the movement is placed on lower third of leg.

Grade 4: Power is sub-normal.
Grade 5: Power is normal.

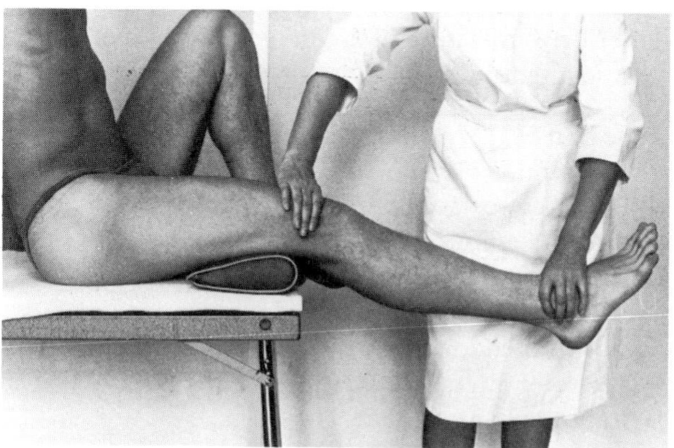

Testing quadriceps with hip flexed.
Grades 3, 4, and 5 may be done in this position.
Rectus femoris is in its shortened position which reduces its action.

Grades 4 and 5 are shown.

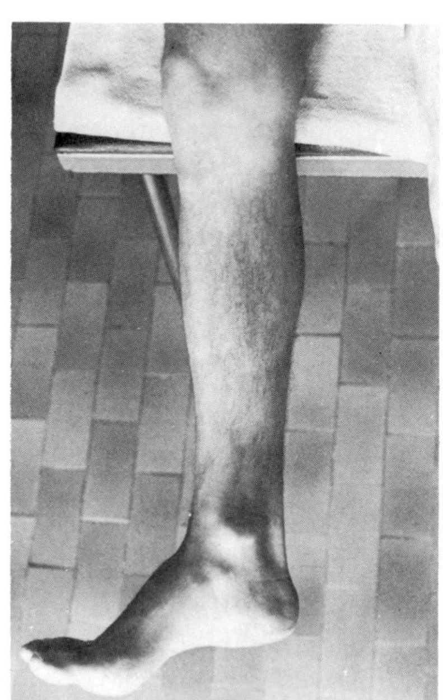

Compensatory external rotation of the thigh (or internal, not shown), a position which minimizes the effect of gravity.

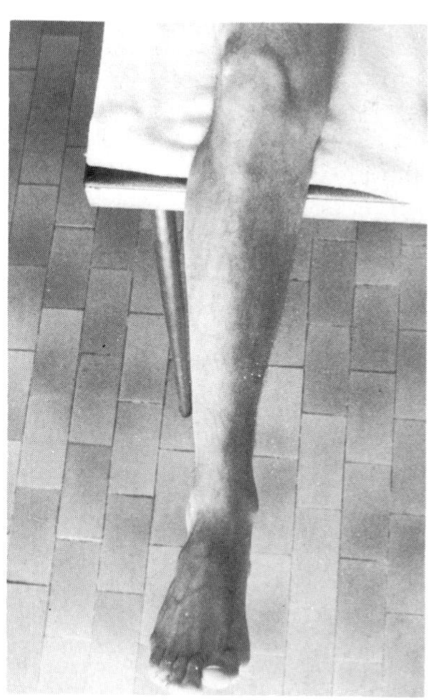

Compensation by pulling knee on to table. Subject thus eliminates part of leg's weight.

HAMSTRING MUSCLES

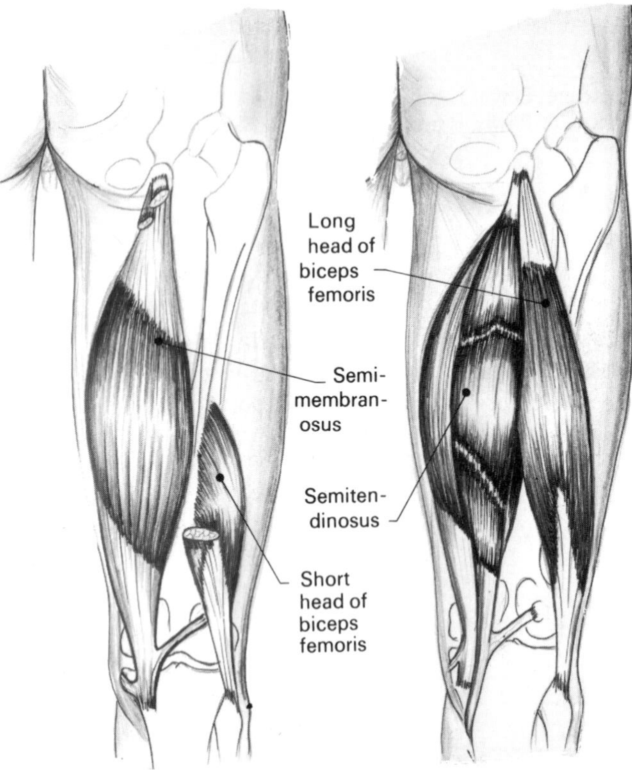

SEMITENDINOSUS

Origin Posterior surface of ischial tuberosity conjoined with tendon of long head of biceps femoris.

Insertion Superior quarter aspect of medial surface of tibia with pes anserinus muscles, below the gracilis and behind the sartorius.

SEMIMEMBRANOSUS

Origin Posterior surface of ischial tuberosity, anteriorly and laterally to conjoined tendon of biceps and semitendinosus.

Insertion

Direct tendon:
—Posterior surface of medial condyle of tibia.
Reflected tendon:
—Anterior surface of medial condyle of tibia.
—Anterior part of infraglenoid margin.
Recurrent tendon:
—Posterior lateral capsular ligament.
—Lateral sesamoid bone.

BICEPS FEMORIS

Origin

Long head:
—Conjoined tendon with semitendinosus on posterior surface of ischial tuberosity.
Short head:
—Lateral lip of linea aspera, and superior part of lateral supracondylar line.

Insertion

—Common on superior lateral aspect of head of fibula and lateral tuberosity of tibia.
—Expansion into aponeurosis of the leg.

Innervation

All three hamstrings are innervated by the sciatic nerve.
—Biceps (L5, S1, s2).
—Semimembranosus (l4, L5, S1, S2).
—Semitendinosus (l4, L5, S1, S2).

Function

On knee joint:
—Flexion of lower leg on thigh.
—Biceps is an external rotator of knee when this joint is in flexion.
—Semitendinosus is an internal rotator assisted slightly by semimembranosus when knee is in flexion.
—Hamstrings stabilize knee in rotary movements.

On hip:
—Extension of thigh on pelvis. Efficiency of hamstrings depends on knee position during this movement. It lessens when knee is flexed.
—They participate in adduction of thigh on pelvis.
—The biceps contracts during external rotation of thigh on pelvis when the knee is in extension.
On pelvis:
—They tilt the pelvis anteriorly.
—They stabilize the pelvis in the sagittal plane.
The hamstring muscles are part of the posterior muscle chain: they contract when a subject rises from a forward leaning position.
They contract in triple flexion (withdrawal) of lower extremity.

Hamstrings, muscle activity during gait cycle

(J. B. Piera and A. Grossiord)

Stance phase

They prevent full knee extension at heel strike, and remain contracted during loading response. According to Scherb, the semimembranosus is contracted and not the semitendinosus. The semitendinosus, a two-joint muscle, wraps around the medial surface of the knee joint with the gracilis and the sartorius, and during weight-bearing opposes the increase of the physiological valgus angle of the knee. Thus it stabilizes this joint and plays the role of an active ligament.

Swing phase

The short head of the biceps femoris, the only single-joint muscle of the hamstring group, contracts in order to control quality and range of knee flexion. The long head of biceps femoris, the semimembranosus, and the semitendinosus contract at the end of movement, controlling the forward swing of the leg, i.e. knee extension of terminal swing before heel strike.
In other words, specialized action of various hamstring muscles seems to exist. The short head of biceps femoris flexes the knee during initial swing, the two-joint hamstrings act principally to restrain knee extension during heel strike (terminal swing). They participate in knee stability at heel strike.

Contracture or tightness See next page.

Deficiency

—Marked limitation of knee flexion, partially, weakly compensated for by gracilis, sartorius, popliteus and gemelli.
—Knee instability: sagittal, transverse and rotational.
—Possibility of genu recurvatum.
—Deficiency of biceps can lead to varus deformity (bow leg).
—Deficiency of medial hamstrings can lead to valgus deformity (knock knee) and external rotation of leg on thigh (predominant activity of biceps femoris).
Paralysis of hamstrings leads to anterior tilting of trunk: the subject compensating by extending hip and trunk, and leaning on the anterior hip ligaments (mainly Bertin's ligament).
Decrease of step length in terminal stance, excessive hip flexion and foot dorsiflexion during initial swing. Fast gait, jumping and other free ranging activities are impossible.

Contracture or tightness

Limitation of hamstring elasticity is quite frequent, even in normal people.

Contracture leads to:

—genu flexum.

—In younger children, it often leads to a posterior glide of tibial tuberosities.

—Contracture of biceps femoris, often associated with that of tensor fasciae latae, encourages external rotation of leg on thigh with valgus deformity of knee.

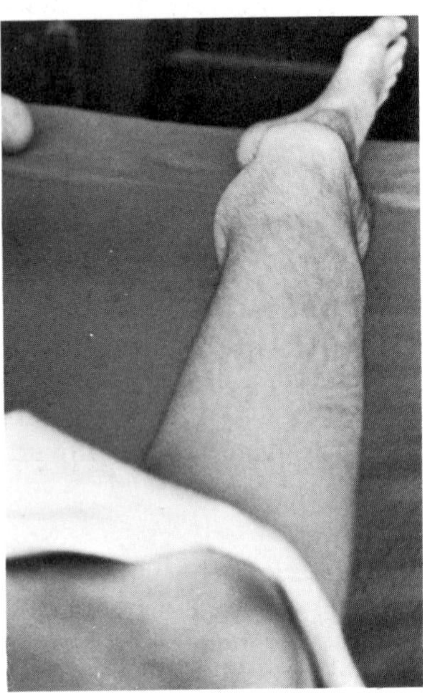

Contracture of biceps and tensor fasciae latae. Note genu flexum and rotation of leg in relation to thigh.

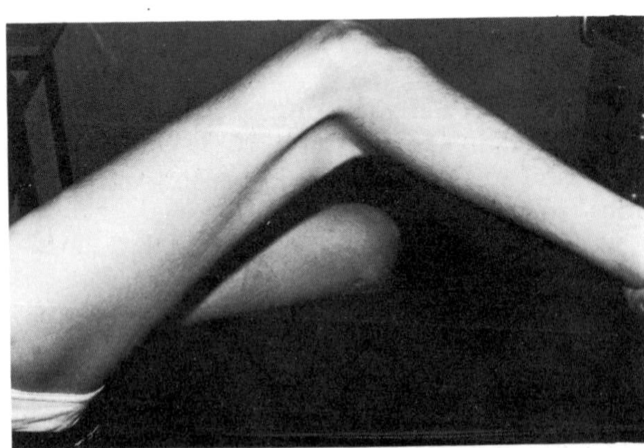

Subject presenting contracture of hamstrings.

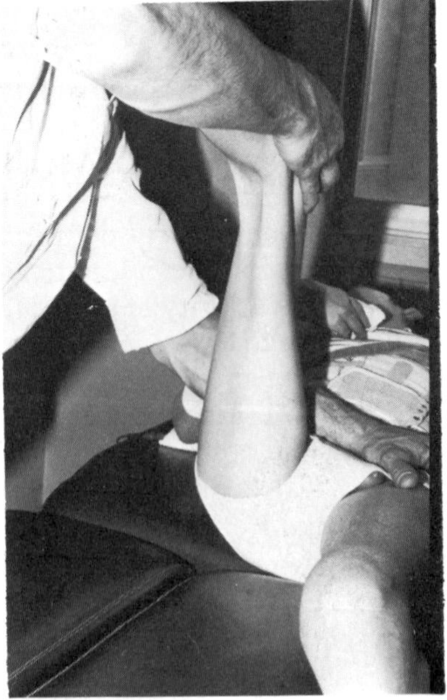

Examination of contracture.

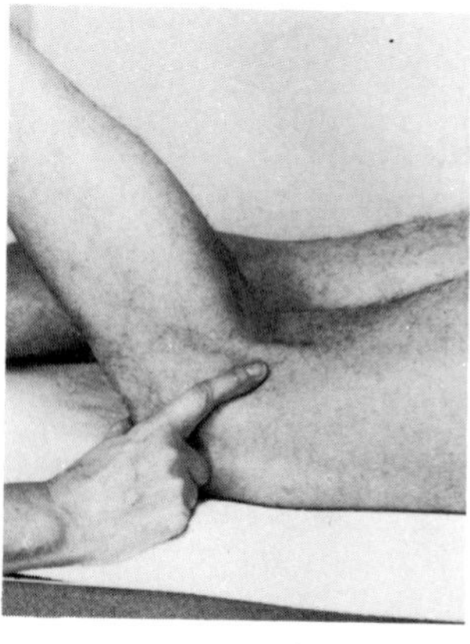

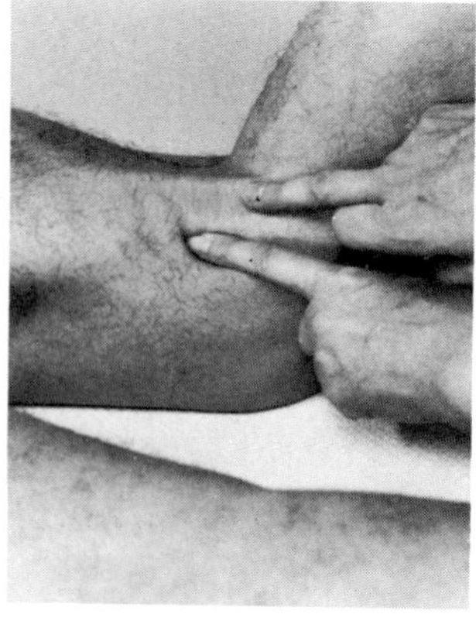

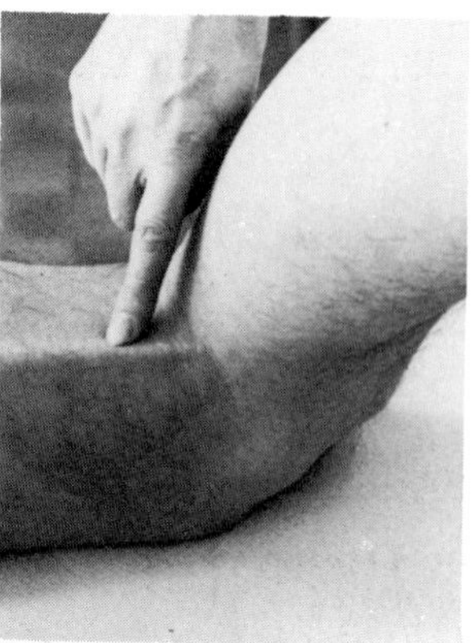

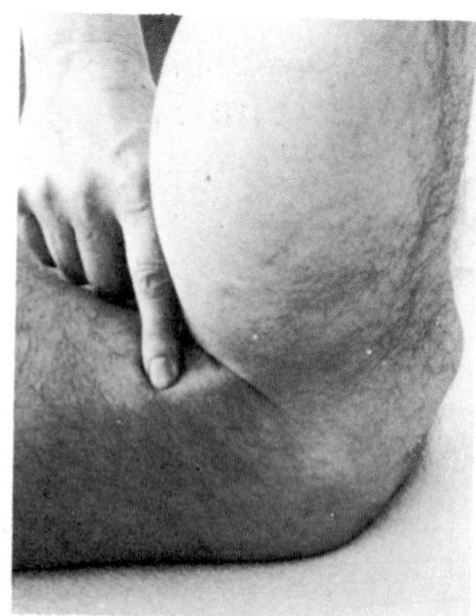

0 and 1 Subject prone, or side-lying (not shown).

Place knee in slight flexion.
Ask subject to flex leg on thigh.
Palpate biceps tendon at posterior lateral aspect of knee.
Palpate semimembranosus and semitendinosus tendons at the posterior medial aspect of knee, semimembranosus is medial in relation to body's axis.
Palpate muscle bellies on posterior aspect of thigh.
Note: Semimembranosus is apparent when knee is flexed at about 70°, semitendinosus appears after 90°.

Grade 0: No tendon movement is felt.
Grade 1: Tendon movement is felt but there is no joint movement.

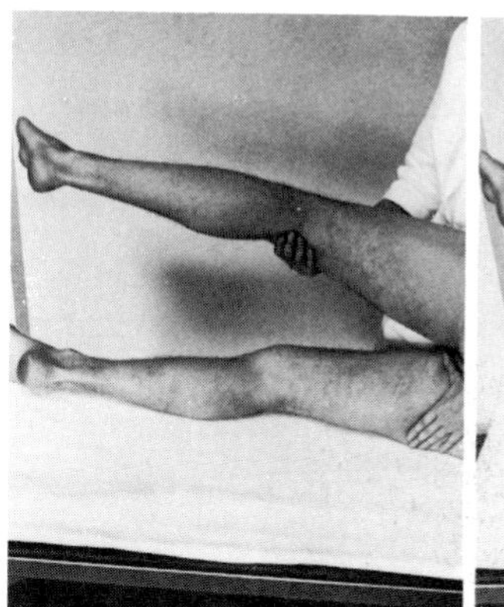

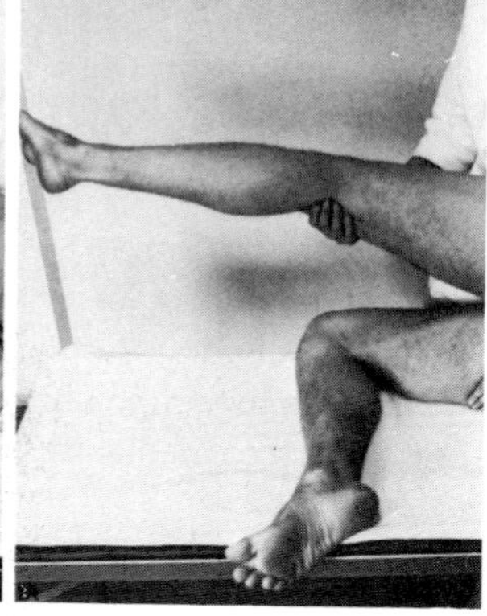

2 Subject side-lying.

Limb to be tested on table, hip straight, knee extended.
Contralateral lower limb supported by examiner.
Stabilize upper portion of posterior aspect of thigh.
Ask subject to flex leg on thigh.
During the movement the foot remains relaxed in plantar flexion, diminishing action of gemelli as much as possible.
Range of movement should be complete.

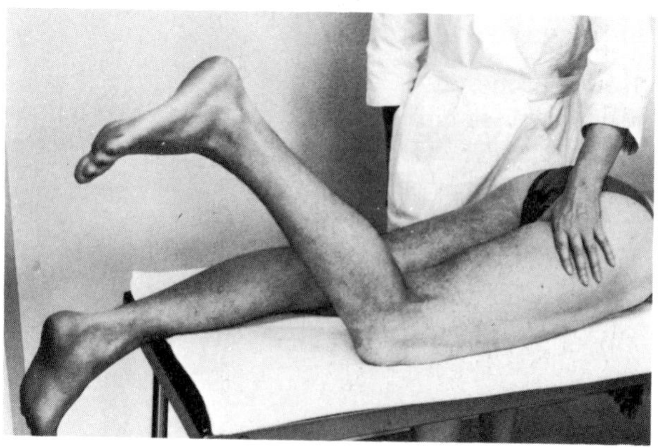

3 Subject prone.

Lower extremities extended.
Stabilize upper portion of posterior aspect of thigh without squeezing muscular bulk. Ask subject to flex leg on thigh.
The movement should be made in a strictly sagittal plane.
From 90°, light resistance to the movement is placed on the lower third, posterior aspect of the leg. Foot should remain relaxed in plantar flexion. External rotation of the leg signifies predominance of biceps, internal rotation signifies medial hamstrings.
Range of movement should be complete.

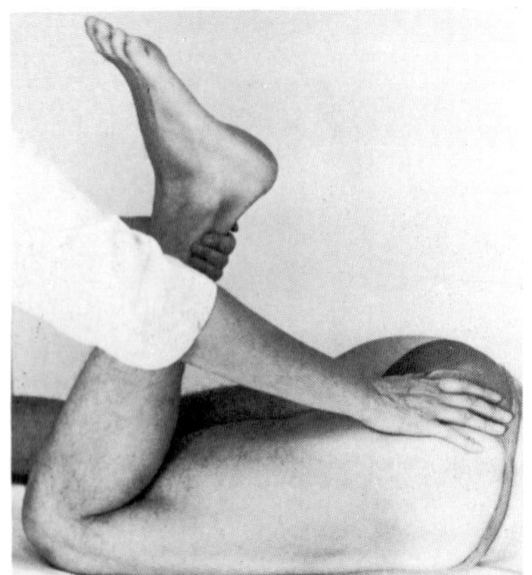

4 and 5 Global test

Same position.
Same stabilization, precautions and movements.
Resistance contrary to the movement is placed on the lower third of posterior aspect of the leg.

Grade 4: Power is sub-normal.
Grade 5: Power is normal.

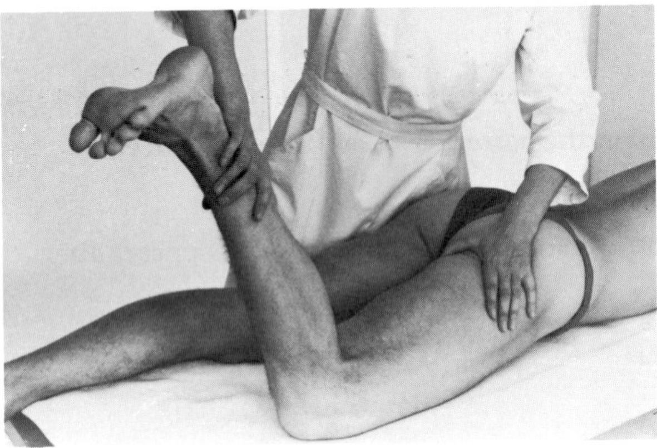

Biceps, individual muscle test

(Grades 4 and 5 shown)
Same position.
Same stabilization.
Ask subject to flex leg on thigh while rotating leg externally.
Resistance to components of movement is placed on lower third of lateral aspect of leg.

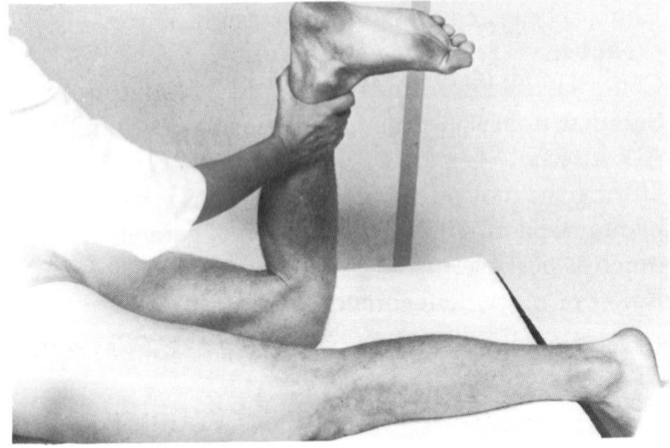

Internal hamstrings, individual muscle test

(Grades 4 and 5 shown)
Same position.
Same stabilization.
Ask subject to flex lower leg on thigh while rotating leg internally.
Resistance to components of movement is placed on lower third of medial aspect of leg.

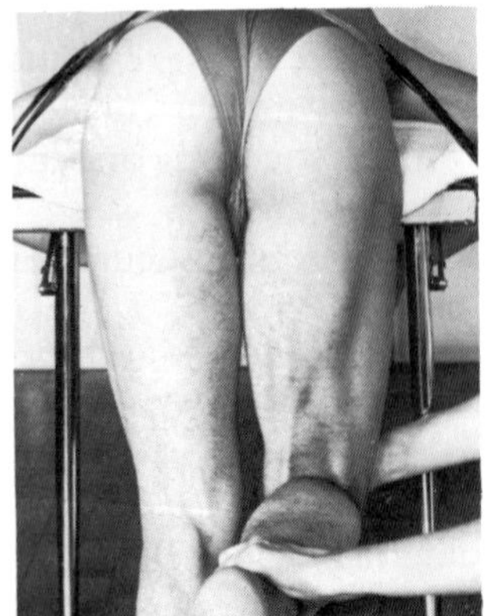

Alternative

Subject prone.
Pelvis at end of table strapped, lower limbs hanging.
Ask subject to flex leg on thigh.
This method may be used for grade 3, and grades 4 and 5 (shown).
Note: Positioning hip in flexion is helpful.

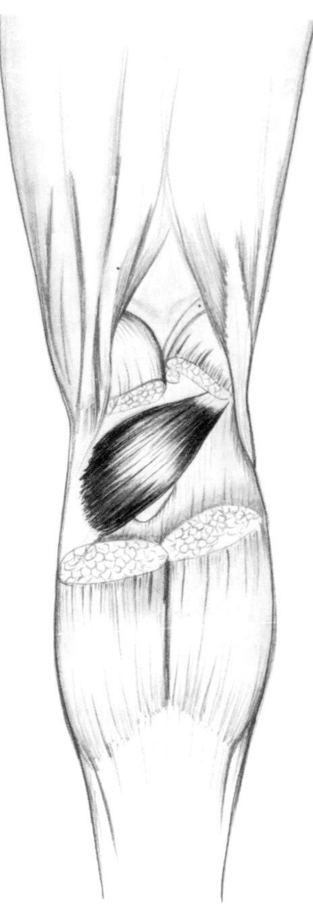

Origin

—Posterior surface of tibia above soleal line.
—Superior lip of soleal line.

Insertion

—Lateral surface of lateral condyle of femur.
—Some aponeurotic fibres mingle with the posterior lateral capsular ligament and arcuate complex.

Innervation Tibial (L5, S1).

Function

—Internal rotation of leg on thigh (knee flexed).
—Flexion of leg on thigh.
It forms an active ligament of the knee joint.
Synergist to the hamstrings, it very weakly substitutes in leg flexion, and can be tested in its function of internal rotator.
Only rough testing is possible. Popliteus cannot be isolated from other internal rotators of the knee joint.
Subject sitting on a stool.
Knee flexed at 90°.
Stabilize lower portion of thigh.
From a position of external rotation of the leg, ask subject for internal rotation.

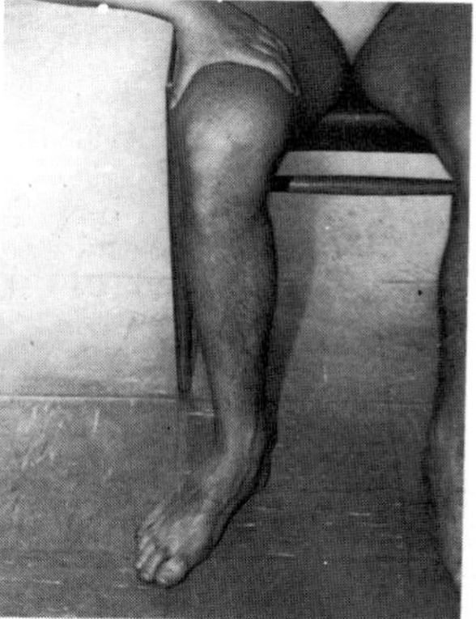

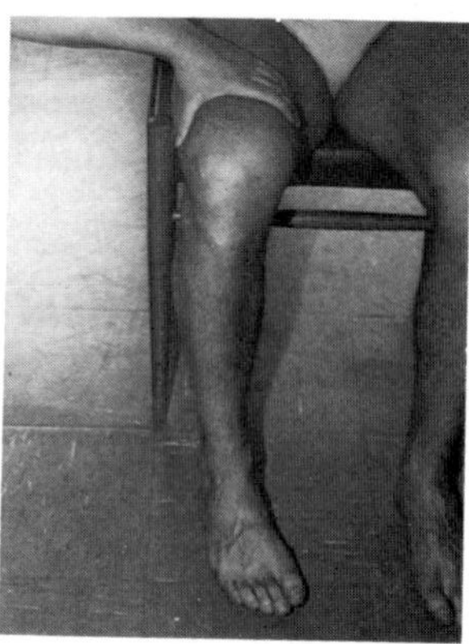

TIBIALIS ANTERIOR

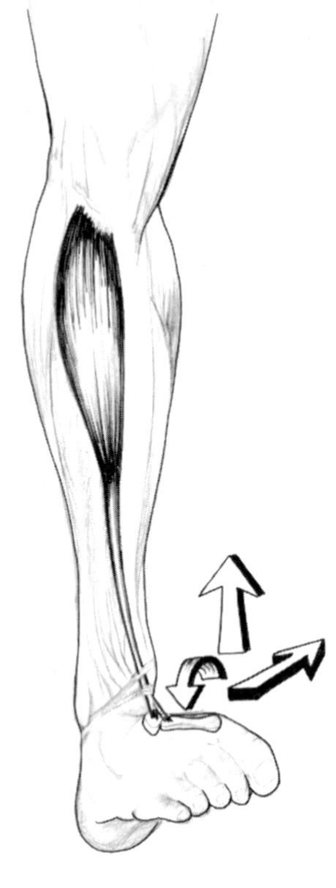

Origin

—Superior two-thirds of lateral surface of tibia.
—Lateral condyle of tibia.
—Anterior surface of interosseus membrane.
—Aponeurosis of leg.
—Intermuscular septum separating it from the extensors.

Insertion

—Medial and plantar surface of medial cuneiform bone.
—Medial and plantar surface of base of first metacarpal bone.

Innervation Anterior tibial (L4, l5).

Function

—Dorsiflexion of foot on leg.
—Adduction and supination of foot.
It draws the dome of the medial arch of the foot upwards.
It participates in the triple flexion pattern of movement of the lower extremity (withdrawal).
Tibialis anterior is the direct antagonist of peroneus longus.

Tibialis anterior, muscle activity during gait cycle (with extensor hallucis longus and extensor digitorum longus).

(J. B. Piera and A. Grossiord)

Stance phase

—Loading response: They play a considerable role in stabilizing the ankle plantar flexion. Tibialis anterior contracts powerfully while it lengthens (dynamic, eccentric contraction).
—Pre-swing: They start contracting to prepare the elevation (dorsiflexion) of the forefoot.

Swing phase

They contract throughout this phase.
Their contraction is dynamic, concentric, quite different from the restraining activity produced during loading response.
Their activity is much less intense in the action of lifting the forefoot than in that of controlling foot plantar flexion.

Contracture or tightness Foot position of dorsiflexion, adduction and supination: talus varus foot.

Deficiency Limitation of foot dorsiflexion. If extensor digitorum longus and peroneus tertius are intact, foot dorsiflexion is combined with abduction and pronation. There is a tendency towards foot valgus. If all the foot dorsiflexors are affected, there is a risk of progressive installation equinus deformity and of a steppage gait: at initial contact the foot slaps down on the ground, and during initial swing, the patient has to pick the lower limb up higher in order to clear the ground.

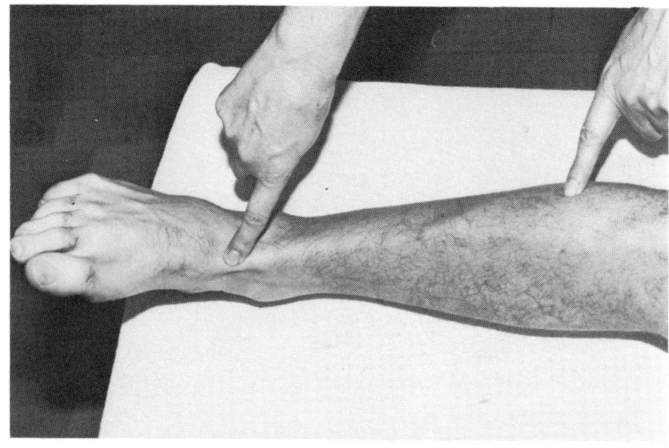

0 and 1 Subject supine.

Lower limbs extended, foot resting on table in neutral position.
Ask subject to dorsiflex the foot on the leg in adduction and supination.
Palpate tendon on dorsal medial aspect of foot instep.
Palpate muscle belly on anterior aspect of leg, laterally as far as tibial crest.
Do not confuse with extensor hallucis longus which is located more laterally.

Grade 0: No contraction or tendon movement is felt.
Grade 1: Contraction is felt but there is no joint movement.

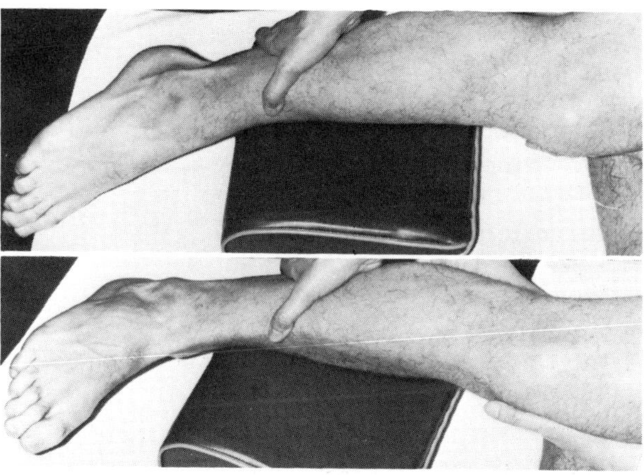

2 Subject side-lying.

Contralateral limb flexed, stabilizing pelvis.
Stabilize limb to be tested at lower third of thigh, knee flexed.
Cushion under lower third of leg, heel free.
Foot in plantar flexion, abduction and pronation.
Stabilize lower third of leg.
Ask subject for foot dorsiflexion on the leg, adduction and supination.
Precaution: toes, particularly the big toe, should remain relaxed in order to avoid substitution of extensor hallucis longus.
Range of movement should be complete.

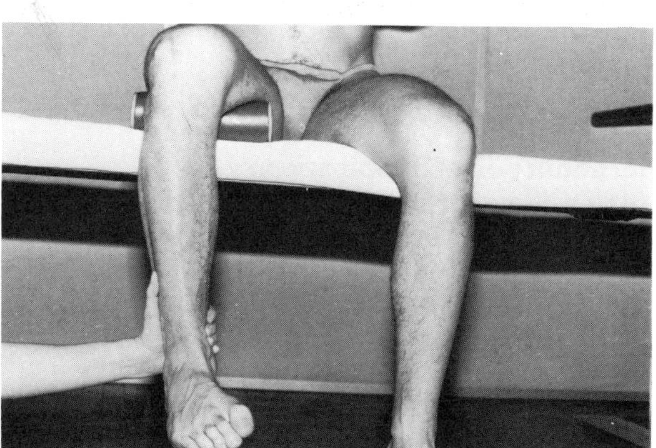

3 Subject sitting at edge of table, leg hanging, foot relaxed.

Cushion under knee.
Stabilize lower third, posterior aspect of leg.
Ask subject for foot dorsiflexion, adduction and supination.
Toes should remain relaxed.
Range of movement should be complete.

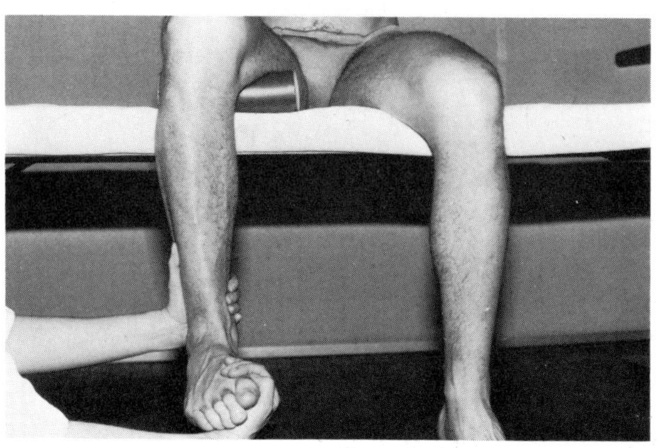

4 and 5 Same position.

Same stabilization, precaution and movement.
Resistance to the different components of the movement is placed on the upper medial portion of the metatarsus.

Grade 4: Power is sub-normal.
Grade 5: Power is normal.

EXTENSOR HALLUCIS LONGUS

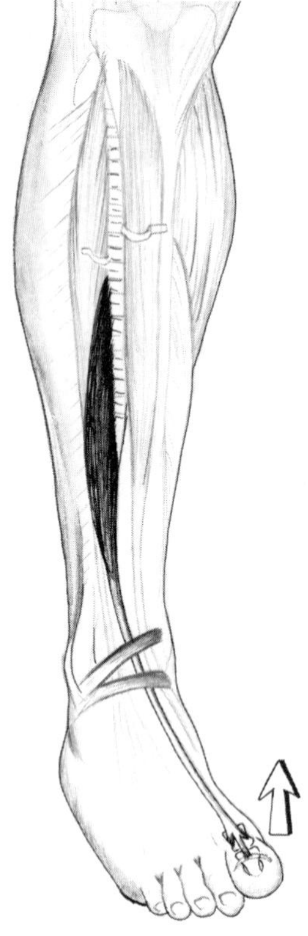

Origin

—Middle third of medial surface of fibula on the preligamentary area.
—Adjacent interosseous membrane.

Insertion

—Dorsal surface of base of distal phalanx of big toe.
—Lateral expansions into proximal phalanx and metatarsophalangeal joint.

Innervation Anterior tibial (l4, L5, s1).

Function

Dorsiflexion of big toe. It acts mostly on extension of proximal phalanx on first metatarsal, the extension of the distal phalanx on the proximal being partly produced by the short plantar intrinsic muscles of the big toe.
During extension of proximal phalanx, there is often a concomitant flexion of the distal phalanx caused by flexor hallucis longus.
Extensor hallucis longus participates in foot dorsiflexion, adduction and supination.
Note: The extensor digitorum brevis also produces the extension of proximal phalanx, and draws the big toe outwards: its role is not to be neglected. It balances the adduction component of the extensor hallucis longus.

Extensor hallucis longus, muscle activity during gait cycle

(J. B. Piera and A. Grossiord)
See tibialis anterior.

Contracture or tightness

Proximal phalanx held in extension, particularly if the short plantar muscles of the big toe are paralysed: claw toe deformity or hammer toe.

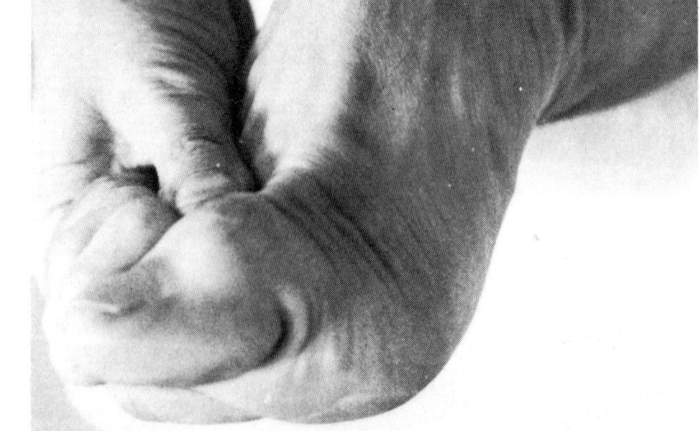

Subject demonstrating contracture of extensor hallucis longus.

Deficiency Inability to extend proximal phalanx if extensor hallucis longus and extensor digitorum brevis are paralysed.
Extension of distal phalanx is possible because of action of intrinsic plantar muscles of the big toe.
If only the extensor digitorum brevis is intact, extension of proximal phalanx is possible but with lateral deviation of toe.
Slight decrease in foot dorsiflexion force.
Extensor hallucis longus and extensor digitorum brevis.

Extensor hallucis longus and extensor digitorum brevis

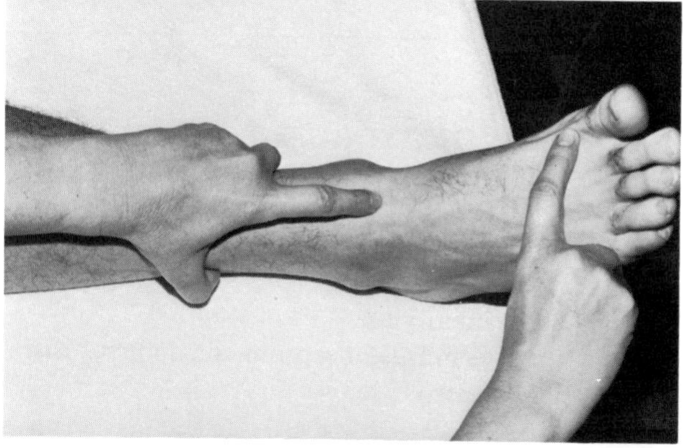

0 and 1 Subject supine.

Lower limbs extended.
Foot in neutral position resting on table.
Ask subject to extend the big toe.
Extensor hallucis longus tendon is palpated on dorsal aspect of proximal phalanx of first metatarsus, or on anterior aspect of the instep laterally to tibialis anterior.
Extensor digitorum brevis is palpated on dorsal lateral aspect of foot laterally to extensor digitorum longus.

Grade 0: No tendon movement is felt.
Grade 1: Tendon movement is felt but there is no joint movement.

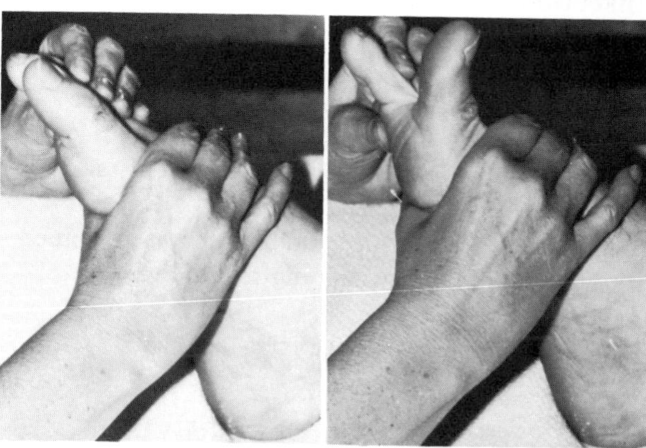

2 Global testing of big toe extension.

Same position.
Stabilize metatarsus.
Ask subject for same movement.
Subject completely extends proximal phalanx.
Extension of distal phalanx is weaker.
If extensor digitorum brevis dominates, big toe deviates outwards.
Range of movement should be complete.
Note: Examiner may test extension of distal phalanx on the proximal by holding sides of proximal phalanx.

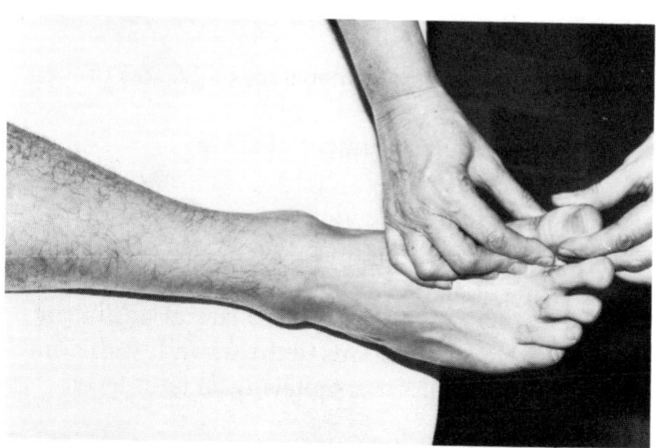

3, 4, and 5 Individual testing of distal phalanx extension.

Same position.
Stabilize sides of proximal phalanx.
Resistance to the movement is placed on sides of the distal phalanx.
Resistance is minimal for grade 3 and increases for grades 4 and 5.

Grade 4: Power is sub-normal.
Grade 5: Power is normal.

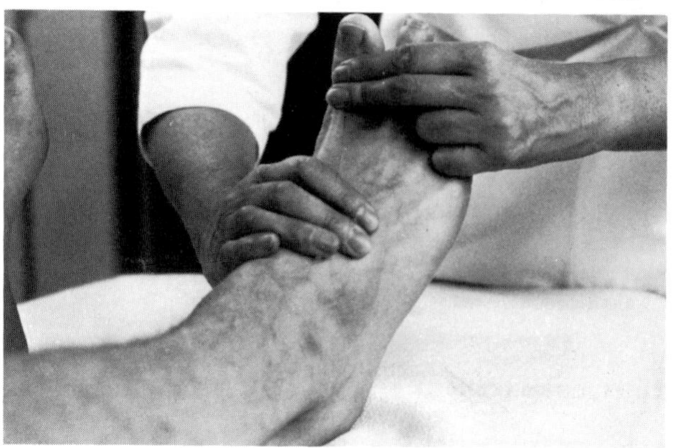

3, 4, and 5 Global testing of big toe extension.

Same position.
Stabilize metatarsus.
Ask subject to extend the big toe.
Resistance to the movement is placed on the dorsal aspect of both phalanges.
Resistance is minimal for grade 3 and increases for grades 4 and 5.

Grade 4: Power is sub-normal.
Grade 5: Power is normal.

EXTENSOR DIGITORUM LONGUS, EXTENSOR DIGITORUM BREVIS, and PERONEUS TERTIUS

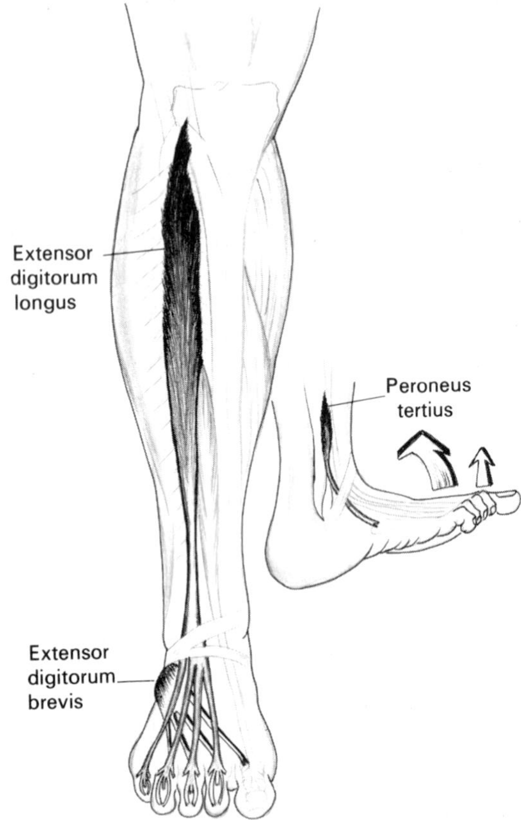

EXTENSOR DIGITORUM LONGUS

Origin

—Upper three-quarters of medial surface of fibula.
—Lateral tuberosity of tibia.
—Interosseus membrane.
—Adjacent intermuscular septum and aponeurosis.
—Leg aponeurosis.

Insertion By four tendons to the last four toes.
—One expansion into lateral sides of proximal phalanx.
—One slip onto dorsal surface of base of middle phalanx.
—One slip onto dorsal surface of base of distal phalanx.
Tendons to second, third and fourth toes are joined by a tendon of the extensor digitorum brevis.

Innervation Anterior tibial (l4, L5, s1).

EXTENSOR DIGITORUM BREVIS

Origin

—Lateral part of superior surface of greater process of calcaneum.
—Anterior annular ligament of ankle.

Insertion By four tendons:
—The first inserts into dorsal surface of base of proximal phalanx of big toe.
—The last three tendons join the lateral side of the extensor longus tendons to the second, third, and fourth toes at metatarsophalangeal joint level.

Innervation Anterior tibial (l4, L5, s1).

PERONEUS TERTIUS (not always present)

Origin

—Lower third of medial surface of fibula.
—Interosseus membrane.
—Anterior intermuscular septum.

Insertion Dorsal surface of tubercular eminence of fifth metatarsal bone.

Innervation Anterior tibial (l4, L5, S1).

Function

Extensor digitorum longus and extensor digitorum brevis

—Extension of toes. Extension takes place principally at level of proximal phalanges on the metatarsal bones.
—Extension of fifth toe is caused by the extensor longus. Its tendon does not receive an expansion from the extensor brevis.
—Extensor digitorum brevis deviates the toes outwards.

Extensor digitorum longus and peroneus tertius

—Peroneus tertius is sometimes considered a part of extensor digitorum longus.
—Both are responsible for:

- Foot dorsiflexion (the extensor digitorum longus is more efficient in this function when the toes are stabilized by the interossei and lumbricales).
- Foot abduction.
- Foot pronation: the lateral side of the foot rises and the plantar surface faces outwards.

Note: When the foot is fixed, the extensor digitorum longus and the peroneus tertius flex the leg on the foot, pulling the tibia forward.

Extensor digitorum longus muscle activity during gait cycle

(J. B. Piera and A. Grossiord)
See tibialis anterior.

Contracture or tightness

Clawing of the toes, placing the metatarsophalangeal joint in hyperextension, especially if the interossei and lumbricales are deficient.
Tendency towards talus valgus foot deformity if contracture is marked.

Deficiency

—If the extensor digitorum longus and brevis are paralysed, extension of the proximal phalanx is impossible. Extension of the middle and distal phalanges may, in certain subjects, be weakly substituted for by the interossei.
—Foot dorsiflexion occurs with adduction and supination by the dominant action of tibialis anterior and extensor hallucis longus.
—Foot dorsiflexion power is decreased.
—When all foot dorsiflexors are paralysed, equinus deformity appears.
During stance, at loading response, the foot slaps the ground.
During swing, the patient must pick his limb up higher for foot clearance: steppage gait.

NOTES

EXTENSOR DIGITORUM BREVIS and EXTENSOR DIGITORUM LONGUS

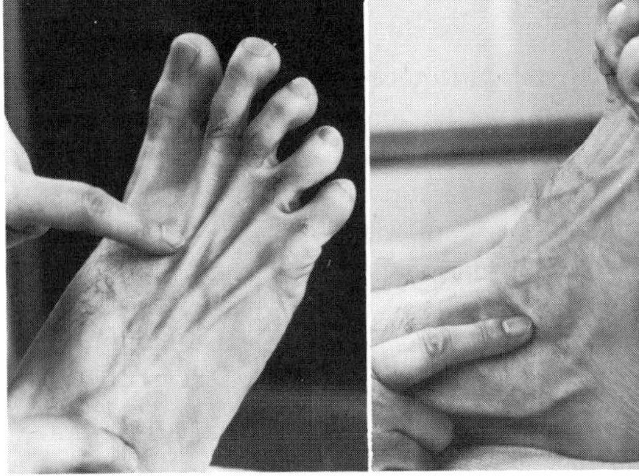

0 and 1 Subject supine.

Lower limb extended.
Foot in neutral position resting on table.
Ask subject to extend last four toes.
Extensor tendons are palpated on proximal phalanges, or on dorsal surface of metatarsus, or on lateral aspect, anterior surface of instep.
Extensor digitorum brevis is palpated on lateral aspect of dorsal surface of foot laterally to extensor digitorum longus.

Grade 0: No tendon movement is felt. No contraction of extensor digitorum brevis is felt.
Grade 1: Tendon movement or muscle contraction is felt but there is no joint movement.

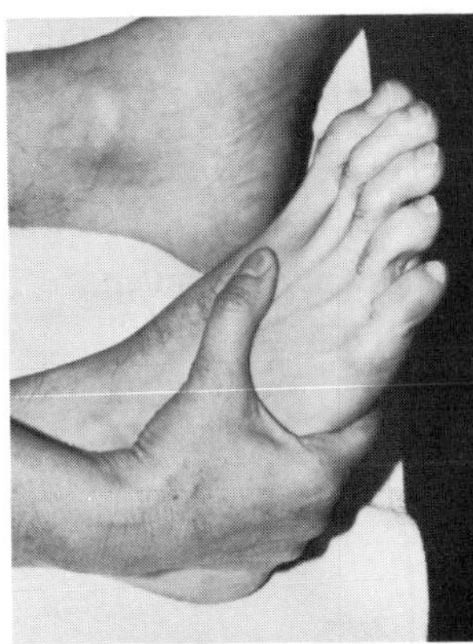

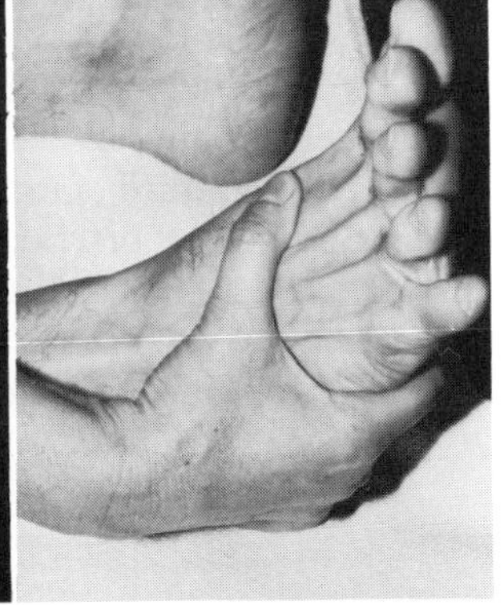

2 Same position.

Stabilize the metatarsus.
Ask subject to extend last four toes.
If the three middle toes are drawn outwards, the extensor digitorum brevis is predominant.
Toe extension takes place primarily at the metatarso-phalangeal joint level.
Range of movement should be complete.

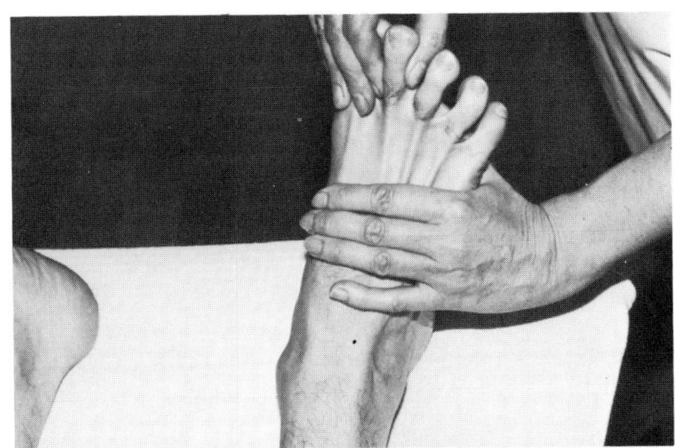

3, 4, and 5 Same position.

Same stabilization and movement.
Resistance to the movement is placed on sides of proximal phalanges of last four toes.
Resistance is minimal for grade 3 and increases for grades 4 and 5.
Toes may be tested individually or together.

Grade 4: Power is sub-normal.
Grade 5: Power is normal.

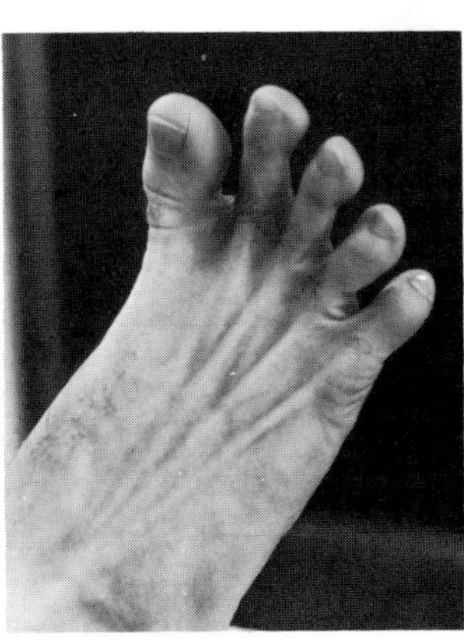

Contraction of extensor digitorum longus.
Note the slight protrusion of extensor digitorum brevis tendons in the second and third intermetatarsal spaces.

EXTENSOR DIGITORUM LONGUS and PERONEUS TERTIUS

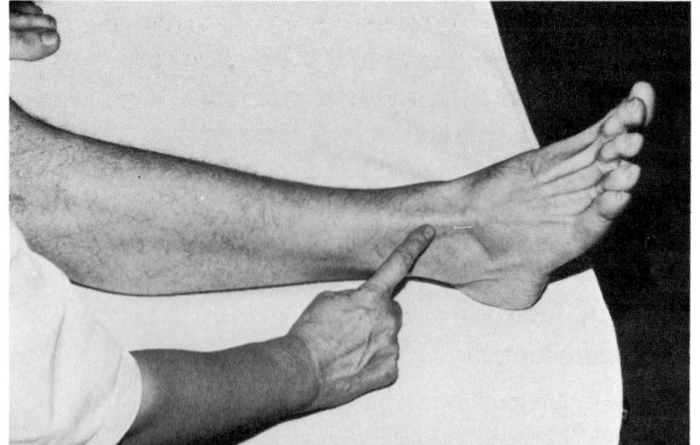

0 and 1 Subject supine.

Foot in neutral position resting on table.
Ask subject for foot flexion with abduction and pronation.
Extensor digitorum longus tendon is palpated on antero-lateral aspect of dorsal surface of instep.
Peroneus tertius tendon is palpated on dorsal surface of base of fifth metatarsus just medially to peroneus brevis.

Grade 0: No tendon movement is felt.
Grade 1: Tendon movement is felt but there is no joint movement.

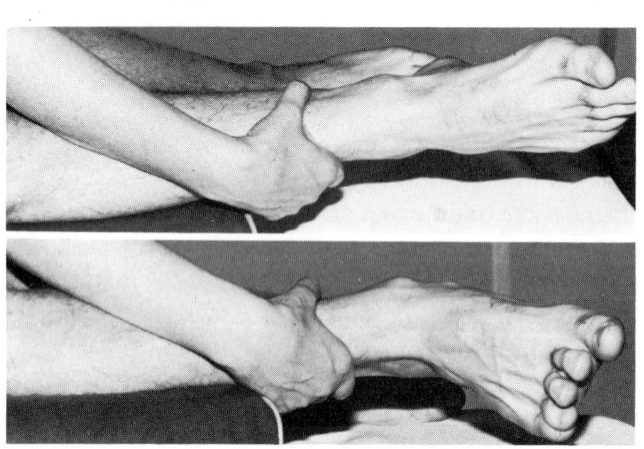

2 Subject side-lying.

Limb to be tested resting on lateral side on a cushion placed beneath the leg.
Stabilize lower third of leg.
From a position of plantar flexion, adduction and supination, ask subject for foot dorsiflexion with abduction and pronation.
Range of movement should be complete.

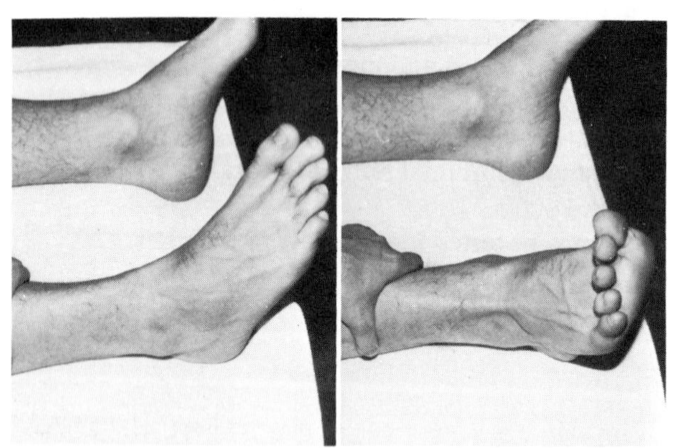

Alternative: Grade 2 may be given with subject supine, foot resting on table.
From a position of plantar flexion, adduction and supination, ask subject for foot dorsiflexion with abduction and pronation.
Range of movement should be complete.

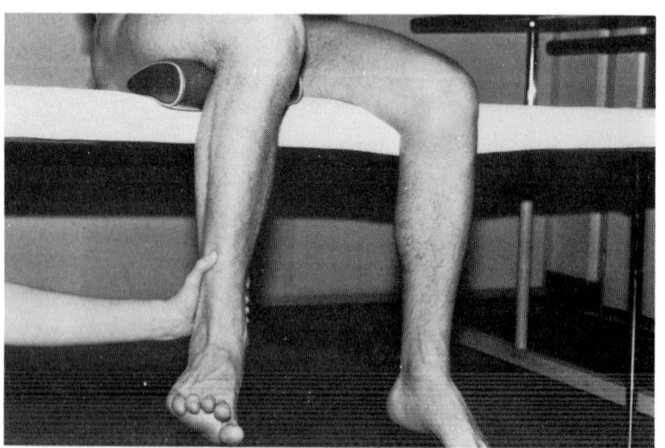

3 Subject sitting on edge of table, legs hanging.

Stabilize lower third of leg.
Ask for same movement.
Range of movement should be complete.

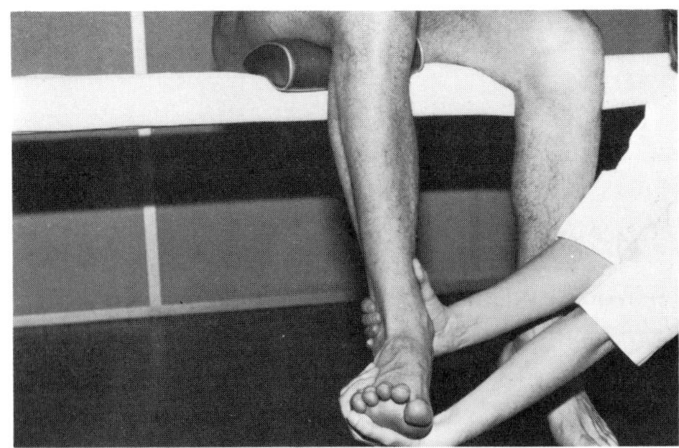

4 and 5 Same position.

Same stabilization and movement.
Resistance to the components of the movement is placed on lateral aspect of dorsal surface of foot.

Grade 4: Power is sub-normal.
Grade 5: Power is normal.

PERONEUS BREVIS

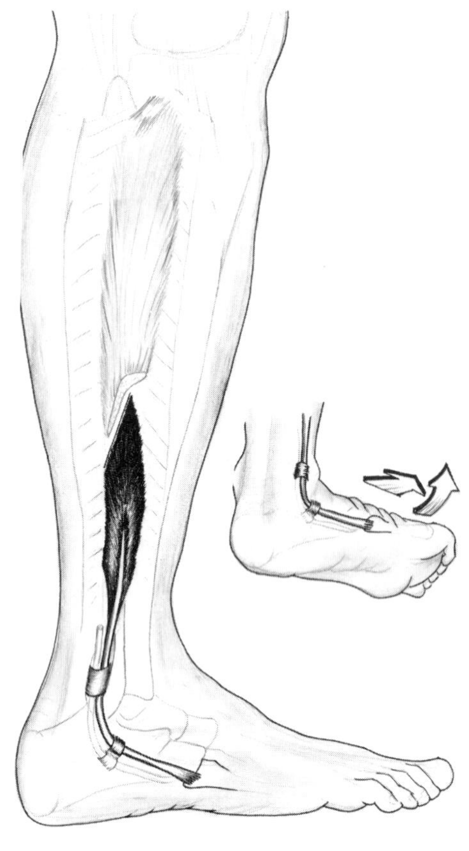

Origin

—Middle portion of lateral surface of fibula.
—Anterior and lateral intermuscular septa.

Insertion

—Styloid tuberosity of base of fifth metatarsal bone.
—Expansion to fourth metatarsal bone.

Innervation Musculocutaneous branch of peroneal (l4, L5, s1).

Function

—Direct and powerful abduction of foot: in this movement it is stronger than peroneus longus.
—Pronation: it raises the lateral side of foot, and sometimes places the fifth metatarsal above the fourth.
—According to Duchenne de Boulogne, it is in direct opposition to tibialis posterior: when the foot is in dorsiflexion, it can plantarflex it; when the foot is in plantar flexion, it can dorsiflex it.
—Acting in synergy with the peroneus longus, it stabilizes the foot laterally.
—It maintains the lateral arch of the foot, preventing downward gaping of the joints.
Note: The movement of abduction and pronation of the foot is the result of synergistic activity of peroneus brevis, peroneus longus, extensor digitorum longus and peroneus tertius.

Peroneus brevis muscle activity during gait cycle

(J. B. Piera and A. Grossiord)
See peroneus longus.

Contracture or tightness Leads to valgus foot deformity.

Deficiency

—Decrease in abduction pronation capacity, partially compensated for by the other abductors and pronators.
—Flattening of the lateral arch of the foot.
—Foot instability.
—Tendency towards varus deformity of the foot if tibialis posterior and tibialis anterior predominate.

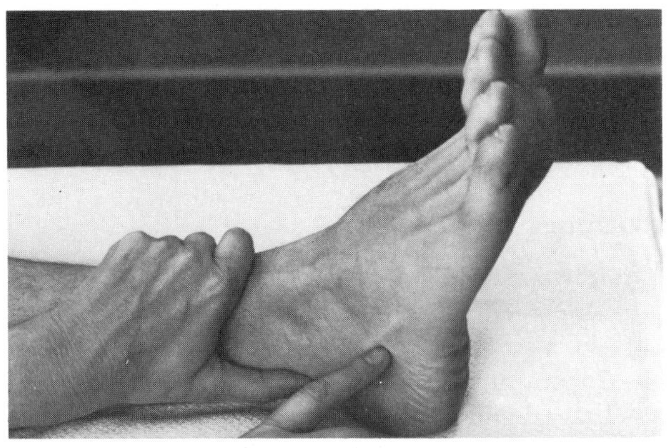

0 and 1 Subject supine.

Lower limbs extended.
Foot in neutral position resting on table.
Ask subject for direct abduction and pronation of foot.
Palpate tendon posteriorly to styloid tuberosity of fifth metatarsal bone.
Do not confuse with peroneus longus and peroneus tertius: peroneus brevis is situated between them.

Grade 0: No tendon movement is felt.
Grade 1: Tendon movement is felt but there is no joint movement.

2 Same position.

Stabilize lower third of leg.
Ask subject for same movement, the foot remaining in neutral position between dorsal and plantar flexion.
Abduction pronation with lowering of the first metatarsal head signifies action of peroneus longus and, with foot dorsiflexion, that of extensor digitorum longus.
Range of movement should be complete.

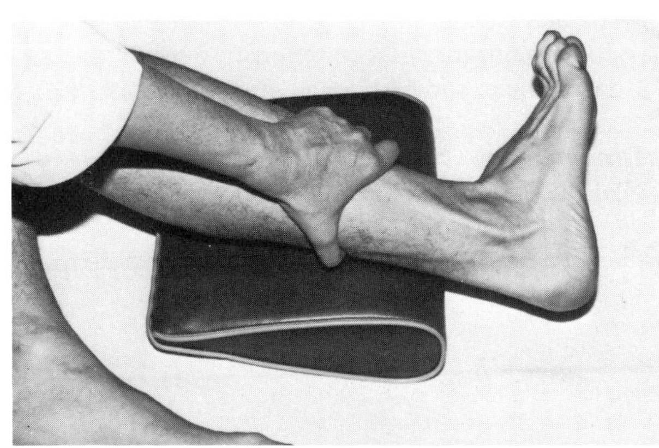

3 Subject side-lying.

Contralateral limb flexed, braced on table stabilizing the pelvis.
Limb to be tested resting on medial side on a cushion leaving heel free.
Stabilize lower third of leg.
Ask subject for the same movement.
Watch for possible trick movements.
Range of movement should be complete.

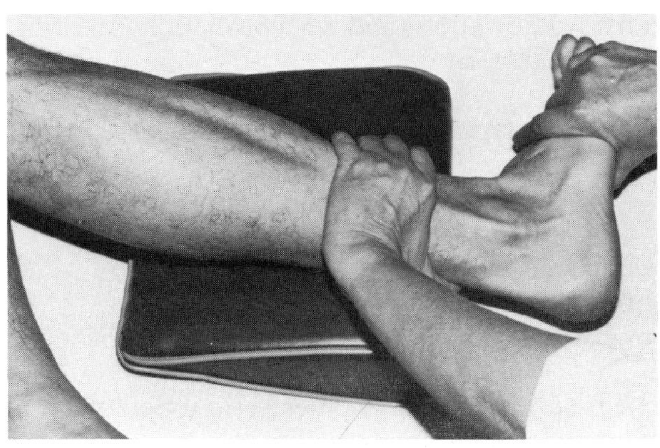

4 and 5 Same position.

Same stabilization and movement.
Same risks of trick movements.
Resistance to the movement is placed on lateral side of fifth metatarsal bone.

Grade 4: Power is sub-normal.
Grade 5: Power is normal.

PERONEUS LONGUS

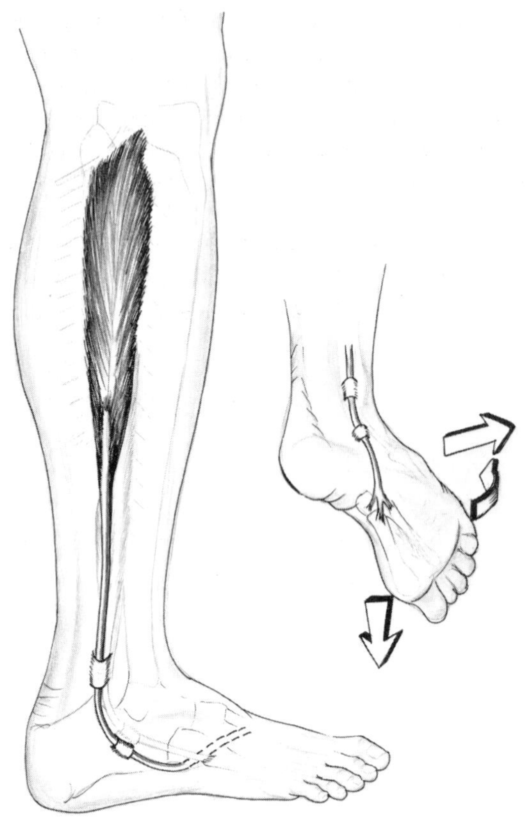

Origin

Epiphyseal head:
—Antero-lateral part of the circumference of head of fibula.
—Superior tibio-fibular ligament.
—Lateral tuberosity of tibia.
—Adjacent intermuscular septa.
—Leg aponeurosis.
Anterior inferior diaphyseal head:
—Upper half of anterior part of lateral surface of fibula.
—Upper half of anterior border of fibula.
—Anterior intermuscular septum.
Posterior inferior diaphyseal head:
—Upper half of posterior part of lateral surface of fibula.
—Upper half of lateral border of fibula.
—Lateral intermuscular septum.

Insertion

—Lateral tuberosity of plantar surface of base of first metatarsal bone.
—Expansions into first cuneiform, to second metatarsal bone, and first dorsal interosseus muscle.

Innervation Musculocutaneous branch of peroneal (l4, L5, s1).

Function

It lowers the first metatarsal and deviates it laterally. Its pull strengthens all the metatarsal bones: thus it enables a better insertion of the action of the gastrocnemius which draws the foot inwards.
The movement continues with foot abduction (foot outward deviation) and with pronation (twisting of the foot so that lateral side rises and medial malleolus protrudes).
It is an accessory plantar flexor.
Because of its contact with peroneal tubercle, it supports the anterior part of the foot and participates in preserving the various arches of the plantar dome:
—Medial arch: lowering of first metatarsal bone.
—Lateral arch: prevents inferior gaping of joints.
—Anterior arch: by its course, it maintains the transversal curve.
It is a lateral foot stabilizer in synergy with peroneus brevis.
Its role is important in the standing position because of ground contact of the first metatarsal bone (foot fixed).
Note: The movement of abduction and pronation of the foot is the result of synergistic action of peroneus longus, peroneus brevis, extensor digitorum longus and peroneus tertius.

Peroneal muscle activity during gait cycle

(J. B. Piera and A. Grossiord)

Stance phase

—Mid stance: Their contraction starts before midstance, when, because of centre of gravity displacement towards the stance limb, a lateral restraining action becomes necessary in order to reduce the range of this movement and oppose the action of tibialis posterior. Peroneus longus, then peroneus brevis therefore contract, and at terminal stance their action predominates over that of tibialis posterior, permitting the raising of the lateral side of the foot, that is to say, the shift of body weight from the head of the fifth metatarsal bone to the head of the first metatarsal and the big toe.

—Pre-swing: They increase their contraction, raise the lateral side of the foot, and thus oppose the inversion component of the triceps surae. The peroneus longus, because of its insertion at the base of the first metatarsal, contributes to prolonging contact of this bone with the ground, enabling it to be the last one to leave it with the big toe.

Contracture or tightness Formation of a high-arched, valgus foot, or flat-foot valgus deformity, depending on state of adjacent muscles.

Deficiency

Decrease of abduction–pronation capacity: this motion can be partly substituted for by the extensor digitorum longus and peroneus tertius, but is associated with dorsiflexion. Flattening of plantar arch: the medial side of the foot collapses. It may lead to flat-foot deformity.
Transverse instability of foot.
Disorder in balance and gait, in spite of presence of triceps surae.
Difficulty in standing on tip toe.

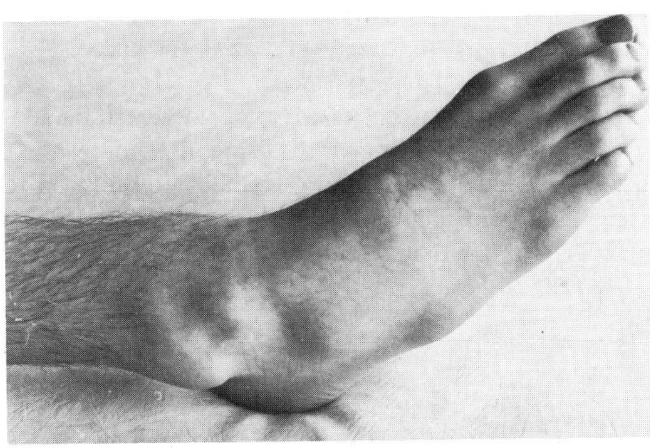

Subject presenting paralysis of peroneals. Predominating triceps surae and tibialis posterior invert the foot.

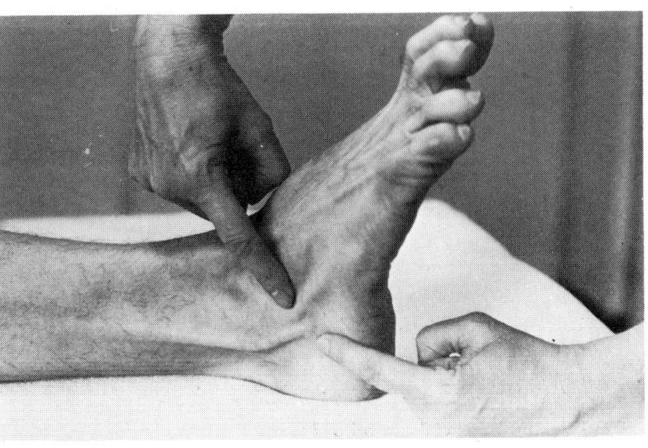

Palpating peroneals on lateral surface of calcaneum.

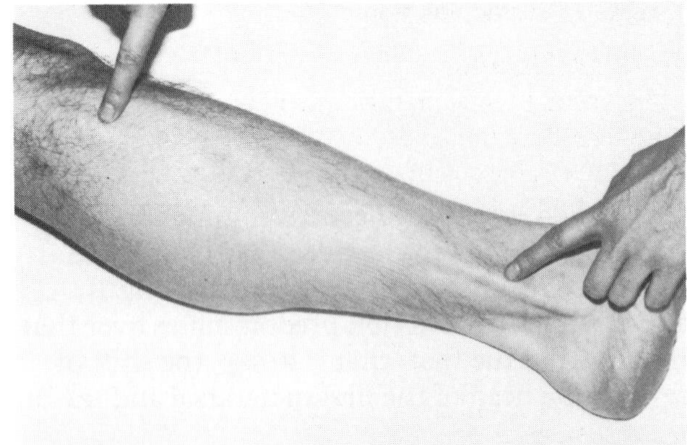

0 and 1 Subject side-lying.

Leg and foot resting on medial side.
Ask subject to lower first metatarsal while abducting and pronating the foot.
The tendon is palpated on lateral surface of the calcaneum beneath the peroneus brevis (not shown), or at posterior border of lateral malleolus.
Examiner may palpate muscle belly anteriorly to and beneath the head of fibula, lateral surface of leg.

Grade 0: No contraction or tendon movement is felt.
Grade 1: Contraction or tendon movement is felt but there is no joint movement.

2 Subject supine.

Foot resting on table.
Stabilize lower third of leg.
From a position of foot dorsiflexion, adduction and supination, ask subject for the same movement.
In muscle's most shortened position, watch for an exaggerated vertical crease in the first intermetatarsal space.
Range of movement should be complete.

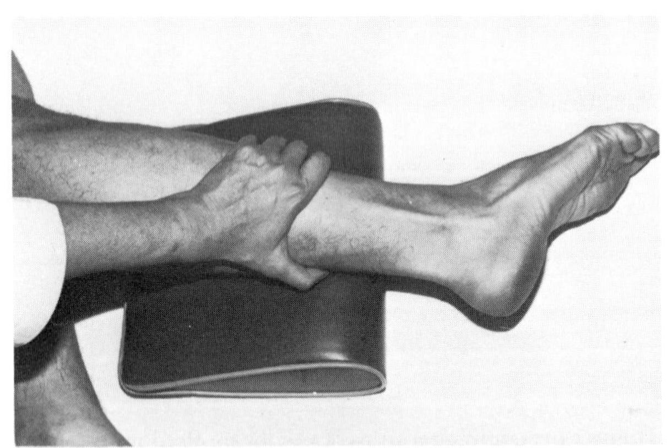

3 Subject side-lying.

Contralateral limb flexed, stabilizing the pelvis.
Cushion under leg to be tested.
From a position of foot dorsiflexion, adduction and supination, ask subject for the same movement.
Watch head of first metatarsal which should lower in an abduction pronation movement of the foot.
Range of movement should be complete.

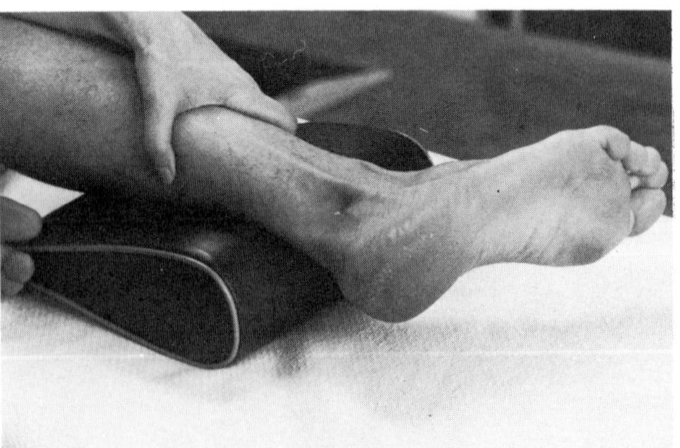

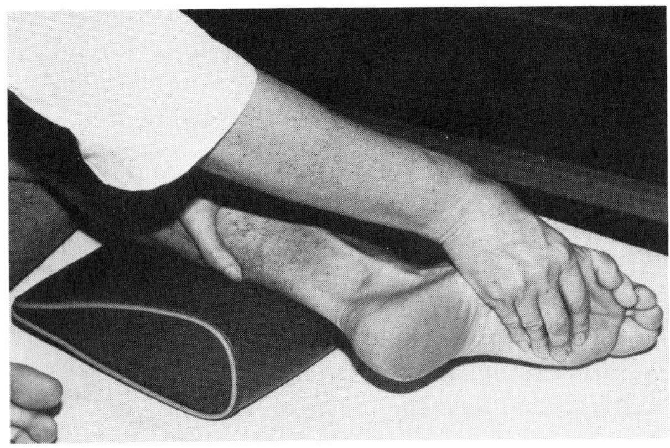

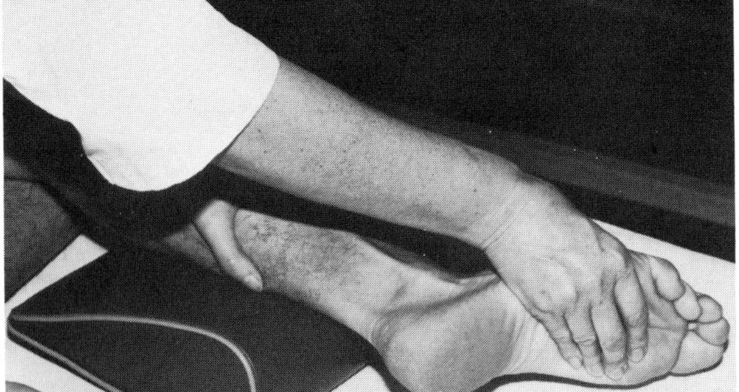

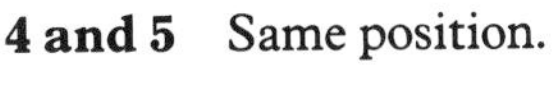

4 and 5 Same position.

Same stabilization.
Ask subject for same movement.
Resistance to the components of the movement is placed on plantar surface of the head of first metatarsal bone first, then on lateral edge of foot by grasping the metatarsus.

Grade 4: Power is sub-normal.
Grade 5: Power is normal.

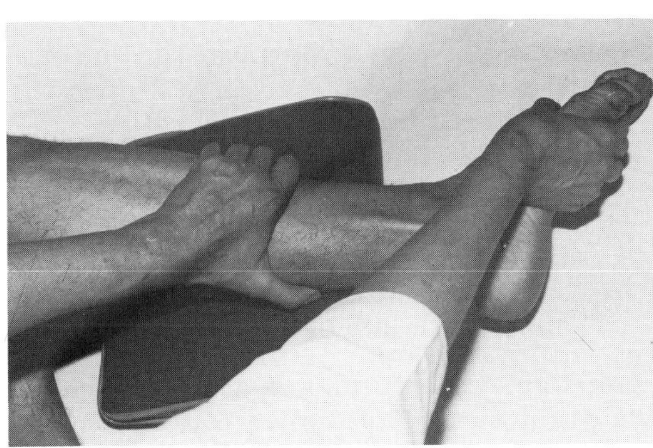

TRICEPS SURAE (GASTROCNEMIUS): GEMELLI and SOLEUS, and PLANTARIS

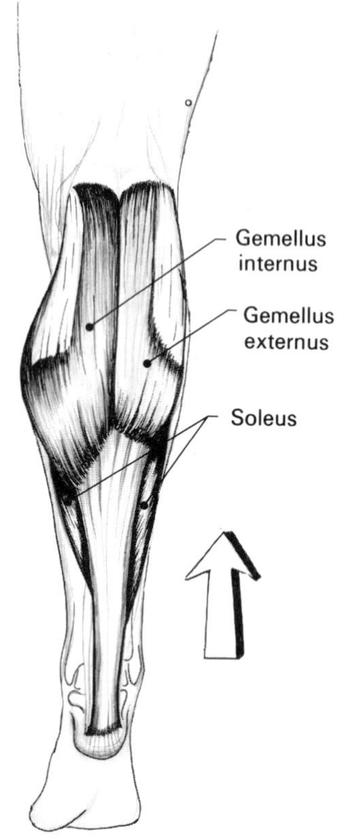

TRICEPS SURAE

Origin

Medial head (gemellus internus):
—Medial surface of medial femoral condyle.
—Inner tuberosity of medial condyle.
—Posterior medial capsular ligament.
Lateral head (gemellus externus):
Same origins, but on lateral side.
Soleus:
—Tibial head:

- Medial four-fifths of posterior oblique line of tibia.
- Middle third of medial border of tibia.

—Fibular head:

- Posterior surface of fibular head.
- Superior quarter of posterior surface of fibula and superior third of its lateral border.
- Lateral intermuscular septum.
- Fibrous arch joining the two heads.

Insertion Common, through Achilles tendon on the posterior and inferior surfaces of calcaneum.

PLANTARIS

Muscle not always present.

Origin Lateral condyle of femur.

Insertion Posterior surface of calcaneum medially to Achilles tendon.

Innervation

Gemelli and plantaris: internal popliteal (S1, S2).
Soleus: posterior tibial (l5, S1, s2).

Function

Plantar flexion of foot on leg. The efficiency of gemelli depends on knee position: it is lessened when knee is flexed.
Ankle plantar flexion caused by triceps is associated with a slight varus of the foot. It is also accompanied by contraction of flexor digitorum brevis and flexor digitorum longus.
Gemelli and plantaris participate weakly in knee flexion.
The triceps intervenes in normal gait, in walking upstairs (concentric contraction), in walking downstairs (eccentric contraction), in jumping, running etc.
The triceps participates in the mass extension pattern of the lower extremity.
With foot fixed, triceps lifts body's weight when rising on to tip toes.

Triceps surae muscle activity during gait cycle

(J. B. Piera and A. Grossiord)

Stance phase

—Loading response: Soleus and flexor digitorum longus contract a little before the end of loading response, when the sole of the foot is flat on the ground. They start their stabilizing activity on the knee by restraining the forward displacement of the leg, acting from the fixed ankle.

—Mid stance: The foot is flat on the ground, the leg is vertical. Forward glide of pelvis causes the forward displacement of the leg, and knee flexion. This tibial forward displacement is controlled by the contraction of triceps surae, in particular by the single-joint muscle, the soleus.

Hence the triceps' contraction is at first a dynamic, eccentric one. At 40% of the gait cycle, the heel is going to rise because of the nearly isometric contraction of triceps surae which does not induce a real plantar flexion but locks the leg on the foot, enabling the heel to rise as the foot pivots around the head of the metatarsal bones (push-off).

Recent works have shown that the various heads of triceps surae may have individual activities:

—Soleus, especially its medial portion, is a powerful stabilizer of the leg on the foot.

—Gemelli do not—or hardly—participate in walking on a flat surface. They act only to give the body a sudden and more powerful thrust, in jumping, walking uphill or running, constituting a sort of reservoir of power.

Let us note the synergistic action of soleus and flexor digitorum brevis which appear to be placed in line. This makes available an increasing force while maintaining precise adjustment of contraction greater than that of a single muscle.

—Pre-swing: The action of triceps surae and toe flexors, very important for propulsion, remains essentially the same as in the preceding phase: they lock the leg on the foot. However, the triceps surae must produce a powerful, dynamic, concentric contraction in order to draw the foot to 15° of plantar flexion.

Contracture or tightness

—Decrease of ankle dorsiflexion.
—Equinus deformity with varus tendency.
—Possibility of relaxing gemelli when flexing the knee.

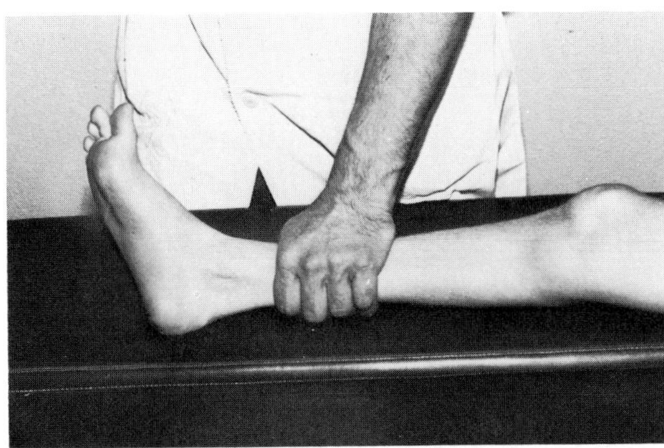

Subject demonstrating contracture of triceps surae. Note varus tendency.

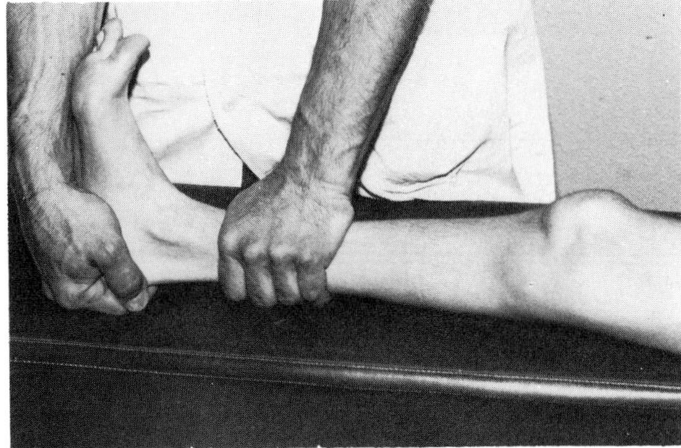

Examining the contracture.

Deficiency

—Inability of rising on toes.
—Loss of push-off during gait, and leg stability, hence a limp: the patient must lift his thigh and increase hip flexion.
—Shortening of step length on sound side in consequence.
—Inability to run or jump.
—Deficiency can be partly compensated by the accessory plantar flexors, the combined force of which cannot, however, come close to replacing the considerable force of the triceps.
—Tendency towards talus cavus foot if the dorsiflexors are predominant.

Note: Triceps surae is tested globally (gemelli and soleus), then the soleus alone.

GLOBAL TESTING OF TRICEPS SURAE

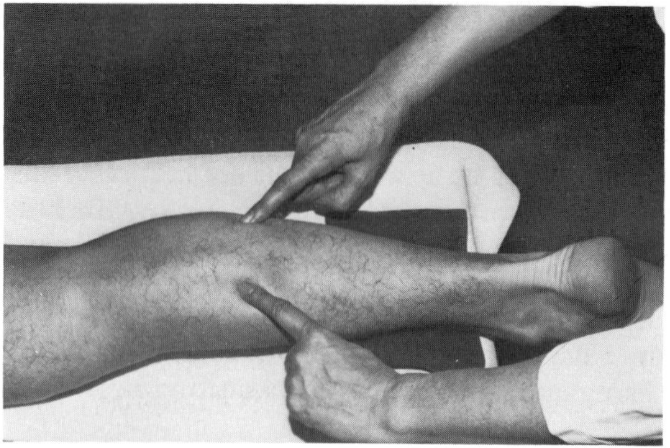

0 Subject prone.

Foot hanging over end of table.
Cushion under lower third of leg.
Ask subject for plantar flexion of foot on leg.
Lateral gemellus is palpated on lateral portion of calf, the medial gemellus, on the medial portion.
Palpation of soleus is covered under individual testing of soleus, page 378.

Grade 0: No contraction is felt.

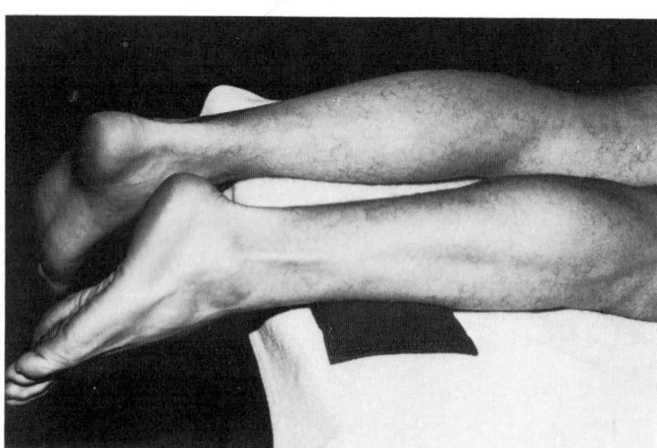

1 Same position.

Ask for the same movement.
Toes should remain relaxed, plantar flexion of the foot taking place at the ankle and not at forefoot level.
Range of movement should be complete.

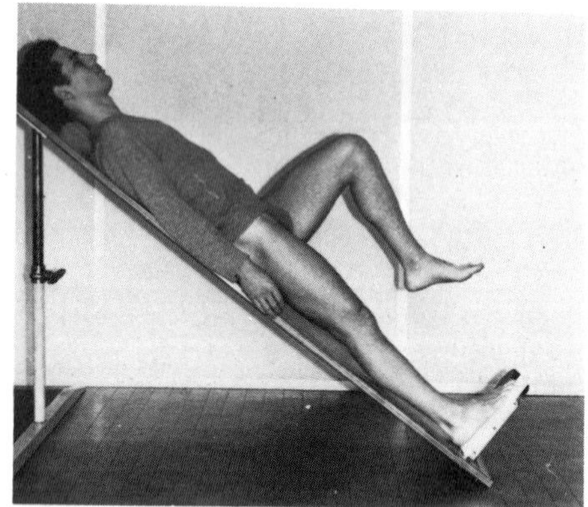

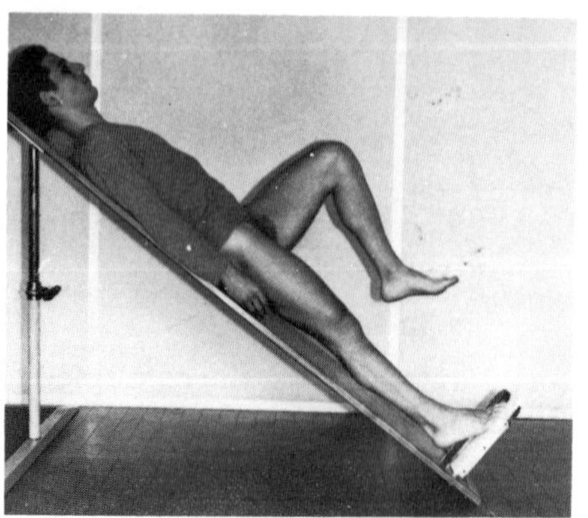

2 Subject supine on tilt table at 45°.

(Dr Maury MD)
Contralateral lower limb flexed but not braced.
Place a block under lateral edge of foot in order to eliminate pressure on first metatarsal bone, thus diminishing activity of peroneus longus.
Ask subject to rise onto toes (foot plantar flexion), knee extended. Body weight should be lifted.
Subject should remain clothed to facilitate sliding on table.
Range of movement should be complete.

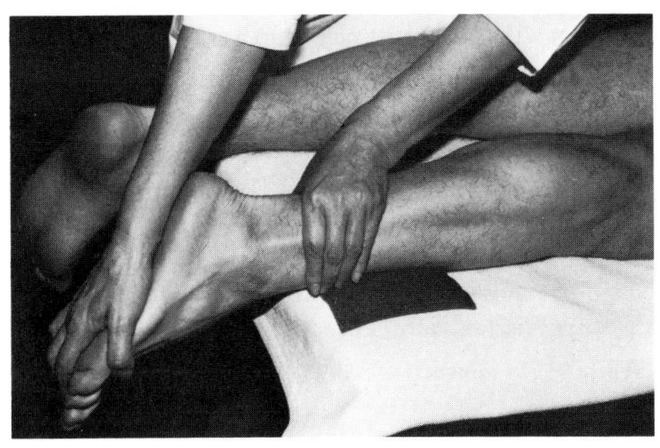

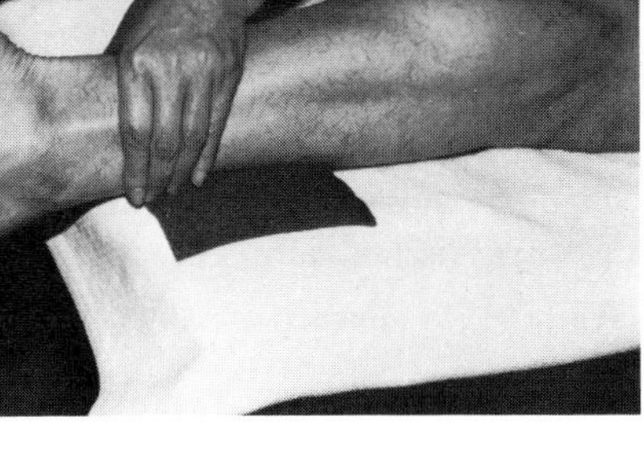

2 *Alternative.*

This position is for patients who cannot be placed in standing position.
Subject prone.
Foot hanging over end of table, cushion under lower third of leg.
Ask subject to plantar flex his foot on leg, knee extended.
Strong resistance to the movement is placed on the plantar surface of the foot.
Range of movement should be complete.
It can be used for grades 3, 4, and 5, considering that from grade 2 on, the triceps' force is practically invincible.

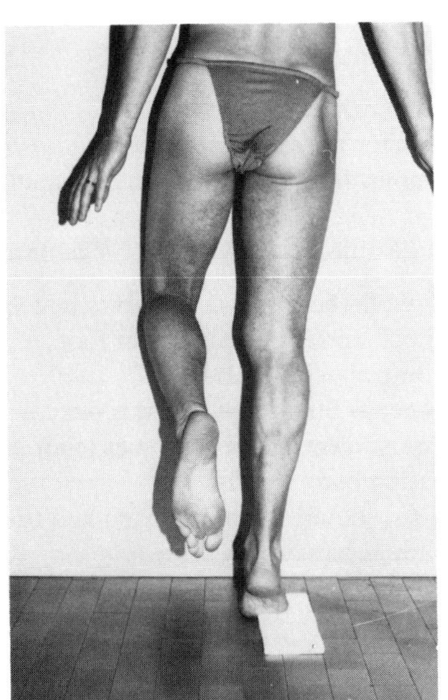

3 Subject standing without hand support.

Place a block under lateral edge of foot in order to eliminate weight-bearing on first metatarsal bone, and to diminish action of peroneus longus.
Contralateral limb flexed but not braced.
Ask subject to rise onto toes (foot plantar flexion), lifting his body weight.
Range of movement should be complete.

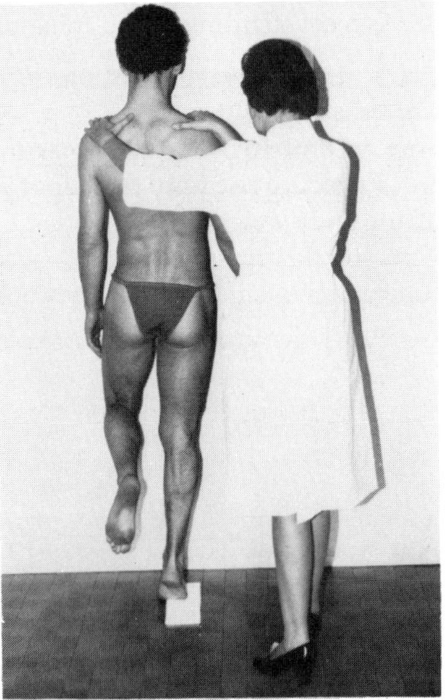

4 and 5 Same position.

Same movement.
Resistance to the movement is placed on subject's shoulders.
Resistance may also be placed on pelvis at iliac crest level (not shown).

Grade 4: Power is sub-normal.
Grade 5: Power is normal.

INDIVIDUAL TESTING OF SOLEUS

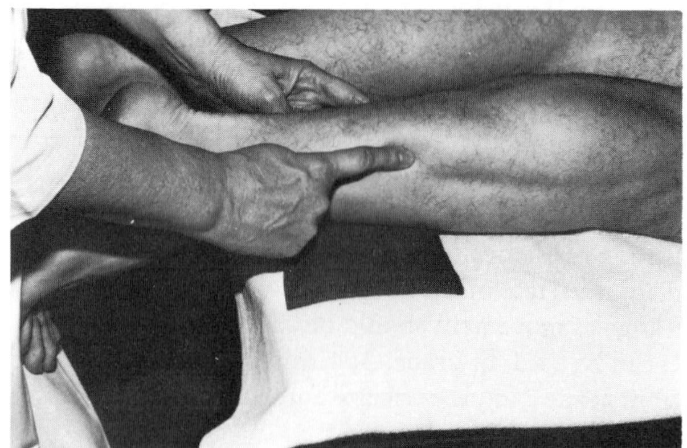

Subjects who cannot be placed in a standing position are tested prone.

0 Subject prone.

Foot hanging over end of table, cushion under lower third of leg.
Ask subject to plantar flex his foot on the leg.
Soleus is palpated on lateral and medial sides of upper portion of Achilles tendon.

Grade 0: No contraction is felt.

1 Same position.

Stabilize knee flexed at 20°.
Ask for the same movement.
Range of movement should be complete.

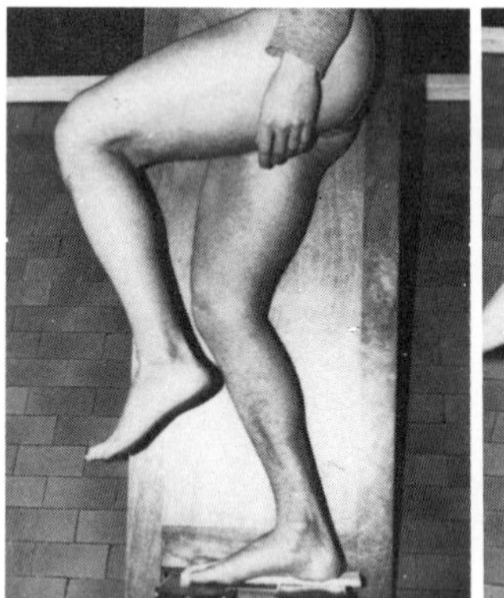

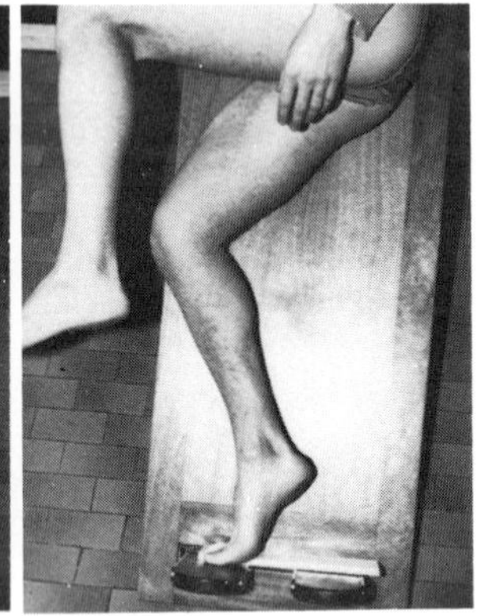

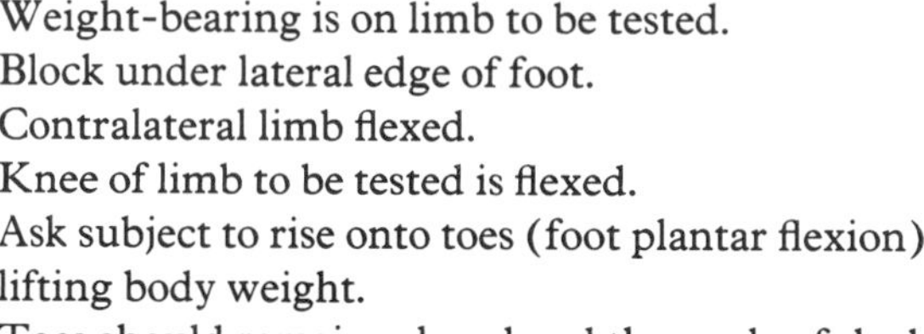

2 Subject side-lying on tilt table at 45°.

Weight-bearing is on limb to be tested.
Block under lateral edge of foot.
Contralateral limb flexed.
Knee of limb to be tested is flexed.
Ask subject to rise onto toes (foot plantar flexion) lifting body weight.
Toes should remain relaxed and the angle of the knee should remain the same.
Range of movement should be complete.

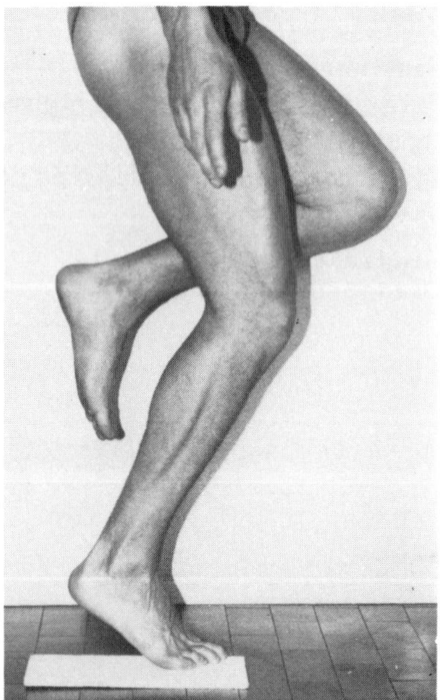

3 Subject standing without hand support.

Block under lateral edge of foot.
Contralateral limb flexed.
Knee of limb to be tested is flexed.
Ask subject to rise onto toes (foot plantar flexion) lifting body weight.
Toes should remain relaxed.
Range of movement should be complete.

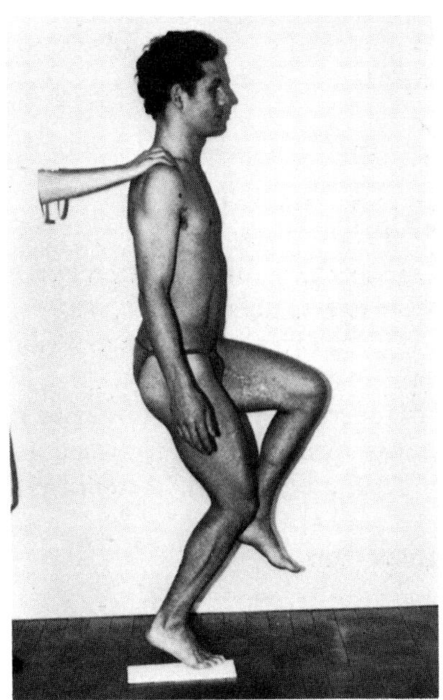

4 and 5 Same position.

Same precautions and movement.
Resistance to the movement is placed on subject's shoulders.
Resistance may be placed on pelvis at iliac crest level (not shown).

Grade 4: Power is sub-normal.
Grade 5: Power is normal.

TIBIALIS POSTERIOR

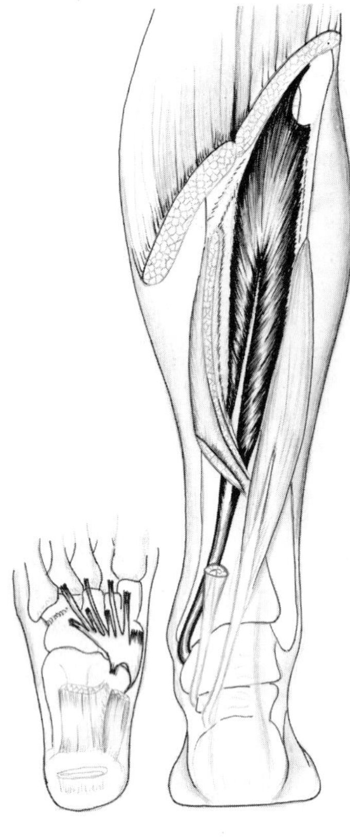

Origin

—Upper two-thirds of lateral part of posterior surface of tibia.
—Upper two-thirds of post-ligamentary aspect of medial surface of fibula.
—Upper two-thirds of posterior surface of interosseus membrane, and adjacent intermuscular septa separating it from flexor muscles.

Insertion

—Tuberosity of navicular bone.
—Expansions onto all bones of tarsus and metatarsus except for the astragalus, the first, and the fifth metatarsal bones.

Innervation Posterior tibial (15, S1).

Function

Adduction and supination of foot.
According to some authors, it participates in plantar flexion: the combination of these three functions causes foot inversion.
Owing to its main insertion and expansions, it plays an important role in plantar arch support.

Tibialis posterior muscle activity during gait cycle (J. B. Piera and A. Grossiord)

Stance phase

—Initial contact: It starts contracting at heel strike to stabilize the foot laterally. Then the tibialis posterior produces a dynamic eccentric contraction to control the valgus thrust. This contraction tends to shift the body weight onto the lateral side of the foot. It also acts as an active medial ligament of the ankle joint.
—Mid stance, terminal stance, and pre-swing: It contracts throughout nearly the entire stance phase, maintaining the medial, transverse balance of the foot.

Contracture or tightness Equino varus foot and metatarsus varus deformity.

Deficiency

Tendency towards foot abduction and pronation.
Collapse of the medial plantar arch.

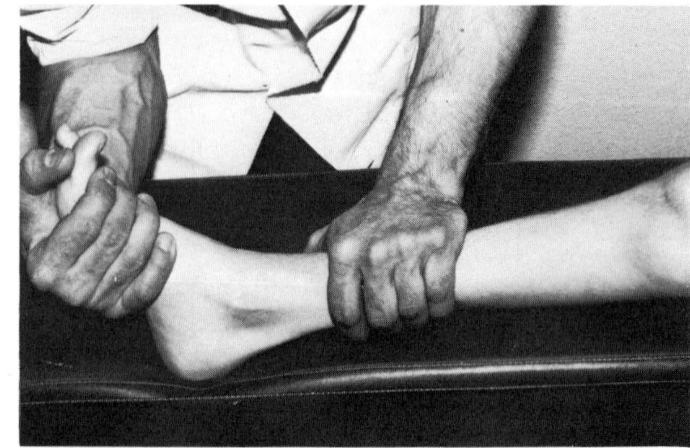

Examining contracture of tibialis posterior.

TESTING TIBIALIS POSTERIOR

Foot placed in plantar flexion, tibialis posterior will be tested in its function of adduction and supination.

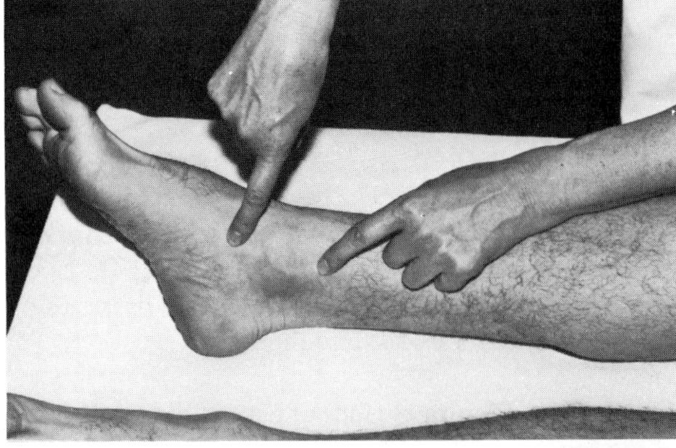

0 and 1 Subject supine.

Foot resting on table in plantar flexion to eliminate action of tibialis anterior.
Ask subject for foot adduction and supination on leg.
The tendon is palpated between medial malleolus and the navicular bone, or posteriorly to the malleolus.
Do not confuse with tendons of flexor hallucis longus or flexor digitorum longus.

Grade 0: No tendon movement is felt.
Grade 1: Tendon movement is felt but there is no joint movement.

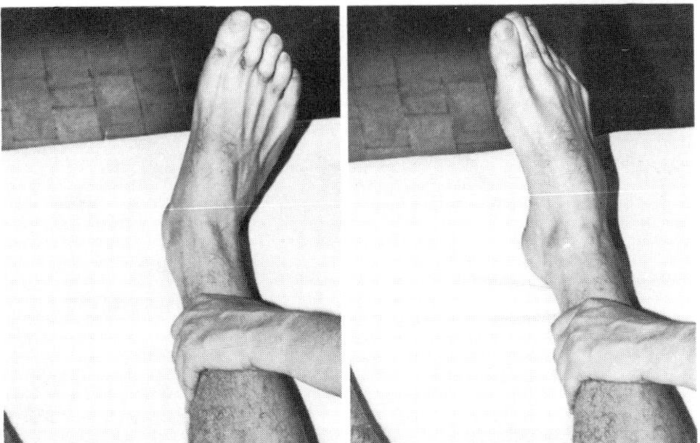

2 Same position.

Stabilize lower third of leg.
Ask subject for the same movement.
Precaution: possibility of compensation by tibialis anterior and flexor hallucis longus.
Range of movement should be complete.

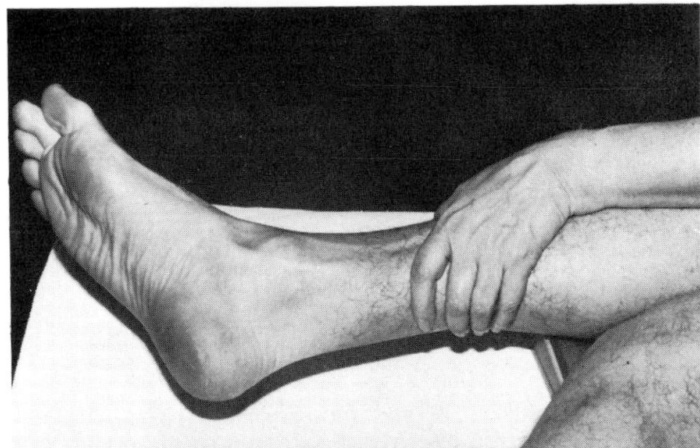

3 Subject side-lying.

Lateral aspect of leg resting on cushion.
Stabilize leg.
Ask subject for foot adduction and supination.
Range of movement should be complete.

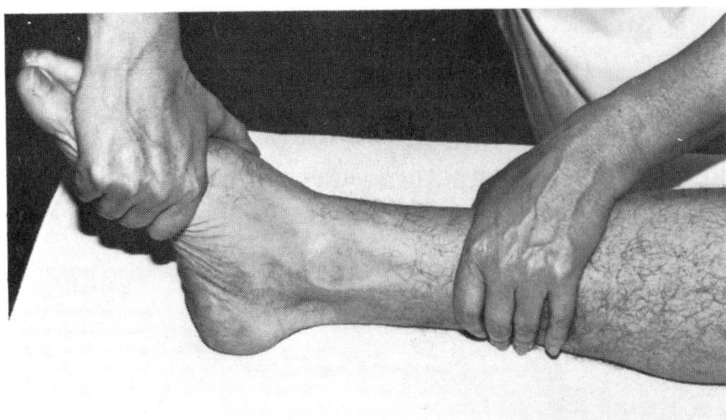

4 and 5 Subject supine.

Stabilize lower third of leg.
Ask subject for foot adduction and supination.
Resistance to the two components of the movement is placed on medial aspect of metatarsus.

Grade 4: Power is sub-normal.
Grade 5: Power is normal.

FLEXOR HALLUCIS LONGUS

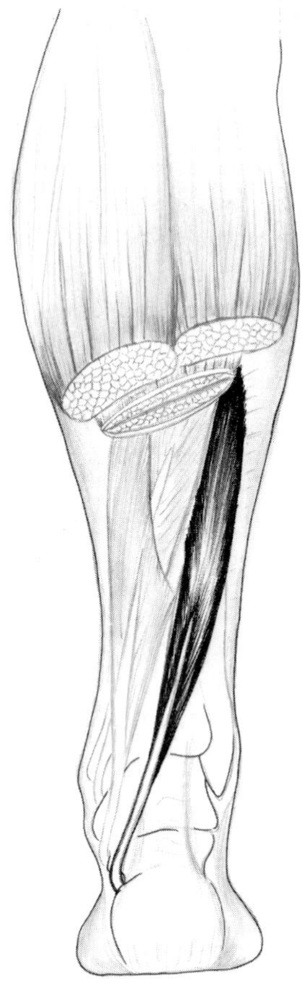

Origin

—Lower three-quarters of posterior surface of fibula.
—Lower posterior surface of interosseous membrane.
—Adjacent intermuscular septa.
Note: At plantar arch level, an expansion passes to flexor digitorum longus tendon.

Insertion Plantar surface of base of distal phalanx of big toe.
See drawing of flexor digitorum, page 384.

Innervation Posterior tibial (l5, S1, s2).

Function

Plantar flexion of distal phalanx on the proximal.
It participates in:
—Plantar flexion of proximal phalanx on first metatarsal bone.
—Adduction, supination and plantar flexion of foot.
Because of the course of its tendon, it plays an important role in supporting the medial plantar arch and in stabilizing the astragalus and the calcaneum.

Flexor hallucis longus muscle activity during gait cycle (J. B. Piera and A. Grossiord)

Stance phase

—Loading response, mid stance, and terminal stance: the flexor digitorum longus quickly followed by the flexor hallucis longus, start firing at loading response in the same manner as the triceps surae from which their activity cannot be dissociated. Furthermore, they stabilize the toes on the ground, and at terminal stance, produce a dynamic eccentric contraction in response to stretch while in dorsiflexion during heel-off.
—Pre-swing: (with triceps surae and flexor digitorum longus). Their action, very important for push-off, remains essentially the same as in the preceding sub-phase. They lock the leg on the foot.

Contracture or tightness Flexed position of distal phalanx.

Deficiency

Difficulty in, or inability of flexing distal phalanx.
Slight instability of gait.

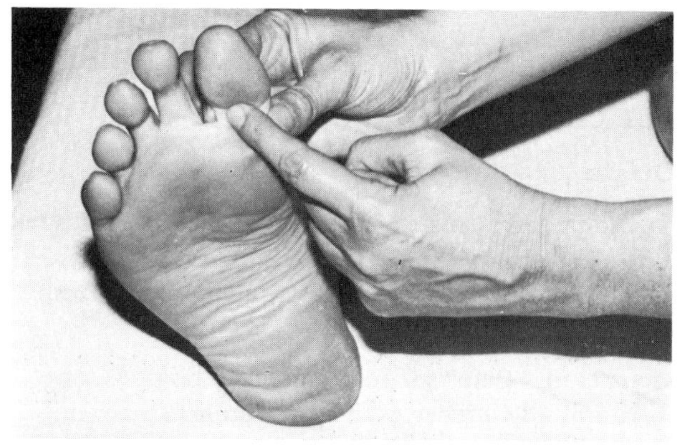

0 and 1 Subject supine.

Foot resting on table in neutral position.
Stabilize sides of proximal phalanx.
Ask subject for plantar flexion of distal phalanx on the proximal phalanx.
The tendon is palpated on plantar surface of metatarsophalangeal joint.

Grade 0: No tendon movement is felt.
Grade 1: Tendon movement is felt but there is no joint movement.

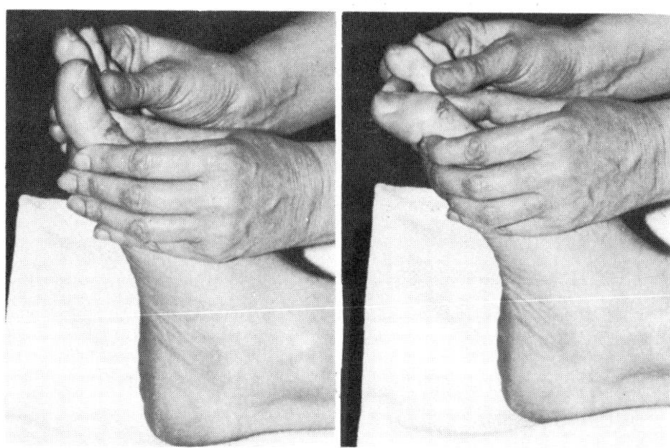

2 Same position.

Same stabilization.
Ask subject for same movement.
Range of movement should be complete.

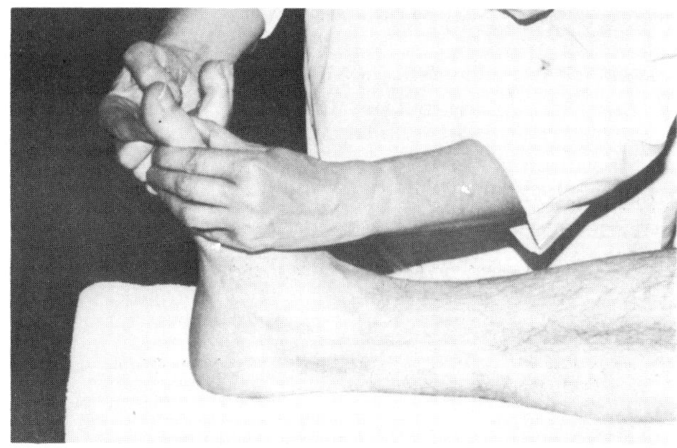

3, 4 and 5 Same position.

Same stabilization.
Ask subject for plantar flexion of distal phalanx on the proximal phalanx.
Resistance to the movement is placed on plantar aspect of distal phalanx.
Resistance is minimal for grade 3 and increases for grades 4 and 5.

Grade 4: Power is sub-normal.
Grade 5: Power is normal.

FLEXOR DIGITORUM LONGUS and QUADRATUS PLANTAE

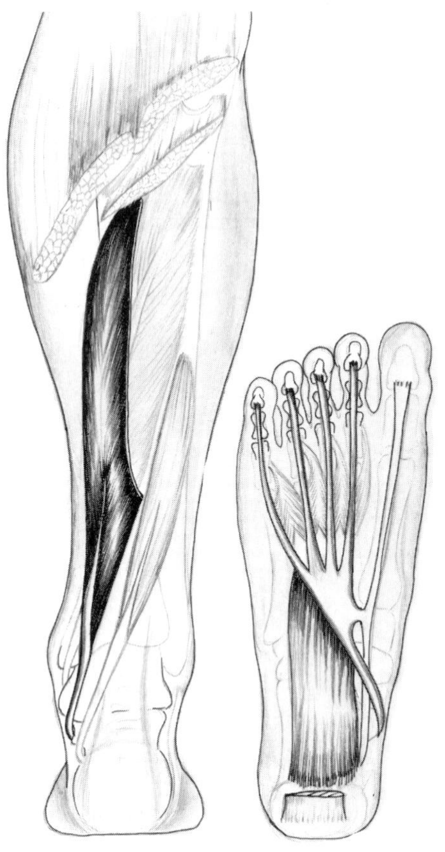

FLEXOR DIGITORUM LONGUS

Origin

—Middle third, posterior surface of tibia, medially to crest which separates it from tibialis posterior.
—Intermuscular septum separating it laterally from tibialis posterior.

Insertion Before dividing into four tendons, on its lateral side it receives the quadratus plantae and sends an expansion to flexor hallucis longus.
The four tendons insert into plantar surface of base of distal phalanges of last four toes.

Innervation Posterior tibial (15, S1).

QUADRATUS PLANTAE

Origin

Lateral head:
—Plantar surface of calcaneum.
—Long calcaneocuboid plantar ligament.
Medial head:
—Medial plantar surface of calcaneum spreading over its medial surface.
—Long calcaneocuboid plantar ligament.

Insertion Deep surface and lateral side of tendon of flexor digitorum longus.

Innervation Lateral plantar (S1, S2).
Medial plantar (15, S1).

Function

Plantar flexion of distal phalanges of last four toes on middle phalanges.
The quadratus plantae stabilizes the flexor digitorum longus by pulling it laterally.
Flexor digitorum longus participates in plantar flexion of middle phalanges on the proximal, and plantar flexion of proximal phalanges on metatarsal bones.
It also participates in adduction, supination and plantar flexion of the foot.
It actively participates in plantar arch support.

Flexor digitorum longus muscle activity during gait cycle

(J. B. Piera and A. Grossiord)

Stance phase

—Loading response: It contracts just before the end of loading response when the sole of the foot is flat on the ground; it starts its action on knee stability by restraining forward displacement of tibia in relation to ankle.
—Mid stance, terminal stance: The flexor digitorum longus, rapidly followed by the flexor hallucis longus, contracts in the same way as the triceps surae from which their action cannot be dissociated. Furthermore, they stabilize the toes on the ground, and at terminal stance produce a dynamic, eccentric contraction, since they are stretched by placement of the toes in dorsiflexion, during heel-off.
—Pre-swing: Toe flexors have a very important action in push-off. Their action remains essentially the same as in the preceding sub-phase; they stabilize the leg on the foot.

Contracture or tightness Flexed position of third phalanges of last four toes leading to toe clawing.

Deficiency Difficulty or impossibility of plantar flexing the third phalanx. Toe instability during gait cycle.

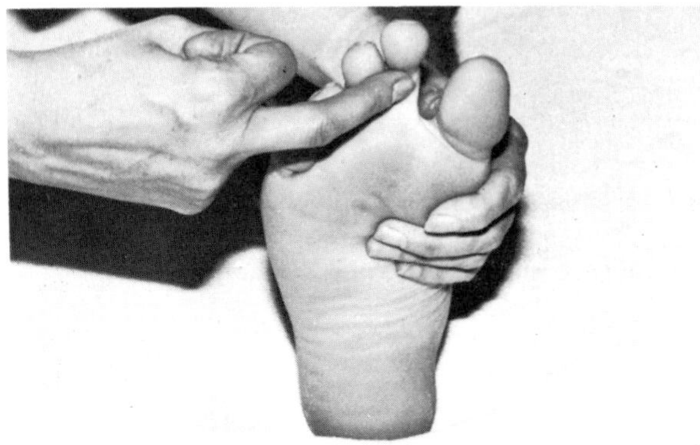

0 and 1 Subject supine.

Lower limb extended.
Foot resting on table in neutral position.
Stabilize sides of middle phalanx.
Ask subject for plantar flexion of distal phalanx on the middle. Tendons are palpated on plantar surface of distal phalanx of last four toes, or between medial malleolus and sustentaculum tali. Do not confuse with tibialis posterior which passes superiorally.

Grade 0: No tendon movement is felt.
Grade 1: Tendon movement is felt but there is no joint movement.

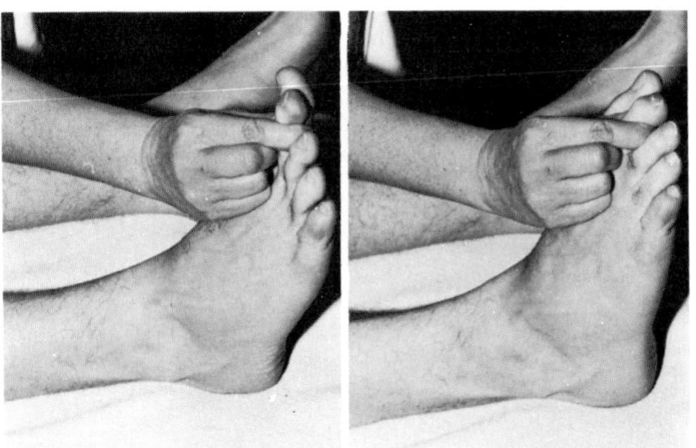

2 Same position.

Same stabilization.
Ask subject for same movement.
Range of movement should be complete.
Note: Toes are examined separately or together.

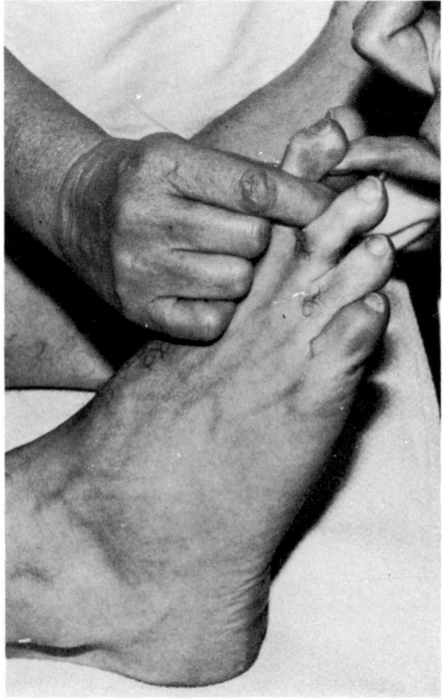

3, 4, and 5 Same position.

Same stabilization and movement.
Resistance to the movement is placed on plantar surface of distal phalanx.
Resistance is minimal for grade 3 and increases for grades 4 and 5.

Grade 4: Power is sub-normal.
Grade 5: Power is normal.

Performance of single joint movement by intrinsic foot muscles is very difficult thus testing of them is unreliable.

ABDUCTOR HALLUCIS

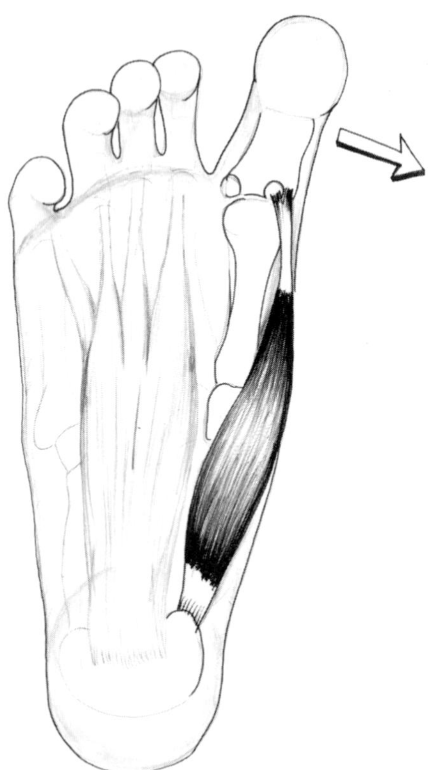

Also called adductor in relation to body axis.

Origin

—Posterior medial process of lower surface of calcaneum.
—Inferior part of medial annular ligament.
—Plantar aponeurosis.

Insertion

—Medial sesamoid bone.
—Plantar surface of base of proximal phalanx of big toe.

Innervation Medial plantar (15, S1).

Function

(Sometimes considered a dorsal interosseus.)
Abducts the proximal phalanx on the first metatarsal bone (in relation to foot axis), and then abducts first metatarsal bone.
Abductor hallucis participates in flexion of proximal phalanx on first metatarsal.
It increases concavity of medial plantar arch.

Abductor hallucis muscle activity during gait cycle

Stance phase

In terminal stance, when the hind part of foot is lifted by triceps surae, it contracts powerfully pressing the big toe against the ground and prolonging the push off motion with the help of the other intrinsics and the flexor hallucis longus.

Contracture or tightness Tendency towards varus deformity of forepart of foot.

Deficiency

Valgus tendency of forepart of foot.
Hallus valgus deformity of big toe.

Range of movement is short. However, on many subjects a well-developed abductor is able to withstand fairly strong resistance.

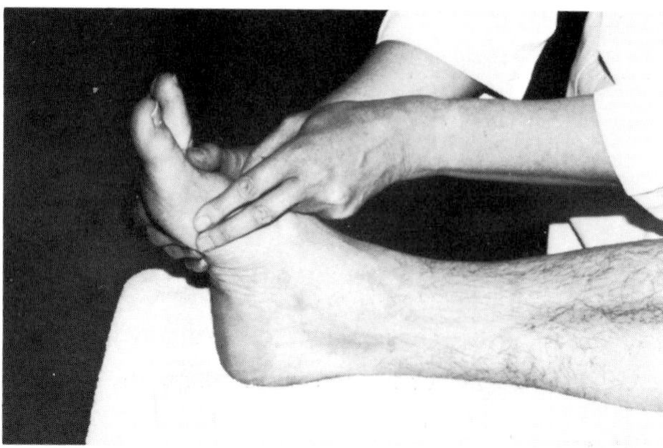

0 and 1 Subject sitting inclined.

Foot resting on table in neutral position.
Stabilize metatarsus leaving first metatarsal bone free.
Ask subject for abduction of big toe.
The abductor hallucis is palpated at medial aspect of first metatarsal bone.

Grade 0: No contraction is felt.
Grade 1: Contraction is felt but there is no joint movement.

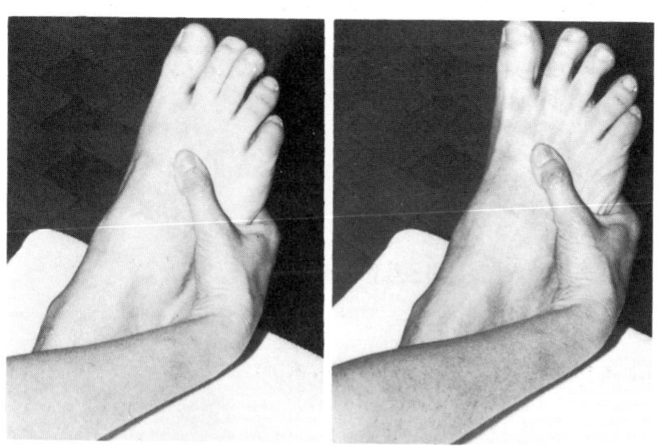

2 Same position.

Same stabilization and movement.
Range of movement should be complete (short in many subjects).

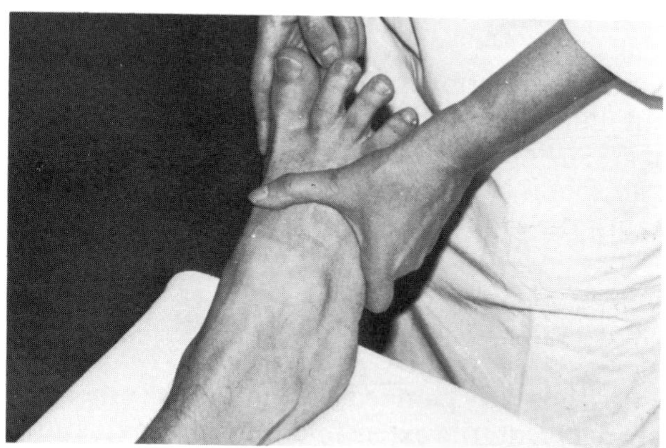

3, 4, and 5 Same position.

Same stabilization and movement.
Resistance to the movement is placed on medial side of proximal phalanx.
Resistance is minimal for grade 3 and increases for grades 4 and 5.

Grade 4: Power is sub-normal.
Grade 5: Power is normal.

FLEXOR HALLUCIS BREVIS

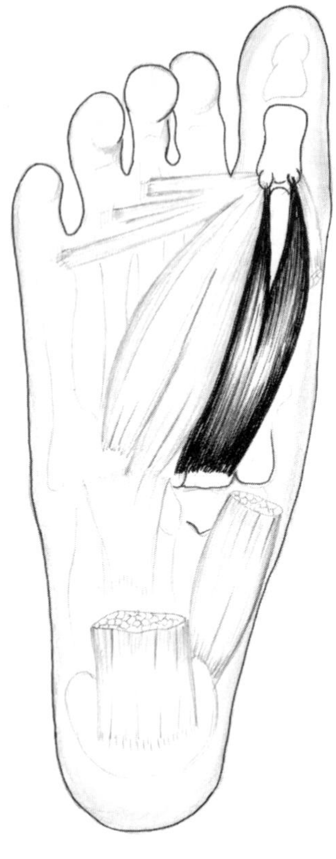

Origin

—Plantar surfaces of second and third cuneiform bones.
—Calcaneocuboid plantar ligament.
—Expansions of tibialis posterior.

Insertion

By two heads.
—Medial or medial sesamoid head, on the tendon of abductor hallucis.
—Lateral or lateral sesamoid head, on the tendon of adductor hallucis.

Innervation Medial plantar (l5, S1).

Function

—Plantar flexion of proximal phalanx on first metatarsal.
—Lateral head lowers the proximal phalanx and pulls it into adduction, but more weakly than the adductor. The medial head pulls it into abduction but more weakly than the abductor hallucis.

Flexor hallucis brevis deepens the medial plantar arch.

During gait, in terminal stance when the hind part of the foot is lifted by the triceps, it contracts with the other intrinsic muscles of the big toe pressing it against the ground, continuing the push-off motion.

Contracture or tightness Marked position of proximal phalanx in plantar flexion.

Deficiency Decrease of proximal phalanx flexion capacity, partly compensated for by the other intrinsics and flexor hallucis longus. If these muscles are deficient, the extensor digitorum longus and brevis can lead to a clawing deformity of the big toe.

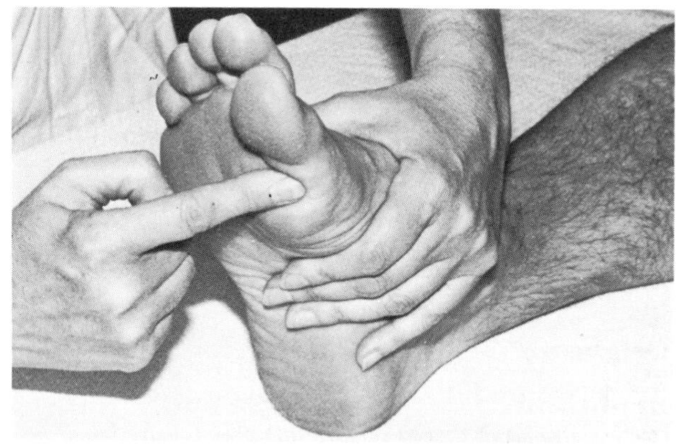

0 and 1 Subject sitting inclined.

Foot resting on table in neutral position.
Stabilize metatarsus.
Ask subject for flexion of proximal phalanx on first metatarsal.
Flexor hallucis brevis of the big toe is palpated on medial portion of plantar surface of first metatarsal bone near metatarsophalangeal joint.

Grade 0: No contraction is felt.
Grade 1: Contraction is felt but there is no joint movement.

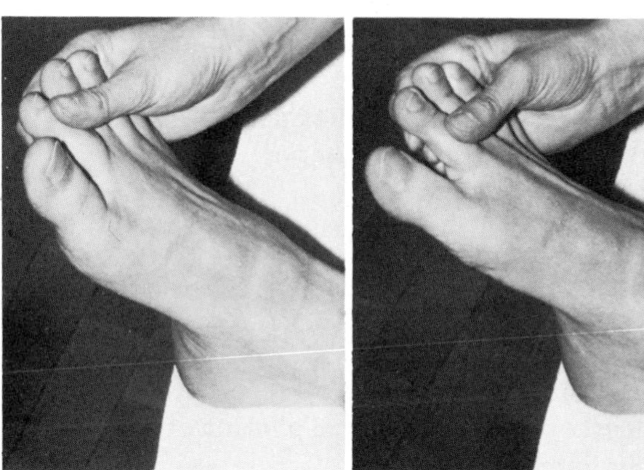

2 Same position.

Same stabilization and movement.
Range of movement should be complete.

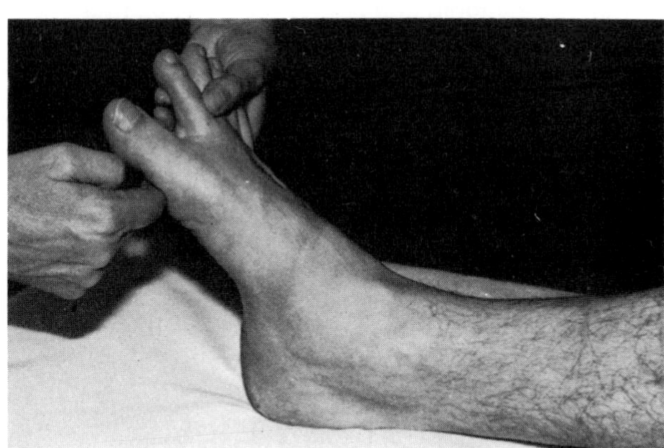

3, 4, and 5 Same position.

Same stabilization and movement.
Resistance to the movement is placed on plantar surface of proximal phalanx.
Resistance is minimal for grade 3 and increases for grades 4 and 5.

Grade 4: Power is sub-normal.
Grade 5: Power is normal.

ADDUCTOR HALLUCIS

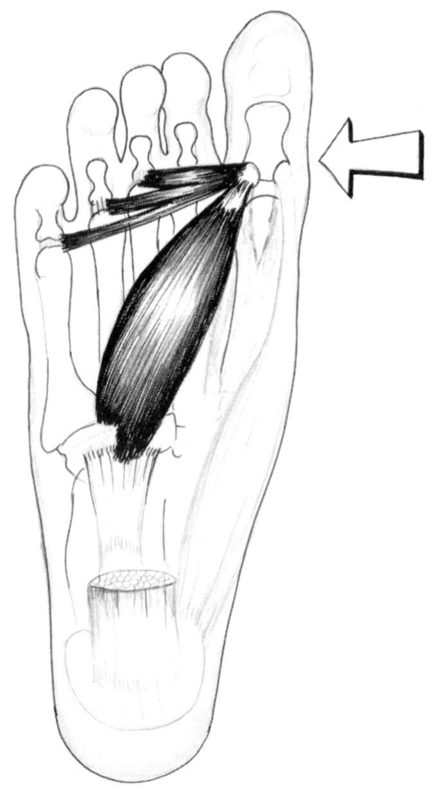

Also called abductor in relation to body axis.

Origin

By two heads.
Oblique head:
—Medial lip of cuboid ridge.
—Plantar surface of third cuneiform bone.
—Plantar surface of base of third and fourth metatarsal bones.
—Calcaneocuboid plantar ligament.
—Expansions of tibialis posterior.
Transverse head:
—Capsules of third, fourth and fifth metatarsophalangeal joints.

Insertion

—Lateral sesamoid and lateral aspect of plantar surface of base of proximal phalanx of big toe.
According to Rouviere:
Oblique head:
—Lateral sesamoid and lateral aspect of plantar surface of base of proximal phalanx
Transverse head:
—Extensor tendon and sheath of flexor hallucis longus.

Innervation Lateral plantar (S1, s2).

Function

Adduction of proximal phalanx on first metatarsal bone, specifically by the transverse head.
It participates in plantar flexion of proximal phalanx and dorsiflexion of distal.
It is the main muscle of the anterior plantar arch which it very actively supports.
During gait, in terminal stance, at heel-off when the hind part of foot is lifted by triceps surae, it contracts with the other intrinsics of the big toe stabilizing it on the ground and continuing push-off.

Contracture or tightness

Marked position of big toe in adduction (abduction in relation to body axis) producing the formation of hallux valgus deformity, especially if the abductor hallucis is paralysed.
Exaggerated hollowing of the anterior plantar arch.

Deficiency

Big toe can be pulled into abduction by the abductor hallucis.
Collapse of the anterior plantar arch.

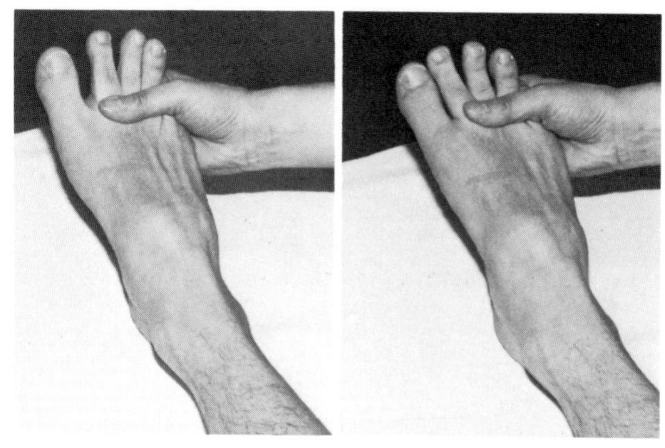

Subject sitting inclined.
Foot resting on table in neutral position.
Stabilize metatarsus leaving first metatarsal bone free.
Ask subject for adduction of big toe.
The adductor is palpated on plantar aspect of first intermetatarsal space near head of first metatarsal bone.
Note if subject is able or unable to perform the movement.
Certain subjects can resist a force placed on lateral side of proximal phalanx. It is difficult to separate it from extensor digitorum brevis, also an adductor (abductor in relation to body axis).

FLEXOR DIGITORUM BREVIS

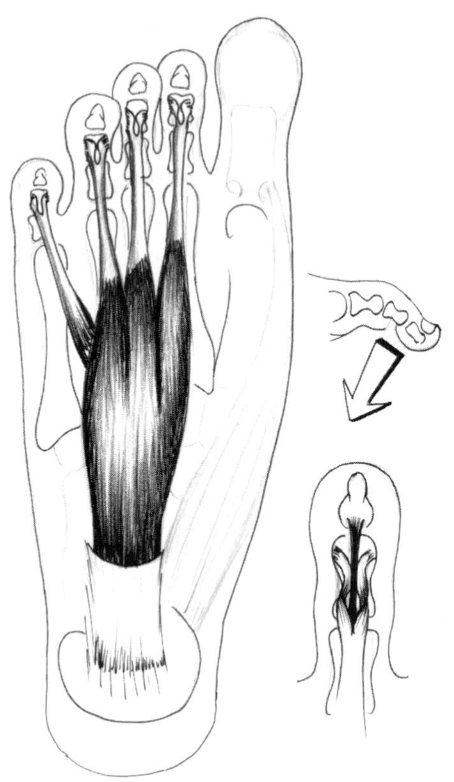

Origin

—Plantar aspect of posterior lateral and posterior medial processes of plantar surface of calcaneum.
—Superficial plantar aponeurosis.
—Adjacent intermuscular septa separating it from other muscles.

Insertion Sides of middle phalanx of last four toes.

Innervation Medial plantar (l5, S1).

Function

Plantar flexion of middle phalanx on the proximal phalanx of last four toes (its function may be compared to that of flexor digitorum superficialis at hand level).
It participates in plantar flexion of proximal phalanx on metatarsal bone.
It plays an important role in supporting and hollowing the plantar arch.
During gait, it contracts from end of loading response to terminal stance.
It constitutes an anatomo-physiological continuation of triceps surae in the foot (see triceps surae muscle activity during gait).

Contracture or tightness

Marked position of middle phalanx in plantar flexion.
If there is an associated deficiency of triceps surae and a predominancy of dorsiflexors, cavus foot deformity may occur.

Deficiency

Decrease of middle phalanx plantar flexion capacity.
Tendency for plantar arch collapse.

Isolated flexion of middle phalanx on the proximal is difficult to achieve.

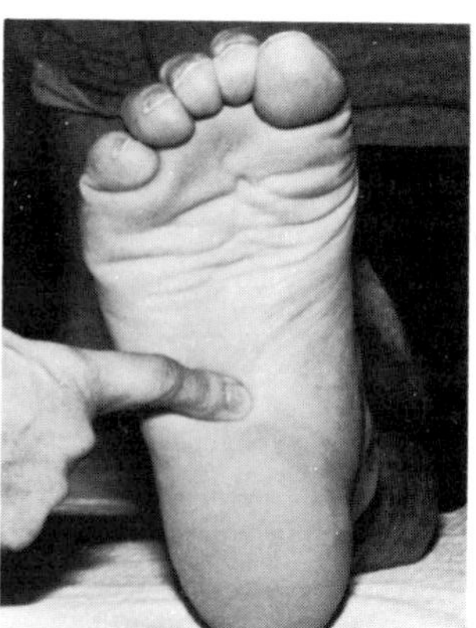

0 and 1 Subject sitting inclined.

Foot resting on table in neutral position.
Ask subject for plantar flexion of middle phalanx of last four toes.
Flexor digitorum brevis is palpated in middle of posterior part of plantar arch.

Grade 0: No contraction is felt.
Grade 1: Contraction is felt but there is no joint movement.

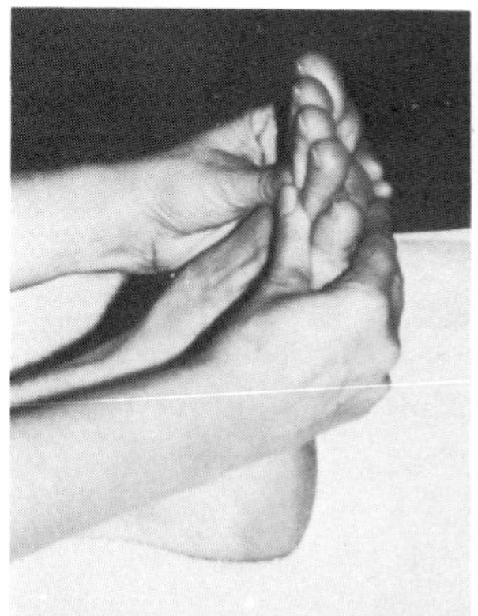

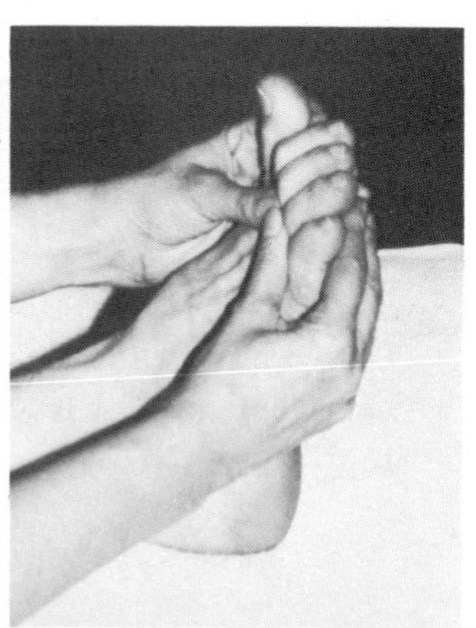

2 Global test.

Same position.
Stabilize proximal phalanges.
Ask subject for same movement, distal phalanx being relaxed.
Range of movement should be complete.
Note: Testing may be global or individual.

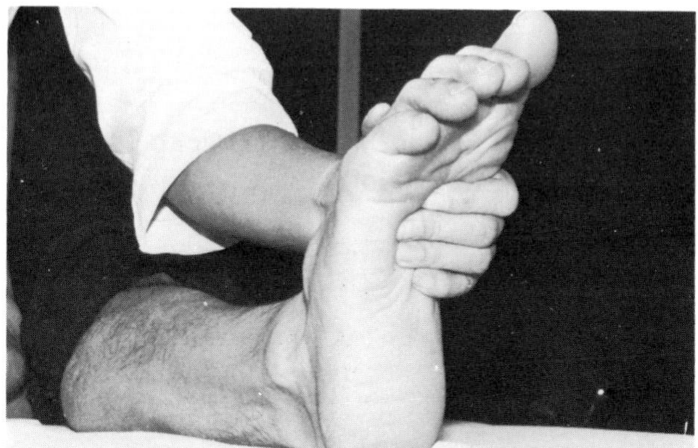

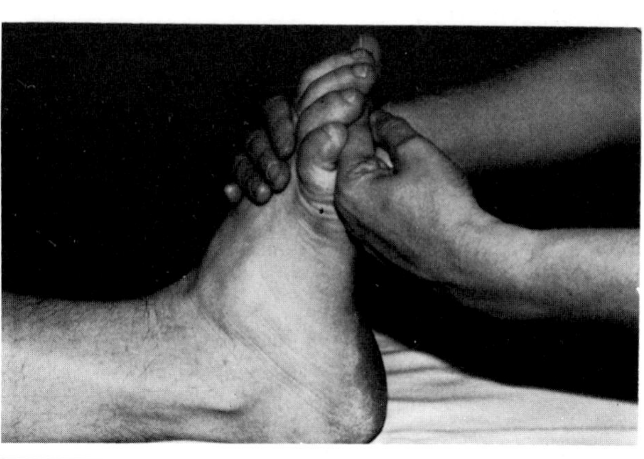

3, 4, and 5 Global test.

Same position.
Individual test is not shown.
Stabilize proximal phalanges.
Ask for same movement.
Resistance to the movement is placed on plantar surface of middle phalanges.
Resistance is minimal for grade 3 and increases for grades 4 and 5.

Grade 4: Power is sub-normal.
Grade 5: Power is normal.

LUMBRICALES PEDIS

They are four in number.

Origin Sides of third, fourth and fifth tendons of flexor digitorum longus except for the first lumbrical which originates from the medial side of flexor tendon to second toe.

Insertion

Medial side of base of proximal phalanx of last four toes.
—Expansion into extensor tendon of same toe.

Innervation

—Medial plantar for the first lumbrical (l5, S1).
—Lateral plantar for second, third, and fourth lumbricales (S1, s2).

Function

Plantar flexion of proximal phalanges of last four toes on metatarsal bones. They stabilize the metatarsophalangeal joints, acting in synergy with the extensor digitorum longus as it dorsiflexes the foot.

Deficiency

Decrease in proximal phalanx plantar flexion capacity.
Clawing position of last four toes due to traction by extensor longus.
They cannot be separated for single muscle testing.

GLOBAL TESTING OF INTEROSSEI AND LUMBRICALES

Subject sitting inclined.
Foot resting on table in neutral position.
Stabilize metatarsus.
Ask subject for plantar flexion of proximal phalanx of last four toes on the metatarsus.
Note if subject is able or unable to perform the movement.
Certain subjects can resist a force placed on plantar surface of proximal phalanges.

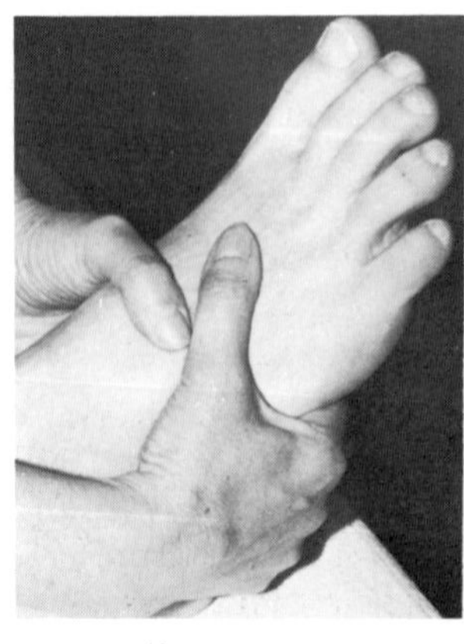

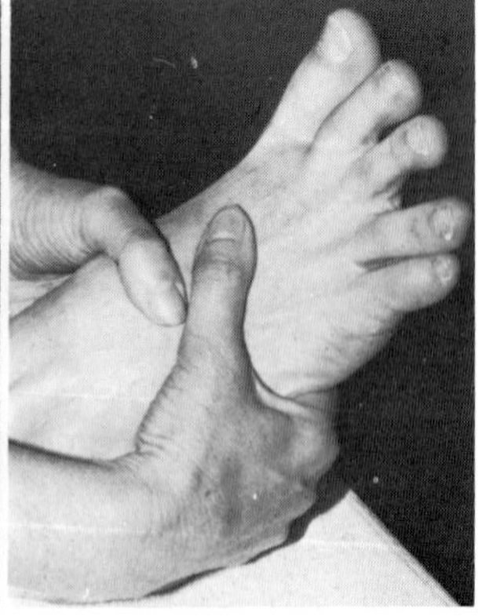

INTEROSSEI PLANTARES

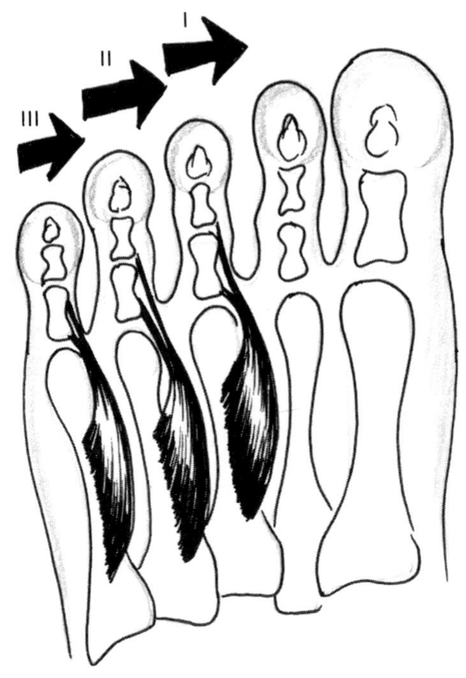

They are three in number.

FIRST PLANTAR INTEROSSEUS

Origin Medial side of third metatarsal bone.

Insertion Medial side of proximal phalanx of third toe.

SECOND PLANTAR INTEROSSEUS

Origin Medial side of fourth metatarsal bone.

Insertion Medial side of proximal phalanx of fourth toe.

THIRD PLANTAR INTEROSSEUS

Origin Medial side of fifth metatarsal bone.

Insertion Medial side of proximal phalanx of fifth toe.
Note: According to Rouviere, the dorsal expansions to the extensor tendons are often absent.

Innervation Lateral plantar (S1, S2, s3).

Function

Adduction of last three toes towards the second.
They participate in plantar flexion of proximal phalanx on corresponding metatarsal, acting with dorsal interossei and lumbricales.
They dorsiflex the last two phalanges weakly.
They stabilize the toes, acting in synergy with the extensor digitorum longus when it dorsiflexes the toes.
During gait they stabilize the metatarsophalangeal joints.

Deficiency

Decrease in plantar flexion capacity of proximal phalanx of last three toes. The proximal phalanx is pulled into dorsiflexion by extensor longus and extensor brevis.
Hammer toe condition.
Decrease in adduction capacity of last three toes.

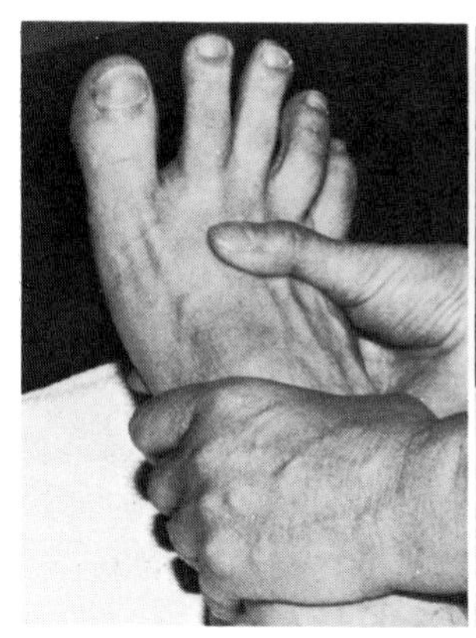

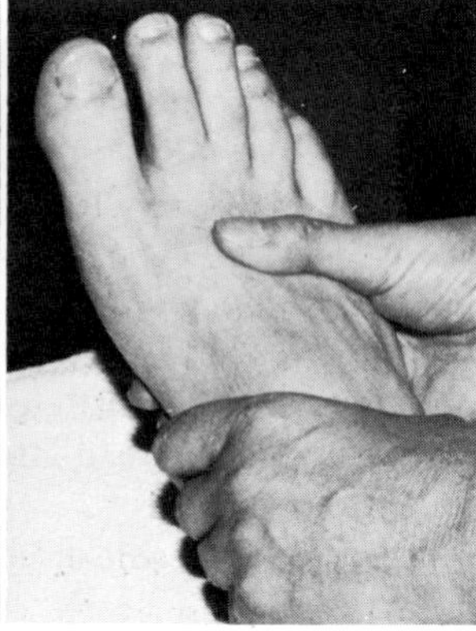

Subject sitting inclined.
Foot resting on table in neutral position.
Stabilize metatarsus.
Ask subject to adduct last three toes.
Palpation is impossible.
Note if subject is able or unable to perform the movement.
Certain subjects can resist a force placed on medial surface of proximal phalanges of last three toes.

INTEROSSEI DORSALES PEDIS

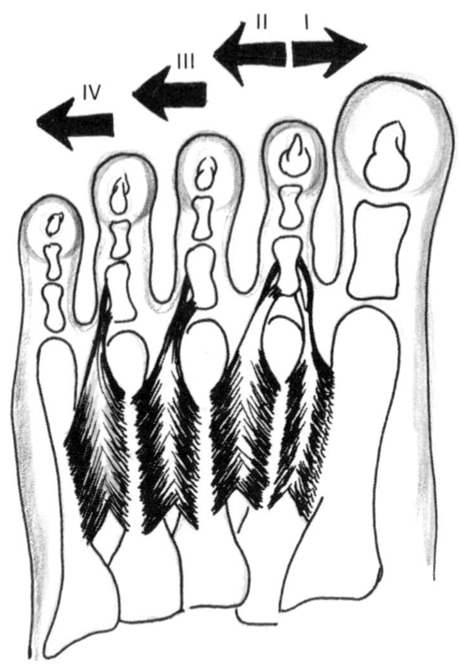

They are four in number.

FIRST DORSAL INTEROSSEUS

Origin In first intermetatarsal space, from adjacent sides of first and second metatarsal bones.

Insertion Medial side of proximal phalanx of second toe.

SECOND DORSAL INTEROSSEUS

Origin In second intermetatarsal space, from adjacent sides of second and third metatarsal bones.

Insertion Lateral side of proximal phalanx of second toe.

THIRD DORSAL INTEROSSEUS

Origin In third intermetatarsal space, from adjacent sides of third and fourth metatarsal bones.

Insertion Lateral side of proximal phalanx of third toe.

FOURTH DORSAL INTEROSSEUS

Origin In fourth intermetatarsal space, from adjacent sides of fourth and fifth metatarsal bones.

Insertion Lateral side of proximal phalanx of fourth toe.

Innervation Lateral plantar (S1, s2, s3).

Function

First dorsal interosseus adducts the second toe.
Second dorsal interosseus abducts the second toe.
Third and fourth dorsal interossei abduct the third and fourth toes.
They participate in plantar flexion of proximal phalanges on the metatarsal bones.
They weakly dorsiflex the last two phalanges.
They stabilize the toes, acting in synergy with the extensor longus in dorsiflexion of the foot.
During gait they stabilize the metatarsophalangeal joints.

Deficiency

Decrease of abduction capacity, partly compensated by extensor digitorum brevis.
Decrease in plantar flexion capacity of middle three toes.
Proximal phalanx is drawn into dorsiflexion by the extensor longus and brevis.
Hammer toe condition.
Subject sitting inclined.
Foot resting on table in neutral position.
Stabilize metatarsus.
Ask subject to adduct and abduct second toe, and to abduct third and fourth toes.
Palpation is possible in intermetatarsal spaces on dorsal surface of foot.
Note if subject is able or unable to perform movement.

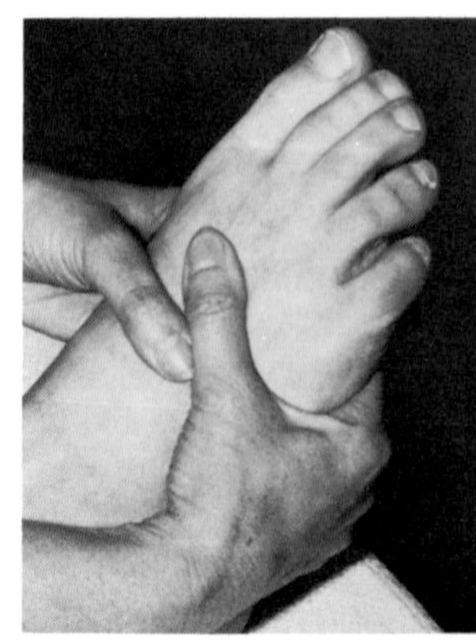

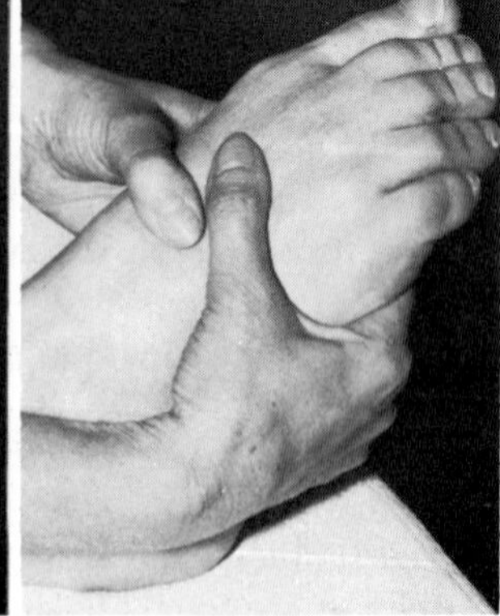

FLEXOR DIGITI MINIMI BREVIS PEDIS

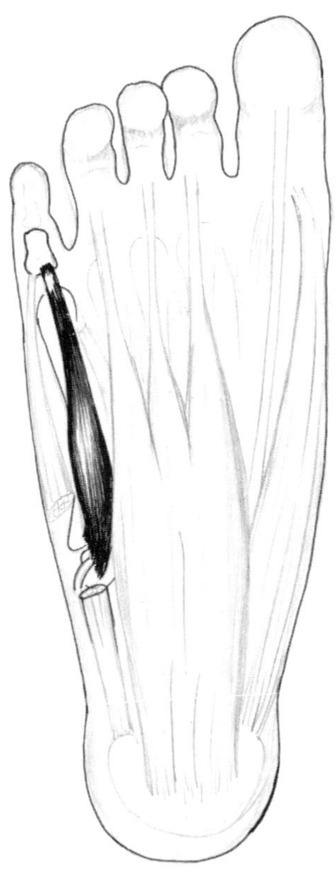

Origin

—Lateral margin of cuboid ridge.
—Sheath of peroneus longus.
—Base of fifth metatarsal bone.

Insertion Plantar surface of base of proximal phalanx of fifth toe.

Innervation Lateral plantar (S1, S2).

Function

Plantar flexion of proximal phalanx of fifth toe on fifth metacarpal bone.
It deviates the proximal phalanx of fifth toe outwards, but more weakly than the abductor digiti minimi.

Subject sitting inclined.
Foot resting on table in neutral position.
Stabilize lateral portion of metatarsus.
Ask subject for plantar flexion of proximal phalanx on fifth metatarsal.
Palpation is possible on median portion of plantar surface of fifth metatarsal near metatarsophalangeal joint.
Note if subject is able or unable to perform the movement.
Certain subjects can resist a force placed on plantar surface of proximal phalanx.

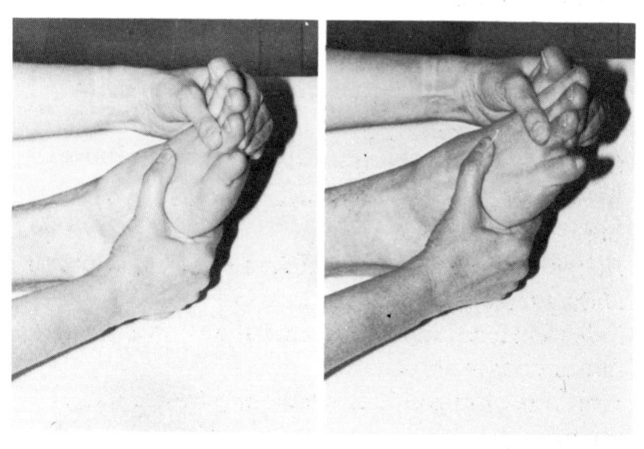

ABDUCTOR DIGITI MINIMI PEDIS

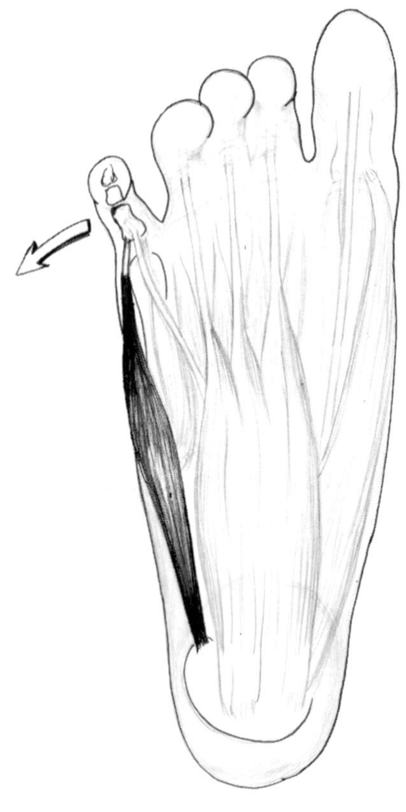

Origin

—Posterior lateral tubercle of plantar surface of calcaneum.
—Plantar aponeurosis.
—Tubercle of base of fifth metatarsal bone.

Insertion Lateral plantar surface of base of proximal phalanx of fifth toe.

Innervation Lateral plantar (S1, S2).

Function

Abducts the proximal phalanx of fifth toe on the fifth metatarsal bone, then slightly abducts the metatarsal bone.
It participates in plantar flexion of proximal phalanx on fifth metatarsal bone.

Subject sitting inclined.
Foot resting on table in neutral position.
Stabilize metatarsus leaving fifth metatarsal free.
Ask subject to abduct his fifth toe.
Palpation is possible on lateral side of fifth metatarsal.
Note if subject is able or unable to perform movement.
Certain subjects can resist a force placed on lateral surface of proximal phalanx.

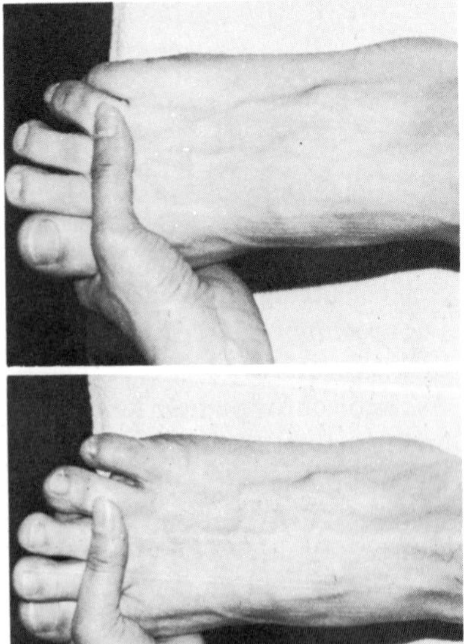

OPPONENS DIGITI MINIMI PEDIS

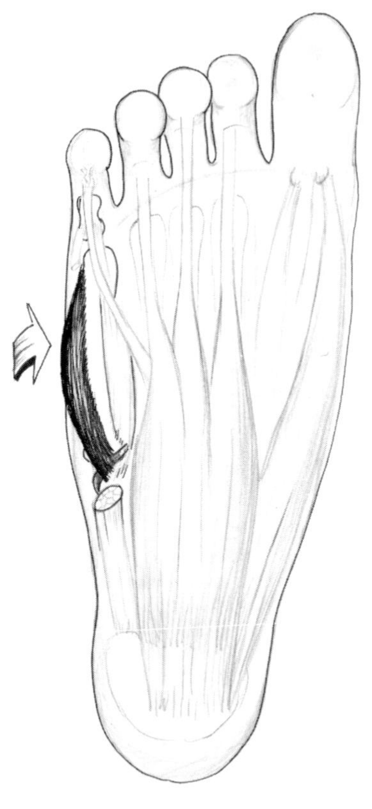

Origin

—Lateral margin of ridge of cuboid.
—Sheath of peroneus longus.

Insertion Lateral side of entire length of fifth metatarsal bone.

Innervation Lateral plantar (S1, S2).

Function

Rotates fifth metatarsal bone from outwards to inwards, and plantar flexes it.

Subject sitting inclined.
Foot resting on table in neutral position.
Stabilize metatarsophalangeal joints, leaving the fifth toe free.
Ask subject to rotate fifth metatarsal bone from outwards to inwards while plantar flexing it.
Palpation is impossible.
Note if subject is able or unable to perform the movement.

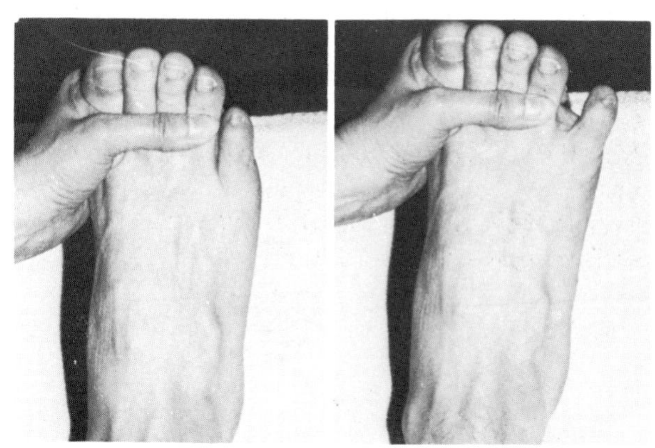

BIBLIOGRAPHY

Benassy J 1963 L'innervation motrice radiculaire des membres. Revue de Rhumatologie

Benassy J 1963 Topographie métamérique de la moelle et de ses racines. Laboratoires Besins

Boubée M 1975 Bilan, analytiques et fonctionnels en rééducation neurologique. Monographie de l'Ecole des Cadres de Bois Larris. (Ed.) Masson, Paris

Boubée M 19— Bilans musculaires. E.M.C. Kinésithérapie, Paris, 26010. A10, C10 et E10 4, 6, 07

Brizon J, Castaing J 19— Les feuillets d'anatomie. Ed. Maloine, Paris

Cambier J, Masson M, Dehin H 1975 Abrégé de neurologie. 2nd edn. Ed. Masson, Paris

Castaing J 1969 Anatomie fonctionnelle de l'appareil locomoteur. Medicorama IV, VI, VII, IX, XII. Ed. E.P.R.I.

Daniels L, Williams M, Worthingham C 1958 Le testing. 2nd edn. Maloine, Paris

Debeyre J 19— Conférences d'anatomie—distribuèes par Vezin. Paris

Dolto B J 1976 Le corps entre les mains. Hermann, Paris

Duchenne de Boulogne 1867 Physiologie des mouvements. Ed. Baillières, Paris

Ducroquet R P, Ducroquet P 1965 La marche et les boiteries. Ed. Masson, Paris

Dumoulin, Bisschop (De) 1971 Electrothérapie. 2nd edn. Maloine, Paris

Efther G 1980 Manuel de technologie de base à l'usage des masseurs-kinésithérapeutes. 2nd edn. Masson, Paris

Fort 1868 Anatomie. Vol 2. 2nd edn.

Godebout (De) J, Ster J, Vidal M 19— Bilans musculaires. E.M.C. Kinésithérapie rééducation fonctionnelle, Paris 26010. A10, C10, D10, E10, L10, M10, P10 318–05. Editions techniques

Godebout (De) J, Ster J, Ster F, Thaury M N, Boussagol H, Grégoire M C, Nicolas C H 1981 Particularités du bilan musculaire dans les atteintes neurologiques traumatiques de l'épaule et du coude. 5e cours de pathologie des nerfs périphériques. May. Montpellier

Godebout (De) J, Ster J, Ster F, Thaury M N, Boussagol H 1981 A propos de l'effet dit 'Steindler'. 5e cours de pathologie des nerfs périphériques. May. Montpellier

Godebout (De) J, Ster J, Ster F, Hubert M N, Boussagol H 1978 Bilan musculaire des grands syndromes neurologiques traumatiques du membre supérieur. 2e cours sur la pathologie des nerfs périphériques. Montpellier. June

Grossiord A, Held J P 1981 Médecine de rééducation. Flammarion, Paris

Hamonet C, Heuleu J N 1978 Abrégé de rééducation fonctionnelle et réadaptation. 2nd edn. Masson, Paris

Held J P, Vernaut J C 19— Paralysie des nerfs périphériques. E.M.C. kinésithèrapie, Paris. 318 0526465, A10

Hinzelin R 1979 Guide pratique d'électromyographie dans les lésions des nerfs périphériques. Ed. Maloine, Paris

Kahle W, Léonard H, Platzer W 1978 Anatomie—Tome 1, Direction. Cabrol C (ed). Flammarion, Paris

Kapandji I A 1972 Physiologie articulaire. Vols 1, 2, 3. Ed. Maloine, Paris

Kendall H O, Wadsworth G E 1974 Les muscles. 2nd edn. Ed. Maloine, Paris

Laplane D 1969 Diagnostic des lésions nerveuse, périphériques. Cahiers Baillières. Ed. Baillières, Paris

Latarjet A 1949 Manuel d'anatomie appliquée à l'education physique. Ed. Doin G (ed)

Lazorthes G 1971 Le système nerveux périphérique. Ed. Masson, Paris

Leliévre J 1961 Pathologie du pied. Ed. Masson, Paris

Levame J 1965 Rééducation des traumatisés de la main. Ed. Archée

Martinez C 1982 Le rachis. Cahiers d'anatomie vivante. Monographie de Bois Larris. Ed. Masson, Paris

Mézière F undated Notes de conference

Perry, Rainaut J J, Lotteau J 1974 Télémetrie de la marche, goniometrie du genou. Revue de Chirurgie Orthopédique 60:97, 107

Piera J B, Grossiord A 19— La marche. E.M.C. Paris, 4 4.02 26013, A10 et A15

Rocher C 1956 Exploration clinique de la fonction musculaire et bilan musculaire. Ed. Masson, Paris

Rouvière H 1962 Anatomie humaine. Vols 1, 2. Ed. Masson, Paris

Scherb R 1952 Kinetisch diagnostische analyse von Gehstorungen, Technik und Resultate der Myokinesigraphie. Ferdinand Enke Verlag, Stuttgart

Tardieu C, Tardieu G, Gagnard L 19— Etude de l'élasticité passive du muscle normal. Applications à certaines raideurs pathologiques. Annales de médecine physique

Trolard 1962 Anatomie humaine. Vol 2: Le tronc. Masson, Paris. Cited by Rouvière, p 509

Trolard 1948 Traite d'anatomie humaine. Vol 1. Doin. Cited by Testut and Latarjet

Viel E, Ogishima H 1977 Rééducation neuro-musculaire à partir de la proprioception. Bases kinésiologiques. Ed. Masson, Paris

Winckler 1962 Anatomie humaine. Vol 2: Le tronc. Masson, Paris. Cited by Rouvière, p 509

Winckler 19— Manuel d'anatomie topographique et fonctionnelle. Masson, Paris

Winckler 1948 Les muscles profonds du dos chez l'homme. Archives Anatomie Histo-ombryologie 31:1

SECTION THREE

Evaluating Motor Function in Disorders of the Central Nervous System

J. P. Bleton and Collaborators

QUALITATIVE EVALUATION OF MUSCLE AND ITS FUNCTION IN PERIPHERAL AND CENTRAL NERVOUS SYSTEM LESIONS IN ADULTS

Muscular atrophy subsequent to a general affection.

GENERAL REMARKS

In all neurological lesions, the criteria of muscular force alone is not sufficient for evaluating muscle.
Muscle efficiency should be observed in its contraction, its release, and in its ability to hold a position. The possibility of reflex, automatic, and voluntary contractions must be seen objectively.

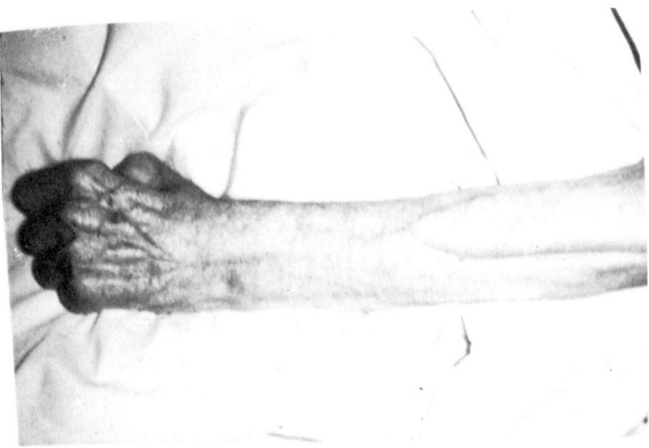

Localized forearm muscular atrophy.

CLINICAL OBSERVATION

Observing the trophic condition

Muscular atrophy is defined by the wasting of muscle mass. This wasting exposes bony relief which is usually invisible, or nearly so. The muscle has lost its curves. When contracted, its morphology is no longer different from that of its resting position. External anatomical characteristics which should be evident, no longer appear.
It is important to relate muscular atrophy to one of its main causes:
—Peripheral neuropathy, either:
- motoneuron lesion
- root lesion
- trunk lesion.

Note: Central lesions do not generally cause important muscular wasting.

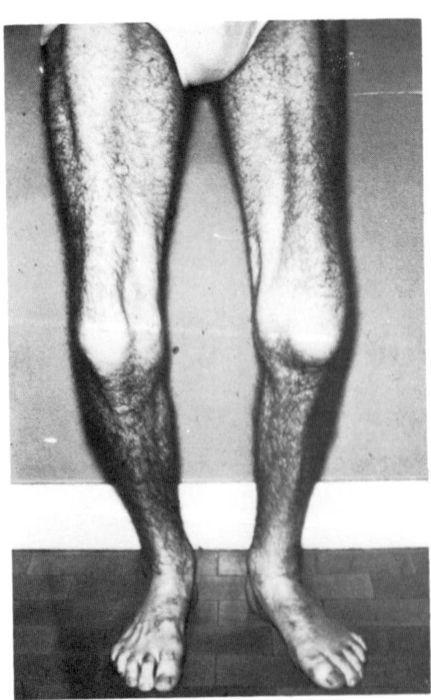

Muscular atrophy localized in vasti muscles.

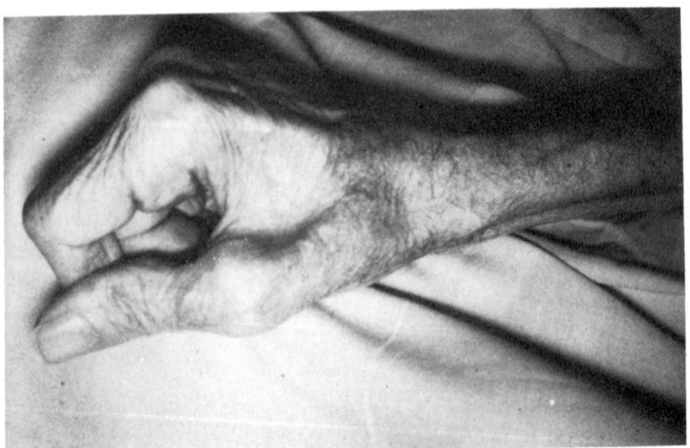

Muscular atrophy of first dorsal interosseus.

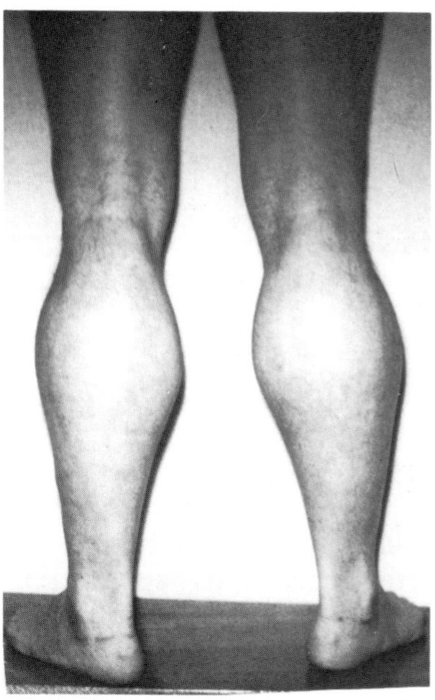

Pseudo-hypertrophic appearance of calves.

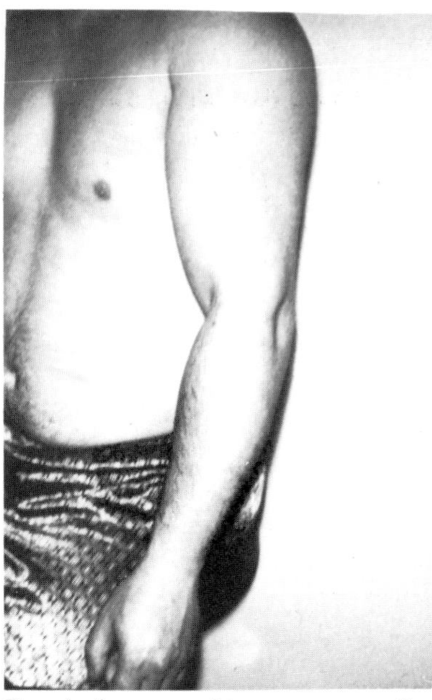

Pseudo-hypertrophic appearance of upper extremity.

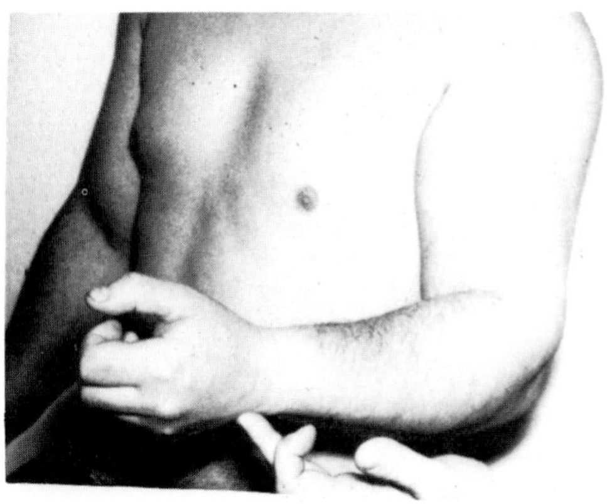

—Disuse, or semi-disuse muscular atrophy caused by immobility or slowing of motor activity capacity (coma, bedridden, Parkinson's disease).
—Loss of body weight secondary to a general affection (progressive disease, metabolic, or other).

Observation is insufficient for evaluating muscular atrophy which may be masked by oedema. Sometimes muscle tissue is infiltrated with cellulo-adipose tissue giving the appearance of pseudo-hypertrophia (as seen in muscular dystrophies and some peripheral neuropathies).

Localized muscular atrophy is quantified by measuring with a tape measure. The measurement obtained is compared with that of the sound side in unilateral lesions. Repeated measuring permits monitoring change.

Observing muscular contraction

A muscle contracts in a homogenous manner when no neurological affection is present. Anarchic and asynchronized contraction of motor units is characterized by fasciculations which give the examiner the impression of brief quivering running along the surface of the muscle. These fasciculations may appear during examination of the contraction (idiomuscular reflex) if the muscle belly is directly tapped, or if a test-tube of iced water is moved along the skin over it. They also appear when scratching the inside of the concha of the ear. These fasciculations usually indicate the presence of pathological motor units mixed with sound motor units (a state encountered in peripheral lesions and particularly in anterior horn cell disease).

The muscle may also be shaken by brief, non-volitional contractions called myoclonia, which may or may not mobilize the part. Myoclonia can be localized in a single muscle group, or diffused and generalized. It is caused by the firing of motoneuron groups, secondary to lesions of various regulatory systems (volitional, automatic or reflex).

Other disturbances of muscular activity:
—choreiform movements
—athetoid movements
—ballic movements
—tremors
will be described later.

Elbow flexion is only possible when assisted.

Observing muscle relaxation

Delay in contraction release often defines a muscular pathology (myotonia, cramps of peripheral neuropathies), or a pathology of motor regulation (hypertonia).

Observing state of fatigue

If repeated contractions lead to fatigue, paresis or myasthenia may be present.

Observing motor performance

Absence of control in displacing a segment may be of several types:
1. Imprecise movement, wandering, undulating from the start to the end of the movement, or going beyond the planned goal. The lesion is probably ataxic.
Paresis of proximal muscles poorly stabilizing the fixators of a limb may cause the same disturbance.
2. Disturbed motor function, where movement may be initiated voluntarily, but control is poor. The muscle with a certain degree of spasticity poorly controls the limb in which movement has been induced. This is evident in a functional activity such as walking where a trigger knee may appear. For example, a hemiplegic's knee during stance phase may be pulled into recurvatum by the hypertonic quadriceps. This characteristic can be demonstrated in the following manner: ask the subject to contract the muscle to be tested so as to mobilize the joint through its complete range of movement, and during this ask him to stop. Involuntary continuation of the movement is proportional to intensity of hypertonia. If a certain degree of spasticity is present, the muscle does not relax immediately, it escapes the patient's control and the movement is continued by spasticity.
3. Speed of displacement may also be uncontrolled. Motor activities may be exaggeratedly slow if the antagonist to the desired movement acts as a brake. This can be observed in Parkinson's disease where both agonist and antagonist muscle groups are hypertonic and when a spastic subject tries to overcome hypertonia. On the other hand movement in the direction of hypertonic reflexes resulting from loss of cortico-spinal inhibition is abnormally fast.

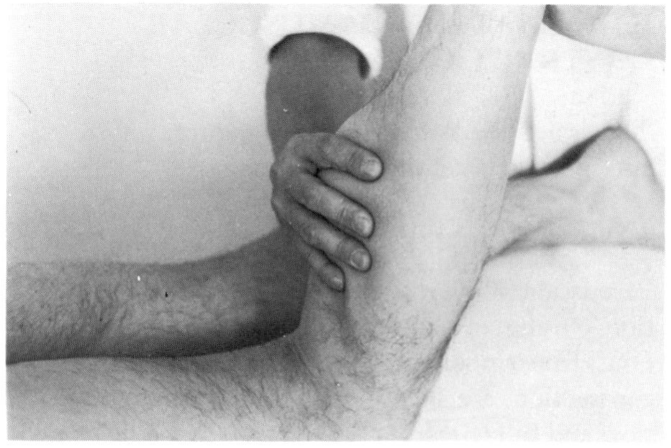

Palpation.

PALPATION

Enables examination of muscular tissue consistency. The hypotonic muscle is easily depressed by digital pressure but it remains passive, giving the impression of being inert when shaking.
The hypertonic muscle, on the contrary, under the examiner's fingers, seems to be in constant contraction and is unaffected by pressure.
Muscular fibrosis eliminates the elastic property of the muscle and transforms its supple consistency into an unextendable, undepressable tissue which has lost its characteristic viscosity.
Pain may appear when muscular tissue is pressed. This is typical of tissues suffering from toxic impregnation (alcoholic polyneuritis).

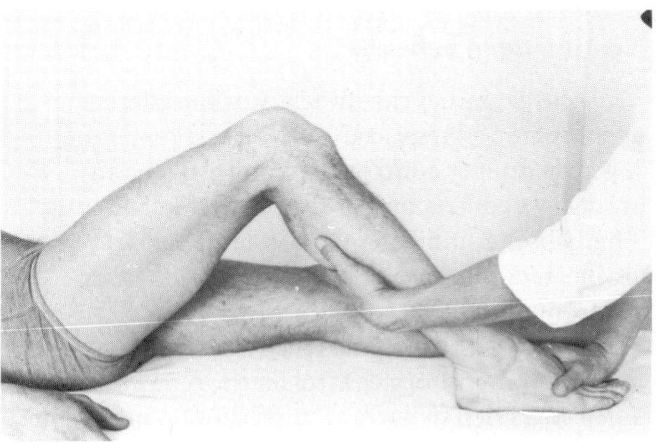

Muscle shaking.

PASSIVE STRETCHING

Demonstrates:
1. The degree of muscle elasticity which permits or limits joint movement, and which is evaluated in muscle stretching tests,
2. Muscle tone.

Hypotonia

Is characterized by:
—Hyperextendability, that is the faculty of the muscle for exaggerated, passive stretching into extreme ranges of motion. It is found in both central and peripheral nerve lesions.
—An agonist-antagonist asynergia. It is demonstrated by shaking the tested limb, inducing exaggerated range of movement of the distal segment. This is typical of cerebellar syndromes.
Peripheral neuropathies demonstrate the two elements of hypotonia: passivity and extendability.

Hypertonia

Is a resistance to passive stretching of the muscle, whether of elastic type as in spasticity, or of plastic type as in rigidity of extra-pyramidal syndromes.

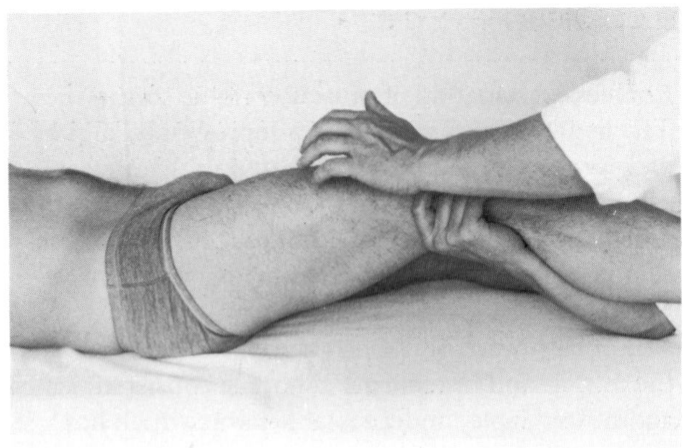

Digital percussion.

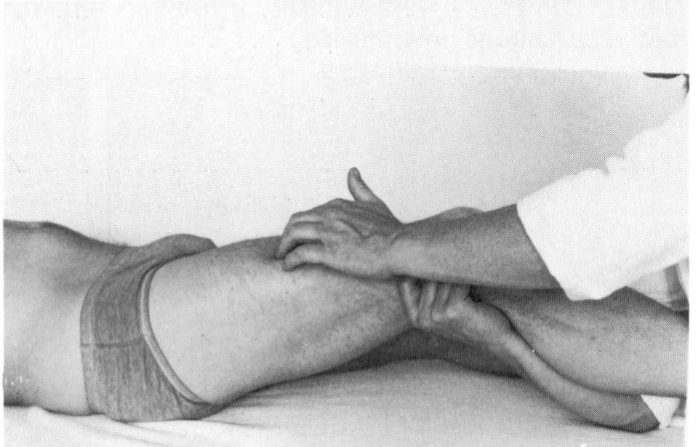

Examination of reflex-muscular contraction.

EXAMINATION OF MUSCULAR CONTRACTION

This is done following different modes: reflex, automatic and voluntary.

Reflex muscular contraction

Percussion of the muscle belly provokes a contraction. On the sound muscle, the response is homogenous. In peripheral nervous lesions reflex-muscular contractions are always exaggerated: they are then slow and full (Sicard's myodiagnostic). Their presence contrasts with muscular atrophy and atonia.

Testing deep reflexes

This test examines the integrity of the afferent-efferent monosynaptic arc as well as qualities of supra medullary control. Tapping the tendon provokes a contraction of the fully stretched muscle. Their absence indicates a break in the sensory-motor arc or predominance of a pyramidal deficiency syndrome.
Their exaggeration signals the release of supra medullary inhibitory control on the reflex arc.
Their inversion denotes an interruption in the reflex arc. In this case the transmission of periosteal vibration leads to contraction of the antagonist, the only muscle capable of responding.
Thus in hemiplegia, percussion of a deficient muscle with abolished reflexes is transmitted to the hypertonic antagonist whose response threshold is lowered.

Automatic contraction

Observed in postural reaction of the body or of a segment.
Tonic adaptations necessary for holding a given posture are automatically regulated.
In certain pathologies muscles may contract voluntarily without fulfilling their stabilizing role. (For example, gluteus medius may voluntarily abduct the hip but not stabilize the pelvis during gait.)
These automatic contractions may be:

—*Either absent or diminished* (hypotonia or paresis): the muscle does not have enough tone to hold the position.

—*Or uncontrolled*: the same postural alignment reaction appears whatever modification is made in posture. Thus in hemiplegia, a single, reflex, inefficient reaction is the response to any change of the centre of gravity, resulting in accentuation of the hemiplegic posture. In other cases the patient is locked in his position and prevented from making new postural adaptations (parkinsonism).

—*Or exaggerated:* postural adaptation goes beyond the patient's needs. Hypermetria which appears during equilibrium reactions creates its own new imbalance (as in cerebellar ataxia).

EXAMINATION OF PRINCIPAL DEEP REFLEXES

Principal deep reflexes	Testing method: area tapped	Spinal level	Contraction obtained
Sternocleidomastoidal reflex	Origin on clavicle	C3–C4	muscle contraction is visible
Coracoid reflex	Delto-pectoral groove	C5	shoulder flexion and adduction
Biceps reflex	Biceps tendon at elbow flexion crease	C5–C6	elbow flexion
Triceps reflex	Tendon above olecranon	C7	elbow extension
Ulnar pronator	Above ulnar styloid process	C8	forearm pronation
Brachioradialis reflex	Brachioradialis tendon on styloid process	C6	elbow flexion
Finger flexors reflex	Tendons at wrist level	C8–T1	finger flexion
Puboadductor reflex	Adductor magnus tendon, medial aspect of knee	L2–L3	hip adduction
Patellar reflex	Patellar tendon	L3	knee extension
Achilles tendon reflex	Achilles tendon	S1	plantar flexion of ankle

DEMONSTRATION OF AUTOMATIC MUSCLE ACTIVITY

Postural reactions provoked by disturbing a patient's balance are a good means of demonstrating the phasic and tonic responses of his muscles.
When the disturbance is slow, a compensatory arching response contrary to the force appears.
When the disturbance is abrupt, free limbs move suddenly towards the force anticipating displacement of the centre of gravity.
When the disturbance is slow and then released, the compensatory arching response is inverted.

Evaluating voluntary contraction

Enables objective appreciation of single muscle activity in manual testing. Muscle functional activity is assessed through global movement or normal gestures: e.g. gait, prehension. In certain conditions such as hemiplegia, the patient may show automatic muscle contraction, but not voluntary. For example, muscular contraction of tibialis anterior may appear in primitive flexion patterns of the lower extremity, but not respond to voluntary command.

Automatic muscular contractions

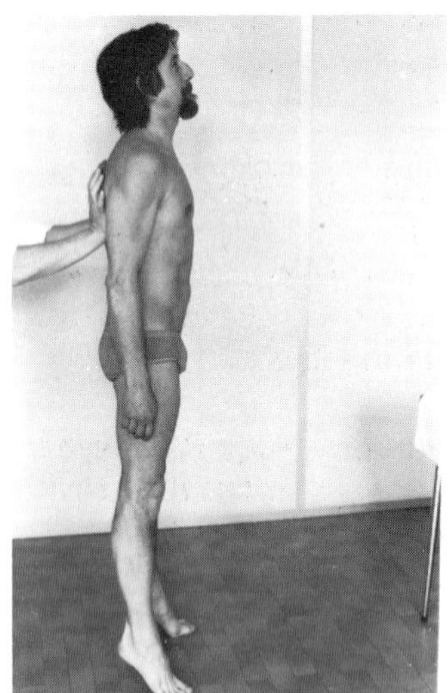

Arching reaction.

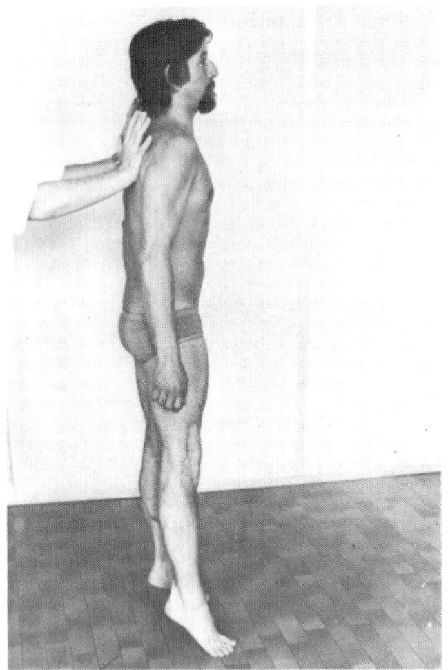

Kinetic reaction of limbs during disturbance of centre of gravity.

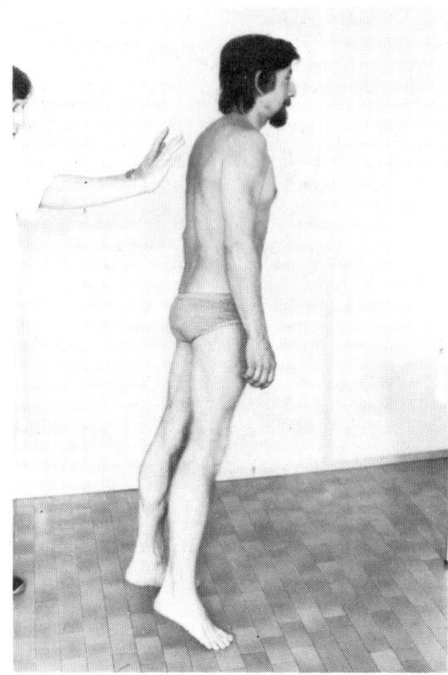

Therapist's push suddenly stopped

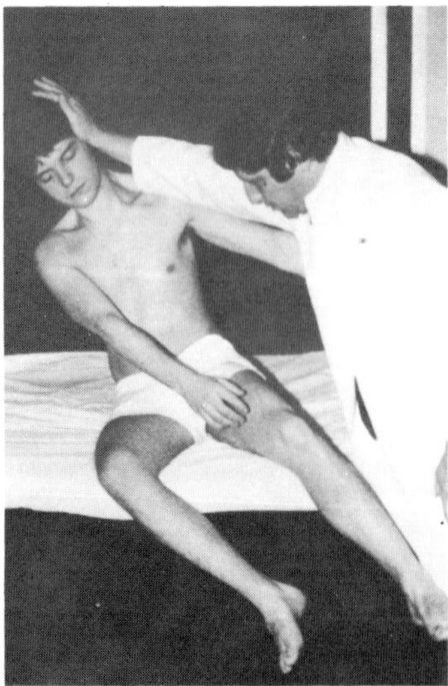

Contraction of antagonists to arching reaction muscle groups.

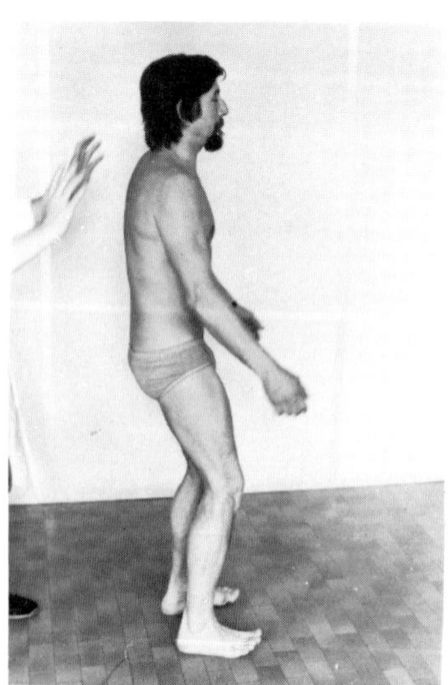

Therapist pushes strongly.

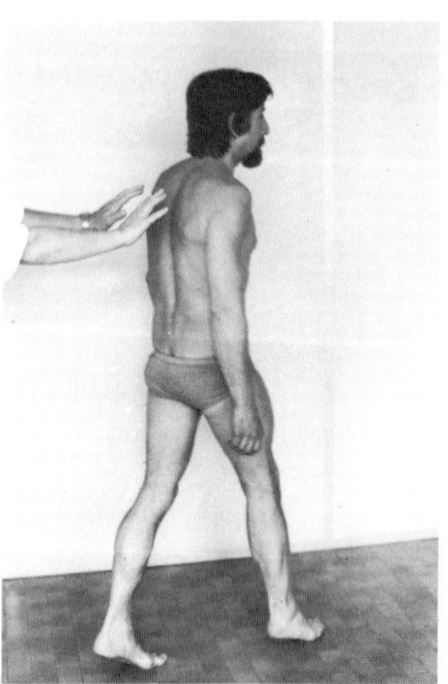

'Parachute' reaction.

EVALUATION OF HAND FUNCTION

GENERAL REMARKS

The use of the hand as a prehensile organ is the result of the evolution of an innate motor activity towards the acquisition of coordinated activity. Babies are born with a reflex prehension: the grasp reflex. Voluntary hand control evolves in two phases:

—*First phase.* During the first six months of life, the ulnar side of the hand is predominant in prehension, which is of the digito-palmar type.

—*Second phase.* During the second ten months of life, objects are drawn towards the radial side of the hand. Manipulation of objects begins at the thenar eminence and progresses towards the distal extremity of the thumb.

90% of all persons are right handed, far outnumbering those left handed. This proportion seems favoured by:

—An anatomical cerebral dominance: the left hemisphere of the brain most frequently dominates.

—An environment of objects of daily use adapted to manipulation by the right-handed.

—Imitation by adults closest to the child who may also force him to use his right hand.

—Cultural environment favours right handedness (Right: right way. Left: left out. Latin root words such as dextra (right hand) gave dexterity, and sinistra (left hand) gave sinister).

Efficiency of the prehensile function requires a certain number of conditions:

—Motor integrity of the hand and the upper extremity.

—Capacity for decoding somaesthetic information transmitted during manipulation. Perception of this sensory input creates stereognosis (recognition of objects by palpation).

—Eye-hand coordination. This enables topo-kinetic pattern organization. The guided course of the hand towards an object follows a preset motor programme founded upon past experience. Once the motor activity is begun, only visual feedback brings in new sensory information likely to alter that course.

—Memorizing hand sensory-motor feedback which enables comparison of manipulation with former experience. This function permits recognition of an object by analogy and adjustment of the grip.

Evaluating the hand with its different grips is a means of appreciating the functional value of contributing muscles.

GRIP BY ACTIVE TENODESIS

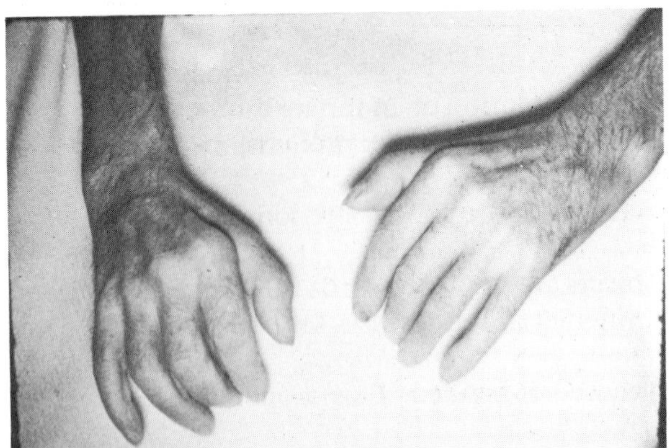

Hands presenting deficiency of flexor muscles.

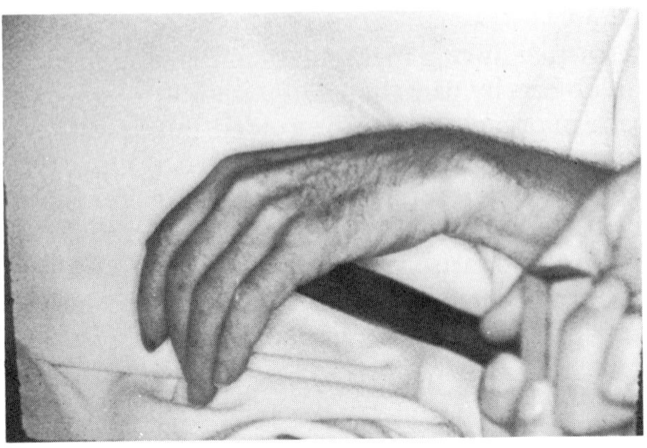

Reach.

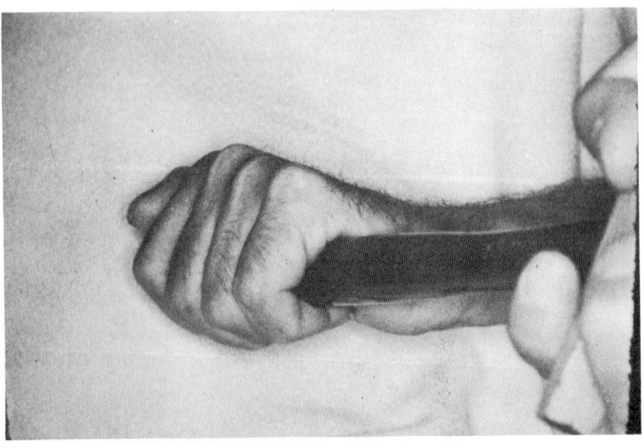

Grip.

PREHENSION MODES FAVOURING THE ULNAR SIDE OF THE HAND

The pseudo-automatic hand

Description: The pseudo-automatic hand enables the achievement of a grip by an active tenodesis effect during certain paralytic disorders of finger motor muscles. If finger flexors present a slight degree of contracture or spasticity, active wrist extension shortens them mechanically drawing fingers into contact with the palmar aspect of the hand.

Examples of use: Pathologies leading to either paralysis of all finger flexors (C8 quadriplegia) or lower brachial plexus lesion (C8–T1 Dejerine-Klumpke), or hypertonia of flexors with no active, volitional, extensor control (spastic hemiplegia or Charcot's disease).

Muscles essential for maintaining the grip:

Wrist extensors:
—Extensor carpi radialis longus and brevis.
—Extensor carpi ulnaris.

Characteristic: Compensatory grip, adaptation to the handicap.

Primary innervation: Radial nerve, C7 root.

Description of reach: The hand hovers over the object and grasps it in a raking sweep. Motor muscles of shoulder and elbow take the hand to object.

Description of release: Gravity flexes the wrist, when contraction of extensors stops, the hand opens automatically by mechanical lengthening of finger flexors.

Note: In certain major hand paralyses, this grip is carried out in supination. The weight of the object in the hand draws the wrist into extension, closing the fingers.

Reach: Hand to object presentation.

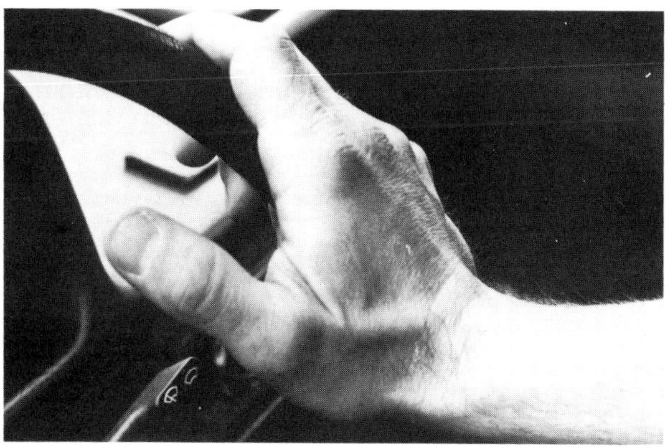

Grip.

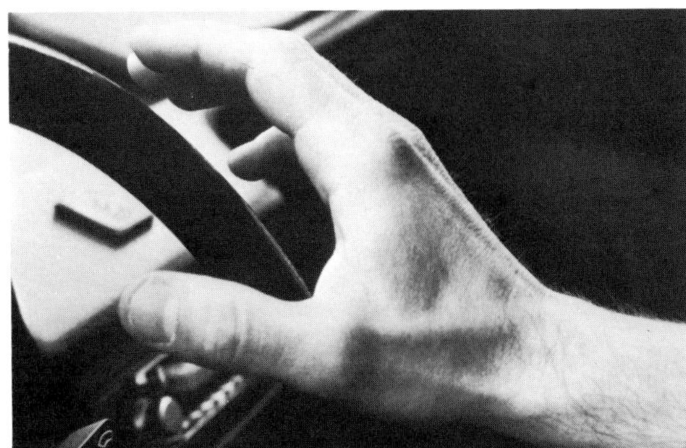

Release.

Digito-palmar grip

Description: Digito-palmar grip enables gripping of a middle-sized object by actively enclosing it between the strongly flexed fingers and the palmar surface of the hand. The thumb does not participate.
Examples of use: Holding the steering wheel of a car.
Muscles essential for maintaining the grip: Flexors of metacarpophalangeal, proximal and distal interphalangeal joints of all digits except thumb: flexor digitorum superficialis and profundus, interossei and lumbricales.
Characteristic: Power grip.
Primary innervation: Ulnar and median nerves, C8–T1 roots.
Description of reach: Extended fingers slide along the object before locking it at the level of proximal phalanges.
Description of release: As flexor activity stops, contraction of extensors of the three phalanges opens the hand. Extensor digitorum communis, interossei and lumbricales.

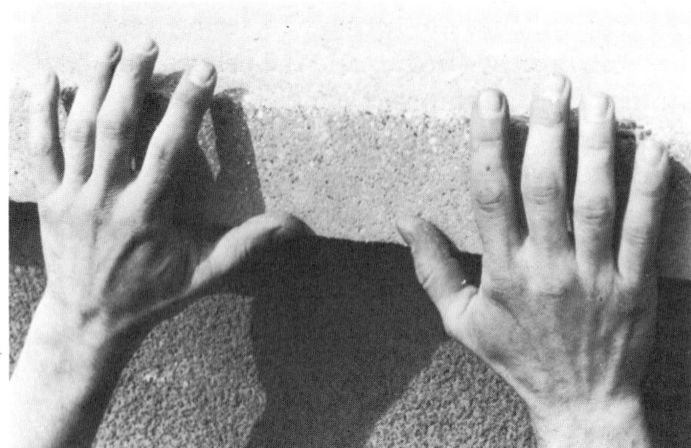

Reach: Hand to object presentation.

Hanging by the hands.

Release.

Hook grip

Description: Hook grip enables hooking the fingers and the last two phalanges in particular, in order to carry heavy loads or to hang by one's hands. Essentially a bilateral activity in which the thumb is not used.

Examples of use:

—Carrying a large, heavy box.
—Hand hold in rock climbing.

Muscles essential for maintaining the grip:

Extrinsic finger flexors:
—Flexor digitorum profundus,
—Flexor digitorum superficialis.
The grip is possible only if all flexors of the upper extremity are in good condition.
Characteristic: Power grip.
Primary innervation: Median nerve with participation of the ulnar, C8–T1 roots.
Description of reach: Slightly flexed fingers approach object so that their fleshy pads make contact with it.
Description of release: As flexor contraction stops, the weight of the load opens the fingers.

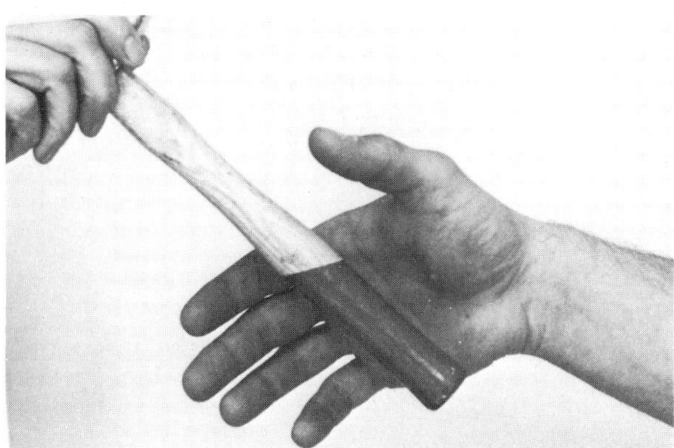

Reach: Hand to object presentation.

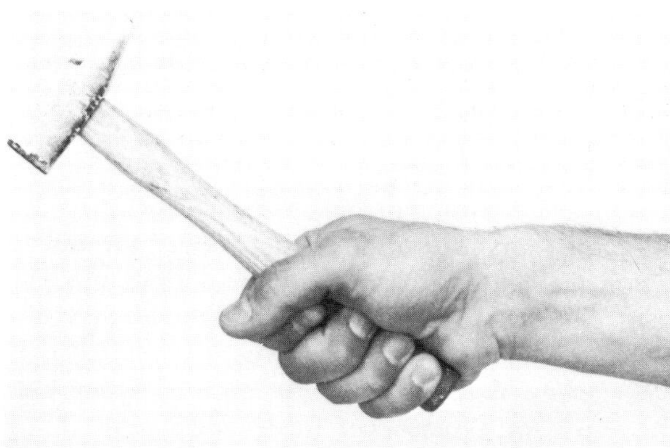

Grip.

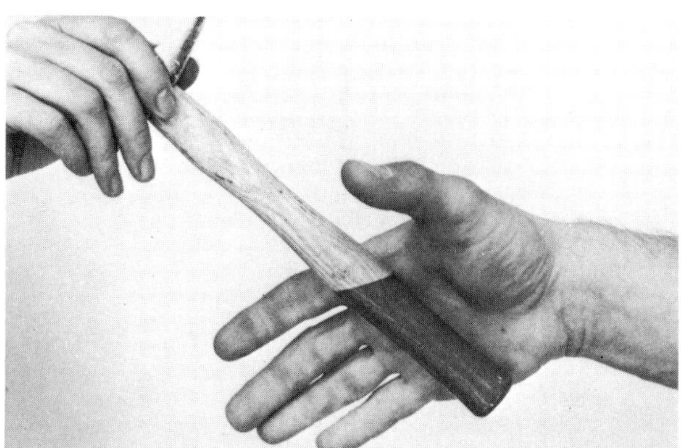

Release.

Tight, full palmar grip

Description: Tight, full palmar prehension enables gripping with the whole hand. The force of the thumb wrapping itself around the object is added to digito-palmar grip to lock the hold.

Examples of use:

—Stabilizes a tool in the hand: sledgehammer, pickaxe handle.
—Hanging by hands from a bar.

Muscles essential for maintaining the grip:

Long fingers:
—Flexor digitorum profundus
—Flexor digitorum superficialis
—Interossei and lumbricales.
Thumb:
—Lateral thenar muscles:
- abductor pollicis brevis
- opponens pollicis
- flexor pollicis brevis, superficial head.

—Flexor pollicis longus.
Characteristic: Power grip.
Primary innervation: Median and ulnar nerves, C8–T1 roots.
Description of reach: Palm of the hand hollows along its vertical axis. Fingers are partially extended by extensor digitorum, interossei and lumbricales. Thumb is extended in palmar abduction by abductor pollicis longus, assisted by extensor pollicis longus and brevis. From this pincer position the hand is ready to grasp the object.
Description of release: As activity of the muscles securing the hold stops, the contraction of finger and thumb extensors releases the object.

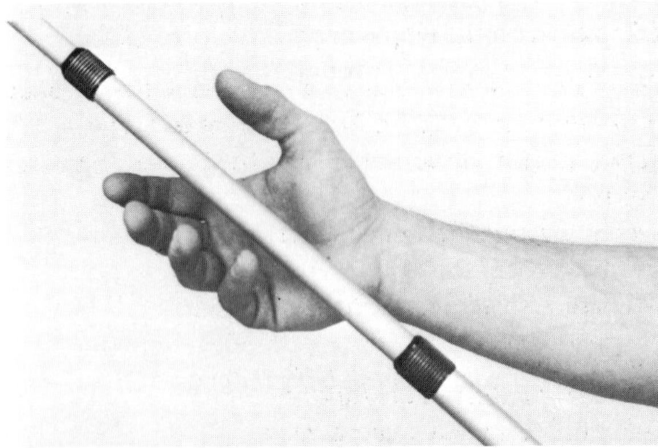

Reach: Hand to object presentation.

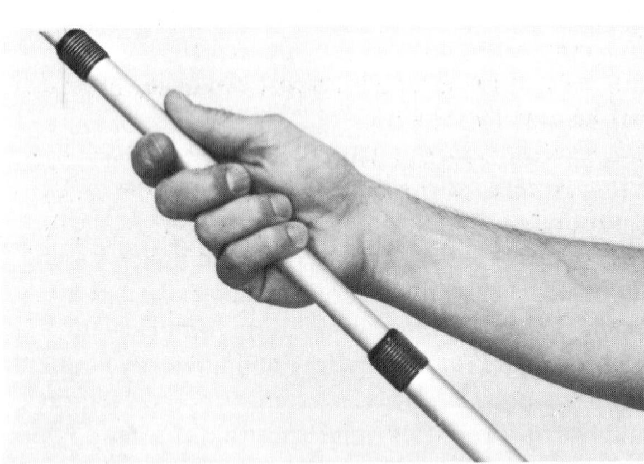

Grip.

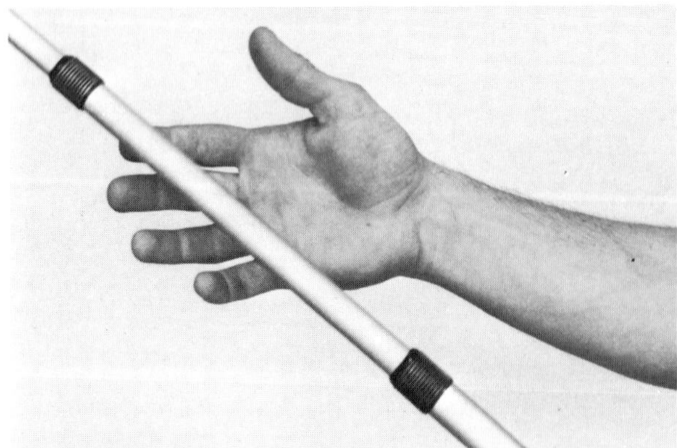

Release.

Directional, full palmar grip

Description: Directional, full palmar grip enables grasp of cylindrical objects upon which the extended thumb, by its palmar surface, transmits the pressure which is directed through the various planes of space.

Examples of use:

—Manipulating a tool with a handle such as a hammer.
—Directing the fisherman's cast with the flick of the wrist.

Muscles essential for maintaining the grip:

—Long fingers:
- Flexor digitorum profundus
- Flexor digitorum superficialis
- Interossei and lumbricales.

—Thumb:
- Adductor pollicis.

Characteristic: Power grip which not only holds the object but also orients it.
Primary innervation: Ulnar and median nerves, C8–T1 roots.
Description of reach: Fingers open partially depending upon size of object to be seized through action of extensor digitorum, interossei and lumbricales. The thumb is drawn into radial abduction by extensor pollicis longus and brevis.
Description of release: As activity of muscles securing the grip stops, extensors of fingers and thumb contract releasing the object.

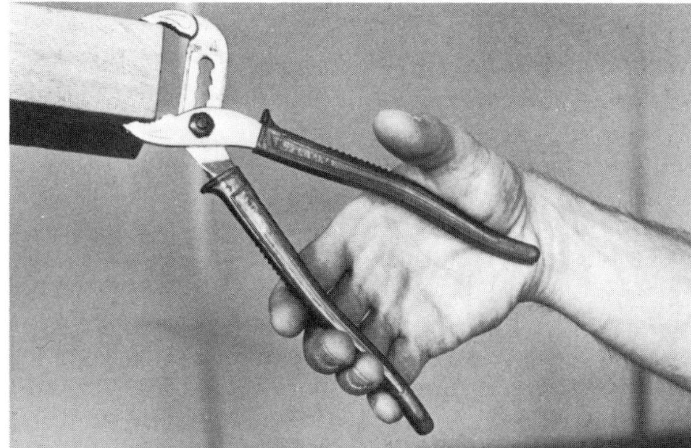

Hand organization in preparation for the grip.

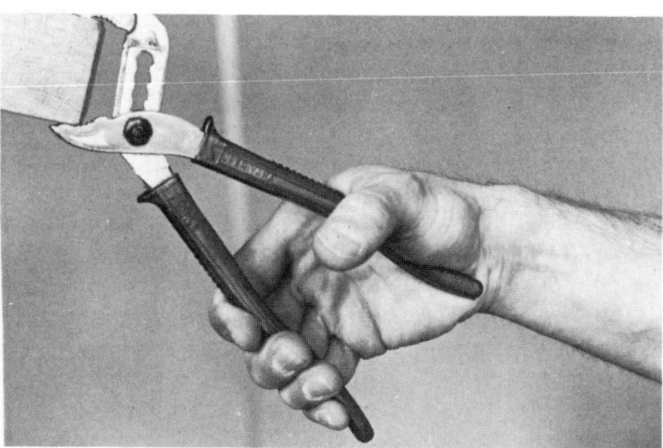

Squeezing.

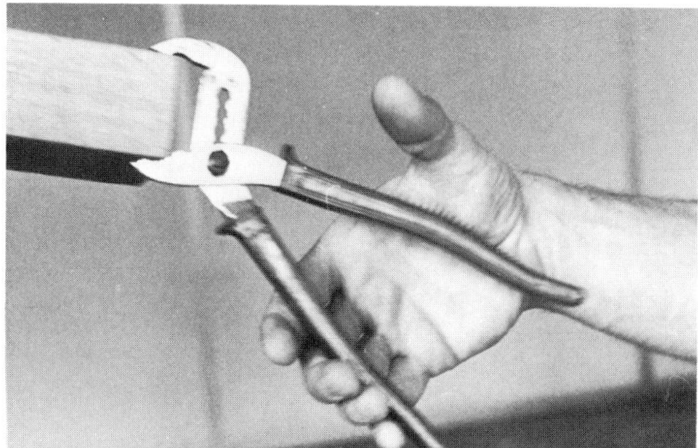

Release.

PREHENSION MODES FAVOURING THE RADIAL SIDE OF THE HAND

Digito-thenar grip

Description: Digito-thenar grip enables exertion of an important compression force on an object held between semi-flexed fingers and the base of the thumb.
Example of use: Handling pliers.

Muscles essential for maintaining the grip:

Long fingers:
—Flexor digitorum profundus
—Flexor digitorum superficialis
—Interossei and lumbricales.
Thumb: Approximation of first metacarpal towards the other digits is accomplished by lateral thenar muscles and particularly abductor pollicis brevis.
Characteristic: Power grip.
Primary innervation: Median and ulnar nerves, C8–T1 roots.
Description of reach: Hand opens, extends, and palm flattens around object.
Description of release: At the end of the power grip, extensors and particularly the extensor digitorum communis open fingers, releasing pressure on the object.

Lumbrical grip (vice grip or 'bec de canard')

Description: Enables extended thumb to approach the other fingers, interphalangeal joints extended, metacarpophalangeal joints flexed. The hand is thus able to mould itself around flat, middle-sized objects.

Examples of use:

—Taking a book from a library shelf.
—Holding a harmonica.

Muscles essential for maintaining the grip:

Thumb: Stability of the length of the thumb and its approximation towards the fingers is carried out by abductor pollicis brevis and opponens pollicis.
Fingers: Metacarpophalangeal joint flexion and interphalangeal joint extension are performed by the interossei and lumbricales.
Characteristic: This is a simple, unrefined, functional grip where the hand seeks neither force nor dexterity, but merely adapts itself to the object.

Primary innervation:

Thumb: median nerve.
Fingers: ulnar nerve.
Roots: C8–T1.
Description of reach: The hand appears open, thumb in palmar abduction, directed towards the fingers. All fingers are slightly flexed.
Description of release: At the end of the hold, the abductor pollicis longus opens the thumb and the extensor digitorum releases finger pressure from the object held.

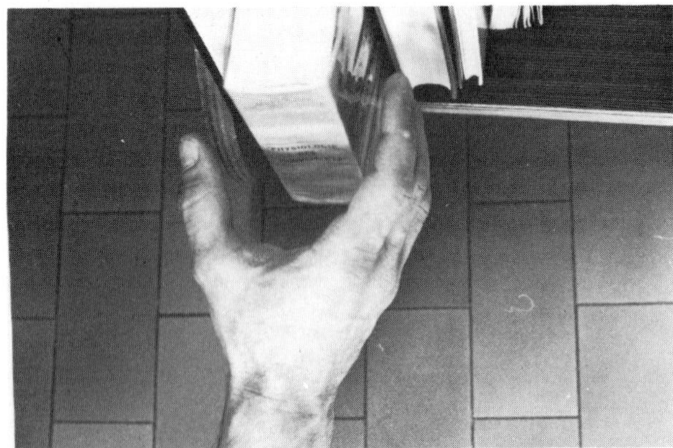

Reach: Hand to object presentation.

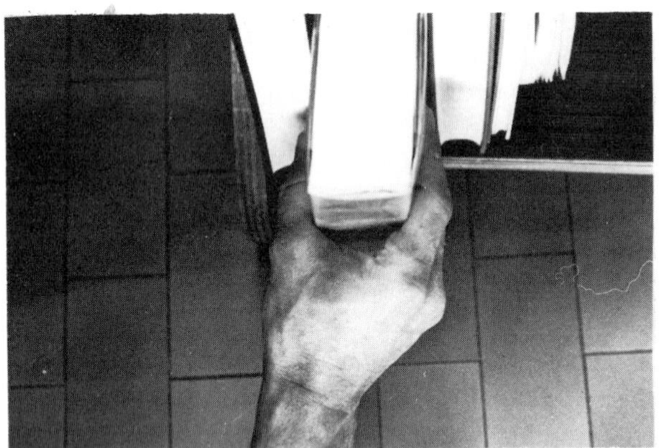

Grip.

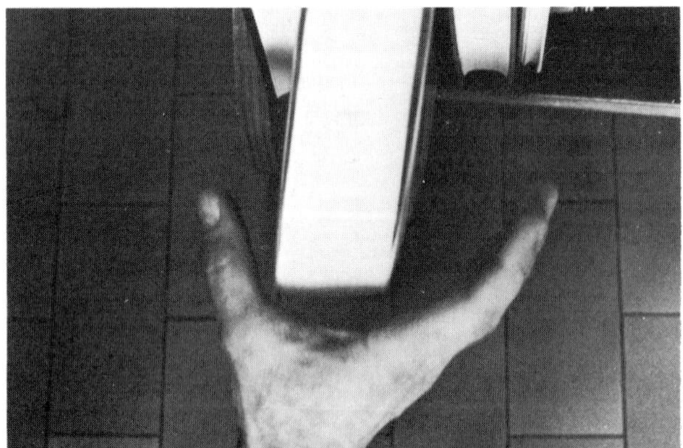

Release.

Reach : Hand to object presentation.

Grip.

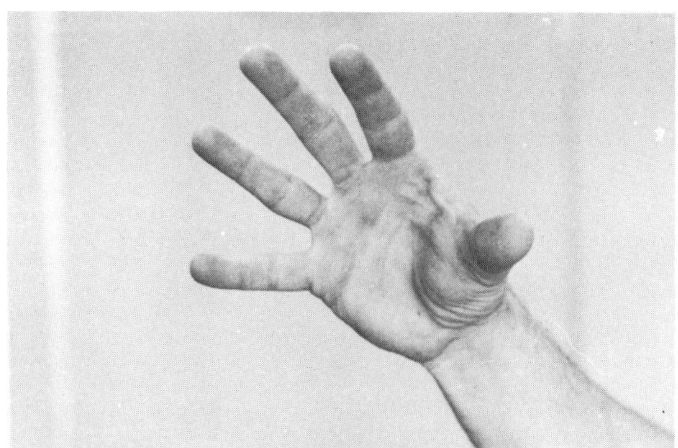

Release.

Spherical palmar grip

Description: Spherical palmar grip enables the hand, fingers spread, to mould itself around a large spherical object.

Examples of use:

—Screwing in a lightbulb.
—Holding a bowl.
—Holding fruit such as an apple.

Muscles essential for holding the grip:

Thumb:
—Lateral thenar muscles, particularly
- abductor pollicis brevis
- opponens pollicis.

Fingers:
—Palmar interossei
—Lumbricales.

Characteristic: Simple, unrefined, circumstantial grip in which the hand cannot develop great power.

Primary innervation:

—Thumb: median nerve
—Fingers: ulnar nerve
—Roots: C8–T1.

Description of reach: The bigger the object, the wider spreads the approaching hand, and the further the fingers open and extend. If the object is particularly voluminous, such as a hand ball, the palm makes contact with it.

Description of release: Hand opening is accomplished by abductor pollicis longus, for the thumb, and by extensor digitorum and dorsal interossei for the fingers.

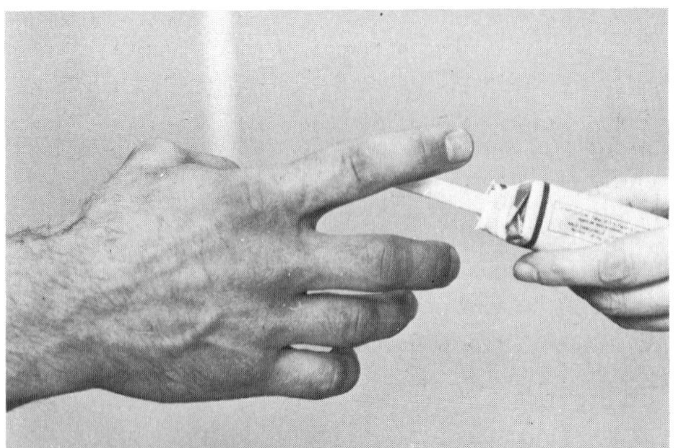

Reach : Hand to object presentation.

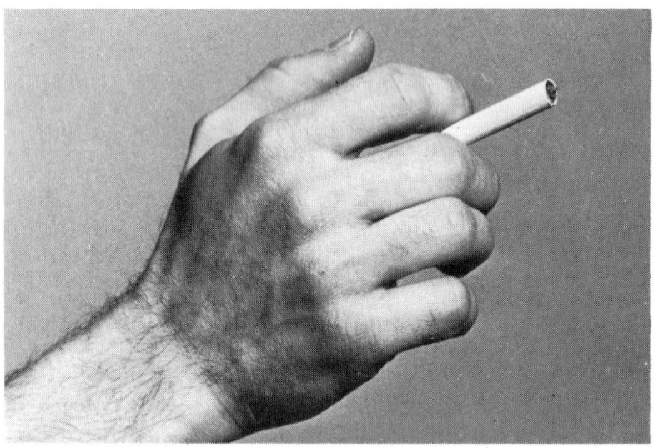

Grip.

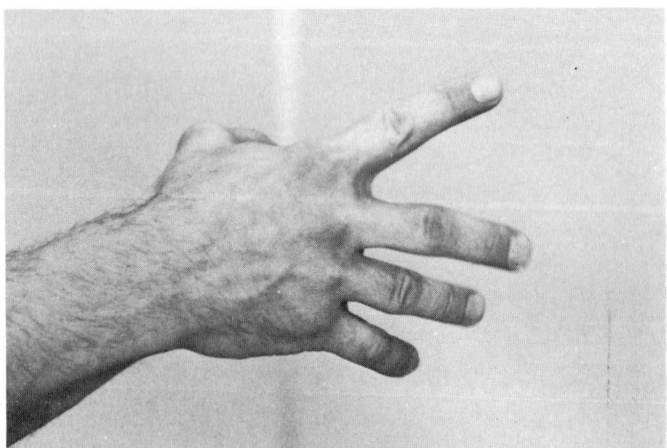

Release.

Interdigital latero-lateral grip

Description: Interdigital latero-lateral grip enables stabilization of an object between the sides of two adjacent fingers. More spontaneously used by the dominant hand between the index and middle fingers, it is nevertheless feasible with other fingers.
Example of use: Holding a cigarette.
Muscles essential for maintaining the grip: Second dorsal and palmar interossei.
Characteristic: Simple, unrefined, hold well adapted to the particular form of the object. It reveals a well-organized sensory pattern. This same hold performed with crossed fingers, the object being held between opposite sides of the adjacent fingers, gives the impression of holding two objects.
Primary innervation: Ulnar nerve, T1 root.
Description of reach: Index and middle fingers present themselves spread like a pair of scissors.
Description of release: Fingers are opened by first and third dorsal interossei.

THUMB TO INDEX GRIP, SUBTERMINO-LATERAL OPPOSITION

Reach: Hand to object presentation.

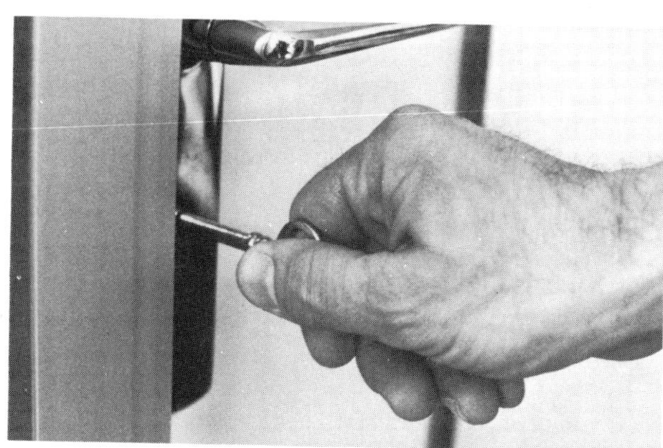

Grip.

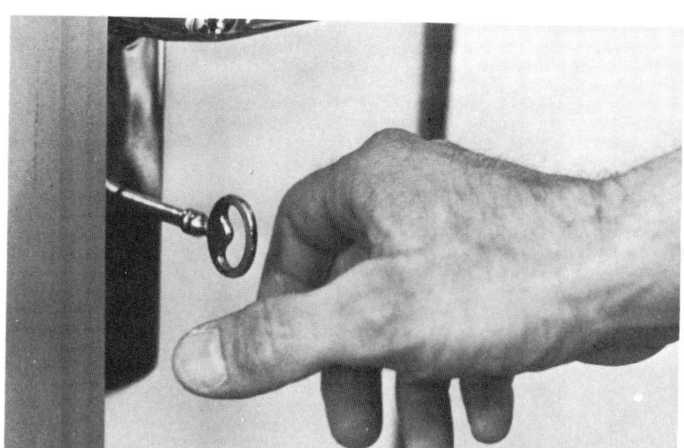

Release.

Thumb to index grip, subtermino-lateral opposition

Description: Enables pressure of pad of thumb against radial aspect of middle phalanx of a slightly flexed index finger. This is stabilized as it presses against the other fingers which are held tightly together. This is an efficient grip for holding flat objects.

Examples of use:

—Turning a key in a lock.
—Cutting a piece of meat.

Muscles essential for maintaining the grip:

—Adductor pollicis.
—Deep head of flexor pollicis brevis.
Characteristic: Thumb to index power grip.

Primary innervation:

—Ulnar nerve, paralysis of which alters this hold. A precision grip then replaces the power grip. Flexor pollicis longus compensates for loss of adductor forming a loop between thumb and index (Froment's sign).
—Roots: C8–T1.
Description of reach: Thumb is in radial abduction, metacarpophalangeal and interphalangeal joints extended. The other fingers remain tightly together, slightly flexed.
Description of release: The object is freed as contraction of the medial thenar muscles stops and the thumb returns to a slightly outspread position.

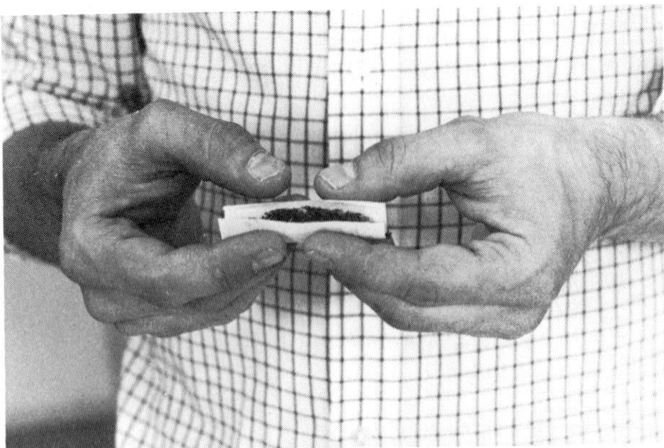

Hand organization in preparation for the grip.

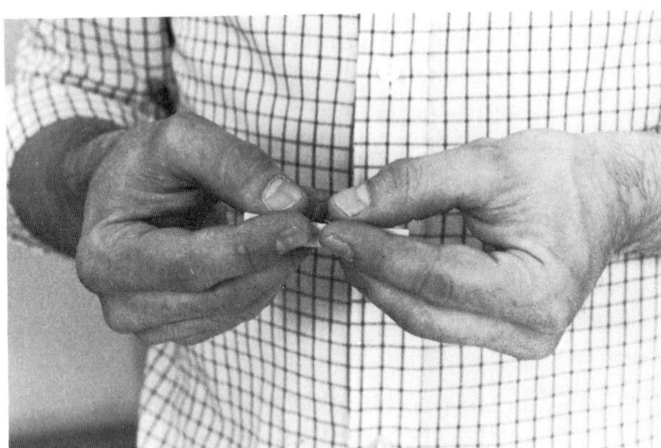

Grip.

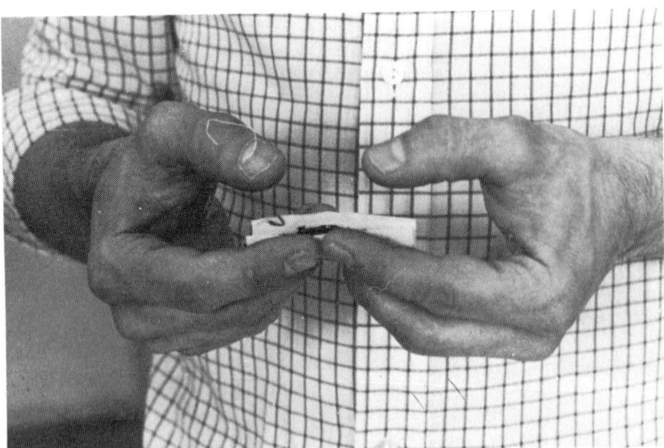

Release.

Tridigital grip (three jaw chuck grip)

Description: Tridigital grip enables thumb opposition to index and middle fingers. Fingers not involved commonly remain folded.

Examples of use:

—Writing, by dominant hand.
—Rolling a cigarette, bimanual activity.

Muscles essential for maintaining the grip:

This grip is performed by simultaneous contraction of: Thumb, lateral thenar muscles:
—Opponens pollicis.
—Abductor pollicis brevis.
—Flexor pollicis brevis, superficial head.
Index and middle fingers:
—Interossei.
—Lumbricales.
Note: During writing, flexor pollicis longus and flexor digitorum superficialis, index and middle fingers flexors, are responsible for pen movements.
Characteristic: Precision grip permitting manipulation in a closed, mobile circumference.

Primary innervation:

—Median nerve is responsible for sensory innervation of cutaneous zone in contact with the object.
—Ulnar nerve.
—Roots: C8–T1.
Description of reach: Thumb is in slight anterior position. Its pad faces the index and middle fingers which are extended in slight abduction.
Description of release: Simultaneous contractions of extensor digitorum, interossei and lumbricales of the index and middle fingers, and extensor pollicis longus separate finger pads from the object.

THUMB TO INDEX GRIP, TERMINAL OPPOSITION

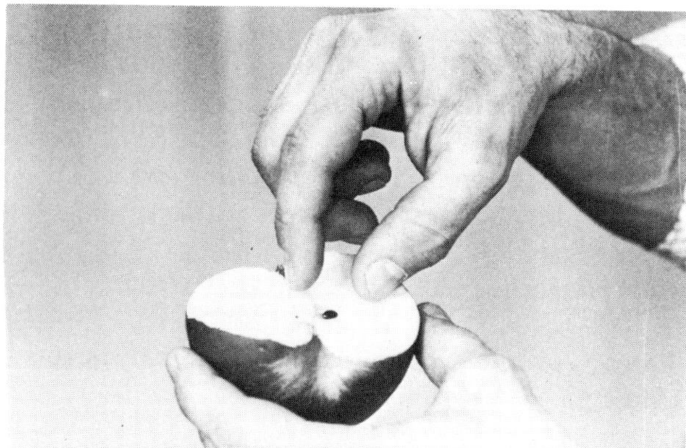

Reach: Hand to object presentation.

Grip.

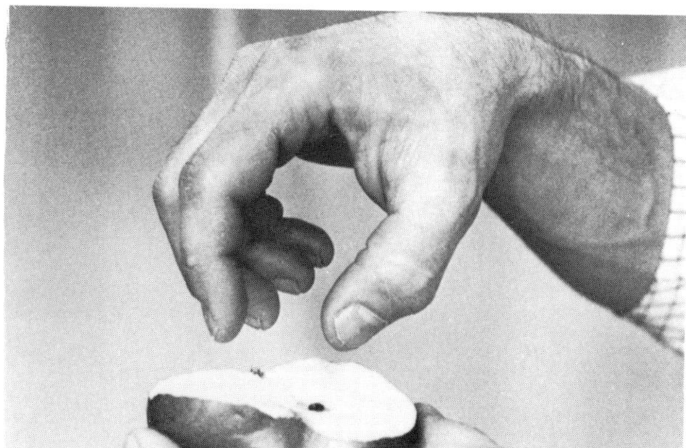

Release.

Thumb to index grip, terminal opposition

Description: Enables holding or picking up minute objects with contact of pads adjacent to nails of thumb and index fingers. In order to place these zones in contact, the two fingers form a nicely rounded loop.
Example of use: Threading a needle.

Muscles essential for maintaining the grip:

Thumb:
—Lateral thenar muscles.
—Flexor pollicis longus.
Index:
—Interossei and lumbricales.
—Flexor digitorum superficialis.
—Flexor digitorum profundus.
Characteristic: The most precise grip.

Primary innervation:

—Median nerve: The quality of this grip indicates the integrity of this nerve.
—Roots: C8–T1.
Description of reach: In order to pick up a very tiny object, the two fingers present themselves with a slight separation between their two distal phalanges. The hold is tightened by flexion of the metacarpophalangeal joint of the index or by adduction of the trapezio-metacarpal joint of the thumb.
Description of releases: Simultaneous extension of index distal interphalangeal and thumb interphalangeal joints releases the object.

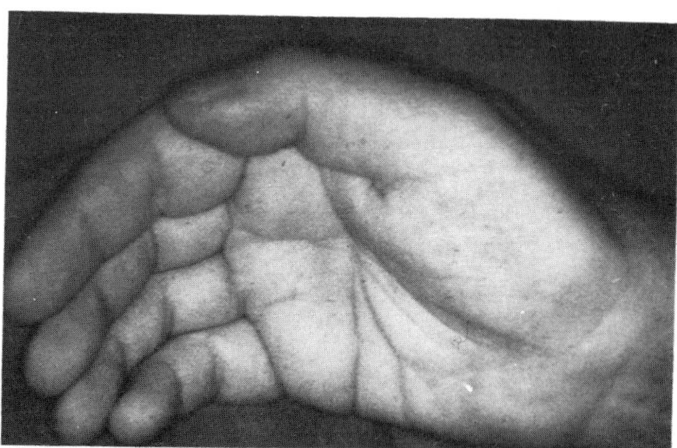

'Ape's hand' Main de Singe: Paralysis of thenar muscles, terminal opposition is impossible.

THUMB TO INDEX GRIP, SUB-TERMINAL OPPOSITION

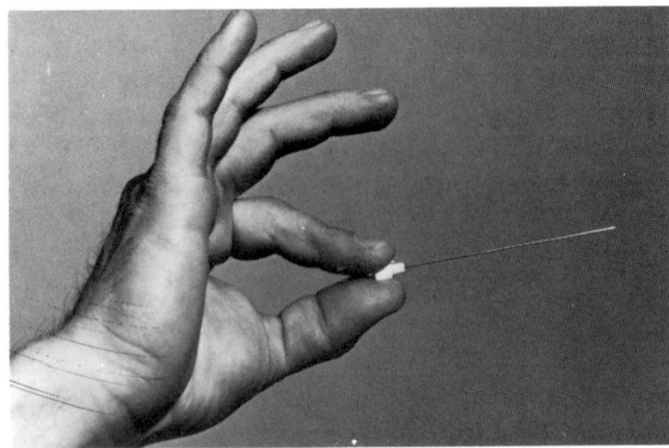

Precision grip of small object.

THUMB TO INDEX GRIP, PADS IN OPPOSITION

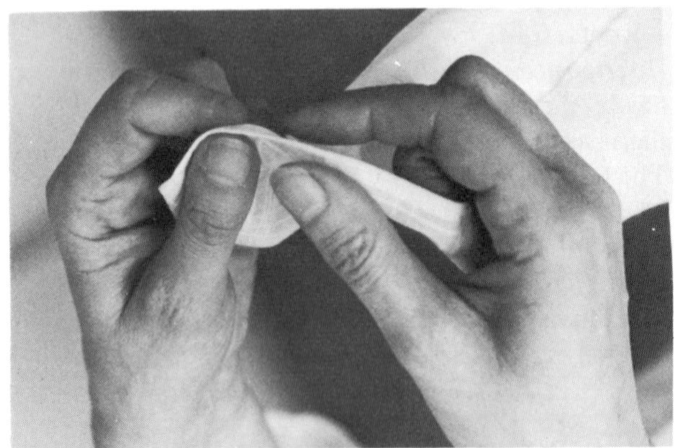

Reach: Hand to object presentation.

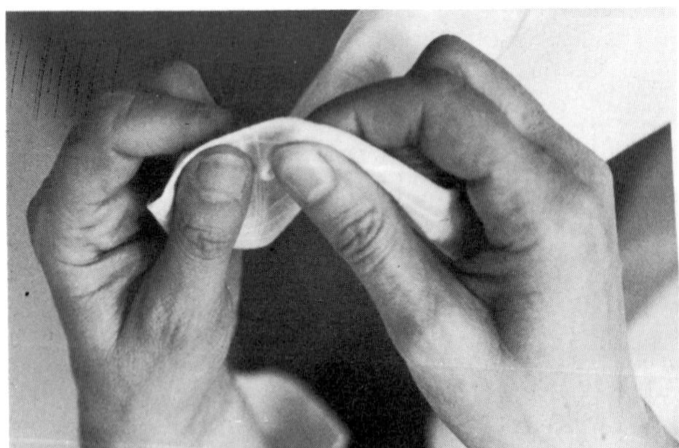

Sliding fingers over object.

Thumb to index grip, sub-terminal opposition

(Note: Variation of thumb to index, terminal opposition.)
Description: A precision grip identical to terminal opposition except for zones of contact. The object is held by the pads close to nails of thumb and index. This implies that distal phalanges are flexed to a lesser degree.
Example of use: Inserting an injection needle.

Thumb to index prehension, pads in opposition

Description: Enables seizing or manipulating small objects between pads of thumb and index. Index distal interphalangeal and thumb interphalangeal joints are extended.

Examples of use:

—Feeling the quality of a piece of fabric.
—Turning the pages of a book.

Muscles essential for maintaining the grip:

—Thumb: Lateral thenar muscles.
—Index: Interossei and lumbrical.
—To a lesser degree, flexor digitorum superficialis, the action of which is secondary.
Characteristic: Precision grip which places the object in contact with zones most adapted for tactile sensation.

Primary innervation:

—Median nerve.
—Roots: C8–T1.
Description of reach: The two pads face each other and are drawn together by simultaneous contractions of flexor digitorum superficialis of the index and abductor pollicis brevis.
Description of release: The object is released as the extension of thumb interphalangeal joint and index proximal interphalangeal joint separate the two pads.

While the hand is the essential organ of manipulation and grip, man is able to accomplish these functions by substituting other parts of his body. These substitutions appear for professional reasons or as socio-cultural habits. They may also compensate for a motor handicap.

Grips by substitution of another body part

a. Mouth area: mouth and dental grips, particularly by the molars enable crushing of objects, carrying a pipe or keeping a cigarette between the lips. With a bilateral affection of the upper extremities, some handicapped people use the mouth for writing or drawing.
b. Chin: may be used to stabilize an armload, preventing objects from falling off.
c. Head to shoulder grip: enables stabilizing a telephone receiver while noting information, or holding a violin while playing it.
d. Arms: objects may be loaded upon the arms, elbows flexed.
e. Thighs and knees: medial aspects of thighs and knees secure a rider upon his horse or aid in everyday activities such as grinding coffee.
f. Feet: manipulating or picking up objects with the feet is an efficient substitution for certain handicapped people who have lost their hands, or congenital amputees who are without upper extremities.

Illustrations on following page.

GRIP BY SUBSTITUTING A PART OF THE BODY OTHER THAN THE HAND

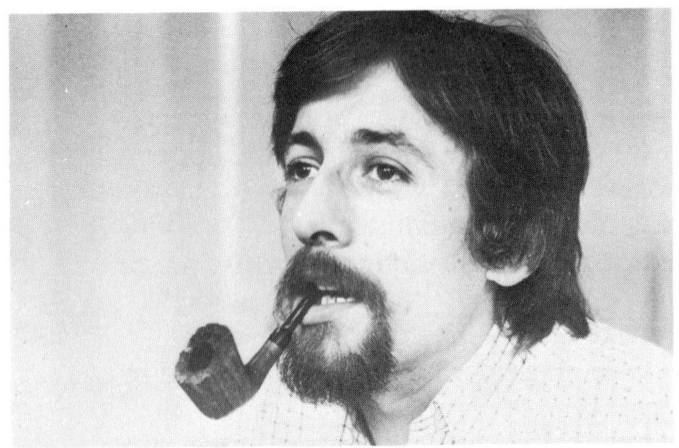

Dental grip.

Head to shoulder grip.

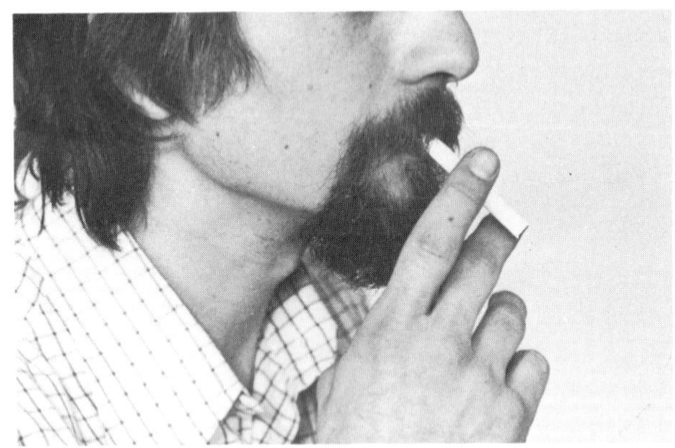

Lip grip.

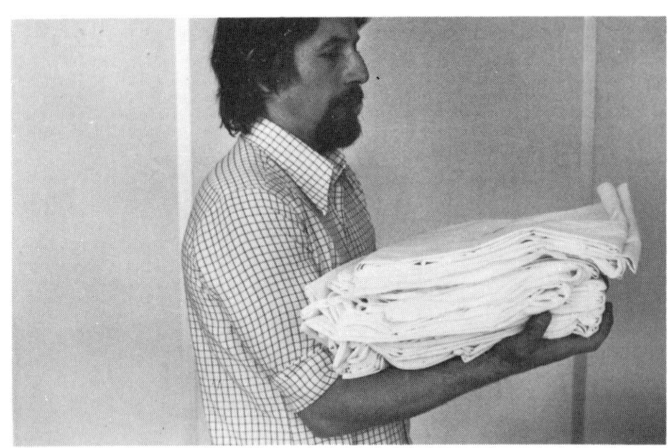

Armload.

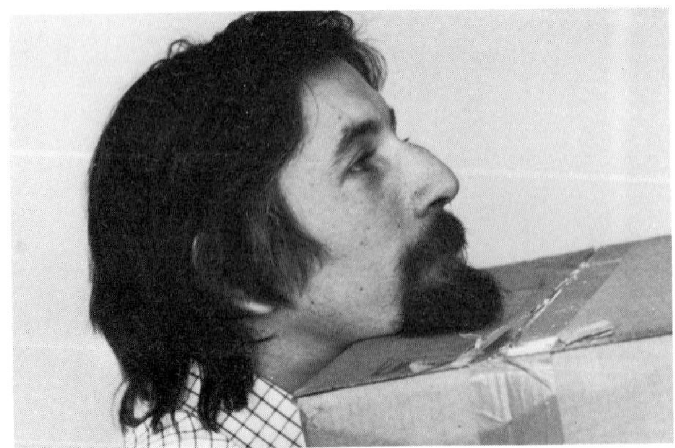

Chin participation in carrying a cumbersome load.

Gripping with thighs and knees.

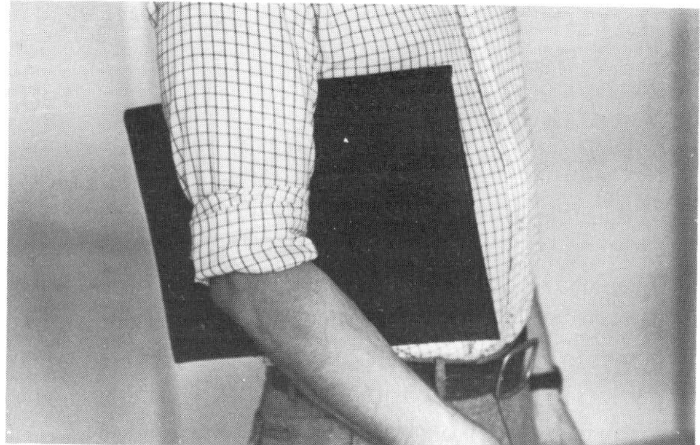

Arm to thorax grip.

Bimanual grip.

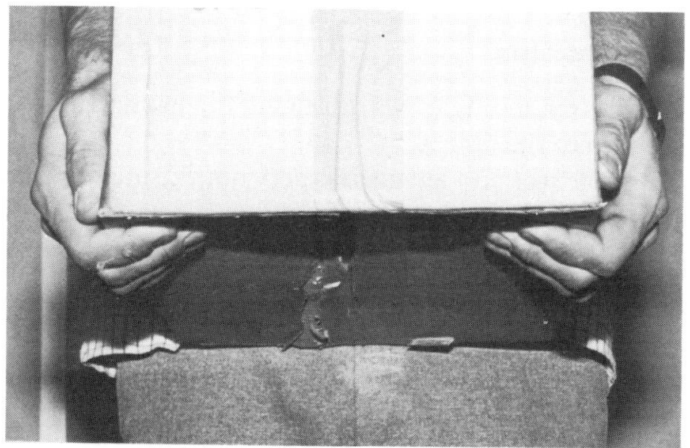

Bimanual grip (Another example).

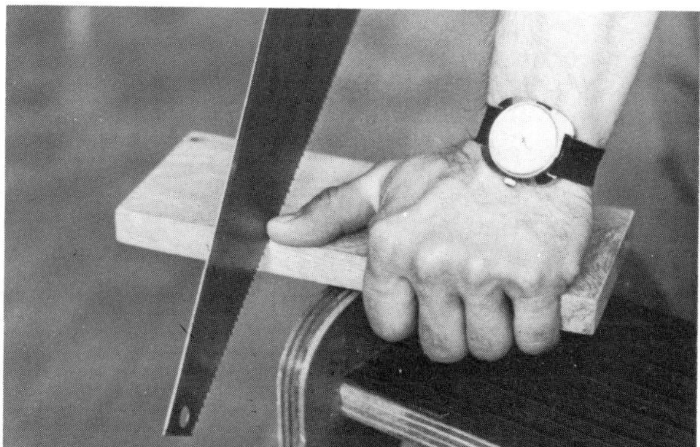

Pressure assistance.

SUPPLEMENTARY GRIPS

Arm to thorax grip

Maintains an object between medial aspect of elbow or arm and lateral aspect of thorax.

Example of use:

Carrying a newspaper or a binder. Requires participation essentially of arm adductors: teres major, latissimus dorsi, pectoralis major.

Bimanual grip

Enables seizing and carrying an object by holding the two hands close to each other.

Example of use:

Bringing a bowl up to the lips. Requires the use of shoulder flexors which place the upper limbs out in front of the trunk, and shoulder adductors which bring the two hands towards each other.
Height adjustment of the hold is determined by:
—Either elbow flexors.
—Or shoulder flexors.

Pressure assistance

Certain handicapped people, such as hemiplegics, hold an object in place with the affected limb in order to work upon it with the sound one.

Example of use:

Holding down a piece of paper in order to write upon it. Making the upper limb rigid is usually sufficient.

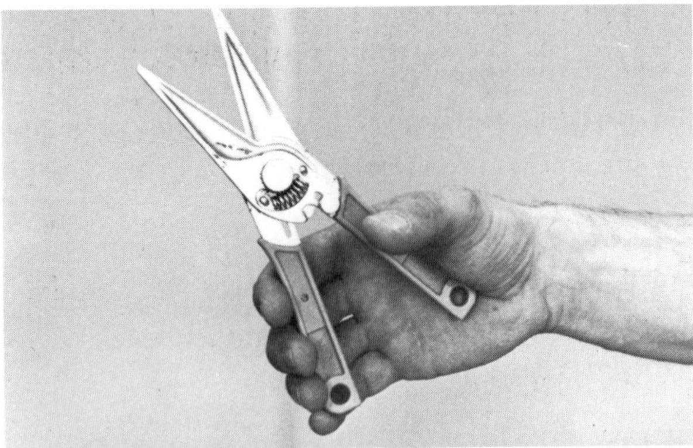

Ulnar deviation.

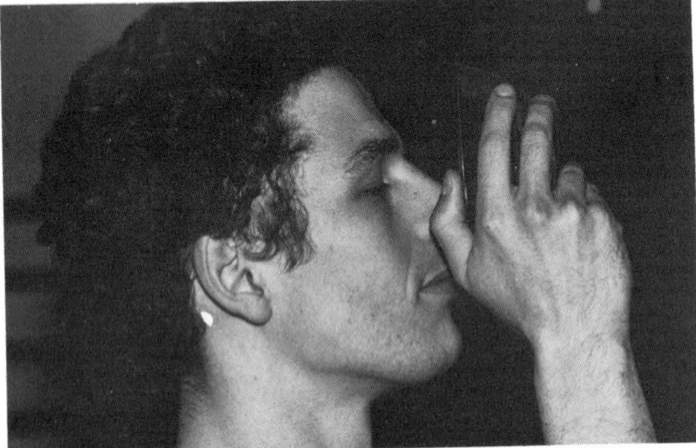

Radial deviation enabling lifting the bowl to the mouth.

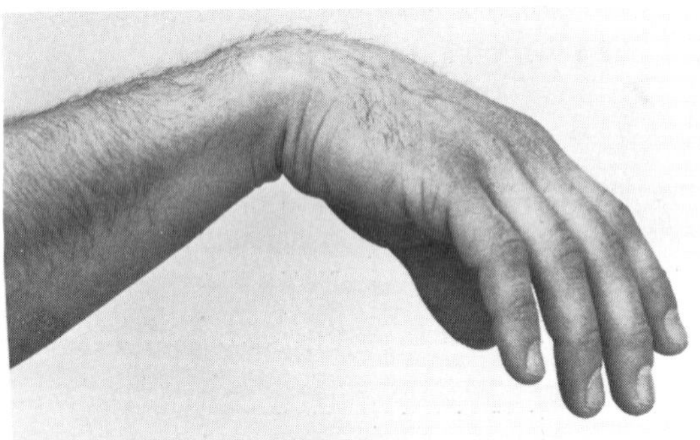

Wrist flexion opens fingers.

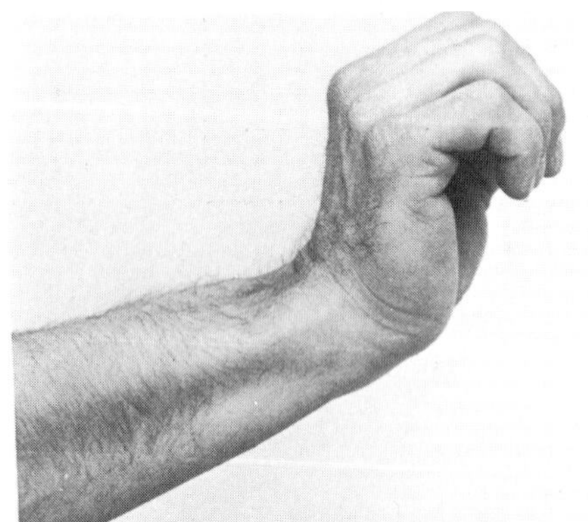

Wrist extension enables strong finger flexion.

THE FUNCTION OF ADJACENT JOINTS ON EFFICIENT GRIP

Wrist action

Wrist adopts:

—Slight ulnar deviation during radial prehension in order to bring thumb into line with forearm axis. This movement is carried out by flexor carpi ulnaris and extensor carpi ulnaris.

—Greater ulnar deviation during power grips. The deviation is proportional to the force used. This movement is also carried out by flexor carpi ulnaris and extensor carpi ulnaris.

—Radial deviation during functional activities which requires the hand to approach the body, e.g. eating with a spoon. Action is carried out by flexor carpi radialis and extensor carpi radialis longus.

—Extension in order to favour power grips. This position lengthens finger flexors. Their activity is thus facilitated. Finger flexor power increases from 1 to 5 as the wrist moves from flexion to extension.

Forearm action

Orients the hand in relation to the object.
Pronation favours prehension and manipulation.
Pronators and supinators stabilize the movement.

Elbow action

Enables extending or withdrawing the hand in relation to the object.
Elbow flexors and extensors carry out the movement.

Shoulder action

Because of its spherical conformation, the shoulder enables the upper limb to move through different planes of space. All muscles of the various joints in the shoulder complex are involved: including stabilizers and prime movers of the scapula and shoulder joint.

Spinal column action

Spinal column furnishes a stable point of attachment for the upper extremities. Its musculature adapts it to changes in the position of the centre of gravity created by gripping or carrying activities. In axial hypotonia, the subject is more concerned with his own balance and thus has diminished upper limb function.

The handicapped person is often forced to use one or both upper extremities for support. Thus manual holds compensate for lack of spinal column stability and manipulation is partially or totally compromised.

GRADING TYPES OF GRIP

Testing different types of grip enables planning socio-professional readaptation possibilities. A grade from 0 to 3 progressively quantifies the seriousness of the handicap.

Grade 0: There is no trouble, grip is easy.
Grade 1: Trouble is minimal, visible, but does not impair the efficiency of the grip (slowness, trembling, stiffness).
Grade 2: Problems are important. Grip is possible but is functionally mediocre.
Grade 3: Subject is unable to perform the manual activity requested.

TYPES OF GRIP	0	1	2	3
Digito-palmar grip				
Hook grip				
Tight, full palmar grip				
Directional, full palmar grip				
Digito-thenar grip				
Lumbrical grip (vice grip or 'bec de canard')				
Spherical palmar grip				
Interdigital latero-lateral grip				
Thumb to index grip, subterminal-lateral opposition				
Tridigital grip (three jaw chuck grip)				
Thumb to index grip, terminal opposition				
Thumb to index grip, sub-terminal opposition				
Thumb to index prehension, pad opposition				
Supplementary grips: —Arm to thorax grip —Bimanual grip —Pressure assistance				
Grip by substitution of another body part				

FUNCTIONAL EVALUATION OF GRIP

The study of different modes of grip is oriented towards function. Motor activity evaluated outside of a concrete, realistic context cannot give a precise reflection of the subject's ability to adapt his motor behaviour to his environment. Testing reflects not the precise action of the muscle but its functional efficiency. A four point grading system defines success or failure from 0 to 3.

Easy 0—Movement requested is accomplished without difficulty.
Disturbed 1—Awkwardness is present but does not impair performance.
Difficult 2—Outside help is needed, either technical aid or a second person, without which the movement would be incomplete or non-functional.
Impossible 3—The function is lost.

Functional activities of the upper extremities fall into the four essential categories of daily living:

—Self-care.
—Dressing.
—Eating.
—Communication and social activity.

SELF-CARE ABILITY	0	1	2	3
Washes face				
Washes hands				
Washes arms				
Washes feet				
Washes legs				
Washes trunk				
Takes bath or shower				
Brushes teeth or dentures				
Grooms hair				
Shaves				
Uses make-up				
Washes intimate parts				
Uses bed pan and urinal				
Uses toilet paper				
Uses bottles or tubes				
Fills and empties a basin				

DRESSING	PUTTING ON				TAKING OFF			
	0	1	2	3	0	1	2	3
Underwear								
Shirt								
Pants								
Sweater								
Jacket								
Brassiere								
Tights, socks								
Shoes								
Splint, brace								
Laces								
Zipper								
Snap fasteners								
Buttons								
Knots tie, scarf								

EATING	0	1	2	3
Drinks from a bowl				
Drinks from a glass				
Drinks from a cup				
Pours a drink				
Uses a knife				
Eats with a fork				
Eats with a spoon				
Prepares a meal				
Aspirates into trachea:				
Liquids	yes		no	
Semi-solids or foods prepared with blender	yes		no	
Solids	yes		no	

COMMUNICATIONS AND FREE TIME	0	1	2	3
Rings bell				
Picks up object from night table				
Smokes without danger of burning				
Holds and reads book				
Writes or uses a typewriter				
Plugs in an appliance or TV				
Uses a telephone				

DEFICIENCY SYNDROME IN PYRAMIDAL LESIONS

CLINICAL EVALUATION

GENERAL REMARKS

A lesion of the pyramidal system leads a disturbance of motor control. It is diagnosed by the appearance of three pathological manifestations:
—Motor deficiency.
—Spasticity.
—Synkineses.
Depending upon its intensity, the motor deficiency is seen in:
—Absence of voluntary movement in one limb.
—Or decrease in power of all motor activities of this limb.
—Or loss or decrease of power in some movement patterns of this limb.

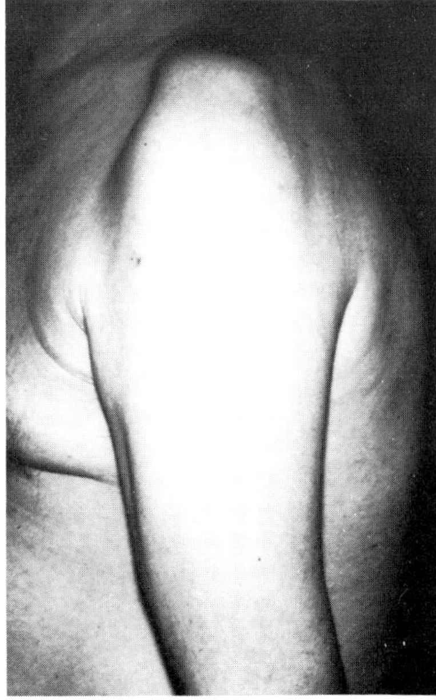

Shoulder subluxation induced by muscular deficiency.

OBSERVATION

Observation of a subject presenting complete deficiency syndrome of one or more limbs reveals:
—Total immobility of affected parts.
—Serious inertia when passively lifting the affected limb which falls heavily back onto the examination table when it is released.
Recovery follows one of two modes:

1. Progressive return of motor command

Examination of the deficiency syndrome must be more precise.
—*Face.* Asymmetry appears in smiling or closing the eyes (sign of Souques's lashes).
—*Neck.* Platysma contracts only on the sound side.
—*Upper extremities.* If the subject holds both upper limbs straight out in front of him, his eyes closed, the arm on the affected side will gradually fall.
—*Lower extremities.* If the subject, prone, holds both

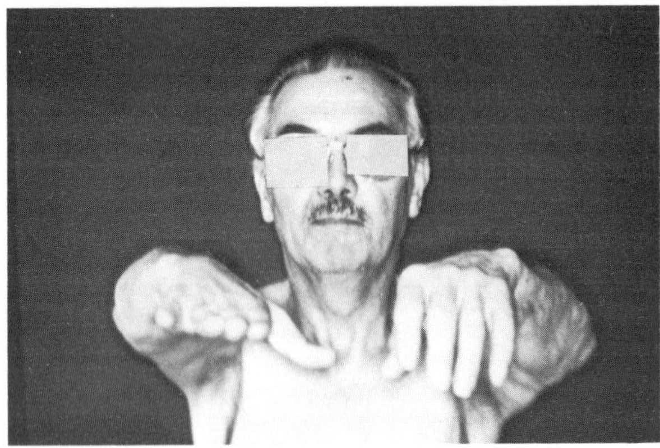
Barre's test with eyes closed.

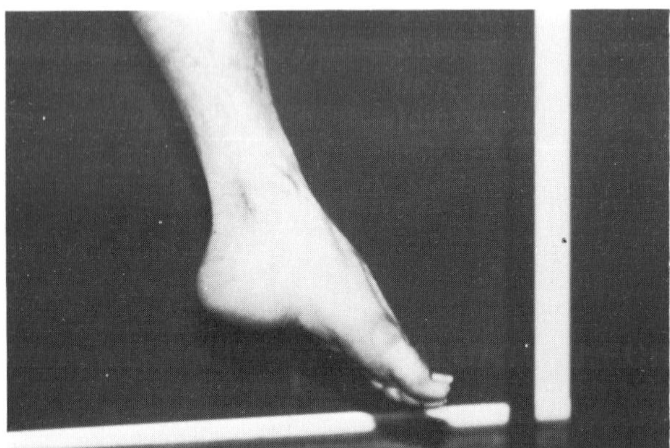
Sequelar hemiplegic foot in relaxed state.

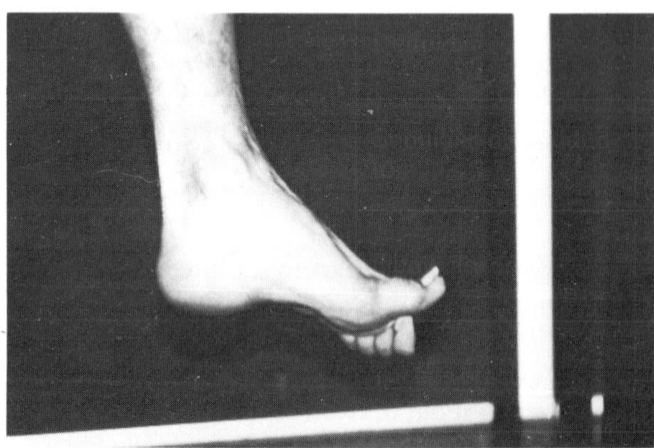
Attempt at dorsiflexion.

knees flexed, the leg on the affected side will gradually fall. If the subject, supine, holds his lower limbs flexed, hips and knees at right angles, the limb on the affected side will fall.

Generally speaking, spontaneous use of the affected side is poorer. The lower limb catches on the slightest obstacle. The hand is not readily used in daily activities.

In order to quantify deficiency syndrome, testing must be adapted. For this, the motor syndrome must be pure, free from associated pathologies such as:

—Deep sensory disorders.
—Visual field deficiency, lateral homonymous hemianopia.
—Praxic and gnosic disturbances.
—Cognition, attention-span deficiencies.

In these circumstances evaluation of peripheral lesions finds a punctilious and exceptional place in central neurological pathologies.

2. Motor command returns accompanied by spasticity.

Hypertonic muscles, more apt to contract, will mask tentative contractions of their deficient antagonists. The dominant pattern of abnormal muscle power is generally that of upper extremity extensors and lower extremity flexors.

The disorder is more serious in distal parts of limbs which are more dependent upon the pyramidal system.

Proximal segments have a more automatic motor control. Testing the power during motor activities remains possible but, in order to follow Held and Pierrot-Deseilligny's grading system, certain criteria must be respected:

—Make quantitative evaluation of the segmental movements and not of the single muscles.
—Consider the angle travelled and not gravity.
—Consider degree of spasticity.
—Note support which makes the movement possible.
—Note coordination synkineses which may appear.
—Consider the patient's position: spasticity increases with postural difficulties. The more significant the spasticity, the more it hinders motor capacity of deficient muscles.

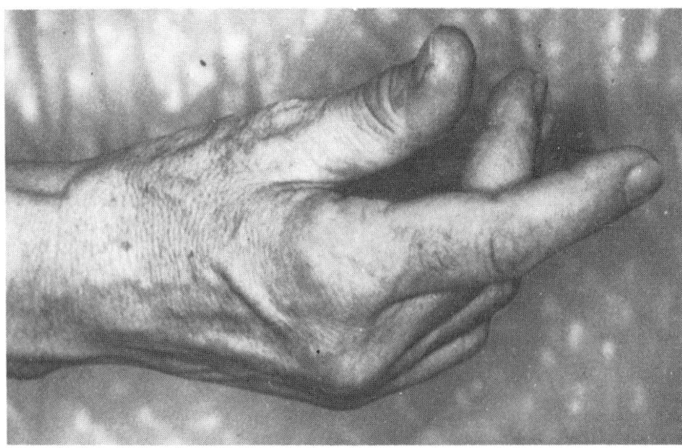

Deficiency of active finger extension although finger closing is efficient.

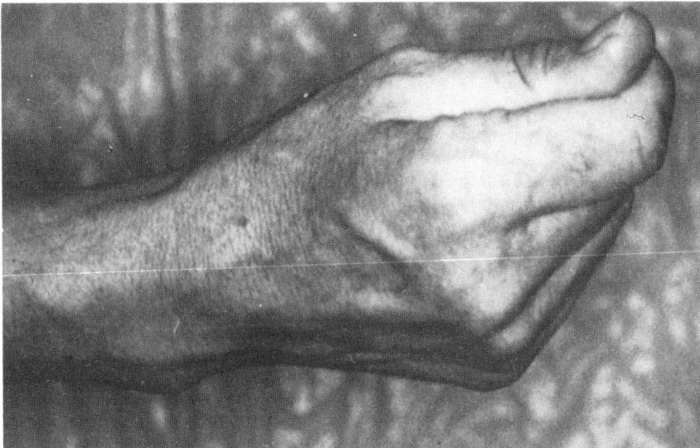

Sequelae of hemiplegia.

Force is estimated by grading from 0 to 5.

0 Contraction is absent.

1 Contraction is perceptible but the segment does not move.

2 The contraction displaces the segment no matter what angle is travelled.

3 The displacement is possible against light resistance.

4 The displacement is possible against more resistance.

5 Force of the movement is identical to that of the sound side.

See charts on following pages.

HEMIPLEGIA: MUSCLE EVALUATION (Held and Pierrot-Deseilligny)

Patient's name:						
		Date: Examiner:				
Muscle	Position *	Force	R.O.M.	Synkineses	Facilitation techniques
Gluteus maximus	Sup				
	P				
Gluteus medius	Sup				
	SL				
Hip adductors	Sup				
Hip external rotators	Sup				
	Sit				
Hip internal rotators	Sup				
	Sit				
Iliopsoas	Sup				
	Sit				
Quadriceps	Sup				
	Sit				
Knee flexors	Sit				
	P				
Tibialis anterior	Sup				
	Sit				
Extensor digitorum long.	Sup				
	Sit				
Extensor hallucis long.	Sup				
	Sit				
Foot abductors	Sup				
	Sit				
Triceps surae	Sup				
	Sit				
Tibialis posterior	Sit				
Quadratus lumborum	Sup				
Abdominals	Sup				
Sternocleido-mastoideus	Sup				

* Sup—supine, P—prone, SL—side-lying, Sit—sitting, R.O.M.—range of movement

Patient's name:				
	Date: Examiner:			
Muscle	Force	R.O.M.	Synkineses	Facilitation techniques
Deltoid: anterior middle posterior				
Shoulder joint adductors: anterior posterior				
Shoulder joint external rotators				
Shoulder joint internal rotators				
Serratus anterior				
Scapular adductors				
Elbow flexors				
Triceps brachii				
Pronators				
Supinators				
Wrist extensors				
Wrist flexors				
Finger flexors: * MCP PIP DIP				
Finger extensors: MCP PIP DIP				
Interossei dorsales				
Interossei palmares				
Thumb: flexor pollicis longus flexor pollicis brevis extensor pollicis longus extensor pollicis brevis abductor pollicis longus abductor pollicis brevis opponens pollicis adductor pollicis				

* MCP—metacarpophalangeal joint,
PIP—proximal interphalangeal joint,
DIP—distal interphalangeal joint.

PYRAMIDAL HYPERTONIA OR SPASTICITY OF CEREBRAL ORIGIN

CLINICAL EVALUATION

GENERAL REMARKS

Spasticity results in an exaggerated myotatic reflex secondary to a lesion of the pyramidal system. Passive stretching of the muscle elicits the hypertonic response.

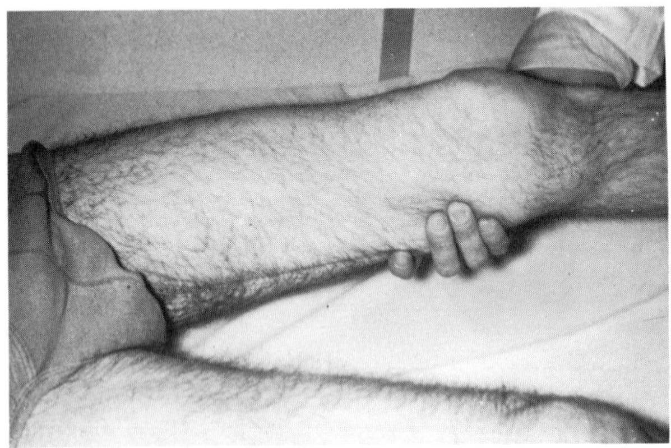

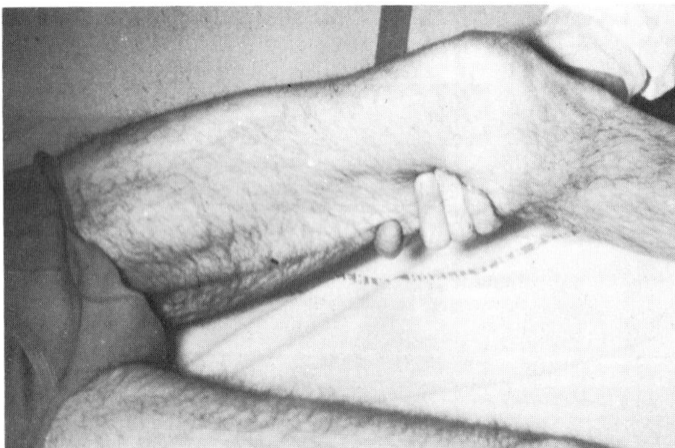

Quadriceps spasticity. Examination technique.

OBSERVATION

During passive muscle lengthening:
First phase: The mobilized segment travels through a free range of movement without meeting any opposition.
Second phase: When the muscle has been stretched to a precise length, an elastic resistance (presence of myotatic reflex) is elicited. This length corresponds to an angle of displacement of the bony segment, the faster the speed of the stretch, the smaller the angle. When spasticity is weak, there may be a threshold speed below which the myotatic reflex does not appear.
In certain cases, short bursts of hypertonia are followed by inhibition. These tonic jerks constitute clonus.
Third phase: If stretching is continued, opposition to passive mobilization grows progressively.
If not, the segment of the limb tends to return to its initial position. By holding the spastic muscle in a maximally stretched position, pyramidal hypertonia is decreased, rapidly when spasticity is weak, more slowly when it is severe (P. Rondot).

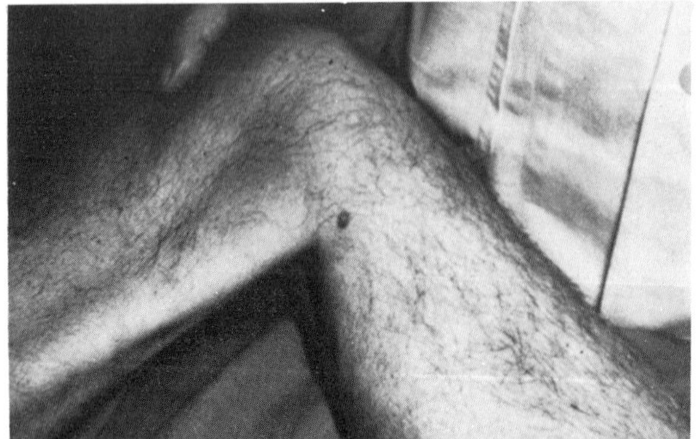

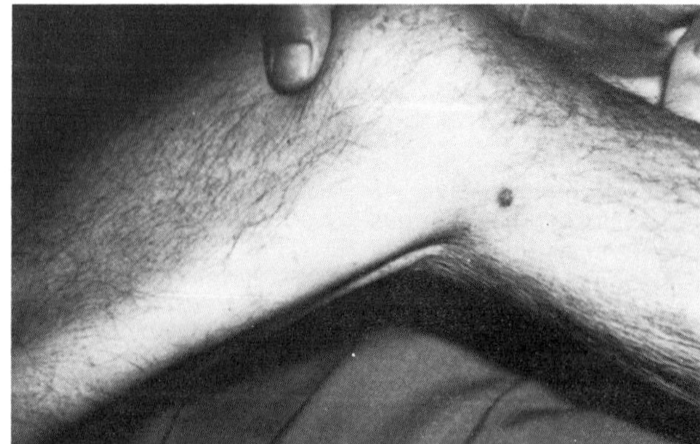

Spasticity of knee flexors. Examination technique.

POSTURAL REACTIONS MODIFYING INTENSIFYING OF SPASTICITY

The subject's posture influences the intensity of pyramidal hypertonia. Minimal when lying down, it increases with sitting, then standing, to attain its maximum when balance is precarious.
The cervical spine position plays a role in muscle tone distribution to the four limbs. These reactions are increased during pyramidal tract conditions. They exist in normal newborn babies:

—Cervical spine extension encourages muscle tone distribution in extension of all four limbs. Inversely, cervical spine flexion encourages flexion.
—Latero-flexion or rotation to one side encourages extension of the two ipsilateral limbs, and flexion of the two contralateral limbs.

All of these pathological postural reactions of cervical origin dominate the upper extremities.
The lumbar spine position also influences muscle tone response of the four limbs but dominates the lower limbs. Lumbar extension encourages extension of the four limbs. On the other hand, lumbar flexion leads to a flexion response. Latero-flexion or rotation to one side induces lower limb extension and upper limb flexion on that side. On the opposite side, lower limb flexion and upper limb extension appear.

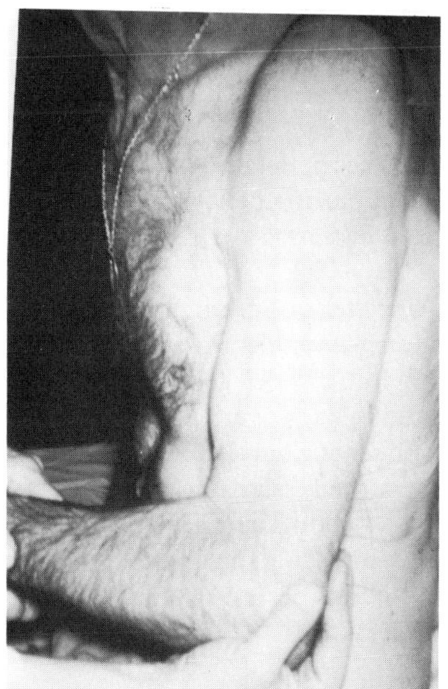

Spasticity of elbow flexors.
Examination technique.

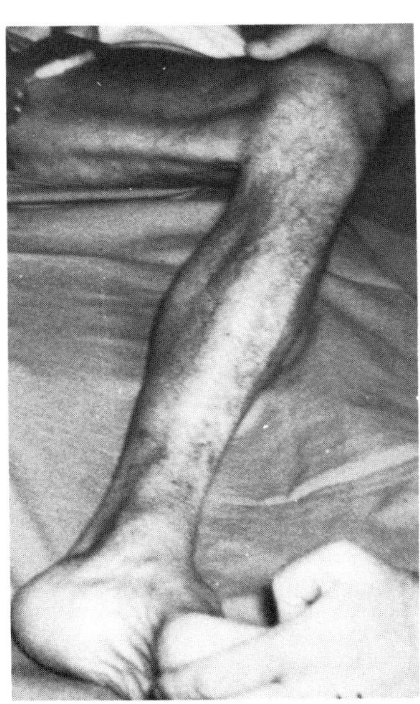

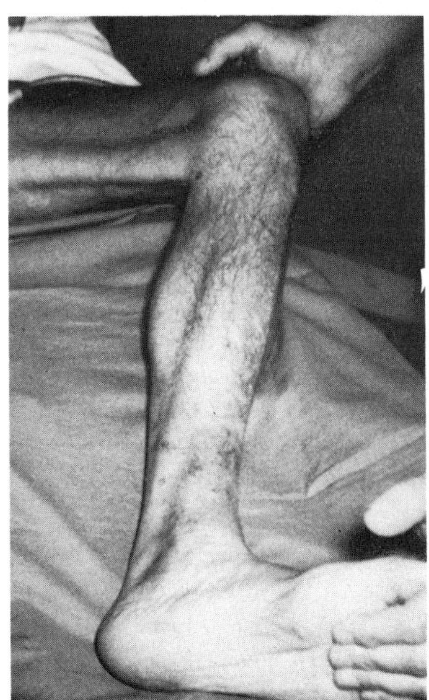

Soleus spasticity. Examination technique.

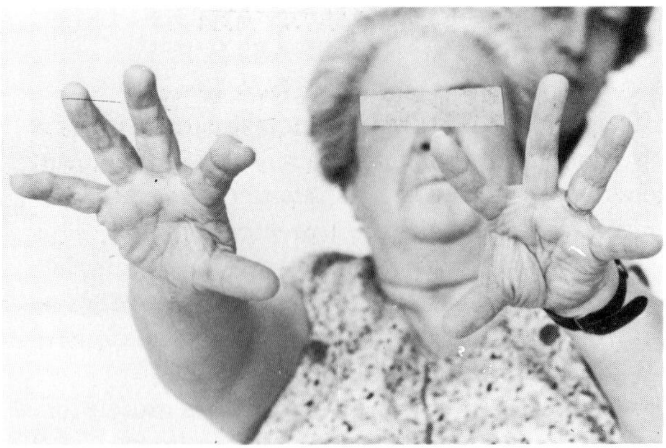

Garcin's hollow hand test.
Left hand spasticity.

FACTORS WHICH INCREASE SPASTICITY

Hypertonia is accentuated by:
—Muscular effort, even at a distance from the hypertonic area.
—Repeated contraction of spastic muscles.
—Effort while holding the breath.
—Pain, of exteroceptive as well as interoceptive origin.
—Emotional reactions (especially anxiety).

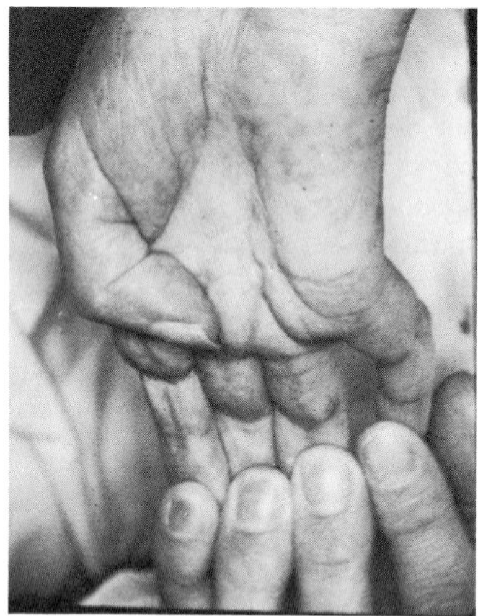

Stretching the fingers increases thumb hypertonia.

FACTORS WHICH INHIBIT SPASTICITY

Hypertonia is diminished by:
—Maintained stretch of the hypertonic muscle.
—Volitional contraction of the spastic muscle in its most lengthened position, which reinforces inhibition compared to its previous tonic state (inhibition rebound).
—Lengthening all muscles belonging to the spastic pattern of a limb, and placing deficient muscles in their most shortened positions (i.e. reflex inhibiting postures).
—Placing the limbs at reflex-inhibiting key points of control: e.g. for the upper extremity, by abducting the thumb and stretching finger flexors.
Note: Positioning hand in back over buttocks facilitates opening of the fingers (Temple Fay).
—For the lower extremity, dorsiflexing the ankle, extending the toes and abducting the big toe.
—Prolonged ice treatment: Cold, localized baths or ice packs greatly reduce hypertonia, though the effect is only temporary.

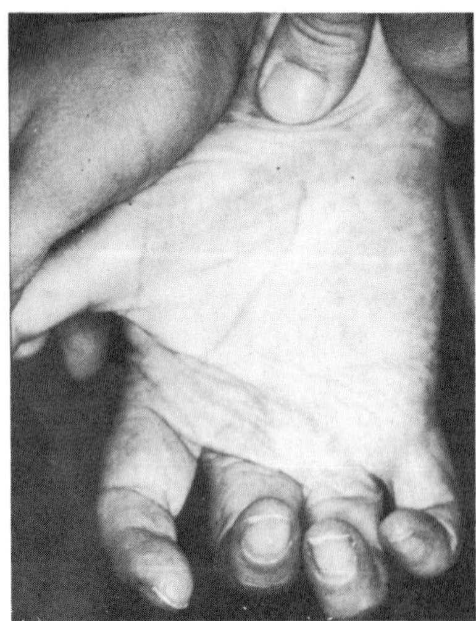

Passive thumb abduction diminishes finger hypertonia.

Shaking of spastic triceps surae is impossible.

QUALITATIVE EVALUATION

Palpation of the hypertonic muscle reveals:
—An abnormally high tone. The muscular tissue is not easily depressed by the examiner's hand and remains difficult to mobilize.
—A loss of 'bounciness' when the examinar shakes the muscle at regular intervals.

Morphological characteristics of the relaxed, hypertonic muscle differ little from those it presents when contracted. Muscular atrophy due to semi-use or disuse appears much later.

—An imbalance among muscular groups is created around the joint. The joint is pulled by the hypertonic muscle which imposes a posture upon it. It is not unusual to see spasticity evolve towards contracture of a muscle tendon unit and a stiffening of joint capsulo-ligamentary structure.

The fairly consistent pattern of spasticity is distributed essentially towards lower extremity extensors and upper extremity flexors.

Observing spontaneous motor behaviour reveals a disturbance in the harmony of the natural gesture. Hypertonic muscles whose force dominates, are more likely to contract. Voluntary movements in their direction are accomplished with exaggerated ease. Those in the opposite direction, when possible, are accomplished slowly and with difficulty.

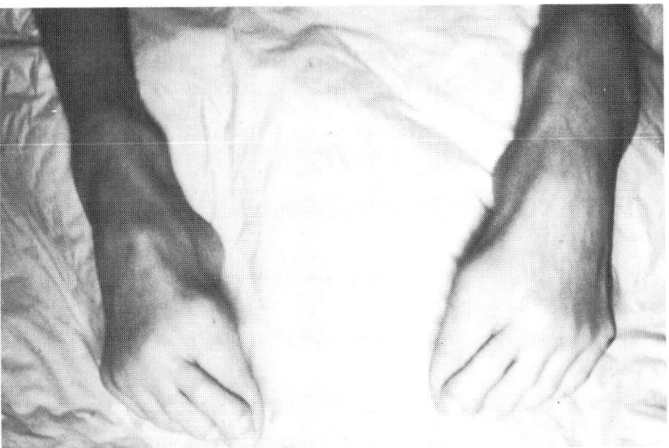

Equinovarus feet due to established spasticity.

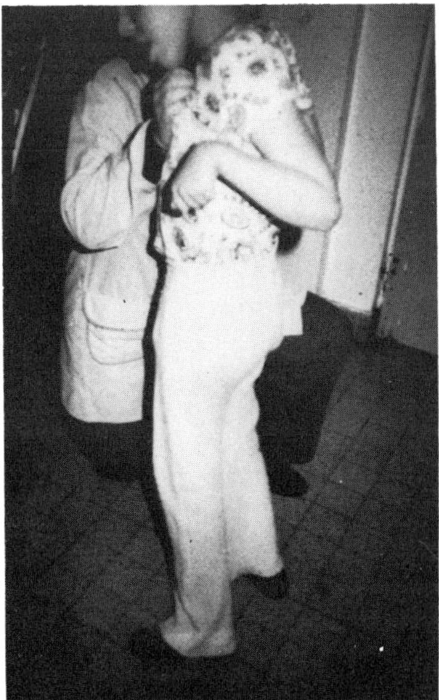

Spastic upper extremity posture with tight fist.

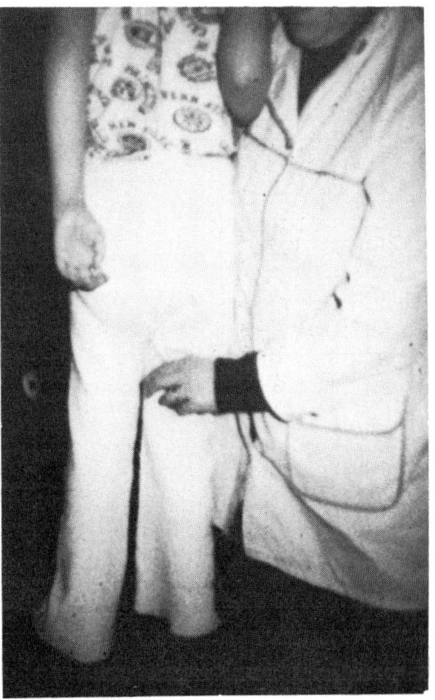

Opening of fingers by placing hand in the back.

TARDIEU AND HELD'S QUANTITATIVE EVALUATION

Based on the three characteristic features of spasticity:

1. Passive stretching speed. Three speeds are differentiated:
   —*V1, slow speed enables examination of tonic response to stretching during passive mobilization.
   —V2, medium speed corresponds to gravity's pull on the segment.
   —V3, fast speed permits examining phasic response to stretching.
2. Angle at which the myotatic reflex appears. The faster the stretch and the more severe the spasticity, the smaller the degree of movement.
3. Response intensity, graded from 0 to 4.
   0 Muscle is sound, there is no sign of hypertonia.
   1 Myotatic reaction is visible or palpable but does not interfere with passive mobilization.
   2 Short pause in the passive mobilization (1 to 3 seconds) is provoked by the myotatic reaction.
   3 Presence of clonic jerks or a longer pause (about 10 seconds) in mobilization.
   4 Established spasticity does not yield to stretching. Differentiation from contracture is possible but only by extreme means (while subject sleeps, nerve block, or general anaesthesia).

Conditions for use:

—The test is begun at slow speed in order to detect eventual contracture.
—Repeated examinations are avoided so as not to facilitate the appearance of myotatic reactions. Tests must be separated by fifteen seconds of rest.
—Subject's position should not change during successive tests. Difficulty in maintaining a posture increases hypertonia.
—Joint positions of a limb should remain identical from one test to another, especially when testing polyarticular muscles.

*V: Speed.

EVALUATION FORM FOR HEMIPLEGIC SPASTICITY (Held and Pierrot-Deseilligny)

Patient's name:													
		Date: Examiner:						Date: Examiner:					
Muscle	Position*	VI		V2		V3		V1		V2		V3	
		Ang	Int	Ang	Int	Ang	Int	Ang	Int	Ang	Int	Ang	Int
Deltoid	Sup												
	Sit												
Shoulder joint adductors	Sup												
	Sit												
Elbow flexors	Sup												
	Sit												
Triceps brachii	Sup												
	Sit												

Patient's name:													
		Date: Examiner:						Date: Examiner:					
Muscle	Position*	VI		V2		V3		V1		V2		V3	
		Ang	Int	Ang	Int	Ang	Int	Ang	Int	Ang	Int	Ang	Int
Pronators	Sup												
	Sit												
Supinators	Sup												
	Sit												
Flexor carpiradialis & palmaris longus	Sup												
	Sit												
Radialis longus & brevis	Sup												
	Sit												
Flexor digitorum super.	Sup												
	Sit												
Flexor digitorum profund.	Sit												
Flexor pollicis longus	Sit												
Gluteus maximus	Sup												
Hip adductors	Sup												
	Sit												
Quadriceps femoris	Sup												
	Sit												
Knee flexors	Sup												
	Sit												
Foot dorsiflexors	Sup												
	Sit												
Triceps surae (lower extrem. ext)	Sup												
	Sit												
Triceps surae (lower extrem. flex)	Sup												
	Sit												

*Position:
Sup—supine
Sit—sitting
Ang—angle

Int—intensity
V1—slow speed
V2—medium speed
V3—fast speed

SPASTICITY OF SPINAL ORIGIN

CLINICAL EVALUATION

GENERAL REMARKS

Spasticity of spinal origin presents itself as spastic hypertonia upon which spinal reflexes have been added. The extension posture produced by spasticity is transformed when nociceptive stimuli are applied, into a posture of lower extremity withdrawal. Spinal spasticity appears with lesions to the spinal cord and is most important in complete transections (traumatic paraplegia). It is the reflection on one hand of corticospinal tract interruption (spasticity in extension), and on the other, of the release of spinal reflexes from supra spinal control (spasticity in flexion).

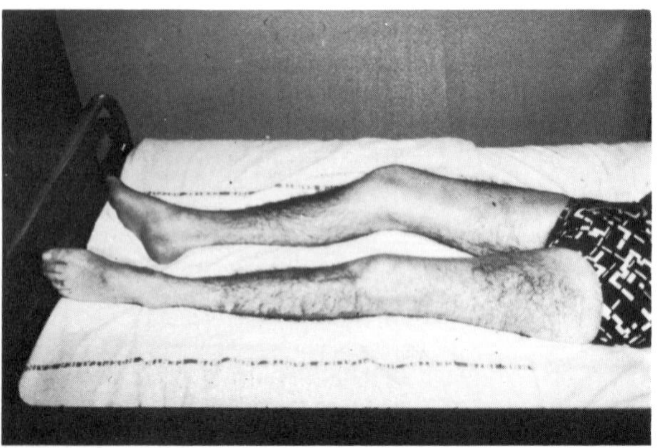

Extension of lower extremities in spasmodic paraparesia.

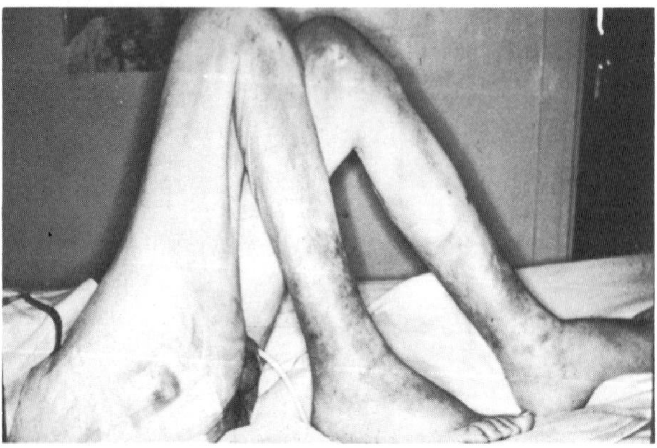

Withdrawal posture of lower extremities.

OBSERVATION

Extension posture of lower extremities is enforced by extensor spasticity. The flexors are deficient. The limb most often appears with:

—Hip in adduction, internal rotation (sometimes external rotation), and extension.
—Knee in extension.
—Ankle in plantar flexion.
—Foot in adduction and supination.
—Toes in plantar flexion.

In complete spinal transections, when the spinal cord below the lesion is intact, posture is generally in extension. Stimulating the skin on the dorsal aspect of the foot by pinching it or by passively flexing the toes, triggers a withdrawal reflex on that side, with occasional contralateral extension.

Flexion posture, schematically:

—Hip in adduction, internal rotation, and flexion.
—Knee in flexion.
—Ankle in flexion.
—Foot in adduction and supination.
—Toes in plantar flexion.

Points of irritation, either visceroceptive (kidney stones, urinary infections) or exteroceptive (pressure sores), may provoke bilateral withdrawal reflexes which can lead to contractures. There are few inhibiting factors and the patient has no control over his spasms. In incomplete spinal transections, spasticity leading to withdrawal posture is much more moderate and sometimes non-existent. As progressive spinal neuropathies evolve (multiple sclerosis, Strumpell-Lorrain's hereditary paraplegia), lower extremity spasticity also occurs, first in extension and then into flexion. This spasticity hinders selective control and the functional quality of lower extremity movement, e.g. gait. A limp appears, progressively worsening until the patient is bedridden.

QUALITATIVE EVALUATION

This consists of noting functioning results of spinal spasticity which may:
—Either encourage function: spasticity in extension may contribute in certain cases to stability in standing and walking with or even without a light brace. Straightening of the limb is assured by extensor spasticity.
—Or hinder function: the appearance of withdrawal reflex at weight-bearing, skin irritation by the brace during gait, abnormal articular stress, impedes motor function and may lead to falling.

QUANTITATIVE EVALUATION

The grading of spasticity, as proposed by Tardieu and Held, is based on three parameters:
—Speed.
—Joint angle at which stretch reflex appears.
—Intensity of response.

It may be used when spinal lesions are incomplete. In complete transections, pathological elements due to autonomic spinal activity add to exaggeration of the myotatic reflex forcing the therapist into using a more superficial approach to the pathology in establishing a therapeutic programme. The therapist tests spasticity by grading from 0 to 3:

0 There is no spasticity.
1 Spasticity is present but does not interfere with re-education.
2 Spasticity interferes with passive mobilization.
3 Spasticity is a functional hindrance in standing with technical aids and in wheelchair activities.

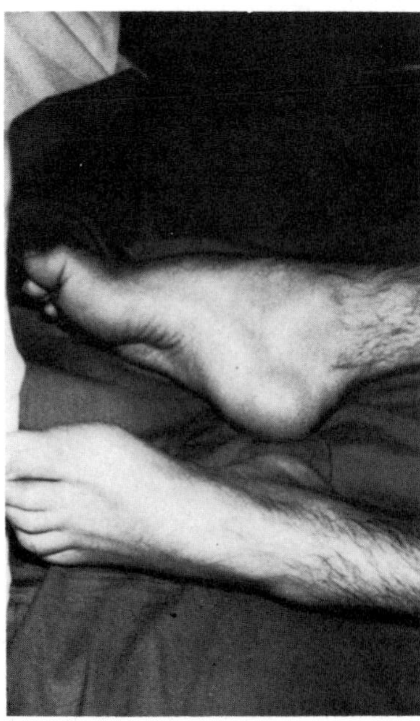

Spastic hollow feet in advanced stage.
Multiple sclerosis.

SYNKINESES OR ASSOCIATED REACTIONS

CLINICAL EVALUATION

GENERAL REMARKS

'Synkineses are involuntary, often unconscious movements which appear at the same time as other movements which are usually voluntary and conscious.' (Pierre Marie and Charles Foix.) They appear primarily after lesions of the pyramidal system and are most frequent in hemiplegia.

CLINICAL OBSERVATION

MASS SYNKINESIS

This is the contraction of all of the muscles of the hemiplegic limb, or even of the whole musculature of that side. In general, it accentuates primitive hemiplegic posture. It appears during effort of the sound or hemiplegic side, and at an increase of postural difficulties or loss of balance.

IMITATION SYNKINESIS

This is the appearance of symmetrical movements, identical and bilateral: motor function appears as if reflected in a mirror.

Different features of imitation synkinesis are:

—Movements of the hemiplegic side which are reproduced on the sound side. This synkinesis is easy to inhibit voluntarily. It seems to be an attempt to substitute for the motor difficulty. It is not systematically linked to corticospinal tract lesions. Thus it may appear on a sound subject. Such is the case when a right-handed person performs a complex motor activity with his left hand. His right hand tends to imitate the movement as if it were helping the left.

—Movements which start on the sound side and automatically appear on the affected side. Rare in hemiplegia, they are more common in choreo-athetoid deficits of the child. When they exist in bilateral pyramidal lesions, they are more often the simultaneous expression of coordination synkineses or mass synkineses in both hemibodies.

—Movements performed simultaneously by upper and lower extremities. If one of the extremities on the affected side moves voluntarily, the other tries to imitate it. This form of imitation synkinesis is frequently the sign of a thalamic lesion.

COORDINATION SYNKINESIS

When one muscular group contracts, the automatic contraction of another muscular group appears instantaneously. This irradiation of motor activity causes pathological synkineses in flexion or extension in hemiplegia. They seem to be created by:

—The release of medullary control.

—The muscle imbalance around a joint where hypertonic and deficient muscles co-exist.

—The action of polyarticular muscles which, faced with postural disorders, act upon all the articulations they cross.

COMBINED SYNKINESES

This is the transformation of a coordination synkinesis into a mass synkinesis. Thus, when actively mobilizing one joint, the patient provokes a coordination synkinesis. The effort needed for this motor activity increases the tonic state. If it is continued, the limb gradually changes its course and ends up locking itself in the mass synkinesis posture.

EVALUATION OF SYNKINESES

Mass synkinesis

Qualitative evaluation

Synkineses appear when the subject has made an effort on the sound or on the affected side. They appear particularly in:
—Gait.
—Loss of balance.
—Precision movements.
—Effort, e.g. while holding breath.
—Resisted exercises.

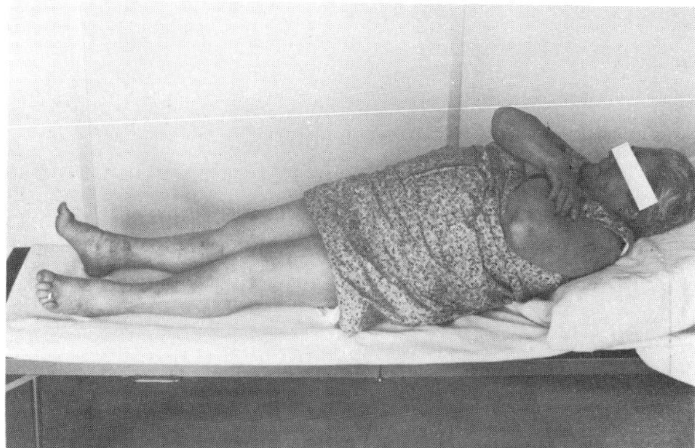

Supine subject.

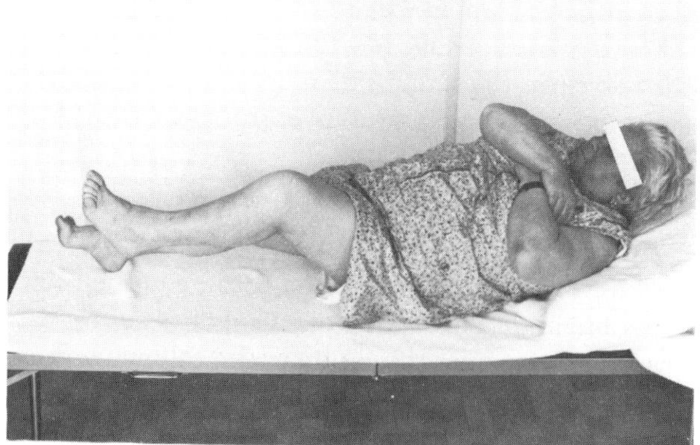

As subject attempts sitting, left lower extremity shows withdrawal.

Babinski gives the following examples:
—The patient sits on a seat which is high enough so that both legs hang and feet do not touch the floor. If he clenches both fists, a movement of knee extension followed by foot equinovarus appears in the lower limb.
—The patient is supine on the treatment table. Lower extremities are relaxed resting on heels, upper extremities are crossed over the chest. If he attempts to rise to a sitting position, the knee extends, the foot goes into equinus, and the hip flexes on the affected side. The limb on the sound side remains immobile.

When mass synkineses appear:
—In the upper limb, the following are accentuated:
- hemiplegic posture, adduction and internal rotation of the shoulder are increased.
- elbow flexion.
- forearm pronation.
- wrist and finger flexion.
- thumb adduction.

Their functional contribution is limited. At best the affected hand or forearm may be used to stabilize an object.
—In the lower limb, the following are accentuated:
- hemiplegic posture, stiffening the limb in extension, increasing hip adduction, extension, and external rotation.
- knee extension.
- ankle equinus.
- foot varus and supination.
- clawing of toes.

Even if these synkineses do simulate the upright position, it is a mere caricature, and they do not permit dynamic balance responses.

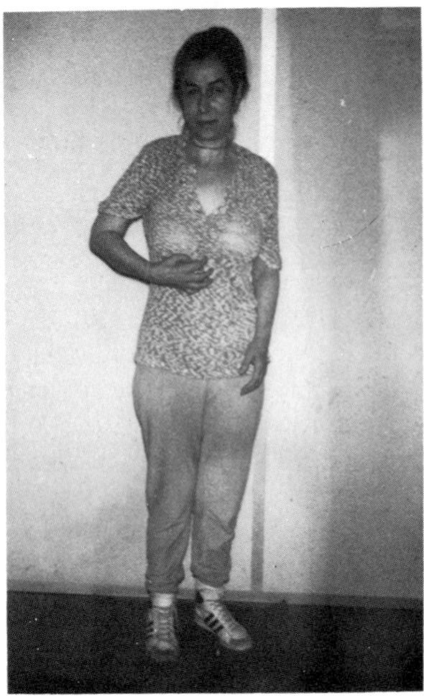

Forearm suppination appears with elbow flexion.

Coordination synkineses.

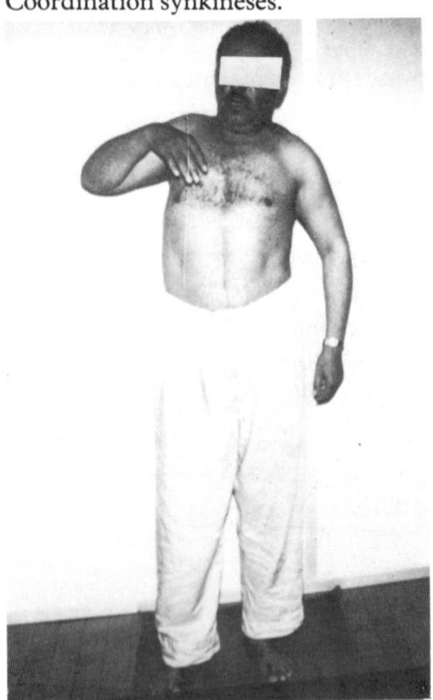

Posterior deltoid contracts and forearm supinates when elbow flexes.

Note: In certain cases, mass synkinesis in flexion appears in the lower limb comprising stance and gait.

Quantitative evaluation

This is graded on a 0 to 3 scale for upper and lower extremities. Grading attempts to identify the intensity of the synkinetic response.

0 Synkinesis is absent.
1 Synkinesis is weak, appearing only during strongly resisted efforts.
2 Synkinesis of medium intensity appears during weakly resisted efforts or postural modifications.
3 Intense synkinesis appears spontaneously at the slightest movement.

Imitation synkineses

Rather rare in the adult, they are frequent in infantile hemiplegia.

Qualitative evaluation

These appear:

—In the affected lower extremity:

- when abduction or adduction of the sound hip is resisted, a symmetrical, automatic movement appears in the affected side.
- during circumduction or to-and-fro movements of the affected wrist, a similar movement may appear in the ipsilateral ankle and foot.

—In the effected upper extremity:

- if the patient tightens his fist on the sound side, the hand on the affected side may close, but this is often only the beginning of mass synkinesis.
- when the patient actively mobilizes his affected ankle, similar movements may appear in the ipsilateral upper extremity.

These synkineses have few functional uses. When they do exist, the sound side may be used to facilitate contraction of certain muscular groups on the affected side.

Quantitative evaluation

Because synkineses are rare and because of their few functional side-effects, the therapist does not need a graded test for quantifying them and following their evolution.

Coordination synkineses

Qualitative evaluation

Coordination synkineses are the result of two voluntary patterns of movement:

—Withdrawal reaction of the limb: The basic flexion synergy appears easily in the upper extremity. It is frequent in the lower limb.

—Lengthening or reaching reaction of the limb: The basic extension synergy is nearly constant in the lower limb. It is not always found in the upper.

When the patient contracts a muscle in one of these basic synergies, he can, depending on the intensity of the contraction and its patterned response, trigger, from one muscle to the next, automatic contractions in corresponding muscle groups participating in the same basic synergy.

Movements composing pathological patterns of motion are found in tables on following pages. The tables are based on the work of Albert, A., 1965; Brunnstrom, S., 1970; Johnstone, M., 1980; and Michels, E., 1959.

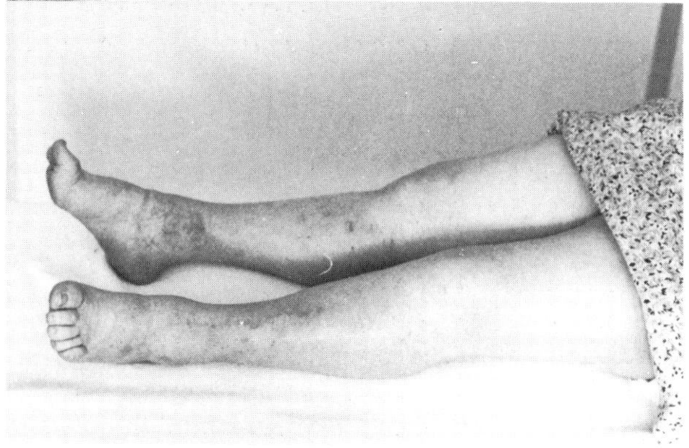

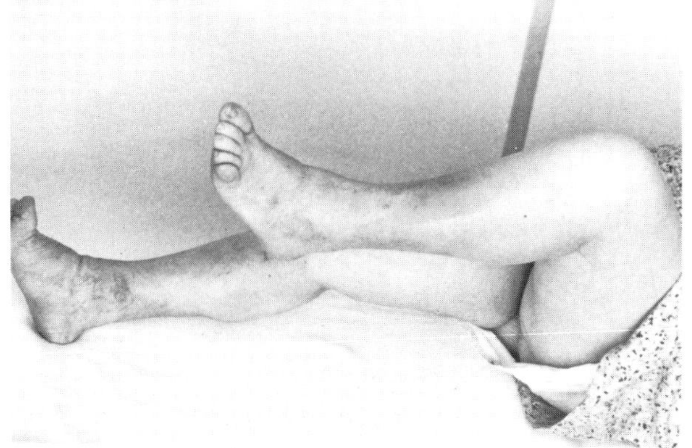

Basic flexion synergy of lower extremity.

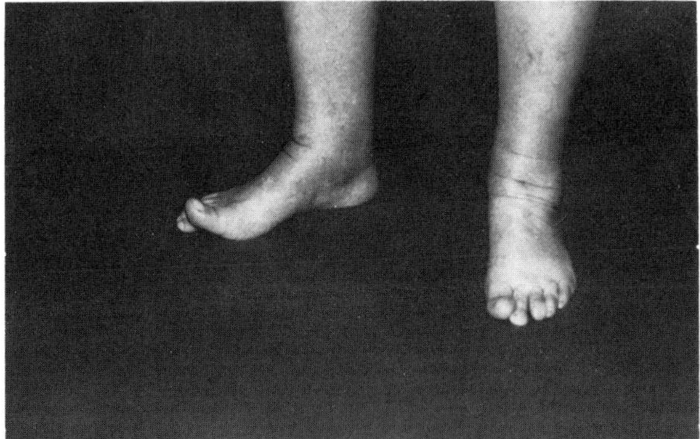

Synkinetic extension of right big toe during weightbearing (rare reaction).

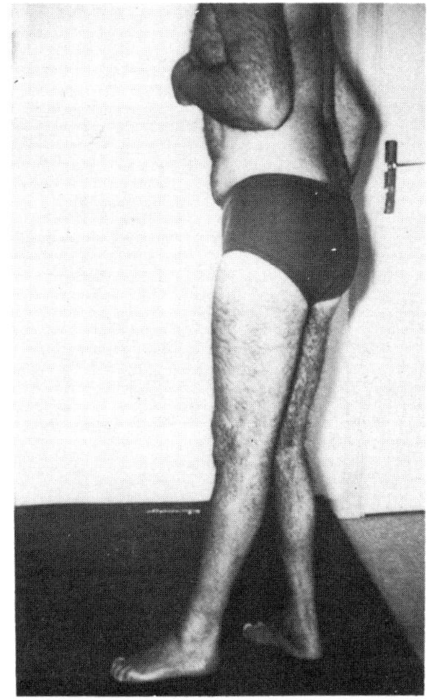

Clawing of the toes during weightbearing (frequent reaction).

LOWER EXTREMITY

Joints mobilized	Hip	Knee	Ankle	Foot	Toes
Basic extension synergy Segmental components:	Extension ⇄ ↓↑ Adduction ↓↑ External rotation	Extension ⇄	Equinus ⇄	Varus ⇄ Supination	Claw
Active muscles	Hamstrings & gluteus maximus ⇄ ↓↑ adductors	Quad'cps femoris & esp. rectus femoris ⇄	Triceps surae & esp. gastrocnemius ⇄	Tibialis posterior ⇄	Extrin. & intrin. flexors
Imbalance causing synkineses	Hamstr'gs act as extensors since knee is locked in exten. Adductors & external rotators dominate gluteus minimus & gluteus medius	Stretched rectus femoris extends knee joints	Stretched gastrocnemius extends ankle	Invertors dominate evertors	Absence of extensors
Basic flexion synergy Segmental components	Flexion ⇄ ↓↑ Abduction ↓↑ External rotation	Flexion ⇄	Flexion ⇄	Varus ⇄ Supination	Claw
Active muscles	Iliopsoas ⇄ ↓↑ Sartorius	Hamstr'gs ⇄	Tibialis anterior ⇄	Tibialis posterior ⇄	Flexors
Imbalance causing synkineses	Hamstr'gs shortened by knee flexion do not balance hip flex. Hip external rotators dominate	Facilitated by hip flex. which relaxes rectus femoris	Gastrocnemius relaxed by knee flexion	Invertors dominate evertors	Flexors dominate esp. the extrinsic stretched by talus posture

UPPER EXTREMITY

Joints mobilized	Scapula	Shoulder joint	Elbow	Forearm	Wrist	Fingers
Basic extension synergy Segmental components	De- ⇄ pression ↓↑ Abduction	Flexion ⇄ ↓↑ Adduction ↓↑ Internal rotation	Extension (often ⇄ incomplete)	Pro- ⇄ nation	Flexion ⇄ or Extension ⇄	Claw Extension abduction
Active muscles	Pectoral. major & minor ↓↑ Teres major	Pectoral. major ⇄	Triceps brachii ⇄ (limit'd by biceps brachii)	Pronator teres & ⇄ Pronator quad.	Flexor carpi ulnaris ⇄ & Radialis & palmar. longus Extensor carpi ⇄ ulnaris	Finger flexors Dorsal interossei
Imbalance causing synkineses	Protractor muscles tract scapula	Pectoralis major imposes arm position	Elbow flexors limit partial triceps action	Pronation favoured by stretching of pronator teres		Flexors dominate or extensors participate (when yawning)
Basic flexion synergy Segmental components	Elevation ⇄ ↓↑ Adduction	Extension ⇄ ↓↑ Abduction ↓↑ Internal rotation	Flexion ⇄	Supination ⇄	Flexion ⇄	Claw
Active muscles	Trapezius ⇄ ↓↑ Levator scapulae ↓↑ Rhomboids	Latissimus dorsi ⇄ ↓↑ Posterior deltoid ↓↑ Teres major	Biceps brachii ⇄ Brachio radialis	Biceps brachii ⇄	Flexor carpi ulnaris ⇄ & Radialis & palmar. longus	Finger flexors
Imbalance causing synkineses	Scapula elevators dominate	Middle deltoid is weak, external rotators absent	Flexors dominate	Pronator teres shortened by flex. position. Supination component of biceps brachii is singled out		Flexors dominate

Head, neck, and lumbar region postures can influence synkinetic reactions and thus may be used as facilitation techniques in providing a basic synergy:

1. Neck posture influences tonic responses leading to certain motor reactions. Most common in infant pathologies, they can also be observed in the adult and are known as Magnus' tonic neck reflex.

Tonic neck reflex:

—Symmetrical: Neck flexion favours upper extremity flexion. Its extension favours extension of these limbs.

—Asymmetrical: Neck rotation or latero-flexion favours extension synergy of ipsilateral upper extremity, flexion of the contralateral.
Influence on adult lower extremities is not well marked. If they do exist, the lower extremity flexes or extends as does the ipsilateral upper extremity.

2. Head posture in space influences tone in the four extremities (Magnus, 1924; Walsh, 1957; Fukuda, 1959).

Tonic labyrinthine reflex:

—Tone of extensors reaches its maximum when the head is positioned so that the line of the oral plane meets the horizontal at a 45° angle which opens upwards and backwards. This position is reached when subject is supine.

—Tone of flexors reaches its maximum when the head is positioned so that the line of the oral plane meets the horizontal at a 45° angle which opens downwards and forwards. This position is reached when subject is prone.

Tonic neck and tonic labyrinthine reflexes, depending on respective positions of the head and neck, combine with or counteract each other as do the terms of an algebraic equation.

3. Lumbar posture also changes tonic reactions of the limbs.

Tonic lumbar reflex:

—Symmetrical: Lumbar flexion favours withdrawal synkineses, lumbar extension, the extension synkineses.

—Asymmetrical: Lumbar rotation or latero-flexion favours the appearance of ipsilateral lower extremity extension synkineses, and contralateral withdrawal synkineses. Its effect on upper limbs is minimal or non-existent. When present, the upper limbs extend with lumbar extension and flex with flexion. Latero-flexion or rotation causes the ipsilateral upper limb to flex, the other to extend.

Quantitative evaluation

In order to examine coordination synkineses, the examiner asks the patient for an active movement of a muscular group. He notes involuntary movements which appear during the patient's volitional movement.

This examination is realized first during free movement, then against gradually increasing resistance. The patient's posture is considered since the more demanding it is, the more easily synkineses will appear. Held and Pierrot-Deseilligny's form may be used for grading. It accurately registers disturbances in hemiplegic single segment movements. This test with its many parameters is long and delicate for everyday use. The therapist is content with simply noting the presence of abnormal motor responses while evaluating global movements of upper and lower extremities. These are quantified from 0 to 3 (see page 434).

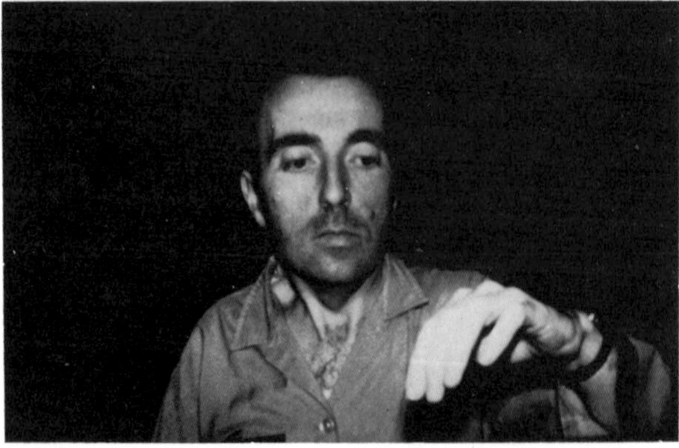

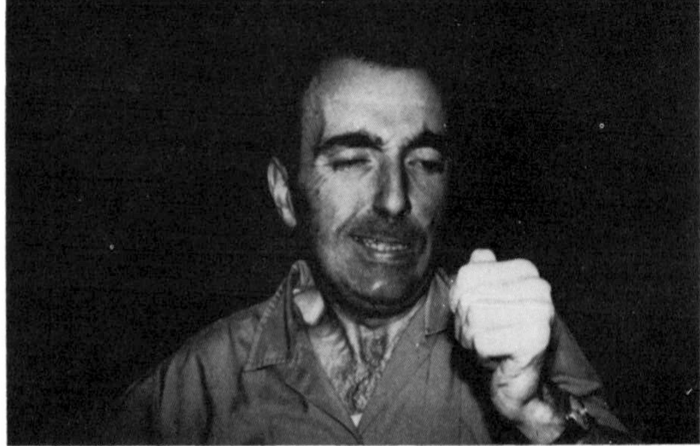

Facial spasm appearing as hemiplegic hand closes.

EVALUATION OF GAIT IN SPINAL CORD DISORDERS

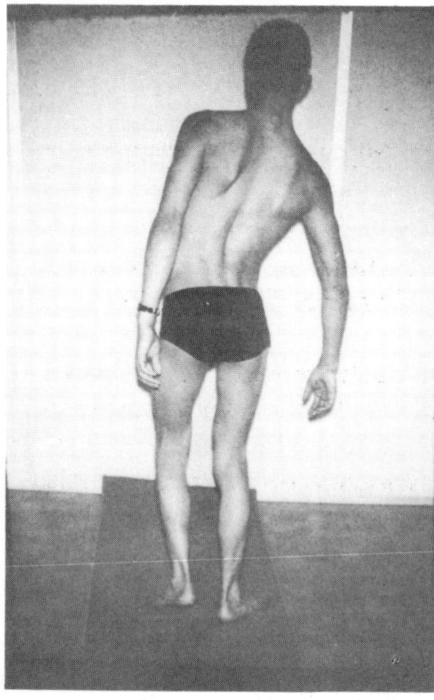

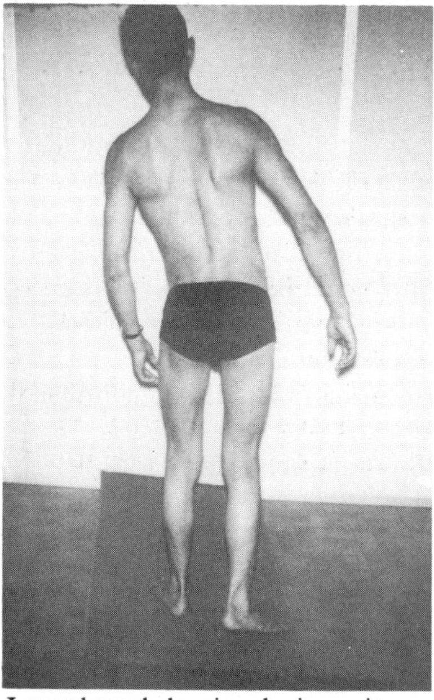

Lateral trunk-leaning during gait (spastic paraparesis).

GENERAL REMARKS

Paraplegia is paralysis of both lower extremities. Depending upon the level of the lesion, the trunk is more or less involved. If the upper extremities are affected, even slightly, paralysis is called tetraplegia. This lesion almost always involves both corticospinal tracts in their spinal course.
Paraplegia may be flaccid:
—In a transitory way after an injury to the spinal cord (spinal shock).
—In a definitive way after destruction of the conus medullaris below the lesion.
But more often it is spastic.
Its evolution may be slow. It appears in the form of a spastic paraparesis which worsens gradually (multiple sclerosis, Strumpell Lorrain familial paraplegia or slow spinal cord compression). Its evolution may be abrupt and definitive as with post-traumatic cord lacerations.
In either case, the essential part of the pathology results in motor symptoms. These are frequently aggravated by an exteroceptive and proprioceptive sensory pathology.

CLINICAL OBSERVATION OF MEDULLAR ORIGIN LIMPING

With incomplete medullar lesions (progressive paraparesis), gait deficiency appears progressively. The different stages of progression generally follow the same pathological pattern:
—Intermittent limp: It is only after the subject has walked a certain distance (a few hundred metres) that the limp appears. Lower limbs go into extension spasticity. The patient lifts his feet from the ground with difficulty. After a period of rest, symptoms disappear.

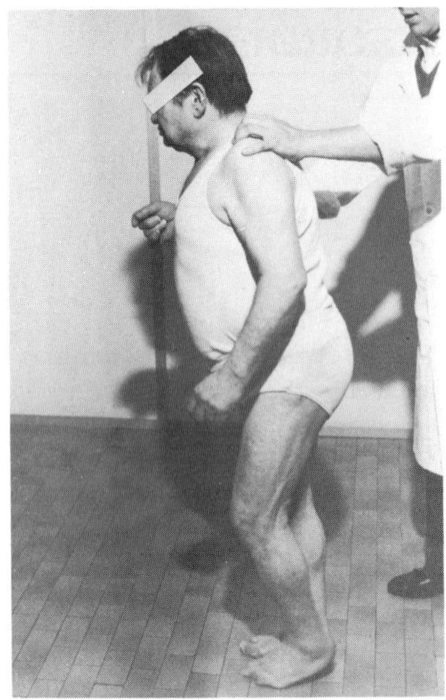

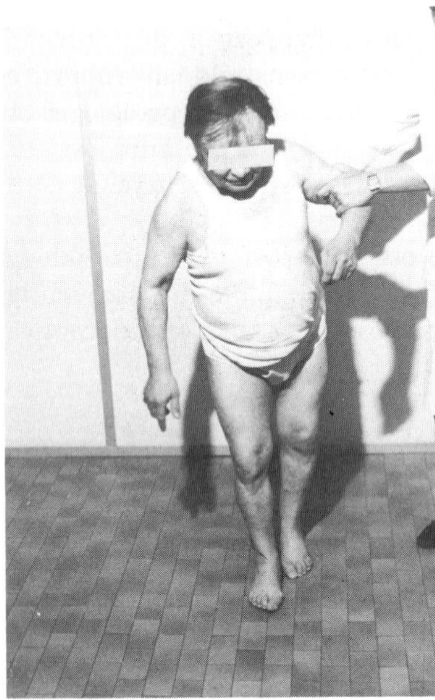

Spastic gait.

As the pathology develops gait deviations increase.

—'Hopping' gait appears when hypertonia becomes more dominant. During swing phase the unweighted spastic triceps surae pulls foot into equinus. During stance, initial contact starts with toes. Loading the lower extremity breaks the spasticity, and the heel touches the ground. This jerking during weight-bearing is accentuated by the knee being pulled into hyperextension by quadriceps hypertone.

—As condition worsens, the equinus position becomes fixed leading to a toe walking gait.

—The next step is the 'chicken gait' (Charcot), when shortening the limb during swing phase becomes impossible. To swing the limb forward, the patient must lean heavily, laterally over stance side. The limb is moved forward by pelvic hitching and hip circumduction. Taking several steps is accompanied by pronounced, compensatory trunk lateral-leaning to each side.

—The last stage of ambulation is possible only with forearm crutches or axillary crutches which permit a dragging gait. The patient leans on his walking aids and drags alternately one limb and then the other.

—After this point the patient is confined to a wheelchair.

EVALUATION

Use of a form which covers the different functional possibilities of patient pathology, identifies the effects of pathology on normal motor function.

A four-point scale quantifies the degree of the lesion:

0 There are no pathological side effects.

1 Minimal pathological disturbances are visible, but functional repercussions are light.

2 Difficulty in performing function tested, or need for outside help.

3 Inability to perform function tested.

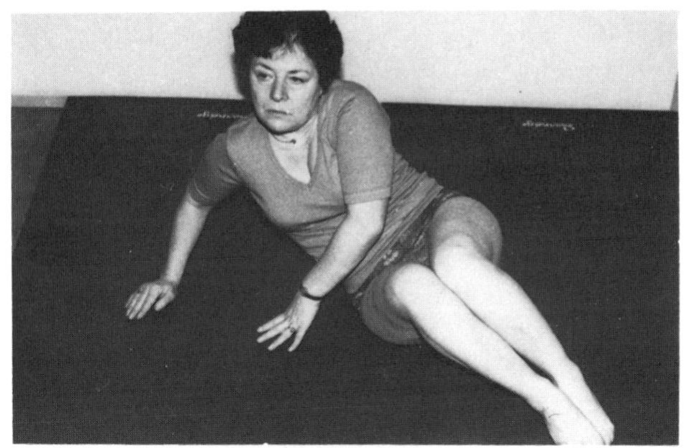

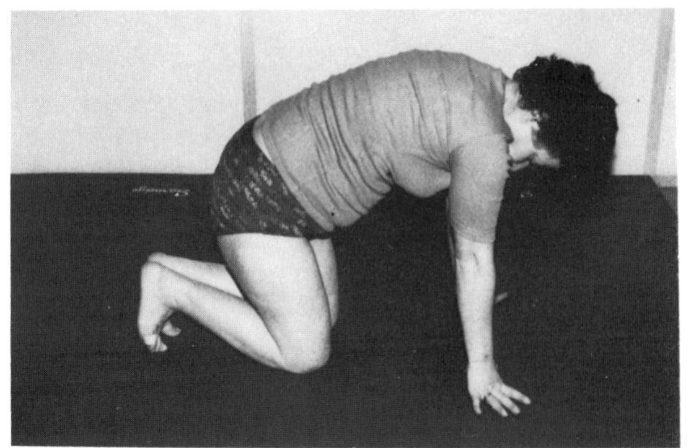

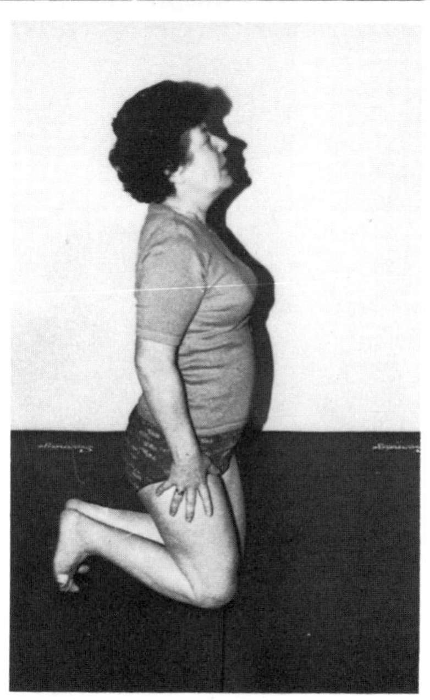

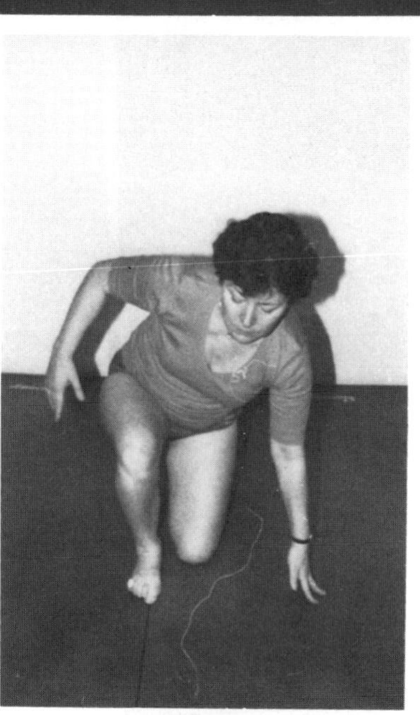

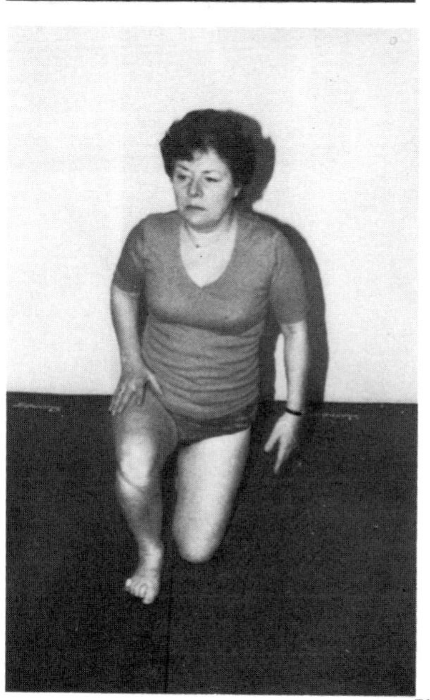

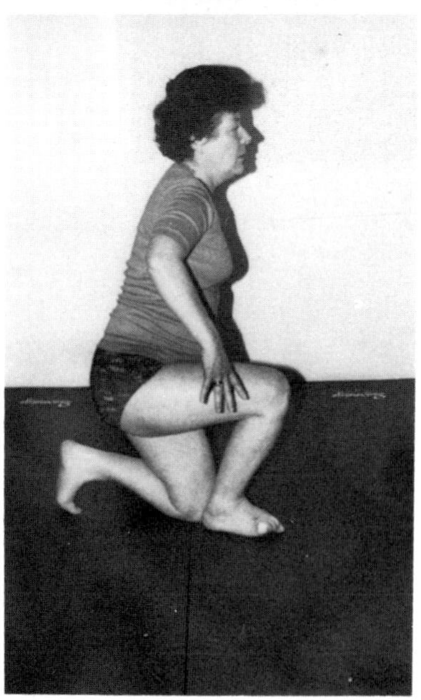

Illustrations from page 455.

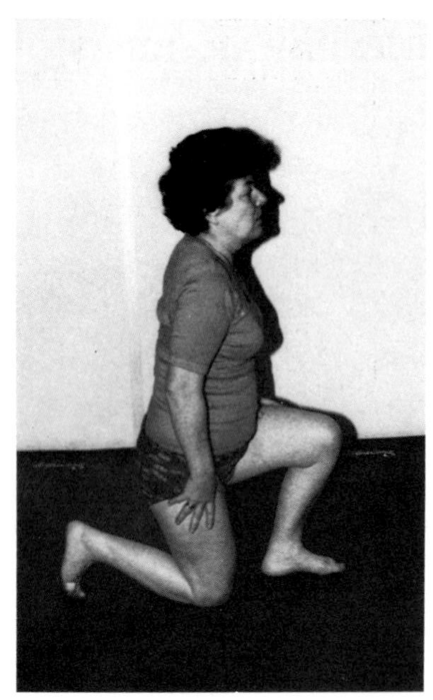

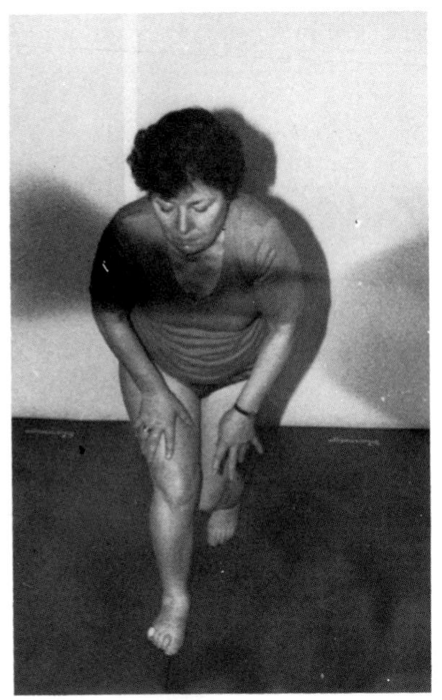

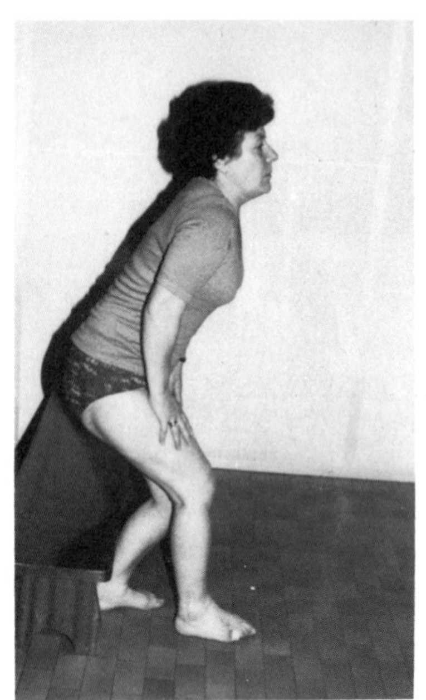

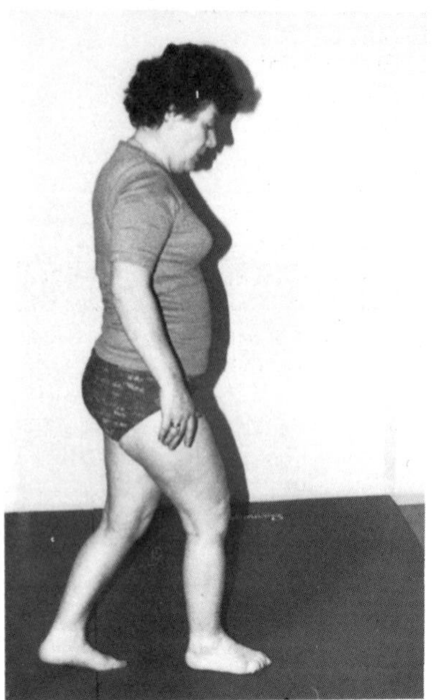

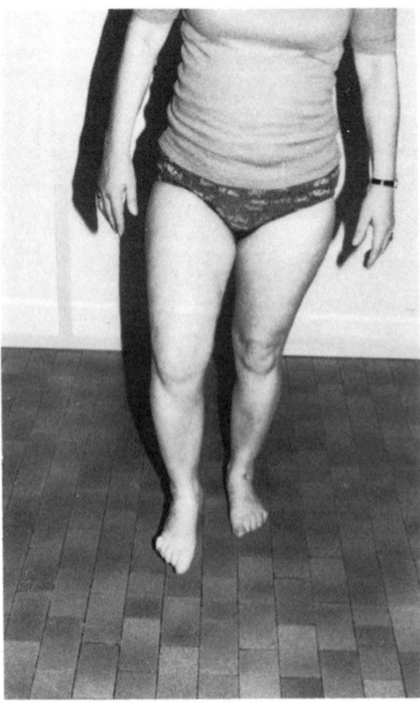

	0	1	2	3
In bed —Lies down —Rolls to right side —Rolls to left side —Rolls to prone position —Sits up in bed —Sits on edge of bed —Takes an object from night stand —Stands up from bed				
On mat —On all fours —Kneels upright on knees —Kneels on one knee • right foot forward • left foot forward —Gets up from floor with support —Gets up from the floor without support —Stands without support				
In a chair —Sits on a high seat —Sits on a low seat —Stands from a high seat —Stands from a low seat —Uses (propels) wheelchair				
At home —Walks without help —Walks with walking aid —Independent mobility in home —Independent in household activities —Independent in bathroom —Independent in toilet				
Outside area —Climbs stairs —Walks downstairs —Steps up over kerb —Steps down from kerb —Walks in street —Walks carrying a load —Walks uphill —Walks downhill —Independent on public transport —Drives a car				
Evaluation of walking perimeter —Note necessary walking aid(s)				

CLINICAL OBSERVATION OF COMPLETE SPINAL LESIONS (PARAPLEGIA)

Standing and ambulation depend upon:

—Level of lesion: Motor function of paralysed muscles must be compensated for by muscles located above the lesion level.
The higher the lesion, the fewer muscles remain to compensate for the deficiency below the lesion.

—Degree of damage: The prognosis is better when part of the spinal cord is intact, that is, an incomplete spinal transection. Depending upon whether the anterior or the posterior portion of the cord is preserved, some motor function or sensation remains.

—The possibility of splinting which requires:
- skeletal alignment.
- absence of severe spasticity which can throw patient off balance.
- absence of pressure sores in skin area in contact with brace.

—State of muscles above the site of the lesion.

—Morphology of patient (obesity is an additional handicap).

—Patient's general condition: Effort of walking with crutches requires good cardio-respiratory adaptation.

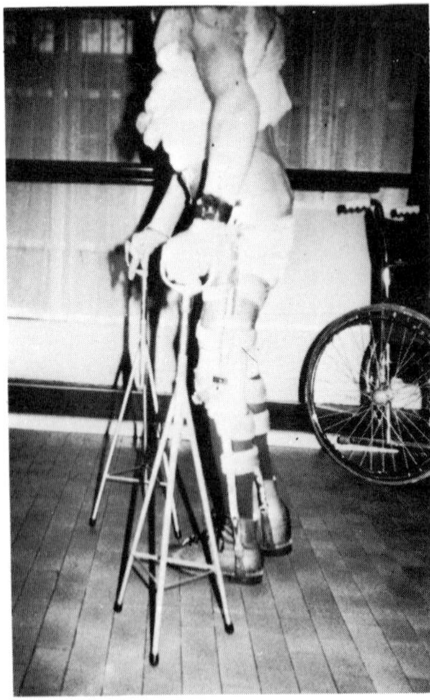

Alternating gait.

DIFFERENT TYPES OF BRACED PARAPLEGIC AMBULATION

They can be classified in two categories:

1. Gaits which require hip flexion.

Two-point gait, possible only in low medullar lesions: the paraplegic advances simultaneously one crutch and the opposite lower extremity.

Four-point gait is similar to the two-stroke, but the patient moves each point of support separately: thus:

—Right crutch then left lower extremity.

—Left crutch then right lower extremity.

Three-point gait is possible when the paraplegic has retained forward pelvic rotation. Function of quadratus lumborum, obliquus externus abdominus, obliquus internus abdominus and latissimus dorsi is necessary.

The patient places both crutches in front, leaning his trunk forward. When one side of his pelvis is hitched, his foot is lifted from the ground. Gravity swings the limb beneath the pelvis, thereby advancing the foot. Then the patient raises the opposite side of his pelvis and the other lower extremity swings forward.

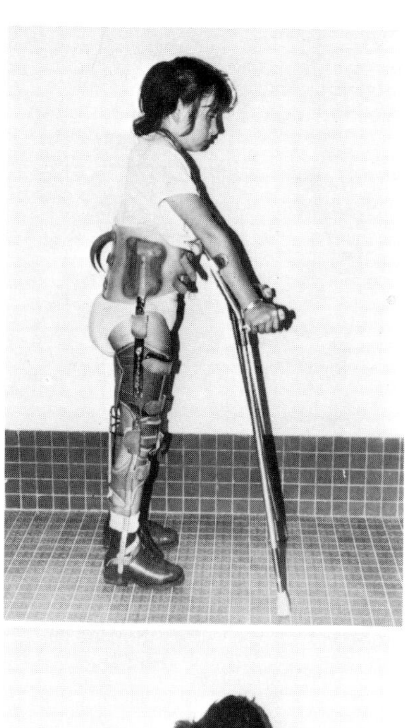

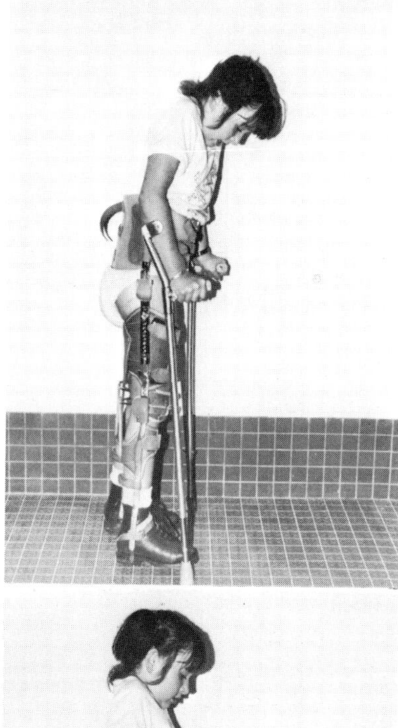

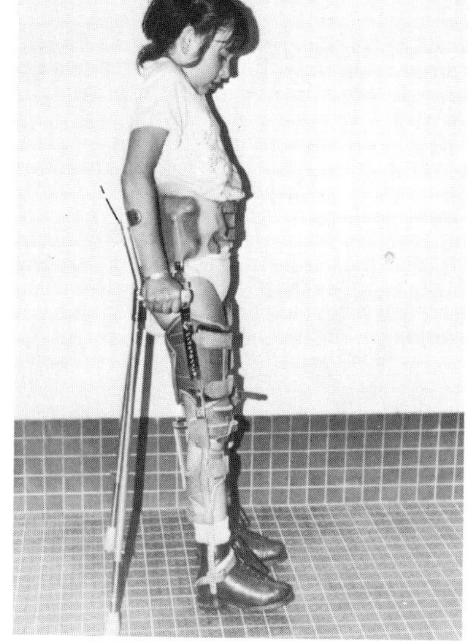

Swing-through gait.

Movement of:
—Two crutches.
—One lower extremity.
—The other lower extremity.
Produce the steps of three-point gait.

2. Swing-through gaits

These are possible whatever the level of the lesion, but they are the only mode of ambulation left to the paraplegic who has no hip flexion.

Two or three-point swing-through gait

The paraplegic leans on his forearm crutches. By extending upper extremities and depressing shoulder girdle, splinted lower extremities clear the ground. By extending his spine, he thrusts them forward. To retrieve balance, he then brings his crutches forward quickly. This is two-stroke swing-through gait. It is a gait which is rather fast but unstable. In order to correct this drawback, the patient strives to increase the number of contacts with the ground and spontaneously invents other gaits. Crutches may be brought forward one at a time for three-stroke swing-through gait.

A semi swing-through gait also exists

It is managed by the paraplegic who has retained a little hip flexion. He uses the same movement as in two-stroke swing-through gait. The difference is in movement of the lower extremities. Having lifted them from the ground, the patient uses hip flexion to perform a scissor-like movement so that the feet land one after the other.

POSSIBLE PARAPLEGIC AMBULATION CHART

Cord lesion	Affected muscles	Consequences	Splinting	Walking aids	Mode of ambulation
S2–S3	Intrinsic foot muscles	Clawing toes	Orthopaedic foot supports		Sub-normal gait (S2–S3, S1)
S1	Gluteus maximus, Biceps femoris, Triceps surae, Toe flexors	Equinus spasticity Flaccid talus	Anti-equinus shoes Anti-talus foot (pes calcaneus)		
L5	Gluteus medius, Extensor hallucis longus, Extensor digitorum longus, Peroneus longus	Lateral pelvic imbalance	same as above	2 standard crutches	Alternative gait (L5 to L2)
L4	Semi-tendinosus Semi-membranosus	Droop foot or equinus	Dorsiflexion assist, Plantar flexion stop	2 standard crutches	
L3	Quadriceps femoris Vastii	Knee may be actively locked Knee cannot be actively locked	Short leg braces Long leg braces	2 standard crutches, or 2 forearm crutches	
L2	Iliopsoas adductors, Rectus femoris	Only slight hip flexion remains	Long leg calipers	2 forearm crutches	
L1	Sartorius	No mobility of lower extremities	Possible addition of pelvic band with hip locks	2 forearm crutches	Swing-through gait (L1 to T5–T2)
T12, T11, T10	Quadratus lumborum, Abdominals, Lower trunk spinal extensors	Absence of pelvic forward rotation	Long leg calipers, Pelvic band or corset	2 forearm crutches	
T10 to T6	Upper abdominals Lower intercostals	Absence of trunk stability	Corset reaching inferior angle of scapula	2 forearm crutches	
T5 to T2	Upper intercostals	No adaptation to effort	Thoracic corset	2 forearm crutches	

USE OF A WHEELCHAIR

Spinal cord injury patients, even if they are able to walk with calipers, tend to use a wheelchair. This means of mobility is practical and perfectly adapted to their handicap.

Two types of wheelchairs should be considered:

—standard wheelchair for paraplegic and low tetraplegic patients.

—electric wheelchair for high tetraplegic patients.

T1 injury level enables use of a lightweight wheelchair allowing full-hand propulsion.

C8 injury level enables use of a wheelchair allowing palmar propulsion (hand rim or modified hand rim) at home and preferably an electric wheelchair for outdoors.

C7 injury level enables the use of an electric wheelchair with hand control.

C6 injury level enables the use of an electric wheelchair with hand control and hand stabilization wrist support.

C5 injury level enables use of an electric wheelchair with chin or head switch control. Patient's position is sitting-inclined because of respiratory distress during effort.

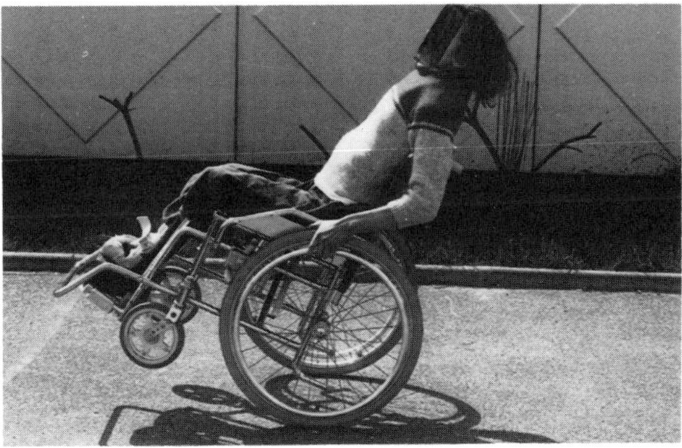

Hand operated wheelchair.

Electric wheelchair.

EVALUATION OF POSTURE AND GAIT IN HEMIPLEGIA

GENERAL REMARKS

Muscle activity of trunk and of lower extremities cannot be dissociated in:
—posture.
—balance reactions.
—gait.
Whether lower extremities or trunk are affected, locomotor function is disturbed. It is therefore this function which can be tested.

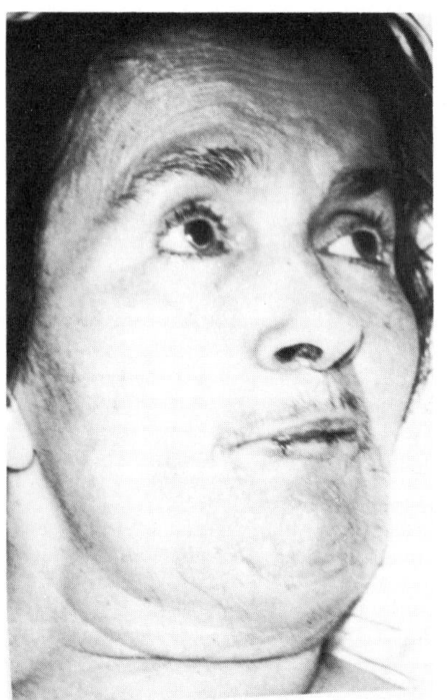

Simultaneous deviation of head and eyes.

CLINICAL OBSERVATION

SUPINE POSITION

Spontaneous posture

Head is most often rotated towards the sound side. 'The patient looks at his cerebral injury' (common deviation of head and eyes). Occasionally the head rotates towards the affected side.
Trunk presents scapular and pelvic girdle rotation, and trunk latero-flexion towards the affected side.

Affected lower extremity

—hip is in adduction, extension and external rotation. Posture in abduction, slight flexion and external rotation is less frequent.
—knee is in hyperextension, sometimes in flexion.
—foot is in equinovarus and supination.
—toes are either in full claw with flexion of all three phalanges, or in hammer with metatarsophalangeal extension and interphalangeal flexion.

Motor behaviour

Trunk partially retains its motor control. Patient can protract his girdles, extend chest on affected side, and perform respiratory movements.
Lower extremity retains in this open-loop position:
• Motor ability in flexion.

This withdrawal is often performed in a froglike manner:

—hip in abduction, flexion and external rotation.

—knee in flexion.

—foot in pes calcaneus (talus), varus and supination.

—clawing of toes.

- Motor ability in extension:
  - — hip in adduction, extension and external rotation,
  - — knee in hyperextension,
  - — foot in equinovarus,
  - — clawing of toes.

SITTING POSITION

Spontaneous posture

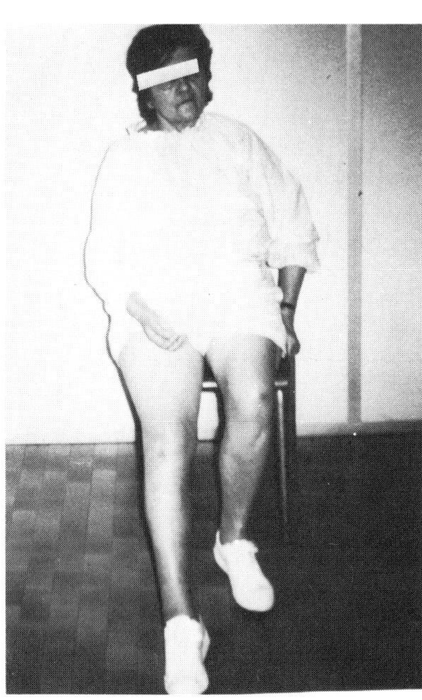

Hemiplegic posture in sitting position.

Trunk deviates towards the hemiplegic side. Bodyweight is placed predominantly on the sound buttock.

Head is held in lateral deviation and rotation generally towards the sound side. This posture is less accentuated in sitting than in supine.

Patient tends to keep affected hand on knees. This posture accentuates pathological posture of upper extremity.

Lower extremity remains in extension, in hip abduction and external rotation. Foot rests on its lateral side. The limb thus retains hemiplegic posture in sitting position.

Motor behaviour

Sitting is a good position for testing upper extremity function.
It enables examination of:
—Trunk postural control by spinal extension.
—Ability for weight transfer on buttocks and scapular girdle de-rotation.
—Postural equilibrium reactions triggered by examiner's stimulation.
Trunk postural adaptation is necessary for use of the upper extremity. Dynamic limb movement is only possible when based on a correctly balanced trunk postural pattern.

STANDING POSITION

Spontaneous posture:

Upper extremity demonstrates its characteristic posture.

Trunk displays:

—Rotation of paralysed side.
—Tendency for latero-flexion towards affected side. This pathological posture is that of the tonic lumbar reflex which leads to lower extremity extension and external rotation posture.

Lower extremity tends to assume:

—Slight hip flexion causing the patient to lean forward a little. If patient holds trunk straight, lower extremity is held in front, thereby partly unloading body weight.

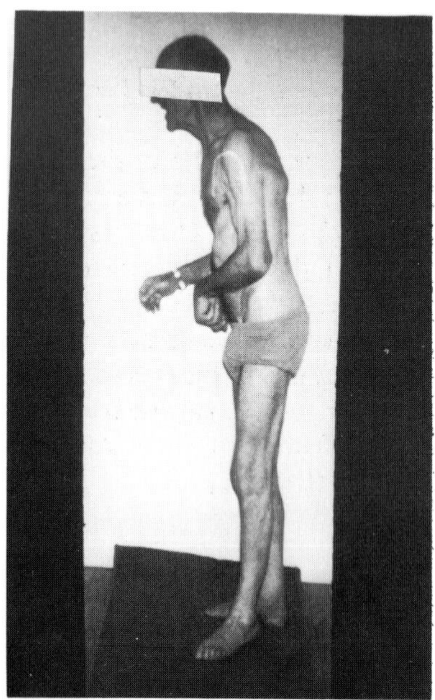
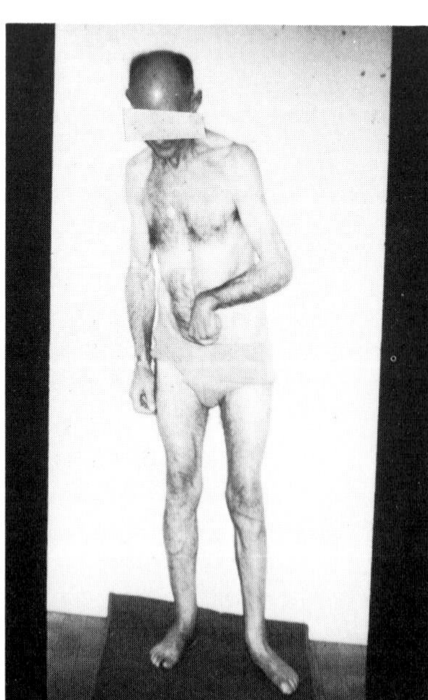
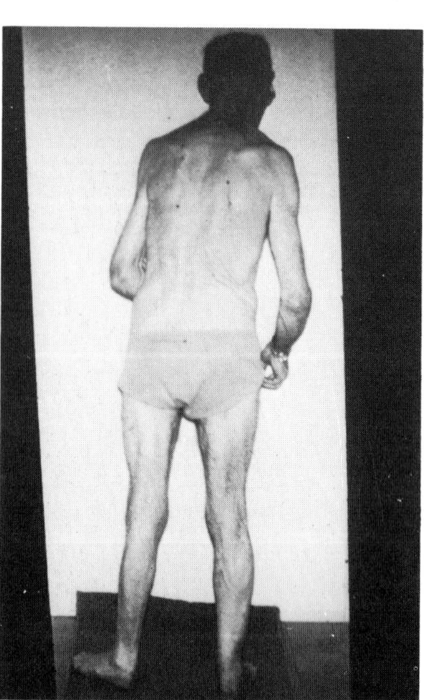
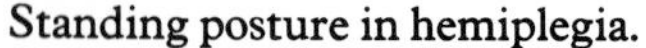
Standing posture in hemiplegia.

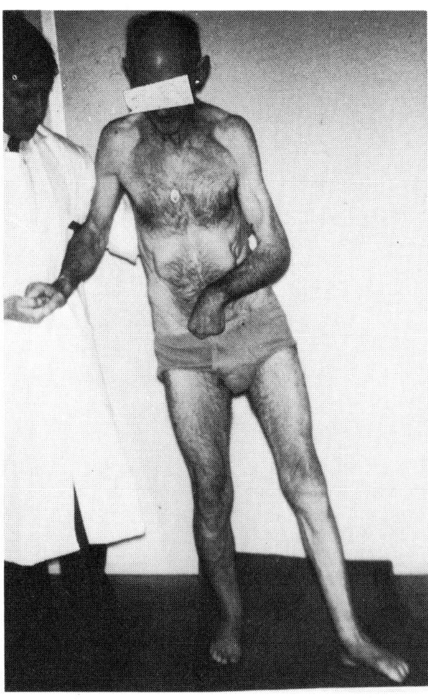

Hip circumduction during swing phase ('fauchage').

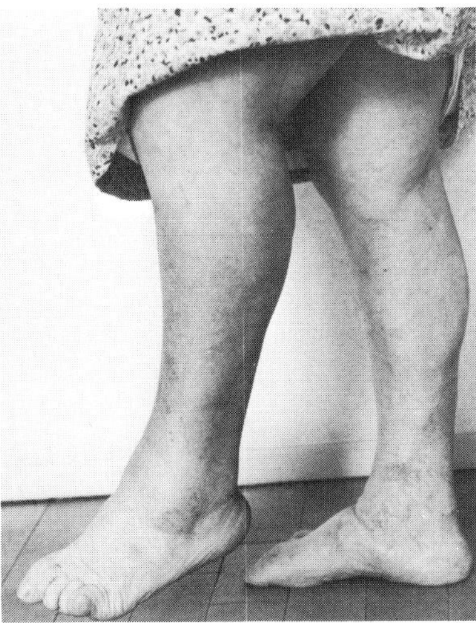

Spasticity of triceps surae during swing phase.

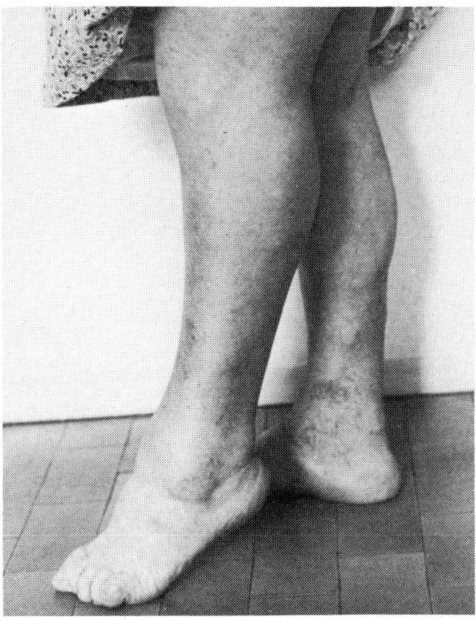

Toes first in initial contact of stance phase.

—External rotation due to pelvic rotation forward on affected side, and posture induced by hypertonia.

—Equinovarus of the foot. The effect of mechanical increase in limb length associated with flexed hip position leads to the 'trigger' knee. The knee joint is abruptly drawn into hyperextension during stance because of two factors:
- quadriceps femoris spasticity,
- pathological hip and foot posture.

—Clawing of the toes. Sometimes pressure on foot plantar surface can trigger a toe dorsiflexion reaction.

—*Note:* Certain types of hemiplegia lead to a posture of lower extremity flexion. Foot sole contact can facilitate limb mass flexion (withdrawal), the lower limb folds up leaving the patient standing on his sound limb like a stork in water.

Motor behaviour during gait

Hemiplegic gait is characterized by hip circumduction ('fauchage'). During swing phase, the patient thrusts his limb outwards and forwards in a semicircle from the hip. This movement is made possible by pelvic hitching and hip circumduction. The resulting limp is due to absence of active hip or knee flexion, and of mechanical limb lengthening by the equinus foot. During foot-clearance, the toes drag on the ground. Initial contact is on the anterior-lateral aspect of foot sole.

During stance on the affected side, the absence of roll-off, precarious single-limb balance and the presence of a 'trigger' knee force the patient to place his unaffected foot on the ground quickly. Step length is shortened. The body leans considerably over the affected side, the sound upper extremity swinging away from the trunk to regain balance.

Other types of gait disturbance are found in hemiplegia:

If there is significant trunk rotation towards the affected side, this side stays behind the unaffected side. During gait, the hemiplegic patient moves his sound lower limb forwards and 'drags' the other from behind without moving it forward freely. If while standing the patient clearly places his affected limb forwards, during gait he moves his sound limb to the level of the other, then, by leaning the trunk backwards, he thrusts the affected limb forwards. The patient 'pushes' his affected side.

QUANTITATIVE EVALUATION

A four point grading system describes hemiplegic postural and locomotor performance:

0 Activity requested is performed without difficulty.
1 Activity requested is performed with difficulty.
2 Activity requested is performed with difficulty, or external help is needed.
3 Activity requested is impossible to perform.

Evaluation form is based on work of Albert, A., 1970; Bobath, B., 1975; and Michels, E., 1959.

	0	1	2	3
1. Lies down: —on right side —on left side				
2. Turns over to prone position, alone: —rolling to right side —rolling to left side				
3. Sits up from supine, leaning on: —right upper limb —left upper limb				
4. Holds himself up on all fours				
5. Rises from prone to kneeling on all fours				
6. Rises from on all fours to sitting: —using right side —using left side				
7. Moves from sitting to kneeling on all fours: —using right side —using left side				
8. Kneels upright on knees				
9. From hands and knees, rises to kneel upright on knees. —From kneeling upright on knees, returns on all fours				
10. Kneels on one knee: —right foot forward —left foot forward				

	0	1	2	3
11. Moves from kneeling upright on knees to one-knee stance: —right foot forward —left foot forward				
12. Returns from one-knee stance, right foot forward, to kneeling on all fours Returns from one-knee stance, left foot forward, to kneeling on all fours				
13. Moves from one-knee stance: —right foot forward, to standing —left foot forward, to standing				
14. From standing position, moves to one-knee stance: —right foot forward —left foot forward				
Standing position 1. Maintains balance				
2. Maintains balance: —affected lower extremity forward —affected lower extremity behind				
3. Actively controls locking and unlocking affected knee				
4. Assumes single limb stance: —on sound side —on affected side				
5. Has protective reactions on affected side when falling: —forward —sideways —backwards				

Gait	0	1	2	3
1. Walks: —without stick —with stick				
2. Exhibits: —complete hip circumduction gait —hip circumduction during terminal swing —hip circumduction during initial swing				
3. Stance phase on affected lower extremity: —loading response —terminal stance (roll-off) —shifts weight and pelvis to affected side —actively controls knee in slight flexion —rolls foot off ground				
4. Swing phase of affected lower extremity: —controls initial swing (hip extension with knee flexion) —controls terminal swing (hip flexion with knee extension)				
5. Equal speed of movement on affected and sound sides				
6. Walks: —sideways —backwards				
7. Walks upstairs: —right foot forward —left foot forward				
8. Walks downstairs: —right foot forward —left foot forward				

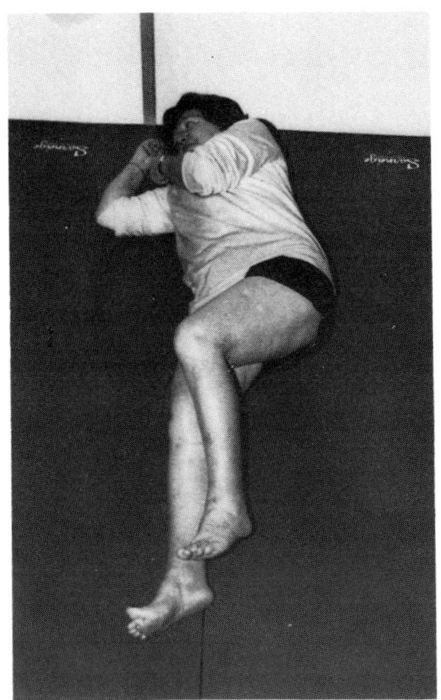

Test 1.

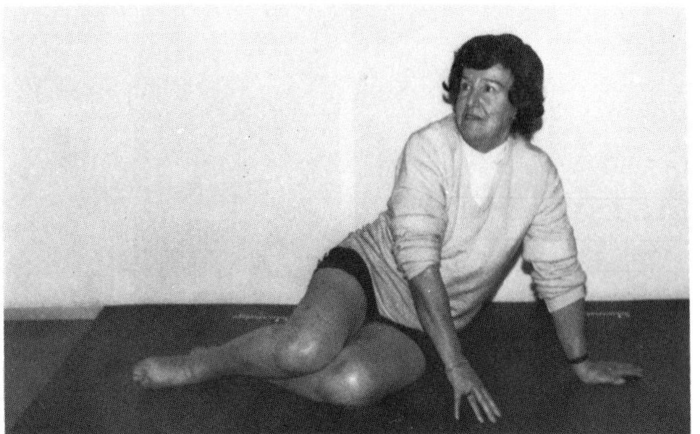

Test 3.

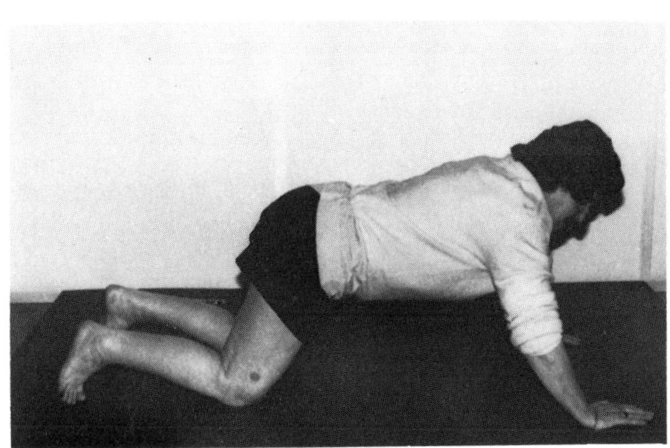

Test 4.

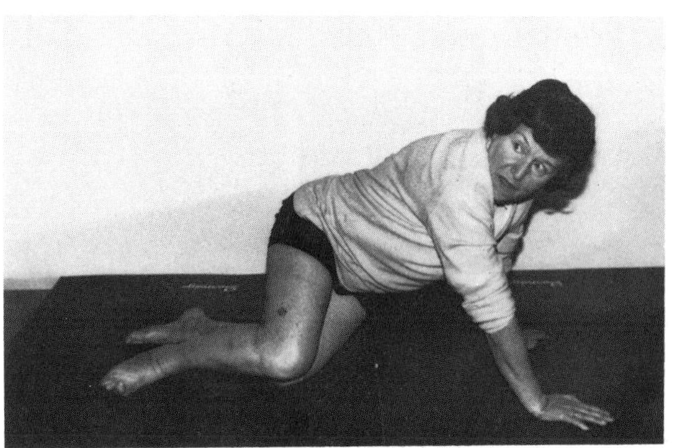

Test 5.

Illustrations from page 464.

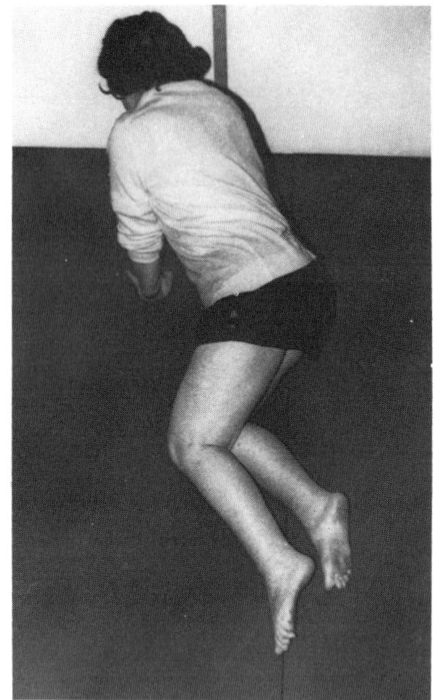
Test 6.

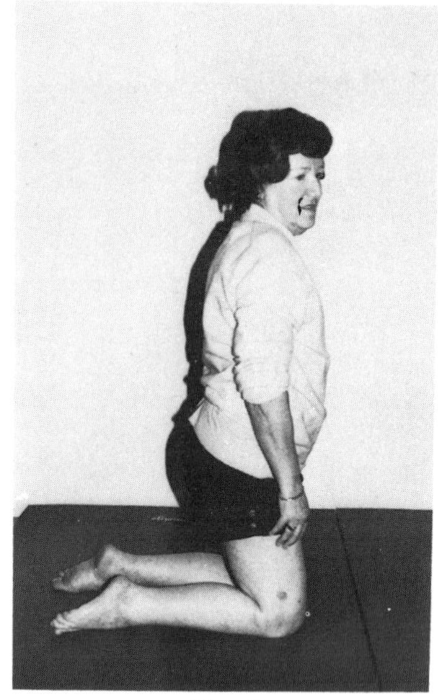
Test 7.

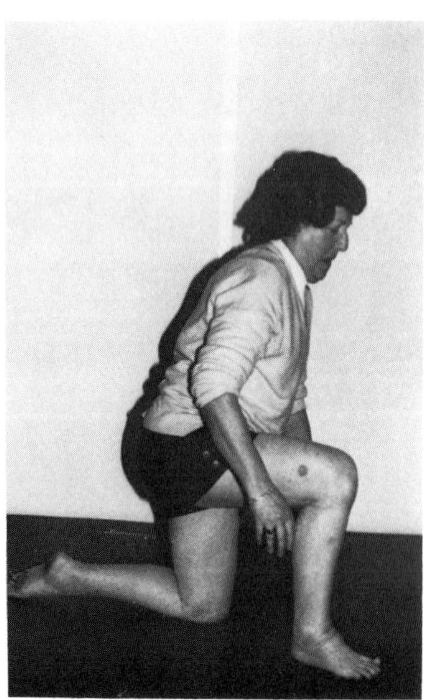
Test 9.

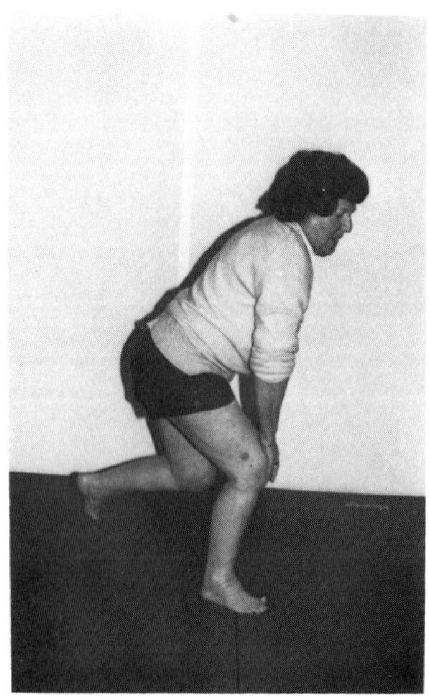
Test 12.

Illustrations from page 464.

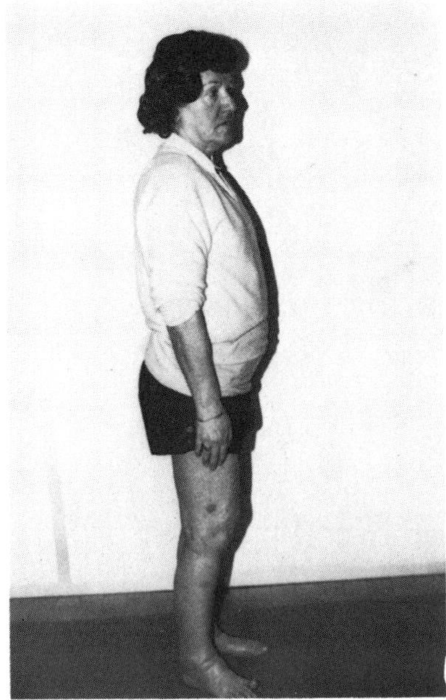

Test 1.

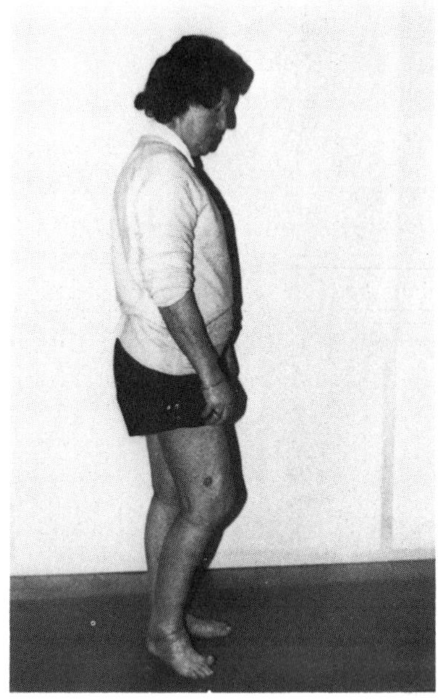

Test 3.

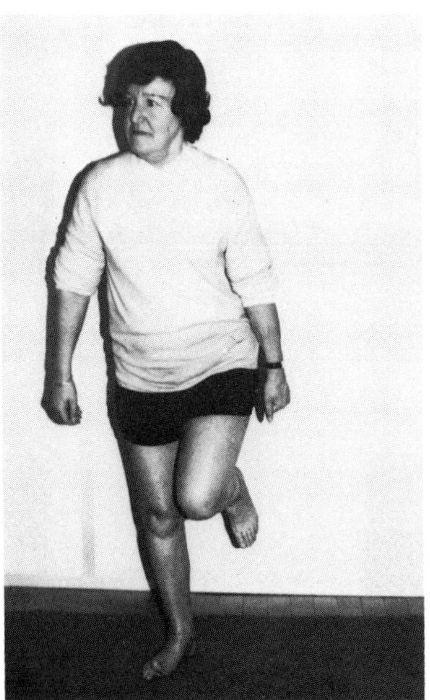

Test 4
on affected side.

Illustrations from page 465.

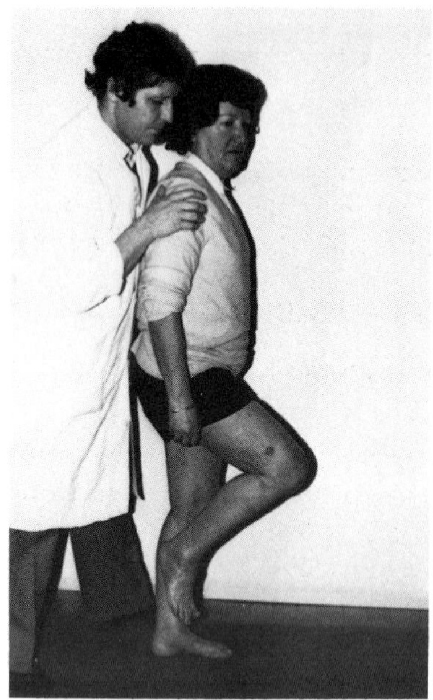

Starting position.
Test 5.

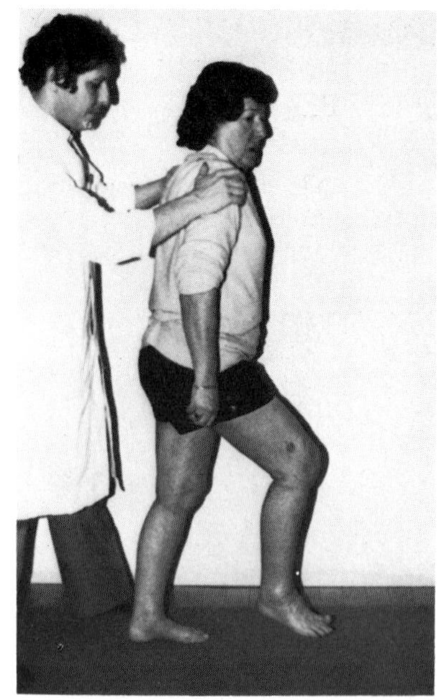

Protective reactions.
Test 5.

Illustrations from page 465.

WALKING

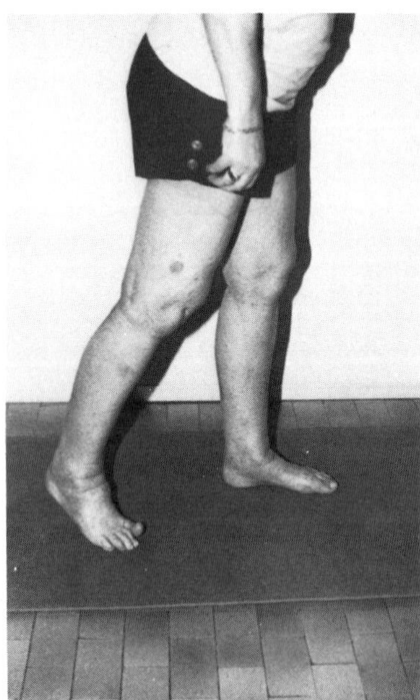

Gait.
Push-off.

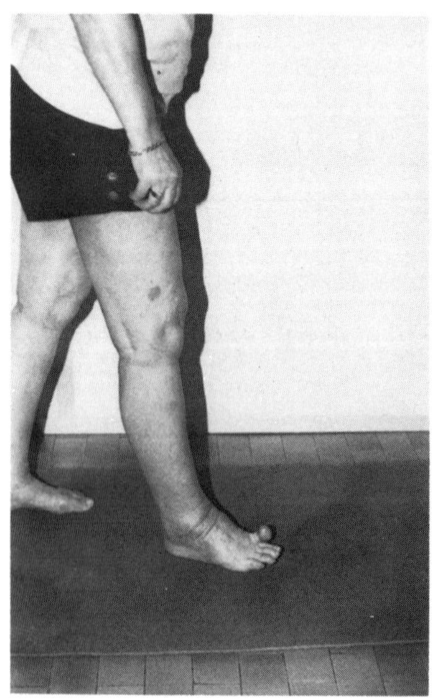

Terminal swing.

Illustrations from page 466.

EVALUATION OF HEMIPLEGIC UPPER EXTREMITY MOTOR FUNCTIONS

CLINICAL EVALUATION

GENERAL REMARKS

Unlike the lower extremity, the hemiplegic upper extremity does not spontaneously recover in a functional position. The patient is left with the problems of a single-handed person. Involvement of the dominant upper extremity further decreases the hope of social and professional function.

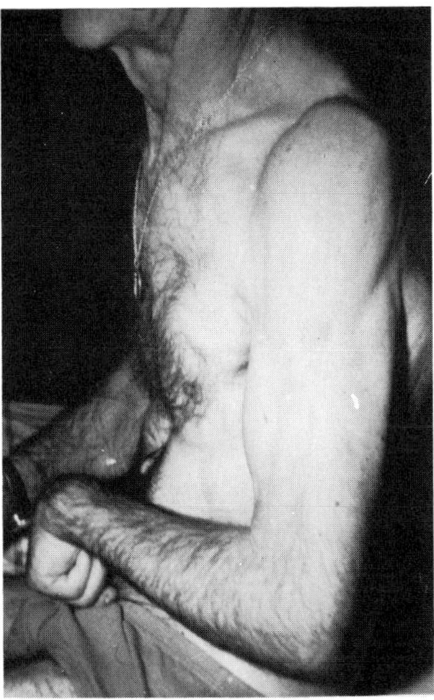

Hemiplegic upper extremity.

CLINICAL OBSERVATION

Most common spontaneous posture for upper extremity

Scapula is retracted in adduction and tilts in downward rotation. Arm clings to body in adduction and internal rotation. Elbow is partly flexed, but elbow extension is not uncommon. Forearm is most frequently in pronation.
Wrist is straight or more or less flexed.
Metacarpophalangeal joints are straight and interphalangeal joints are in flexion.
Thumb tends to be positioned in the palm in slight adduction. The thumb interphalangeal joint is either in extension or in slight flexion.

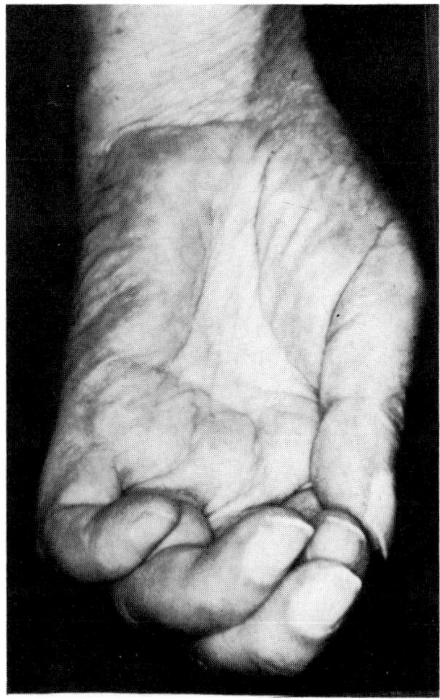

Hemiplegic hand.

Hemiplegic upper extremity, forearm supination posture.

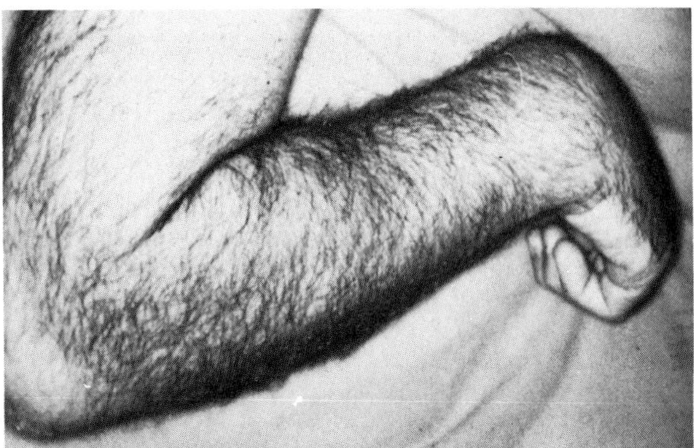

Forearm pronation posture.

Spontaneous motor behaviour

Distribution of:
—muscular deficiency,
—spasticity,
—synkineses,
lead to motor difficulties which result in the lack of selective movement.

In the most severe cases, patient's upper extremity is almost completely immobile. Sometimes only limited movement of the shoulder girdle, by trapezius, is present. During effort or gait, massive reactions appear in the upper extremity which accentuate the pathological posture, stressing the extent of the mass synkineses.

When partial recovery is gained, the patient acquires voluntary control of flexion and extension synergies. Disturbance produced by coordination synkineses leads to little functional use.

When selective motor control appears, it follows a proximal to distal progression. Selective control first appears in the shoulder, the muscles of which are mostly under automatic control.

These centres are generally unaffected in corticospinal tract lesions.

However, the voluntary motor tracts mostly contol the distal portions of the extremities, and so the hand.

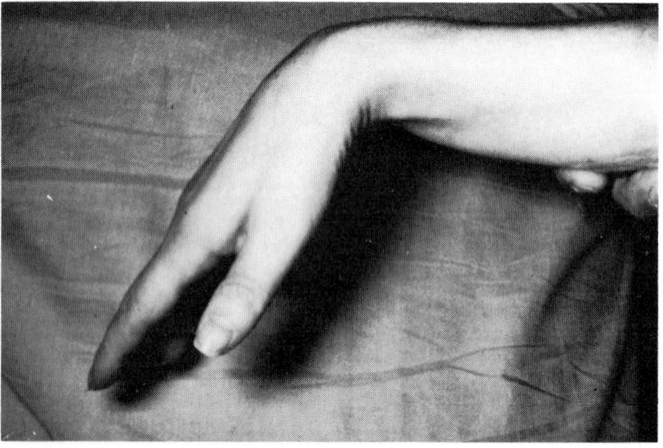

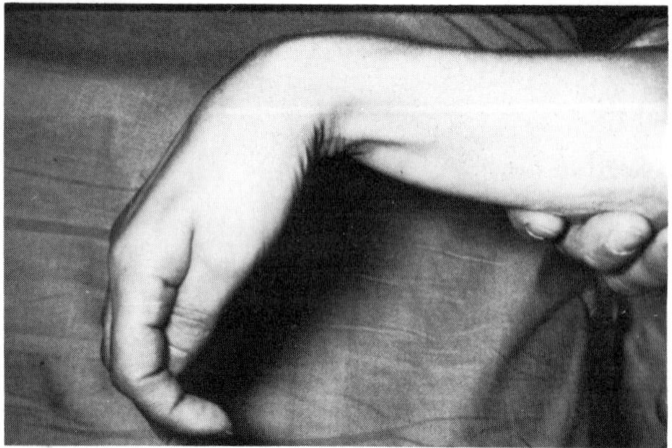

Grip difficulty in hemiplegia.

QUALITATIVE EVALUATION

Complete grading of functional ability may be applied to the hemiplegic hand depending on the severity of the lesion. From the most severe to the least, the effects of impaired hand control may be:
—Loss of grip of any type.
—Use of limb as a fixator which holds paper or objects firmly while the other hand manipulates.
—Use of limb for arm to trunk grip.
—Use of tenodesis grip. Active wrist extension closes the fingers against the palm by pulling on finger flexors already shortened by either contracture or spasticity. Releasing wrist extension opens the fingers.

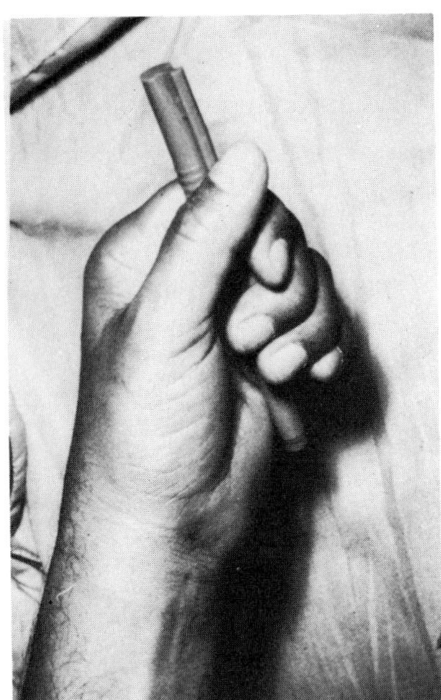
Digito-palmar grip of a hemiplegic hand.

—The ability to achieve digito-palmar grip: synkineses dominate finger flexors preventing them from moving independently of each other. Fingers open and close together.
—Use of thumb with more or less precision.
—The ability to perform two-handed activities without coordination. Motor and sensory feedback from the sound side dominates and overshadows information coming from the affected side. In activities requiring the use of both hands, the patient has the tendency to forget the affected side.

In spontaneous activity, movements of the impaired hand slow and stop.

Except for light, transitory hemiparesis, complete recovery of hand function is rare. Depending upon the lesion, the following may persist:
—Slight hypertonia of flexor muscles hindering hand suppleness and dexterity.
—Postural instability.
—Increased tone when precision is required.
—Slight ataxia.

These different disturbances may be aggravated by deep sensory impairment, or apraxia resulting from cortical lesions.

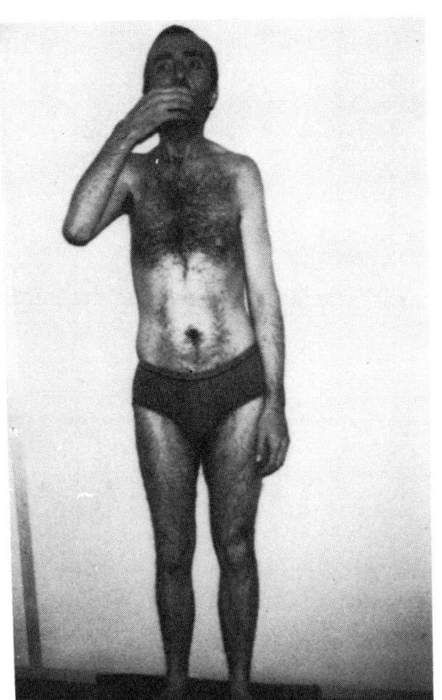
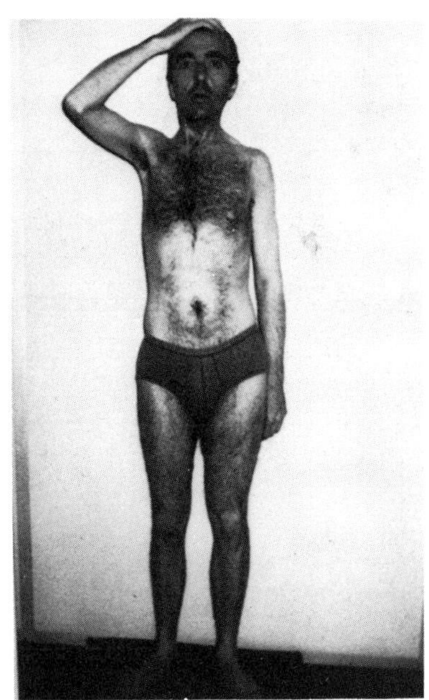
Hemiplegic upper extremity: correct control of proximal joints.

QUANTITATIVE EVALUATION

Follows a form which explores independence of hand activity, and considers different positions of upper limb in space.

A four point grading scale is used:

0 Function tested is performed without difficulty.
1 Function tested is performed although moderate pathological signs are present.
2 Function tested is performed with difficulty or incompletely; it is not spontaneous. Major pathological signs are present.
3 Function tested is impossible to perform.

	0	1	2	3
Spontaneous use of: —affected limb to stabilize objects —arm to trunk grip				
Digito-palmar grip: —reaches —holds —releases				
Hand to mouth, forearm in supination: —opens fingers —closes fingers —thumb to index grip • lateral opposition • pad opposition				
Hand to mouth, forearm in pronation: —opens fingers —closes fingers —thumb to index grip • lateral opposition • pad opposition				
Upper extremity in forward horizontal position: Forearm in supination: —opens fingers —closes fingers —thumb to index grip: • lateral opposition • pad opposition				
Forearm in pronation: —opens fingers —closes fingers —thumb to index grip: • lateral opposition • pad opposition				

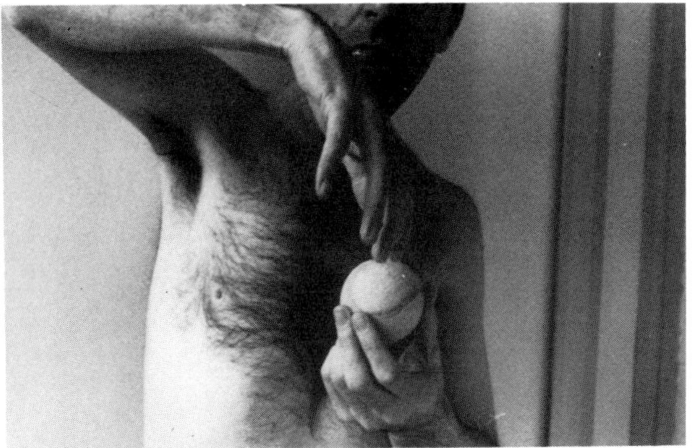

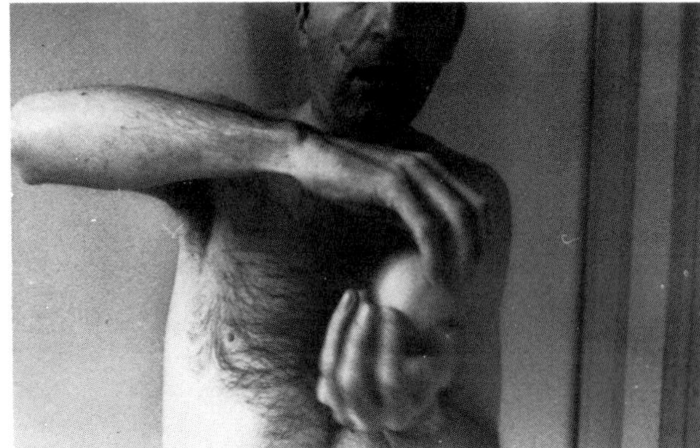

Digito palmar grip: both grip, release.

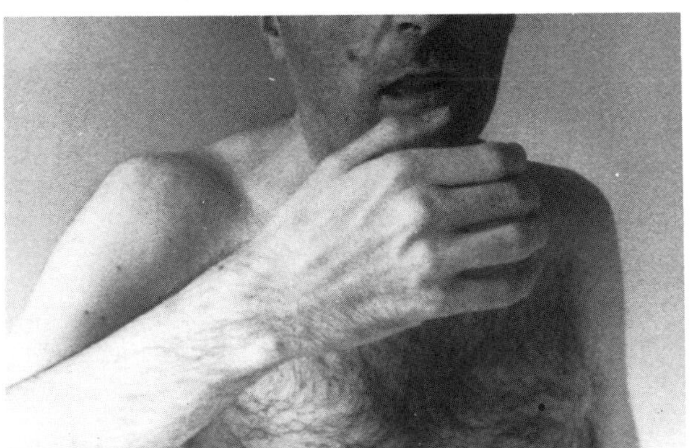

Hand to mouth: Forearm in supination.

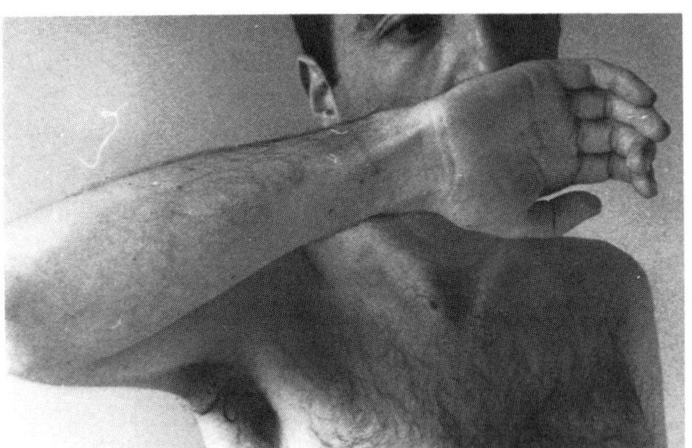

Hand to mouth: Forearm in pronation.

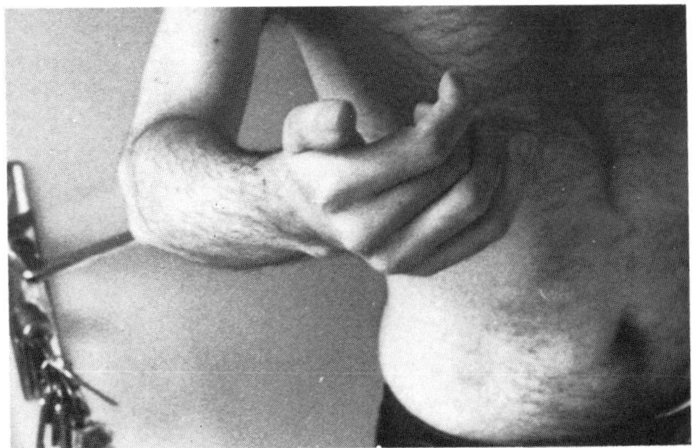

Upper extremity in forward horizontal position:
Opens fingers with forearm in supination.

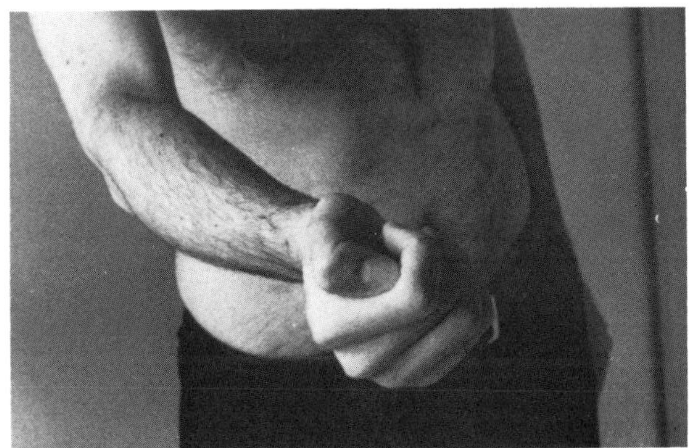

Closes fingers.

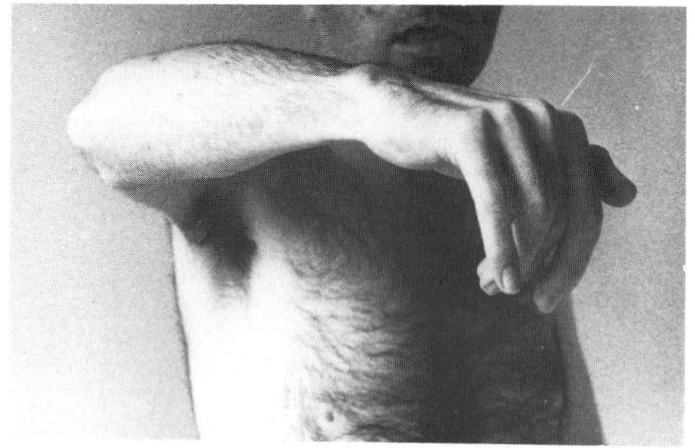

Upper extremity in forward horizontal position:
Opens fingers with forearm in pronation.

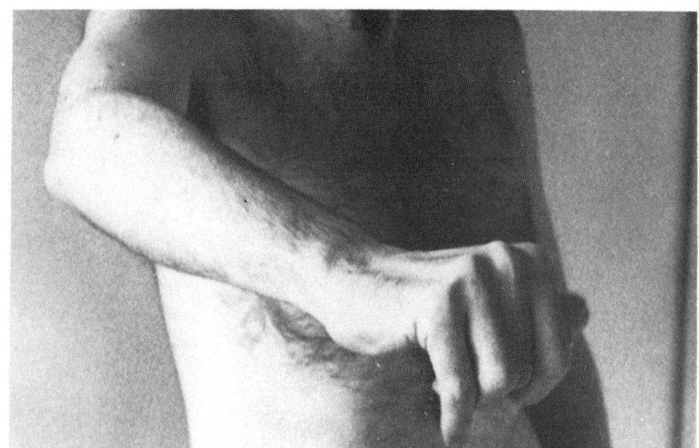

Closes fingers.

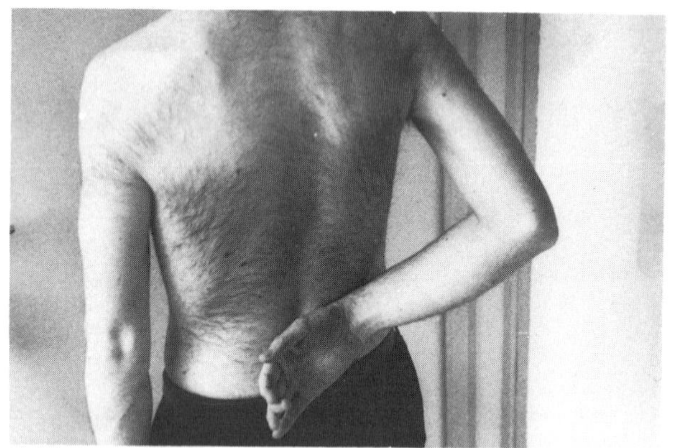

Hand behind back:
Opens fingers.

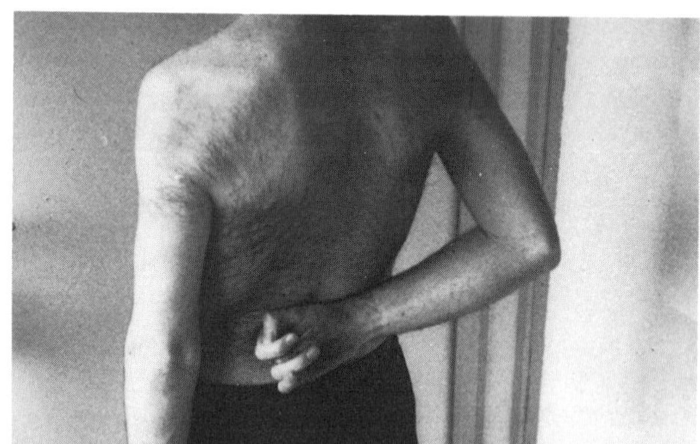

Closes fingers.

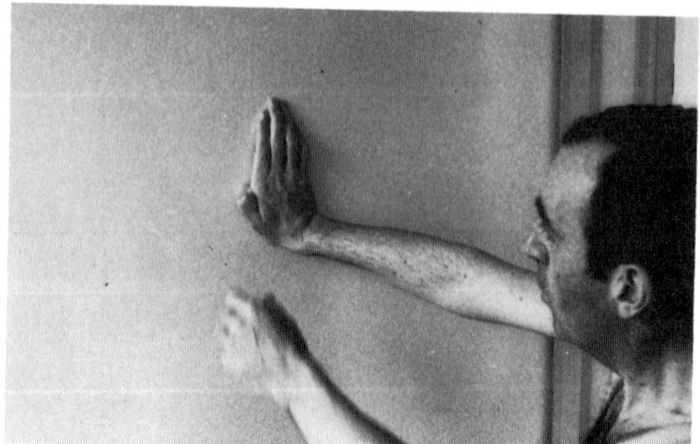

Leans against a vertical surface.

	0	1	2	3
Hand behind the back: —opens fingers —closes fingers —thumb to index grip: • lateral opposition • pad oppositon				
Individual finger extension (flick of the finger): —index —middle finger —ring finger —little finger				
Coordinated bilateral activities: —knots a scarf —knots a string —beats time on table alternating both hands				
Leans on palm, upper extremity straight in external rotation: —on a horizontal surface —on a vertical surface				
Protective reaction with affected limb when falling: —forward —sideways				
Upper limb participation in speech				
Upper limb swings automatically during gait				

MOTOR FUNCTION EVALUATION IN PARKINSONIAN SYNDROMES.
THE EFFECT ON MANUAL DEXTERITY

GENERAL REMARKS

Classical paralysis agitans described by James Parkinson is characterized by:
—Resting tremor.
—Rigidity.
—Akinesia.
Only the latter two signs affect voluntary motion.

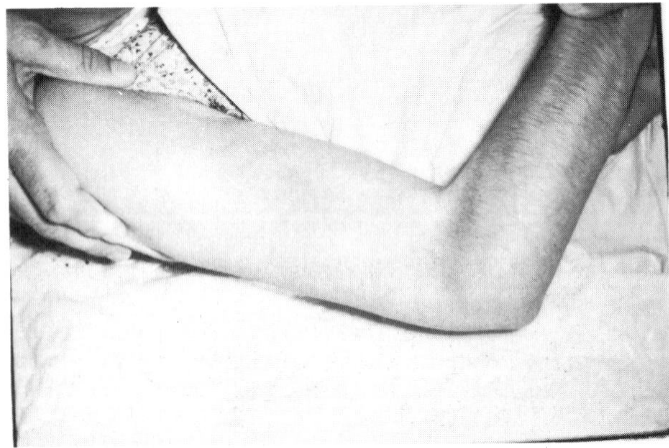

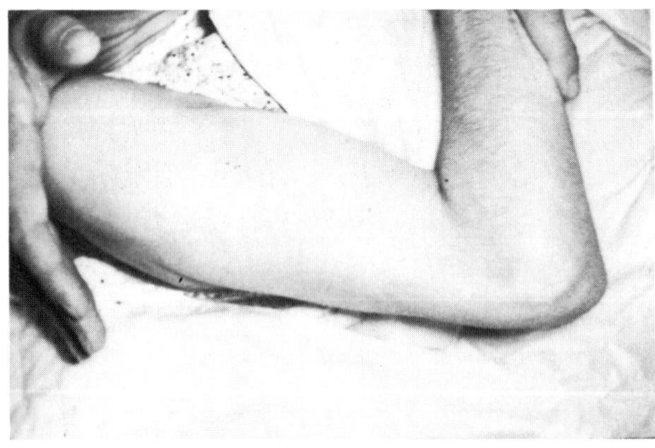

Agonist, antagonist rigidity.

CLINICAL OBSERVATION

Rigidity

An extra-pyramidal type of hypertonia, which becomes obvious during passive movement. It is characterized by a resistance which is apparent from the very beginning of movement.
This resistance is either:
—Constant throughout passive movement, called 'lead pipe' resistance.
—Yielding by steps, a 'cog-wheel' resistance.
Rigidity affects both agonist and antagonist muscle groups with a greater effect on flexors. It is composed of two distinct elements:
1. A stretching reaction on the agonist muscle which overreacts to lengthening. This reaction appears at the beginning of passive elongation and remains constant during passive mobilization.
Consequently, voluntary isometric muscle force is preserved while dynamic force diminishes since the agonist muscle must overcome resistance of its antagonist.
2. A shortening reaction: the muscle, of which origin and insertion are passively brought together, contracts to hold the position. This leads to an increase in postural reactions which predominate in flexor muscles creating the characteristic parkinsonian stooping posture.

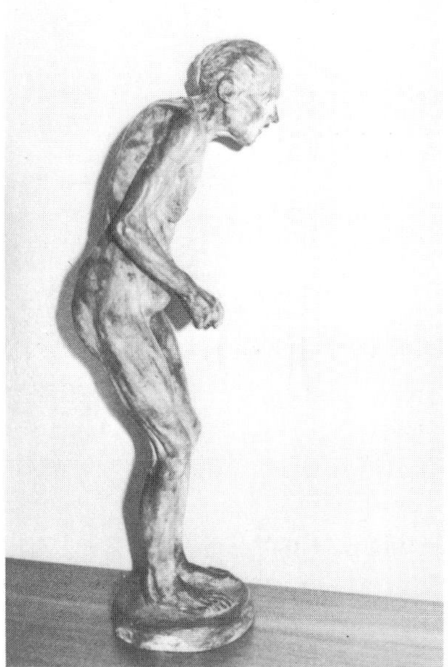

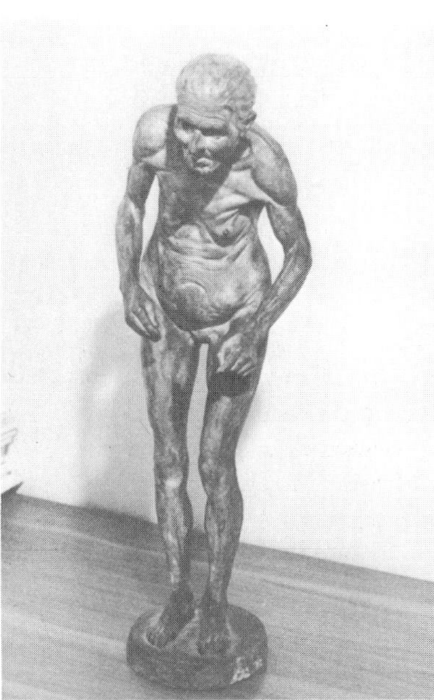

Statuette sculpted by Paul Richer (Pr. Rondot collection).

Rigidity is variable:

—There is a relationship between the position of the eyes and rigidity of the upper extremities (Quix's gaze).

—Parkinsonian rigidity decreases when the attention of the patient is distracted from the active area of his body. A true 'phenomenon of chasing' hypertonia occurs. Rigidity which was significant in one part of the body decreases suddenly while hypertonia increases in another part with no apparent reason or logic.

This conditions the posture of the parkinsonian patient who presents the following:

—Trunk:
- upon which the head protrudes,
- back is rounded by kyphosis which may be marked.

—The upper extremities:
- arms are along sides of body,
- slightly flexed,
- hands at hip-joint level in a characteristics posture (the most common of which is the 'writer's hand') (i.e. thumb and forefinger in apposition).

—The lower extremities:
- held tightly together,
- slightly flexed,
- feet and ankles are often in equinus.

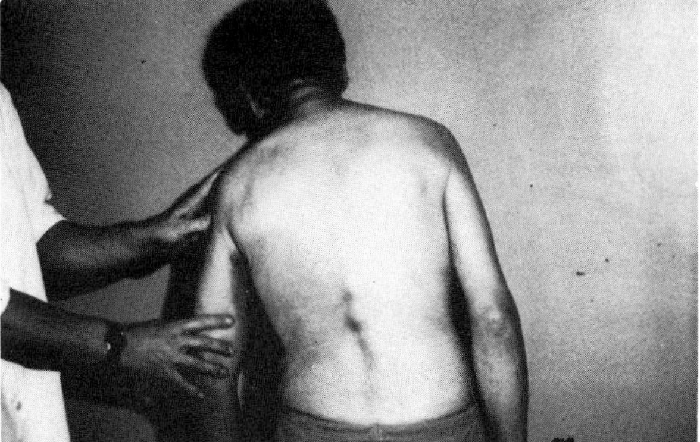

Trunk latero-flexion. Parkinson's disease predominant on left side.

PARKINSONIAN HANDS

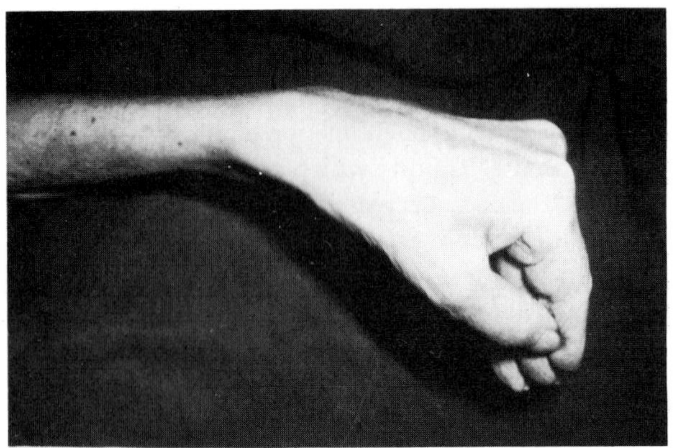

Writer's hand.

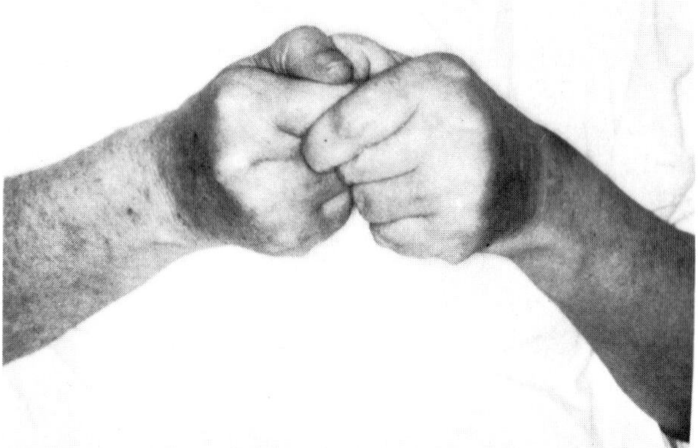

Magician's hands, pill-rolling tremor.

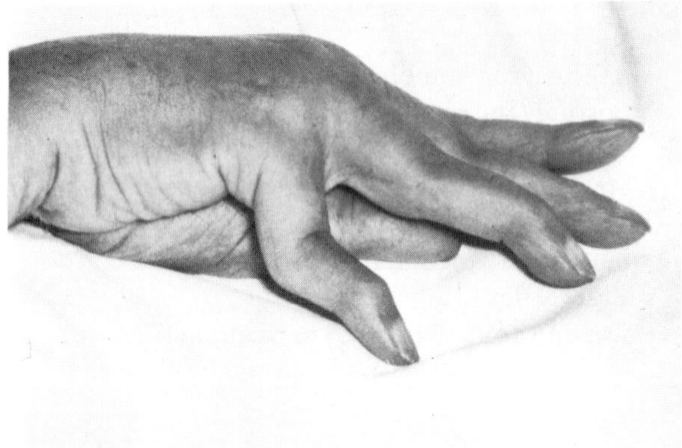

Pseudo-rheumatic hand.

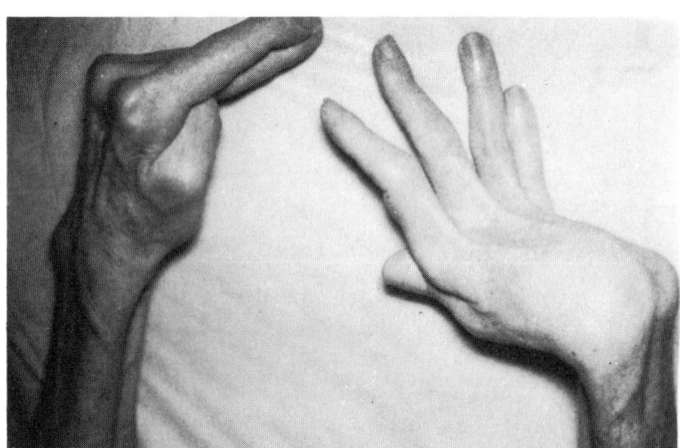

Hand described by Dejerine (at right).

PARKINSONIAN FEET

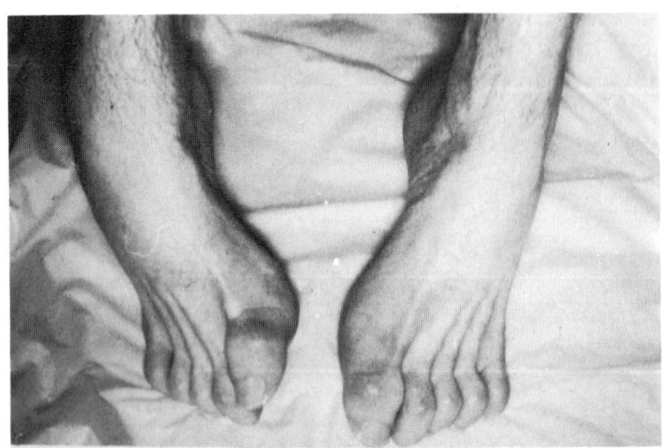

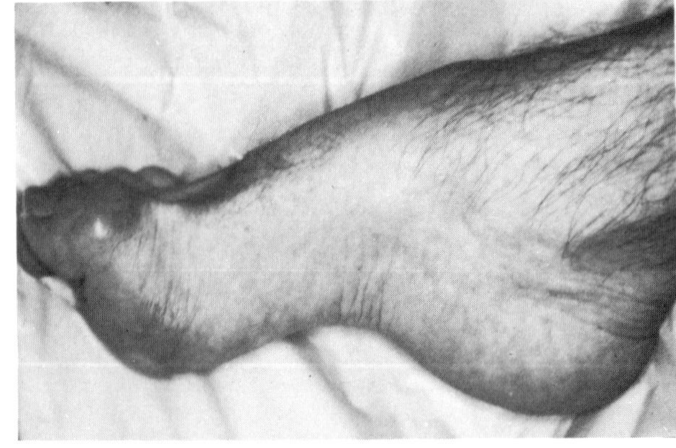

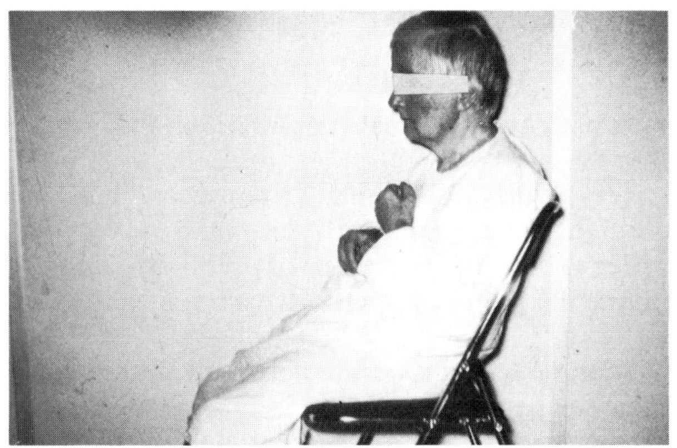

Parkinsonian posture.

Akinesia

This is a disturbance in the initiation and performance of movement associated with the loss of automatic functioning.
Gestures become slow and rare. The patient needs to think each move out. Functional activity, composed of various patterns of movement whose organization is normally under automatic control, becomes difficult or impossible to perform.
The parkinsonian patient must voluntarily reconstruct the patterns and reassemble the different sequences.
Akinesia is responsible for:
—Loss of upper extremity swing during gait
—Loss of facial expression
—Loss of automatic equilibrium reactions
—Inability to perform two activities simultaneously (such as speaking while cutting meat; when he speaks, the patient must concentrate on the discussion and his hands remain motionless)
—Difficulty in organizing complex motor functions (such as rising from a chair).

QUALITATIVE EVALUATION

The two major disturbances of this pathology are interwoven. When examining motor function it is difficult to differentiate precisely which loss is directly related to rigidity and which is related to akinesia.
Passive mobilization demonstrates rigidity. It is graded in upper and lower limbs.

	Absent Grade 0	Slight Grade 1	Severe Grade 2	Uncontrollable Grade 3
Rigidity				

Akinesia becomes apparent with loss of movement: the incapacity to move spontaneously and to alter posture.

	Absent Grade 0	Slight Grade 1	Severe Grade 2	Completely immobile Grade 3
Akinesia				

QUANTITATIVE EVALUATION

Akinesia and rigidity are evaluated simultaneously.
Increase in flexed posture can be graded by measuring the patient's height at regular intervals.
Slowing of upper extremity movement is evaluated by:

—The fist test where the patient opens and closes his fingers several times during a ten second period. The number of completed movements is noted for the right hand, then for the left hand.

—The prono-supination test: first done bilaterally in order to determine the side of predominant motor disturbance. Then the examiner counts the number of prono-supinations each forearm makes individually within a ten second period.

—Two-handed dexterity is demonstrated with manipulation tests such as, time necessary for tying a knot, linking objects, arranging objects in a precise fashion.

Note: Tremor appears sometimes during precision movements of the hand. The motor activity is then disturbed.
Delay in performing movements is difficult to quantify clinically without resorting to electronic measuring devices. Evaluating delay in performance remains subjective.
The examiner also notes the speed of stopping a given motor activity. It happens that at the end of a movement, the patient needs a certain time, sometimes several seconds, to release the object, which is evidence of tonic carry-over.
Objective evaluation of pathological responses in upper extremity activities is limited to testing activities of daily living:

—Hygiene, self-care,
—Feeding,
—Dressing, handling shoe laces, buttons, . . .
—Communication and leisure activities.

(See pages 428, 429, and 430.)

EVALUATION OF GAIT IN PARKINSONIAN SYNDROMES AND FUNCTIONAL CONSEQUENCES

GENERAL REMARKS

Associated directly with disturbance in the motor state, particularly that of rigidity and akinesia of the lower extremities, and frequently aggravated by secondary effects of immobility (parkinsonian rheumatism), the characteristic gait is an exterior sign of the evolution of the disease.

Stepping on the spot during a reversed direction of movement.

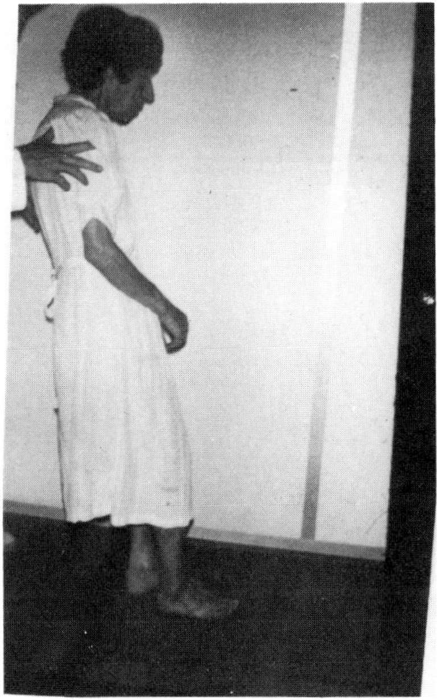

Retropulsion.

CLINICAL OBSERVATION

Each phase of movement may be disturbed and requires specific analysis.

Initially the pathology is manifested by stepping on the spot, a sort of motor activity jamming. The patient appears unable to programme elements composing automatic activity of gait.
This transient, hectic, astasia-abasia (R. Garcin) only lasts one to two seconds.
Gait proceeds:
—Spontaneously.
—Or through the use of facilitation techniques such as exaggerated hip flexion, stepping over obstacles placed on the ground, rising on toes.

Forward movement Consists of quick, short steps, 'marche a petit pas'. Sometimes speed increases during forward ambulation. The patient seems to be running after his own centre of gravity (Lhermitte). This is festination so typical of parkinsonism.
Gait is performed without automatic swinging of the arms and counter rotation of the girdles. Trunk latero-flexion is absent. During forward movement, the characteristic stooped posture is accentuated.

Reversed movement Reversing direction is often difficult. The function of one side is usually dominant. Stepping on the spot frequently reappears.

Stopping The trunk tends to lean backwards. This retropulsion causes the patient to take a step back with one foot in order to avoid falling. When stopping, the patient moves out of his centre of gravity.
In severe cases, the patient supports himself on a wall or a piece of furniture to slow down or to stop.

Risk of falling This is common in more advanced states demonstrating disturbance in equilibrium. This risk is increased by events such as the ringing of the telephone while the patient is walking.

EVALUATION OF GAIT DISTURBANCES

Global appraisal

When a patient is undergoing treatment, modifications in pathological states during the circadian cycle lead to a confused pathological picture.
In some cases a single functional test of whether or not walking is possible must suffice.

	Easy Grade 0	Troubled Grade 1	Very difficult Grade 2	Impossible Grade 3
Walking				

Gait analysis

When walking is graded 1 or 2 and if the pathology is sufficiently stable, an analysis of factors which contribute to the disturbance will describe the handicap more precisely.

Qualitative evaluation

	Easy Grade 0	Difficult but possible Grade 1	Very difficult or just possible Grade 2	Impossible Grade 3
Start Forward movement Right about turn Left about turn Ability to stop: —spontaneous —on verbal command Swinging: —of right arm —of left arm Unilateral stance, right limb Unilateral stance, left limb Protective reactions present in: —right lower extremity —left lower extremity Movement: —backwards —laterally to right —laterally to left				

Note:

	yes	no
Risk of falling		
Consistency of performance throughout the day		

Absence of equilibrium reactions on the right side of a hemi-parkinsonian patient.

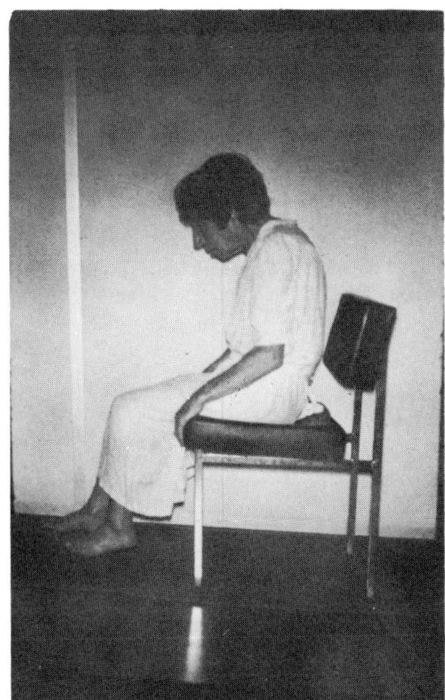

Attempt at standing up.

Quantitative evaluation

Requires the assessment of objective criteria such as measuring step lengths in centimetres.

—Gait is easy if step length is over 50 cm and difficult if it is limited to 15 cm.

—It is important to establish how many steps are necessary in walking a measured distance of 15 metres.

—How much time is needed to walk that distance?

A notion of the consequence of akinesia is obtained by using a stopwatch to time the latency period between the command to start walking and its initiation.

Functional consequences

Qualitative evaluation

As the disease develops, independence of the parkinsonian patient decreases progressively.

First he loses his capacity to go out, then he becomes less independent domestically until he is finally confined to his bed and an armchair. landmarks of the disease's progression are noted:

—Gait: Described earlier.

—Rising from a chair: The patient tries to stand several times and falls back into the chair before finally succeeding or failing in the effort. He no longer automatically organizes the motor sequences which combine to form functional movement. Difficulty is increased if the seat is low.

—Sitting down: With the loss of automatic equilibrium reactions, the patient lets himself drop into the armchair.

—Rolling over in bed: Particularly at night, often in agitated sleep, the patient finds difficulty in rolling over onto one side or the other, or onto his stomach (prone). These elderly patients have difficulty in getting up to go to the bathroom.

—Walking up or downstairs is often easier than walking on a flat surface. Increased knee flexion required for reaching the steps facilitates motor activity.

These characteristic elements of the disease lead the therapist to specific appraisal of functional locomotor possibilities of the parkinsonian patient.

Quantitative evaluation

See evaluation form page 455.

ATAXIA. EVALUATION OF MOTOR COORDINATION AND EQUILIBRIUM DISTURBANCES

GENERAL REMARKS

The quality of postural stability, equilibrium reactions and motor organization of movement depends upon:

- *Somataesthetic and proprioceptive feedback* which indicates:
  —sensation and pressure exerted by bodyweight on the sole of the foot.
  —angles between bony segments.
  —joint movement.
  —positioning of cervical and lumbar regions, particular areas of postural organization.
- *Labyrinthic feedback*:
  —saccules and utricles give 'head-position-in-space' information.
  —semicircular ducts give information about movement of head displacement in the three spatial planes.
- *Visual feedback*:
  —macular information enables subject to relate to his surroundings.

• *Cerebellar control*: analyses peripheral information sent during the course of postural alignment and motor action. It compares this information with patterns of posture and motion organized at the central level. It orders the corrections necessary for relating action performance to motor command. Depending upon information feedback to the corrective mechanisms, three types of ataxia can appear.

- *Ataxia of peripheral origin* caused by lesion to peripheral pathways of deep sensitivity such as:
  —tabes dorsalis.
  —Dejerine and Sottas infantile disease.
  —Guillain-Barré syndrome.
  —polyneuritis.
- *Ataxia of mixed origin*, peripheral and spinal, which appears as a posterior cord syndrome, affecting both peripheral (posterior roots) and central (Goll and Burdach fasciculi) tracts of proprioceptive conduction, as in:
  —multiple sclerosis.
  —Friedreich's disease.
- *Ataxia of central origin* which affects the encephalon centres of posture and movement control:
  —cerebellar, as in hereditary cerebellar ataxia.
  —labyrinthine, by lesion to vestibular and labyrinthine systems, or to their sensory conduction tracts.

Oscillation of left upper extremity.

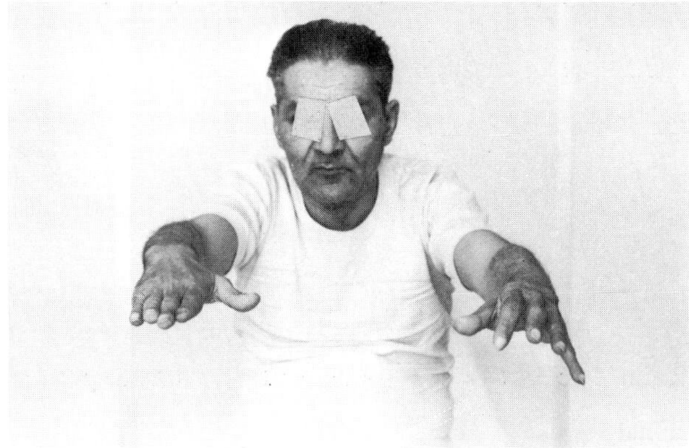

Arms extended.

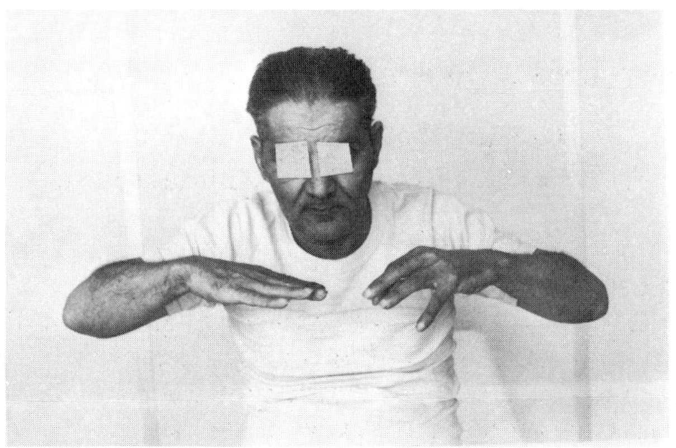

Hands to clavicles.

CLINICAL OBSERVATION

PERIPHERAL ATAXIA

Static ataxia

When sitting without support or standing, the patient experiences continuous visual oscillations. With eyes closed, he cannot remain upright; he falls no matter what position the head is in. Extremities show the same ataxic reactions. It is impossible for the patient to hold an entire limb, or part of one, immobile. Thus a supine subject cannot raise a limb from the treatment table without ataxic movement. For example, constant displacements of the index finger are seen as the patient tries to hold it on his nose.

Kinetic ataxia

Initially, this may be observed in movements (such as picking up a coin or a pin). Then the disturbance quickly becomes apparent in everyday and professional gestures. This pathological factor is characterized by dysmetria, a difficulty in maintaining the direction of movement and attaining the planned target.
For example: if the patient wants to drink, grasping the glass is too forceful and deliberate with a movement pathway which is irregular and deviant. Often the drink is spilled.
The following tests confirm kinetic ataxia.
—Heel to knee.
—Finger to nose.
—Touching the palm of the examiner's hand with the foot or the index finger.
—Blind prehension. With eyes closed, the patient tries to grab his thumb, which the examiner has placed in a given position, with his other hand.
Kinetic ataxia is increased by trunk ataxia. The limb has no stable point from which to perform an intended movement.

Locomotor ataxia

During gait the patient abruptly thrusts the lower extremities forward, knees are usually straight except in polyneuritic disabilities when pathology is further aggravated by motor deficiency resulting in altered gait. Foot strike causes a so called 'heeling gait'.
Movements are so jerky that balance is lost at each step. When eyes are closed, clinical signs increase. Gait is often impossible.

Illustrations on next page.

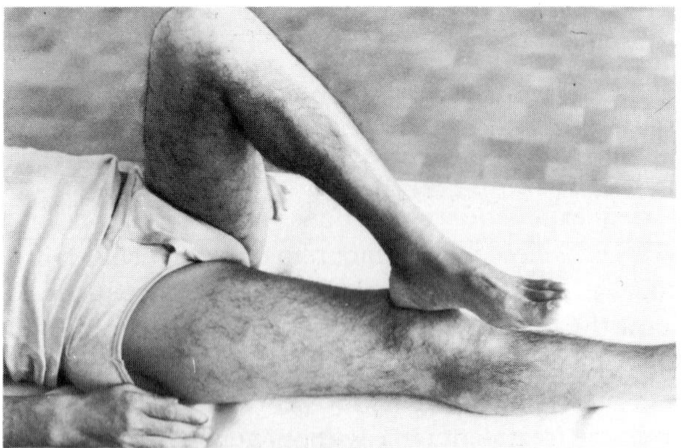
Heel to knee test, starting position.

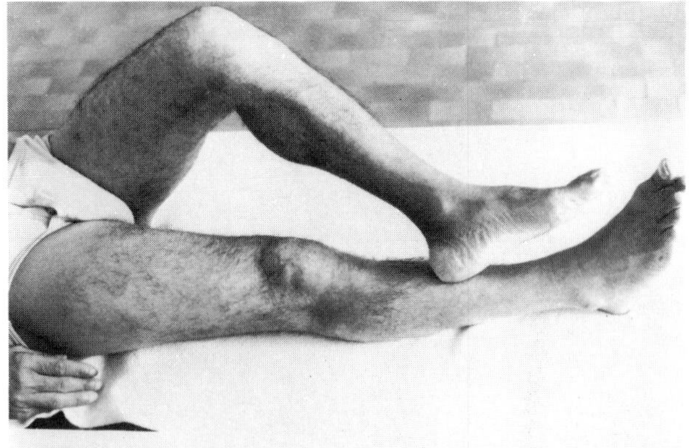
Final position.

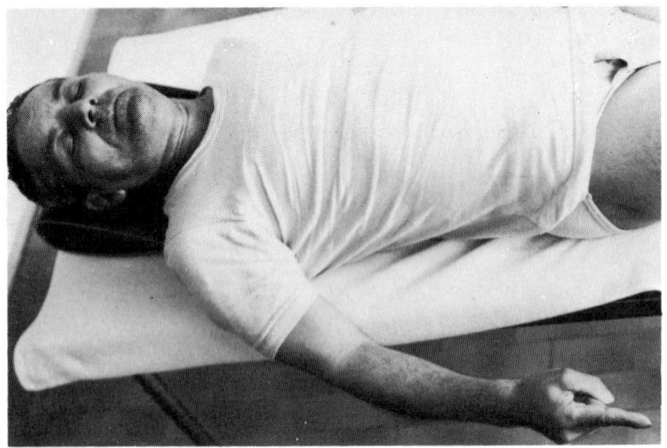
Finger to nose test, starting position.

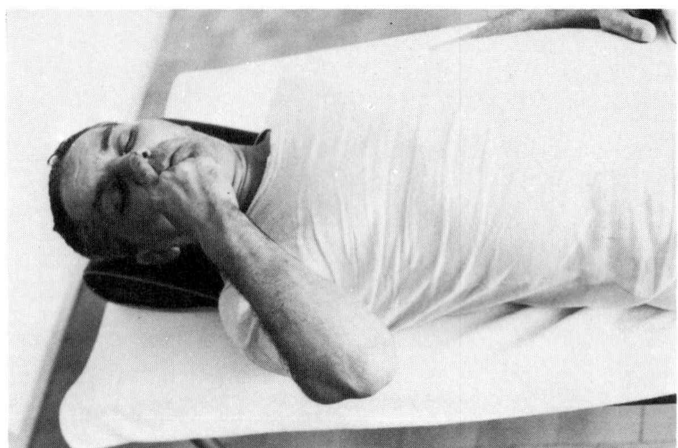
Final position.

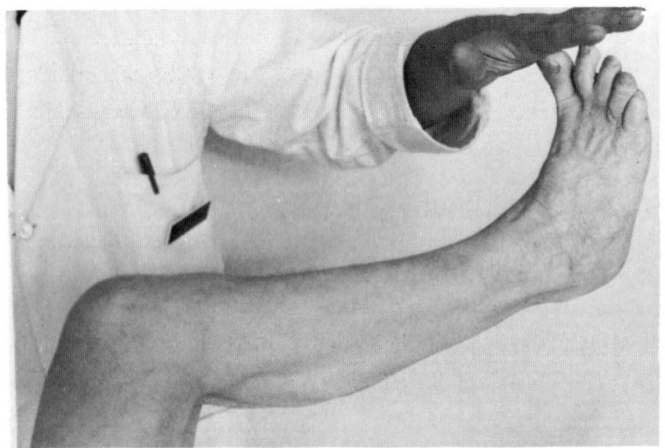
Patient touches examiner's palm with big toe.

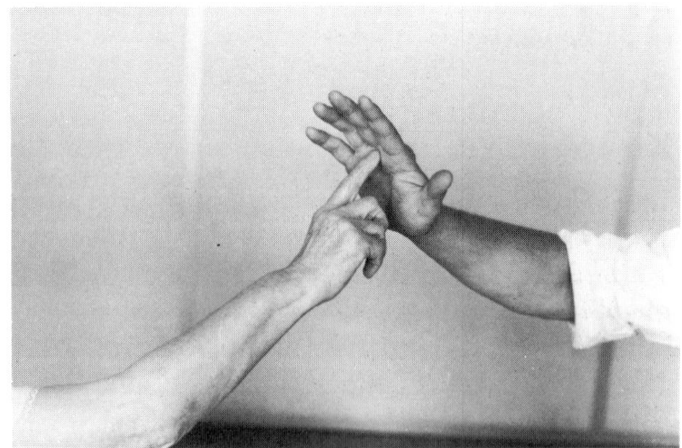
With index finger.

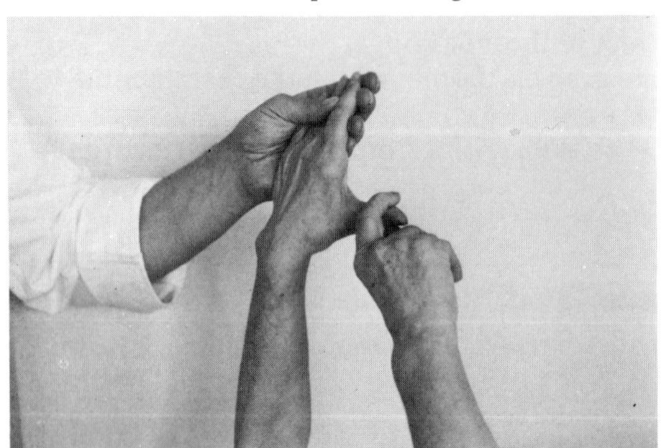
Blind prehension.

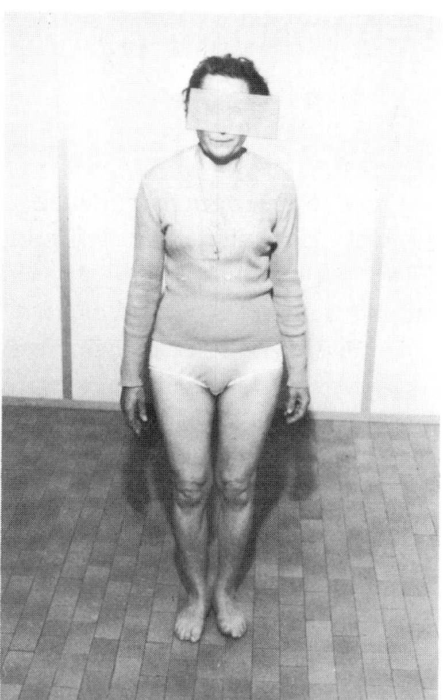

Standing with eyes open.

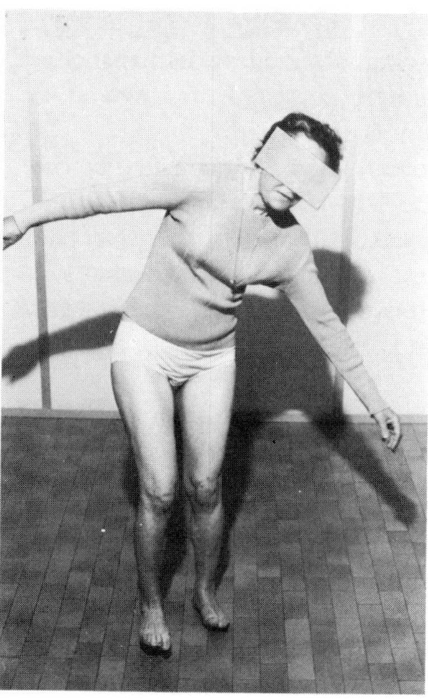

Balance is lost when eyes are closed.

Associated motor disturbances

—Muscles are hypotonic.
—Deep tendon reflexes are abolished.
—Postural reactions have disappeared.
—Equilibrium reactions are abolished or inadapted.

ATAXIA OF MIXED ORIGIN

Disturbances often are seen in lower extremities, impairing standing position and equilibrium, and inducing joint collapse (knees frequently give way into flexion). Clinical signs are often increased by closing the eyes.

Static ataxia

In standing position, patient's head and trunk oscillates.

Kinetic ataxia

Disturbances resemble those of peripheral ataxia. The movement stays roughly on its course, but often overshoots the target.

Locomotor ataxia

Gait is called tabetocerebellar. The patient staggers and cannot walk in a straight line, he zigzags as in cerebellar ataxia. He thrusts out his lower extremities and heels as in peripheral ataxia. Pathological gait is slightly increased when eyes are closed.

Associated motor disturbances

—Muscles are hypotonic.
—Deep tendon reflexes are abolished.
—Postural reactions have diminished or disappeared.
—Equilibrium reactions are diminished and inefficient.

ATAXIA OF CENTRAL ORIGIN.

A. CEREBELLAR ATAXIA

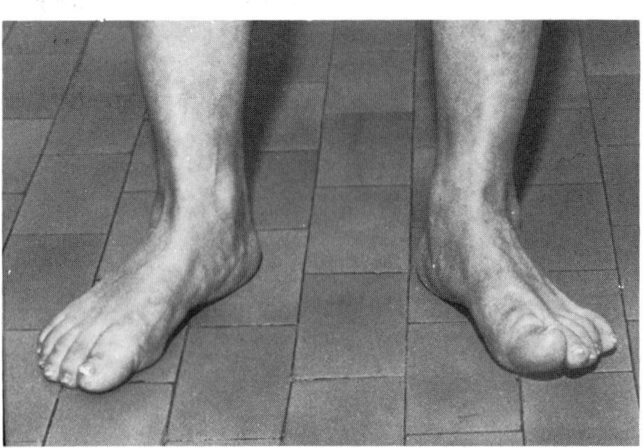

The 'jig of the tendons'.

Static ataxia

In a standing position, foot support area is widened. The body oscillates slightly: cerebellar ataxia without falling.
Muscles of postural adjustment are constantly active, particularly at instep level where alternating contraction of tibialis anterior and extensor digitorum longus do a 'jig of the tendons'.
Movement from sitting to standing is very unstable.
Closing the eyes does not increase postural disturbance.

Kinetic ataxia

'In cerebellar patients, movements are performed without measuring time and space; they are too fast, too abrupt, and they go beyond their targets; initial impulsion is too strong, speed too fast, stopping too late' (André Thomas).
Dyschronometria is characterized by disturbances of the measure of movement in time. Initiating and stopping of movement are retarded.
These preceding problems are demonstrated by the finger-to-nose and heel-to-knee tests.
Hypermetria is characterized by disturbance of the measure of movement in space. If the movement remains on course, it overshoots the planned target. The target is achieved only after a corrective manoeuvre.
Asynergia is characterized by difficulty or impossibility of coordinating all motor components participating in complex movement. Contraction of fixator muscles and agonists is no longer coordinated and synchronized.
Asynergia becomes obvious when passing from the supine to sitting position. When the supine subject, his arms crossed over his chest, tries to sit, he cannot. Instead it is the lower extremities which rise. The synergy of trunk flexion by abdominal muscles and hip extension by gluteal muscles no longer exists.
Asynergia forces the patient to break down a complex movement into each of its basic components. This analysis destroys the natural harmony of motion.
Adiadochokinesia is characterized by the failure to perform rapidly alternating movements.

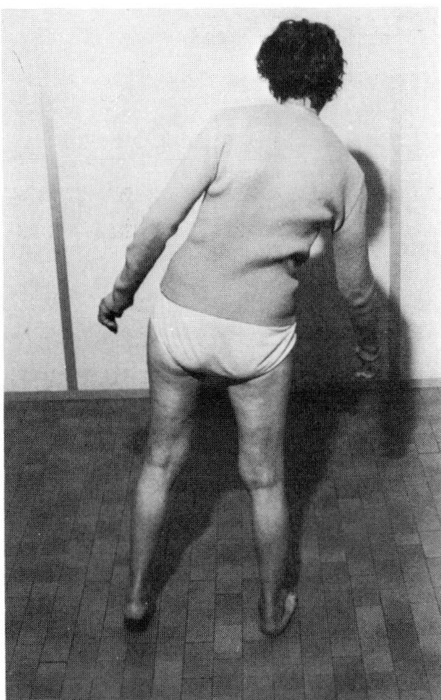

Increased foot support area.

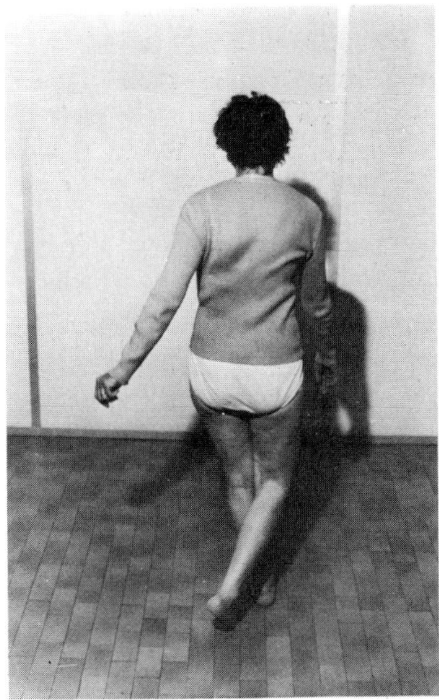

Staggering gait.

It is demonstrated by:
—Testing of simultaneous opening and closing the hands.
—Testing of bilateral prono-supination, proving that is is impossible to perform hand-puppet-like movements.

On a segmental level, kinetic ataxia induces a movement of which:
—the impulse force is too strong,
—the speed is too great,
—the stopping is too late.

Example: If the patient wishes to drink.
—He opens his hand wider then is necessary to grasp the glass.
—He grabs the glass too quickly.
—Hand displacement is directed towards the mouth but it zigzags along the way. The glass quickly arrives in region of the mouth, but the movement needs correcting to make it reach the lips.

Locomotor ataxia

Gait is accomplished, the lower extremities abducted, producing a wide base. Arms are held away from the body for balance.

The patient cannot walk in a straight line but zigzags as though 'tipsy'. He manages a pseudo-drunken or staggering gait.

Asynergia appears when trunk movement are no longer coordinated with those of the lower extremities, leading to a serious risk of falling.

Hypermetria is seen by excessive hip and knee flexion during swing, and increased foot contact during stance.

Eye closure does not aggravate this pathological gait deviation.

Associated motor disturbances

Major muscular hypotonia is reflected in passive manoeuvres. Shaking of a limb by the examiner is exaggerated, joint excursion of the swinging limb being excessive.

Deep tendon reflexes are pendulous.

Postural reactions are disturbed.

Holding a position or voluntary movement triggers an intention tremor which does not exist in the relaxed state.

Equilibrium adjustment reactions are delayed and of exaggerated amplitude.

B. LABYRINTHINE ATAXIA

From an anatomical point of view, this type of ataxia may be considered of peripheral origin. However, it presents clinical signs which resemble those of cerebellar ataxia. It becomes a central lesion only in disorders of nuclei and vestibular pathways.

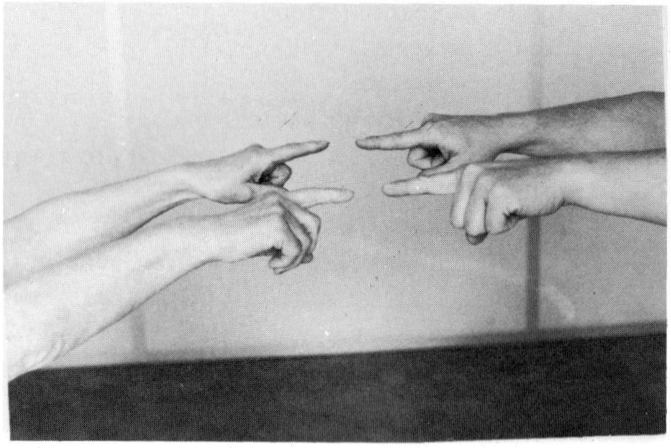

Looking for the sign of index deviation.

Static ataxia

The patient, when standing with feet close together, tends to fall towards the side of the affected labyrinth. This sign is accentuated by closing the eyes. Head position plays a role in this lateral movement. Posture of upper extremities is also disturbed. If the patient stands, pointing both index fingers, upper extremities held straight before him, the index fingers will deviate towards the affected side upon closing the eyes.

Kinetic ataxia

Dynamic voluntary movement is barely disturbed. Only head movements may produce an impression of dizziness.

Locomotor ataxia

In unilateral lesions, gait is characterized by a tendency to lean towards the impaired labyrinth. The patient corrects the deviation, giving the impression of a zigzagging gait. The affected side is confirmed by the blind walking test during which the patient is asked to close his eyes and walk repeatedly forwards and backwards. When the lesion is unilateral, gait traces the outline of a star. When walking, the patient moves himself forward and laterally to the affected side. When he goes back, he moves himself backwards and laterally towards the affected side. When the disorder is bilateral, the patient staggers.

QUALITATIVE EVALUATION

Evaluation of ataxia attempts to determine the effect of the pathology on:
—Standing position and gait,
—Disturbances of postural adjustment and of upper extremity dynamic movement.

Examining standing position and gait

This is done with the patient's eyes open first of all, then closed.
The examiner notes if the test requested is:

—Possible without disturbance:	Grade 0
—Possible with moderate disturbance (tremor, hesitation, oscillation):	Grade 1
—Very difficult, disturbances are major:	Grade 2
—Impossible to perform:	Grade 3

Static examination	Eyes open	Eyes closed
Standing position (note eventual Romberg's sign or tendency to fall towards a preferred side)		
Measure widening of foot support area in centimetres		
Weightbearing predominance on one side (use two bathroom scales)		
Preferred weight-bearing area on foot plantar surface		
Need to extend arms for balance when standing		
Presence of efficient protective reaction when losing balance: —right side —left side		
Evaluate the ability to correct static posture		
Standing with heels together: —Arms out —Arms along sides of trunk —Hands on hips —Hands on shoulders —Arms up in the air		
Unilateral stance: —Right —Left		
Standing with head in different positions: —Flexed —Extended —Rotated to right —Rotated to left —Right latero-flexion —Left latero-flexion		

Static-dynamic examination	Eyes open	Eyes closed
Sitting without back support: —Hands on knees —Hands on hips —Hands on shoulders —Arms up in the air		
Squatting on heels: —Hands on knees —Hands on hips —Hands on shoulders —Arms up in the air		
Kneeling upright: —Arms along sides of trunk —Hands on hips —Hands on shoulders —Arms up in the air		
Kneeling on one knee, Right foot forward: —Arms along sides of trunk —Hands on hips —Hands on shoulders —Arms up in the air		
Kneeling on one knee, Left foot forward: —Arms along sides of trunk —Hands on hips —Hands on shoulders —Arms up in the air		

Note ability to get up off floor: —With a support —With no support	
Note effects on gait: —Determine area of base while walking —Describe spontaneous gait: • heeling • festinating • zigzagging	
Note capacity for walking along a line drawn on the floor.	
Note capacity for: —Right about turn —Left about turn	
Note ability for walking upstairs: —Using the banister —Without using the banister	
Note ability for walking downstairs: —Using the banister —Without using the banister	

Evaluation of postural adjustment and upper extremity dynamic movement

This is done in a sitting position. Examiner notes if patient is able to:
—Maintain this posture, eyes open and eyes closed. Note if posture has been maintained:
—In a stable manner: Grade 0
—Disturbed by postural tremor: Grade 1
—With difficulty, or is greatly disturbed: Grade 2
—Not at all: Grade 3
—Organize dynamic movements.

	0	1	2	3
Postural attitudes (eyes open, closed)				
Is able to hold right upper extremity:				
—Straight in forward horizontal position				
—Straight in side horizontal position				
—Flexed, hand over clavicle				
Is able to hold left upper extremity:				
—Straight in forward horizontal position				
—Straight in side horizontal position				
—Flexed, hand over clavicle				
Is able to hold both upper extremities simultaneously:				
—Straight in forward horizontal position				
—Straight in side horizontal position				
—Flexed, hands over clavicle				
Dynamic movements (prerequisite: trunk must remain immobile.)				
With right index finger, is able to touch a stationary object placed:				
—In front of patient				
—To his right				
—To his left				
With left index finger, is able to touch a stationary object placed:				
—In front of patient				
—To his right				
—To his left				
With right index finger, is able to touch the examiner's hand which:				
—Moves slowly in front of him				
—Moves rapidly in front of him				
With left index finger, is able to touch the examiner's hand which:				
—Moves slowly in front of him				
—Moves rapidly in front of him				

	0	1	2	3
Is able to follow with his own hand and without touching the examiner's hand, which moves about in space (mirror image movements): —With right upper extremity —With left upper extremity				

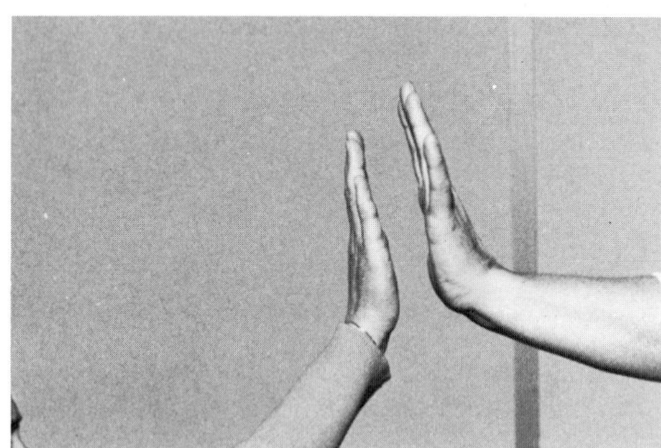

Mirror image movement test.

ABNORMAL MOVEMENTS

CLINICAL EVALUATION

GENERAL REMARKS

Abnormal movements are caused by involuntary, uncontrollable muscular contractions which appear independently of paretic or ataxic disturbances. Localized or generalized, they lead to modifications of resting posture or muscular activity. They disappear during sleep.

CLINICAL OBSERVATION

Abnormal movements are characterized by:

- *Location.* They may be localized:
  - —In a muscle group
  - —In a part of the body
  - —Or generalized in a symmetrical or non-symmetrical manner.
- *Form.* They impose a form of movement on part of, or the whole body.
- *Conditions favouring their presence:*
  - —Motor:
    - —resting position.
    - —maintaining posture.
    - —dynamic movement.
  - —Sensory: noise.
  - —Vegetative: cold.
  - —Psychological fatigue, emotion, concentration.
- *Rhythm:*
  - —Regular or irregular.
  - —Spaced (remission periods may be several minutes long).
  - —Intermediate (remission periods last only a few seconds).
  - —Frequent (abnormal movements produced without interruption).
- *Amplitude*
  - —Slight: muscles jerk without displacing segments.
  - —Medium: localized segmental displacement is visible.
  - —Severe: creation of complex pathological movements.

TREMORS

'They are characterized by involuntary, rhythmic oscillations induced by part of or the entire body around its position of balance' (Dejerine).

With the exception of physiological trembling due to:

—emotion
—cold chills
—fatigue
—intense muscular activity

tremor is a pathological neurological manifestation.

Parkinsonian tremor

Location

Unilateral at the beginning of the disease, it usually becomes symmetrical as the pathology evolves. It is predominant at distal portions of extremities:
—Hands.
—Feet.
—Lower portion of the face.
It may become generalized in the later stages of the disease.

Form

In upper extremities:
—In the hands the tremor is 'pill rolling' (patient crumbles bread or counts coins).
—If tremor is severe the patient beats his chest or 'flaps his wings'.
In lower extremities: Alternating flexion and extension movements give the impression that the patient is beating time to music.
Face: Tremor predominates in lips giving the impression of chewing.

Conditions favouring their presence

At the onset of the disease, tremor can appear when certain positions are held, but more often it comes with fatigue or emotion.
Once the disease is well established, tremor characteristically appears when resting and disappears with volitional movement:
—If the subject thinks about it, it disappears
—If he forgets it, it reappears.
Tremor increases with mental activity (mental calculating).
In final stages it can prevent normal posture and fine finger movements.

Rhythm. Rhythm is relatively slow, from 4 to 7 cycles per second. Although it appears to be regular, bursts of tremor are followed by relative remission periods.

Amplitude. Is therefore variable.

Evaluation. Intensity can be graded on a four-point scale:
0 Tremor is absent.
1 Slight tremor appearing only under stress, e.g. mental calculating.
2 Tremor is visible and constant when subject is resting.
3 Tremor is severe when subject rests. It may also disturb posture and movement. It has a tendency to be generalized. In more severe forms the patient is quite shaken by the tremor.

Cerebellar tremor

Location

It predominates in:
—Upper extremities where it becomes obvious when the patient straightens his arms out in front.
—Head and trunk.
It also appears when subject is seated without support and when he stands.

Form. Brief muscular contractions lead to unstable posture and disjointed motion.

Conditions favouring their presence. Absent while the patient rests, it appears with muscular activity irrespective of whether the contraction is postural or kinetic.
It is an action or intention tremor.
It is severe at the beginning and at the end of the movement.
This characteristic is demonstrated with tests:
—Finger to nose.
—Heel to knee.

Rhythm. Slow and irregular.

Amplitude. Oscillations are large.

Examination

Clinically, the therapist is limited to testing consequences of the tremor on functional activities.

Attitude tremor

Often familial, this tremor is considered an exaggeration of physiological postural reactions. Most of its characteristics are contrary to those of the parkinsonian tremor.

Location. It concerns essentially the extremities, more rarely the head.

Form. It is seen by fine finger movements and minimal oscillations of the head.

Conditions favouring their presence. It appears when the subject maintains posture. It increases with emotion and fatigue. Calmed by mental calculating, it disappears when subject moves or rests.

Rhythm. Rapid oscillations.

Amplitude. Slight.

Evaluation. Clinical observation notes the appearance of trembling when patient holds:
—Upper extremities extended.
—Fingers spread apart and extended.
When performing certain graphic tests (examples below) tremor is demonstrated with the irregularity of the lines.

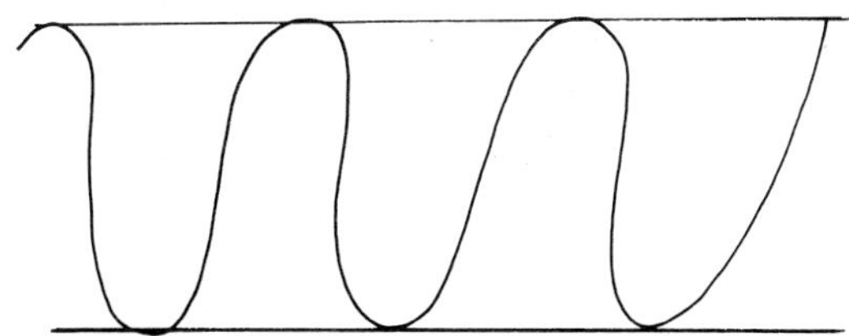

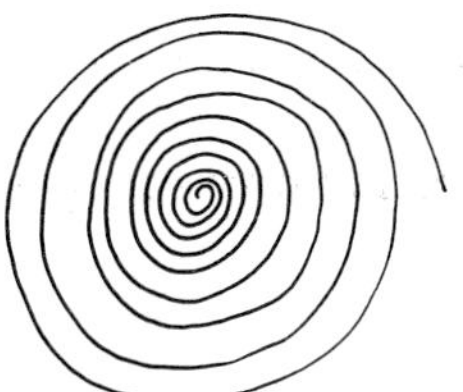

Athetosis.

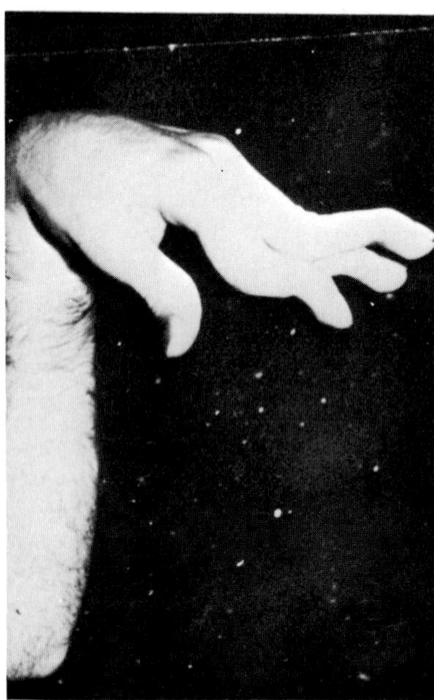

Athetoid hand.

ATHETOSIS

Is characterized by the inability of maintaining a resting position or any fixed position.

Location

—Either unilateral: hemiathetosis is a complication of infantile hemiplegia,
—Or bilateral: as in congenital double athetosis.

It concerns the face and the limbs where it predominates in the extremities.

Upper extremities are affected more severely than the lower.

Form

—On the face, it creates distortions, grotesquely imitating laughter or sadness. The lower part of the face is particularly active in elaborating these grimaces although muscles of the eyes and the frontalis may also participate. Tongue protrusions and difficulties in phonation are frequent, hindering capacity for oral communication.
—Upper extremities: Worm-like movements appear, similar to those of an octopus' tentacles. They are often compared to gestures of Javanese of Borneo Island dancers.
—Lower extremities: Movements affect especially the toes, feet and ankles. To a lesser degree they resemble those of the upper extremity.

Conditions favouring their presence

The patient cannot stay still when requested to do so. Uncontrollable muscle contractions occur in this basic state: Tardieu's B factor.

Muscular effort tends to spread athetoid movement. It takes the form of an 'immense and generalized synkinesis' (Forster).

Posture, voluntary movement, standing position and gait, even if they are preserved, are disturbed.

Pathological reactions are accentuated by:
—Intellectual activity.
—Speech.
—Anxiety.
—Sudden noises.
—Threats.
—Postural efforts.
—Voluntary movements.

Cutaneous stimulation by superficial massage of the skin modifies the segmental positions induced by athetosis. Thus the wrist goes into extension when the dorsum of the hand is rubbed, and into flexion when the palm or the anterior aspect of the forearm are rubbed.

Rhythm

Athetosis is composed of slow movements. When an agonist contracts, the antagonist becomes active but to a lesser degree and slows down the movement.

The pathological attitude may be fixed for a few seconds. Agonist–antagonist co-contraction provokes a real spasm (muscular rigidity) which can become permanent in evolved forms.

Movements are:

—Arrhythmic.

—Irregular.

—Unpredictable.

—Variable in speed and direction.

Amplitude

Is more or less marked involving notable displacement of the segments. For a given joint, pathological movement is often not apparent until reaching a certain degree of range of motion, which is always the same. Thus selective movement initiated voluntarily, is rudely disrupted by pathological motor activity beyond the patient's control.

Evaluation

Consists of describing:

- In resting position
  —involuntary contractions and jolts which appear
  —tonic modifications and involuntary movements which are triggered by a loud noise (hands clapping) or a threat (sting).
  The following are noted:
  —intensity of the abnormal movements
  —increase in their rhythm and amplitude
  —time necessary for returning to original resting position.
- Assessing repercussions of the disturbance on functional activities and movement.

CHOREA

Chorea is the name given to various disorders characterized by involuntary, continuous jerking movements of the affected muscles.

—Sydenham's chorea or Saint Guy's dance is an infantile disease linked to and regarded as one of the manifestations of streptococcic rheumatic fever (acute disorder).

—Huntington's chorea is a chronic, progressive, hereditary disease occurring in adults and accompanied by mental deterioration.

Location

Choreiform movements are diffused, bilateral and symmetrical. They predominate at attachments of limbs, but also affect the face and neck.

Tongue and pharynx can be involved, impairing speech and swallowing.

Certain automatic movements may be affected (cardiac or respiratory movements).

Form

Choreiform movements appear in a patient whose musculature is hypotonic, the face grimaces imitating laughter, anxiety or terror.

Constant neck motion resembles that of a 'bird on the watch'. At upper extremity level:

—Shoulders shrug (elevation, protraction).

—Elbows flex and extend.

—Finger activity is ceaseless, particularly thumb activity.

These movements are like those of a puppet being manipulated on strings.
Lower extremities are generally less involved. However, in sitting position, incessant activity is apparent, crossing, uncrossing, spreading, drawing together. When standing, hopping and oscillations created by involuntary movements make gait awkward, sometimes impossible. Postural adjustment reactions are decreased.

Conditions favouring their presence. Choreiform movements are more or less severe during the waking state, they are accentuated by:
—Physical effort.
—Intellectual effort.
—Fatigue.
—Emotion.
They are decreased by rest.

Rhythm. Movements are:
—Arrhythmical.
—Disorderly.
—Involuntary.
—Fast.
—Explosive.
Movements are uncontrolled by antagonists which explains their irrepressive and brutal character.

Amplitude. Joint movement is marked.

Evaluation. Consists of counting the number of involuntary movements occurring over a given period of time:
—One minute when rhythm is rapid.
—Three minutes when it is slow.
The count is done:
—With patient in a resting position: seated comfortably.
—With patient in a position facilitating pathological movements: standing with upper and lower extremities spread apart, fingers in abduction and tongue sticking out.

BALLIC MOVEMENTS

Symptom is rare, prognosis is poor.

Location. Generally unilateral (hemiballism), it affects upper and lower limbs with emphasis on the upper.
Sometimes face and neck are involved.
Biballism concerns both sides of the body.

Form. These movements tend to move the limbs into internal rotation. Stereotyped, they repeat the same pattern of movement for several minutes.
They occur despite decreased tonic reactions and reduced muscle strength.

Conditions favouring their presence. Decreased during rest, they increase with voluntary movement, emotion and speech.

Rhythm. If their form resembles that of choreiform movements, they are different in their intensity (their violence may easily cause traumatic lesions).
They appear in an outburst, they are illogical, rapid and even explosive.
Rhythm is asynchronous.

Amplitude. Segmental movements induced are marked.

Evaluation. Is limited to noting location and intensity of the movements.

Spasmodic torticollis.

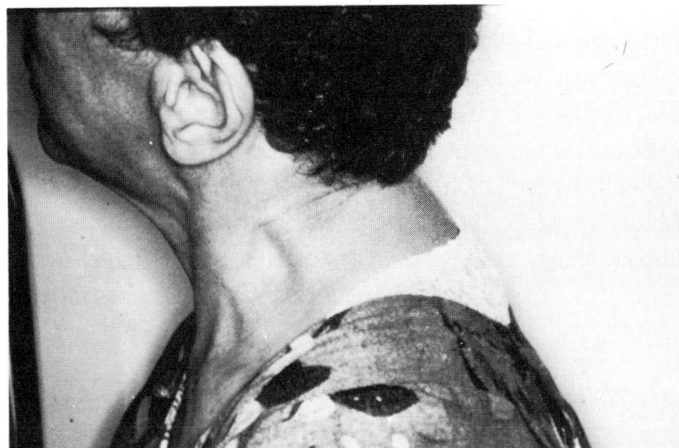

Spasmodic sternocleidomastoideus and splenius capitis.

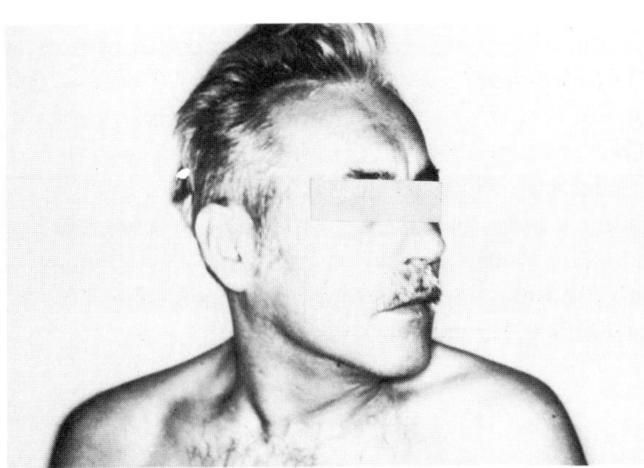

Torticollis.

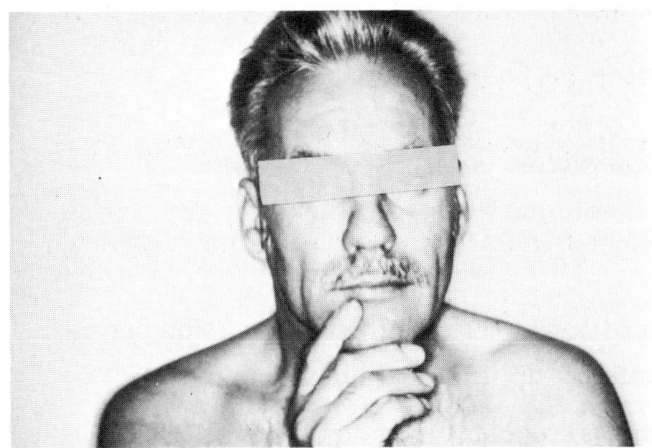

Antagonistic movement.

SPASMS

These are localized, transient contractions which repeat themselves regularly in a same muscle group. Example:

Spasmodic torticollis

This is a tilted neck posture which may appear regularly (clonic torticollis), or which may be fixed by a continual, uncontrollable spasm (tonic torticollis).

Location

It affects the neck muscles. In more severe forms, the shoulder can be elevated by trapezius contracture.

Forms

—Torticollis: is seen as a head rotation caused by activity of splenius capitis on one side and sternocleidomastoideus on the other. These two muscles form a rotating torque.

—Laterocollis: is seen as neck latero-flexion caused by the simultaneous activity of:

- sternocleidomastoideus.
- splenius capitis.
- superior trapezius on the same side.

—Antecollis: is seen as an anterior head projection caused by simultaneous activity of both sternocleidomastoidei.

—Retrocollis: is seen as posterior head positioning caused by activity of both trapezii or of both splenii.

Conditions favouring their presence

The different forms of torticollis are not well marked when subject is lying down, except for retrocollis. They become more evident when subject sits, increasing when he stands to reach a maximum during gait.

If torticollis seems to disappear clinically during sleep, electromyographic recordings are not as conclusive. A slight activity may persist (P. Rondot).

The different forms are attenuated and sometimes completely disappear when antagonistic or remedial manoeuvres are used, such as placing the hands behind the head or one hand to the chin.

Fatigue and public appearances aggravate torticollis.

Rhythm

The 'muscular convulsion' may be constant. The patient retains his pathological attitude: tonic spasmodic torticollis.

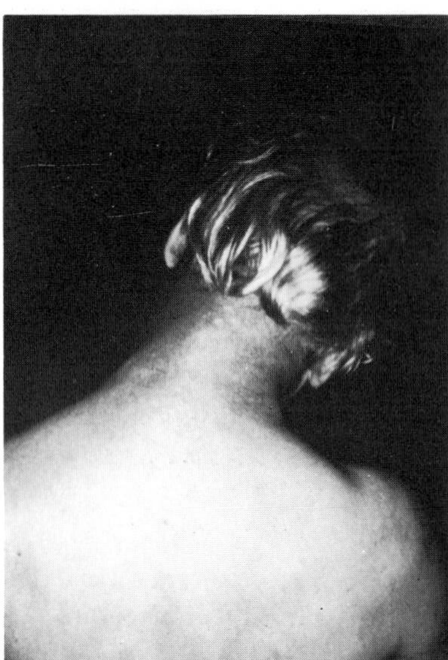

Laterocollis.

Pathological position may be reduced voluntarily, but as soon as the patient stops trying to correct the pathological attitude, it reappears: tonico-clonic, spasmodic torticollis.
Torticollis may come in the form of uncontrollable, involuntary movements appearing with quicker or slower rhythm. This form is a real 'neck tic' (Duchenne de Boulogne). Its rhythm may be fast, causing an actual jerking of the head, or slower, appearing every 20 to 30 seconds: clonic, spasmodic torticollis.

Amplitude

Amplitude may be slight when there is a mere trembling of the head.
It is often marked, including extreme head and neck movement.

Evaluation

The pathological attitude is noted with a tridimensional measurement of the head's position in space.
The measurement is taken in spontaneous position and during attempts at correction.
The number of spasms is counted for a given time:
—One minute when rhythm is rapid,
—Three minutes when it is slow.
Feedback instrumentation, especially myofeedback devices, enable a precise spasm count.
Spasm intensity appears in microvolts on the machine's voltmeter.

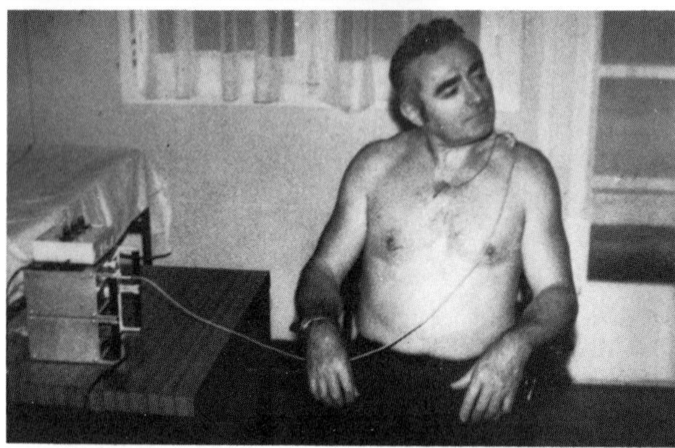

Torticollis: evaluation of spasm intensity with a myofeedback device.

TICS

'The tic is a convulsive, habitual and conscious movement resulting from the involuntary contraction of one or several muscles of the body, reproducing often but in an untimely way, some of the reflex or automatic gestures of everyday life' (G. Guinon).

Location. Locations are innumerable and migratory, from the simplest movement (blinking the eyes or shrugging shoulders) to complex, organized movements such as dance steps. The most frequent are located in:
—face.
—neck.
—fingers.
Sometimes they are vocal noises, of eructation.

Form. Tonic or clonic, they create unharmonious, illogical motor activity.

Conditions favouring their presence. Though at times under voluntary control, when they appear tics present a liberating aspect.
They are increased by fatigue and activities in public.

Rhythm. They appear without reason, generally in series of outbursts spaced with remission periods.

Amplitude. Depending upon location and movement created, amplitude is quite variable.

Evaluation. Is limited to clinical study of their characteristics and identification of the muscles responsible for the tic.

APRAXIA

GENERAL REMARKS

It would be incomplete to assess movement only by the abnormality of one or more segments. Every voluntary gesture results from an intended action. Thus, evaluating movement also involves judging its performance and success in relation to the planned goal. Praxia corresponds to the perfect actualization of a planned act, signifying its skilfulness. Conversely, apraxia has been defined in a negative manner since Dejerine as:

> A disorder of intentional, gestural activity in a subject whose performance apparatus is intact (absence of paralysis, ataxia and of choreo-athetosis) and who fully understands the act to be accomplished (absence of gnosic disorders and of global intellectual deficiency).

CLINICAL OBSERVATION

Several types of apraxia are described depending upon the forms of motor disorganization presented by subjects.

Ideomotor apraxia

Disorder lies in a simple motor act which the patient cannot execute upon request. He conceives the required gesture but cannot implement the voluntary act. The concept of actualization is preserved, but the execution phase is disturbed. On the other hand, this same act appears spontaneously when he is faced with the real situation. Thus the patient cannot show the examiner how one says, 'Good-bye', with the hand, but does it automatically from the window when his family leaves.

Complex motor activities are often disturbed because, even though the patient is able to formulate the strategy, he is not always able to implement the elementary sequences which permit its actualization.

Motor failures caused by ideomotor apraxia appear as either

—the absence of the response,

—the attempt alone,

—inadequate synkinetic movements,

—or even as the repetition of the movement previously studied, thus leading to perseveration.

In a right-handed subject, ideomotor apraxia may be bilateral or unilateral on the left side.

Ideational apraxia

Disorder appears during the performance of complex acts which require an elaborate strategy. The patient cannot put together the various elementary sequences which lead to its success. However, each single phase considered individually is performed correctly.

The tests which demonstrate this form of apraxia are given with objects whose use forces the patient to formulate an adequate plan, such as lighting a candle with a match placed in its box. Hecaen (1978) considers ideational apraxia as a disorder of the association between a concept, in this case the one of an object, and the act which corresponds to its use.

The gestural disorganization may present itself in different ways:

—the patient may stop right away with the first sequences of the complex action he is to accomplish.

—or he may perform an act which resembles but is different from the one requested.

—or he may invert the order of the phases.

Ideational apraxia affects the whole body and cannot be unilateral.

EVALUATION

Ideomotor apraxia

Is assessed first for the right hand, then for the left hand.
This demonstrates the unilaterality or bilaterality of the disorder. The possibility, difficulty or impossibility of performing elementary gestures are noted after:
—verbal command.
—visual imitation (reproduces the examiner's gesture).
—kinaesthesic imitation (the examiner passively performs the gesture for the patient who should then reproduce it).
Orders are given from most difficult to easiest: verbal command, visual imitation and kinaesthesic imitation.
Different medical teams when evaluating apraxia have their own tests which take into account local habits as well as the patient's socio-cultural level. Nevertheless, the test must register different levels of complexity in the gestures:
—gestures of conventional symbolism.
—expressive gestures.
—descriptive gestures.
—imitations of movements deprived of finality and significance.
The proposed tests belong to the standard examination of Unite 111 of I.N.S.E.R.M. of Paris. For each movement tested the examiner gives verbal commands as in the following example:
—on verbal command: 'With your right hand, give a military salute'.
—on visual imitation: 'Do what I am doing'.
—on kinaesthesic imitation: 'Do what I just made you do'.
The examiner notes if the patient is able or unable to perform the requested movement.

	Right hand	Left hand
Conventional symbolism		
1. Military salute. —verbal command —visual imitation —kinaesthesic imitation		
2. How do you say, 'Pay attention' with your finger? —verbal command —visual imitation —kinaesthesic imitation		
3. Thumb your nose. —verbal command —visual imitation —kinaesthesic imitation		
4. Cross yourself. —verbal command —visual imitation —kinaesthesic imitation		

	Right hand	Left hand
Expressive gestures		
1. Threat.		
—verbal command		
—visual imitation		
—kinaesthesic imitation		
2. Foul odour.		
—verbal command		
—visual imitation		
—kinaesthesic imitation		
3. Plead.		
—verbal command		
—visual imitation		
—kinaesthesic imitation		
4. Cold.		
—verbal command		
—visual imitation		
—kinaesthesic imitation		
5. Fear.		
—verbal command		
—visual imitation		
—kinaesthesic imitation		

	Right hand	Left hand
Descriptive gestures (imitation without the object)		
1. Twirl the ends of your moustache.		
—verbal command		
—visual imitation		
—kinaesthesic imitation		
2. Scratch your nose.		
—verbal command		
—visual imitation		
—kinaesthesic imitation		
3. Smoke.		
—verbal command		
—visual imitation		
—kinaesthesic imitation		
4. Comb your hair.		
—verbal command		
—visual imitation		
—kinaesthesic imitation		
5. Eat.		
—verbal command		
—visual imitation		
—kinaesthesic imitation		
6. Turn the key in the lock.		
—verbal command		
—visual imitation		
—kinaesthesic imitation		
7. Cut with a pair of scissors.		
—verbal command		
—visual imitation		
—kinaesthesic imitation		
8. Hammer a nail.		
—verbal command		
—visual imitation		
—kinaesthesic imitation		
9. Sharpen a pencil.		
—verbal command		
—visual imitation		
—kinaesthesic imitation		
10. Strike a match.		
—verbal command		
—visual imitation		
—kinaesthesic imitation		

Simultaneous examination of both sides	Right hand	Left hand
Imitation of movements without significance		
1. Hold your fingers together. —visual presentation —kinaesthesic presentation		
2. Form interlocking rings with thumbs and index fingers. —visual presentation —kinaesthesic presentation		
3. Make a butterfly with your hands. —visual presentation —kinaesthesic presentation		
4. Pair of horns. —visual presentation —kinaesthesic presentation		

Ideational apraxia:

Is evaluated through the manipulation of simple objects known to the patient, after having asked him, 'How do you use this?'
The test objects used by the medical team of the Unite 111, I.N.S.E.R.M. of Paris, directed by Professor P. Rondot, are the following:

1. a pair of scissors.
2. a key.
3. pour water into a glass.
4. light a cigarette.
5. fold a letter and put it into an envelope.

Note: In order to complete the evaluation of upper level functions, it is important to know the other forms of apraxia which exist.

The melody kinetic apraxia

Which perturbs 'movement's melody' (Luria). It is described by Kleist as a pure performance apraxia. It affects only certain muscle groups. The agonist-antagonist function, hence the harmony of the performance of the movement, is disturbed.
It manifests itself through an inability, or difficulty, in doing rapid, precise movements in a series, such as table top piano playing, or pressing and releasing a switch upon command. This form of apraxia is difficult to differentiate from a loss or a clumsiness of muscular origin.

Gait apraxia

In which the patient can neither alternate nor coordinate correct movements of the lower extremities during gait, when there is no muscular deficiency, nor ataxia.

Dressing apraxia

In which the patient is unable to put his clothes on correctly or to dress himself. He remains confused over his jacket, which he turns over and over again without knowing how to put it on. He no longer knows how to tie his shoe-laces or his tie.

Bucco-linguo-facial apraxia

In which the patient is unable to make voluntary expressions of the face or of the buccal cavity (tongue, pharanx), such as clucking his tongue, imitating a kissing noise, swallowing, whereas automatic responses are retained.

Constructional apraxia

In which the patient is unable to structure space. He cannot reproduce geometrical figures such as a triangle, a square or a cube by drawing or with sticks. Similarly, he cannot copy a more complex drawing such as a house or a bicycle. This apraxia results from a performance difficulty in the visuo-spatial sector.

BIBLIOGRAPHY

Albert A 1969 Rééducation neuromusculaire de l'adulte hémiplègique. Masson, Paris

Albert A 1965 Bilan fonctionnel de l'adulte hémiplégique. Kinesitherapie 94 : 13–33

Albert A 1978 Kinésithérapie de la spasticité. Annales de Kinésithérapie. April : 134–137

André-Thomas 1940 Equilibre et équilibration. Ed. Masson, Paris

Barrie M L 1970 Kinésithérapie de la maladie de Parkinson. Maloine, Paris

Bauer H J 1972 La spasticité : causes et signification clinique. Birkmayer W (ed.) (Vienna). Masson, Paris

Bessou P 1978 Psychologie humaine. Le système nerveux. Vol 2 Simep editions. Villeurbanne

Bird, Catalo 1978 Experimental analysis of E.M.G. feedback in treating dystonia. Annals of Neurology 3 : 310–315

Bobath B 1973 Anomalies des réflexes de posture dans les lésions cérébrales. Maloine, Paris

Bobath B 1975 Hemiplégie de l'adulte, bilans et traitement. Masson, Paris

Bonafos J P, Calliaguet P 1974 Bilan neuro-musculaire de l'adulte hémiplégique par le kinésithérapeute. Annales de Kinésithérapie 1 : 143–154

Bru C 1978 Attitude du reèducateur devant la sclérose en plaques au cours de son évolution. Cahiers de kinésithérapie. no. 73 : 5–25

Brunnstrom S 1970 Movement therapy in hemiplegia : a neurophysiological approach. Harper and Row, New York

Cahuzac M 1977 L'enfant infirme moteur d'origine cérébrale. Masson, Paris

Cahuzac M, Commanay F B, Montaud B 1976 A propos d'une fiche d'examen du handicapé moteur en vue d l'orientation professionnelle. Revue Réadaption no. 234 : 19–23

Conraux C 1974 Anatomie et physiologie de l'appareil vestibulaire. Revue du Praticien 24 : 13–29

Cruchet R 1907 Traité des torticolis spasmodiques, spasmes, tics, rythmies du cou, torticolis mental. Masson, Paris

Dejerine J 1977 Sémiologie des affections du système nerveux (1914). Ed. Masson, Paris

Delprat J, Mansat M 1981 Exploration de la sensibilité. 5e cours, Pathologie des nerfs périphériques. Montpellier

Duchenne (de Boulogne) G B 1872 De l'electrisation localisée et son application à la pathologie et à la thérapeutique. Baillière, Paris

Duchenne (de Boulogne) G B 1867 Physiologie des mouvements (1967). Annales de Medecine physique. Ed. Baillière, Paris

Durand J 1981 Bilan musculaire. In Grossiord A, Held J P (eds) Médecine de Rééducation. Flammarion Medecine Sciences, Paris

Duval-Beaupère C, Maury M 1966 Les Appareils de marche dans les infirmités neurologiques. Masson, Paris

Finnie N R 1979 Abrégé de l'éducation à domicile de l'enfant infirme moteur cérébral. Masson, Paris

Flament F 1975 Coordination et prévalence manuelle chez le nourrisson. Editions du Centre National de la Recherche Scientifique, Paris

Forster 1980 Cited by de Recondo J (1980)

Gagnard L, Le Metayer M 1979 Rééducation des infirmes moteurs cérébraux. Expansion Scientifique Française, Paris

Gaheric Y, Legallet E 1980 Influence de la situation posturale sur la programmation de la motricité des membres, lors des stimulations du cortex chez le chat chronique, p 345–353. Anticipation et comportement, publié sous la direction de Jean Requin. Editions du Centre National de la Recherche Scientifique, Paris

Garcin R 1933 Les Ataxies. Congrès des aliénistes et neurologistes. Rabat

Garcin R 1970 Les troubles de la démarche dans les affections neurologiques. Gazette Médicale de France 77 : 2359–2369

Gennevième 1977 Maladie de Ménière. Provence Médicale 45 : 3–22

Goudot-Perrot A 1973 Cybernétique et biologie. 'Que sais-je ?' P.U.F., Paris

Gribenski A, Caston J 1973 La posture et l'équilibration. 'Que sais-je ?' P.U.F., Paris

Guibert A, Guibert C 1974 Du symptôme à la maladie en neurologie. 2nd edn. Maloine, Paris

Guinon G 1888 Tic convulsif. Dictionnaire encyclopédique des sciences médicales

Hecaen H 1972 Introduction à la neuro-psychologie. Larousse

Hecaen H, Jeannerod M 1978 Du contrôle moteur à l'organisation du geste. Masson, Paris

Held J P, Pierrot-Deseilligny E 1969 Rééducation motrice des affections neurologiques. Ed. Baillière, Paris

Held J P, Pierrot-Deseilligny E 19–– Rééducation de l'hémiplégie. Encyclopèdie médico-chirurgicale (Rééducation Fonctionnelle) (26455 A 10)

Held J P, Mansa T M, Delprat J 1981 Rééducation de la sensibilité. In Grossiord A, Held J P (eds) Médecine de Rééducation. Flammarion, Paris, p 250–259

Hoehn M M D, Yahr M D Parkinsonism : Onset progression and mortality. Neurology 17 : 247–442

Johnstone M 1980 Home care for the stroke patient : living in a pattern. Churchill Livingstone, Edinburgh

Kapandji I A 1975 Physiologie articulaire. Membre supérieur. Fascicule 1, 4e edition. Maloine, Paris

Kleist K 1907 Corticale (Innervatorische) Apraxie. Journal de Psychiatrie 28

Kleist K 1911 Der Gang u. d. gegenwärtige Stand d. Apraxie Forschung. Ergebnisse der Neurol. 1

Lacert P 1978 Rééducation des paraplégiques. Expansion Scientifique Française, Paris

Lamontagne Y 1976 La rétro-action biologique : instrument du mesure et outil thérapeutique en thérapie comportementale. Psychologie Medicale 8 : 21–90

Luria A R 1966 Human brain and psychological processes. Harper and Row, London and New York

Marchand J 1972 Le médecin practicien et l'examen vestibulaire. Le Concours Médical 94–7 : 1165–1183

Marduel Y N 1980 L'épaule de l'hémiplégique. Kinésithérapie-Scientifique no. 182 : 45–52

Marie P 1911 La pratique neurologique. Masson, Paris

Marie P, Foix C 1916 Les Syncinésies des hemiplégiques. Revue Neurologique 1 : 3–27 ; 2 : 145–162, 620

Massion J 1980 Interrelations entre posture et mouvement, p 355–364. Anticipation et comportement, publié sous la direction de Jean Requin. Editions du Centre National de la Recherch Scientifique, Paris

de Meurisse G 1979 Evaluation de la motricité après un accident vasculaire. Acta Neurologica Belgica 79 : 5–21

Michels E 1959 Evaluation of motor function in hemiplegia. Physiol. Ther. Revue 39 : 589–595

Morel-Maroger A 1975 Aspects neurologiques de la maladie des tics. Encyclopèdie Medico-Chirurgicale (Neurologie) 17011 B 1, 1st edn, 7

Nahon E, Piera J B 1977 Rééducation de la préhension et autonomie du tétraplègique-traumatique complet. Kinésithérapie-Scientifique no. 153 : 39–51

Rademaker G G J 1935 Das stehen, 1931. Réactions labyrinthiques et equilibre. Ed. Masson, Paris

Ratinaud J P 19–– Une forme d'évaluation de la fonction motrice chez l'hémiplégique. Journal de Kinésithérapie no. 183

Recondo (de) J 1978 Principaux syndromes neurologiques. 2nd edn. Laboratoires Roussel

Recondo (de) J 1980 Les mouvements anormaux. Soins T 25 (19) : 3–10

Rocher C, Rigaud A 1964 Fonctions et bilans articulaires. Masson, Paris

Rondot P 1968 Les contractures, étude clinique et physiopathologique. Extrait de l'encéphale Nos 3 and 4 : 185–242 ; 187–332

Rondot P, Jedynak C P 1976 Dystonies de fonction. Encyclopédie Medico-Chirurgicale. 9 : Neurologie 17007 C 20. Paris

Rondot P, Jedynak C P, Ferrey G 1981 Le torticolis spasmodique. Rapport de Neurologie. Congrès de psychiatrie et de neurologie de langue française, Colmar. Ed. Masson, Paris
Rondot P, Recondo (de) J 1976 La maladie de Parkinson. Baillière J B (ed). Paris
Rondot P, Recondo (de) J 1978 Troubles de la préhension manuelle sous le contrôle de la vue. Masson (ed). Annales de Médecine Interne nos 8–9 : 487–492
Signoret J L 1975 Le cerveau neuropsychologique. Cahiers intégrés de Médecine (neurologie XI, 1° partie). Masson, Paris
Signoret J L, North P 1979 Les apraxies gestuelles. Rapport de neurologie présenté au congrès de psychiatrie et de neurologie langue française. Ed. Masson, Paris
Tardieu G 19-- Le dossier clinique de l'infirmité motrice cérébrale. Cahier du C.D.I. no. 39 (2nd edn refondue)
Tardieu G 1971 Athètose. Encyclopédie Médico-Chirurgicale (Neurologie). September : 17010 D 10
Tinel J 1916 Les blessures des nerfs. Ed. Masson, Paris
Zucman E 19-- Rééducation des encéphalopathies infantiles selon la methode Temple Fay. Thèse de docteur en Médecine

SECTION FOUR

Motor Points

MOTOR POINTS AND MORPHOLOGICAL LANDMARKS

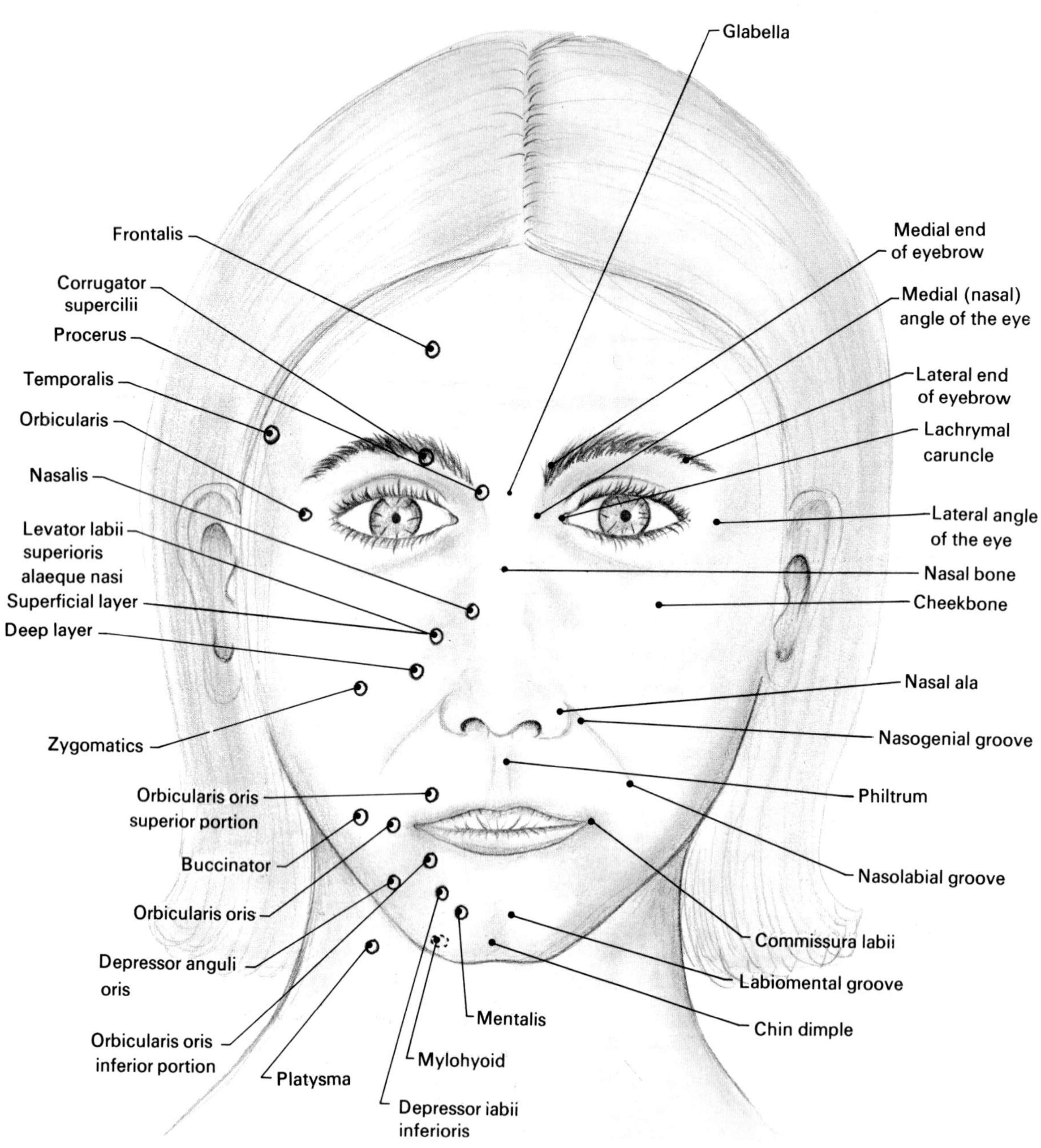

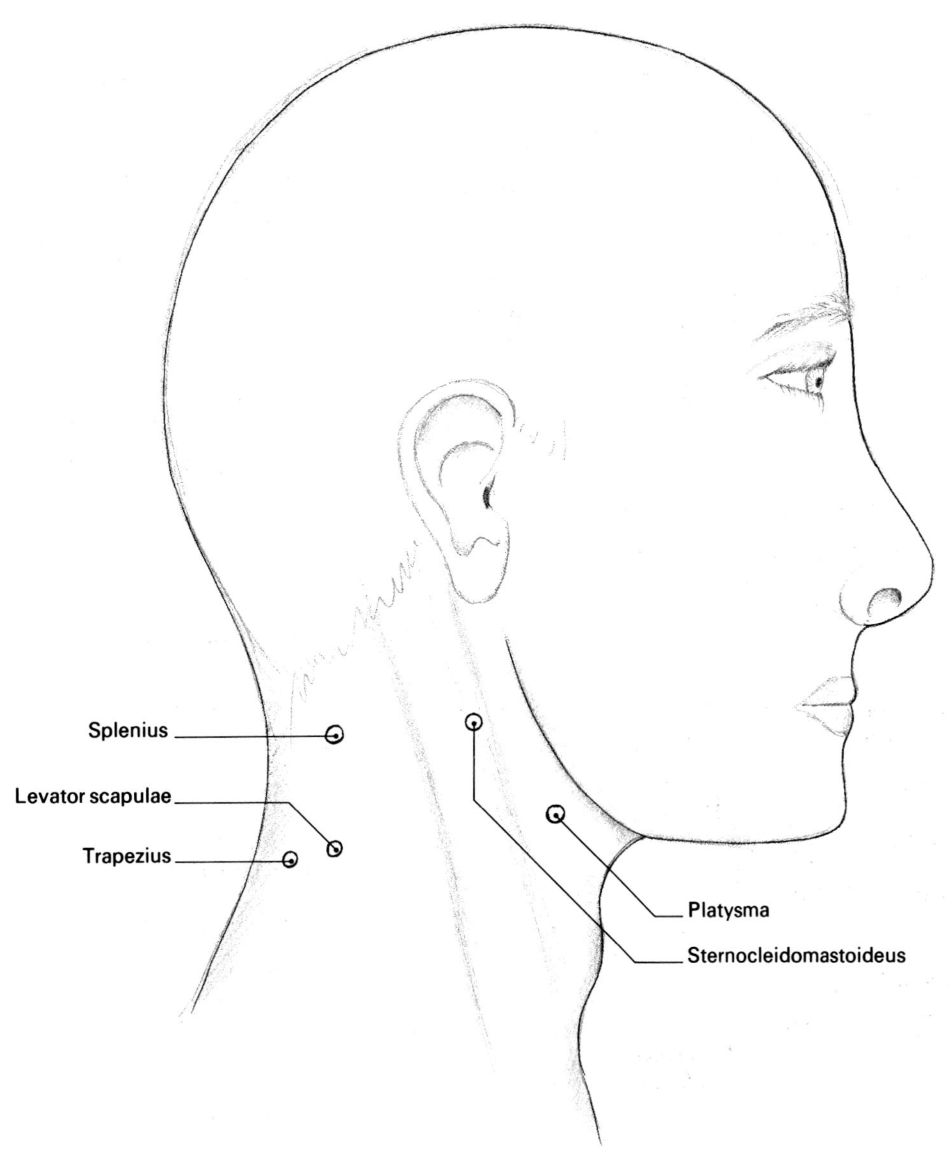
Splenius
Levator scapulae
Trapezius
Platysma
Sternocleidomastoideus

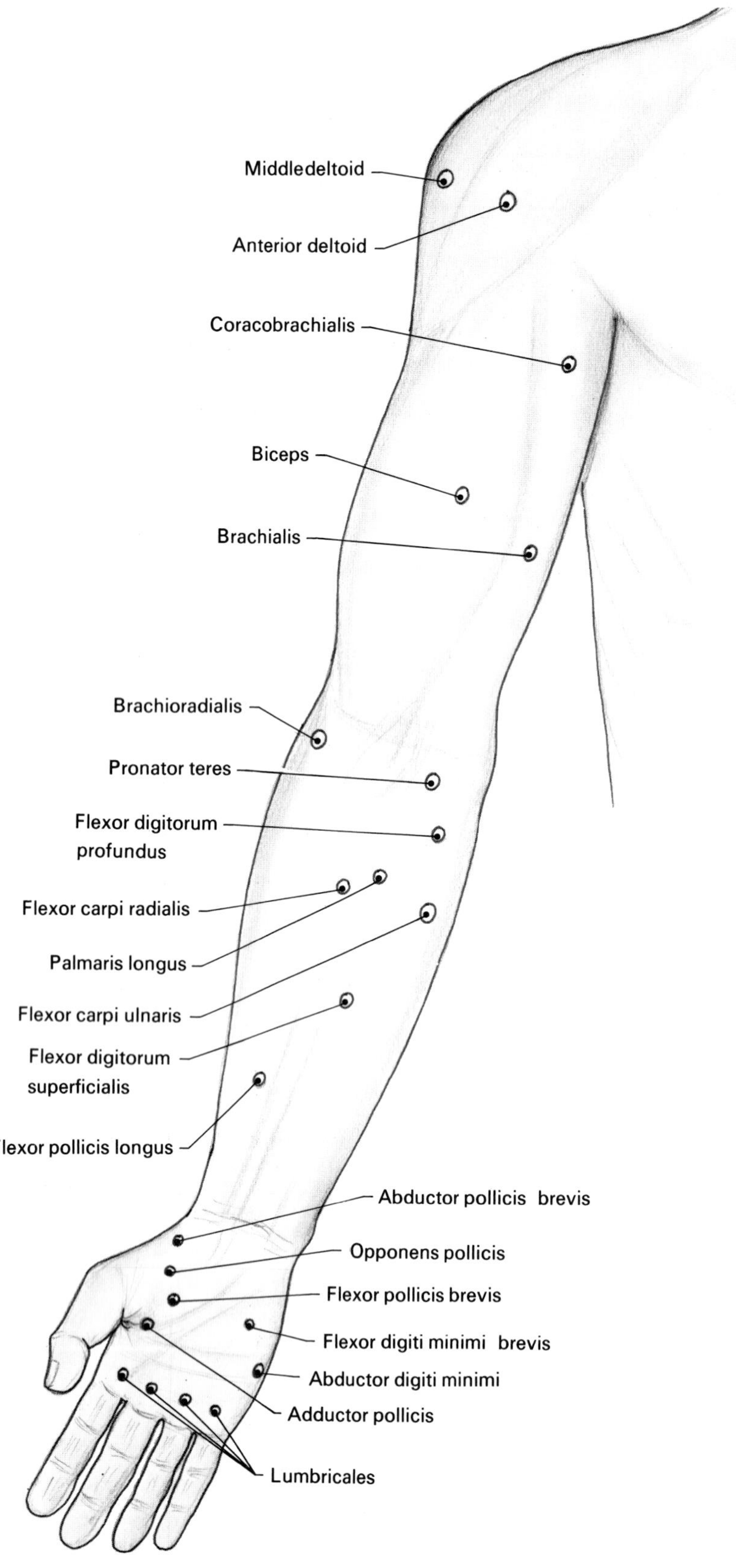
Middledeltoid
Anterior deltoid
Coracobrachialis
Biceps
Brachialis
Brachioradialis
Pronator teres
Flexor digitorum
profundus
Flexor carpi radialis
Palmaris longus
Flexor carpi ulnaris
Flexor digitorum
superficialis
Flexor pollicis longus
Abductor pollicis brevis
Opponens pollicis
Flexor pollicis brevis
Flexor digiti minimi brevis
Abductor digiti minimi
Adductor pollicis
Lumbricales

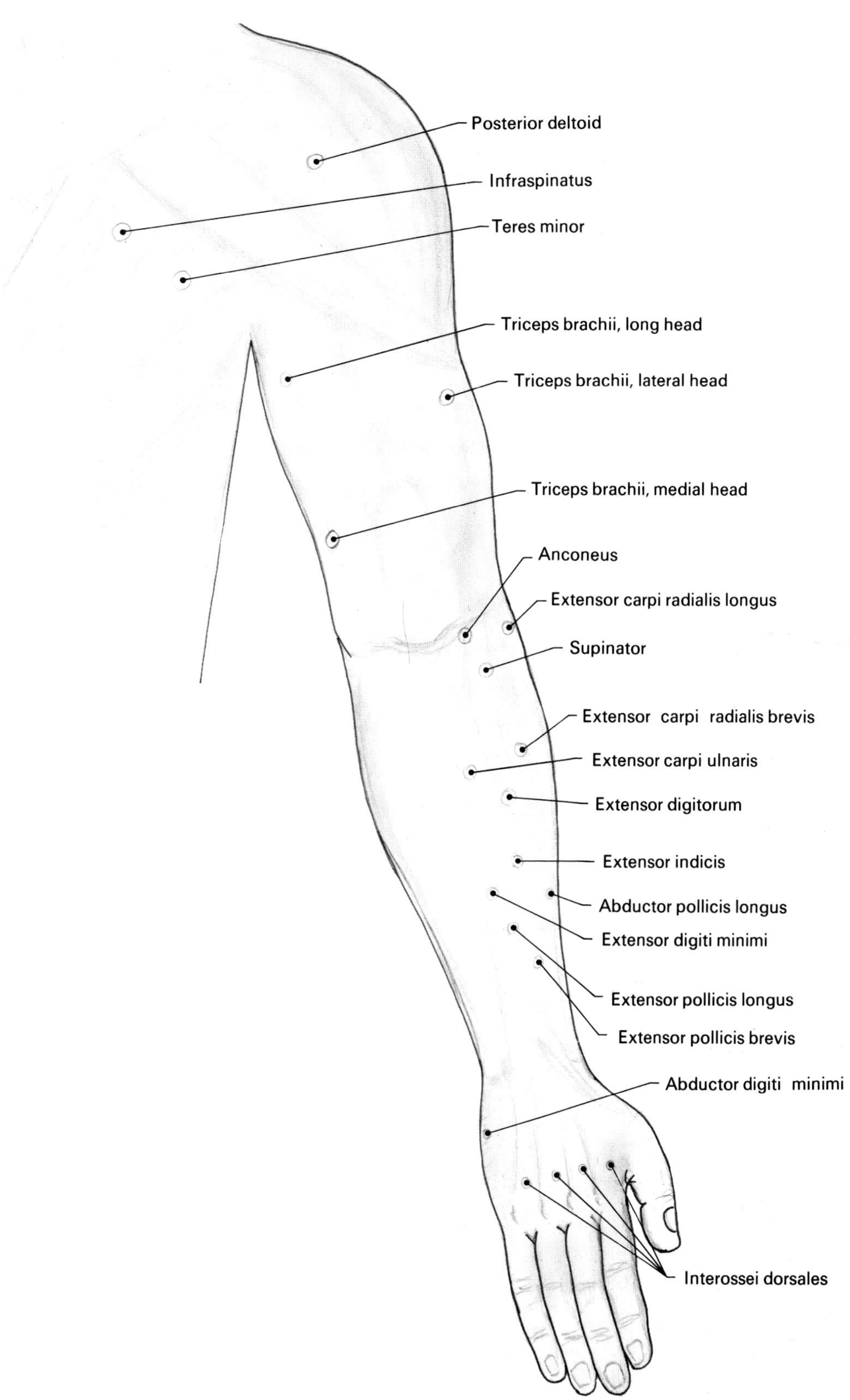
Posterior deltoid
Infraspinatus
Teres minor
Triceps brachii, long head
Triceps brachii, lateral head
Triceps brachii, medial head
Anconeus
Extensor carpi radialis longus
Supinator
Extensor carpi radialis brevis
Extensor carpi ulnaris
Extensor digitorum
Extensor indicis
Abductor pollicis longus
Extensor digiti minimi
Extensor pollicis longus
Extensor pollicis brevis
Abductor digiti minimi
Interossei dorsales

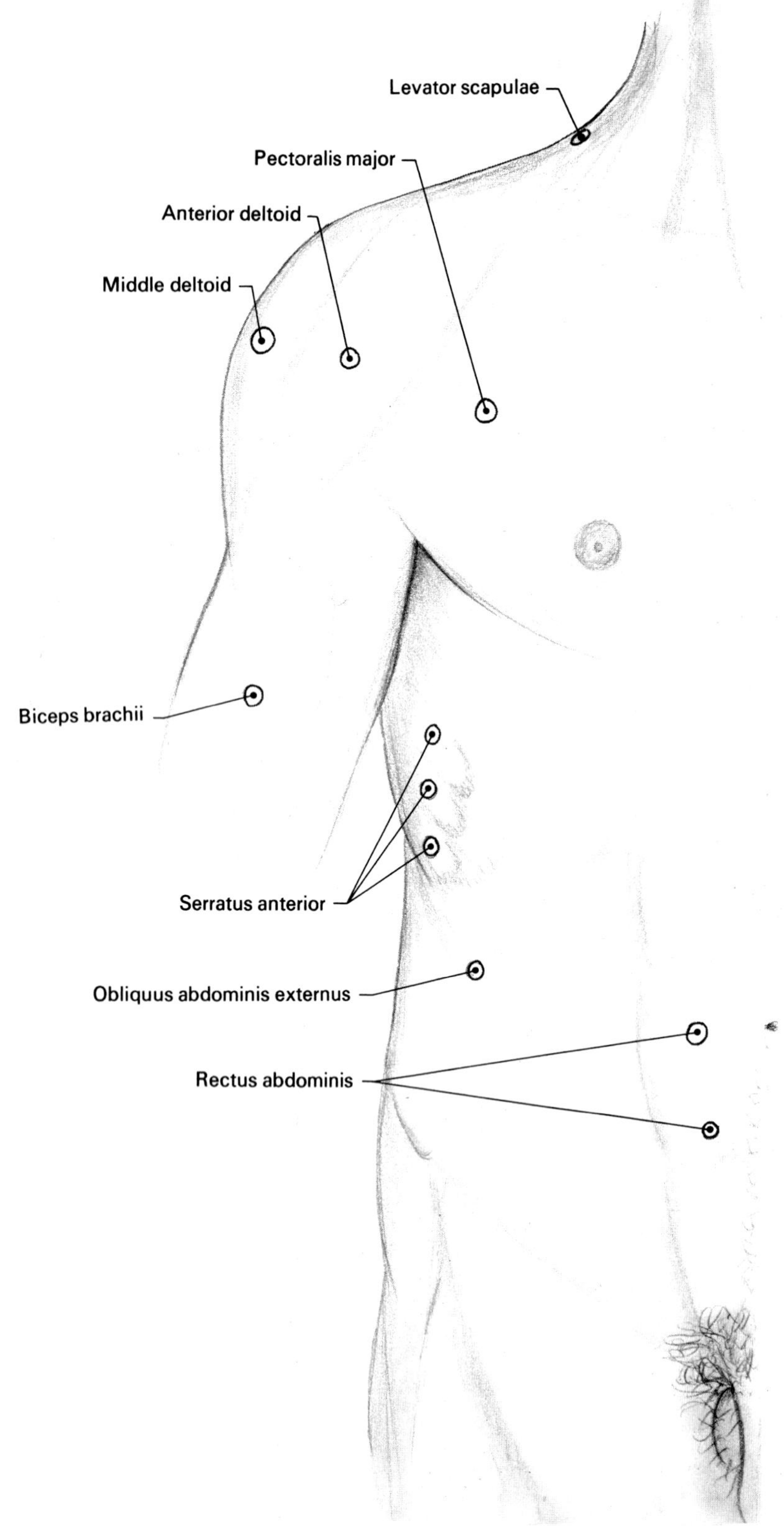
Levator scapulae
Pectoralis major
Anterior deltoid
Middle deltoid
Biceps brachii
Serratus anterior
Obliquus abdominis externus
Rectus abdominis

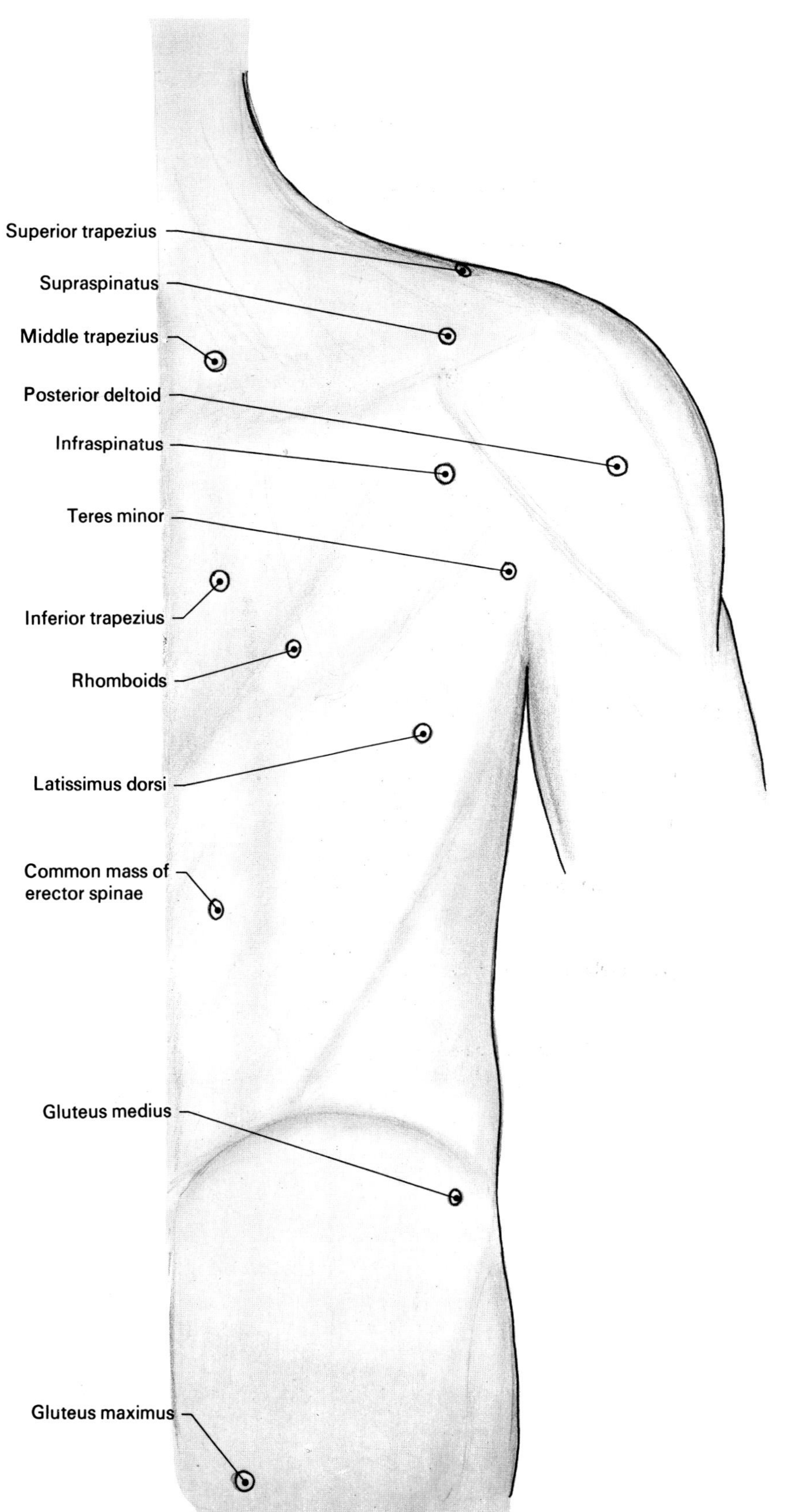
Superior trapezius
Supraspinatus
Middle trapezius
Posterior deltoid
Infraspinatus
Teres minor
Inferior trapezius
Rhomboids
Latissimus dorsi
Common mass of
erector spinae
Gluteus medius
Gluteus maximus

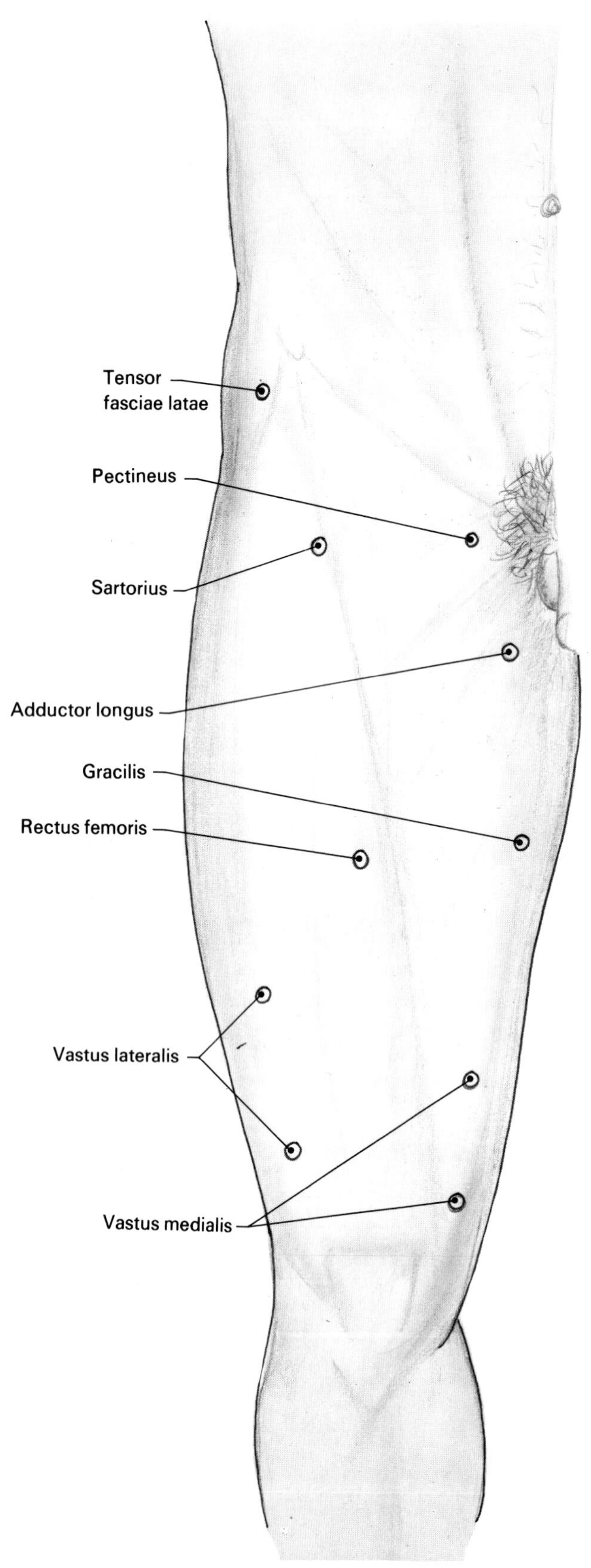
Tensor
fasciae latae
Pectineus
Sartorius
Adductor longus
Gracilis
Rectus femoris
Vastus lateralis
Vastus medialis

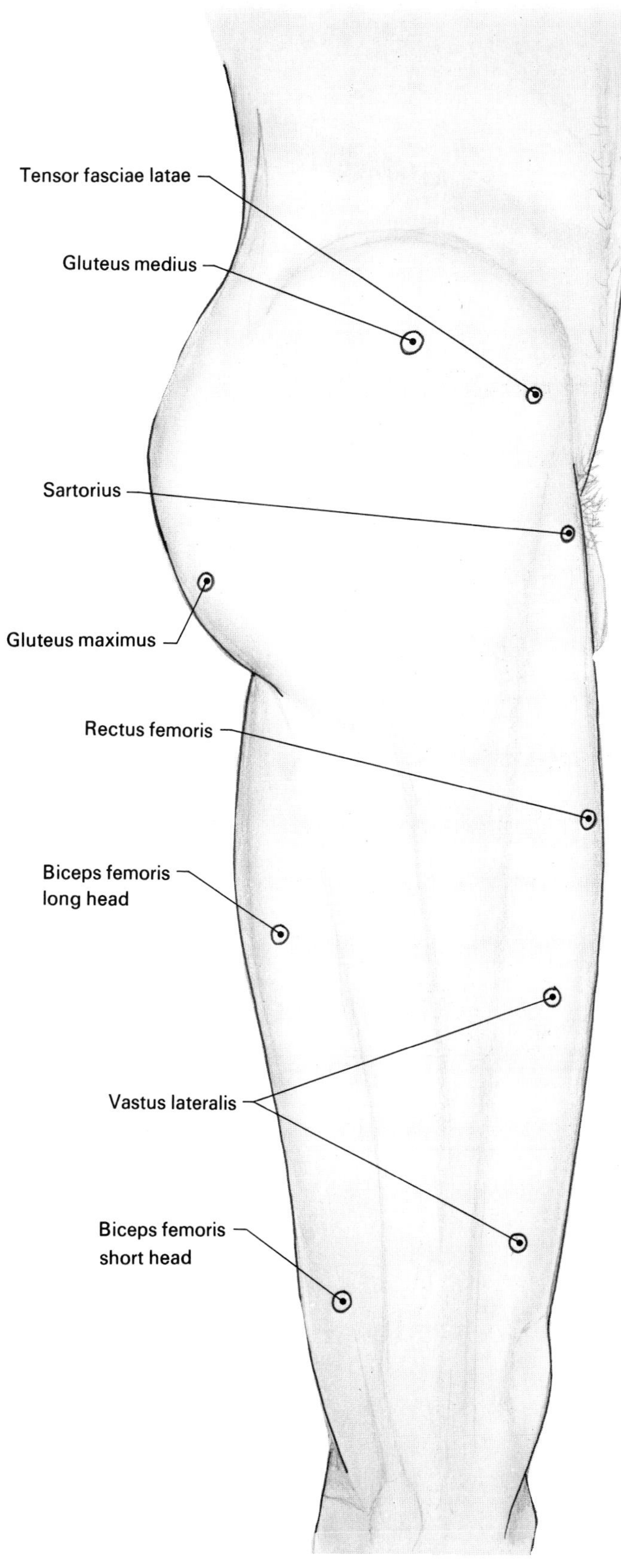
Tensor fasciae latae
Gluteus medius
Sartorius
Gluteus maximus
Rectus femoris
Biceps femoris
long head
Vastus lateralis
Biceps femoris
short head

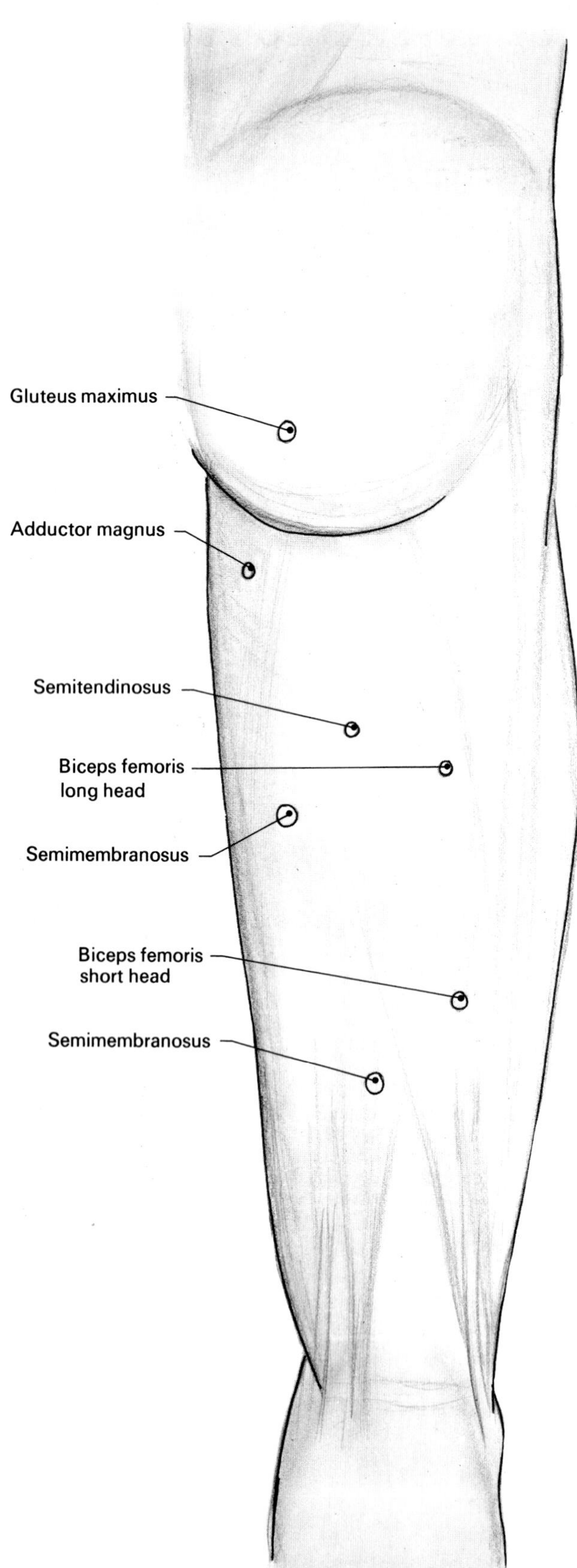
Gluteus maximus
Adductor magnus
Semitendinosus
Biceps femoris
long head
Semimembranosus
Biceps femoris
short head
Semimembranosus

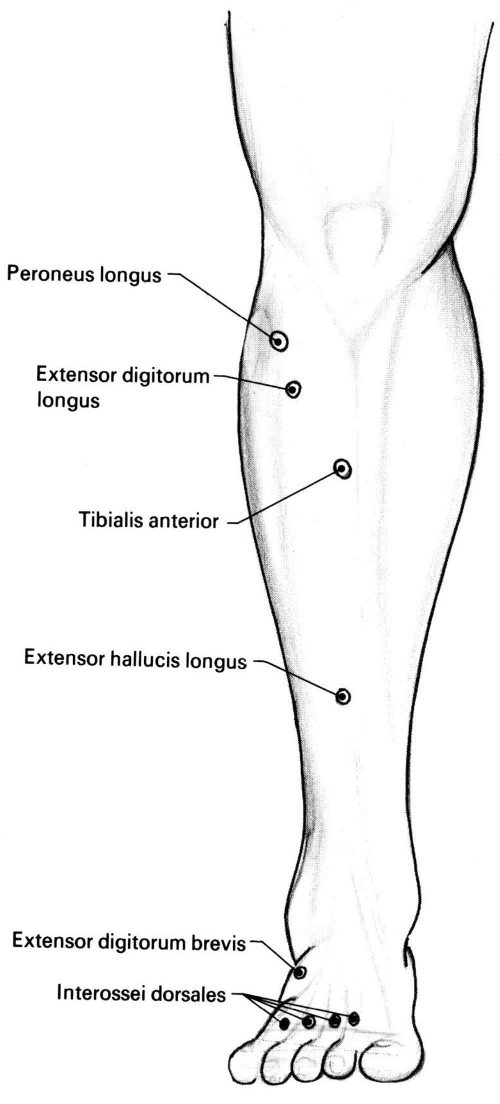
Peroneus longus
Extensor digitorum longus
Tibialis anterior
Extensor hallucis longus
Extensor digitorum brevis
Interossei dorsales

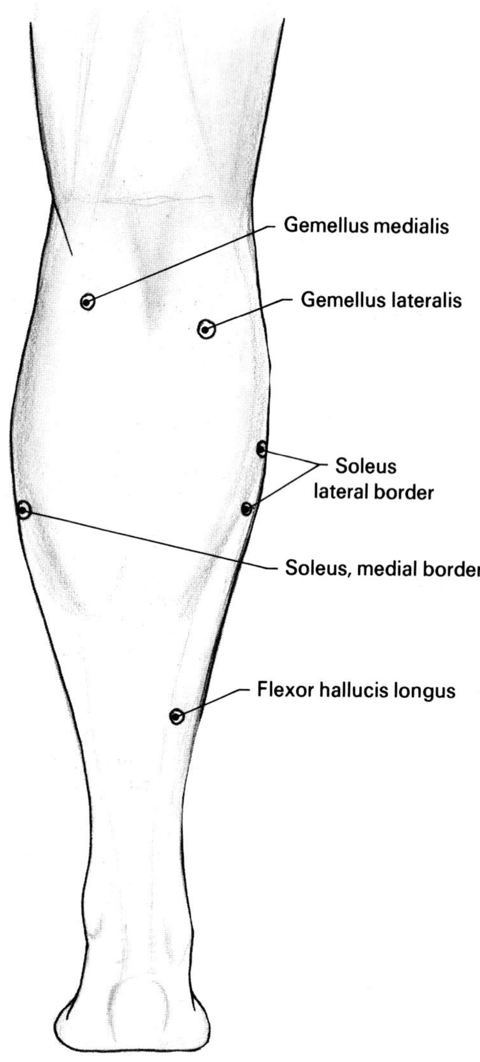
Gemellus medialis
Gemellus lateralis
Soleus lateral border
Soleus, medial border
Flexor hallucis longus

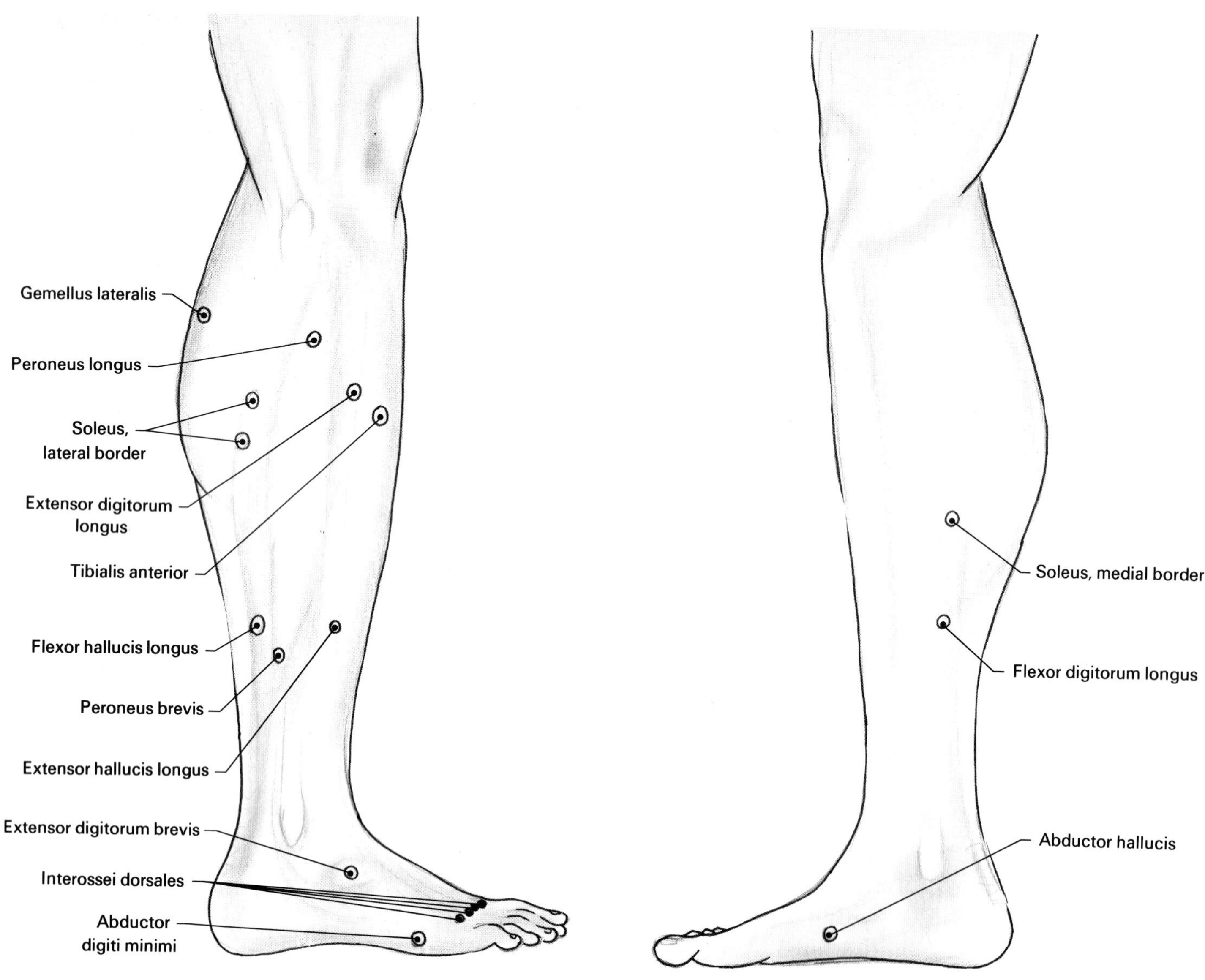

Gemellus lateralis
Peroneus longus
Soleus,
lateral border
Extensor digitorum
longus
Tibialis anterior
Flexor hallucis longus
Peroneus brevis
Extensor hallucis longus
Extensor digitorum brevis
Interossei dorsales
Abductor
digiti minimi
Soleus, medial border
Flexor digitorum longus
Abductor hallucis

SECTION FIVE

Innervation: Schematic Nerve Distribution

M. Lacôte and A. Miranda

PLEXUS AND NERVES

SCHEMATIC FORMATION OF BRACHIAL PLEXUS

1 MUSCULOCUTANEOUS NERVE
2 CIRCUMFLEX NERVE
3 RADIAL NERVE
4 LATERAL HEAD OF MEDIAN
5 MEDIAL HEAD OF MEDIAN
6 MEDIAN NERVE
7 ULNAR NERVE
8 Medial brachial cutaneous (sensory innervation of anterior medial part of arm and of anterior and posterior medial part of forearm)
9 Lesser medial brachial cutaneous (sensory innervation of axillary fossa and of medial aspect of arm)
10 **Nerve to pectoralis major (C5 C6 C7 C8 T1)**
11 LATERAL CORD
12 ANSA PECTORALIS
13 **Inferior subscapular nerve (C5 C6)**
14 **Nerve to teres major (C5 C6 C7)**
15 **Nerve to latissimus dorsi (C6 C7 C8)**
16 POSTERIOR CORD
17 MEDIAL CORD
18 **Nerve to pectoralis minor (c7 C8 T1)**
19 **Nerve to subclavius (C5 C6)**
20 **Superior subscapular nerve (C5 C6)**
21 SUPRASCAPULAR NERVE
22 **Nerve to supraspinatus (C5 C6)**
23 **Nerve to infraspinatus (C5 C6)**
24 **Nerve to levator scapulae (C4 C5)**
25 **Nerve to rhomboids (C4 C5)**
26 UPPER TRUNK
27 MIDDLE TRUNK
28 LOWER TRUNK
29 **Long thoracic nerve (C5 C6 C7)**

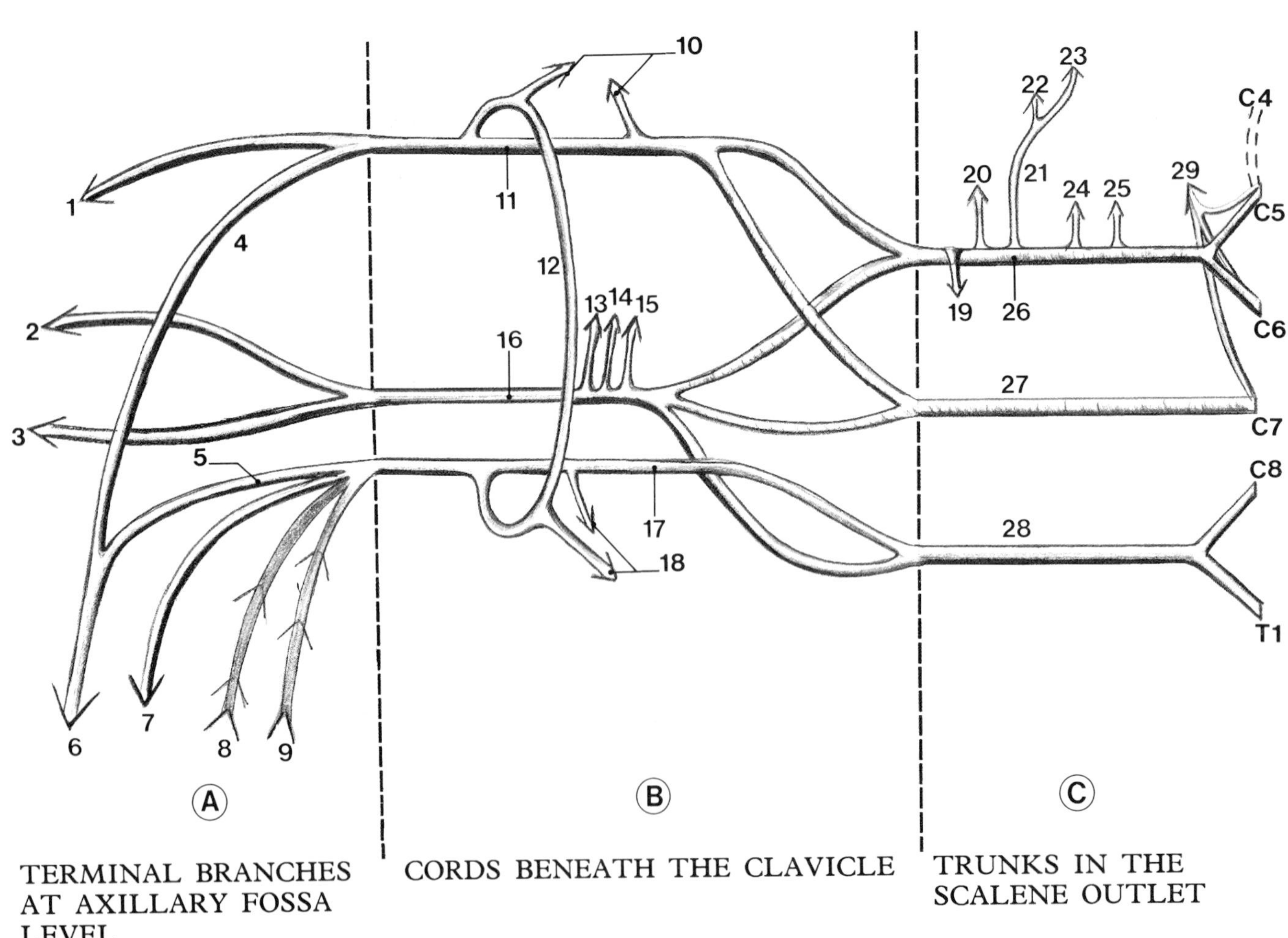

TERMINAL BRANCHES AT AXILLARY FOSSA LEVEL

CORDS BENEATH THE CLAVICLE

TRUNKS IN THE SCALENE OUTLET

MUSCULOCUTANEOUS NERVE

LATERAL CORD (C5 C6)
1 **Superior nerve to coracobrachialis (C6 C7)**
2 *Nerve to humeral shaft*
3 *Vascular nerves (axillary and humeral arteries)*
4 **Nerve to long head of biceps brachii (C5 C6)**
5 **Nerve to short head of biceps brachii (C5 C6)**
6 LATERAL BICIPITAL TROUGH
7 Posterior sensory branch (sensory innervation of posterior lateral part of forearm)
8 Anterior sensory branch (sensory innervation of anterior lateral part of forearm and of upper part of thenar eminence)
9 *Articular nerve to wrist joint*
10 *Coracobrachialis muscle*
11 **Inferior nerve to coracobrachialis (C6 C7)**
12 **Nerves to brachialis, four or five rami (C5 C6)**
13 *Antebrachial aponeurosis*

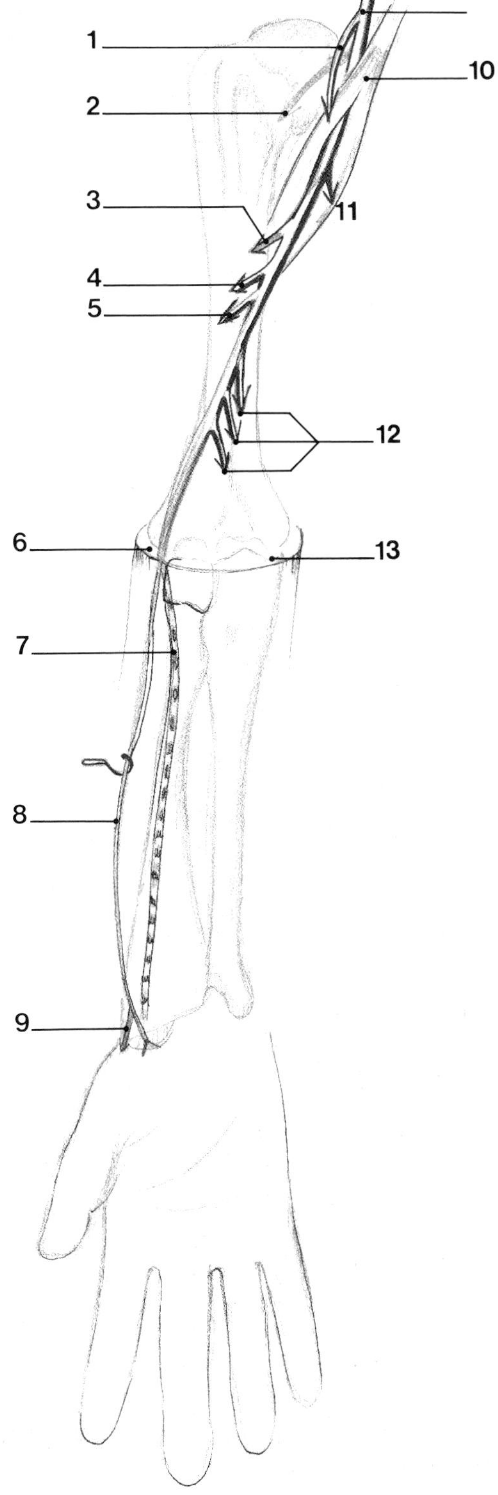

MEDIAN NERVE

MEDIAL CORD AND LATERAL CORD (C6 C7 C8 T1)

1 *Vascular ramus (humeral artery)*
2 *Nerve to humeral shaft*
3 *Pronator teres muscle*
4 **Inferior nerve to pronator teres, ulnar head (C6 C7)**

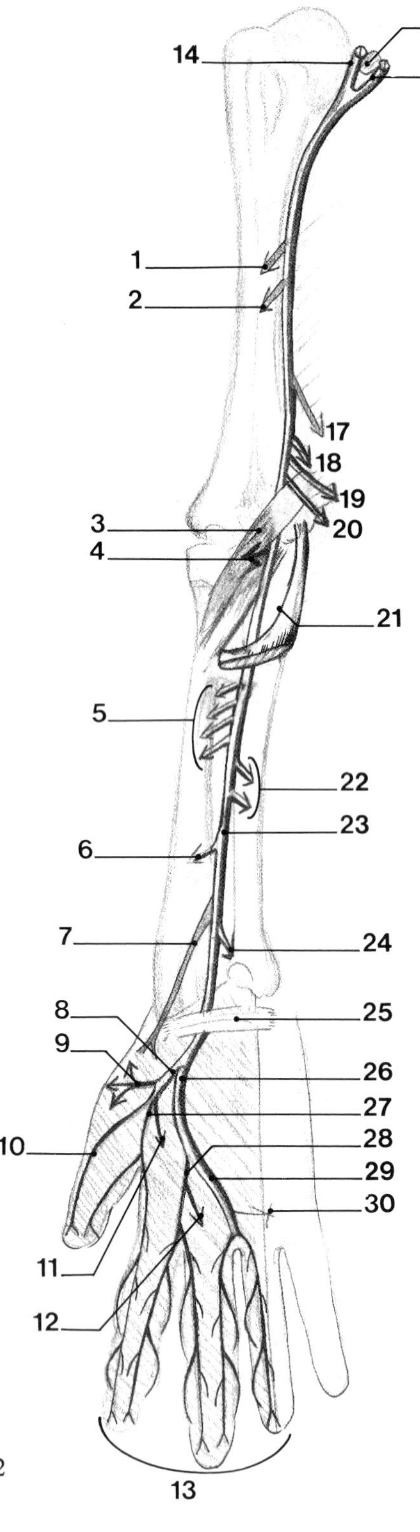

5 **Nerves to flexor digitorum superficialis (c7 C8 T1)**
6 **Nerve to flexor pollicis longus (c7 C8 T1)**
7 Palmar cutaneous nerve (sensory innervation of lateral part of palm of hand
8 *Lateral branch*
9 **Thenar branch (C8 T1)**
— nerve to abductor pollicis brevis
— nerve to opponens pollicis
— nerve to superficial head of flexor pollicis brevis
10 Lateral palmar collateral of thumb
11 **Nerve to first lumbrical (C7 C8 T1)**
12 **Nerve to second lumbrical (C7 C8 T1)**
13 Palmar rami (sensory innervation of palmar surface of proximal, middle and distal phalanges)
Dorsal rami (sensory innervation of dorsal surface of middle and distal phalanges)
14 *Lateral head*
15 *Axillary artery*
16 *Medial head*
17 *Articular nerve of elbow joint*
18 **Superior nerve to pronator teres, humeral head (C6 C7)**
19 **Nerve to palmaris longus (C6 C7 c8)**
20 **Nerve to flexor carpi radialis (C6 C7 c8)**
21 *Arch of flexor digitorum superficialis*
22 **Nerves to flexor digitorum profundus, two lateral heads (c7 C8 T1)**
23 Anterior interosseus nerve
24 **Nerve to pronator quadratus (c7 C8 T1)**
25 *Anterior transverse carpal ligament*
26 *Medial branch*
27 Nerve to first interdigital space
28 Nerve to second interdigital space
29 Nerve to third interdigital space
30 Anastomosis to ulnar nerve
31 Dorsal rami

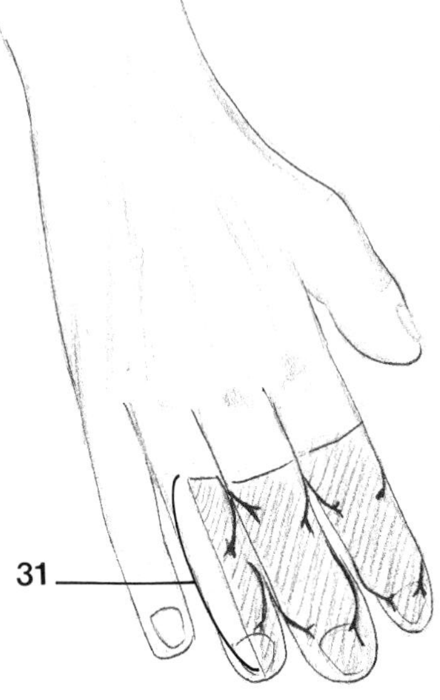

ULNAR NERVE

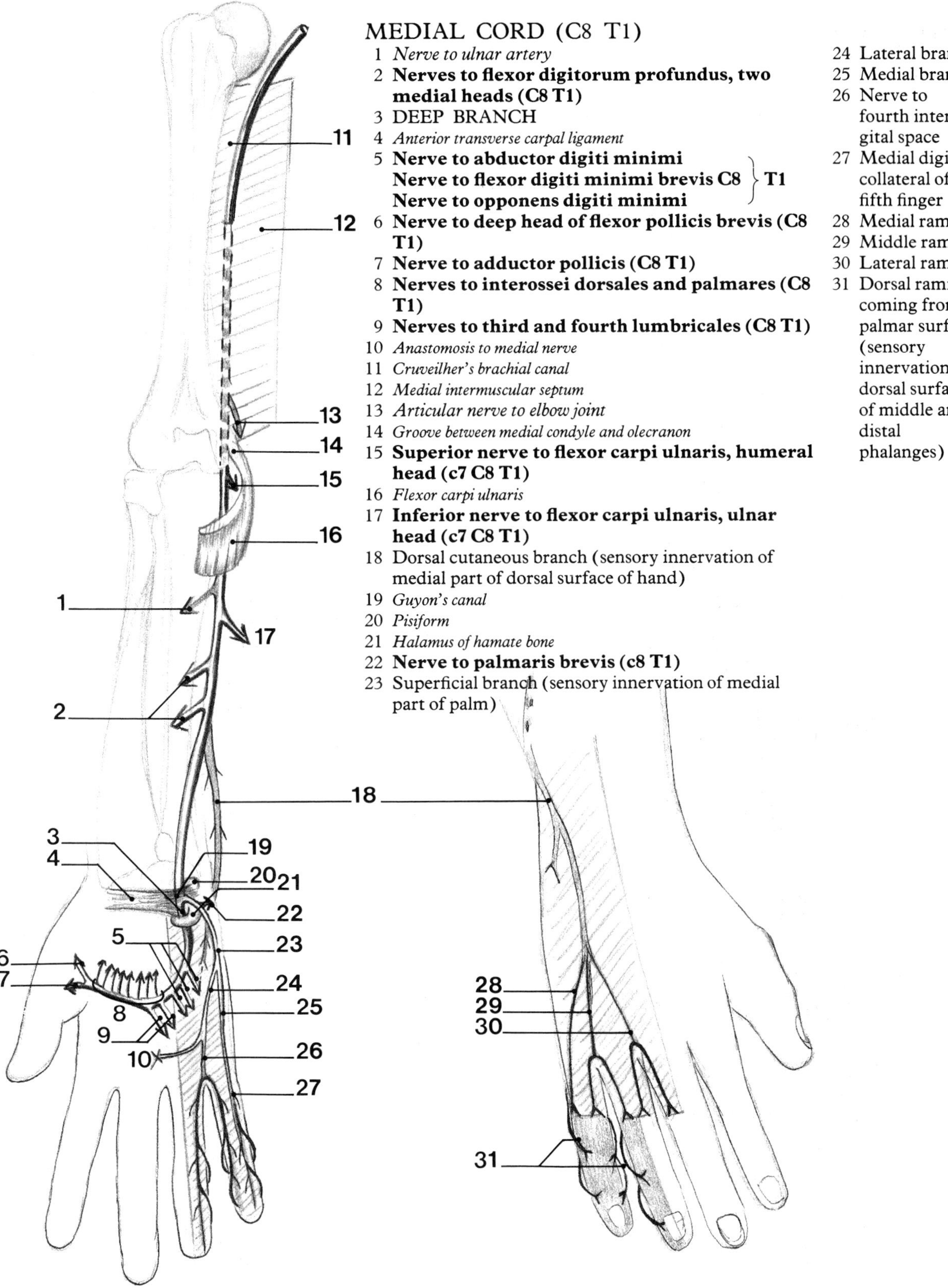

MEDIAL CORD (C8 T1)

1 *Nerve to ulnar artery*
2 **Nerves to flexor digitorum profundus, two medial heads (C8 T1)**
3 DEEP BRANCH
4 *Anterior transverse carpal ligament*
5 **Nerve to abductor digiti minimi**
Nerve to flexor digiti minimi brevis C8 } **T1**
Nerve to opponens digiti minimi
6 **Nerve to deep head of flexor pollicis brevis (C8 T1)**
7 **Nerve to adductor pollicis (C8 T1)**
8 **Nerves to interossei dorsales and palmares (C8 T1)**
9 **Nerves to third and fourth lumbricales (C8 T1)**
10 *Anastomosis to medial nerve*
11 *Cruveilher's brachial canal*
12 *Medial intermuscular septum*
13 *Articular nerve to elbow joint*
14 *Groove between medial condyle and olecranon*
15 **Superior nerve to flexor carpi ulnaris, humeral head (c7 C8 T1)**
16 *Flexor carpi ulnaris*
17 **Inferior nerve to flexor carpi ulnaris, ulnar head (c7 C8 T1)**
18 Dorsal cutaneous branch (sensory innervation of medial part of dorsal surface of hand)
19 *Guyon's canal*
20 *Pisiform*
21 *Halamus of hamate bone*
22 **Nerve to palmaris brevis (c8 T1)**
23 Superficial branch (sensory innervation of medial part of palm)
24 Lateral branch
25 Medial branch
26 Nerve to fourth interdigital space
27 Medial digital collateral of fifth finger
28 Medial ramus
29 Middle ramus
30 Lateral ramus
31 Dorsal rami coming from palmar surface (sensory innervation of dorsal surface of middle and distal phalanges)

LATERAL BRANCH OF POSTERIOR CORD (C5 C6)

1 Cutaneous nerves to shoulder (sensory innervation of posterior and anterior lateral parts of shoulder girdle)
2 *Articular nerves*
3 **Nerves to deltoideus, anterior, middle, and posterior rami (C5 C6)**
4 *Clavicle*
5 RADIAL NERVE
6 **Nerve to subscapularis (C5 C6)**
7 **Nerve to teres minor (C5 C6)**
8 *Velpeau's quadrilateral space*
9 *Long head of triceps brachii*
10 *Teres major and latissimus dorsi*

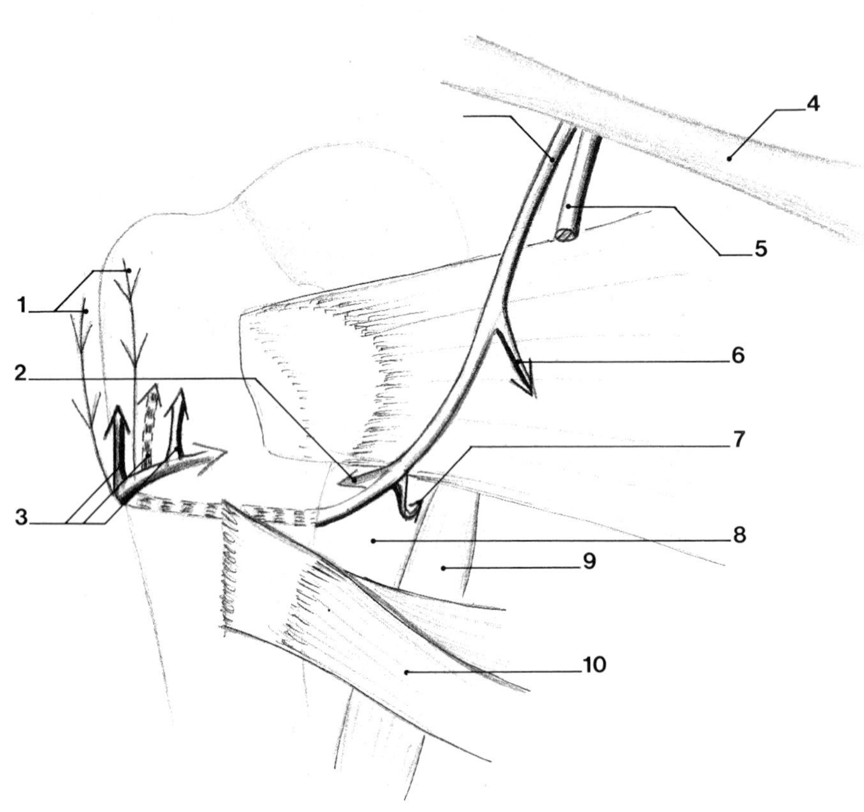

MEDIAL BRANCH OF POSTERIOR CORD (C6 C7 C8 T1)

1 Medial cutaneous branch (sensory innervation of posterior lateral part of arm, anastomosed to medial cutaneous brachial nerve).
2 **Nerve to long head of triceps brachii (c6 C7 c8)**
3 *Musculo spiral groove (posterior surface of humerus)*
4 **Nerve to brachioradialis muscle (C5 C6)**
5 Lateral cutaneous ramus (sensory innervation of posterior part of forearm
6 **Nerve to extensor carpi radialis longus (C6 C7)**
7 **Nerve to extensor carpi radialis brevis (c6 C7 c8)**
8 **Nerve to supinator (C5 C6 c7)**
9 POSTERIOR MUSCULAR BRANCH
10 SUPERFICIAL POSTERIOR BRANCH
11 **Nerve to extensor digitorum (c6 C7 c8)**
12 **Nerve to extensor digiti minimi (c6 C7 c8)**
13 **Nerve to extensor carpi ulnaris (C7 C8)**
14 Lateral branch
15 Thenar twig
16 Middle ramus
17 Nerve to first interdigital space
18 Lateral collateral of thumb
19 *Teres major and latissimus dorsi muscles*
20 *Long head of triceps brachii muscle*
21 *Gap between humerus and triceps brachii*
22 **Nerve to medial head of triceps brachii (c6 C7 c8)**
23 **Nerve to medial head of triceps and anconeous (c6 C7 c8)**
24 **Nerve to lateral head of triceps brachii (c6 C7 c8)**
25 **Nerve to brachialis (C6) inconstant**
26 *Articular nerve to elbow joint*
27 *Supinator muscle*
28 Anterior sensory branch. Division of radial nerve is at radial head level or sometimes more proximally
29 DEEP ANTERIOR BRANCH
30 **Nerve to abductor pollicis longus (c6 C7 c8)**
31 **Nerve to extensor pollicis brevis (c6 C7 c8)**
32 **Nerve to extensor pollicis longus (c6 C7 c8)**
33 **Nerve to extensor indicis (c6 C7 c8)**
34 Medial ramus
35 *Anastomosis to the ulnar nerve at dorsal surface level*
36 Nerve to second interdigital space

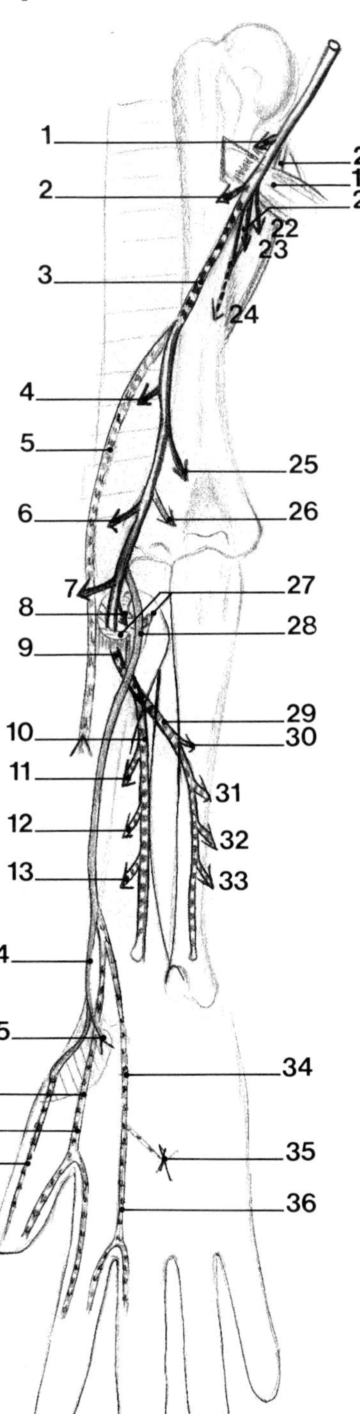

DORSAL VIEW

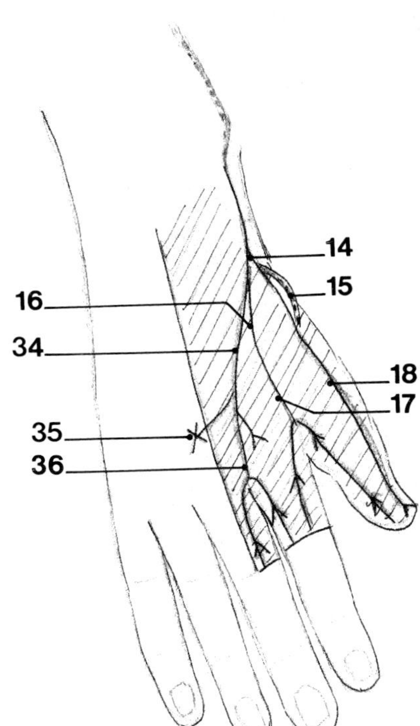

SCHEMATIC FORMATION OF LUMBAR PLEXUS

TWO COLLATERAL BRANCHES
SIX TERMINAL BRANCHES

1 ILIO-HYPOGASTRIC NERVE, **muscular innervation of abdominals**, SENSORY INNERVATION OF UPPER PART OF BUTTOCK, OF ABDOMEN, OF SUPERIOR MEDIAL PART OF THIGH, GENITAL ORGANS.
2 ILIO-INGUINAL NERVE, **muscular**, AND SENSORY INNERVATION SAME AS ILIO-HYPO GASTRIC NERVE
3 GENITO-CRURAL NERVE: **muscular innervation of genital organs**, SENSORY INNERVATION OF GENITAL ORGANS AND OF SUPERIOR MEDIAL PART OF THIGH (SCARPA'S TRIANGLE)
4 Lateral femoral cutaneous nerve, sensory innervation of lateral aspect of buttock, anterior and posterior lateral aspects of thigh
5 FEMORAL NERVE (COMING FROM POSTERIOR PARTS)
6 OBTURATOR NERVE (COMING FROM ANTERIOR PARTS)
7 **Nerve to quadratus luborum (t12 L1 L2)**
8 **Nerve to psoas and psoas minor (L1 L2 L3 L4)**
9 LUMBO-SACRAL TRUNK

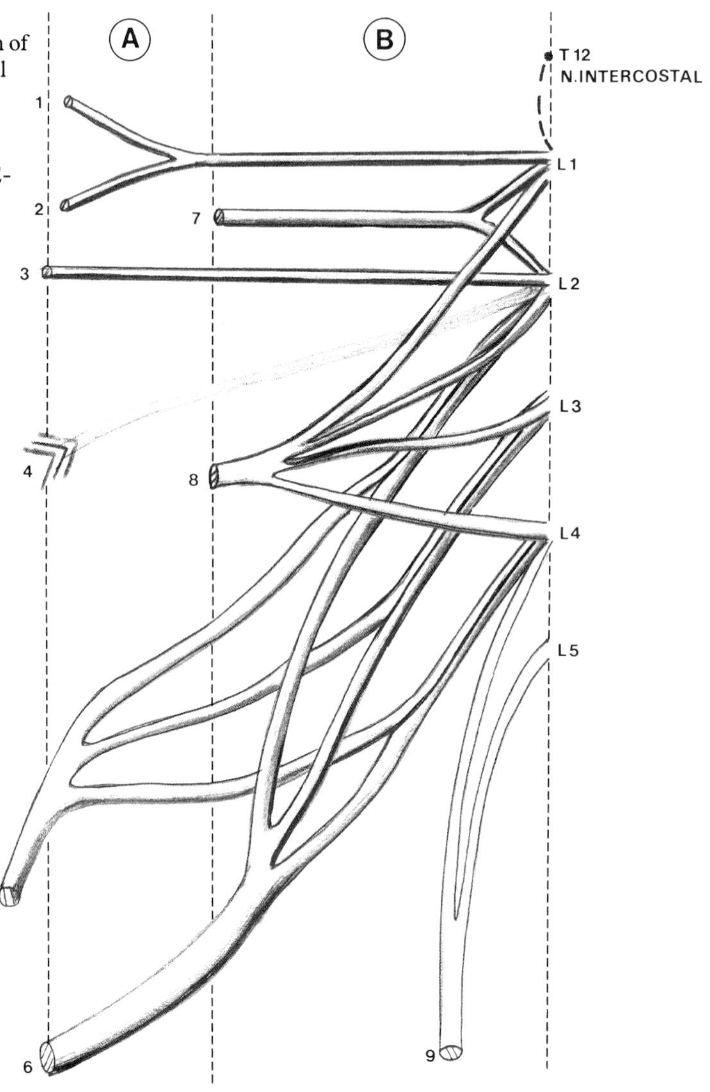

TERMINAL BRANCHES   COLLATERAL BRANCHES

OBTURATOR NERVE (anterior parts)

1 FEMORAL NERVE (POSTERIOR PARTS)
2 *Obturator foramen*
3 **Nerve to obturator externus (L3 L4)**
4 **Nerve to pectineus (L2 L3 L4)**
5 **Nerve to gracilis (L2 L3 L4)**
6 **Nerve to adductor longus (L2 L3 L4)**
7 Cutaneous ramus (sensory innervation of medial part of knee, anastomosis to medial saphenous and third perforating nerves)
8 **Nerves to adductor brevis (L2 L3 L4)**
9 *Articular nerves to hip joint*
10 **Inconstant nerve to obturator externus (Rouvière)**
11 *Articular nerve to hip joint*
12 **Inconstant nerve to adductor brevis (Brizon and Castaing's)**
13 **Nerve to adductor magnus, first and second heads (12 L3 L4)**

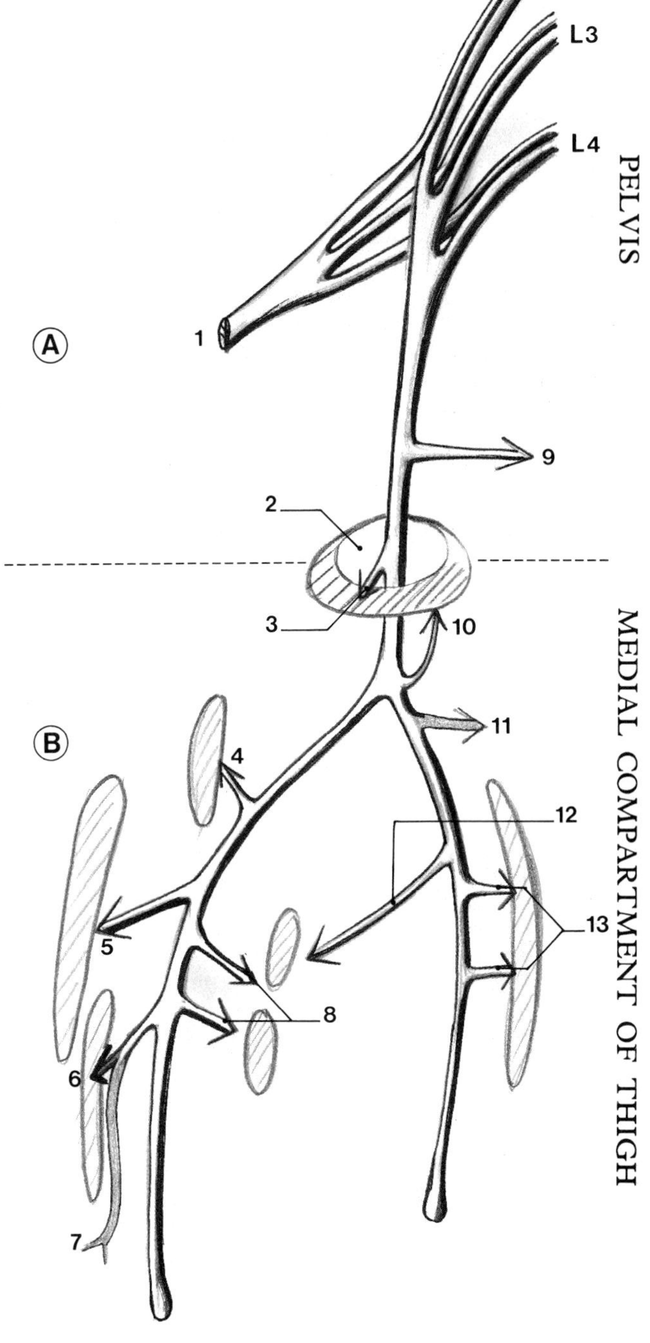

FEMORAL NERVE (posterior parts)

1 **Nerves to iliacus (L2 L3 L4)**
2 *Nerve to femoral artery (Schwalbe's nerve)*
3 *Poupart's ligament*
4 **Nerves to sartorius (L1 L2 L3)**
5 First superior perforating nerve (sensory innervation of anterior part of thigh and knee)
6 Second middle perforating nerve (same territory as the first)
7 Third inferior perforating nerve, accessory of medial saphenous nerve: anastomosis to medial saphenous and obturator cutaneous ramus (sensory innervation of medial part of knee
8 **Nerve to rectus femoris (L2 L3 L4)**
9 **Nerve to vastus lateralis and vastus intermedius (L2 L3 L4)**
10 **Nerve to vastus medialis, vastus intermedius, and subcrureus (L2 L3 L4)**
11 **Nerve to vastus intermedius (L2 L3 L4)**
12 Tibial cutaneous ramus (sensory innervation of medial aspect of calf)
13 Terminal branch to leg (sensory innervation of medial part of leg, of medial malleolus, of instep and medial side of foot)
14 *Articular nerve to ankle joint*
15 Femoral cutaneous ramus (sensory) innervation of medial and inferior part of thigh and of medial surface of knee
16 *Articular nerve to knee joint*
17 Patellar terminal branch (sensory innervation of medial part of knee below the obturator)
18 OBTURATOR NERVE (ANTERIOR PARTS)
19 Anterior lateral femoral cutaneous nerve (Valentin's): anastomosed to femoral cutaneous nerve (sensory innervation of lateral part of thigh)
20 *Articular nerve to hip joint*
21 Cutaneous rami (sensory innervation of superior medial part of thigh)
22 **Nerves to adductor longus (L2 L3 L4)**
23 **Nerve to pectineus (L2 L3 L4)**

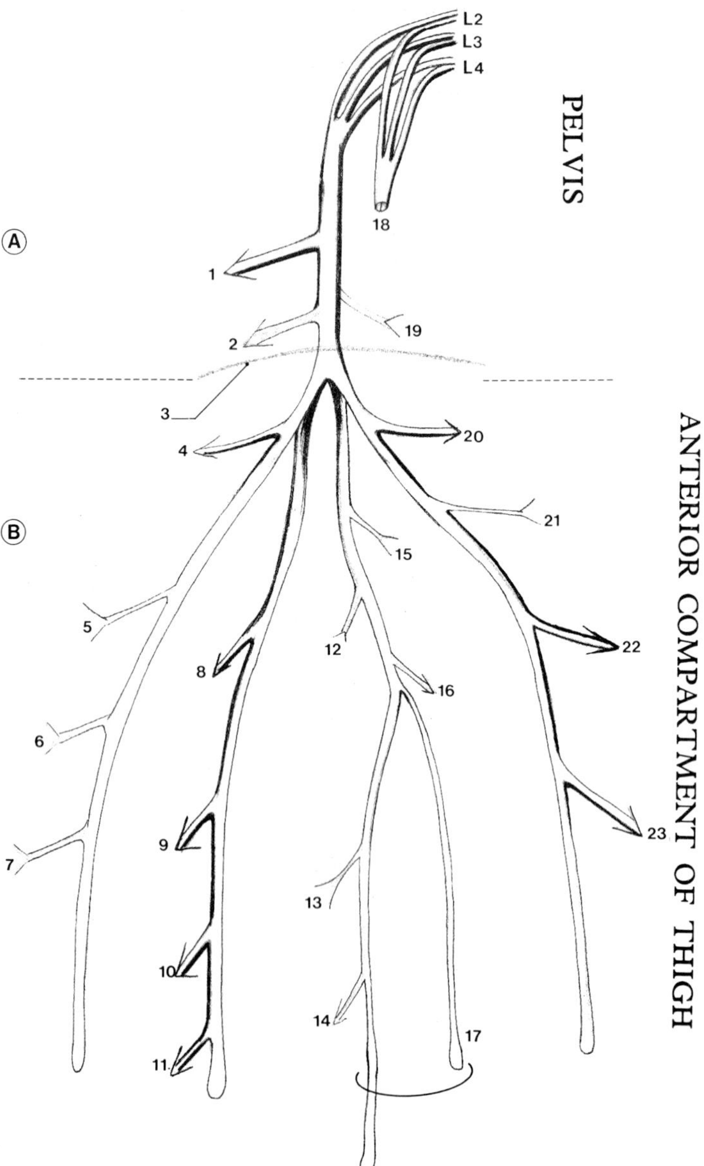

SCHEMATIC FORMATION OF SACRAL PLEXUS: FIVE COLLATERAL BRANCHES ONE TERMINAL BRANCH

1 SUPERIOR GLUTEAL NERVE (L4 L5 S1)
2 *Superior branch*
3 *Inferior branch*
4 **Nerve to gluteus medius and gluteus minimus (L4 L5 S1)**
5 **Nerve to tensor fascia latae (L4 L5 S1)**
6 **Nerve to obturator internus and gemellus superior (L5 S1 S2)**
7 **Nerve to quadratus femoris and gemellus inferior (L4 L5 S1)**
8 **Nerve to piriformis (s1 S2)**
9 SMALL SCIATIC NERVE (L4 L5 S1 S2)
10 INFERIOR GLUTEAL NERVE (L5 S1 S2)
11 **Nerve to gluteus maximus (L5 S1 S2)**
12 Posterior cutaneous nerve (S2), sensory innervation of perineum, of genital organs, of posterior surface of thigh, of posterior surface of upper third of leg. Anastomosis to lateral saphenous nerve
13 *Anastomosis to pudental plexus*
14 GREAT SCIATIC NERVE (COMPLETE MUSCULAR INNERVATION OF THIGH, POSTERIOR COMPARTMENT OF LEG AND OF FOOT)

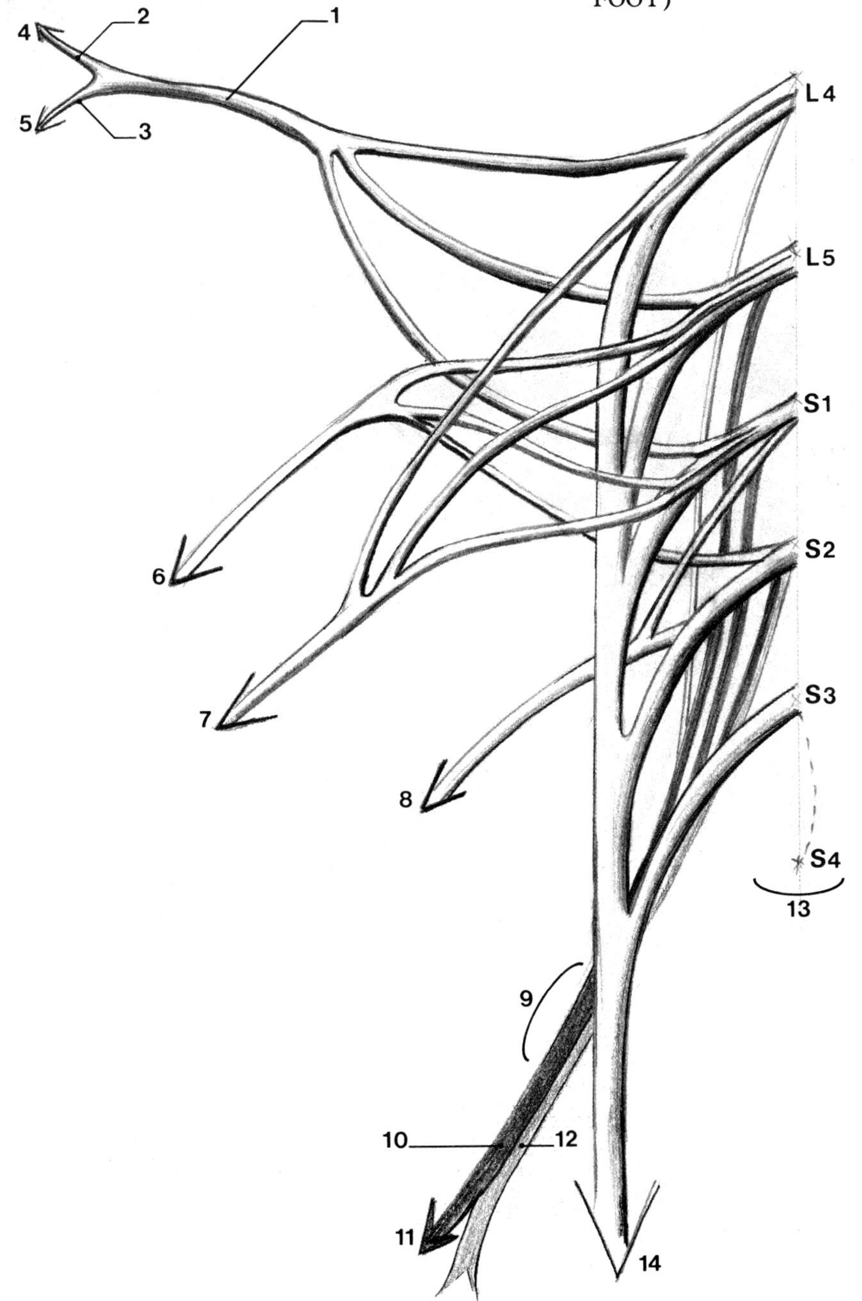

GREAT SCIATIC NERVE

1 **Nerve to long head of biceps femoris (l5 S1 s2)**
2 **Nerve to short head of biceps femoris (L5 S1 s2)**
3 *Articular nerve to knee joint*
4 **Superior nerve to semitendinosus (l4 L5 S1 s2)**
5 **Inferior nerve to semitendinosus (l4 L5 S1 s2)**
6 **Nerve to semimembranosus (l4 L5 S1 s2)**
7 **Nerve to lower portion of adductor magnus (l4 L5 S1)**

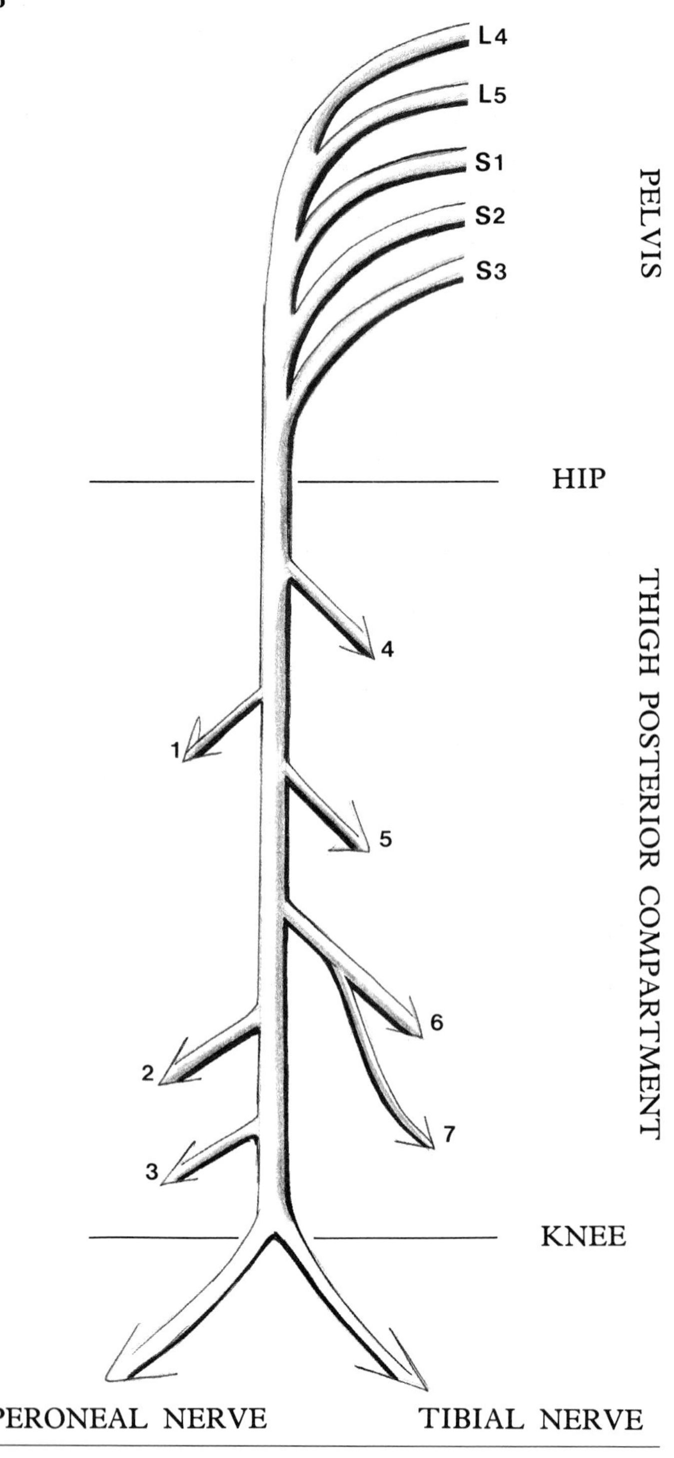

COMMON PERONEAL NERVE (EXTERNAL POPLITEAL NERVE)
MUSCULOCUTANEOUS NERVE
ANTERIOR TIBIAL NERVE

1 *Articular nerve to knee joint*
2 TIBIAL NERVE (INTERNAL POPLITEAL)
3 Peroneal lateral saphenous (accessory lateral saphenous), sensory innervation of lateral malleolus of lateral lower part of leg and of lateral part of heel.
4 Peroneal cutaneous nerve (sensory innervation of upper lateral part of leg and of knee)
5 **Superior nerves to tibialis anterior (L4 L5)**
6 **Nerve to peroneous longus, epiphyseal head (14 L5 s1)**
7 **Nerve to peroneous longus, anterior inferior diaphyseal head (14 L5 s1)**
8 **Nerve to peroneous longus, posterior inferior diaphyseal head (14 L5 s1)**
9 **Nerve to peroneous brevis (14 L5 s1)**
10 **Nerve to peroneous tertius (14 L5 s1)**
11 Cutaneous ramus (sensory innervation of lower lateral part of leg and lateral malleolus)
12 Medial cutaneous branch
13 Lateral cutaneous branch
14 First interdigital space nerve
15 Second interdigital space nerve
16 COMMON PERONEAL NERVE (EXTERNAL POPLITEAL)
17 *Lateral surface of neck of fibula*
18 *Extensor digitorum arch*
19 **Superior nerve to tibialis anterior (14 L5 s1)**
20 **Nerve to extensor digitorum longus (14 L5 s1)**
21 **Nerve to extensor hallucis longus (14 L5 s1)**
22 **Inferior nerve to tibialis anterior (14 L5 s1)**
23 **Nerve to peroneus tertius (not always present) 14 L5 s1**
24 *Vascular ramus*
25 *Articular nerve to ankle joint*
26 Lateral branch
27 **Nerve to extensor digitorum brevis (14 L5 s1)**
28 Medial branch (sensory innervation of first interdigital space)
29 Medial collateral nerve of big toe

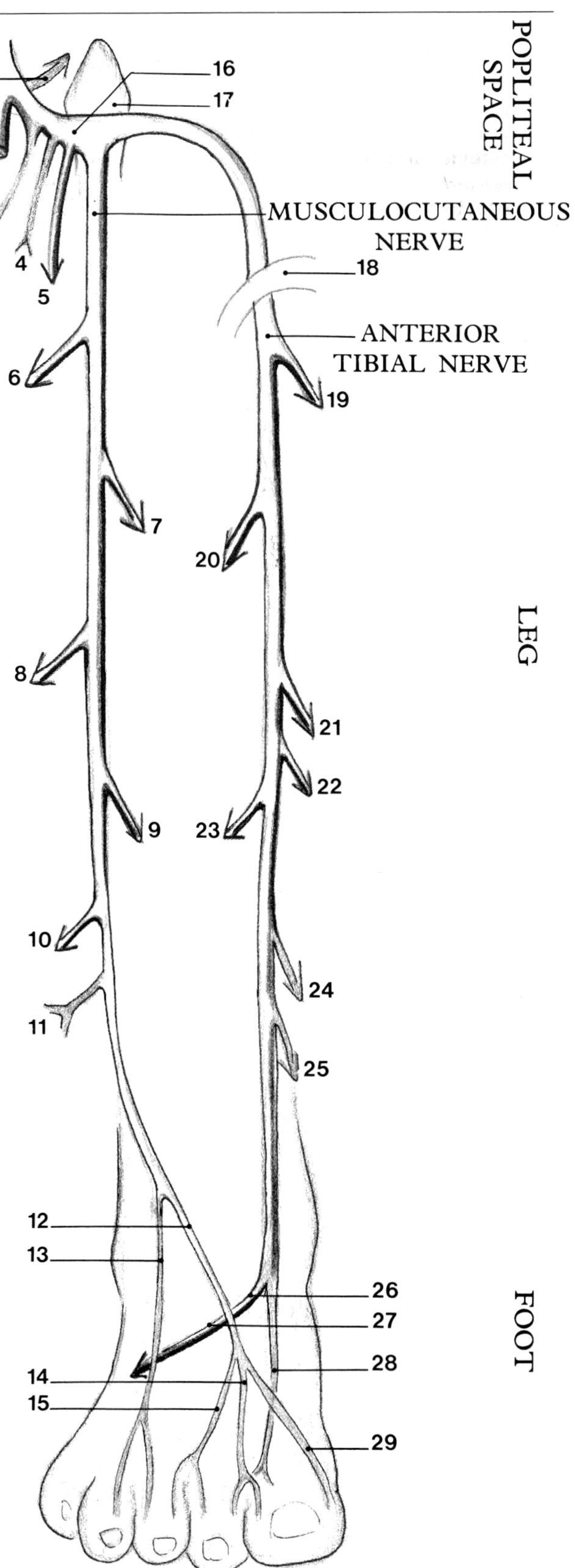

TIBIAL NERVE (INTERNAL POPLITEAL NERVE) POSTERIOR TIBIAL NERVE

1 GREAT SCIATIC NERVE
2 **Nerve to lateral gemellus (S1 S2)**
3 **Nerve to medial gemellus (S1 S2)**
4 **Nerve to plantaris (S1 S2)**
5 **Nerve to popliteus (14 L5 S1)**
6 **Posterior nerve to soleus (15 S1 s2)**
7 *Soleus arch*
8 **Anterior nerve to soleus (15 S1 s2)**
9 **Nerve to tibialis posterior (L5 S1)**
10 **Nerve to flexor digitorum longus (15 S1)**
11 **Nerve to flexor hallucis longus (15 S1)**
12 COMMON PERONEAL NERVE (EXTERNAL POPLITEAL)
13 *Articular nerve to knee joint*
14 Lateral saphenous nerve (tibial saphenous), sensory innervation of lower lateral part of leg, of instep, of fifth toe, and lateral half of fourth toe.
15 *Vascular nerve to popliteal artery*
16 *Articular ramus to ankle joint*
17 Medial calcaneal nerve (sensory innervation of medial malleolus of posterior part of heel and of foot plantar surface

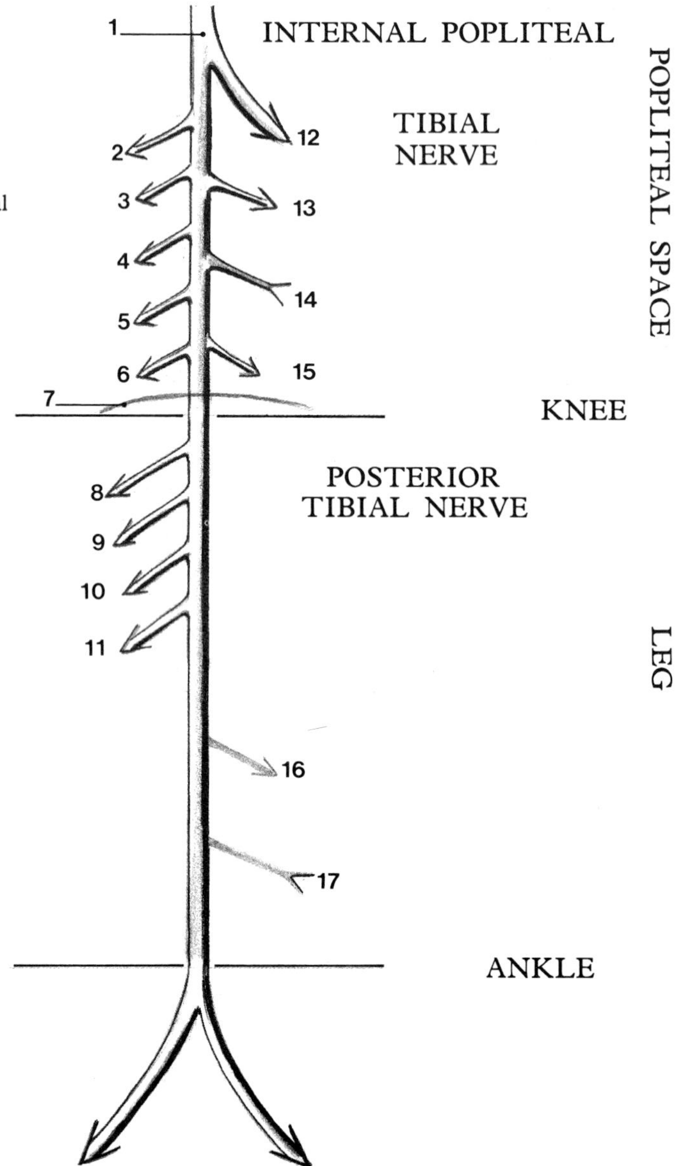

MEDIAL PLANTAR NERVE
LATERAL PLANTAR NERVE

1 POSTERIOR TIBIAL NERVE
2 *Articular nerves*
3 Cutaneous rami (sensory innervation of medial part of heel and of foot plantar surface)
4 **Nerve to flexor digitorum brevis (l5 S1)**
5 **Nerve to abductor hallucis (foot axis) l5 S1**
6 **Nerve to quadratus plantae (l5 S1)**
7 **Nerve to flexor hallucis brevis (l5 S1)**
8 *Division at level of navicular bone*
9 *Medial branch*
10 *Lateral branch*
11 **Nerve to flexor hallucis brevis (l5 S1)**
12 Nerve to first interdigital space
13 Nerve to second interdigital space
14 Nerve to third interdigital space
15 **Nerve to first lumbrical (l5 S1)**
Nerve to second lumbrical (l5 S1)
16 Medial collateral of great toe
and
17 (Sensory innervation of plantar part of proximal and distal phalanges and dorsal part of distal phalanx).
18 Plantar collateral nerves of toes (sensory innervation of
and plantar part of proximal, middle and distal phalanges
19 and dorsal part of distal phalanx)
20 **Nerve to abductor digiti minimi (S1 S2)**
21 **Nerve to adductor hallucis, transverse and oblique heads (foot axis) (S1 s2)**
22 **Nerves to interossei plantares and dorsales (S1 S2 s3)**
23 **Nerves to third and fourth lumbricales (S1 s2)**
24 *Division at level of fifth metatarsal*
25 *Deep branch*
26 *Superficial branch*
27 *Lateral branch*
28 *Medial branch*
29 **Nerve to abductor digiti minimi (S1 S2)**
30 **Nerve to opponens digiti minimi (S1 S2)**
31 **Nerve to flexor digiti minimi (S1 S2)**
32 Nerve to fourth interdigital space
33 Lateral collateral of fifth toe
34 *Anastomosis, medial and lateral plantar nerves*

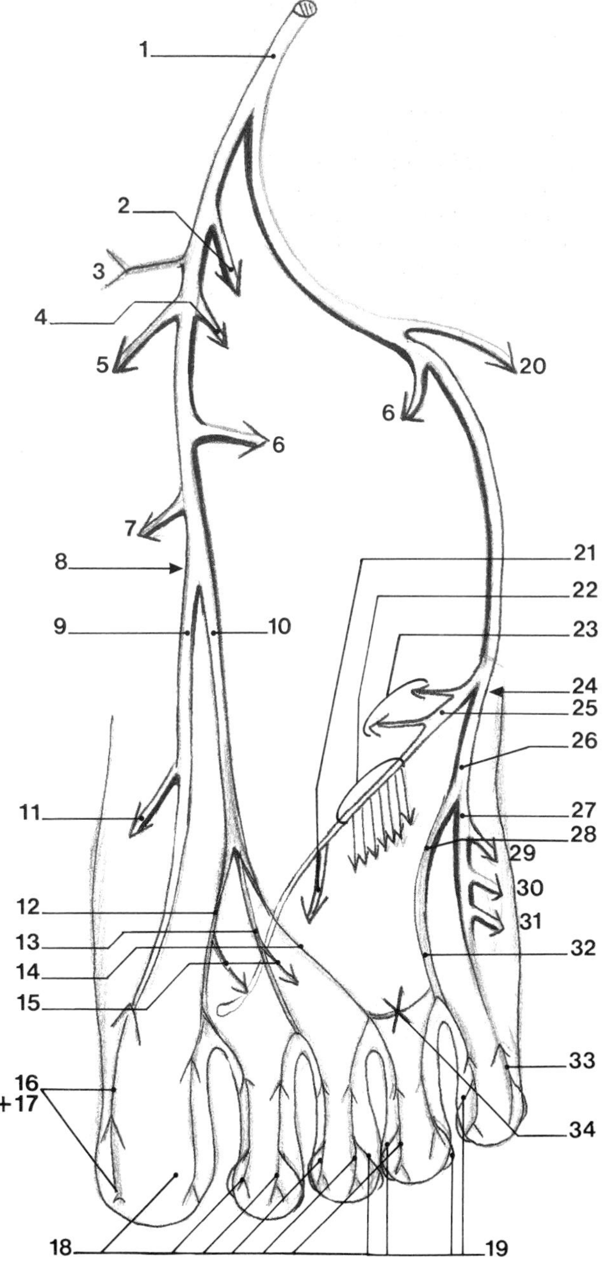

SUMMARY OF RADICULAR NERVE DISTRIBUTION

Region	Muscle	C1	C2	C3	C4	C5	C6	C7	C8	T1
SHOULDER	anterior head and neck flexors	■	■	■	■	■	■	■	■	■
SHOULDER	neck extensors	■	■	■	■	■	■	■	■	■
SHOULDER	sternocleidomastoideus		■	■						
SHOULDER	trapezius		½	■	■					
SHOULDER	levator scapulae				■	■				
SHOULDER	rhomboids				■	■				
SHOULDER	deltoideus					■	■			
SHOULDER	supraspinatus					■	■			
SHOULDER	infraspinatus					■	■			
SHOULDER	teres minor					■	■			
SHOULDER	subscapularis					■	■			
SHOULDER	teres major					■	■	½		
SHOULDER	serratus anterior					■	■	■		
SHOULDER	pectoralis major					■	■	■	■	■
SHOULDER	coraco brachialis						■	■		
SHOULDER	latissimus dorsi						■	■	■	
SHOULDER	pectoralis minor							½	■	■
WRIST	biceps brachi					■	■			
WRIST	brachialis					■	■			
WRIST	brachioradialis					■	■			
WRIST	supinator					■	■	½		
WRIST	pronator teres						■	■		
WRIST	triceps brachii & anconeus						½	■	½	
WRIST	pronator quadratus							½	■	■
ELBOW	extensor carpi radialis longus						■	■		
ELBOW	flexor carpi radialis palmaris longus						■	■	½	
ELBOW	extensor carpi radialis brevis						½	■	½	
ELBOW	extensor carpi ulnar							■	■	
ELBOW	flexor carpi ulnaris							½	■	■
HAND	extensor digitorum						½	■	½	
HAND	extensor indicis propius						½	■	½	
HAND	extensor digiti minimi						½	■	½	
HAND	abductor pollicis longus						½	■	½	
HAND	extensor pollicis brevis						½	■	½	
HAND	extensor pollicis longus						½	■	½	
HAND	flexor pollicis longus							½	■	■
HAND	flexor digitorum superficialis							½	■	■
HAND	flexor digitorum profundus							½	■	■
HAND	lumbricales							■	■	■
HAND	abductor pollicis brevis								■	■
HAND	opponens pollicis								■	■
HAND	flexor pollicis brevis								■	■
HAND	adductor pollicis								■	■
HAND	interossei dorsales								■	■
HAND	interossei palmares								■	■
HAND	flexor digiti minimi brevis								■	■
HAND	abductor digiti minimi								■	■
HAND	opponens digiti minimi								■	■

	Muscle	L1	L2	L3	L4	L5	S1	S2	S3
HIP	iliopsoas	■	■	■	■				
HIP	sartorius	■	■	■					
HIP	tensor fasciae latae				■	■	■		
HIP	gluteus minimus				■	■	■		
HIP	gluteus medius				■	■	■		
HIP	adductor brevis		■	■	½				
HIP	adductor longus		■	■	½				
HIP	pectineus		■	■	■				
HIP	gracilis		■	■	½				
HIP	adductor magnus		½	■	■	■	½		
HIP	obturator externus			■	■				
HIP	gemellus inferior				■	■	■		
HIP	quadratus femoris				■	■	■		
HIP	obturator internus					■	■	■	
HIP	gemellus superior					■	■	■	
HIP	piriformis						■	■	
HIP	gluteus maximus				½	■	■	■	
KNEE	quadriceps		■	■	■				
KNEE	semi tendinosus				½	■	■	½	
KNEE	semi membranosus				½	■	■	½	
KNEE	biceps femoris					■	■	■	½
KNEE	popliteus					■	■		
FOOT	tibialis anterior				■	■			
FOOT	extensor hallucis longus				½	■	½		
FOOT	extensor digitorum longus				½	■	½		
FOOT	peroneus tertius				½	■	½		
FOOT	extensor digitorum brevis				½	■	½		
FOOT	peroneus longus				½	■	½		
FOOT	peroneus brevis				½	■	½		
FOOT	soleus					½	■	½	
FOOT	gemelli & plantaris						■	■	
FOOT	tibialis posterior					■	■		
FOOT	flexor hallucis longus					■	■	■	
FOOT	flexor digitorum longus & quadratus plantae					■	■	■	
FOOT	abductor hallucis					■	■		
FOOT	flexor hallucis brevis					■	■		
FOOT	flexor digitorum brevis					■	■		
FOOT	adductor hallucis						■	■	
FOOT	lumbricales					½	■	½	
FOOT	interossei dorsales						■	■	½
FOOT	interossei plantares						■	■	½
FOOT	muscles of V						■	■	

	C3	C4	C5	C6	C7	C8	T1	T2	T3	T4	T5	T6	T7	T8	T9	T10	T11	T12	L1	L2	L3	L4	L5	S1	S2	S3	S4
Diaphragm	■	■	■																								
Intercostales							■	■	■	■	■	■	■	■	■	■	■										
Erector spinae							■	■	■	■	■	■	■	■	■	■	■	■	■	■	■	■	■	■	■	■	
Rectus abdominis													■	■	■	■	■	■									
Obliquus externus abdominis													■	■	■	■	■	■	■								
Obliquus internus abdominis															■	■	■	■	■								
Transversus abdominis													■	■	■	■	■	■	■								
Quadratus lumborum																		■	■	■							
Muscles of the perineum																							■	■	■	■	■

SENSORY DISTRIBUTION

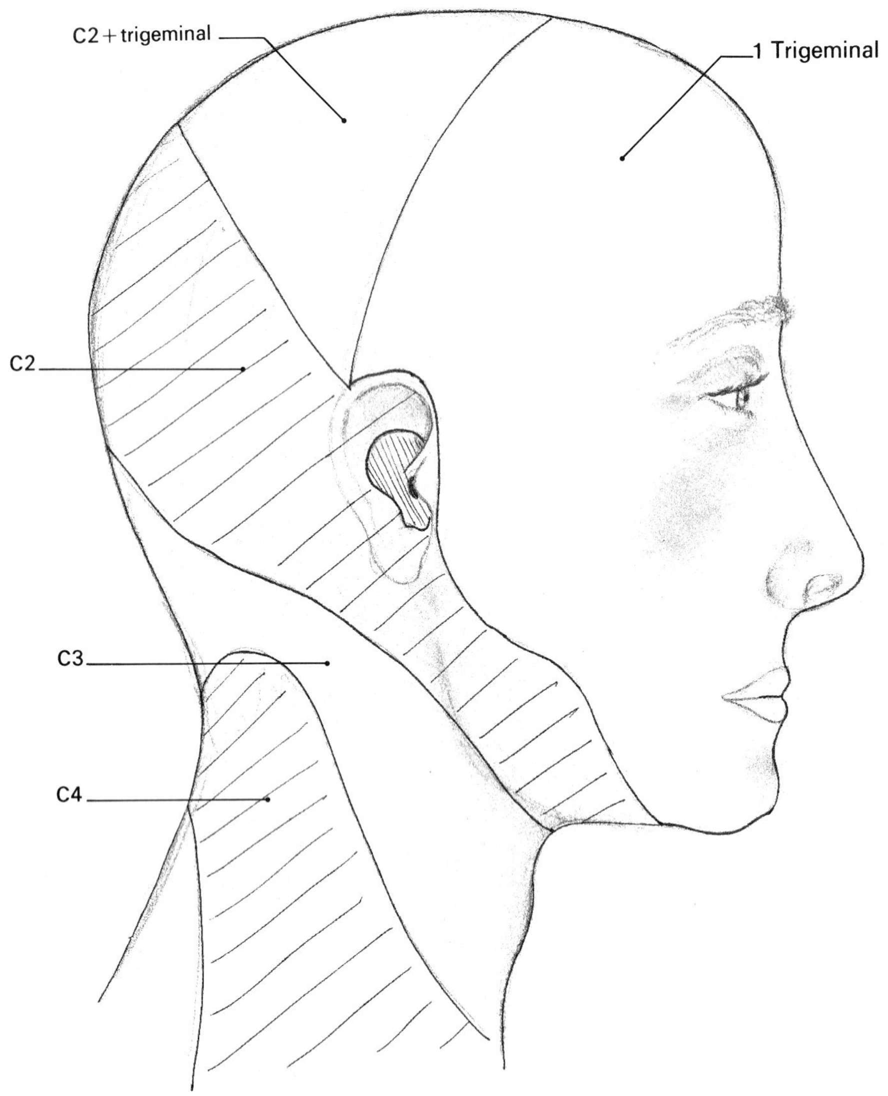
C2 + trigeminal
1 Trigeminal
C2
C3
C4

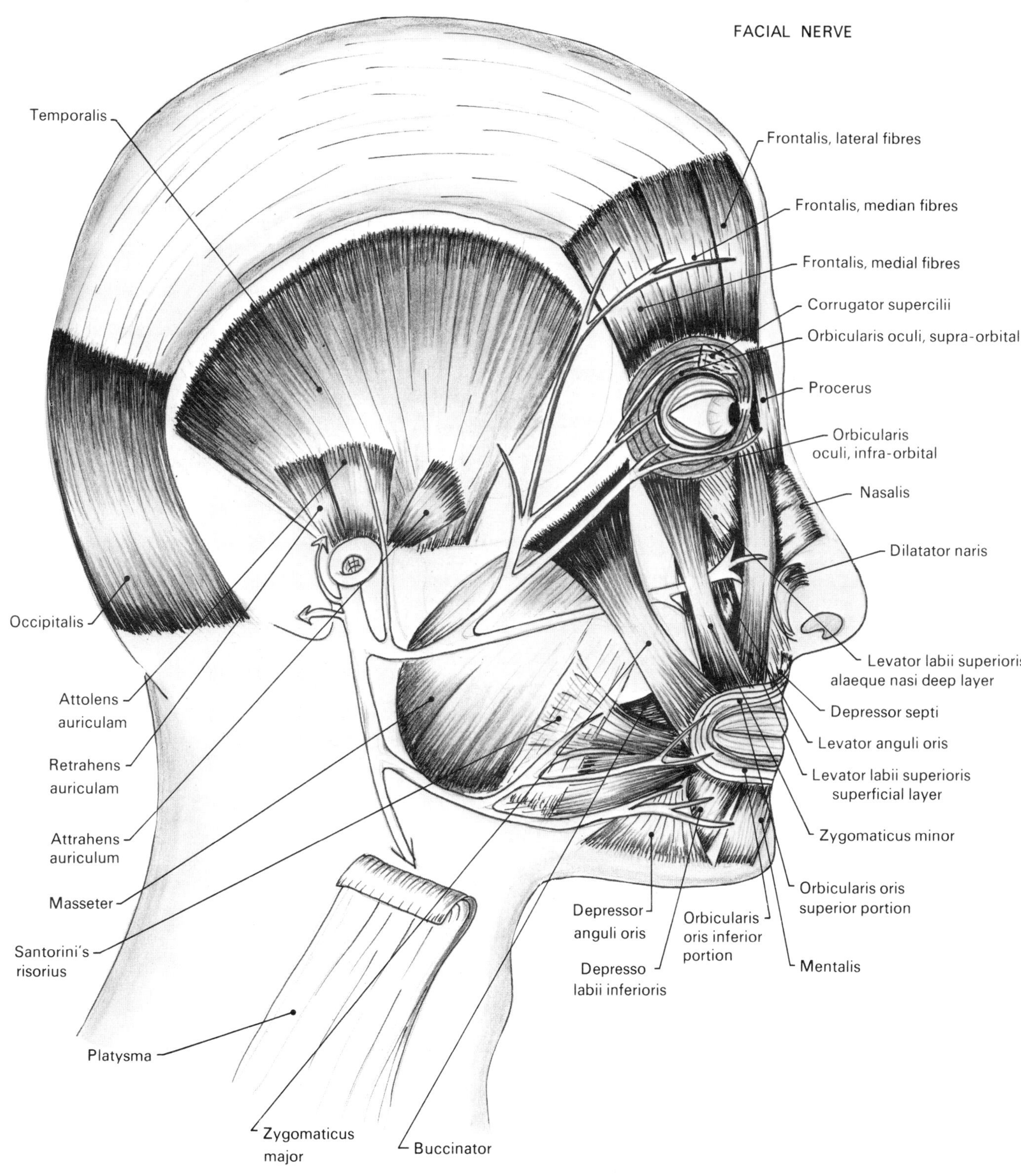
FACIAL NERVE
Temporalis
Frontalis, lateral fibres
Frontalis, median fibres
Frontalis, medial fibres
Corrugator supercilii
Orbicularis oculi, supra-orbital
Procerus
Orbicularis oculi, infra-orbital
Nasalis
Dilatator naris
Occipitalis
Levator labii superioris alaeque nasi deep layer
Attolens auriculam
Depressor septi
Levator anguli oris
Retrahens auriculam
Levator labii superioris superficial layer
Attrahens auriculum
Zygomaticus minor
Masseter
Orbicularis oris superior portion
Depressor anguli oris
Orbicularis oris inferior portion
Santorini's risorius
Depresso labii inferioris
Mentalis
Platysma
Zygomaticus major
Buccinator

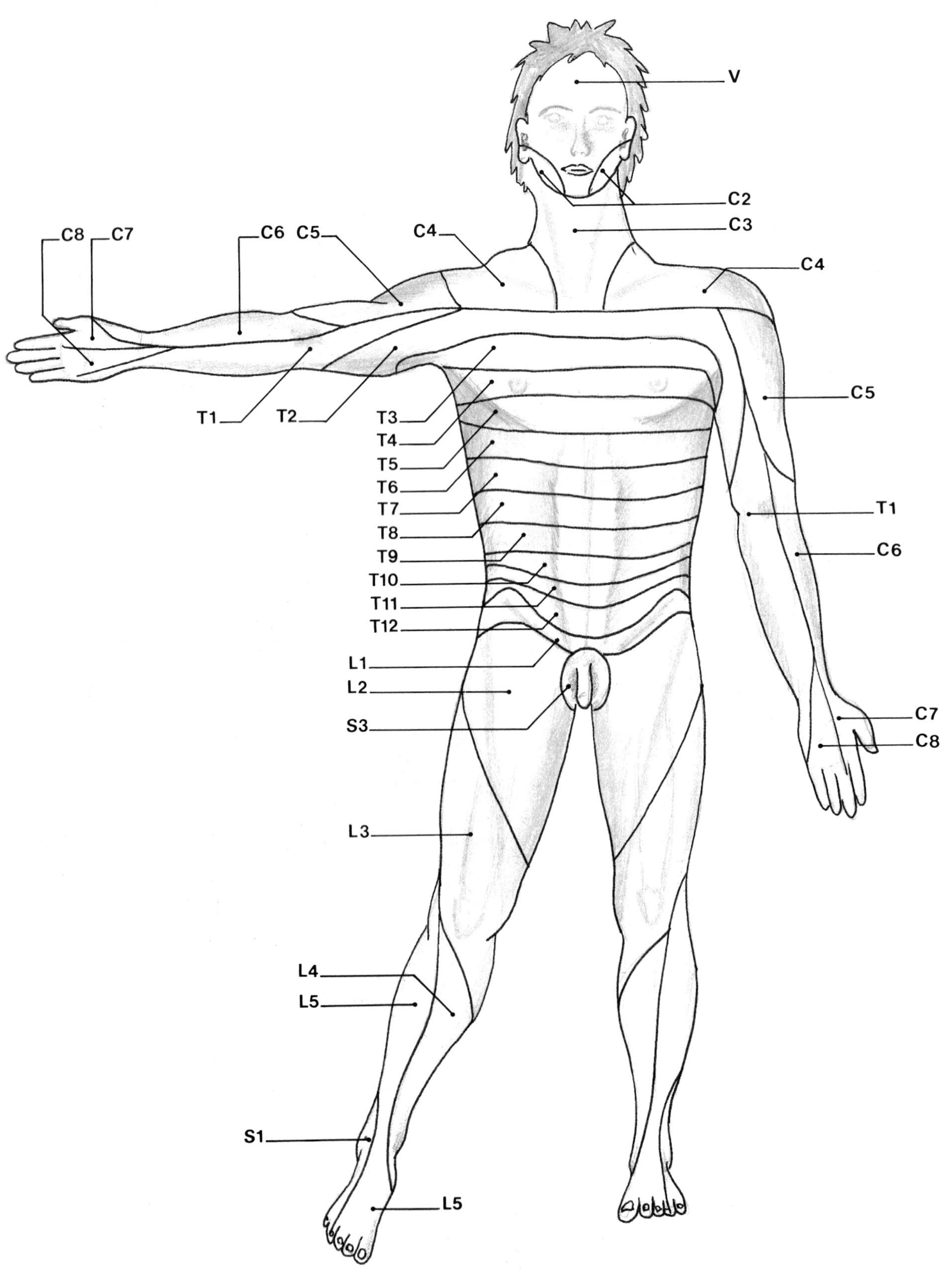
V
C2
C3
C8
C7
C6
C5
C4
C4
C5
T1
T2
T3
T4
T5
T6
T7
T8
T9
T10
T11
T12
L1
L2
S3
T1
C6
C7
C8
L3
L4
L5
S1
L5

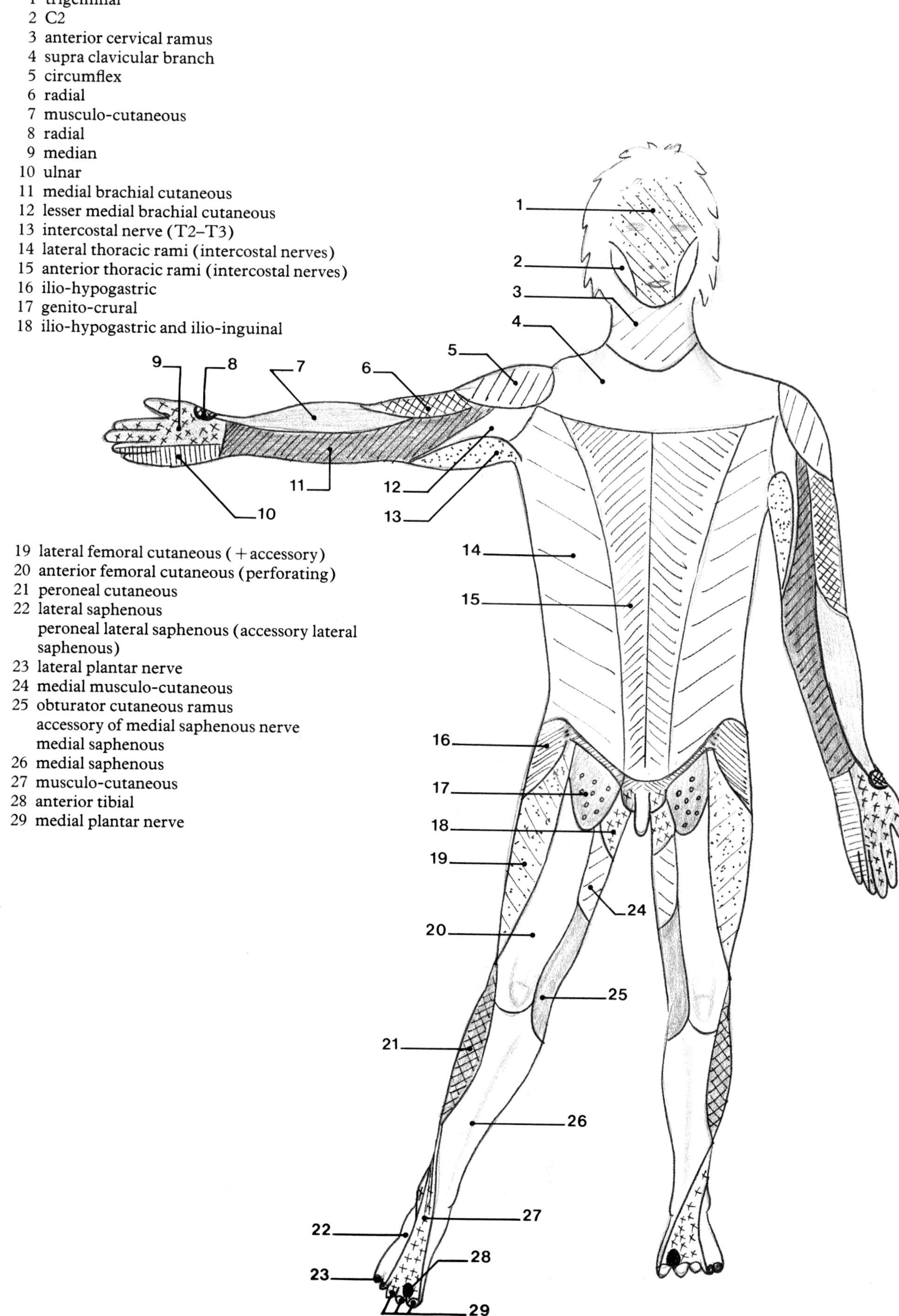
1 trigeminal
2 C2
3 anterior cervical ramus
4 supra clavicular branch
5 circumflex
6 radial
7 musculo-cutaneous
8 radial
9 median
10 ulnar
11 medial brachial cutaneous
12 lesser medial brachial cutaneous
13 intercostal nerve (T2–T3)
14 lateral thoracic rami (intercostal nerves)
15 anterior thoracic rami (intercostal nerves)
16 ilio-hypogastric
17 genito-crural
18 ilio-hypogastric and ilio-inguinal
19 lateral femoral cutaneous (+accessory)
20 anterior femoral cutaneous (perforating)
21 peroneal cutaneous
22 lateral saphenous
peroneal lateral saphenous (accessory lateral saphenous)
23 lateral plantar nerve
24 medial musculo-cutaneous
25 obturator cutaneous ramus
accessory of medial saphenous nerve
medial saphenous
26 medial saphenous
27 musculo-cutaneous
28 anterior tibial
29 medial plantar nerve

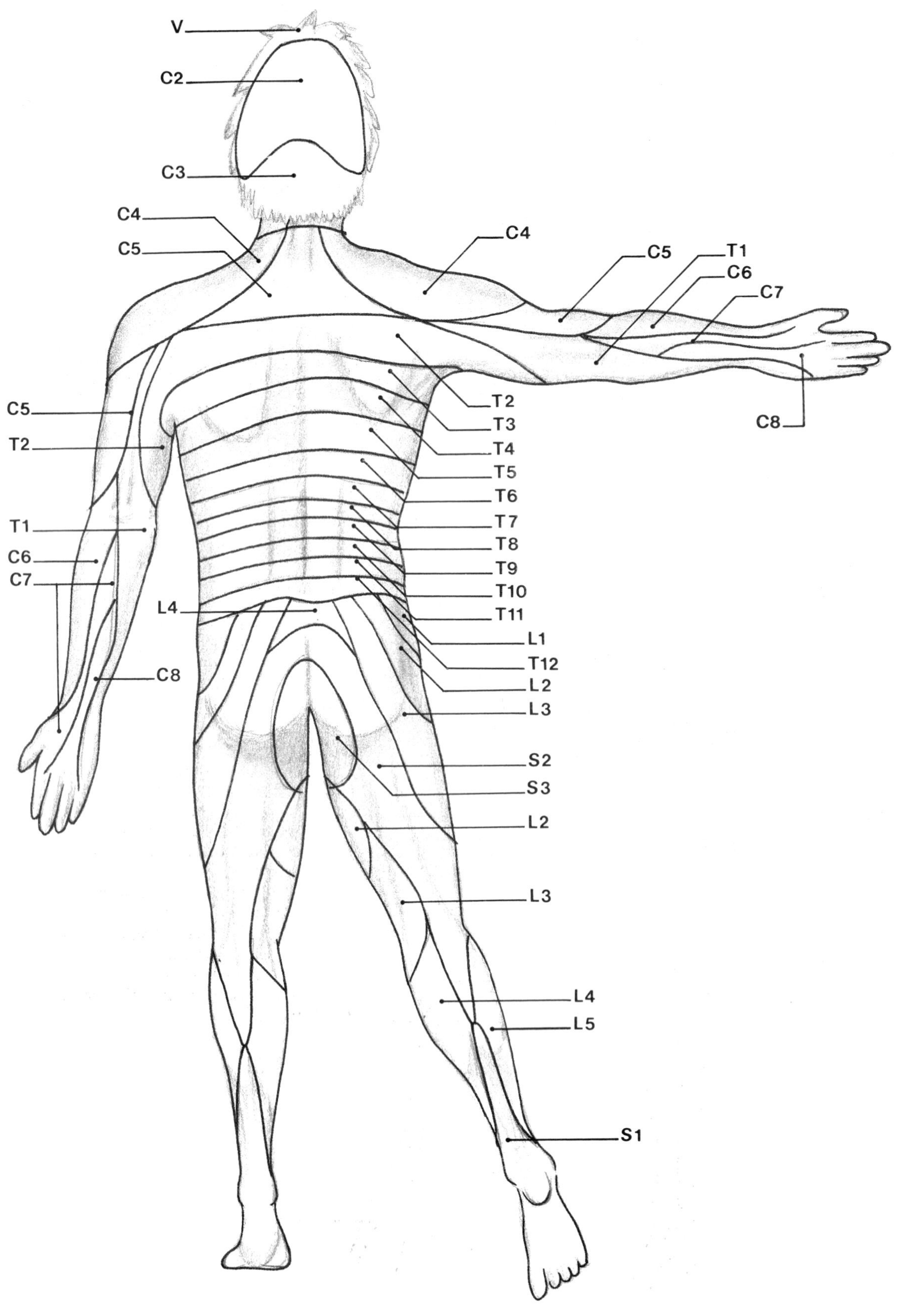
V
C2
C3
C4
C5
C4
C5
T1
C6
C7
C8
C5
T2
T1
C6
C7
T2
T3
T4
T5
T6
T7
T8
T9
T10
T11
L4
L1
T12
L2
L3
C8
S2
S3
L2
L3
L4
L5
S1

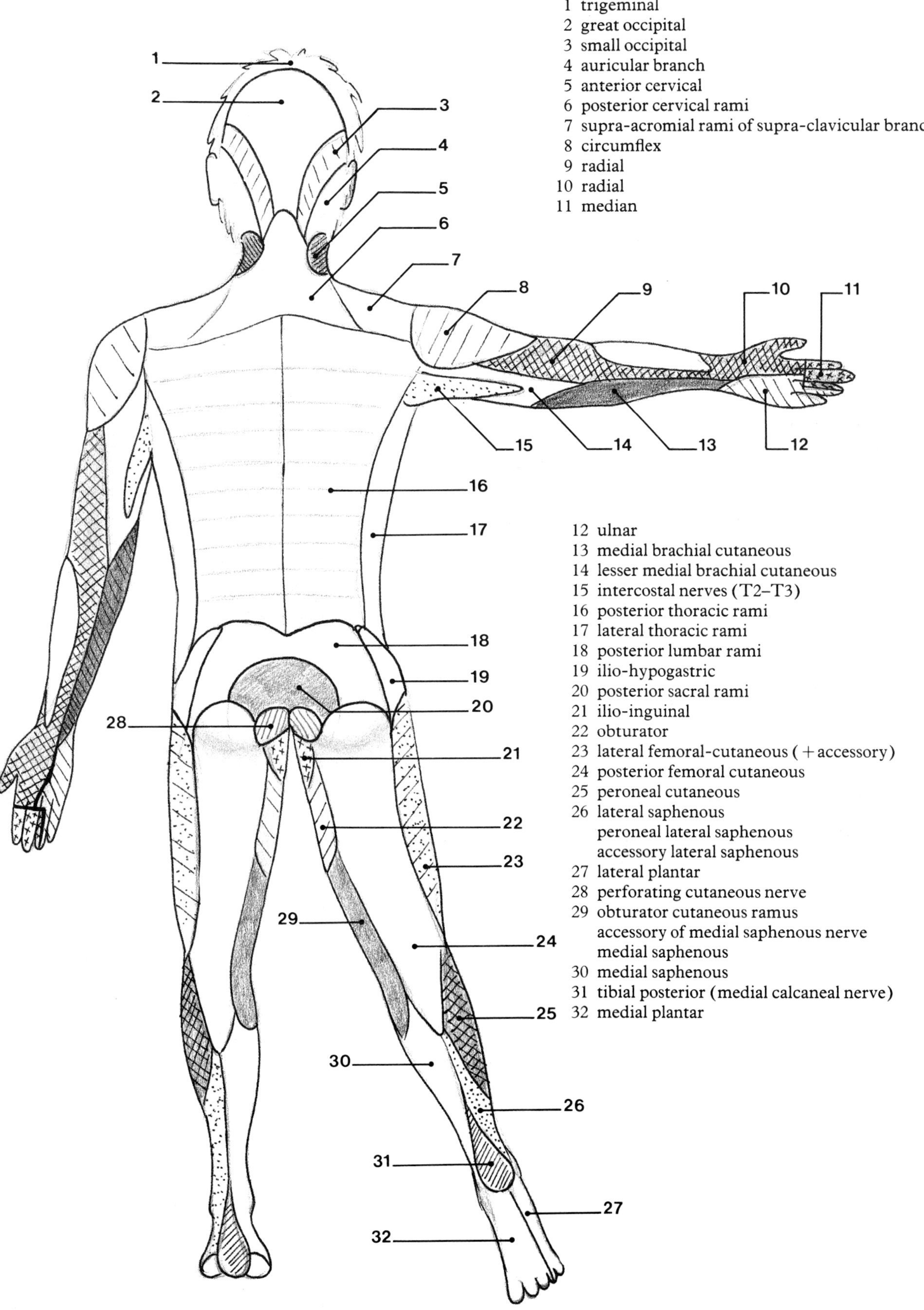
1 trigeminal
2 great occipital
3 small occipital
4 auricular branch
5 anterior cervical
6 posterior cervical rami
7 supra-acromial rami of supra-clavicular branch
8 circumflex
9 radial
10 radial
11 median
12 ulnar
13 medial brachial cutaneous
14 lesser medial brachial cutaneous
15 intercostal nerves (T2–T3)
16 posterior thoracic rami
17 lateral thoracic rami
18 posterior lumbar rami
19 ilio-hypogastric
20 posterior sacral rami
21 ilio-inguinal
22 obturator
23 lateral femoral-cutaneous (+accessory)
24 posterior femoral cutaneous
25 peroneal cutaneous
26 lateral saphenous
peroneal lateral saphenous
accessory lateral saphenous
27 lateral plantar
28 perforating cutaneous nerve
29 obturator cutaneous ramus
accessory of medial saphenous nerve
medial saphenous
30 medial saphenous
31 tibial posterior (medial calcaneal nerve)
32 medial plantar

1 musculo-cutaneous
2 radial (anterior territory : Lejar's surface)
3 radial (posterior territory)
4 median
5 ulnar

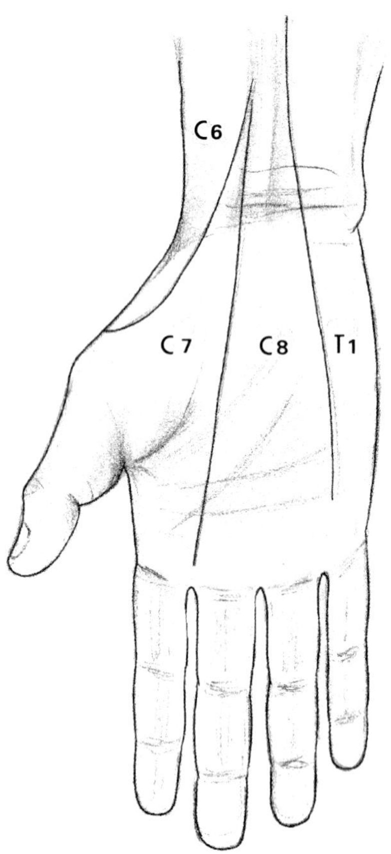

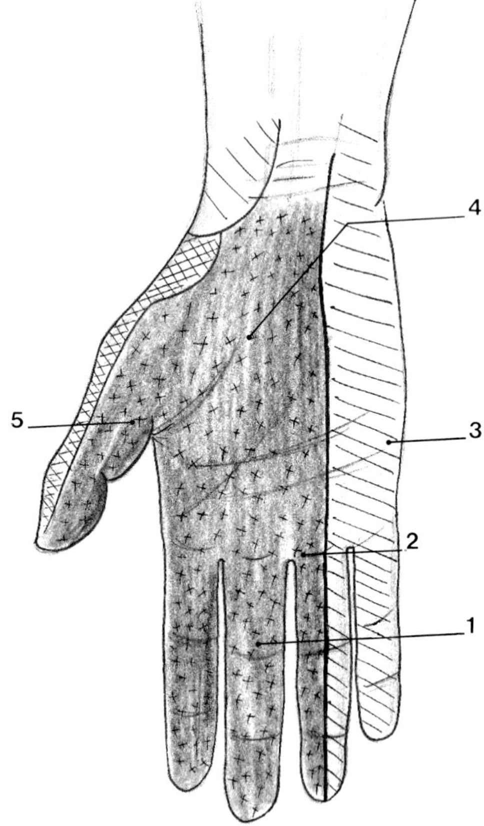

1 ulnar (dorsal cutaneous branch)
2 ulnar (rami coming from palmar surface)
3 radial
4 median (palmar territory)
5 median (rami coming from palmar surface)

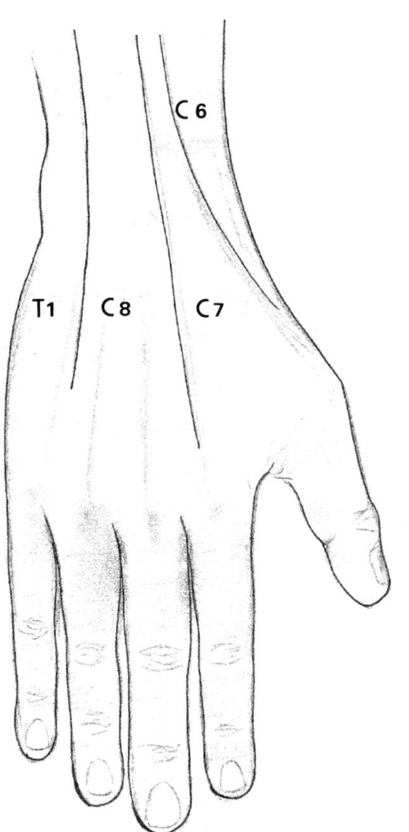

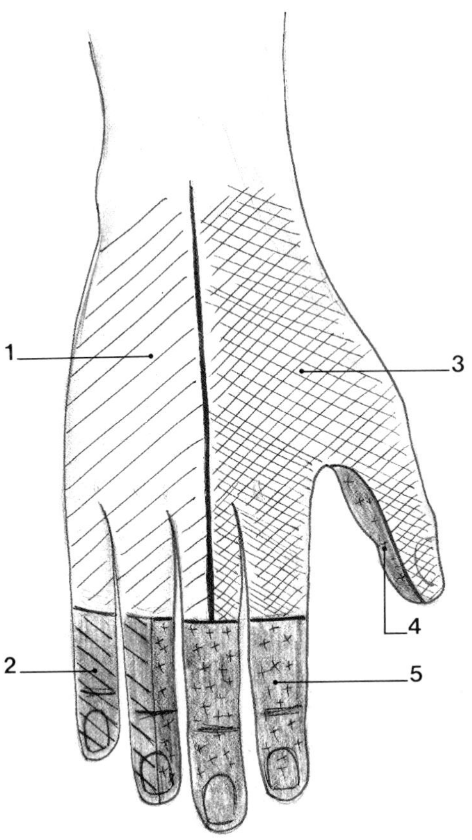

1 genito-crural
2 posterior femoral cutaneous
3 medial pudental
4 iliohypogastric and ilio-inguinal
5 femoral cutaneous
6 medial musculocutaneous
7 lateral musculocutaneous
8 medial saphenous
9 musculocutaneous
10 anterior sacral cossygeal branch and anal cutaneous nerve
11 posterior sacral branch
12 perforating cutaneous nerve
13 medial pudental
14 posterior femoral cutaneous
15 obturator
lateral musculocutaneous
medial saphenous
16 posterior tibial
17 medial plantar nerve

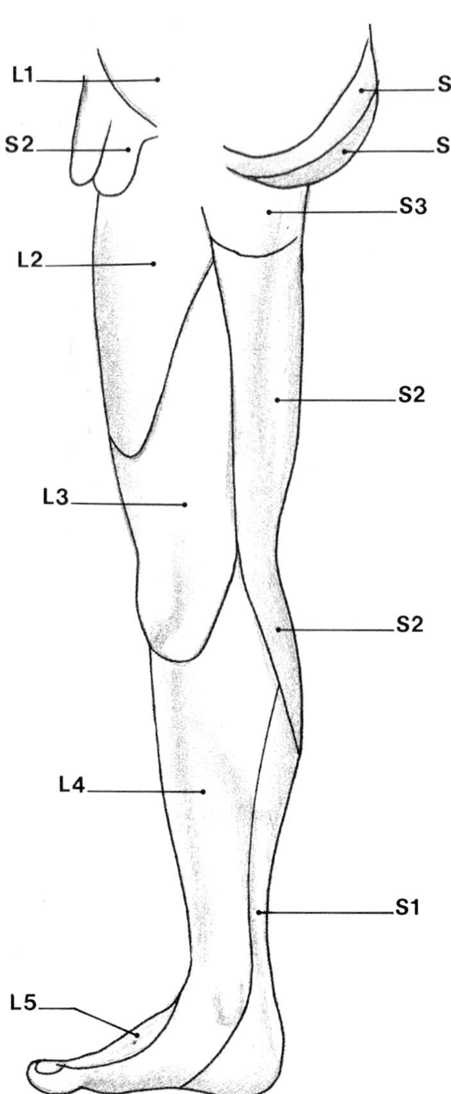

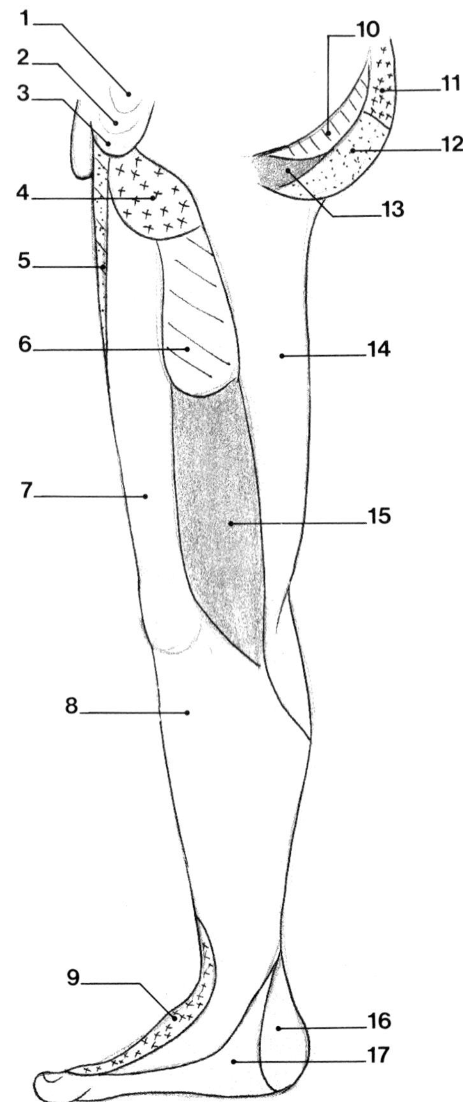

1 musculocutaneous
2 lateral saphenous
3 lateral plantar (dorsal surface of distal phalanx of fifth toe and lateral half of fourth toe)
4 medial plantar (dorsal surface of distal phalanx of all toes)
5 anterior tibial (deep peroneal)
6 medial saphenous

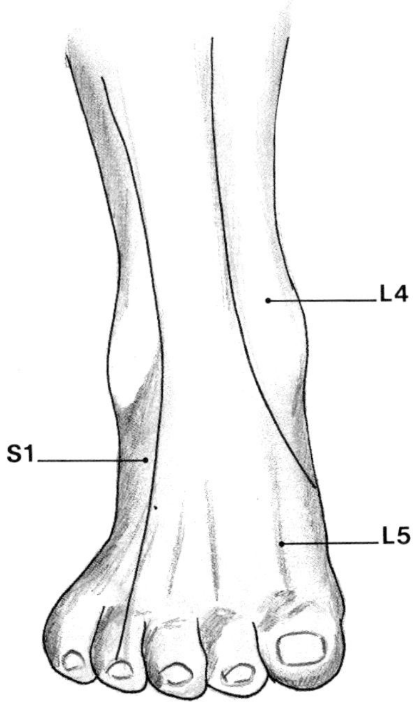

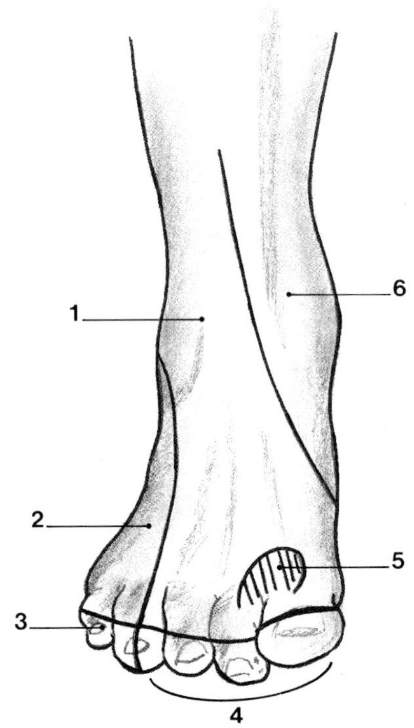

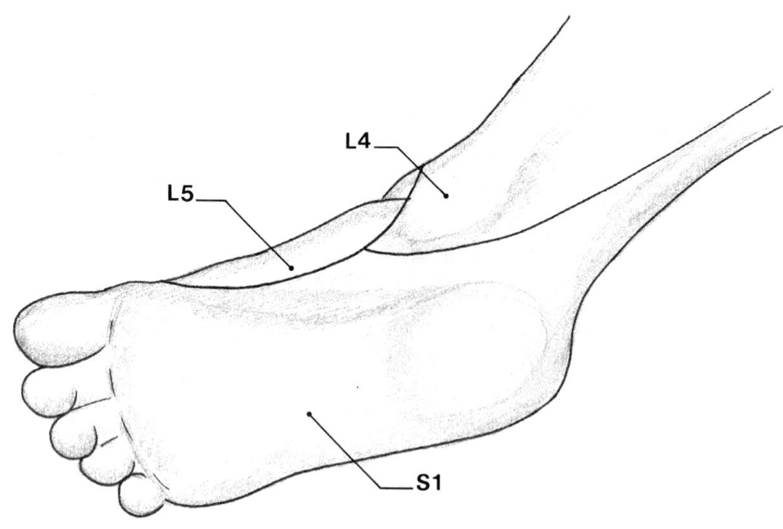

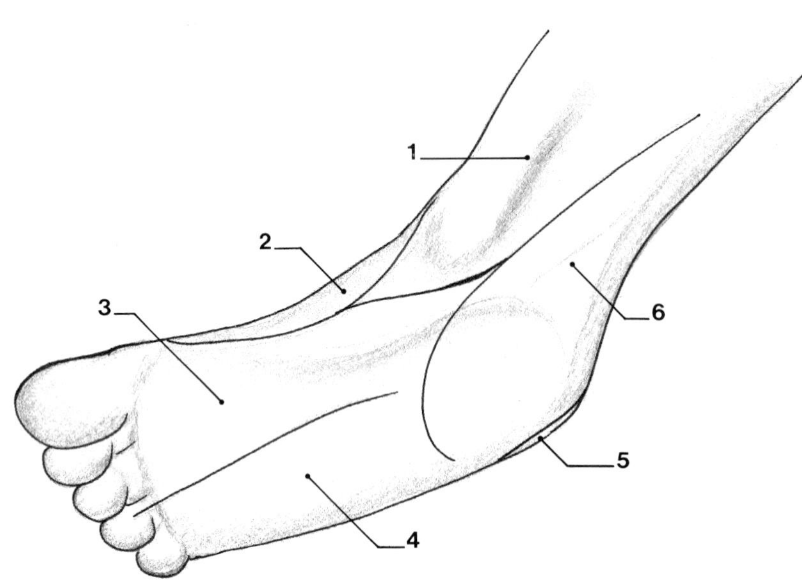

1 medial saphenous
2 musculocutaneous
3 medial plantar
4 lateral plantar
5 lateral saphenous
6 posterior tibial

BIBLIOGRAPHY

Basmadjian J V 1977 Anatomie. Maloine, Paris
Benassy J 1963 Innervation motrice radiculaire les membres. Revue de Rhumatologie
Benassy J 19-- Topgraphie métamérique de le moelle et de ses racines. Laboratoires Besins
Cambier J, Masson M, Dehin H 1975 Abrégé de Neurologie. 2nd edn. Ed. Masson, Paris
Chouard C, Charanchon R, Morgon A, Cathala H P 1972 Le Nerf Facial (anatomie pathologie en chirurgie). Masson, Paris
Chusid J G 1970 Correlative neuroanatomy and functional neurology. 14th edn. Lange Medical Publications, Los Altos, Calif.
Held J P, Vernaut J C 19-- Paralysie des nerfs périphériques. E.M.C. Kinesitherapie, Paris. 3180526465 A 10
Kendall H O, Kendall F P, Wadsworth G E 1974 Les Muscles Bilan et Etude Fonchonnelle. Maloine, Paris
Laplane D 1969 Diagnostic des lesions nerveuses périphériques. Cahiers Baillière. Ed. Baillière, Paris
Lazorthes G 1971 Le système nerveux périphérique. Ed. Masson, Paris
Rouvière H 1962 Anatomie humaine. Vols 1, 2. Ed. Masson, Paris

NERVES INDEX

Nerve Distribution to Head.
—Nerves, 551
—Sensory territories, 550, 553, 555
—Metameric sensory patterns, 550, 552, 554

Nerve Distribution to Upper Extremity.
—Plexus and nerves, 530, 531, 532, 533, 534, 535, 546
—Sensory territories, 553, 555, 556, 557
—Metameric sensory patterns, 552, 554, 556, 557

Nerve Distribution to Trunk.
—Plexus and nerves, 530, 536, 548
—Sensory territories, 553, 555
—Metameric sensory patterns, 552, 554

Nerve Distribution to Lower Extremity.
—Plexus and nerves, 536, 537, 538, 539, 540, 541, 542, 543, 547
—Sensory territories, 553, 555, 558, 559, 560
—Metameric sensory patterns, 552, 554, 558, 559, 560

MUSCLES INDEX

A
Abductor digiti minimi, 266
Abductor digiti minimi pedis, 398
Abductor hallucis, 386
Abductor pollicis brevis, 252
Abductor pollicis longus, 246
Adductor brevis, 342
Adductor hallucis, 390
Adductor longus, 342
Adductor magnus, 342
Adductor pollicis, 258
Amygdaloglossus, 94
Anconeus, 218
Attolëns auriculam, 96
Attrahens, 96

B
Biceps brachii, 212
Biceps femoris, 352
Brachialis, 214
Brachioradialis, 216
Buccinator, 67, 108 109, 111, 114, 115

C
Cleido-mastoideus, 160
Cleido-occipital, 160
Complexus, 168
Compressor labiorum, 85
Coracobrachialis, 192
Corrugator supercilii, 31, 110, 112
Crureus (see vastus intermedius)

D
Deltoideus
Deltoideus anterior portion, 192
Deltoideus posterior portion, 196
Depressor anguli oris, 78, 113, 115
Depressor labii inferioris, 73, 108, 109, 115
Depressor septi, 52, 108
Diaphragm, 294
Digastricus, 104, 105, 164
Dilatator naris, 50, 113, 115

E
Erector spinae, 304
Extensor carpi radialis brevis, 234
Extensor carpi radialis longus, 232
Extensor carpi ulnaris, 236
Extensor digiti minimi, 242
Extensor digitorum, 242
Extensor digitorum brevis, 362
Extensor digitorum longus, 362
Extensor hallucis longus, 360
Extensor indicis, 242
Extensor pollicis brevis, 248
Extensor pollicis longus, 250
External rotatores of the hip, 338

F
Flexor carpi radialis, 228
Flexor carpi ulnaris, 226
Flexor digiti minimi brevis, 268
Flexor digiti minimi pedis, 397
Flexor digitorum brevis, 392
Flexor digitorum longus, 384
Flexor digitorum profundus, 240
Flexor digitorum superficialis, 238
Flexor hallucis brevis, 388
Flexor hallucis longus, 382
Flexor pollicis brevis, 256
Flexor pollicis longus, 244
Frontalis (see occipito-frontalis)

G
Gastrocnemius, 374
Gemelli, 374
Gemellus inferior, 338
Gemellus superior, 338
Genioglossus, 92
Geniohyoideus, 104, 105
Gluteus maximus, 346
Gluteus medius, 330
Gluteus minimus, 334
Gracilis, 342

H
Hamsting muscles, 352
Hip adductors, 342
Hyoglossus, 92
Hyoid muscles, 105

I
Iliocostalis, 305
Iliocostalis cervicis, 305
Iliocostalis thoracis, 305
Iliocostalis lumborum or sacrolumbalis, 168, 305
Infra-hyoid muscles, 106, 114
Infraspinatus, 200
Intercostales externi 298
Intercostales interni, 300
Intercostales intimi, 300
Interossei dorsales, 264
Interossei dorsales pedis, 396
Interossei palmares, 262
Interossei plantares, 395
Interspinalis, 305
Interspinalis cervicis, 168
Intertransversarii, 163
Ischiococcygeus, 314, 315

L
Latissimus dorsi, 208
Levator anguli oris, 59, 112, 113, 115
Levator ani elevating portion, 314
Levator ani sphincter portion, 314
Levatores costarum, 298
Levator labii superioris alaeque nasi, 56, 109, 113, 114
Levator palpabrae superioris, 44, 88, 111
Levator scapulae, 182
Longissimus capitis, 168
Longissimus cervicis, 168
Longissimus thoracis, 305
Longitudinalis linguae inferior, 93
Longitudinalis linguae superior, 94
Longus capitis, 162
Longus colli, 162
Lumbricales, 260
Lumbricales pedis, 394

M
Masseter, 99
Masticators, 101, 113
Mentalis, 75, 108, 109, 113, 115
Middle deltoid, 194
Multifidus cervicis, 167
Multifidus spinae, 304
Mylohyoideus, 164

N
Nasalis pars transversa, 46, 113, 114, 115
Neck extensors, 166

O
Obliquus capitis inferior, 167
Obliquus capitis superior, 167
Obliquus externus abdominis, 282
Obliquus internus abdominis, 286
Obliquus occuli inferior, 88, 90, 91
Obliquus occuli superior, 88, 91
Obturator externus, 338
Obturator internus, 338
Occipito-frontalis, 26, 110, 114
Omohyoideus, 104, 106, 164
Opponens digiti minimi, 270
Opponens digiti minimi pedis, 399
Opponens pollicis, 254
Orbicularis occuli, 38, 109, 111, 112, 113, 114, 115
Orbicularis oris, 82, 108, 112

P
Palatoglossus, 93
Palmaris longus, 230
Pectineus, 342
Pectoralis major, 202
Pectoralis minor, 186
Peroneus brevis, 368
Peroneus longus, 370
Peroneus tertius, 362
Piriformis, 338
Pharyngoglossus, 93
Plantaris, 374
Platysma, 80, 113, 115, 164

Popliteus, 357
Procerus, 35, 113, 114, 115
Pronator quadratus, 224
Pronator teres, 224
Psoas iliacus, 318
Pterygoideus lateralis, 98, 100
Pterygoideus medialis, 98, 100

Q
Quadratus femoris, 339
Quadratus lumborum, 310
Quadratus plantae, 384
Quadriceps femoris, 348

R
Rectus abdominis, 276
Rectus capitis anterior, 163
Rectus capitis lateralis, 163
Rectus capitis posterior minor, 167
Rectus capitis posterior major, 167
Rectus externus, 88
Rectus femoris, 348
Rectus inferior, 88
Rectus internus, 88
Rectus lateralis, 90
Rectus medialis, 90
Rectus superior, 88, 89
Retrahens auriculam, 96
Rhomboideus major and minor, 182
Risorius, 70, 109, 111, 115

S
Sartorius, 322
Scalenus anterior, 163
Scalenus medius, 163
Scalenus posterior, 163
Semimembranosus, 352
Semitendinosus, 352
Serratus anterior, 188
Serratus posterior inferior, 300
Serratus posterior superior, 299
Soleus, 374
Splenius capitis and cervicis, 169
Sphincter ani externus, 316
Spinalis cervicis, 168
Spinalis thoracis, 305
Sternocleidohyoideus, 104, 107, 164
Sternocleidomastoideus, 104, 160
Sternohyoideus, 104, 164
Sternomastoideus, 160
Sternooccipital, 160
Sternothyroideus, 104, 107
Stylohyoideus, 104, 106, 164
Styloglossus, 93
Subclavius, 298
Subcrureus, 348
Subscapularis, 198
Supinator, 222
Supra-hyoid muscles, 105
Supraspinatus, 194

T
Temporalis, 99
Tensor fasciae latae, 326
Teres major, 206
Teres minor, 200
Thyrohyoideus, 107, 164
Tibialis anterior, 358
Transversus abdominis, 290
Transversus linguae, 94
Transversus perinei profundus, 315
Transversus perinei superficialis, 315
Transversus thoracis, 301
Trapezius inferior, 178
Trapezius medius, 174
Trapezius superior, 114, 166
Triceps brachii, 218
Triceps surae, 374

U
Upper trapezius (see trapezius superior)

V
Vastus intermedius, 348
Vastus lateralis, 348
Vastus medialis, 348
Verticulis linguae, 92

Z
Zygomaticus major, 64, 109, 111, 112, 114, 115
Zygomaticus minor, 62, 113, 114, 115